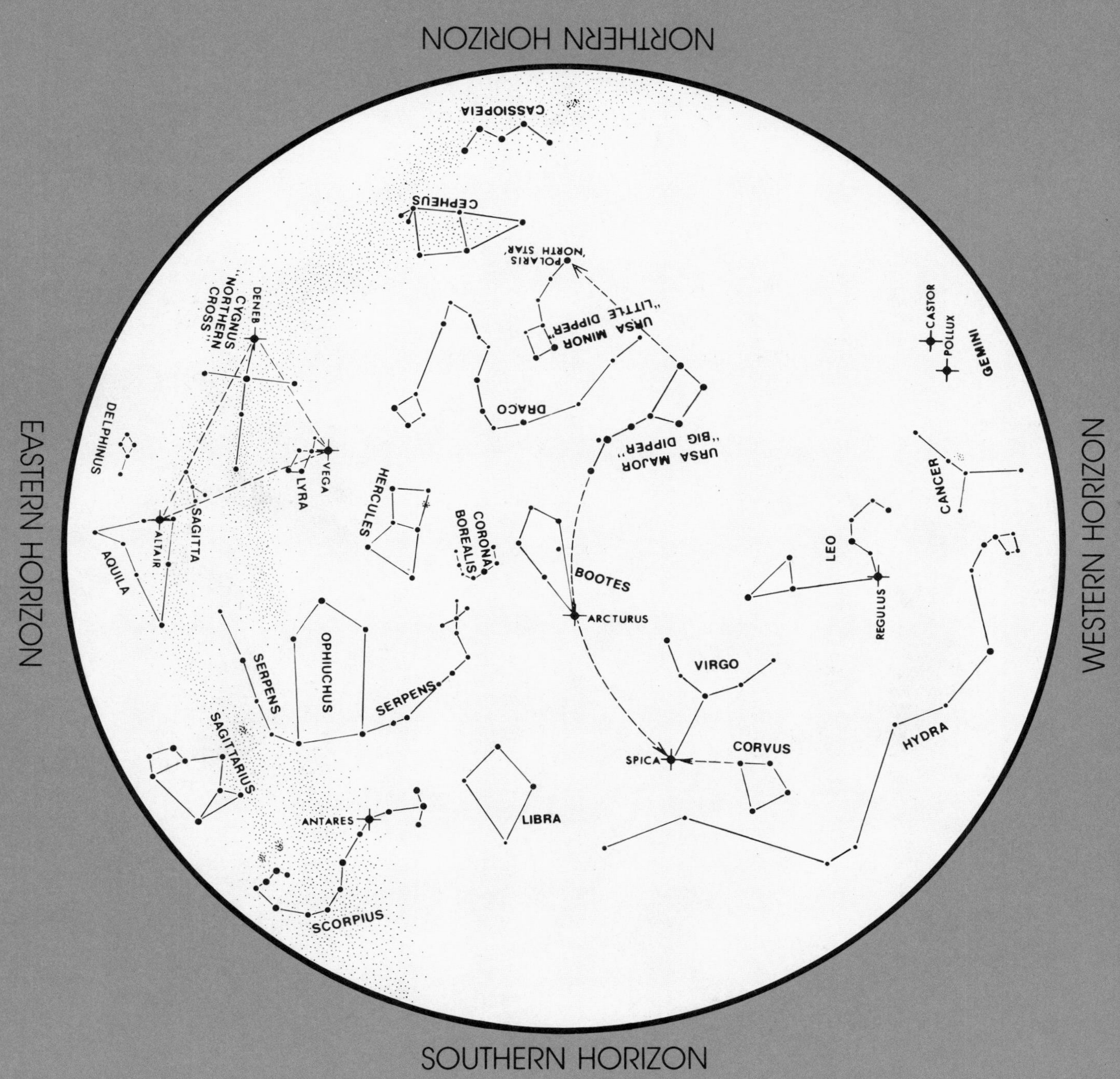


# THE NIGHT SKY IN JUNE

# ASTRONOMY

Michael W. Friedlander
Washington University
St. Louis, Missouri

# ASTRONOMY

## From Stonehenge to Quasars

Prentice-Hall, Inc., Englewood Cliffs, New Jersey 07632

*Library of Congress Cataloging in Publication Data*

FRIEDLANDER, MICHAEL W.
Astronomy, from Stonehenge to quasars.

Bibliography: p.
Includes index.
1. Astronomy. I. Title.
QB45.F84 1985 520 84-22862
ISBN 0-13-049867-X

Editorial/production supervision: Virginia Huebner
Interior design: Judith A. Matz-Coniglio
Cover design: Judith A. Matz-Coniglio
Manufacturing buyer: John B. Hall
Page layout: Charles Pelletreau and Peggy Finnerty
Cover photograph: High Altitude Observatory National Center for Atmospheric Research, Colorado (Sponsored by the National Science Foundation)



Printed in the United States of America

10 9 8 7 6 5 4 3 2 1

ISBN 0-13-049867-X 01

Prentice-Hall International, Inc., *London*
Prentice-Hall of Australia Pty. Limited, *Sydney*
Editora Prentice-Hall do Brasil, Ltda., *Rio de Janeiro*
Prentice-Hall Canada Inc., *Toronto*
Prentice-Hall Hispanoamericana, S.A., *Mexico*
Prentice-Hall of India Private Limited, *New Delhi*
Prentice-Hall of Japan, Inc., *Tokyo*
Prentice-Hall of Southeast Asia Pte. Ltd., *Singapore*
Whitehall Books Limited, *Wellington, New Zealand*

For Jessica

# Contents

# 6

## Galileo 67



# 7

## Newton: Mechanics and Gravity 75



# 8

## The Signals of Astronomy 86



# 9

## Telescopes and Radiation Detectors 109



# 10

## Space Exploration: New Opportunities for Astronomy 141



# 11

## The Solar System—a Summary 156

## 12

## The Earth—Our Home *170*



## 13

## The Moon *187*



## 14

## The Inner Planets: Mercury and Venus *208*



## 15

## Mars *220*



## 16

## The Jovian Planets *239*

# Preface

Astronomy courses provide a popular option each year for thousands of college students as part of their general education. This is a good time to study or teach astronomy, for there is a steady flow of new results from ground-based observatories as well as from satellites and long-range space probes. Many of the discoveries are reported in newspapers and magazines, on radio and television. There is as a result a widespread awareness and a curiosity that we can draw on in our introductory courses.

Our astronomy courses can, however, do more than simply fill a slot in the curriculum. We have, I believe, an opportunity and a responsibility to do more than entertain our students with a picture of this attractive science as we catalog its discoveries and successes. Few of our students will go on to pursue careers in science, and an even smaller number will do so in astronomy. Directly or indirectly, however, most will use the products of science through technology, and they will be asked to support scientific research, both for its anticipated later benefits and, to some extent, for its own interest. Our science will benefit from having more people know about our work, and our survey courses can play an important role if they show that a general understanding of astronomy need not be confined to experts.

Perhaps of even greater importance, with potentially longer-lasting results, is the chance that we have to show something of the internal structure of science. Considering how much we benefit from science, it is ironical that science is still so widely mistrusted, its operation not understood. We can use our astronomy courses to demonstrate how the scientific method has worked with so much success—the endless interplay between observations and theory, the logical application of the laws of science that have been discovered by experiment, and the gradual construction of an impressive understanding of our universe. Long after the factual details of our course have faded or become out of date (and this means almost any time after the end of the semester), it is the method of science that might be remembered. This is a major contribution that we can make as teachers and an important understanding that our students can gain.

There is always far more material available than can be included in an astronomy course or textbook, and the final selection and balance must show, to some extent, the personal preference of author and instructor. In this text, I have tried to combine the content and the method so that the wonder of discovery is often supported by answers to the questions "what is the evidence for . . . ?" or "how do we know that . . . ?" I hope that some of the flavor and excitement of a vigorous science will come across from these pages. Enjoy your course.

## SOME COMMENTS ON THE USE OF THIS TEXT

This text is quantitative but not heavily mathematical. Our ability to describe and predict in science is severely limited if we do not use numbers, some as small as the size of the atom and others as large as the dimensions of galaxies. Without, however, using any more mathematics than high school algebra (and very little of that) we can examine a large number of fascinating topics. To allow this text to be used at a variety of levels, many of the more mathematical discussions have been confined to Boxes that appear in most chapters. These can usually be omitted without loss of continuity or included as desired for further detail. Similarly, there are both qualitative and quantitative questions at the end of each chapter. Some of the questions have been designed to test the knowledge or understanding of a single concept or fact, others to demonstrate some numerical relationship. There is overlap between some questions, allowing an instructor to assign more or less detailed questions covering the same material.

A bibliography has been provided (Appendix O). It has been my experience in teaching this survey course that few students have the time to do much additional reading. The lists are, therefore, intended to provide only an introduction to what is an enormous literature on astronomy and its related sciences. I have tried to list books that are up-to-date, well written, interesting, and often available in paperback editions.

This text can be used for courses that extend for a whole year or for no more than a single semester or quarter. Whatever the course length, most of Chapters 7, 8, and 9 should be included to provide the basic knowledge for understanding the material in later chapters. These three chapters cover mechanics, light, optics and telescopes, and the detection of astronomical signals. Individual instructors will surely omit or select some sections, depending on the degree of detail they propose to use for the rest of the course.

A one-semester course that focused on the solar system might then include Chapters 11 to 18. As an alternative, a course that concentrated on the stars and galaxies might include Chapters 7, 8, 9, and then Chapters 19 to 31. Another variation is a semester course that is intended as a general survey, and this might include Chapters 7, 8, 9, 11, 14, 16, 18, 19 to 21, 23, 24, 26, 28, and 31. Chapter 1 provides an introduction, and Chapter 2 is useful for understanding the easily observed features in the sky. I hope that Chapters 3 and 4 are often used. Chapter 3 describes some of the earliest astronomy, and Chapter 4 shows how some of the old observations are being put to use with current observations. Chapters 5 and 6 provide an abbreviated history through the time of Newton.

Although there have been moves to standardize the units used in science textbooks, most of the astronomical literature employs the cgs units, and accordingly I have used them too. This has the advantage that students can more easily consult other sources without facing the problem of converting the units.

## ACKNOWLEDGMENTS

This text has its origin in the astronomy courses I have taught, in the popular lectures given, and in discussions with friends and colleagues. I am grateful to the many people who have helped in providing illustrations (acknowledged individually and separately) and in granting permission to quote from their books. I am particularly indebted to the following for their comments on various sections of the manuscript and their advice on specialized topics: Geoff Davies, Marty Israel, Jonathan Katz, Steve Margolis, Robert Walker, and Clifford Will. I would like to take this occasion to thank the $B^2FH$ authors for their photograph and the editors of the *Astrophysical Journal* and *Astronomy and Astrophysics* for permission to reprint some illustrations.

I am truly grateful to the readers who gave so much care to their reviews of several iterations of the manuscript. They include Robert L. Chasson, James A. Earl, Lowell R. Doherty, Darrell Hoff, Ernest G. Reuning, John L. Safko and three anonymous readers. It is customary, but no less heartfelt, to thank them for catching many errors and for pointing to the rough and unclear passages. The errors, howlers, and mixed metaphors that surely remain are clearly my responsibility. I would appreciate hearing from readers who find errors that have so far eluded all of us; my thanks, in anticipation.

It is my pleasure to thank Nadja Wilson and Ruth Benedett for their endless efficiency and patience in typing the manuscript, and Vicki Barnard-Ellis for similar assistance with the Index and the Instructor's Manual. Another acknowledgment goes to the unknown inventor of the word processor.

It is also my pleasure to thank my editors at Prentice-Hall, Doug Humphrey and Virginia Huebner, for their considerable assistance in the shaping of this book and in the endless details of its production. Thanks also to the talented art staff at Prentice-Hall, Judith Matz-Coniglio, designer, and Charles Pelletreau and Peggy Finnerty, layout artists, for a very attractive book.

Finally, a special note of appreciation for my family, for their constant interest and support—my deepest thanks.

## CREDITS FOR ILLUSTRATIONS

I am indebted to many who have helped by providing photographs, and especially to the following for their assistance in locating requested items: Linda Brenton, National Space Science Data Center; Agnes Paulsen, Kitt Peak National Observatory; Margaret Weems, National Radio Astronomy Observatory; Sue Glover, NASA Marshall Space Flight Center; Robbie Score, NASA Johnson Space Center; Sandra Preston, McDonald Observatory; Roselyn Arsenault, National Astronomy and Ionosphere Center; Garred Giles, Arecibo Observatory; Lynn Albaugh, NASA/Ames; Helen Horstman, Lowell Observatory; and Richard Dreiser, Yerkes Observatory.

Many individuals have generously supplied me with photographs and diagrams or tabular material. Most are credited at the places of display, but I am glad to have this chance to again express my thanks to the following, for their additional assistance: George Assousa, Owen Gingerich, Ray M. Klebesadel, Craig Leff, Alexander Marshack, O. W. Nuttli, Juan Roederer, Alar Toomre, Joseph Veverka, Jill C. Tarter, Betty Weiss, Ewen Whitaker. Abbreviated credits have been included at each illustration. Further identification is provided here. I am grateful for permission to use these illustrations. The Apollo, Mariner, Viking, and Voyager photography have been supplied by the National Space Science Data Center, Greenbelt, Maryland.

Some illustrations have been obtained direct from Observatories and NASA Centers:

NASA/Ames: Ames Research Center, Mountain View, California

NASA/GSFC: Goddard Space Flight Center, Greenbelt, Maryland

NASA/JSC: Johnson Space Center, Houston, Texas

NASA/Marshall: Marshall Space Flight Center, Alabama

AFGL: Air Force Geophysics Laboratory, Cambridge, Massachusetts

Arecibo: Arecibo Observatory, National Astronomy and Ionosphere Observatory, operated by Cornell University under contract with the National Science Foundation.

Big Bear Observatory: Big Bear Solar Observatory, California Institute of Technology

Center for Meteorite Studies: Center for Meteorite Studies, Arizona State University, Tempe, Arizona

CTIO: Association of Universities for Research in Astronomy, Inc. (AURA) Cerro Tololo Inter-American Observatory

HAO: High Altitude Observatory, National Center for Atmospheric Research, Colorado (Sponsored by the National Science Foundation)

KPNO: AURA, Inc., Kitt Peak National Observatory

Mt. Wilson and Las Campanas Observatories, Carnegie Institution of Washington.

MMT: M.M.T. Observatory, a joint facility of the University of Arizona and the Smithsonian Institution

NRAO: National Radio Astronomy Observatory, operated by Associated Universities Inc., under contract with the National Science Foundation.

NRL: Naval Research Laboratory, Washington, D.C.

Palomar Observatory, California Institute of Technology

Sacramento Peak Observatory: AURA, Inc., Sacramento Park Observatory

TRW: TRW Electronics and Defense Inc.

---

Permission is gratefully acknowledged for the reproduction of the diagrams whose copyright is held as indicated.

Ch. 4, Fig. 1: from "Historical Eclipses," F. R. Stephenson in *Scientific American*, October 1982.

Ch. 5, Fig. 9: from "Copernicus and Tycho," Owen Gingerich, in the *Scientific American*, December 1973.

Ch. 12, Fig. 11: Trapped radiation belts, from James Van Allen, Radiation Belts, from Lerner, Trigg, *ENCYCLOPEDIA OF PHYSICS* © 1981, Addison-Wesley, Reading, Massachusetts.

Ch. 13, Fig. 34: from *Icarus*, vol. 15, p. 368 (1971), copyright 1971 Academic Press.

Ch. 16, Fig. 32: from *Science*, vol. 217, p. 243 (1982), copyright 1982 American Association for the Advancement of Science.

Ch. 18, Fig. 7: from "The Formation of the Earth from Planetesimals," G. W. Wetherill, in the *Scientific American*, June 1981.

Ch. 21, Fig. 10: from "Double Stars," copyright 1978 D. Reidel Publishing Co.

Ch. 22, Fig. 14: from "Ultraviolet Astronomy" Leo Goldberg, in the *Scientific American*, June 1969.

Ch. 24, Fig. 11 & 12: from *Annual Reviews of Astronomy and Astrophysics*, vol. 5, copyright 1967.

Ch. 25, Fig. 27: from *Nature*, vol. 276, p. 476 (1978), copyright 1978 Macmillan Journals Limited.

Ch. 25, Fig. 24: Smithsonian Institution Photo No. 79-6163

Ch. 26, Fig. 5: from *Astronomy and Astrophysics*, vol. 105, p. 164, (1982), article by H. A. Mayer-Hasselwander, K. Bennett, G. F. Bignami, R. Buccheri, P. A. Caraveo, W. Hermsen, J. L. Masnou, J. A. Paul, K. Pinkau, B. Sacco, L. Scarsi, B. N. Swanenburg, and R. D. Wills

Ch. 26, Fig. 23: from *The Structure and Evolution of Normal Galaxies*, copyright 1981 Cambridge University Press.

Ch. 29, Fig. 2: from *Nature*, vol. 197, p. 1038 (1963), copyright 1963 Macmillan Journals Limited.

Ch. 29, Fig. 12: from *Nature*, vol. 290, p. 366 (1981), copyright 1981 Macmillan Journals Limited.

Ch. 30, Fig. 5: from *Science*, vol. 199, p. 377 (1978), copyright 1978 American Association for the Advancement of Science.

Color Plate 26: from *Nature*, vol. 302, no. 5909 (1983), copyright 1983 Macmillan Journals Limited.

# ASTRONOMY

# CHAPTER 1

# The Realm of Astronomy

Astronomy is fascinating and exciting. It can be a wonderful hobby and an exhilarating area of research. For millenia, man has watched the heavens and recorded their appearance, noting those aspects that seem unchanging and those that recur with regularity, and yet again those that are transient. By now, astronomical ideas have penetrated so deeply into our society, into our ways of thinking, and into our language that we often do not notice the extent of this dependence. These days, it is black holes that capture our imagination; in earlier times it was the observation of an eclipse or the passage of a comet that carried with it the threat of cosmic forces only vaguely perceived. Astronomy has provided the concepts that were essential to the construction of what we now call "the scientific method" and the phenomena essential to the development and testing of powerful scientific theories such as Newton's laws and Einstein's theory of relativity.

Astronomy is accessible to us in more ways, at more levels, than many other parts of science. With very modest expense we can build or buy our own telescopes and take our own photographs of the Moon. As our interests deepen, we can probe the mathematical side of astronomy. Whatever our level of interest, we can find a comfortable niche and a steady supply of books that can keep us abreast of the most exciting of the recent discoveries.

What are these discoveries? What can remain to be discovered after centuries of watching? We make discoveries because we learn to look in new ways or learn how to reanalyze older data. Until quite recently, astronomy depended on only the visible part of the spectrum. Then came the first infrared observations (of the Sun), then radio astronomy, and now with balloons, rockets and satellites, we can detect the ultraviolet and X-rays and gamma rays that do not pass through the atmosphere to our observatories. With each new part of the spectrum being examined, we have found new kinds of celestial objects, some radiating far more strongly in the infrared, others emitting X-rays.

Where are the cosmic frontiers? Black holes should probably be listed first. We think that we know how to describe them: They are the collapsed remnants of some stars, so dense and compact that their gravitational attraction prevents even their own light from escaping—and because none of their light reaches us from their surfaces, we call them "black."

For most people, black holes will probably remain as a bizarre and hazy notion, but the exploration of our solar system has gained an immediacy and a reality through television. No complicated mathematics, special effort, or understanding are needed in order for us to be awed by the sight of men on the lunar surface, or by the bleak landscapes of Mars as seen by the Viking landers, or the swirling atmospheric patterns of Jupiter seen by Voyager on its way to Saturn and its intricate rings. These are more than simply spectacular sights—we are in the process of truly discovering more, far more, about the solar system and its history.

In later chapters more detail of these discoveries will be given: distant galaxies, quasars, supernovae, new moons for Jupiter, rings for Uranus, and the list goes on. We shall do more than simply describe them: We are now in the position to understand and explain many of our observations and fit them into logical patterns. This organizing and theorizing can occur only because there is a secure base of modern physics and chemistry. Without those fundamental sciences, astronomy would be little more than a catalog and a set of geometrical rules as it was for thousands of years until the application of the telescope and spectroscope. We shall, therefore, at all stages try to answer the question "how do we know . . .?," for without this questioning, our astronomical surveys will not rise above the level of an entertaining pastime.

We have no record of the earliest systematic observations of celestial happenings. Even the oldest Babylonian clay tablets with their wedgelike indentations are already detailed and completed tabulations of projected lunar and planetary positons that could only have been calculated after the painstaking accumulation of sightings for very many years and the testing and rejection of erroneous ideas. Alexander Marshack has suggested that certain inscriptions on bones dating even further back, around 25,000 years ago to the end of the last ice age, might represent the record of months and lunar phases. If this interpretation is correct, then these bones would be the

**Figure 1.1** Bone tool and chipped quartz point found at Ishango, Zaire. The marks on these bones are in groups, as though being used to keep record of some counting (© Alexander Marshack).

**Figure 1.2** Clay tablet with cuneiform writing describing the total solar eclipse of 136 B. C. (Reproduced by courtesy of the Trustees of the British Museum).

oldest written records. Somewhere between that distant time and the development of settled areas such as those in the Middle East, enough care was taken to be able to draw the distinction between the stars and the planets. In that hazy region of prehistory, the classic planets were identified: Mercury, Venus, Mars, Jupiter, and Saturn. Neither cuneiform tablet nor hieroglyph proclaims the discovery of a new planet—those five were already known more than 5000 years ago, and the next planet would not be added until 1781, with the accidental discovery of Uranus.

As with other branches of science, astronomy serves society's needs. If an emperor was uneasy at the prospect of a bad omen in the heavens, to whom should he turn for assurances, for predictions (so that appropriate prayers could be said), and for the detailed night-by-night records? When a developing society needed reliable information for the planting of crops, to whom were the calendar and the observations entrusted? For the primitive oceanic societies, who should preserve and transmit the astronomical traditions that permit navigation well beyond sight of land? Although much of this might not qualify as astronomy in the modern sense, it required the systematic observing of the heavens—the sort of systematic observation from which all science stems.

Some ancient needs have persisted. An accurate calendar and time service are essential today, and the U.S. Naval Observatory collaborates with the Royal Greenwich Observatory to produce the annual publication *The Astronomical Almanac*. The Naval Observatory provides an accurate time service. For 50¢, you can call 900-410-TIME from anywhere in the United States, and listen to the time signals.

This is a good place at which to pause and to recognize a very fundamental limitation in astronomy. Unlike most other sciences, astronomy is largely observational rather than experimental. In experiments, we plan every detail in order to isolate those factors that interest us. When we find unanticipated results, we at least know which quantities were being kept fixed and which were being changed, and by how much.

**Figure 1.3** Isaac Newton's telescope. This telescope, now in the possession of the Royal Society of London, was thought to have been made by Newton in 1671. Recent examination indicates that the 2 in. diameter primary mirror, a copper-tin alloy with a trace of arsenic, is probably original. The rolled and pasted paper tube was probably made much later, perhaps following descriptions of the original instrument (Royal Society of London).

We cannot do this in astronomy. We cannot, for example, construct a new planet and observe its evolving atmosphere. We cannot change the temperature at the center of the Sun and then examine the changes in the solar spectrum. We cannot invent a new universe. The best we can do is to examine many stars, looking for regularities and trends, assuming that some stars are older than others. We can *construct* hypothetical stars and follow their aging through computer calculations, making many assumptions and testing our results against actual observations. We can be technically sophisticated and inventive as we build our telescopes and spacecraft, but we remain observers.

As astronomers sought to describe the universe they saw, they developed their own language, giving specialized meanings to everyday words and inventing new words where no convenient or precise term could be found. The Glossary at the back of this book contains some terms that we shall need, but a few examples may be of interest at this time.

Even a casual glance at the sky shows that different objects are very different in their brightness. Astronomers classify stars by their **magnitudes**, a term dating back to Hipparchus, who compiled one of the earliest star catalogs around 130 B.C., and to Ptolemy (about 250 years later) through whose writings Hipparchus's listing have survived. Ptolemy's catalog contained about 1000 stars, between first and sixth magnitudes.[1] It is still convenient to refer to the bright stars as first magnitude, fainter stars as second or third or twenty-third magnitude. On this scale, sixth or seventh magnitude is about the faintest star that can be seen by the unaided eye, but by using a large telescope we can detect objects fainter than twenty-fifth magnitude.

Distance measurements play a critical role in astronomy, but it was not until 1838 that reliable values were obtained for the distance of any star other than the Sun. The astronomical unit of distance now widely used is

[1] Shih-shen (fourth century B.C.) is said to have compiled a list of stars and constellations, but only fragments have survived through other authors and many identifications are ambiguous. Ptolemy's catalog is the oldest accessible catalog that we have.

**Figure 1.4** Lick Observatory on Mt. Hamilton, California. Observatories now often comprise a cluster of telescopes located well away from major cities (Lick Observatory).

**Figure 1.5** Cerro Tololo Inter-American Observatory in Chile. Located in the southern hemisphere, these telescopes can observe parts of the sky not visible from the U.S. The largest dome houses the 4 m telescope (CTIO).

**Figure 1.6** Radio telescopes in the Very Large Array (VLA) located near Socorro, New Mexico. Twenty-seven of these dishes, each with a diameter of 25 m can be moved along three tracks arranged in the form of a large Y (NRAO).

the **parsec** (pc), a compound of **parallax** (the method used) and **second** (a small fraction of 1 degree of angle).[2] A more picturesque unit is the **light-year** (LY)—the distance travelled by light in a year. (One parsec is about 3.26 LY). Both of these distance units are relatively new. Whereas the pc and LY find most use when we refer to stars, other distances are so much larger or smaller that different units are more convenient. Thus, megaparsec (million pc or Mpc) and kiloparsec (thousand pc or kpc) come in handy for very distant objects. Closer to home, within the solar system, a smaller and more convenient unit of length is provided by the Earth's orbit around the Sun. This orbit is not circular but oval, and its longest diameter is defined as 2 astronomical units (AU). There are 206,265 AU in 1 parsec.

Parsecs and astronomical units are part of the working language of astronomers and are often far more convenient than their equivalents in centimeters or kilometers. It is obviously much easier to say that Jupiter is 5 AU from the Sun than to say that it is 65,000,000,000,000 cm. But that allows us to introduce the matter of **scientific notation**. The numbers used are indeed astronomic, and no one would want to write 65 and a string of zeros on a regular basis. There *is* a method that is not only more convenient but far less liable to error when it comes to calculations. Scientific notation involves powers of 10, and thus we would write Jupiter's distance as $6.5 \times 10^{13}$ cm or $6.5 \times 10^{8}$ km. This topic may be familiar, to many of you, in which case you may take this as a reminder. However, because of its importance, it is covered separately in Appendix B, for those of you to whom it is new or unfamiliar. Scientific notation will be used extensively.

What is the goal of astronomy? A driving force behind much scientific research is curiosity. We would like to know, in detail, how stars form and evolve, how our solar system formed, how stars group into galaxies, and

[2] We still use the system invented by the Babylonians, dividing each degree into 60 min and each minute into 60 sec. In Latin, these parts were called **partes minutae primae** and **partes minutae secondae** from which present terms are derived. Decimals would be so much easier for angles, but that innovation of the French Revolution did not survive.

**Figure 1.7** Satellite HEAO-2 ("Einstein") being assembled at TRW for launch in 1978. X-ray telescopes must be carried above the Earth's atmosphere in rockets or satellites (TRW Electronics and Defense).

**Figure 1.8** Research balloon during inflation with helium. This balloon carried cosmic-ray detectors to an altitude of about 25 mi (Herb Weitman, Washington University).

**Figure 1.9** Press conference to announce results of the Voyager 2 encounter with Saturn. Space probes and television have changed the way in which scientific discoveries now reach the public (NASA).

even how life began. The other major aim is more practical; with whatever knowledge we can accumulate, we would like to regulate our activities or exert more control over more of our environment. In this sense, there is little difference between watching for the annual appearances of the brightest star, Sirius, in order to predict the flooding of the Nile and searching for possible correlations between the solar activity and the weather on Earth.

In the following chapters, we shall systematically trace out the impressive successes of astronomy and its related sciences, and we shall point to still-unsolved puzzles. We shall see how a systematic application of scientific methods has allowed us to assemble a most remarkable understanding of our universe, even as we recognize the need for yet more research to solve those puzzles. We can also expect our activities to lead to the recognition of new puzzles and the formulation of new research programs.

## PROBLEMS

**1.** Astronomy and astrophysics now require us to deal with distances as small as the atomic nucleus and as large as millions of light years. What is the range of dimensions that you have personal experience with: How small an object have you seen, and how great a distance have you travelled?

**2.** Review of scientific notation: Write the following quantities in scientific notation with powers of 10 (check appendix B for review).

a. population of the United States, two-hundred and forty million

b. planned defense expenditure for the next 5 years, one trillion dollars

c fraction that 1 ft is of 1 mi

d. number of seconds in a day.

**3.** The speed of light is often listed as 186,000 mi per sec. Convert this to miles per hour and cm per sec. How many seconds are there in a year of $365\frac{1}{4}$ days? How far will light travel in 1 year?

**4.** Make a list of astronomical terms that are used for everyday items: Quasar TV, comet cleanser, . . . (check the index or Glossary for additional terms). Have other sciences been used in similar ways?

**5.** Why has astronomy been termed an observational science rather than an experimental science? How and to what extent has this been changed by space exploration?

**6.** List three astronomical objects that you have come across within the past 6 months on radio, TV, newspapers, or magazines. Check the index to this book to see whether they are included. If so, read those sections. Can you follow the descriptions? Come back to this question at the completion of this course.

**7.** We often take familiar things for granted. What evidence do you have that we don't have a new Sun every day? How do you know that the Sun does not cease to exist after sunset? Try to put yourself in the position of someone a few thousand years ago, without access to telescopes, radio, etc.

**8.** Astronomy has attracted much attention in recent years, but can you name five or more twentieth-century astronomers?

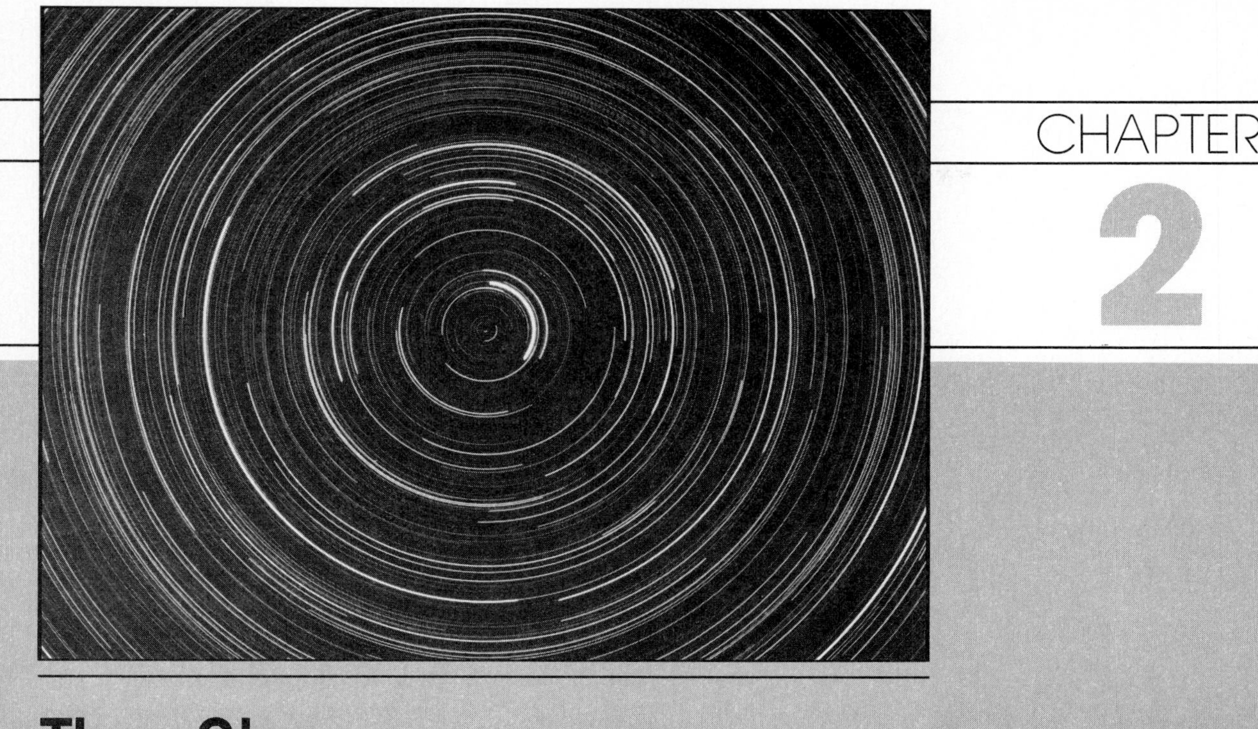

CHAPTER

2

# The Sky

## 2.1 SIMPLE OBSERVATIONS

In this chapter we will review those features of the heavens that you can observe for yourself, without the need for a telescope or even binoculars. You can repeat some of the simple observations made by prehistoric man and look for the patterns and regularities that have been recognized for so long. In Chapter 3, we will have a look at elaborate structures such as Stonehenge, some that were erected as far back as 4000 years ago. Recent surveys have suggested that many of these ancient sites were laid out with great care, incorporating a knowledge of the heavens and their changes that we now find surprising. In principle although you can repeat all of the observations required for laying out a Stonehenge, some require patient observation over many years, far longer than the length of your introductory astronomy course. Accordingly, in the present chapter, we shall describe those happenings that you can easily see and those that you could note if you too made records for many years.

How many of the following questions can you answer? Which bright stars or constellations will be visible in the south at midnight tonight? What is the phase of the moon tonight—new, quarter, or full? Where on the horizon will the Sun and Moon rise tomorrow? Unless astronomy has been your hobby, you will probably not be able to answer all of these.

We no longer use the stars and constellations to tell us the date, nor do we need the Sun or Moon for timekeeping—we have wristwatches whose accuracy is comparable to that of national time standards of not so many years ago. We do not need to keep track of the phase of the Moon, for nighttime travel is no longer concentrated into the few days around full

Moon, as was that of the biblical Israelites, whose major festivals and gatherings were carefully tied to a lunar calendar.

Indeed, we no longer *need* to be aware of these phenomena, for we have, over the centuries, devised far better and more convenient alternatives to such an extent that we seldom look up at night for any reason other than wanting to see whether it is overcast and likely to rain. (You can even avoid looking at the sky for that purpose—most towns have phone numbers you can call for time, temperature, and weather forecasts. But how many towns have phone numbers to call for astronomical information?)

And so, as in so many other ways, we have become less observant. The growth of cities and industry has not helped. The combination of urban pollutants and more and brighter street lights makes it much harder or impossible to see faint objects from some locations. In a city, look at the sky on a typical night, and try to compare its appearance after a rain or a crisp frost when the air is temporarily clean. In addition, urban horizons tend to be obscured by trees and buildings and haze. If you have not already done so, look at the heavens from a place well away from towns. You can then begin to understand why, even beyond their practical uses, these heavens have held such a fascination for man and provided the basis for some of his myths and beliefs.

Life might be very different without the Moon and its influence on the tides, and we could get along perfectly well without the other planets, but without the Sun and its energy we would not be here. During daylight hours, the Sun is so bright that even the small fraction of its light that is scattered in our atmosphere effectively blots out almost all other heavenly objects,

**Figure 2.1** Trails produced by stars close to the celestial equator. These paths are very large circles but appear to be straight when observed over only a short distance (Lick Observatory).

**Figure 2.2** Star trails around the north celestial pole. The North Star, Polaris, is not precisely at the pole, and so it too traces out a circular path each night (Lick Observatory).

with the exception, of course, of the Moon and at times near sunrise or sunset the planets Mercury and Venus. But, after the Sun sets, a very different sky presents itself. Watch the sky for 2 or 3 hours around sunset. At first, only a few bright stars will be visible, then gradually fainter and fainter stars come into view, as the Sun moves further below the horizon and the sky darkens. In some parts of the sky, there seem to be no bright stars, in others there are enough to form patterns.

Repeat your observations over many nights and you will begin to see changes in the sky. Once the Sun has set and the sky has darkened, note the positions of some bright stars in the east or south, and look for them again 3 or 4 hours later. They will have moved, following curving paths that arch from east to west, but as they sweep across the sky they keep their spacing, and they do so night after night. As they rise from the eastern horizon, they seem to follow diagonal paths that show up clearly in a time exposure. Later, as the stars approach the western horizon, you will observe a similar effect (Fig. 2.1). Some stars, however, follow different paths: if you look north (in the northern hemisphere), some stars never set but follow counterclockwise paths around a point in the sky (Fig. 2.2). (Time exposures show us patterns that cannot be seen directly by our unaided eyes: stars that are too faint to be seen unaided, and the orderly stellar paths that we would otherwise have to infer from repeated measurements.) Some stars slightly further from the pole have paths that rise and set, dipping below the northern horizon for only a few hours.

The **celestial pole** is that point around which the stars seem to be moving, and it lies along the extension of the Earth's own axis of rotation. There is no celestial marker precisely at the pole. The star Polaris is the closest star, slightly less than 1° from the pole, providing a useful indicator for north on a clear night. Over the centuries, the Earth's axis moves slowly, tracing out a circle on the sky and taking 26,000 years to complete its circuit. This **precession** has been known for a long time; one effect of the precession is that around 3000 B. C. the pole was at the star Thuban; many thousand years to come it will be at the bright star Vega, and for long periods there is no bright star near the pole. Take advantage of its while it lasts!

**BOX 2.1**
*Polaris and the Big Dipper*

Polaris is quite easy to locate, and you can do this at the same time that you begin to recognize the celestial landmarks (skymarks?) of the brightest stars and most prominent constellations. Look for the group of stars shown in the figure comprising the Big Dipper, known in England as the Plough (Fig. 2.3a). The Dipper rotates around the pole with the two pointers about 5° apart, 30° from the pole, and always pointing to it. This group makes a complete circuit every 24 hours. The figure shows how the Dipper will appear early in the evening at different seasons (Fig. 2.3b).

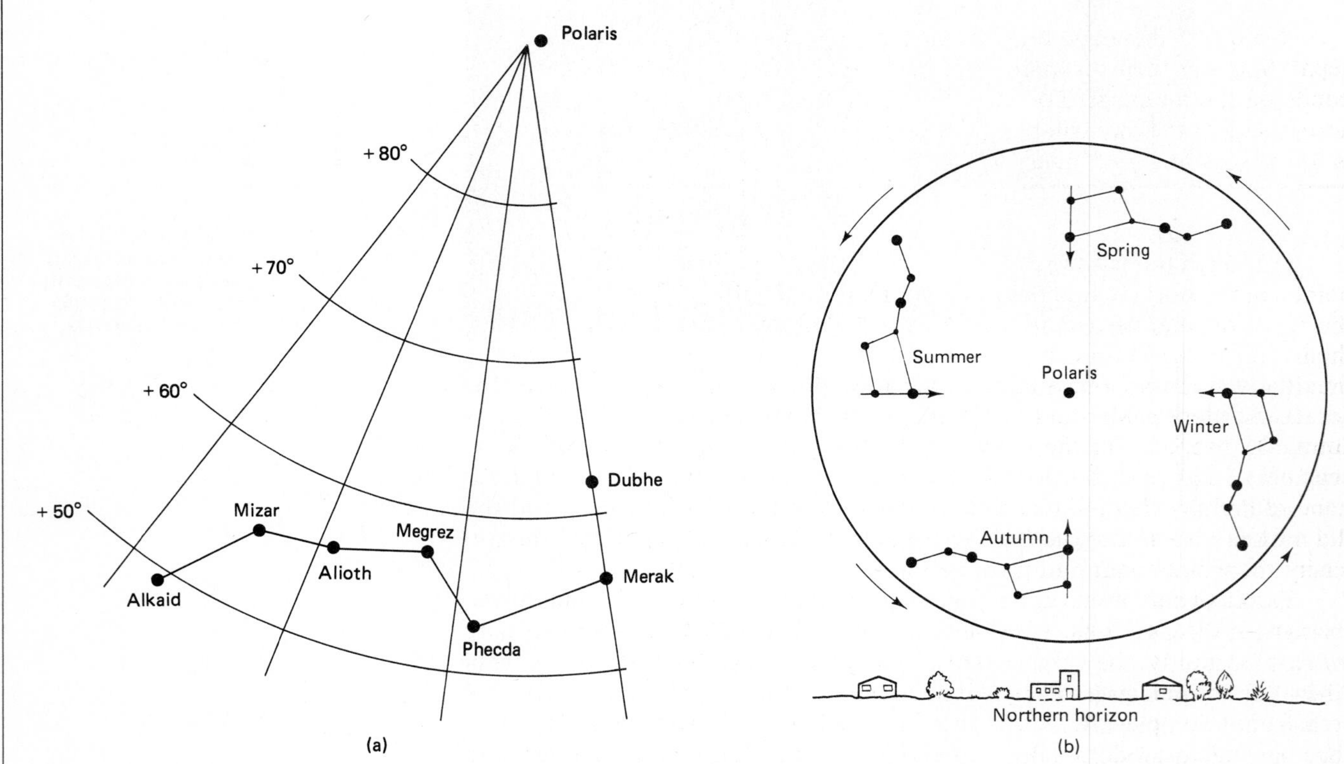


**Figure 2.3** (a) The Big Dipper. This well-known group of stars is easily located on a clear night. (b) The position of the Dipper changes each night and also with the seasons. This diagram shows the position of the Dipper around 9 P.M. for each season.

**BOX 2.2**
*Angles in the Sky*

You will find it easier to locate objects in the sky if you can estimate angles, even very roughly. A typical ball point pen has a diameter of about $\frac{3}{8}$ in. At an arm's length, 25 to 30 in. from your eye, the angular diameter of the pen is about $\frac{3}{4}$°. This is slightly larger than the $\frac{1}{2}$° diameter of the Sun and the Moon. A finger at an arm's length is about 2° wide. A 3 × 5 in. file card provides another useful measure: Its long dimension covers about 10° at an arm's length (Fig. 2.4).

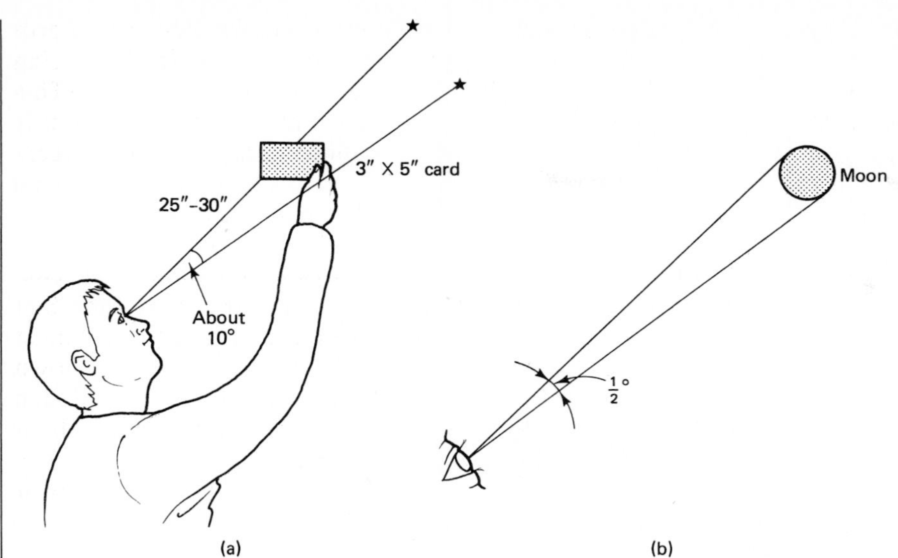


**Figure 2.4** (a) You can estimate angular distances in the sky by using a 3 × 5 in. file card as a 10° standard. (b) Both of the most familiar objects, the Sun and Moon, have angular diameters of close to half a degree.

Note that the angle between two objects depends on both their actual separation and their distance from you. Similarily, an angular diameter depends on the actual size of the object and its distance. Thus an object 1 ft across and 25 ft away will have the same angular size as another object that is 3 ft across and 75 ft away.

Although the patterns of the stars remain fixed, their rising and setting points on the horizon and their maximum altitudes above the horizon during the night depend on where we are on the Earth's surface when we view them—move further north and you will find that some stars that did not rise much above the southern horizon are now not visible at all. Move further south and those same stars rise higher while new stars now appear near the southern horizon. On the equator, all parts of the sky can be seen at all seasons of the year, but for every other latitude of observation, there is a zone of the sky that is invisible. Growing up in the southern hemisphere, I did not see Polaris and the Big Dipper, but the Southern Cross was our most easily recognized star pattern.

Make observations of the sky at intervals over a year and you will see changes and repetitions. Five of the stars twinkle less than the others, and more importantly, they change positions against the background of the stars. These are the **planets** (from the Greek word for *wanderer*). Their wanderings, regular but complex, were observed for millenia and aroused the curiosity that has led to modern astronomy. The starting point was the *recognition* that these planets differ from the stars. There must have been careful recording of their positions night after night not originally with the thought of constructing a theory but out of curiosity. It is difficult to put ourselves in the position of ancient man, when the different nature of the planets was first dimly suspected. At that time, there was no clear distinction between the different sorts of events that occurred well above the Earth's surface. With unquestioning confidence today we distinguish the planets from the stars and have no hesitation in keeping these apart from comets, meteors, auroras, lightning, and clouds. This was not always the case and originally "meteorology" covered *all* "things on high", for there was no way of telling just how far above the Earth's surface these were.

This ancient confusion in the heavens has been well recorded. The Astronomical Bureau in China was headed by the Grand Historian who had

**Figure 2.5** Constellations around the Zodiac, as shown in woodcuts in the 1520 *Compilations* of Leopold of Austria (Yerkes Observatory).

among his other functions the recording of "guest stars" of all types. His office, which existed under different names for 2000 years (until 1911), accumulated records that are now gold mines to modern astronomers seeking records of ancient comets and exploding stars. Great care must be taken, however, in transcribing these records, for positions in the sky were given in terms of Chinese constellations whose boundaries are not precisely known. As is the case with many guests, some guest stars are more interesting than others.

And so, night after night, the stars in their patterns drift across the sky. Some of the patterns lend themselves to fantasy and imagination, providing a twinkling framework for fanciful and extended figures of people and animals—the **constellations** (Fig. 2.5). Products of a more mystical age, the constellations have survived, providing signposts by which we still label the regions of the sky and the stars in them. The oldest constellations in western astronomy are only in those parts of the sky that can be seen from about 35° north latitude, that is, the latitude of Cyprus, Syria, and northern Iran—ancient Babylon falls in this region. The Egyptians, living much further south, could see other stars and constellations and gave them names that have not survived. The southern hemisphere contellations that are found in modern sky maps date back only to the centuries of exploration by the navigators from Europe.

## 2.2 SIMPLE OBSERVATIONS—THE SUN

What could be more obvious than the Sun's daily passage from east to west? But even this seemingly trivial observation can be used to illustrate how we draw conclusions. As far back as we have any records, the Sun has risen in the east. We would be mightily surprised to find the Sun setting in the

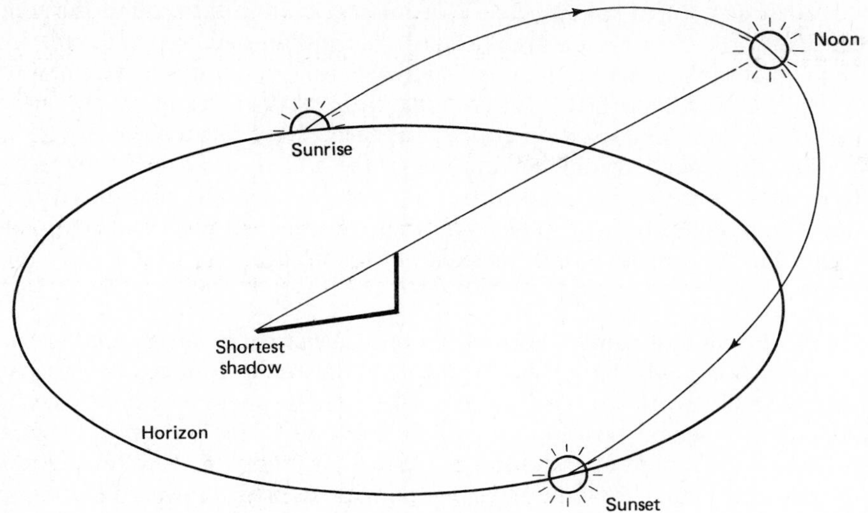


**Figure 2.6** Noon each day is the time when the Sun is highest in the sky, as shown by the shortest shadows. From one noon to the next marks the passage of a solar day.

west one day, then, after 12 hr. of night, rising in the west and setting in the east and thereafter oscillating from east to west. This seems such an absurd possibility as I write this paragraph that you may wonder why I should persist in including it. We all *know* that this scenario is impossible. But is it? A modern explanation might first discuss the rotation of the Earth and then go on to invoke rotational momentum and the law of gravitation—very impressive, but quite unknown to ancient man. All that *he* knew was that the regularity of rising and setting, east to west, came to be expected and accepted. For a long time, this was simply an article of faith although somewhat fragile.

The regularity of the Sun's reappearances is the basis of our familiar measure of time. Each day, the Sun reaches its maximum height above the horizon at noon; a **solar day** is the time interval from one noon to the next (Fig. 2.6). Although the length of the solar day is almost constant (there are small irregularities due to variations in the Earth's rotation rate and orbital speed), the division between light and dark varies markedly through the year, especially at far northern and southern latitudes.

BOX 2.3
*The Day*

*Noon* marks the time of the highest point of the Sun in the sky each day. The time interval from one noon to the next is easily observable and defines the **solar day**. We might think that this interval of time corresponds to one rotation of the Earth on its axis, but when we look at a diagram that shows the positions of the Sun and Earth we find that this is not the case (Fig. 2.7).

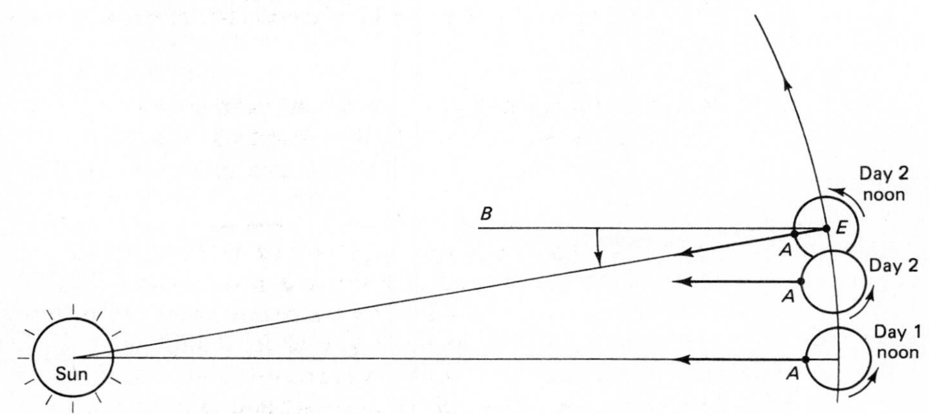


**Figure 2.7** Because of the Earth's orbital motion, the solar day corresponds to more than one complete rotation of the Earth on its own axis. On day 1, it is noon at some place (*A*) on the Earth's surface. After precisely one complete rotation of the Earth on its own axis, A has moved through 360° (day 2), but it will not be noon at *A* until the Earth has rotated through an additional small angle (as shown).

While the Earth rotates, it is also moving in its orbit around the Sun. As a result, the Earth must make slightly more than one complete rotation with respect to the stars before it will again be noon at our observation point. Because the Earth's orbit is not circular, the length of the solar day varies through the year, and its average value is termed the **mean solar day**.

The time needed for one rotation of the Earth, in space, defines the **sidereal day**, and this is found to be $23^h56^m$ in solar hours and minutes. It is most conveniently measured by noting each night the passage of stars through the local north–south meridian of longitude.

As the time of sunrise gets earlier and earlier each spring and the sunsets correspondingly later, the position of the rising Sun on the horizon moves steadily north (as seen in the northern hemisphere). At noon the Sun stands at its highest position in the sky each day. The extremes of these positions are reached at the **summer solstice** (around June 21); for the next 6 months the trend is reversed until at the winter solstice (around December 21) the sunrise and sunset points are farthest south, the duration of the daylight is shortest, and the maximum height of the Sun at noon is least. Midway between the solstices are the **equinoxes** (spring and autumn, around March 21 and September 21, respectively) when the Sun rises due east and sets due west, and day and night are of almost equal length (Fig. 2.8). These daily and seasonal changes can easily be noted with a sundial, perhaps the

**Figure 2.8** The rising and setting points of the Sun move along the horizon through the year. This diagram shows the sunrise positions through the year computed for 40° north latitude, typical for most of the continental U.S.

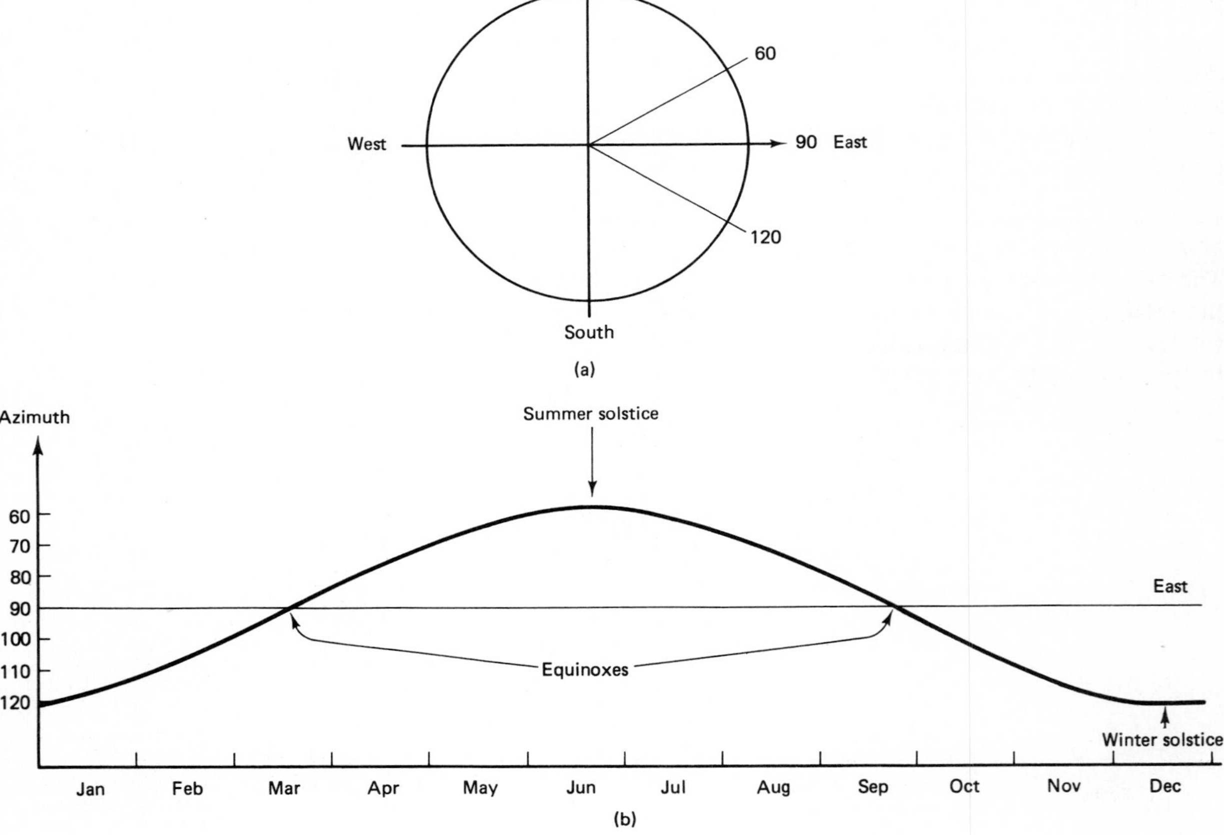

oldest scientific instrument. A pointer casts a shadow whose position indicates the time of day.

Because the daytime sky is so bright, it is necessary to look at night for the slow seasonal drift in the stars that are visible. After 1 year, we will find the same bright stars and their patterns back again in the same places at the same time of night. (See for example Fig. 2.3 showing the seasonal shift in the position of the Big Dipper.)

We can also look for the brightest stars that are visible at twilight in the east shortly before sunrise or in the west soon after sunset. Although we cannot see the Sun's starry background at noon, we can see those stars that are fairly close to the Sun when it is just below the horizon. In this way we will find that the Sun seems to move against the background of stars slowly by about 1° each day completing its circuit in a year.

Some diagrams (Fig. 2.9) show how the relative positions of the Earth, Sun, and stars determine what part of the sky we can see at different times of the year. The path apparently followed by the Sun through the stars is the **ecliptic**. In the modern (correct) view, we would say that it is the Earth's

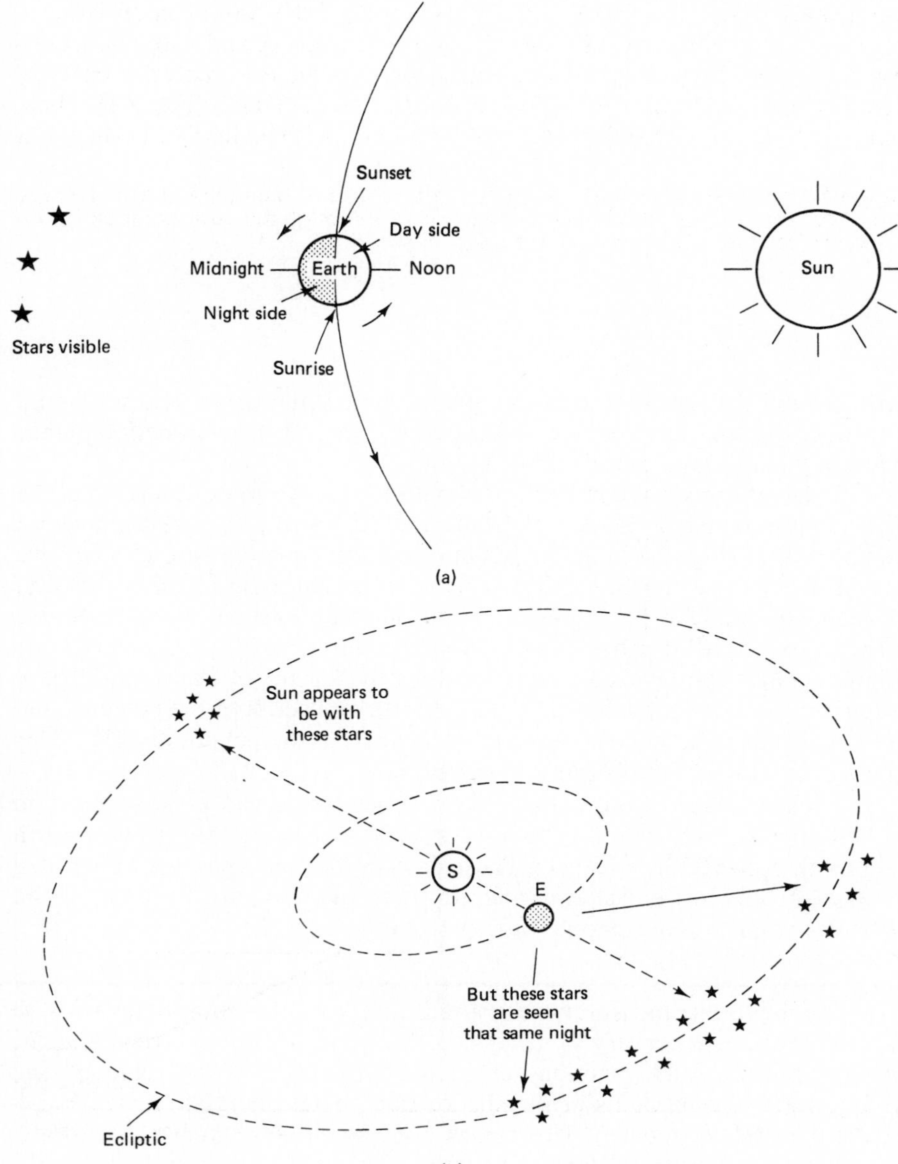


**Figure 2.9** Because of the Sun's brightness, stars cannot usually be seen in the sky on the daylight side of the Earth. Shortly before sunrise and shortly after sunset those stars closest to the Sun can be seen, but most viewing of stars is at night in directions away from the Sun.

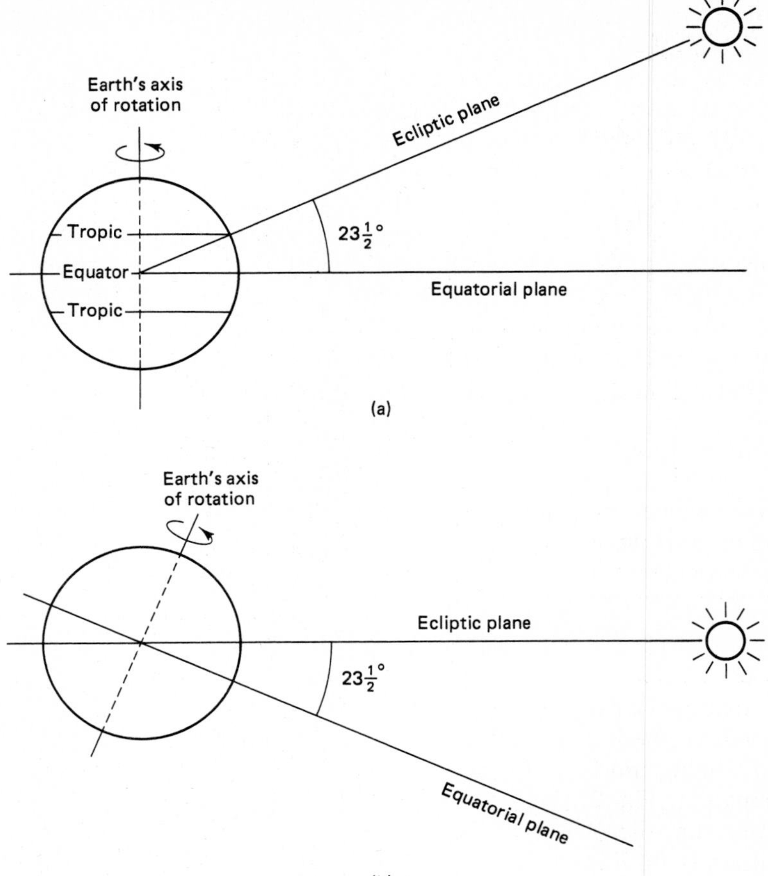


**Figure 2.10** The Earth's orbit around the Sun defines the ecliptic plane. The Earth's equatorial plane is tilted at 23½° with respect to the ecliptic. The Sun will appear overhead only at places that lie between 23½° latitude north and south but will never be overhead at places further from the equator.

orbit around the Sun that defines a plane, the **ecliptic plane**, and we would also note that the orbits of the other planets around the Sun define planes that are mostly quite close to the ecliptic.

Some of the apparent motion of the Sun that we have described is the direct consequence of the ecliptic plane and the Earth's equatorial plane not being parallel (Fig. 2.10). If the ecliptic and the equator were parallel, we would always see the Sun following the same daily path across the sky. Instead, we see seasonal changes. At the summer solstice, the noon Sun is directly overhead at those places along the Tropic of Cancer, and at the winter solstice the favored places lie along the Tropic of Capricorn. These tropics are lines of latitude that are 23½° north and south of the equator, and we could thus infer that the equator is inclined to the ecliptic by 23½°. This angle is known as the **obliquity** of the ecliptic.

People closer to the equator than either tropic can observe the Sun directly overhead at noon on two days each year as the Sun travels north and south between the solstices. These days of the **zenith passage** were noted by the Mayans in Central America, and were as important to them as the solstices were to other people further from the equator.

### BOX 2.4
*Measurement of the Obliquity of the Ecliptic*

You can make this measurement using a simple method that dates back to at least the fourth century B. C. (Fig. 2.11). Set a tall pole vertically in the ground, and note the positions of its noon shadows at the solstices and equinoxes. Measurements of the shadow lengths at noon allow you to calculate the angles indicated. This can be done by either making a scale drawing or using trigonometry for a calculation. The angle between the two sols-

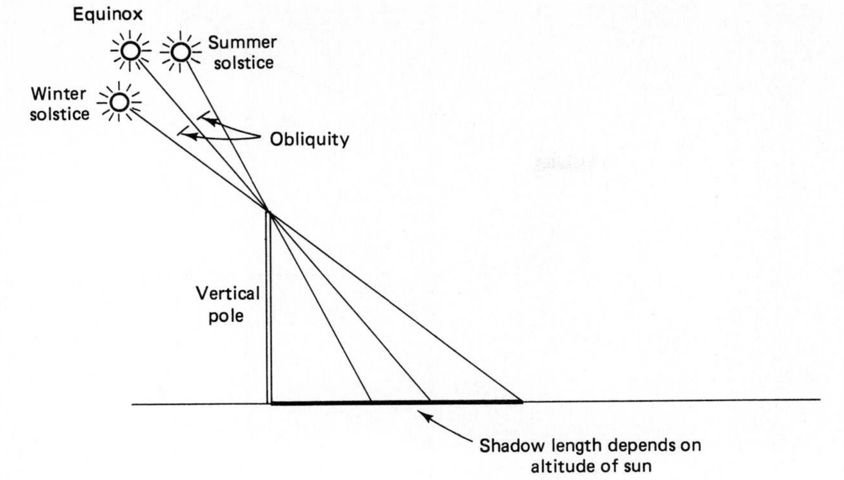


**Figure 2.11** The angle between the ecliptic and the Earth's equatorial plane is the obliquity of the ecliptic, and it can be measured by noting the positions of the longest and shortest shadows cast at the winter and summer solstices, respectively.

tice shadows is twice the obliquity, but this requires that you make measurements in midsummer and midwinter. It may be more practicable to make your measurements at the equinoxes and the winter solstice.

## 2.3 SIMPLE OBSERVATIONS—THE MOON

We are so accustomed to seeing the Moon that we can too easily overlook the immense achievement of realizing that the Moon shines because it reflects the Sun's light, and that the phases of the Moon have a simple geometrical explanation (Fig. 2.12).

Unlike the Sun, the Moon *can* be seen at the same time as the stars, and its movement is sufficiently rapid to be seen within even a single night. If you see the Moon close to a bright star, look at them again a few hours later. Their separation should have changed noticeably. Seen against the background stars, the Moon moves about 13° each day and takes 27.3 days to return to the same place in the sky. This period is known as the **sidereal** month. There is also another way of defining the month, based on phases of the Moon. From one new Moon to the next is a **synodic month**, 29.5 days.

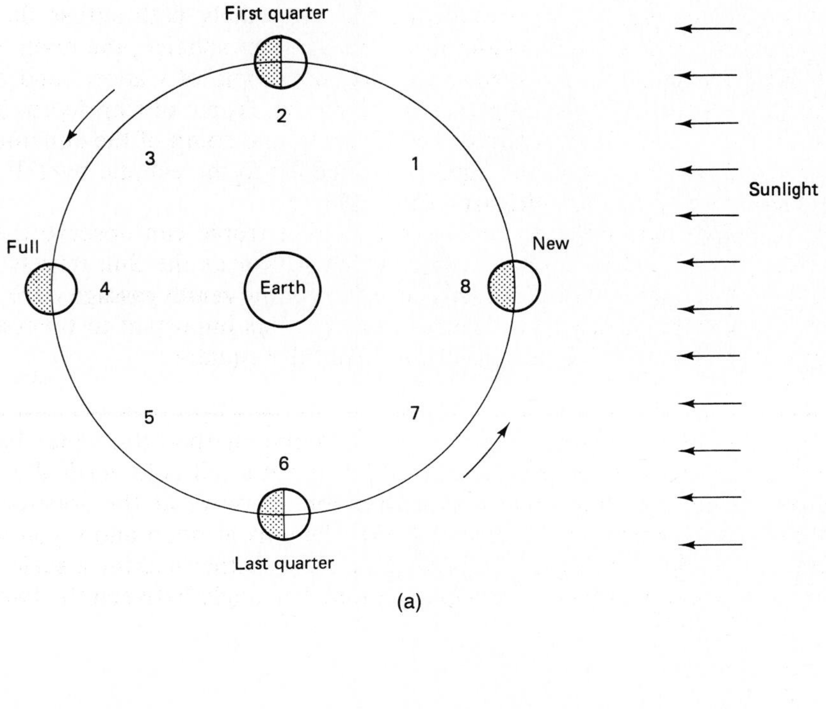

**Figure 2.12** (a) Phases of the Moon. The numbers around the circle correspond to the sequence of photographs shown.

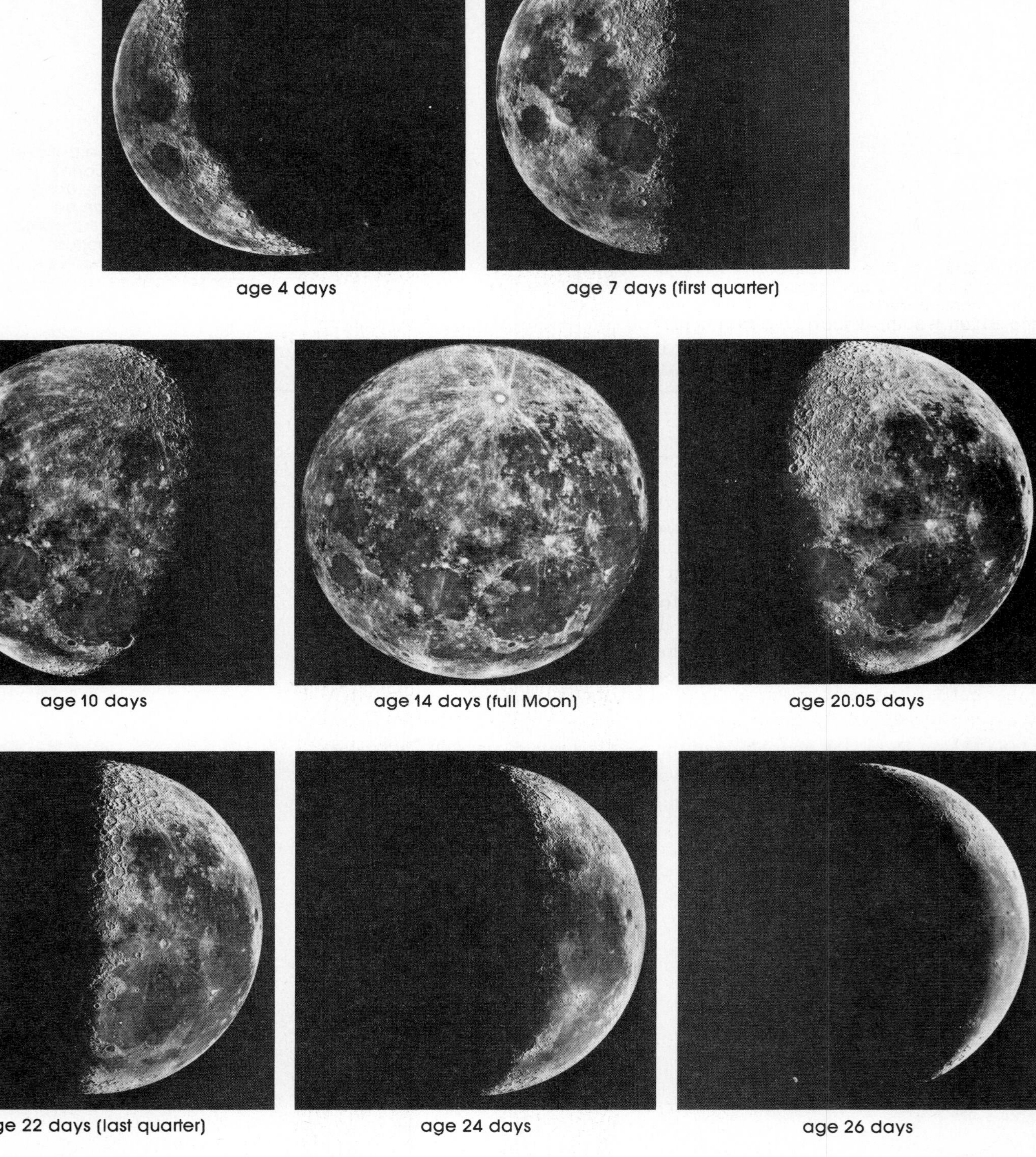


(b)

**Figure 2.12** (b) Sequence of photographs through the lunar month: (Lick Observatory) Many of the features can be seen when heavily shadowed as around first and last quarter. (Note: Because telescopic images are often inverted, photographs may sometimes be mirror images of what would be seen by eye or with binoculars.)

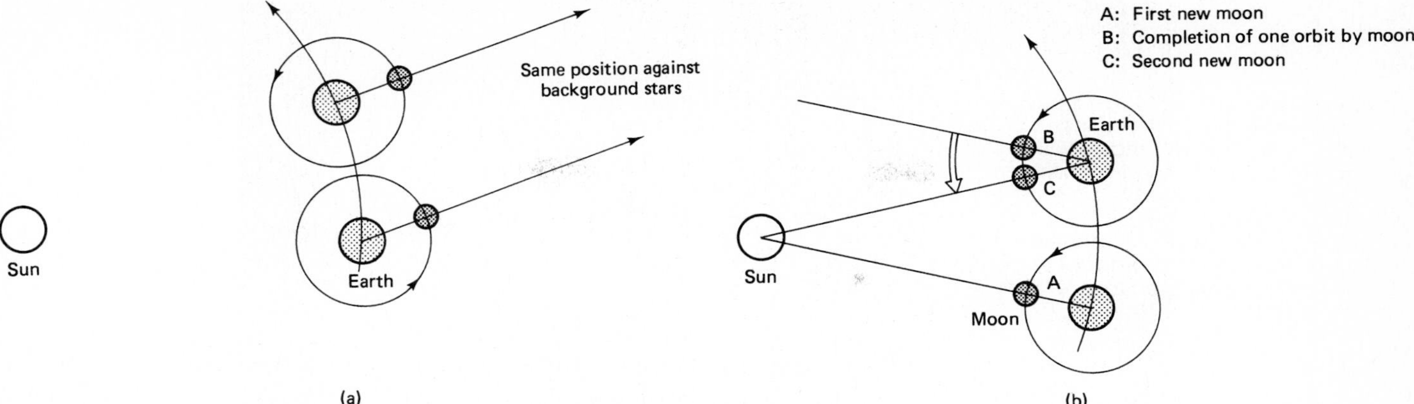


**Figure 2.13** (a) One complete orbit of the Moon around the Earth brings it back again to the same background stars. This defines one sidereal month. (b) One lunar month is defined by the time from one new Moon to the next; this requires more than a complete orbit around the Earth.

The reason for the difference between these two ways of defining the month is that the Earth is not stationary but is moving around the Sun. The sidereal month depends on the completion of exactly one lunar orbit in space (Fig. 2.13). In contrast, between successive *new* positions that define the (longer) synodic month, the Moon has to travel *more* than one complete circuit of its own orbit.

## 2.4 ECLIPSES

Moving in its orbit around the Sun, the Earth casts a long conical shadow into space where we do not see it. Sometimes the Moon's path takes it through this shadow and we have a **lunar eclipse**. The eclipse can be partial or full, depending on the precise position of the Moon (Fig. 2.14). Because the Moon is much smaller than the shadow of the Earth, the duration of a lunar eclipse can be as much as about 1½ hr. As the Sun and Moon must be on opposite sides of the Earth for a lunar eclipse, this can only occur at full

**Figure 2.14** (a) The Earth's shadow has one region (the umbra) that is completely shielded from direct rays from the Sun and a larger region (the penumbra) where only some of the Sun' rays are excluded. (b) When the Moon passes through the umbra, the eclipse is total, but sometimes the Moon's orbit takes it only partially through the umbra or perhaps only through the penumbra.

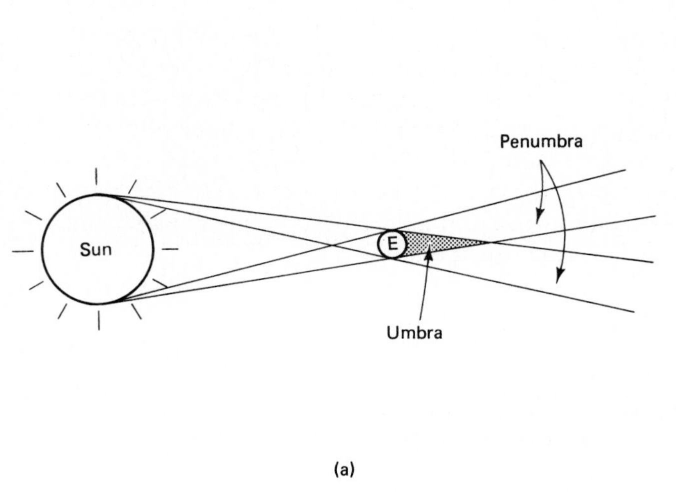


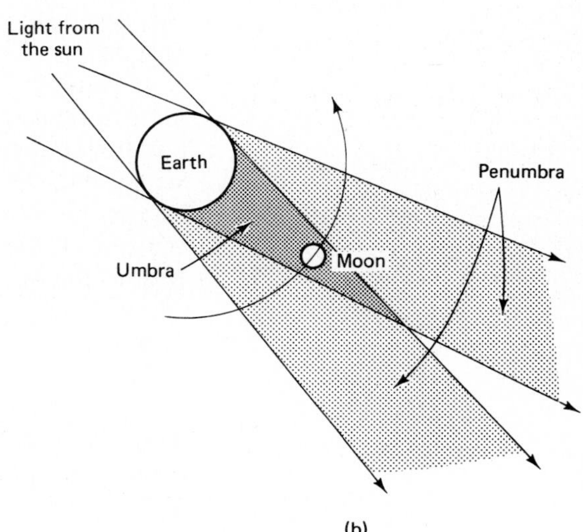

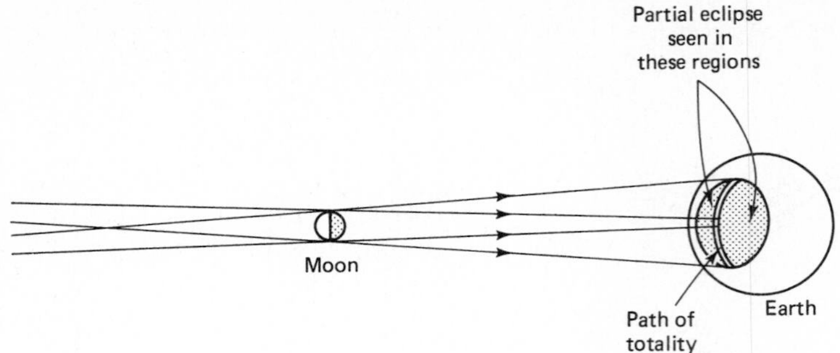


**Figure 2.15** Solar eclipses can be seen when the Earth passes through the Moon's shadow. The movement of the umbra on the Earth's surface marks out a narrow path of totality in which the total eclipse can be seen. Surrounding this path is a much broader region in which a partial eclipse will be observed.

Moon. However, because the Moon's orbit around the Earth is inclined at 5° to the ecliptic, an eclipse does not occur at every full Moon: Usually the Moon will then be slightly above or below the ecliptic.

The Sun, Moon, and Earth can also line up with the Moon in the middle—and we may then see a **solar eclipse**. In this, it is the Moon's shadow that falls on the Earth (Fig. 2.15), and again the eclipse can be full or partial. However, the Moon's shadow on the Earth is small and the duration of the total solar eclipse is therefore quite short. It is set by the size of the Moon's shadow and the speed with which the shadow sweeps across the Earth; $7\frac{1}{2}$ min. is the longest duration at any place on Earth.

Because neither the Earth's orbit around the Sun nor that of the Moon around the Earth is precisely circular, the distances, shadow dimensions, and eclipse durations will vary from one eclipse to the next. Another consequence of these variable distances is the form of the solar eclipse. When the Moon is farthest from the Earth, the full lunar shadow does not even touch the Earth, and observers on the central line of the eclipse can still see a thin ring of the edge of the Sun, an **annular eclipse**.

Lunar eclipses through 1992 are listed in Table 2.1 and the paths of total solar eclipses through 1990 are shown in Fig. 2.16.

*Table 2.1* LUNAR ECLIPSES 1985–1992

| Date | | Type |
|---|---|---|
| 1985 | May 4 | Total[a] |
| | October 28 | Total[a] |
| 1986 | April 24 | Total[a] |
| | October 17 | Total[a] |
| 1987 | October 7 | Partial |
| 1988 | August 27 | Partial |
| 1989 | February 20 | Total |
| | August 17 | Total |
| 1990 | February 9 | Total[a] |
| | August 6 | Partial[a] |
| 1991 | December 21 | Partial |
| 1992 | June 15 | Partial |
| | December 10 | Total |

[a] At mideclipse these cannot be seen from anywhere in the continental United States.

There is an important distinction between solar and lunar eclipses. When the Moon is eclipsed, the event can be seen from almost anywhere on the night side of the Earth. In marked contrast, a solar eclipse can be seen *only* within a very narrow band. The Moon's shadow sweeps along the **path of totality** always less than about 300 km (180 mi) wide. Within this strip, a total eclipse will be seen; in a small adjacent region on each side, the eclipse will be seen as partial—but, outside these parts *nothing*. Because of this great difference in visibility of the solar and lunar eclipses, whenever a lunar eclipse occurs it will be seen by many people, but solar eclipses are rare events at any particular place on the Earth and will be seen by relatively few people (Fig. 2.16).

Another important difference between the solar and lunar eclipses is in their appearances. During a total lunar eclipse, the Moon can still be seen faintly illuminated by sunlight that has been bent by the Earth's atmosphere so that the central shadow is not totally dark. During a total eclipse of the Sun on the other hand, the Moon appears as a dark featureless disk hiding the Sun whose brilliant halo (or corona) can be seen streaming out from behind the Moon. Add this to the dramatic reduction in sunlight, and one can well appreciate ancient fears for the Moon's appetite and the reason for sacrifices to ensure the safe return of the Sun. It is no wonder that solar eclipses should have been so well recorded.

An unusual view of an eclipse was obtained on April 24, 1967. On that day a lunar eclipse was seen as the Moon entered the Earth's shadow. As seen by the lunar probe Surveyor III from the lunar surface, however, this appeared as a solar eclipse (Fig. 2.17). Viewed from the Moon the Earth

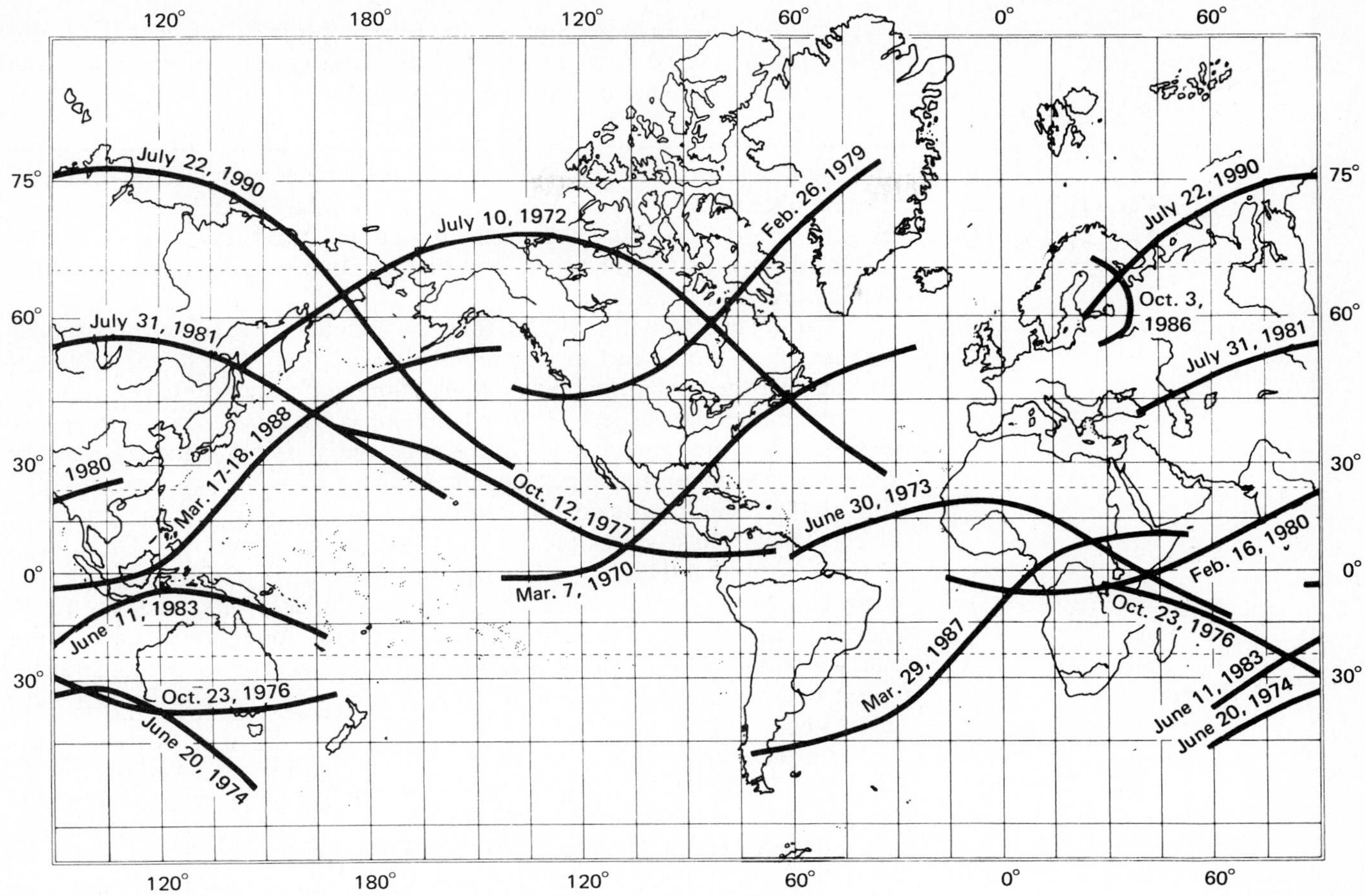

**Figure 2.16** Calculated tracks of total solar eclipses, for the years 1970–1990. No total eclipse will be visible from the U.S. between 1979 and 1990 (NASA).

**Figure 2.17** In April 1967 there was a total eclipse of the Moon as seen on Earth. Viewed from the Moon, however, this appeared as a solar eclipse. These photographs were taken from the Surveyor III on the surface of the Moon, using a wide-angle camera and various colored filters. Filters—(a) and (d): red; (b) and (e): blue; (c) and (f): green (NASA).

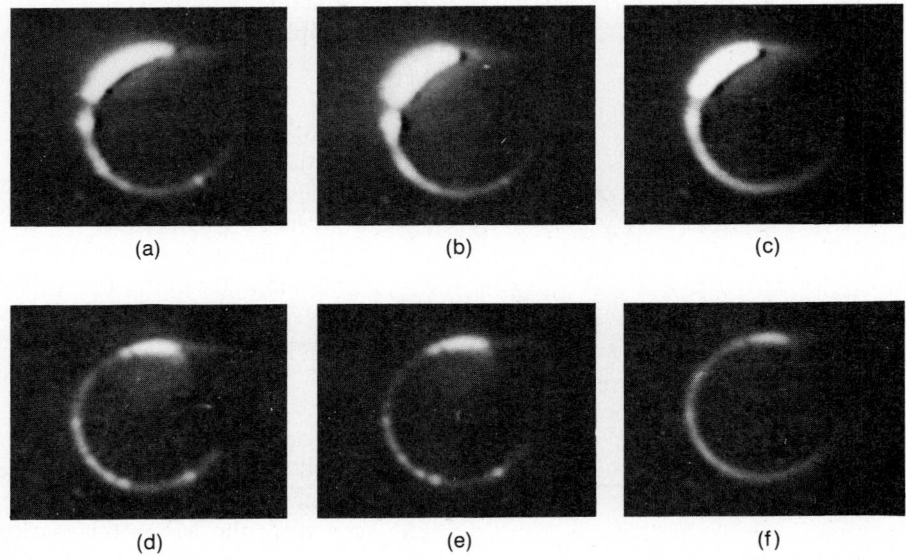

seems much larger than the Sun and the region of totality covered the entire Moon. Refraction of sunlight through the Earth's atmosphere made the eclipse seem somewhat like an annular eclipse.

## 2.5 SIMPLE OBSERVATIONS—THE PLANETS

The classic planets—Mercury, Venus, Mars, Jupiter, and Saturn—draw attention to themselves when repeated observations are made over even a few months. Not only do they change their positions, but they are usually conspicuously as bright as the brightest stars, their brightness changes, and they twinkle less than the stars.

Careful checks will show that their movements are in some ways similar to those of the Moon: east to west nightly but usually with a steady eastward drift against the background stars along paths that are close to the ecliptic. By far the strangest aspect of their behavior is their occasional **retrograde motion**, when they move westward against the stars for some time before apparently reversing and once again adopting the more normal eastward drift (Fig. 2.18). The duration of the retrograde motion is not a constant for a given planet, and it varies greatly from one planet to another. These days we have no great difficulty in understanding this bizarre motion as the natural consequence of our watching the planets in their orbits from an Earth that is itself also in orbit around the Sun. But as long as we think of the Earth as standing still, the behavior of the planets certainly seems strange, and it took many years before the true workings of the solar system were understood.

Further careful observing leads to a differentiation among the planets. Mercury and Venus never seem to venture far from the Sun. The greatest angular separation (**elongation**) between the Sun and Mercury is about 28°, that of Venus about 47° (Fig. 2.19). These two planets are visible for rela-

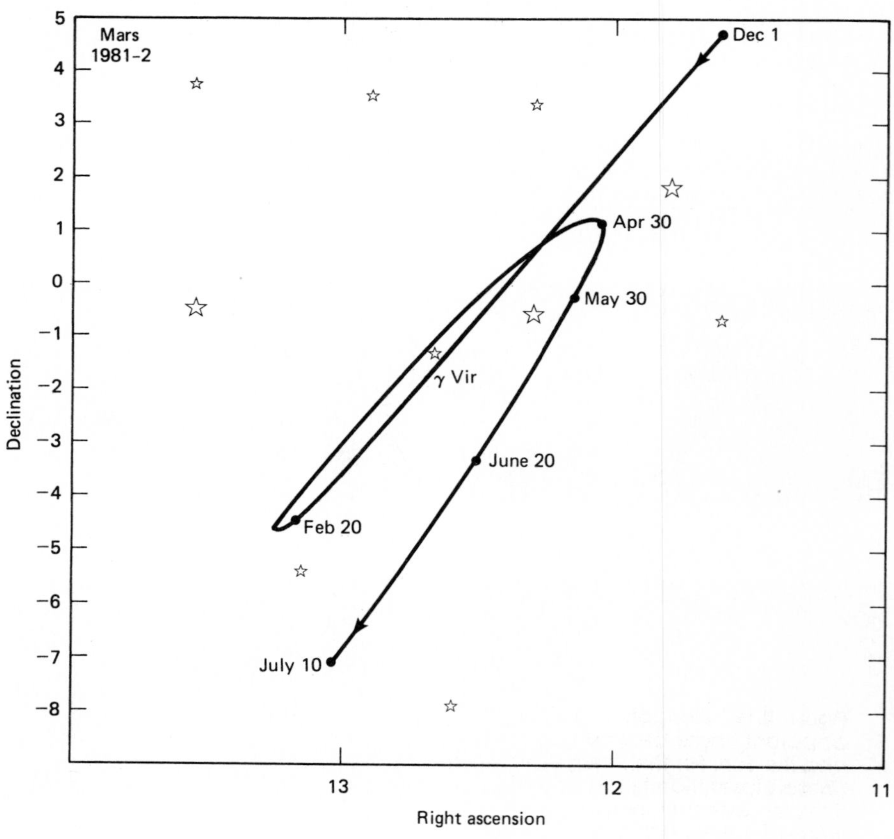


**Figure 2.18** Retrograde loops of Mars and Venus. (a) In 1982 Mars appeared to move back for 2 months before resuming its regular direction of motion.

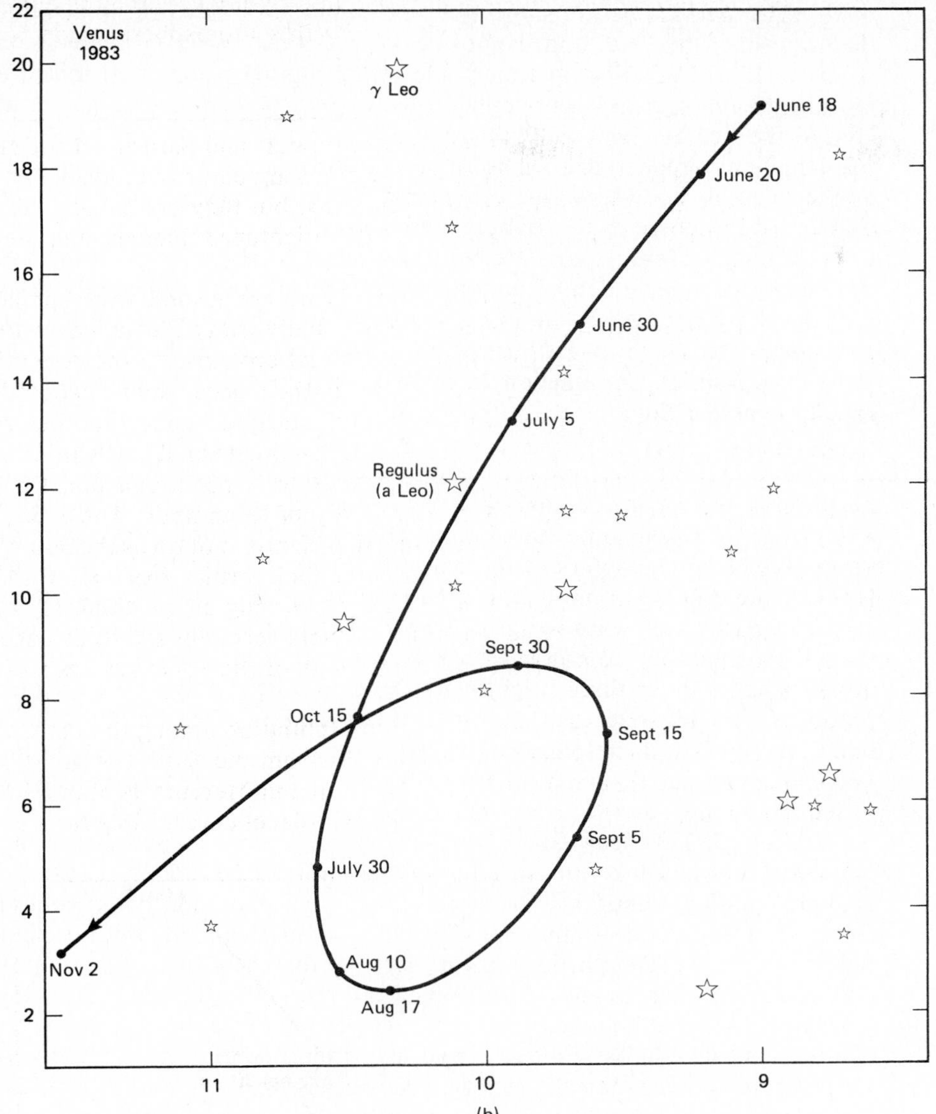


**Figure 2.18** (b) For Venus in 1983 the loop was larger, but the retrograde sector took only about a month and a half. During part of this loop, Venus was too close to the Sun to be seen.

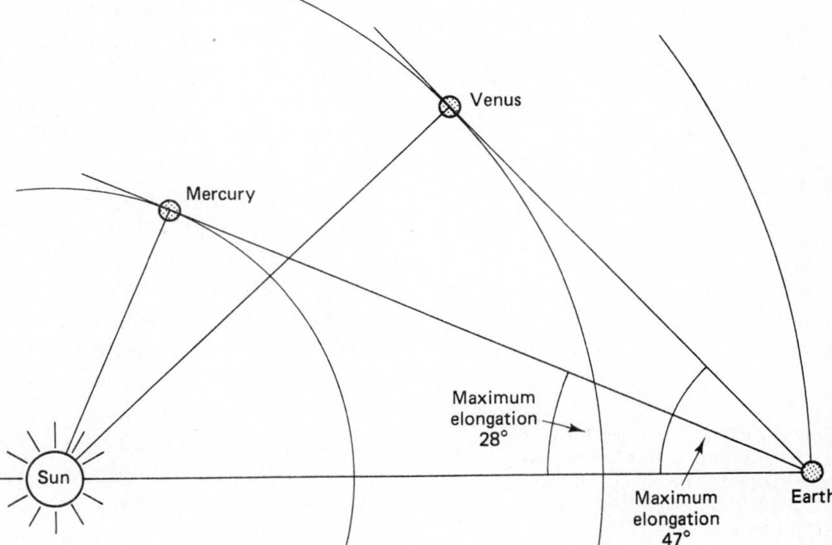


**Figure 2.19** Elongation is the apparent angle between a planet and the Sun. For the inferior planets (those closer than the Earth to the Sun), the elongations can never exceed values that are set by the relative sizes of the orbits.

tively short periods before sunrise or after sunset. As they drift in position, they approach the Sun and cannot be seen against its bright glare in the daytime sky. Depending on which side of the Sun they are, they appear as morning or evening *stars*, but you will never see them at midnight. We should admire the achievement of Pythagoras (around 530 B. C.), who may have been the first to realize that the morning star Phosphorus was the same object as the evening star Hesperus, our planet Venus. (There is some evidence that this was known to the Babylonians about 1600 B. C.)

Although Mercury and Venus never seem to stray far from the Sun, they do move independently, and one may be visible in the morning on days that the other appears in the evening. All of the other planets can be seen well away from the Sun. When a planet is 180° away from the Sun, it is said to be in **opposition**. **Conjunction** occurs when a planet is at the same celestial longitude as the Sun.

## 2.6 SIMPLE OBSERVATIONS—THE BRIGHTEST STARS

Look up at the night sky, and what do you see? Scattered over the sky is a profusion of stars, some very bright, others barely visible, but where are those legendary characters of the constellations—the Twins, the Fish, the Hunter, and others? In these days of television, we all share the same standardized images and we have grown unaccustomed to exercising our individual imaginations. Perhaps with practice and some perseverance, you might imagine those three bright stars do form a belt, and with a bit more prodding, you might accept that bright star as the foot of the hunter (hunter? which hunter?) and the reddish star as marking his shoulder. But, even if you can conjure up this view of Orion, this will not be reinforced by a culture in which the legends take life in the heavens.

After you have learned to recognize the brightest stars and their groupings, you too may take pleasure in delving into the associated stories, but at the beginning identifying the stars seems quite daunting, comparable to flying at night over an unfamiliar city: many twinkling lights but no identifiable landmarks. The purpose of this section, therefore, is to give you just enough information to allow you to pick out the very brightest stars. Once you know them well enough to amaze your friends, you should refer to one of the many detailed sky charts. You may then find it useful to contact a local amateur astronomers' society, whose members are enthusiastic and skillful and more than willing to advise and guide you especially if you should want to purchase or make your own telescope. Our objectives here are far more modest: to provide a guide for the simplest observations.

First, a word on celestial **coordinates**: coordinates are numbers that are used to designate the location of some object or place. For example, 7400 Pershing Avenue specifies a precise position, but for this information to be of use, you must know where the numbering starts and which way it runs. In St. Louis, that is easy—the Mississippi River provides a convenient starting point or origin for the coordinates. The same location could just as accurately but less conveniently be found by specifying 38.6° north latitude, 90.3° west longitude, provided that it is remembered that the origin of longitude is through Greenwich (England, not Connecticut) and that of latitude is the Equator. How then do we specify the positions of the stars and planets?

The easiest system to use is termed the horizon system—we measure **elevation** angle or *altitude* above the horizon, and **azimuth** on a horizontal circle that runs from zero at north through 90° due east, 180° due south, and 270° due west. (This system is often referred to as the alt-azimuth system.) (Fig. 2.20). The immediate problem encountered in using this system is that the coordinates of stars change throughout the night as the Earth (or sky) turns. To be of use, the additional coordinate of time is needed, but even

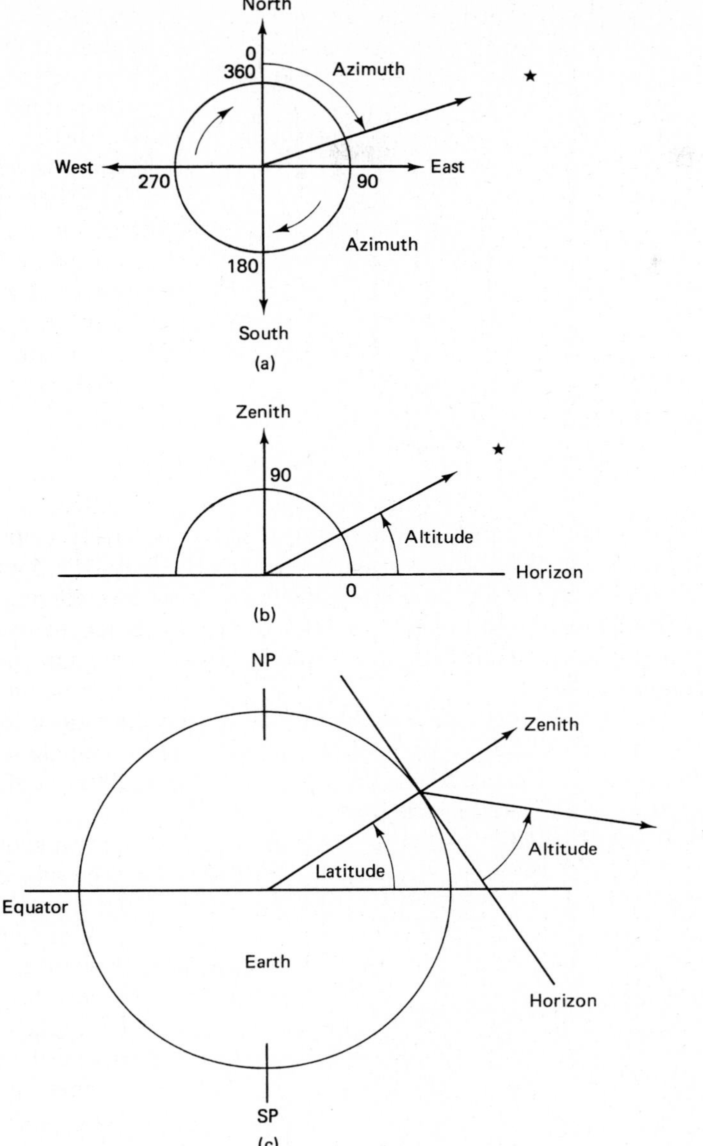


**Figure 2.20** (a) Azimuth is measured around a horizontal circle with zero defined as the direction of North. (b) Altitude (or elevation) is the angle measured up from the horizon. (c) The altitude of a star depends not only on the position of that star but also on the location of the observer on the Earth's surface.

then the numerical value of each of these coordinates depends on your latitude. This is highly inconvenient in astronomy. We need a coordinate system that will allow astronomers everywhere, to know immediately whether they are looking at the same object. The universally-adopted **equatorial** coordinates provide a system of celestial latitude and longitude.

The celestial analog of latitude is easy to imagine (Fig. 2.21). Simply extend the Earth's equatorial plane outwards, and measure directions above or below that celestial equator as either positive (north) or negative (south) **declination**.

**Right ascension** is the celestial version of longitude. The choice of Greenwich for the zero of longitude was arbitrary—many other places on Earth could have served as well, but Greenwich was the choice of convenience. Analogous to the Greenwich meridian is a line of celestial longitude through an arbitrary but agreed point on the celestial equator.

That point is the position of the Sun at the spring equinox. Precession (produced by movement of the Earth's axis) shifts the spring equinox position of the Sun relative to the stars, and as a result both the right ascension

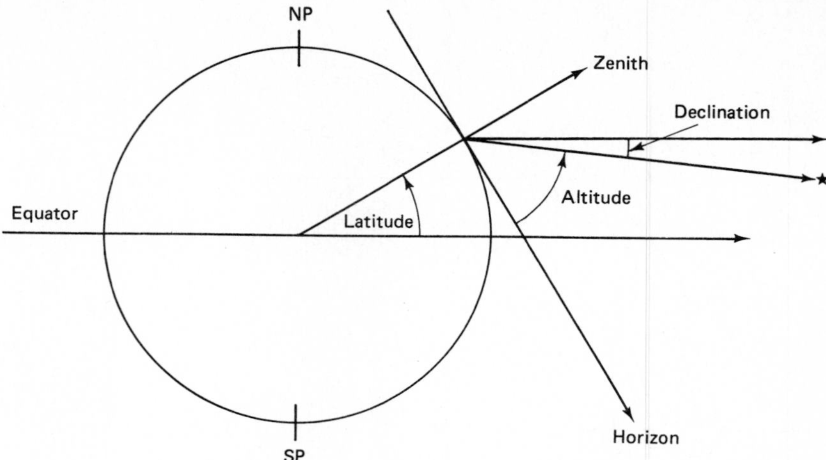


**Figure 2.21** Declination is defined relative to the Earth's equatorial plane, and it can be calculated from the measured altitude and azimuth and the known latitude of the place of observation.

and declination values for any object change slowly. Tables of stellar positions for general use are usually updated every 50 years and more frequently when greater positional accuracy is needed. Right ascension (R.A.) is usually specified in units of time (hours, minutes, seconds) rather than in degrees. Thus, for example, the R.A. of the brightest star, Sirius, is listed as $6^h43^m$.

Although R.A. and declination (dec) may be convenient for more advanced astronomers, the horizon coordinates of altitude and azimuth are still best for beginners. Accordingly, in Table 2.2, I have listed the sixteen brightest stars visible from northern midlatitudes. These are easiest for you to locate. In addition to the R.A. and declination, three other items have

*Table 2.2* STAR POSITIONS[a,b]

| *Star* | *Right Ascension (Hrs, Min)* | *Declination (Degrees)* | *Maximum Altitude above Horizon (Degrees)* | *Month* | *Azimuth at Rising (Degrees from North)* |
|---|---|---|---|---|---|
| Aldebaran | $4^h35^m$ | 16 | 68 | January | 69 |
| Rigel | 5 14 | −8 | 43 | January | 101 |
| Capella | 5 15 | 46 | 97 | January | 23 |
| Orion's Belt (center) | 5 34 | −1 | 50 | January/February | 92 |
| Betelguese | 5 54 | 7 | 59 | February | 81 |
| Sirius | 6 44 | −16 | 35 | February | 112 |
| Procyon | 7 38 | 5 | 57 | February/March | 83 |
| Pollux | 7 44 | 28 | 79 | February/March | 53 |
| Castor | 7 33 | 32 | 83 | February/March | 47 |
| Regulus | 10 07 | 12 | 63 | April | 74 |
| Spica | 13 24 | −11 | 40 | May | 104 |
| Arcturus | 14 15 | 19 | 71 | June | 65 |
| Antares | 16 28 | −26 | 25 | July | 125 |
| Vega | 18 36 | 38 | 90 | August | 37 |
| Altair | 19 50 | 8 | 60 | September | 79 |
| Deneb | 20 41 | 45 | 97 | early September | 25 |
| Fomalhaut | 22 57 | −29 | 22 | October | 129 |

[a] The values of altitude and azimuth have been calculated for St. Louis (latitude 38°39′) and will be correct to within a few degrees for most of the continental United States. The largest errors will be for Capella and Deneb, whose motions appear circumpolar when viewed from places north of about New York City.

[b] Because of precession, coordinates change slowly. The values listed here are for 1980, and may differ slightly from values you will find in other tables. I have rounded the values to the nearest minute or nearest degree—quite accurate enough for our present purposes.

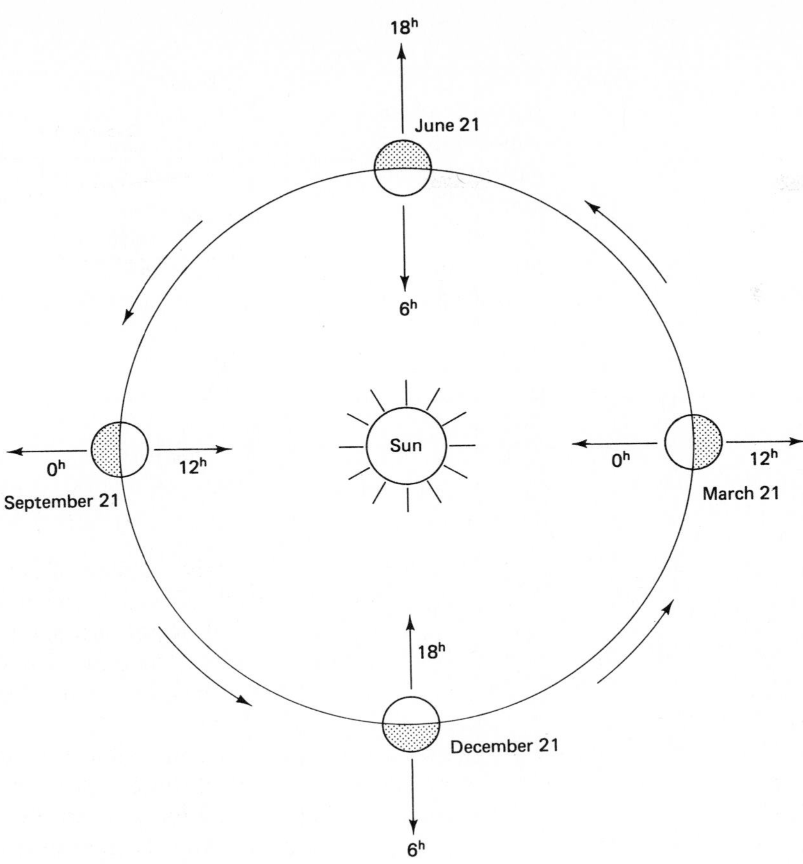


**Figure 2.22** The right ascension (R.A.) astronomical coordinate is widely used and appears on many astronomical charts and in tables of star positions. With this diagram, one can find which values of R.A. will be on the night side of the Earth through each month of the year, and one can check against star charts to see which stars will be visible.

been listed to help you. The maximum elevation (altitude) is the maximum angle of the star above the southern horizon, and the month is listed when the star reaches that altitude around 9 P.M. local time. The final column lists the azimuth at which each star rises, along the eastern horizon. The values range from 23° (north–east) through 90° (east) to 125° (nearly south–east).

## CHAPTER REVIEW

**Celestial pole:** point about which the sky appears to rotate each night. Use Big Dipper to locate the pole star, Polaris.

**Sun:** east to west movement across sky each day. Length of day and maximum height of Sun at noon vary with season. Slow drift of Sun against background stars, about 1°/day, completing circuit in a year.

**Solar Day:** time from one noon to the next.

**Sidereal day:** time for one rotation of the Earth relative to the stars.

**Moon:** east to west movement across the sky each day. Rises about 50 min. later each day, so sometimes seen by day, sometimes at night. Makes orbit around the Earth each month, showing changing phases.

**Sidereal month:** 27.3 days, time for one rotation of the Moon relative to the stars.

**Synodic month:** 29.5 days, time from one new moon to the next.

**Eclipses:**

— lunar eclipse; passage of the Moon through the Earth's shadow.

— solar eclipse: passage of the Earth through the Moon's shadow.

**Planets:** five are visible to the naked eye, three more through binoculars or telescope only. East to west movement across the sky each night, with west to east drift. Occasional reverse (retrograde) motion. Brighter than the stars, also twinkle less.

**Stars:** east to west movement across the sky each night. Rise about 4 min. earlier each day. Positions and patterns appear fixed, but different parts of the sky are visible at different times of the year because of the movement of the Earth relative to the Sun.

## NEW TERMS

altitude
angular diameter
annular
azimuth
Big Dipper
celestial pole
conjunction
coordinates
declination
ecliptic
elevation
elongation
equatorial
equinox
lunar eclipse
obliquity of ecliptic
opposition
path of totality
planet
Polaris
precession
retrograde motion
Right Ascension
sidereal day, month
solar day, eclipse
solstice
synodic
tropics
zenith passage

## PROBLEMS

**1.** On a clear night locate the Big Dipper and note its position relative to Polaris. Make observations (at the same time each night) about every four weeks through the semester, and check the positions of the Dipper against the diagram in the text.

**2.** The Sun and Moon each have an angular diameter of about half a degree. How large an object is half a degree across when it is 2 ft (60 cm) from your eye? Make a cardboard cutout of that size and check this against the Moon.

**3.** Test your general astronomical awareness:

a. What time was sunset the day before you read this question?
b. What phase was the moon last night?
c. Check your answers by observations during the next few clear days.

**4.** Knowing the geometry involved (see Fig. 2.7), calculate the length of the sidereal day in terms of hours of the solar day.

**5.** What sort of observation is needed in order to show that the Earth's axis is tipped at 23.5° to the plane of its orbit around the Sun?

**6.** The Moon takes 27.3 days for a single orbit around the Earth relative to the stars. If you observe the Moon on two successive nights at the same time, how many degrees will the Moon have appeared to have moved?

**7.** From where on the Earth's surface can only half the sky ever be seen?

**8.** From where on the Earth's surface can all of the stars be seen, at some time?

**9.** The obliquity of the ecliptic is $23\frac{1}{2}$°. Suppose that it were 50°. Where would the tropics be located? How would astronomical observations be affected?

**10.** If you wanted to look for the Moon a week before an eclipse, where in the sky should you look shortly after sunset?

**11.** Sketch the positions of the Sun, Moon, and Earth as they need to be for (a) a solar eclipse and (b) a lunar eclipse.

**12.** On September 21 and March 21, the Sun will be directly overhead at the equator. Measure the length of the shadow of a tall pole close to either of those dates. Use your measurement to calculate the latitude of your location.

**13.** The following cities are located at the listed latitudes:

Greenwich, England: 51° north
St. Louis: 38° north
Calcutta: 23° north
Quito: 0°
Rio de Janeiro: 23° south.

At which location(s)

a. is the Sun overhead on June 21?
b. is the Sun overhead on December 21?
c. is Polaris at the northern horizon?
d. will stars of declination +67° never set?
e. will Polaris not be visible?
f. will stars of declination +51° pass overhead?

Illustrate your answers with diagrams.

**14.** Suppose the Earth slowed in its rotation so that a solar day lasted 48 of today's hours. How long would a sidereal day then be (in today's hours)?

**15.** Keep a daily log of the position of the Moon for about 5 weeks. Note its position relative to the Sun, its phase, and whether you are seeing it by day or at night. Combine your observations into a diagram showing the Moon in its orbit around the Earth relative to the Sun.

**16.** On an 8 × 11 in. sheet of paper draw a large circle to represent the horizon. Mark out the cardinal points. Through the duration of this astonomy course, note the setting positions of the Sun each week and try to make the effort to note some sunrise positions too. Check that the sunrise/set positions are indeed where they should be at the equinox (or solstice, if this is a summer school course).

**17.** Check when Venus and Mercury are visible

shortly after sunset. You will need to refer to the *Astronomical Almanac*, the *Handbook of the Royal Astronomical Society of Canada, Sky and Telescope,* or *Astronomy*. Identify these planets in the evening sky.

**18.** On a particular evening just at sunset Venus can be seen in the southwestern sky, about 20° above the horizon. Draw a diagram to show the position of the Sun, Earth, and Venus for this situation.

**19.** What are the Sun's declinations at the summer and winter solstices?

**20.** The Sun's maximum declination is +23° and that of the Moon is +29°. How will this difference show up in the positions of the corresponding sunrise and moonrise?

**21.** Refer to Appendix I, which lists the brightest stars. Identify those that are visible during this course, and keep a record as you find them.

**22.** How would solar and lunar eclipses be changed if the Moon were half its present size but still in the same orbit?

CHAPTER

# 3

# Early Astronomy: The Evidence

## 3.1 INTRODUCTION

In the previous chapter we surveyed the most prominent objects that can be seen in the sky without using a telescope and we took note of their various movements. We might reasonably expect the ancients to have seen these same objects and also taken note—but what evidence do we have? We can try to list the uses to which we think ancient man might have put his astronomical awareness.

It is not easy, however, to imagine how we would react were we living 5000 or more years ago. Of course, we can picture ourselves living in caves or simple shelters, and try to imagine what life was like without the many comforts we take for granted—automobiles, electricity, secure shelter, adequate food and clothing, medical care, and electric can openers. What is far more difficult is for us to exchange our beliefs and fears for those of the ancients. Unless we can truly imagine how *we* would react to *their* pressures and needs and fears (real or imagined), we may well do little more than create a vision of ourselves on a stage or movie set still reacting in accordance with modern knowledge and following patterns of response that were not available to our ancestors.

What evidence do we have for ancient astronomical knowledge? Evidence can take several different forms. We have written evidence and must then deal with the problems of transcription, translation, editing, selection, and perhaps even forgery before we can take the records that have survived as accurate reflections of ancient knowledge. For some of the older written evidence, we have had the initial problem of decipherment of scripts no longer used. We have oral evidence: folklore—sometimes illuminating, sometimes enigmatic. We have customs and traditions whose origins we might surmise even if we cannot date them with precision; various calendars fall into this category. Finally, we have structural evidence, like Stonehenge, that will be the main subject of our present chapter.

## 3.2 STONEHENGE: MEGALITHIC ASTRONOMY

As the best known and most extensively studied ancient astronomical site, Stonehenge deserves our initial attention. The study of such sites is now sometimes called **megalithic** astronomy, sometimes **archaeoastronomy**, although the latter term is not restricted to large stone structures.

Stonehenge lies on the Salisbury Plain of southern England not far from the old cathedral city of Winchester. Set on a gentle hill, its location appears very ordinary. As you approach, the first impression is one of scale—the stones seem so much smaller than you might have expected. Standing next to them, you realize how large they are, but we are so accustomed to skyscrapers that the real Stonehenge, as seen in Fig. 3.1, seems something of an anticlimax.

The stones are in the central area of a circular arena, bounded by the outer ditch about 5 ft deep that provided the Earth for its adjacent bank, thought to have been about 6 ft high originally but now eroded to a gentle hump. Just inside the bank is the circle of fifty-six Aubrey holes, named after John Aubrey, the seventeenth-century English antiquary. These holes lie accurately around a circle with a radius of 141.8 ft (43.2 m).

Amidst the Aubrey holes there are additional markers whose significance was long unsuspected. These are the four station stones, usually designated as numbers 91, 92, 93, and 94 on the official maps, proceeding clockwise and starting with the stone on the east side of the circle. Two of these stones (91 and 93) are still in their places, but evidence for the other two is provided by remnant holes with a small ditch around each overlying the Aubrey holes.

**Figure 3.1** General view of Stonehenge. The Heelstone is upright almost at the road, and the Slaughter Stone slightly closer in flat on the ground and close to the circle of Aubrey holes (white markers). The large arches (the Trilithons) stand around a U-shaped arc opening to the northeast where the summer solstice sunrise can be seen sighting from the center of the circle across the Heelstone. The smaller arches are all that remain from the originally complete ring of thirty arches that stood around a 97 ft diameter circle. Two station stones can be seen also on the Aubrey circle (British Crown Copyright, reproduced with permission of the Comptroller of Her Britannic Majesty's Stationery Office).

**Figure 3.2** Close-up view of some of the standing stones with a Trilithon arch at the left and two of the Sarsen arches (in the center). The horizontal stones on the arches (lintels) were curved to follow the arcs along which the uprights were set out. The careful shaping of the stones is well illustrated by these lintels (British Crown Copyright, reproduced with permission of the Comptroller of Her Britannic Majesty's Stationery Office).

In the central area is the structure that is evoked by the name *Stonehenge*. As constructed, there was a circle of thirty vertical *sarsen*[1] stones capped with lintels to form a continuous ring, 97 ft in diameter and 13 to 14 ft tall. Today, only sixteen of the sarsens are upright and only six of the lintels are in place. Parts of some stones lie on the ground. Over the centuries others were removed, broken by villagers who carted away the pieces to use for buildings.

Within the sarsen circle are the largest stones, the trilithons. Each has two massive uprights capped with a lintel and reaching a maximum height of about 17 ft. The trilithons are not connected, as were the sarsen arches, but originally stood around a horseshoe-shaped curve open to the northeast. Three of the original five trilithons are still standing. Between the trilithons and the sarsen circle there was once another horseshoe of bluestones but fewer than half of the original nineteen stones are still there.

Some of the stones have acquired special names. At the center is the Altar stone, and at the northeast on the Aubrey circle is the Slaughter stone. Neither name is anything other than colorful. There is no evidence to link the ancient Druids and their ceremonies with Stonehenge; there is no evidence of slaughtering at the one stone or the use of the other as an altar, and neither of these need concern us further. To the northeast and just beyond the ditch is another massive stone standing about 16 ft tall and known since around the middle of the sixteenth century as the Heelstone or Friar's Heel stone. The Heelstone is not quite in the center of what is termed the Avenue, two parallel ditches and low banks that run in the northeast direction from just outside the main circular ditch. The Avenue runs straight for over a quarter of a mile, then bends and runs downhill until it reaches the river Avon. It is thought that the Avenue might have been a ceremonial path along which the giant sarsens were brought to Stonehenge, but for our purposes the most important aspect is its initial direction. Atkinson, the archaeologist who is the recognized authority on Stonehenge, surveyed the Avenue in 1978 and obtained a value of 49°54′40″ for the azimuth of the initial 509 m. We shall return to this.

[1] Sarsens and bluestones are descriptive terms like granite or sandstone for the types of stone that were used.

What do we know of the age of Stonehenge? Some of the evidence is archaeological, some from measurements of radioactivity of carbon in bone or charcoal. Stonehenge was built in stages, extending over nearly a thousand years. Stonehenge I consisted of the ditch and bank and the Aubrey holes, the Heelstone, and the station stones. This construction phase has been dated between 2500 and 3000 B. C. Stonehenge II (about 2100 B. C.) had the bluestone circle, neither completed then nor to be seen today. Stonehenge III concluded with the dismantling of the bluestone circle and the erection of the structures we see today, and it is dated at about 2075 B. C.

Despite some uncertainties in the calibration of the radiocarbon method, the Stonehenge dates are probably secure enough for us to say that parts of Stonehenge predate the Babylonian cuneiform records and that Stonehenge I is roughly contemporary with the oldest pyramids.

We know something about the construction; what do we know about the astronomy of Stonehenge? For a long time, it has been known that Stonehenge was aligned on the position of sunrise at the summer solstice. Stand at the center and look through one of the remaining sarsen arches in the northeast direction towards the Heelstone. On midsummer's day the Sun will rise above the Heelstone. Well, almost.

Sir Norman Lockyer, one of England's leading astronomers at the end of the nineteenth century, realized that the precession of the Earth's axis would slowly change the azimuthal position at which the Sun rises at the solstice. As a result, an alignment that was correct when Stonehenge was built would be slightly in error today. Conversely, from the measurable error today, we can calculate back to find how long ago the alignment was perfect, and thus determine Stonehenge's age. Lockyer applied his ingenious idea and deduced a date of 1480 to 1880 B. C. If we now use Atkinson's new measurement of the azimuth of the Avenue, we find a date close to 2000 B. C.

Alignment on the solstices had been pointed out in 1846 by a Reverend Edward Duke, who had shown that the summer solstice sunrise was well marked by a line from station stone 92 to stone 91, whereas the line from 94 to 93 aimed at the winter solstice sunset. Thereafter, apart from Lockyer's calculation, there was no advance in our understanding of Stonehenge until 1963 when C. A. Newham, an amateur archaeologist, and Gerald Hawkins a professional astronomer, independently arrived at similar conclusions, showing that there were other alignments of major astronomical importance.

What Newham and Hawkins did was to move from the obvious alignments to the geometrical, from those that can be easily seen today at Stonehenge to those that emerge when the plans are examined. If we stand today at stone 91, we can look only to 92, *but* before Stonehenge III existed there were also clear sight lines from 91 to 93 and 94 and similarly from 92 to 93

**Figure 3.3** Sunrise positions as seen from Stonehenge through the year. Each initial track of the Sun makes an angle of 31° to the horizon, and the extreme positions occur at the solstices. The moonrise positions vary in a more complicated way between extreme northern and southern points, completing a cycle every 18.61 years.

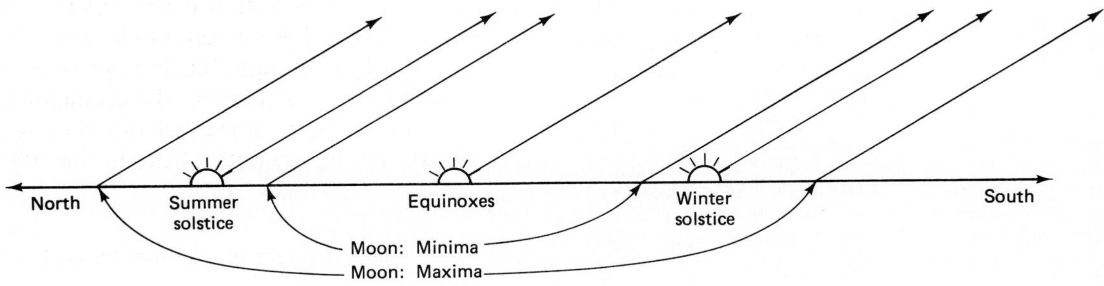

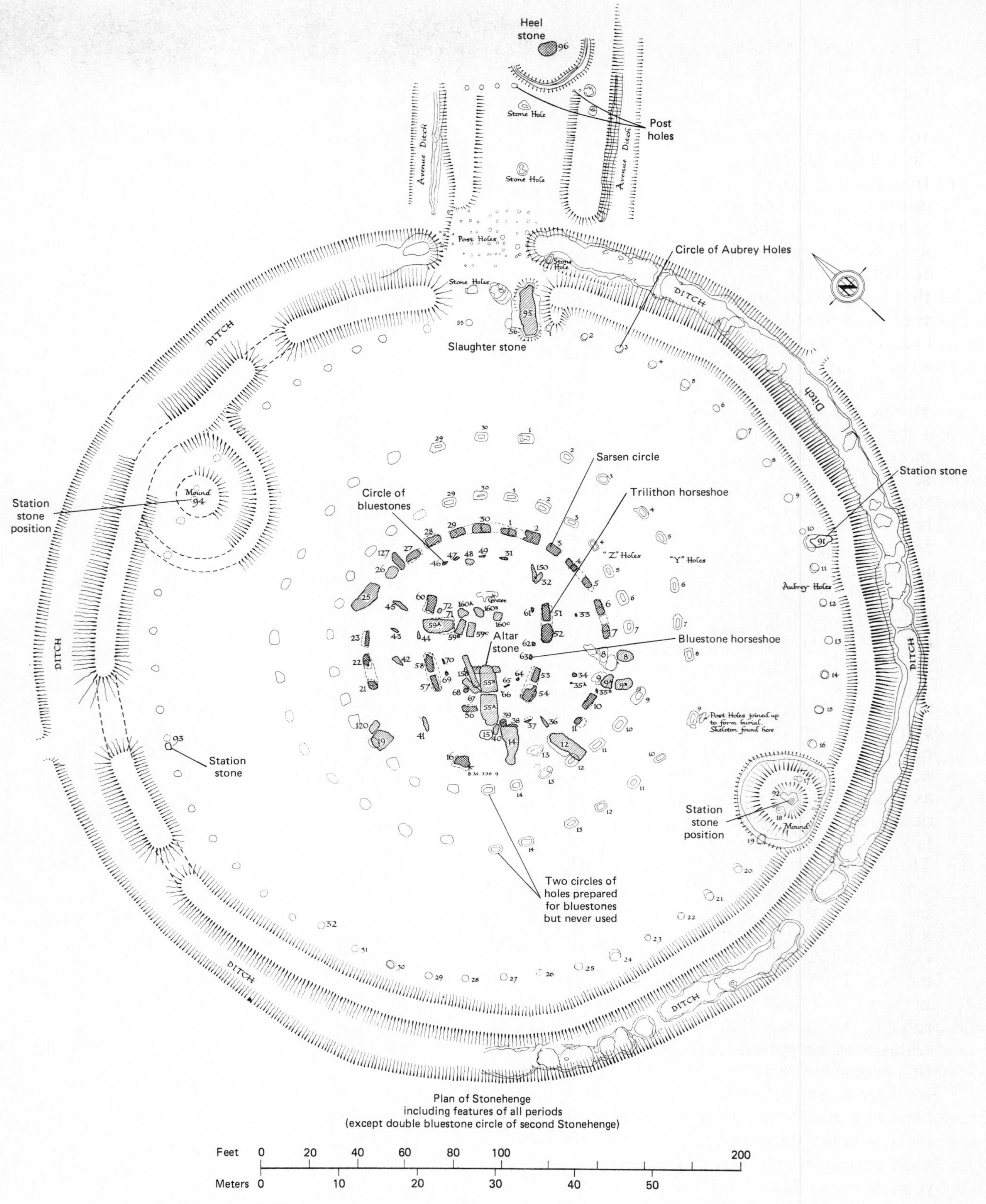


**Figure 3.4** The four station stones (around the Aubrey circle) define a rectangle. The sides and diagonals of this rectangle mark out sightlines to important rising and setting points of the Sun and Moon. Some of these sightlines are duplicated by lines between various arches but for clarity are not shown here. (British Crown Copyright, reproduced with permission of the Comptroller of Her Britannic Majesty's Stationery Office).

and 94. Today those lines are obstructed by the sarsens and trilithons, but we can reconstruct them on paper—with quite unanticipated results.

To appreciate the astronomical significance of these lines, we need to consider the Moon's movements in more detail than we did in Chapter 2. It helps to compare its movements to those of the Sun. The Earth's equatorial plane is inclined at 23.5° to the ecliptic plane. As a result, the Sun's position as seen from the Earth goes through an annual cycle. The positions of sunrise and sunset move north and south regularly along the horizon, between extremes at each solstice. The actual horizon positions depend on the latitude of the place of observation.

The motion of the Moon is much more complicated. Its orbit around the Earth is in a plane inclined at 5.2° to the plane of ecliptic, and the relative positions of these two planes change steadily. This causes the extreme northern and southern horizon positions at midsummer and midwinter moonrise and moonset to go through a complex cycle that repeats every 18.61 years.

A line from 91 to 94 turns out to mark the setting point of the Moon at its extreme midwinter position (corresponding to a declination of 29°), whereas the diagonal line from 91 to 93 marks out the other extreme midwinter moonset (declination = 19°). From 93 to 92 locates the midsummer moonrise (declination = −29°) and 93 to 91 the other moonrise extreme (declination = −19°).

There are many other lines that can be drawn on a Stonehenge map. Some between adjacent stones provide too short a baseline for accurate work, but Hawkins used a computer to do the tedious geometrical calculation involved in determining azimuths and hence declinations. In this way the solstice lines appeared, not only as anticipated from the center but also for sight lines between various sets of sarsen and trilithon arches. The extreme lunar positions were also duplicated in this way. So, although the construction of Stonehenge III obscured the major sight lines across the diagonals between the station stones, some of the new arches replaced them. No clear alignments turned up to any of the planets or brightest stars. Given the precision of its construction, Stonehenge has clearly been set out with a purpose, and we are probably entitled to take those solar and lunar alignments as deliberate and not simply as an accident that we are now attempting to overinterpret.

Hawkins has also proposed a theory to explain the use of the Aubrey holes. These do not seem to have held posts. Excavation has revealed wood ash in one, flakes of flint in another, and cremated human bones in most. Hawkins has suggested that they were used for eclipse prediction, with the fifty-six holes representing a 56-year eclipse cycle, and the tally being kept by moving six marker stones around the circle in an elaborate and carefully specified way. Fred Hoyle, also an astronomer, has devised a somewhat different theory for eclipse prediction also using moving markers in the holes. It is probably fair to say that neither of these innovative approaches has found widespread acceptance, and we still have no satisfactory understanding of this prominent feature of the site.

But there can be little doubt about the intentional solar and lunar alignments. It is one thing to mark out the positions of sunrise and sunset: These recur every year and the drift due to precession is imperceptible in a lifetime. The Moon's movements, though, present a challenge. To note a cycle of 18.61 years requires that several cycles must have been followed and somehow recorded simply to recognize that there was a regularity. Then careful measurements or alignments must have been made for many years to establish the extreme positions. None of this necessarily requires that the length of the cycle need to be known, but only that extreme positions are revisited (every 18.61 years). There is no trace at all of any written records—

no writing, no rock paintings, no tangible aids to memory—yet the construction must have been preceded by observations that extended over several generations.

It has been the traditional view that western civilization diffused outwards from its ancient origins in Egypt and Babylonia. If Stonehenge is correctly dated and interpreted, we have evidence for the presence of a developed community contemporary with the oldest in the middle east. This implies a major revision in our view of the pattern of growth and diffusion of civilization and it explains why the fruits of archaeoastronomy are of such great interest to many archaeologists.

One final comment of Stonehenge is the order at this time. Stonehenge has sometimes been termed a neolithic observatory, but that is misleading for Stonehenge was not an observatory in the modern sense. It was designed for the celebration of existing knowledge rather than for investigations to add to that knowledge. It was and remains an impressive summation of astronomical knowledge of that time.

## 3.3 OTHER MEGALITHIC SITES

Our survey of other major megalithic sites should start with the Pyramids, subjects of much published nonsense. All sorts of fanciful theories have been constructed, often based on contorted manipulation of the Pyramids' dimensions. Some of these theories are naive, others seem simply designed to sell books.

The sides of the Great Pyramid at Giza are aligned very closely north-south (within 4′) and east-west (within 2′), and although it seems to be agreed that these deviations are larger than would be expected for building errors, no accepted explanation has been found. Similarly, the orientations of various passages and chambers in the Pyramids have been subjects of contested claims that they were aligned on certain stars, but no convincing case has been made.

Scattered over the British Isles, there are remnants of hundreds of rings of stones. Some are accurately circular, others are oval, and others appear to be composed of irregular curves. Their diameters range from about 10 ft to over 150 ft. From careful surveys, it appears that some of these rings were probably set out to indicate astronomical alignments, but for none is the evidence as overwhelming as at Stonehenge. There are, also, many other megalithic sites in Europe, but their astronomical content is even less well established.

## 3.4 ARCHAEO-ASTRONOMY IN THE AMERICAS

What evidence do we have for the role of astronomy in the Americas? There is a growing body of quite diverse information, and we therefore broaden our discussion from megalithic- to archaeoastronomy and also do not confine ourselves to the extremely ancient. In this section we shall review only one representative site in detail, Cahokia, in Illinois near St. Louis.

St. Louis, Missouri is situated in a very low-lying area a few miles south of the junction of the Mississippi and Missouri Rivers and rather further north of the point where the Ohio River joins them.

With its favorable position for transportation and trade, the area became well settled. Estimates of population around A. D. 1000 run as high as 30,000. This Indian population constructed more than a hundred mounds of earth including Monks Mound, the largest in America north of Mexico, located a few miles east of St. Louis. After A. D. 1300 the population declined, and the white settlers found no inhabitants.

Working about half a mile due west of Monks Mound, Warren Wittry has uncovered evidence for what he has termed a "woodhenge," remnants

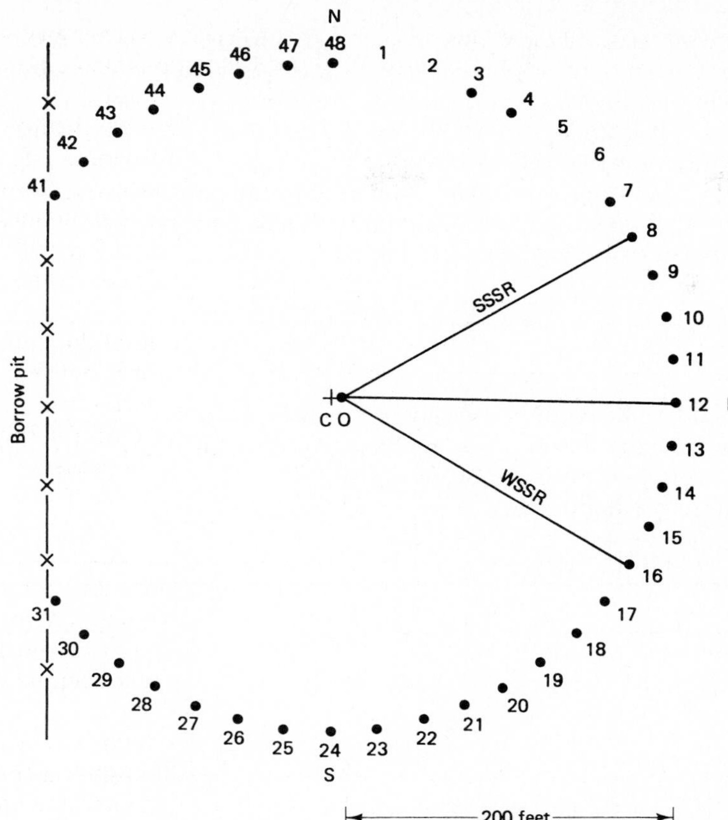


**Figure 3.5** Schematic diagram of the circle of post holes at Cahokia, Illinois. There is a posthole at *O*, a few feet to the east of the geometrical center of the circle (at *C*). From *O*, sightlines to three of the post holes mark out the sunrise positions for the solstices and equinoxes. Comparable sunset positions cannot be found because of damage to the western part of the circle (Warren Wittry and *Archaeoastronomy*).

of a very accurately laid out circle of red cedar posts, each around 18 in. diameter. As the site was excavated, a stump of one post was found in 1978, and the impressions of others long gone could still be seen in other post holes. Wittry has good evidence for three interlocking circles, with Circle 2 being the best defined with a diameter of 205 ft. There is no marker at the geometric center, but 5 ft east there is a hole that at one time held a very large post.

From this woodhenge, the solstice sunrise positions are along a low line of bluffs, about 4 mi to the east, and the equinox sunrises are over the southern tip of Monks Mound. Viewed from the nearcentral position of Circle 2 across appropriate post holes on the circumference, these sunrise positions are well defined. Wittry's final analysis of his field data and the accurate specification of the positions of post holes are not yet complete. A preliminary map was published in 1964 and an updated map after his 1977 dig. This revealed post holes close to the rising and setting points of Capella, which is not circumpolar at this latitude. Bisecting Capella's rising and setting positions could have been used for determining celestial north, but this remains an interesting conjecture, and there is no post hole due north. The western sector of the circles has suffered damage, and it will not be possible to complete the reconstruction of the main circle.

Wittry has had a local utility company place posts at the near center and at the three sunrise positions. On several occasions he has clambered up the observation post to observe and photograph the sunrise. The alignments seem very accurate. We must remember, though, that we have no knowledge of the height of the original posts nor of the actual method of observing, but the reconstruction is impressive. The site is perhaps unique in its easy access from a major city and is worth a visit.

Numerous other sites have now been examined, in the U.S., Mexico and central America. In many, there is evidence for astronomical alignments involving the sun, Venus and some of the brightest stars. From carved stones and from the few Mayan writings that survived the Conquest, we know that the Mayans had a considerable knowledge of astronomy.

It is clear that people in widely separated parts of the Earth created structures that reflected their own astronomical knowledge and needs. Further work may elucidate some of the remaining puzzles, but many sites are in such ruinous condition that little can be done apart from preventing further spoilage or misguided restoration.

## CHAPTER REVIEW

**Stonehenge** construction shows that builders knew of seasonal variations in positions of sunrise and sunset and also knew of the 18.6-year lunar cycle. No clear evidence for use of monument for eclipse prediction.

**Cahokia, Illinois:** circle of markers dating from around A. D. 1000. Positions of solstice and equinox sunrises were marked.

## NEW TERMS

**archaeoastronomy**
**megalithic astronomy**

## PROBLEMS

**1.** We can try to imagine how a primitive people might have kept astronomical records by seeing what anthropologists have found about Bushmen, the Aborigines, and isolated tribes in America and Africa. Consult your reference library for one of these and see what is known.

**2.** Construct a circular diagram after the fashion of Stonehenge but for your location. Mark the azimuthal positions of sunrise for the solstices and equinoxes. Mark also the extreme lunar positions.

**3.** Suppose that you wished to mark out a circle for astronomical sightings, such as at Cahokia or Stonehenge. If the radius of the circle is 100 ft, what accuracy is needed for positioning a marker on the circumference if the alignment is to be good to 1°?

**4.** The four station stones at Stonehenge are by no means the most striking feature of that site, but their positions indicate some important astronomical awareness by its builders. What was that?

**5.** If the directions of moonrise and moonset have been correctly identified at Stonehenge and the builders did indeed know of the 18.61-year lunar cycle, why is this an important but puzzling fact?

**6.** Most of Stonehenge was constructed before 2000 B. C. What events do we know of for about the same time in Egypt and the Middle East?

**7.** Which astronomical directions appear to be marked at Cahokia and also at Stonehenge?

**8.** Examine a modern street map. Does it contain any features that might be considered astronomical alignments if the ruins of the city were rediscovered in A. D. 4000. What evidence might be found in layout or in the architecture of buildings for astronomical knowledge on the part of today's inhabitants?

CHAPTER 4

# The Modern Use of Ancient Records

## 4.1 THE NATURE OF THE RECORD

In the previous chapter we saw how astronomical knowledge was incorporated by some early societies into large and impressive monuments. Perhaps *monuments* might appear a strange choice of word, but in some important ways Stonehenge and the other sites are indeed monuments for they have served to memorialize for their users and for us much later the achievements of their builders. As we study these ruins, we are learning about those builders and their societies, but we do not expect this research to contribute to the solution of current problems in astronomy. In an unexpected contrast, some of the written astronomical records *are* turning out to be a unique and invaluable resource for modern astronomy.

Not all ancient records lend themselves to this particular use. Many, such as the cuneiform tablets from Babylon and the Dresden Codex of the Maya, contain information on regular cycles in the solar system. They list the appearances of Venus or the positions of the Moon and seem to have played a role similar to our *Astronomical Almanac*. We can use these and other records to trace the development of astronomical knowledge and theories. In this chapter, however, we shall look at the ways in which the old records are making contributions to modern astronomy but should first point to some intrinsic differences between structural and written records.

The structures could be planned after the repeated observation of regularly occurring risings and settings. The careful placing of markers would identify special alignments, but we do not know how or even whether such information was transferred from one site to another, nor do we know when these exploratory observations were made, for we can date only the structures (if at all) and not the prior discoveries. Many of the written records, however, can be dated quite accurately, for they refer to kings or dynasties or the occurrence of events that can be pinpointed from other evidence.

As the rest of this chapter shows, accurately dated records can be combined with modern observations in different ways, yet all relate to a central theme in astronomy today: the evolution over time of our universe and its components. This is an important change from the view that prevailed in medieval Christian Europe when doubts about the unchanging nature of the heavens invited persecution. We now know that we can improve our understanding of a particular star, for example, if we have reliable information on variations in its brightness at different times. If we can extend the period for which we have reliable data, even to times before the use of telescopes, we may be able to eliminate some theories or strengthen others. All of this presupposes that the ancient information is trustworthy, and we need to be alert to the notorious unreliability of some records and to the pitfalls that go with the careless use of some materials.

## 4.2 ECLIPSES

In Chapter 2 we mentioned an important feature of total solar eclipses. Each solar eclipse can be seen only within a narrow band of totality on the Earth's surface; people close by but outside that band will see a partial eclipse or if further away no eclipse at all. The eclipse of April 15, 136 B. C. was reported by Babylonian astronomers, who also noted that Mercury, Venus, Mars, Jupiter, and many stars could be seen during totality. With our present accurate knowledge of the positions and speeds of the Earth and Moon in their orbits and with the use of computers, we can calculate the positions of all of these bodies at various times both in the future and in the distant past. We can in a sense run the celestial movie backwards and tabulate those occasions when the Sun, Moon, and Earth lined up for an eclipse (Figs. 4.1 and 4.2). The accuracy of these calculations is quite sufficient for our purposes. (This calculation *cannot* be pushed infinitely far back in an attempt to see how the solar system started, for over very long periods various errors will accumulate. But for at least 10,000 years the results are taken to be reliable.) We can therefore confirm from the calculations whether an eclipse actually did occur in 136 B. C.—it did. Computing *where* on the Earth the band of totality fell introduces a new quantity: the rotational speed of the Earth on its own axis. All that the orbital calculations can reveal is whether the three bodies lined up, but the rotational speed is needed to determine which places on the Earth's surface happened to fall into the shadow.

By calculating back using the Earth's present speed of rotation, we find that the eclipse should have been visible across parts of western Europe, but not in Babylon, where it *was* actually seen. The explanation lies in the changing speed of rotation of the Earth, that is, in the changing length of the day.

From the difference in longitude between the region where the shadow actually fell and the region where it was calculated to have been, we can now compute what the change in the Earth's rotation rate must have been, between 136 B. C. and now. Over this span of twenty-one centuries, the length of the day has been increasing by only 1.8 thousandths of a second per century. The day was almost 40 milliseconds shorter then than now.

This type of calculation can be repeated for any accurately dated eclipse, and the most recent tabulation of such eclipses is that of Stephenson and Clark. They list thirty-nine eclipses that were reliably recorded before the use of telescopes. For completeness it should be pointed out that the calculations are more complicated than we have described here, involving additional quantities such as variations of the Moon's speed. Of course, the

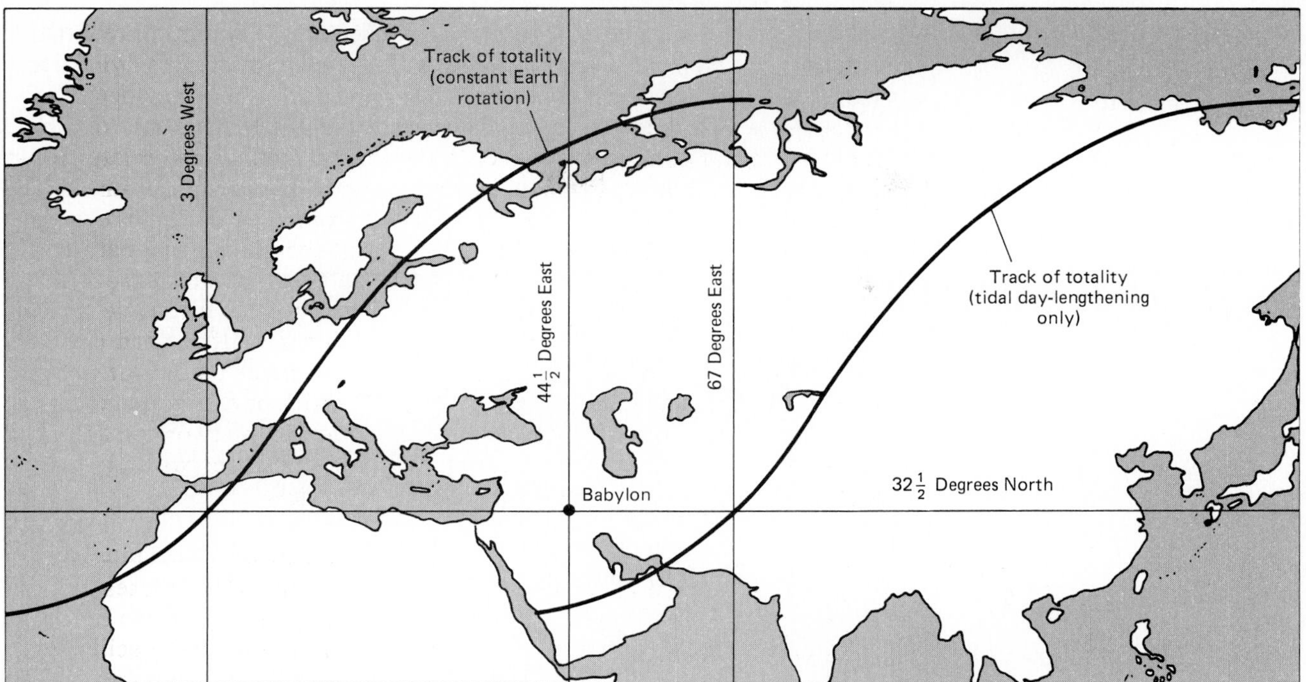


**Figure 4.1** Tracks of totality computed for the eclipse of 136 B. C. If the Earth's rotation speed had remained constant over the past two millenia, then the path of totality would have crossed present Finland and Germany. If, however, allowance is made for the slow lengthening of the day because of tidal friction, then totality would have been seen well east of Babylon. Well-documented observation of the eclipse in Babylon points to the operation of an additional acceleration in the Earth's rotational speed (F. Richard Stephenson and *Scientific American*).

**Figure 4.2** Clay tablet from Babylon with inscription describing the total solar eclipse of 136 B. C. (Reproduced by courtesy of the Trustees of the British Museum).

later and more recent use of telescopes and clocks permits much greater accuracy, but the old records allow us to follow changes in the Earth's rotational speed over many centuries. Changes in this speed are also related to the Moon's distance from the Earth and to the tides so that the whole problem is quite complex, but there is no mistaking the real effect. From reliable reports of past eclipses we can draw important conclusions that are of current interest.

## 4.3 SUNSPOTS

As part of the Church's doctrine of celestial perfection, the Sun was long considered to be without blemish. We now study a variety of phenomena that relate to solar irregularities, with sunspots the most easily observed. Despite the past theological denial of the presence of these spots, there are many records of their observation long preceding the use of telescopes. For a long time they were ignored as being due to atmospheric effects, but Galileo's observations with his new telescope in 1610 showed that they were indeed on the Sun (Fig. 4.3). Sunspots are regions on the Sun's surface of lower than average temperature, and their structure is extremely complex. Systematic counts of sunspots have been maintained only since about the middle of the nineteenth century, and their numbers have been found to increase and decrease in a fairly regular way with a cycle of a little more than 11 years (Figs. 4.4 and 4.5). Two recent studies have drawn on old records of sunspots and shown that this 11-year periodicity is more complicated than previously suspected.

John Eddy, a solar astronomer at the High Altitude Observatory in Colorado, has collected a wide range of data showing that very few sunspots were observed between about 1645 and 1715—it seems that the Sun was unusually quiet. (This reduction in the number of sunspots had also been noted by Maunder in the 1890s but subsequently forgotten. The period of low sunspot activity is now termed the Maunder Minimum.) Could Eddy

**Figure 4.3** Illustration from *Rosa Ursina* (1630) by Christopher Scheiner. A solar image was projected onto a screen where measurements of sunspots could be safely made. The caption may be translated as "Spots and faculae are confirmed by various methods of observing." (Owen Gingerich and The Houghton Library, Harvard).

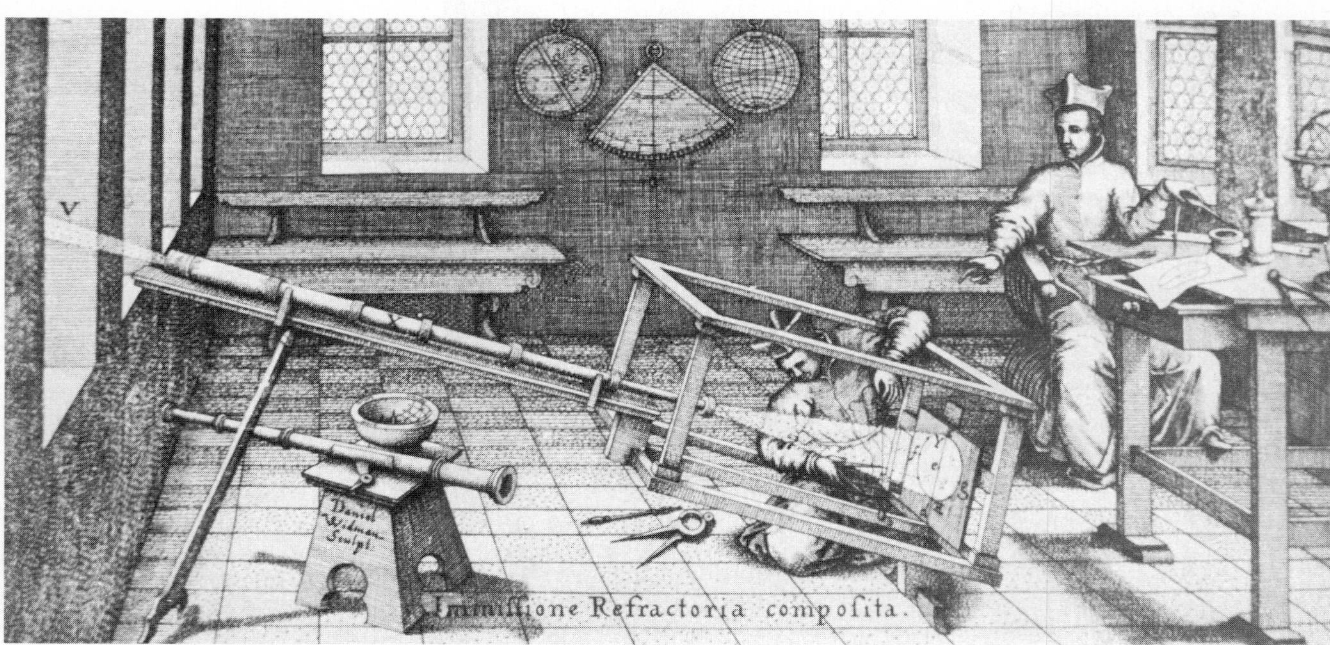

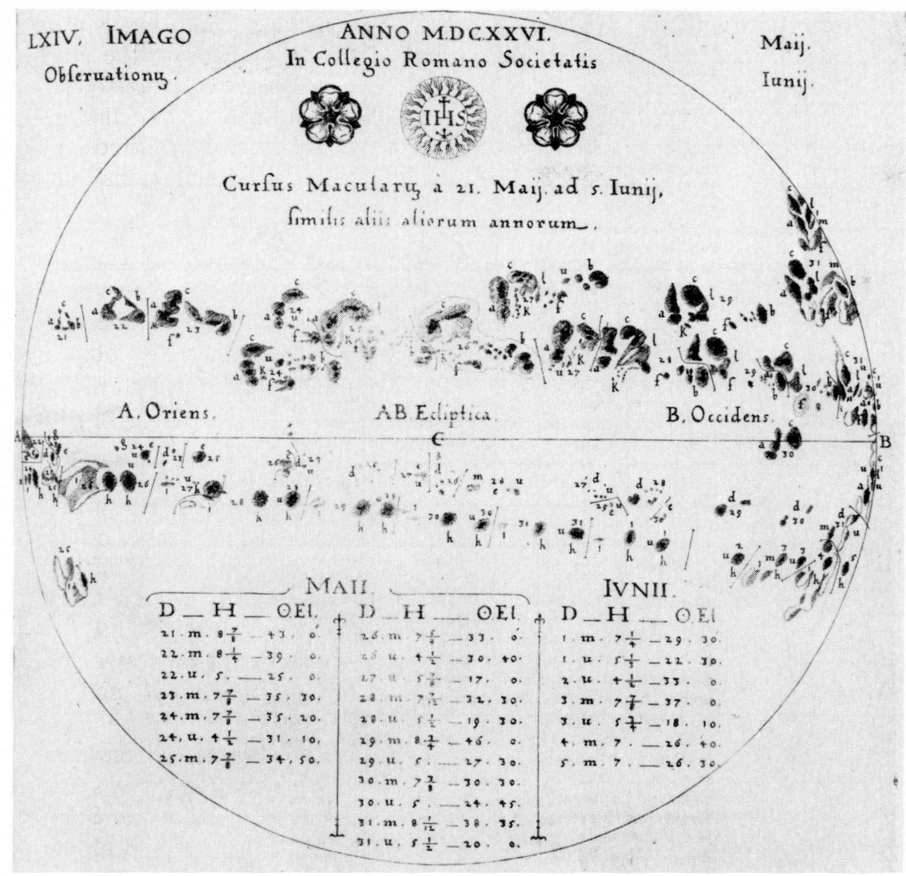


**Figure 4.4** Diagram from the *Rosa Ursina* (1630) showing daily positions of sunspots between May 21 and June 5, 1626. (Original diameter of the circle is 22 cm.) Rotation of the solar disc is clearly shown by this and other drawings, and the rotation speed is found to vary (Yerkes Observatory).

(or Maunder) be wrong? Could it simply be that (for some unknown reason), spots were seen but not recorded?

Eddy has assembled other evidence that strongly supports the proposition that solar activity was significantly reduced during those years. For example, very few auroras were seen in those years. Auroras are brilliant displays of atmospheric lights, often covering large parts of the sky and generally seen only at northern latitudes; these are now known to be closely associated with increased solar activity as signaled by sunspots.

The average temperature on Earth is also an indicator of solar activity and is reflected in the growth of glaciers. Eddy has shown that the records of these apparently unrelated phenomena are consistent with the diminished solar activity suggested by the sunspot data, but no good explanation is available for the underlying causes. In a later chapter when we discuss the structure of the Sun, we will see that this Maunder minimum adds to the growing puzzle of the Sun's behavior.

Some sunspot data can be put to a rather different use. The total number of sunspots is a simple but poorly understood index of solar activity. The spots that we tally for keeping track of the 11-year cycle appear, grow, and shrink until they finally disappear, all the while being carried across the visible face of the Sun as it rotates. The equatorial region of the Sun takes about 25 days to rotate once, and we can see some spots disappear around the one edge (*limb*) and reappear later at the other limb. Spots can therefore be used to measure the rotational speed of the Sun. Some old records included accurate sketches showing groups of spots and their changing positions day after day. It has emerged, from the analysis of these records, that the Sun's rotational rate was slower in 1612 than it is now, and that it

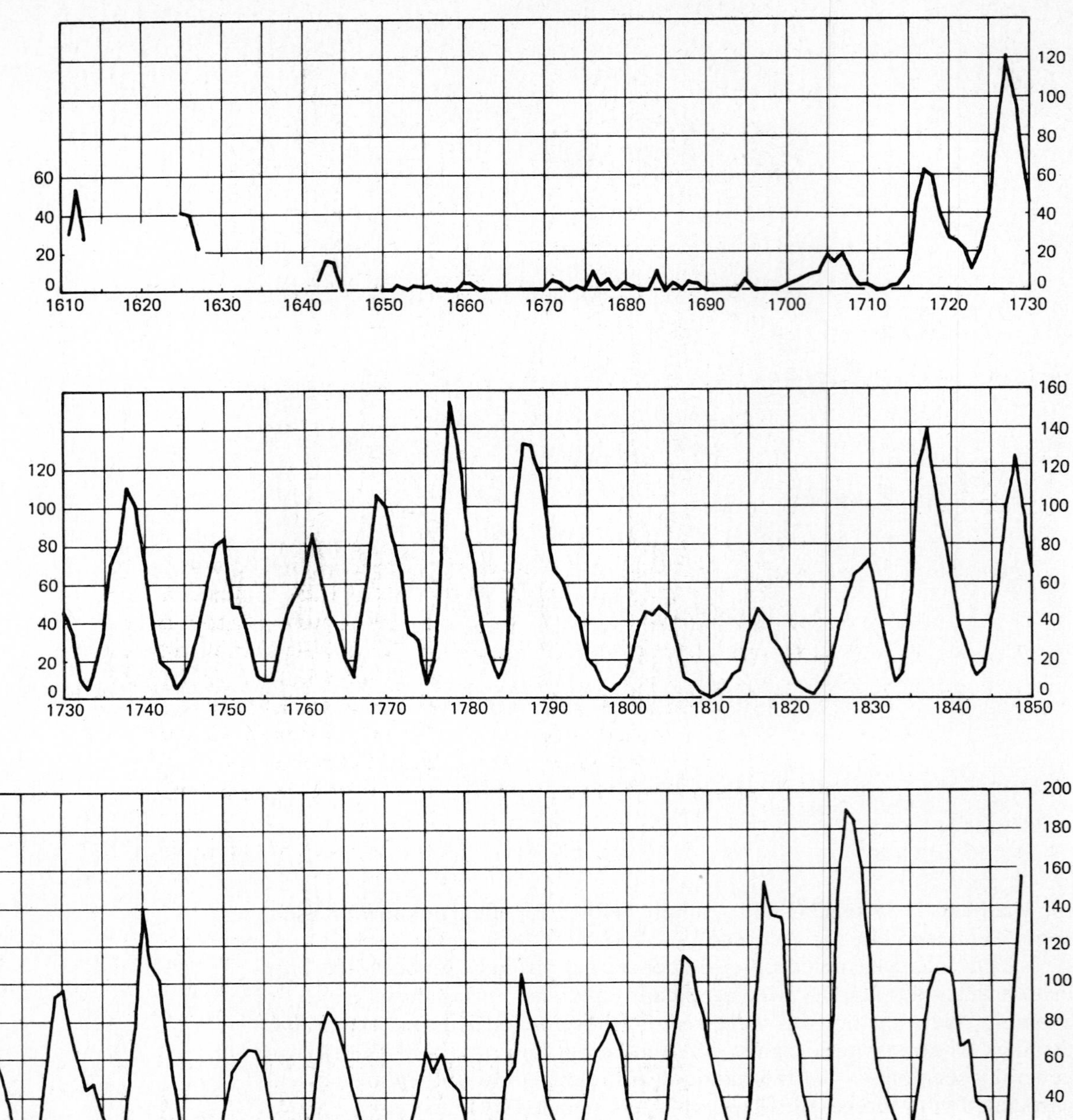


**Figure 4.5** Annual mean sunspot numbers from 1610 to 1979. Note the 11-year periodicity since about 1717 and the very few sunspots between about 1645 and 1715, the period now termed the Maunder Minimum (John A. Eddy, High Altitude Observatory).

accelerated in the years leading to the Maunder Minimum. As with the Maunder Minimum, we have no explanation for these apparent changes in the solar rotational speed.

## 4.4 SUPERNOVAE

Probably the best known use of ancient records involves novae—violently explosive stars whose unpredictable increases in brightness were so sudden that to the early astronomers it seemed as though new stars virtually appeared from nowhere. These were called new stars, *stellae novae,* now abbreviated to **novae** and, in a few cases of exceptional brightness we term them **supernovae**. The historical record is fascinating for its inclusions and

**Plate 1** Observatories on the summit of Mauna Kea, Hawaii. Large domes, from the left: the Canada-France-Hawaii 3.6m reflector, the University of Hawaii's 2.2m reflector, the United Kingdom's 3.8m IR reflector, and the 3m IR Telescope Facility. The smaller domes house 60 cm reflectors (Institute for Astronomy, University of Hawaii).

**Plate 2** Atmospheric refraction distorts the appearance of the Moon, as it approaches the horizon. Viewed from Skylab (NASA/Marshall).

**Plate 3** The 3m IRTF reflector (Institute for Astronomy, University of Hawaii).

**Plate 4** Arecibo radio telescope (Arecibo Observatory).

**Plate 5** Arecibo radio telescope: view up, towards the antenna (Arecibo Observatory).

**Plate 6** Earth, as seen from Apollo 17. Visible in this view are Africa, the Arabian Peninsula, Madagascar (off the east coast of Africa), and the eastern Antarctic ice sheet. Most of the earth's surface is covered by water (NASA).

**Plate 7** Apollo 11 Lander (NASA).

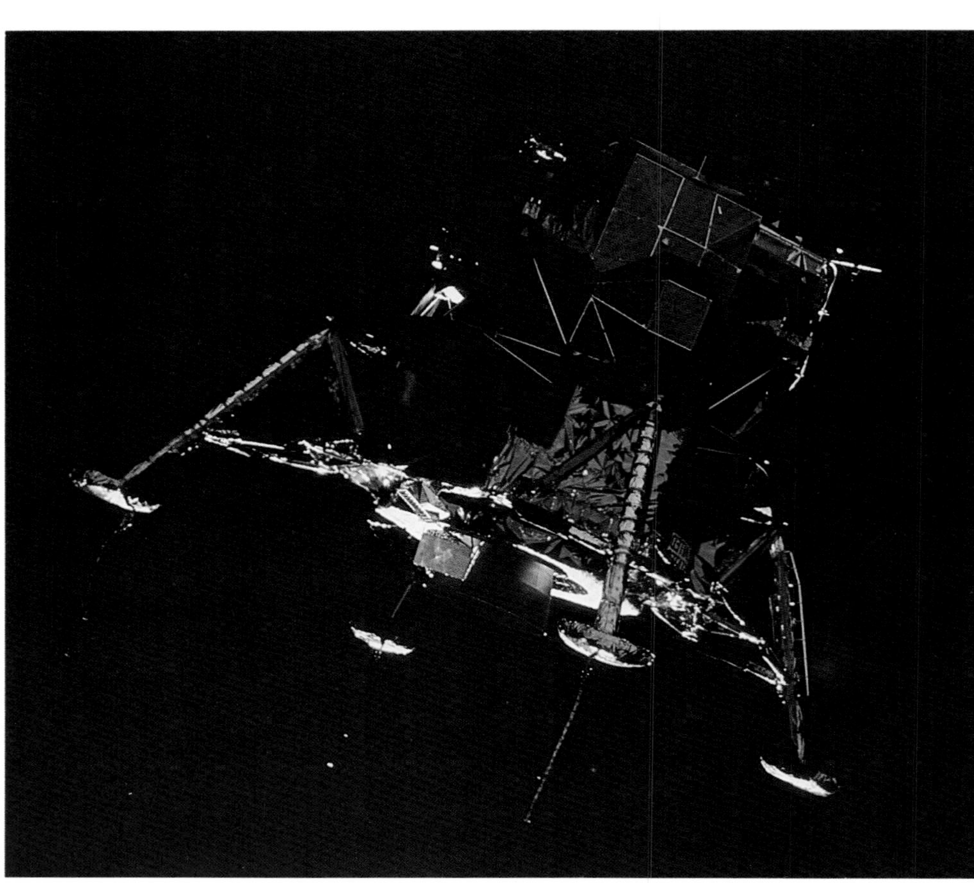

**Plate 8** Astronaut Harrison Schmitt shown using the rake to collect a sample of rocks ranging from 1 to 4 cm in diameter during the Apollo 17 mission (NASA).

**Plate 9** Lunar sample collected during the Apollo 15 mission. (NASA).

**Plate 10** Solar Maximum Mission (SMM) spacecraft (HAO/NCAR).

**Plate 11** High Energy Astronomical Observatory (HEAO)-1 (TRW Electronics and Defense).

**Plate 12** High Energy Astronomical Observatory (HEAO)-3 (TRW Electronics and Defense).

**Plate 13** Small Astronomical Satellite (SAS)-2 (Carl Fichtel, NASA).

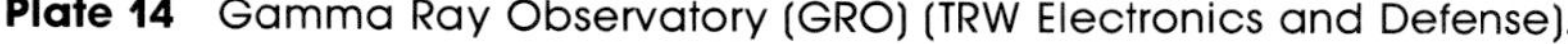

**Plate 14** Gamma Ray Observatory (GRO) (TRW Electronics and Defense).

**Plate 15** False-color map of Venus from IR observations (NASA).

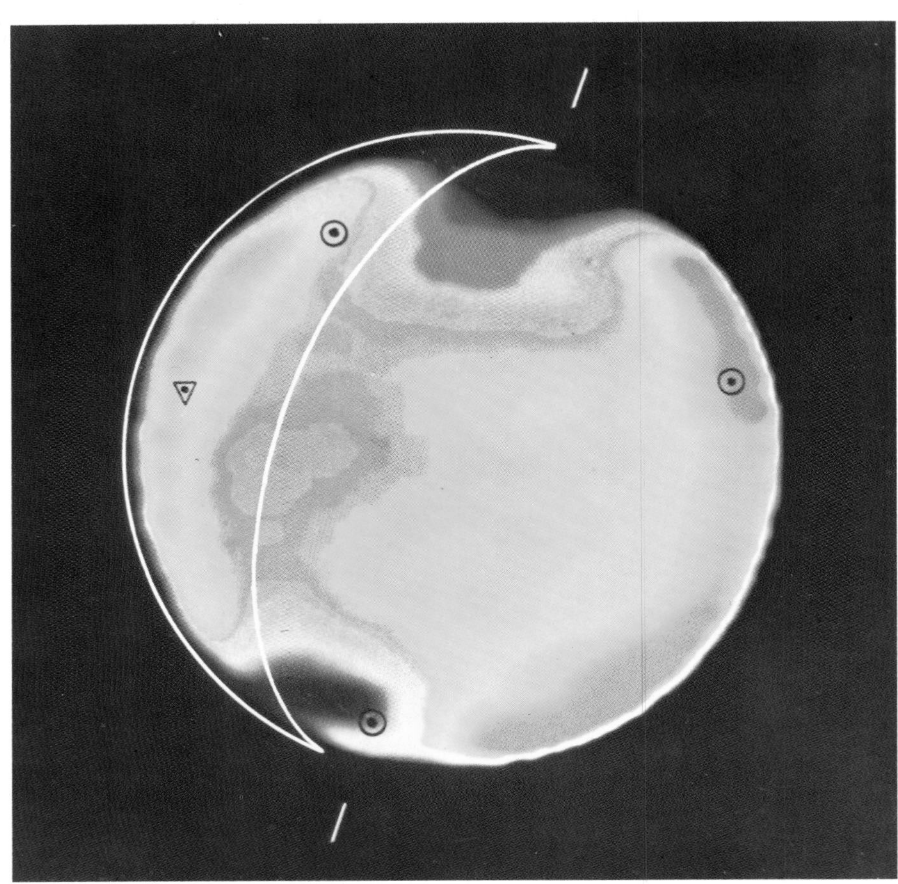

**Plate 16** View from the Viking Lander 2 on the surface of Mars. Color of the sky is produced by scattering of sunlight by dust particles in the atmosphere (NASA).

omissions. There are many reports in the Chinese, Korean, and Japanese annals, stretching as far back as several centuries B. C. In China new arrivals in the heavens were reported in different categories: **K'o-hsing** ("guest stars"), **po-sing** ("rayed stars" or "bushy stars"), and **Hui-sing** ("brook stars"). Comparison with other records shows the occasional subjectivity of these classifications, but generally the *k'o* referred to starlike objects that are now identified with novae or supernovae, the *po* to tailless comets (identified by their movement), and the *hui* to comets with tails.

No supernova has been seen in our galaxy since 1604, before Galileo invented the astronomical telescope. As a result, we have had no very close look at the features and changes as a supernovae evolves, although we have been able to observe supernovae in other galaxies where the distance limits our investigation. We do, however, have very good naked eye descriptions of the two most recent galactic supernovae from Tycho Brahe (in 1572) and Johannes Kepler (in 1604). From these and from the modern (extragalactic) observations, we have developed a clear enough idea of the early appearance of a supernova that we can pick out the supernovae from amongst the ancient records of guest stars. What use can we make of them?

First is the fairly obvious calculation, the frequency of occurrence of supernovae in our own galaxy. Modern stellar theory covers the formation, evolution, and aging of stars. Supernovae are thought to occur very late in the lives of certain types of stars. We know something of the number of stars in the galaxy and the rate at which new stars are still forming; knowledge of the rate of supernovae explosions could complement this. Making an estimate of the rate for the entire galaxy on the basis of the observed supernovae is not quite so simple, for there are large regions of the galaxy that we cannot see. We can now probe those parts with radio waves but not visible light, and we know that the visual record must be incomplete. Other sources of bias can be allowed for, and the rough overall figure is about one supernovae per fifty years. This number is used again later.

Beyond the statistical evidence, we are interested in the individual supernovae. Our understanding of this stage in the evolution of a star is still very tentative in many aspects. The amount of energy that is released in the explosion is indicated by the brightness, and the way in which the brightness changes in the days after the outburst tells us something about the expansion of the stellar debris and the changes taking place. A graph that displays the observed brightness, as it changes with time, is termed a **light-curve**. Tycho Brahe's records were accurate enough to be used by modern astronomers to reconstruct the light curve for his supernova (Fig. 4.6). Brahe compared

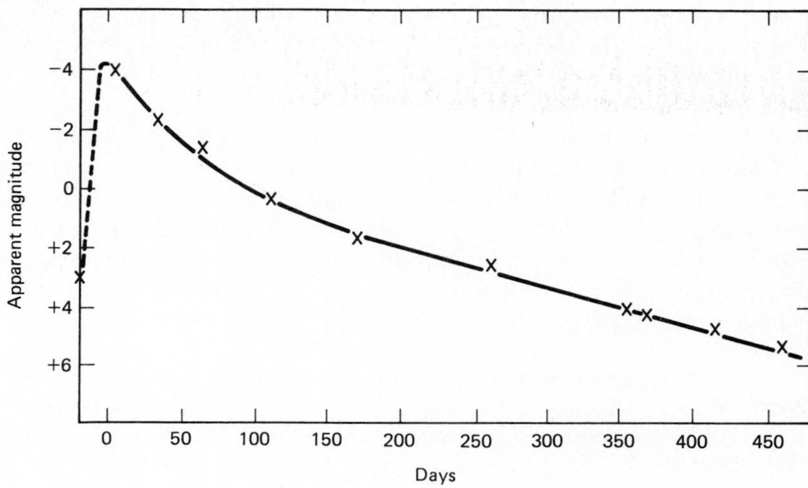


**Figure 4.6** The light curve reconstructed from descriptions of the A. D. 1572 supernova. Apparent magnitudes have been estimated from comparisons that were made between the "nova" and planets and well-known stars (D. H. Clark and F. R. Stephenson, *The Historical Supernovae*, Pergamon Press).

the new star with the brightness of planets and well-known stars nearby so that even though he had no light measuring device, his data are reliable enough to be used today.

The star was visible for about a year and a half, and it is interesting to quote from Brahe's original description:

> . . . behold, directly overhead, a certain strange star was suddenly seen, flashing its light with a radiant gleam and it struck my eyes. When I had satisfied myself that no star of that kind had ever shown forth before, I was led into such perplexity by the unbelievability of the thing that I began to doubt the faith of my own eyes, and so, turning to the servants who were accompanying me, I asked them whether they too could see a certain extremely bright star when I pointed out the place directly overhead. They immediately replied with one voice that they saw it completely and that it was extremely bright. . . . immediately I got ready my instrument. I began to measure its situation and distance from the neighbouring stars of Cassiopeia, and to note extremely diligently those things that were visible to the eye concerning its apparent size, form, colour, and other aspects[1].

This supernova was also observed in the Far East, where it was recorded in the **Ming-shih-lu**.

> (6th year in the Lung-ch'ing reign period, 10th month), (day) *hsin-wei*. . . . Previously in the 10th month on the 3rd day *ping-ch-en* at night a guest star was seen at the north-east; it was like a cross-bow pellet. It appeared beside *Ko-tao* in the degrees of (*Tung*)-*pi* lunar mansion. It gradually became fainter. It emitted light in the form of pointed rays. After the 19th day, *jen shen* at night the said star was orange in color. It was as large as a lamp and the pointed rays of light came out in all directions. . . . It was seen before sunset. . . . At the time the Emperor saw it in his palace. He was alarmed and afraid, and at night he prayed in the open air on the Vermillion Steps[2]

No star can be seen today at the location of Brahe's nova, but the careful analysis of Brahe's observations by Walter Baade in the 1940s stimulated a search by radio astronomers that led to the discovery of the supernova remnants in the 1950s. These remnants emit radio waves and show up as a wispy shell of gas, still expanding at a very high speed and forming a nearly circular ring centered on the place so well described by Tycho. Very recently, these remnants have also been found to be a source of X-rays.

Although the Eastern records are rich in their reports of comings and goings in the sky, there is an almost total silence in the West. Stephenson and Clark list seventy-five novae and supernovae that were sighted before the invention of the telescope. Apart from the sightings of Brahe and Kepler, *only* the giant supernova of A. D. 1006 was recorded in the West, and that in a very brief note in the monastery of St. Gallen in Switzerland. That supernova was possibly the most spectacular ever seen, but its location in the sky was too far south to be seen from much of Europe. For the rest there is silence, and our knowledge of those events rests entirely upon Eastern sources. This silence is generally attributed to the rigid position of the Church, within which the perfection of the heavens and the centrality of the Earth could be questioned only at a great personal risk. In the case of comets, which were consigned to the atmosphere, there was no problem, and the

[1] D. H. Clark and F. R. Stephenson, *The Historical Supernovae* (Oxford, Pergamon Press, 1977) p. 174.

[2] D. H. Clark and F. R. Stephenson, *The Historical Supernovae* (Oxford, Pergamon Press, 1977) p. 175.

**Figure 4.7** The A. D. 1066 appearance of Halley's comet as recorded in the Bayeux Tapestry. The legend reads "Isti mirant stella," ("They stand in awe of the star.") Made between 1073 and 1083 to commemorate Harold's victory at the Battle of Hastings. The appearance of the comet was taken as an omen (Yerkes Observatory).

terrifying appearance of Halley's comet in A. D. 1066 has been perpetuated in the Bayeaux Tapestry (Fig. 4.7).

The great event of A. D. 1054, now identified with the Crab nebula, provides another interesting example of this selective silence (Fig. 4.8). There is not one report from Western Europe and only one from the Middle East, but amongst many in the Chinese annals we find

> 1st year of the **Chih-ho** reign period, 7th month, 22nd day . . . Yang Wei-te said, 'I humbly observe that a guest star has appeared; above the star in question there is a faint glow, yellow in color. If one carefully examines the prognostications concerning the Emperor, the interpretation is as follows: The fact that the guest star does not trespass against *Pi* and its brightness is full means that there is a person of great worth. I beg that this be handed over to the Bureau of Historiography[3].'

[3] D. H. Clark and F. R. Stephenson, *The Historical Supernovae* (Oxford, Pergamon Press, 1977) pp. 141–142.

**Figure 4.8** The Crab Nebula present remnant of the A. D. 1054 supernova explosion. Now too faint to be seen by the naked eye, the supernova outburst was bright enough to be seen by day for 23 days (Lick Observatory).

By now, the Crab nebula has probably been studied more than any other object in the sky outside the solar system, and it is fortunate that its early phases have been so well documented (see Chap. 25 for further discussion). Unlike Tycho's nova, the Crab does have a visible remnant that was listed in catalogs long before its full significance was appreciated. In the eighteenth century, Charles Messier was one of the foremost authorities on comets. Hunting for comets involves looking for objects that suddenly appear and which move against the fixed pattern of the stars. Comets will not have the pointlike appearance of stars, but they can be confused with other hazy objects known as **nebulosities**. Messier found that some nebulosities turned up repeatedly in his comet searches. He therefore compiled a list of the most prominent of these noncomets so that other astronomers could save time by not repeatedly having to check on well-known objects. The Messier catalog has 103 items, with M1 being the Crab and M31 the Andromeda nebula. It has turned out to be a list of great interest, both visually and scientifically, although Messier himself had no idea of the nature of his number 1.

An important use of the old far Eastern records lies in allowing us to identify supernovae whose remnants we can still see, and in establishing their ages very firmly so that their evolution can be followed. Radio astronomers have found other supernova remnants (SNR) for which we have no records. In many cases their present characteristics suggest that the explosions took place a very long time ago. Without knowing the ages of these SNR, some of our analysis is limited, but their present properties are of considerable interest.

## 4.5 IN GENERAL

An important new field is being developed by scholars who are combining language skills and astronomical knowledge. We have already mentioned Baade's pioneering studies relating to the Crab nebula and its description in the Chinese records. More recently in the 1950s Iosef Shklovsky of the Sternberg Astronomical Institute in Moscow utilized his contacts with the Chinese to encourage an expansion of these studies into a systematic search of their ancient records for reports of other novae. Several tabulations and detailed analyses have appeared. The old records are in a sense being recycled to good effect.

## CHAPTER REVIEW

Some historic astronomical observations can be used because of their reliable dating in the present study of changes in the behavior of the Earth, Sun, and stars.

Eclipse records have been used to check on variations in the Earth's rate of rotation.

Sunspots records show that the Sun's 11-year cycle has not always been regular. In addition, the sunspot records have been used to measure the rate of solar rotation and this, too, shows changes.

Supernovae are violent outbursts of massive stars. The ancient record of supernovae explosions can be combined with present-day observations in the development of theories of stellar structure and evolution.

## NEW TERMS

**light curve**
**limb (of the Sun)**
**nebulosity**
**nova**
**supernova**
**sunspots**

## PROBLEMS

**1.** Why can historic records of solar eclipses but not lunar eclipses be used to provide information on changes in the speed of rotation of the Earth?

**2.** Which pieces of information are needed in order to be able to predict where on the Earth's surface the next total solar eclipse will be visible?

**3.** Why does a solar eclipse have only a few minutes of totality but a lunar eclipse last for much longer?

**4.** For the eclipse of 136 B.C., totality was seen about 47° east of where it would have been if the Earth's rotation speed had not changed. How long does it take the Earth to rotate through 47°?. This difference has accumulated over the course of about 2100 years. What is the average change per day?

**5.** The solar diameter is $1.4 \times 10^{11}$ cm, and the lunar diameter is $3.5 \times 10^{8}$ cm. Draw a diagram to show the umbra in the Moon's shadow, and calculate the length of the umbra. Compare this to the average Earth—Moon distance. Refer to Appendix E and G for dimensions of the orbits of the Earth and Moon.

**6.** It has been suggested (but not proven) that the circle of Aubrey holes at Stonhenge was used for eclipse prediction. Suppose that it had been used for predicting solar eclipses. The Aubrey holes were probably laid out before 2500 B.. C. If the Earth's rotation speed had changed steadily, as indicated by the 136 B. C. eclipse, by how much would the longitude of an eclipse prediction be in error over the time span since Stonehenge?

**7.** The Earth's rotation speed actually changes in an irregular and unpredictable fashion. As a result, leap seconds are added every year or six months. Check with a local observatory to find out when the next second will be added—the extra second will be included in the recorded time service from the U.S. Naval Observatory.

**8.** You can observe sunspots safely by projecting a solar image onto a screen. There are several safe ways of observing sunspots. If you can observe them, do this over a period of about a month to see their rotation across the solar disc. You can also count them and see whether the number changes through the duration of your astronomy course.

**9.** Use the data shown in Fig. 4.5 to calculate the average length of the solar sunspot cycle. Use this to predict when the next maximum will occur.

**10.** Why are historic observations of supernova outbursts important for modern astrophysics?

CHAPTER

# 5

# From Observations to Theories: The Triumph of Geometry

## 5.1 THE EARLY STAGES

Scientific progress is made at a very uneven rate. Sometimes a line of development is halted by the lack of tools—conceptual, mathematical, instrumental—and for many years the practitioners may sense a need and a frustration as they realize that they are unable to make progress but cannot see how to take the next steps or even in which direction to go.

We can place a critical step in astronomy's progress in the fifth century B. C. Before that time we find only observers—those who took note, sometimes kept records, and sometimes saw order amidst their records. Since that time science has developed around theories constructed by those who sought unifying descriptions and underlying causes beneath the observed patterns.

Those patterns might have been seen in the regular recurrence of events such as eclipses, the appearance of Venus as a morning or evening star, or the cycle of phases of the Moon. To whatever extent these cycles led to reliable predictions, they were wholly arithmetical and implied no overall geometrical view of construction of the solar system. There were of course cosmological ideas—combinations of beliefs and myths handed down as legends in which the positions of the Earth, heavens, and nether regions were described and the activities of the gods heroically set out. But these provided no basis for the astronomical tabulations, no ordering of the planets in relation to one another and to the Sun. They were musings, speculations, elaborately embroidered but of no help to predictive astronomy.

Today, when we describe our solar system, we think of each planet moving in its orbit, as the Earth moves around the Sun, with the Moon being carried along in its own orbit around the Earth. We have a fairly clear spatial conceptualization—we call it a model.

Models are now used almost everywhere in science but were conspicuously absent from astronomy before about 500 B. C. Then quite suddenly, we find models being introduced and elaborated until Ptolemy's model (about A. D. 125) appeared to provide such a complete solution to the celestial puzzle that it remained in use for nearly 1500 years. The ingredient that led to this scientific revolution was the development of geometry—the intellectual tool that provided the means of describing the motions in the heavens. Geometry was unknown to the Babylonians; without it they kept score but could not explain the layout of the field.

## 5.2 THE GEOMETRICAL REVOLUTION

Over a period of several hundred years, the Greeks devised geometrical models for the heavens. They combined astronomical observations with geometry to obtain rough values for the relative distances and sizes of the Sun and Moon. Aristarchus developed a sun-centered model for the solar system but there was insufficient evidence for it to prevail against the philosophically-supported model in which the Earth was at the center of the solar system and of the universe.

In the progress towards a realistic understanding of the scale of dimensions in astronomy, the size of the Earth is important. No reliable value existed until the measurement by Eratosthenes (276–195 B. C.). By this time the extension of the empire of Alexander the Great had seen the shift in the center of astronomical learning to the new city of Alexandria, where Eratosthenes was one of the first directors of the Library. The town of Syene (now Aswan), lay on the Nile near the first cataract and almost exactly on the tropic. At noon on the summer solstice, the Sun stood overhead and cast no shadow from a vertical post (Fig. 5.1). On the same day in Alexandria the length of the shadow of another vertical post indicated that the Sun was almost 7° (close to $\frac{1}{50}$ of a circle) south of the vertical. Accordingly, Eratosthenes reasoned that Alexandria must be $\frac{1}{50}$ of the Earth's circumference north of Syene. Since the distance between these two towns was known to be 5000 **stadia**, the full circumference of the Earth had to be $500/(\frac{1}{50})$ or 250,000 stadia. There is uncertainty today about the size of the Greek unit of length, the **stadium** that Eratosthenes used. Units of length were not rigorously standardized at that time, and sometimes different lengths carried the same name. The Egyptians used both cubits and Royal cubits, and the stadium is known to have varied between $7\frac{1}{2}$ and 10 to the Roman mile, with 157 m (515 ft) being considered a likely value for Eratosthenes to have used.

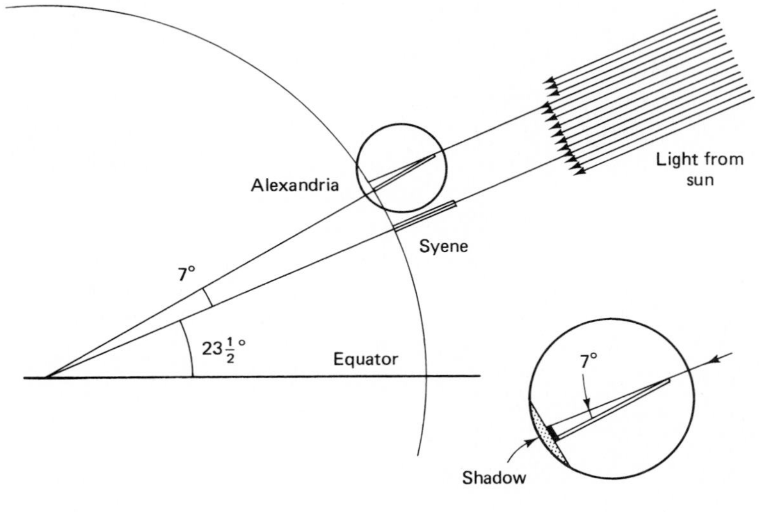

**Figure 5.1** The first good determination of the size of the Earth came from Eratosthenes' observations of the length of shadows in Alexandria on the day of the summer solstice. On that day the Sun was directly overhead at the southern town of Syene (Aswan). The distance between Alexandria and Syene was known allowing the Earth's circumference to be calculated.

The value of 5000 stadia is itself somewhat uncertain. It was measured by the time taken for the king's messengers to run it. Despite the final value being a bit fuzzy (in our critical view) a reasonably accurate value was obtained for the size of the Earth.

Contributions to astronomical knowledge came from many of the Greek philosophers, but probably the most important and certainly the longest-lasting came from Ptolemy. We do not know the exact dates of Ptolemy's birth and death; however, observations made by him between A. D. 127 and 151 are known. His great work in astronomy comprised the *Thirteen Books of Mathematical Composition* (the *Mathematike Syntaxis*). In its Arabic translations it became known as *Al-majasti* (the Greatest), from which comes its best known name, the *Almagest*.

In the *Almagest* Ptolemy elaborated on a model that originated with Hipparchus. The starting point was familiar: The stars were on a sphere rotating on a fixed axis. At the center was the Earth, also a sphere.

To explain the observed motions of the heavens, he built on the philosophical principle of uniform motion—steady speed in a circle. This was clearly sufficient to account for the nightly motion of the stars. But, as noted many times, the planets display their retrograde motions and also show variations in speed. Variations in speed could be incorporated into the model by the ingenious introduction of an **epicycle** (Fig. 5.2). Here, a planet was carried uniformly on a circle (**the epicycle**) whose center was in turn moving uniformly around the Earth on another circle, the **deferent**. The rotations on the two circles were synchronized so that each took the same time for one revolution. At one part of the orbit, the planet would be closer to the Earth and moving backwards as required.

With an array of geometrical tools at his disposal, Ptolemy set out to describe the motion of the Sun, Moon, and planets using whichever construction gave the results in closest agreement with observation. To account for the Moon's motion, he used a large epicycle, but this resulted in such large variations of the Moon's distance from the Earth that changes in the apparent size of the Moon should have been far larger than were actually observed. When he wanted to account for the known changes in the Moon's apparent diameter, he simply used a different construction just for that effect.

Ptolemy knew that the orbits of the planets did not lie exactly in the ecliptic plane but that each orbit was inclined at a slightly different angle.

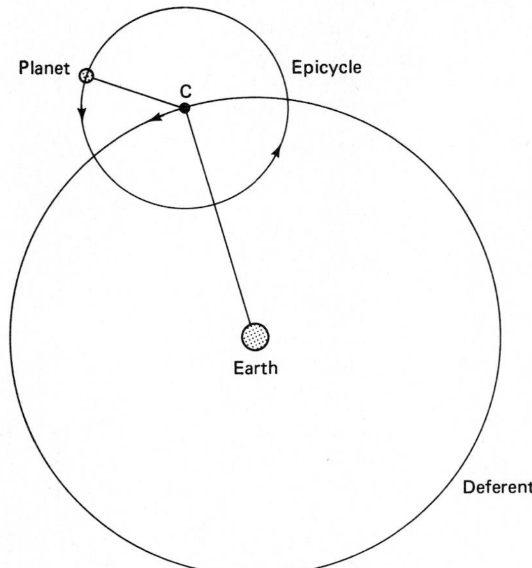


**Figure 5.2** Retrograde motion could be reproduced if a planet were carried on one circle (the epicycle), and the center of this circle itself moved around another circle (the deferent). All circular motions could still be uniform, and the retrograde motion could be simulated by appropriate choices of the radii of the circles and the speeds of the planets and cycles.

Therefore, more calculations were needed to describe the movements in celestial latitude (above and below the ecliptic plane).

The aim of Ptolemy's geometrical choices was the accurate description of the complicated motions in the sky. In this, he succeeded magnificently by great cleverness. What he did *not* seek was a unified picture of these motions. No relation was shown between the movements of the different planets, for example, and it was apparently quite acceptable to produce a different explanation for each planet even though the motions of the planets appeared superficially similar. All that was demanded of the theory was that it gave the right results—and, indeed, this scheme survived for nearly fifteen centuries. As Ptolemy himself put it:

> I do not profess to be able thus to account for all the motions at the same time, but I shall show that each by itself is well explained by its proper hypothesis.[1]

This is an achievement we should honor.

## 5.3 THE COPERNICAN REVOLUTION

There was a very long and silent period after Ptolemy, before the astronomical revival that we usually associate with Copernicus. No major advances took place in the understanding of the workings of the solar system, and there were no improvements in the methods of computation. The Romans, despite the extent of their empire, made no major contributions. The Eastern Roman Empire centered in Byzantium was not tolerant of novel ideas, while in the West Greek science was equated with Paganism and as such was out of favor. Science prospered for awhile in Baghdad under the Caliphs, and astronomy was a beneficiary. The Greek works were translated into Arabic. In 829 a new observatory was built in Baghdad that flourished as a center of mathematics and astronomy for 200 years. To this period, we can trace the names of many stars. The bright star Aldebaran was al-dabaran, the Follower (of the Pleiades); Betelgeuse started as yad al-jawza, the hand of al-Jawza (Orion).

Arabian astronomers made extensive use of the **astrolabe**, which can be admired in some museums today. These fine instruments consisted of elaborately inscribed brass discs with moving pointers and were used for calculating star positions (Fig. 5.3).

After about A. D. 1000 the Arabic translations spread from the Islamic countries to Europe where they were translated into more accessible languages. During this period there was the continuing need for tables of celestial positions, especially to permit the prediction of close approaches (conjunctions) between planets. These conjunctions and their associated omens could be favorable or unfavorable and could influence decisions—when to start on a long journey, whether to marry, whether to enter into a treaty.

The most influential and longest-used tables were those produced at the court of Alfonso X of Leon and Castile in the thirteenth century. By the middle of the fifteenth century, however, errors as large as several degrees were being found between observations and the predicted positions, and the need for improved tables was clear.

Nicolaus Copernicus was born in 1473 in Torun, Poland to a German family that had migrated there when the area was ruled by the Knights of the Teutonic Order. Copernicus' interest in astronomy had started in his student days in Krakow. His copy of the Alfonsine Tables is now in the University of Uppsala along with other volumes from his personal library,

[1] J. L. E. Dreyer, *History of Astronomy from Thales to Kepler* (Cambridge, UK, Cambridge University Press, 1905) p. 201, (reprinted by Dover, 1953).

**Figure 5.3** Persian astrolabe, 18 cm in diameter and made of brass. This astrolabe was made in the early eighteenth century, but these instruments were already widely used by the twelfth century. The network of lines on the base (the mater) represent celestial coordinates. The open-pattern disc (rete) rotates on the base about a pivot point that represents the north celestial pole. Decorated points on the rete represent stars whose positions at different times can thus be computed (St. Louis Art Museum).

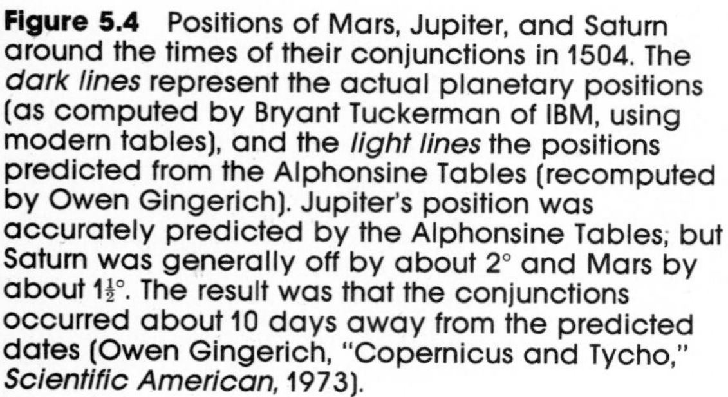

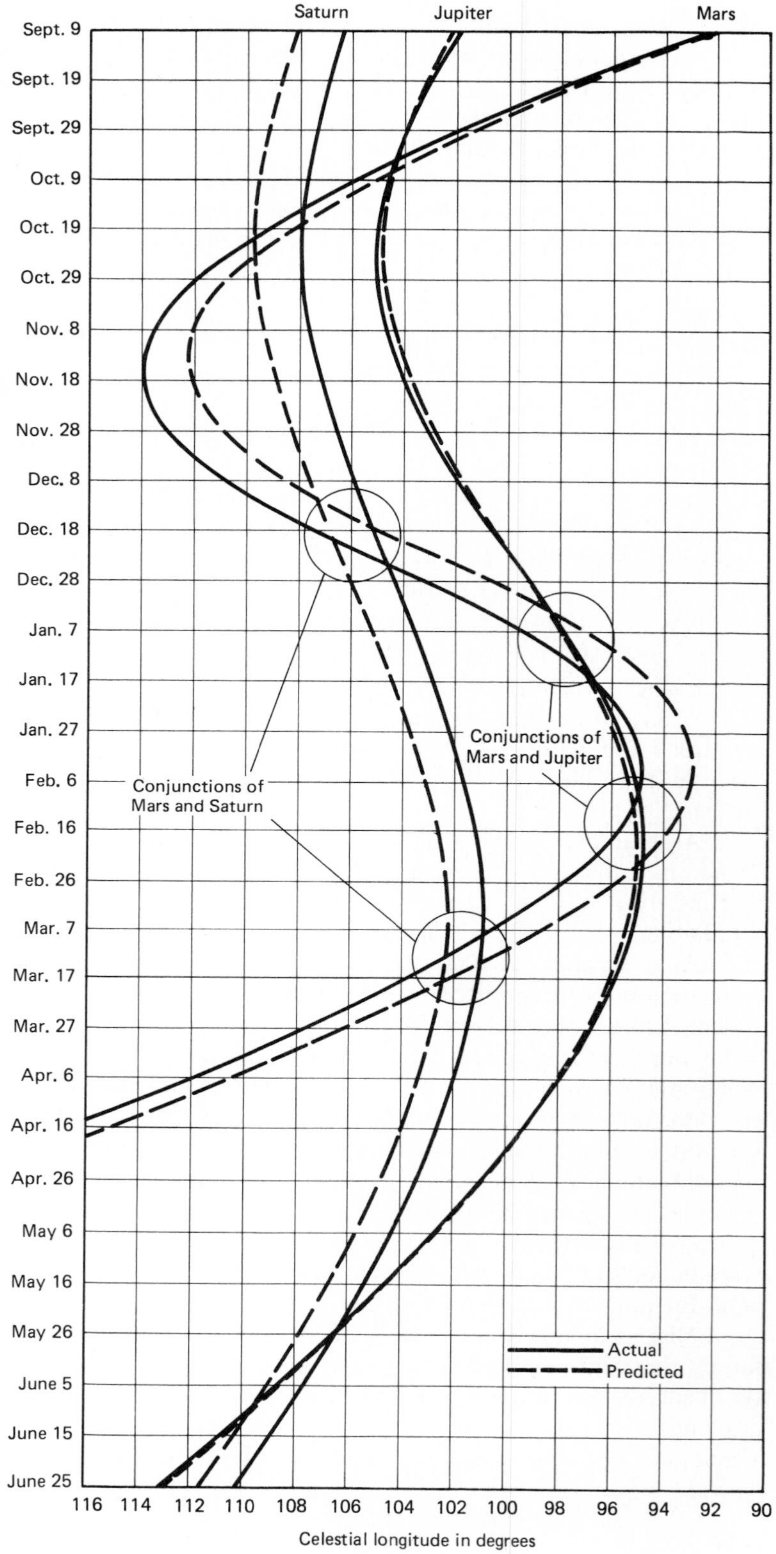

**Figure 5.4** Positions of Mars, Jupiter, and Saturn around the times of their conjunctions in 1504. The *dark lines* represent the actual planetary positions (as computed by Bryant Tuckerman of IBM, using modern tables), and the *light lines* the positions predicted from the Alphonsine Tables (recomputed by Owen Gingerich). Jupiter's position was accurately predicted by the Alphonsine Tables; but Saturn was generally off by about 2° and Mars by about $1\frac{1}{2}$°. The result was that the conjunctions occurred about 10 days away from the predicted dates (Owen Gingerich, "Copernicus and Tycho," *Scientific American*, 1973).

sent to Sweden by Gustavus Adolphus in 1627 during the Thirty Years' War. The Alfonsine volumes have many of Copernicus' own notes including, on the last page (in Latin):

> "Mars surpasses the numbers by more than two degrees; Saturn is surpassed by the numbers by $1\frac{1}{2}$ degrees[2]."

Owen Gingerich, of Harvard University, has shown that these enigmatic notes refer to the **conjunctions** between Saturn, Jupiter, and Mars of 1504. Gingerich used modern computer calculations to check on the Alfonsine Tables that Copernicus had used. The conjunctions actually occurred about 10 days away from the times predicted in the Alfonsine Tables. At the times that the Tables had listed for the conjunctions, Mars was indeed 2° off from its predicted position, and Saturn $1\frac{1}{2}$°. Jupiter's positions were accurately given by the Tables. These deviations may well have served to confirm Copernicus in his view that the Ptolemaic system, the basis of the Alfonsine Tables, needed revision.

Perhaps Copernicus felt that the scheme he was about to propose would permit easier computation, but the major motivation behind his theory was strictly theoretical. He described it in this way:

> Having become aware of these defects, I often considered whether there could be found a more reasonable arrangement of circles, from which everything would move uniformly about its proper center, as the rule of absolute motion requires.[3]

Copernicus then set out his assumptions. The Sun was at the center of the solar system, and the Earth and planets were in orbit around the Sun. In addition, the Earth rotated on its own axis. These were certainly revolutionary proposals.

The system has an immediate intellectual appeal. The troublesome retrograde motions were now described as natural consequences of differences between the speeds of the Earth and the planets. The daily apparent motions of the heavens were now seen as the result not of their own movement but of the Earth's rotation; the extreme distance of the Earth from the stars explained the apparent lack of any movements among the stars.

### BOX 5.1
*Retrograde Motions*

Ptolemy solved the problem of the retrograde motions by introducing epicycles. A planet would appear to move backwards at certain times when seen from the Earth, and the radii of the circles could be adjusted to fit the observations (Fig. 5.5).

When Copernicus moved the center of the solar system to the Sun and put the Earth and planets into motion about this new center, he also suggested that the orbital speeds diminished as the distances increased. The figure shows illustrative positions of the Earth and Mars at equally spaced time intervals in their respective orbits. While the Earth completes one orbit around the Sun, Mars moves across only a small fraction of its orbit. The **apparent** positions of Mars seen from the Earth are then given by the lines joining corresponding positions in the two orbits. The appearance of the retrograde loop occurs when the Earth and Mars are closest together (**opposition**) moving in the same direction and Mars appears brightest. During

[2] Owen Gingerich, "Copernicus and Tycho," *Scientific American* Vol. 229 (1973) 89, (reprinted by permission of W. H. Freeman and Company).

[3] E. Rosen, *Three Copernican Treatises*, 3rd ed., (New York, Farrar, Straus, & Giroux 1971) pp. 57–59.

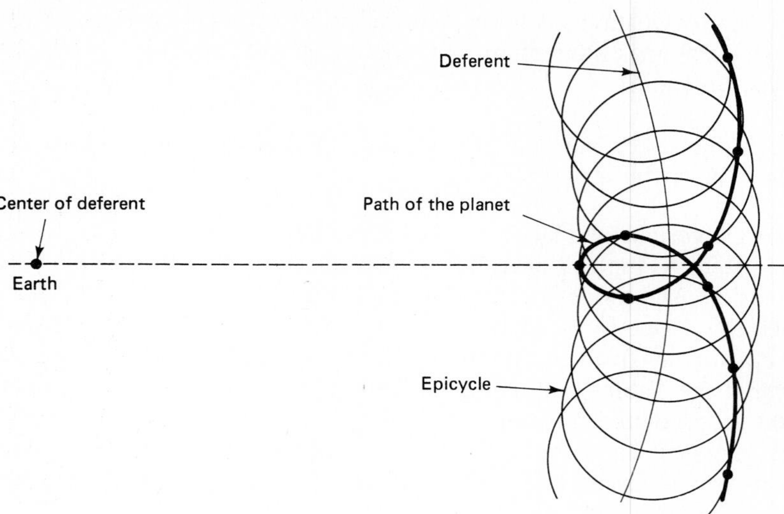


**Figure 5.5** In the Ptolemaic system, the retrograde motions came from the combinations of movements around the epicycle as it moved along the deferent, which had the Earth at its center.

this brief period the Earth is overtaking Mars (positions F to I). Direct motion is resumed when the Earth moves almost at right angles to Mars and then in the opposite direction (positions I to M) (Fig. 5.6).

With those planets between Earth and the Sun, that is, Mercury and Venus, it is the Earth that moves more slowly but the same sort of effect occurs, as Copernicus put it,

> not from their motions but from the Earth's . . .

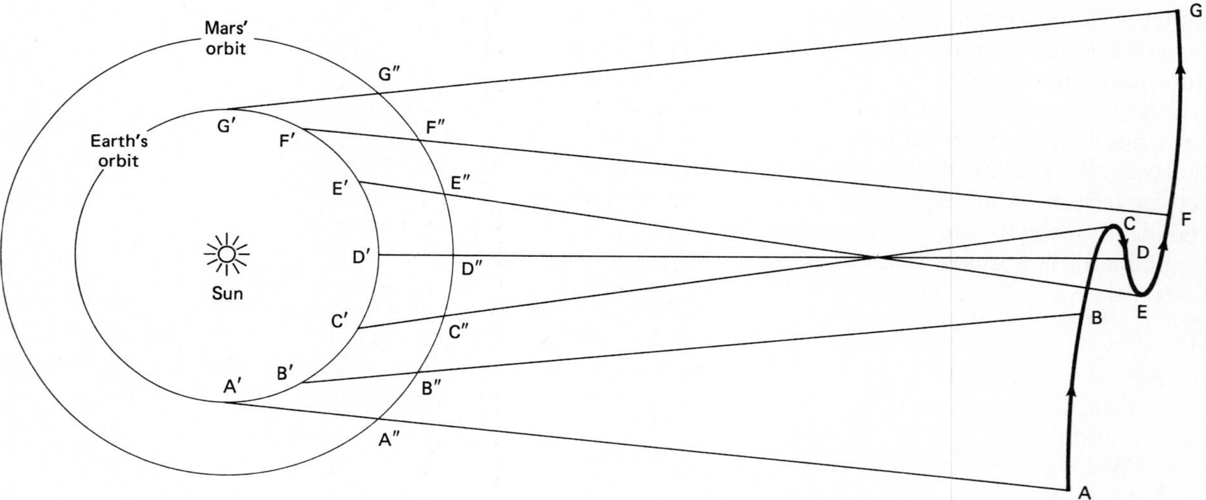


**Figure 5.6** Retrograde motion of Mars, as understood with the Copernican system. The Earth takes 1 year to complete one orbit around the Sun, but Mars takes 1.88 years. As a result the Earth overtakes Mars whose apparent positions (against the background of the distant stars) are shown. Mars moves in its direct (forward) path while the Earth moves from *A′* to *C′*, but then the Earth's movement carries it ahead of Mars which seems to be moving back (*C* through *D* to *E*). By this stage, the Earth's path is curving rapidly (*E′*, *F′*, *G′*), and Mars appears to resume its direct movement (*E, F, G*).

Copernicus now had a simple explanation for the sequence of the planetary distances and speeds. With the heliocentric system and the observed synodic periods of the planets, he was able to calculate their sidereal pe-

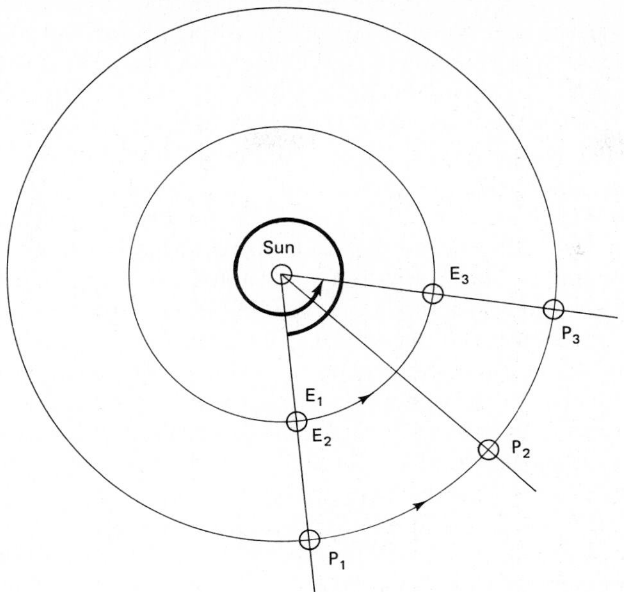


**Figure 5.7** Sidereal and synodic periods of planetary motion. The sidereal period of a planet (*P*) is defined by the completion of one orbit around the Sun. The synodic period is defined by the alignment of the Sun, Earth, and planet. Thus the Earth, Sun, and planet *P* are initially aligned (at some place $E_1$ for Earth and $P_1$ for the planet). After 1 year, the Earth has completed one orbit (and is back again at the starting point, $E_2$). In this example, the planet *P* is farther from the Sun and so moves slowly so that it has progressed only as far as $P_2$ in 1 year. The faster-moving Earth continues and catches up so that the next alignment occurs at $P_3$ with the Earth at $E_3$. The time between alignments is a synodic period for planet *P*, and its relation to the sidereal period is easily calculated.

riods—and the planets immediately fell into the correct order with the slowest being furthest from the Sun and with the two fastest within the Earth's own orbit.

### BOX 5.2
*Sidereal and Synodic Periods of the Planets*

In an earlier section, we saw how the length of the day could be defined in different ways because of the Earth's own motion. We defined the sidereal day (based on the Earth's rotation in space) and the solar day (based on the apparent rotation relative to the Sun). We find a similar situation when we come to describe the orbital periods of the planets.

The sidereal period of a planet is the time it takes to make one orbit around the Sun. As viewed from the moving Earth, the planet appears to complete one orbit each time the Sun, Earth, and planet line up. This does not define exactly one orbit in space.

There is a simple relation between the **synodic period** (S years) and the **sidereal period** (P years) for each planet. For planets whose orbits lie between the Earth and the Sun

$$\frac{1}{P} = 1 + \frac{1}{S} \qquad (\text{``inferior'' planets})$$

For planets whose orbits lie further from the Sun,

$$\frac{1}{P} = 1 - \frac{1}{S} \qquad (\text{``superior'' planets})$$

For example, Saturn has a synodic period, S, of 378 days or 1.035 years. Its sidereal period is thus,

$$\frac{1}{S} = 1 - \frac{1}{P} = 1 - 0.966 = 0.034$$

$$S = 29.6 \text{ years.}$$

This value differs slightly from the value listed in the table of planetary data, because we have used a value of P rounded off to the nearest day.

The new theory had an elegant simplicity in this design, but did it work? Copernicus did not have at his disposal any significant body of reliable, recent data. Worse yet, since he was still fully committed to uniform circular motion, his model had to have its full share of epicycles.

The number of circles in the Ptolemaic system has been subject to gross exaggeration. It has sometimes been said that after Ptolemy, astronomers kept adding circles to try to improve the accuracy of the predictions. The 1969 edition of the *Encyclopedia Brittanica* asserts that forty to sixty circles were needed for *each* planet. This is simply not correct. Gingerich has shown that the Alfonsine Tables can be recomputed today using only one circle and a single epicycle for each planet, and he has also shown that if one compares the models of Copernicus with that of Ptolemy, counting circles just for the longitude of the planets, the score is Copernicus 18, Ptolemy 15. Economy of circles was simply not attained. In addition, in order to get the right answers, Copernicus had to use different circles for his calculations of planetary longitude from those needed for latitude. Finally, again in order to improve the agreement with observation, Copernicus was forced to shift the Sun so that the center of his new system was the geometrical center of the Earth's orbit with the Sun off to one side. In summary: Copernicus' theory had some attractive features, but in simplicity and accuracy in actual use it offered no clear advantage over Ptolemy's.

## 5.4 TYCHO BRAHE

Tycho Brahe played a critical role in the progress of astronomy, but today tends to be somewhat neglected, sandwiched as he was between Copernicus and Kepler. Born in Denmark in 1546, he observed the conjunction of Jupiter and Saturn in 1563 finding the Alfonsine Tables off by a month. He first attracted attention, though, through his observation of the 1572 nova (Chap. 3). In 1576 Frederick II of Denmark established the Uraniborg Observatory for him on the island of Hveen, where he remained until 1597. Later he settled in Prague, where he died in 1601.

Brahe considered that some of the discrepancy between the predicted and observed positions of the planets might have come from the use of inaccurate measuring instruments. He therefore systematically set about improving their designs, having them more rigidly built and mounted with differing designs depending on the task at hand. One specific example will suffice. The depiction of Tycho with his great mural quadrant is well known (Fig. 5.8). The scale for the quadrant was set out along a wall that was accurately aligned in the north-south meridian. What is usually overlooked in this depiction of the quadrant is the scale of degrees. What looks like a simple geometric pattern is actually a clever device for increasing the accuracy of the measurements. The quadrant had a radius of just over 2 m, so that each degree division was 3.5 cm along the perimeter. To read angles to 1 arc minute accuracy would have required the subdivisions to be spaced at intervals of almost half a millimeter, which would have been impossible to scribe or read at that time. Brahe adopted a simple but ingenious expedient: transverse divisions.

It is relatively easy to set out rows of dots, with ten dots per row, six rows per degree, as shown in Fig. 5.9. Brahe could thus read his scales to the nearest arc minute by seeing with which dot his movable pointer was lined up. Brahe's earlier measurements are generally good to better than 10′, but recent studies have shown that many of his later positions are reliable to better than 1′. His precision was well known during his lifetime.

As might have been expected, Brahe also developed a model of the solar system. He was fully familiar with Copernicus model but was critical of it on very reasonable grounds. He had difficulty, of course, in reconciling

**Figure 5.8** Tycho Brahe's great mural quadrant. Brahe is shown both as an observer and also in his study against a background of his other instruments. The actual quadrant had a radius just over 2 m, and each degree division was 3.5 cm (just over an inch) along the quadrant. The elevation of a planet or star was measured by sighting from the movable aperture through the fixed slot in the wall (upper left). The transversals along the quadrant are shown in detail in Fig. 5.9 (Yerkes Observatory).

the model with the Scriptures, not a negligible problem in Lutheran Denmark. There were scientific objections, too. The ''heavy and sluggish'' Earth seemed an unlikely object to be whirling rapidly in space, and the distance between Saturn and the sphere of the stars seemed unreasonably vast. The motion imputed to the Earth was hard to believe, and Brahe seemed persuaded by the standard argument that a stone dropped from a tower should fall far from the base if the Earth were truly rotating. In addition, if the Earth did move in an orbit around the Sun, then the positions of some stars should appear to change, but this had never been seen.

Brahe proceeded to construct his own model, a very sensible blend of the models of Ptolemy and Copernicus. The Earth was left at the center with the Moon in its usual orbit. Further out in orbit around the Earth was the Sun, but now the planets revolved around the Sun, and thus (indirectly) still around the Earth (Fig. 5.10).

This model computationally was as good as the Copernican model, and it represented a conservative compromise that had at least as much scientific validity as that of Copernicus.

Viewed in the context of his time, Brahe's resistance to the Copernican model was probably just what a prudent astronomer should have shown. He had more accurate data than Copernicus had, and his approach was more critical. No clear refutation of Brahe's model was possible on the basis of the evidence then available.

**Figure 5.9** Transverse divisions. Each degree on Brahe's quadrant was divided into sixty parts using an ingenious idea of marking out the small divisions along six transverse rows of dots. The alignment of the moving slide (Fig. 5.8) could then be read to the nearest 1/60 of a degree.

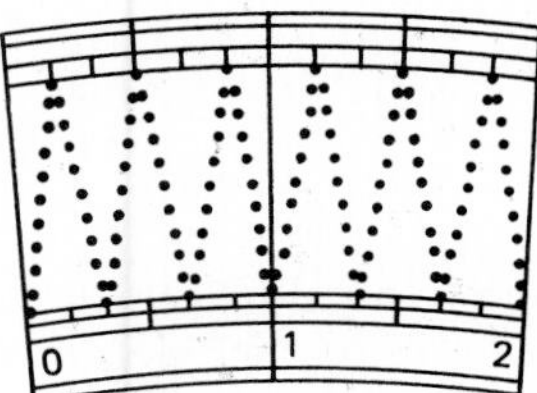

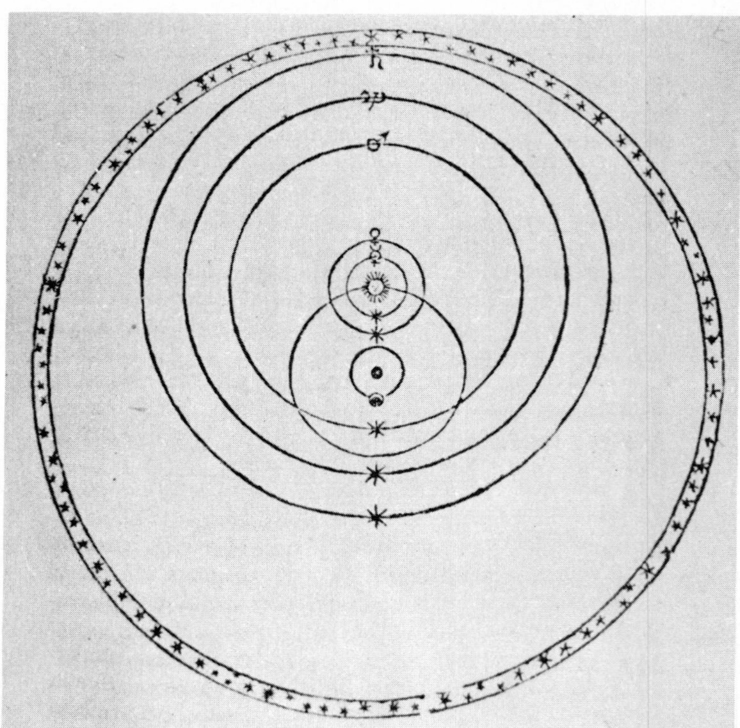

**Figure 5.10** Brahe's model of the solar system. Like Copernicus, Brahe considered the five planets to be in orbit around the Sun. Unlike Copernicus, however, he considered the Sun (with its attendant planets) to be in orbit around the Earth (Owen Gingerich and The Houghton Library, Harvard).

## 5.5 JOHANNES KEPLER

Johannes Kepler was born in 1571 in Weil der Stadt, near Stuttgart. He first learned of the Copernican theory from his teacher, Michael Maestlin. His early attempt to construct a geometrical model for the planetary orbits yielded only fair agreement with the known orbital sizes. Kepler wanted to improve the model and at this stage realized that he was limited by the accuracy of the data available. He knew that Brahe was the only astronomer with data of the needed accuracy, and he was able to take up a position with Brahe in Prague in 1600, starting work on the problem of the determination of the orbit of Mars. After Brahe died in 1601, Kepler inherited all of the data despite some dispute with the family. Kepler continued his work on Mars, and his first discovery related to the speed of that planet in its orbit. This has become known as his Second Law, but it was in fact discovered first and was the necessary ingredient for his subsequent discoveries.

In trying to describe the geometrical shape of Mars' orbit, Kepler was guided by what we would call physical reasoning, that is, he postulated physical causes for the motion and searched for effects that these causes might produce. His notions of the causes were wrong, but in his errors he arrived at the correct answer. In his *Mysterium Cosmographicum* (1596), Kepler had made the suggestion that a force emanating from the Sun was responsible for the motion of the planets, and this feeling was strengthened by the appearance of William Gilbert's *De Magnete* in 1600. Kepler felt that the effect of any magnetic force from the Sun should decrease with distance. The planets that were farthest from the Sun traveled slowest, and the orbits were all very nearly in the same plane (the ecliptic). It was thus logical for Kepler to assume that as the solar force had to spread out to cover the larger orbital circumference, the force at any place had to decrease in inverse proportion to the distance from the Sun. Thus the force would be one-fifth as strong at Jupiter as it was at the Earth.

This is precisely the type of reasoning used today, except that the influence is considered to spread out in all three dimensions and not in a single plane. If Kepler had not made this error, he might have anticipated

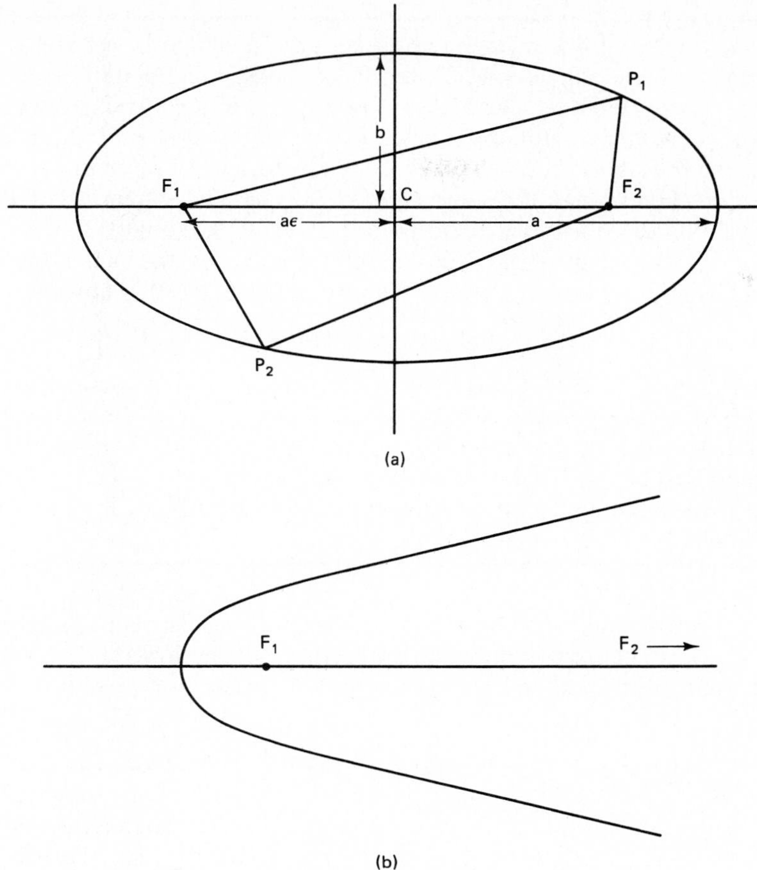


**Figure 5.11** (a) An ellipse has two foci ($F_1$ and $F_2$) located equidistant from the center $C$. The combined distances $F_1P + F_2P$ are constant for all orbits around the ellipse. Thus $F_1P_1 + F_2P_1 = F_1P_2 + F_2P_2$. In a circle, all diameters are equal. In an ellipse, there is one diameter that is longest (the major axis, $2a$) and one diameter that is shortest (the minor axis, $2b$). The distance from either focus to the center of the ellipse is $\epsilon a$, where $a$ is the semimajor axis, and $\epsilon$ is defined as the eccentricity of the ellipse. (b) An ellipse with $\epsilon = 0$ is a circle. As the eccentricity $\epsilon$ approaches the value 1, the ellipse becomes more and more elongated. One focus moves to infinity, and the ellipse becomes a parabola.

Newton's discoveries, but the strong concentration of the planets in the ecliptic was misleading.

In any event Kepler proceeded to examine the observed variation of Mars' speed at different parts of its orbit.

Through a series of calculations that included plausible assumptions but also errors that later cancelled, Kepler arrived at his correct (second) law: That the time taken to cover an arc of an orbit is proportional to the area swept out. The Second Law contains the well-known observation: Planets move more quickly when closest to the Sun (at perihelion) and slowest when farthest (at aphelion).

With this discovery and with observations taken at perihelion and aphelion, Kepler recomputed Mars' orbit and found apparent deviations as large as 8′ between Brahe's data and the theory. As Kepler then observed:

> after the divine goodness had given us in Tycho Brahe so careful an observer, that from his observations the error of calculation amounting to 8 minutes betrayed itself, . . . we should take the pains to search out at last the true form of the heavenly motions[4].

The variations in the speed and the 8′ discrepancy forced Kepler to reconsider the shape of the orbit, something that the Second Law does not mention explicitly. Kepler tried various noncircular orbits, including an oval and an epicycle. Although none of these worked, Kepler had clearly departed from the use of circles. He then computed twenty-two distances of Mars from the Sun using a new method of his invention and found that an excellent fit to the data was obtained with an ellipse with the Sun at one focus.

[4] M. Caspar, *Kepler,* New York, Collier Books, 1959) p. 133.

**BOX 5.3**
*The Geometry of Orbits*

An ellipse looks like a flattened circle, and its shape is specified by its **eccentricity**. Unlike the circle, center $C$ is no longer equidistant from all points on the circumference. The distance between $C$ and $P$ varies between a maximum value $a$ and a minimum value $b$. The largest diameter, $2a$, is the major axis and the smallest, $2b$, is the minor axis. Instead of referring to the radius, we therefore have $a$, the semimajor axis and $b$, the semiminor axis. On the major axis there are two focal points, $F_1$ and $F_2$, equally spaced from the center $C$. The shape of the ellipse is defined such that the sum of the distances from $F_1$ and $F_2$ to any point on the circumference ($P$) is always constant.

$$P_1F_1 + P_1F_2 = \text{constant} = P_2F_1 + P_2F_2$$

The distance of each focus from the center is then defined as $a\epsilon$, where $\epsilon$ is the **eccentricity**. If the two foci ($F_1$ and $F_2$) coincide, the radii $a$ and $b$ are equal, and $\epsilon = 0$; the ellipse then becomes a circle. As $\epsilon$ approaches the value 1, the ellipse become more and more elongated. When $\epsilon = 1$ precisely, one focus recedes to infinity and the ellipse opens up so that we have a parabola.

A crowning touch was Kepler's demonstration that a single curve—an ellipse—sufficed to describe correctly both the longitudes and latitudes provided the planet's ellipse was inclined slightly to the ecliptic.

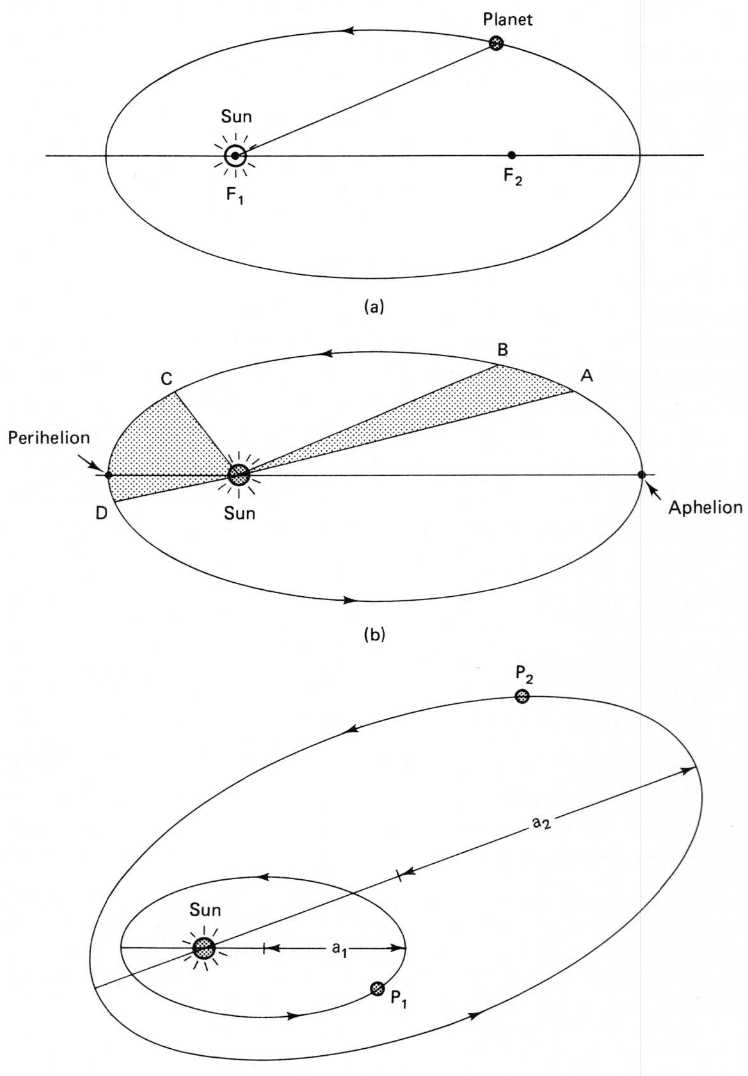


**Figure 5.12** (a) Kepler's first law: Each planet follows an orbit that is an ellipse with the Sun at one focus. (b) Kepler's second law: as a planet moves in its orbit, the line joining it to the Sun sweeps out equal areas in equal times. The planet moves slower when farther from the Sun, and thus it travels only through distance *AB* in the same time that it takes to travel from *C* to *D* when closer to the Sun. The shaded areas, however, are equal. (Closest approach to the Sun in termed perihelion and farthest is aphelion.) (c) Kepler's third law: The sizes of the orbits of different planets are related to the times they take to complete their orbits. Thus for two planets $P_1$ and $P_2$, in orbits with semimajor axes $a_1$ and $a_2$, by Kepler's third law we have $P_1^2/P_2^2 = a_1^3/a_2^3$

This magnificent synthesis for Mars was decribed by Kepler in his *Astronomia Nova*. It was truly a celestial physics leading to a new astronomy even though the foundations included erroneous assumptions. Kepler's magnetic model differed from the modern view of gravity in another important respect besides the variation of the force with distance. Kepler assumed that the solar force acted along the circumference, whereas it is now known that the direction of the gravitational attraction lies along the radius. But his theory was still the first modern scientific attempt to incorporate causes into astronomy.

The *Astronomia Nova* deals only with Mars, but in his popular *Epitome astronomiae* (1618), the structure of the solar system was set out fully for the first time, and the first two laws were extended to the other planets.

Kepler's Third Law relates the size of the orbit of a planet to the time it takes to complete one orbit, that is, its sidereal period. This law appeared in 1619.

**BOX 5.4**
*Kepler's Laws*

The three laws that Kepler derived can be summarized as follows:

1. Each planet follows an orbit that is an ellipse with the Sun at one focus.
2. As a planet moves in its orbit, a straight line joining the Sun to the planet sweeps out equal areas in equal times. Thus when the planet is closest to the Sun (perihelion) it is moving fastest, and when it is furthest from the Sun (aphelion), it is moving slowest.
3. The squares of the sidereal periods ($P$) of the planets are directly proportional to the cubes of the semimajor axes ($a$) of their orbits. This law can be written as

$$P^2 = ka^3$$

where the value of the constant $k$ depends on the units of length and time used for $a$ and $P$.

This law takes a particularly simple form when we measure the sidereal periods in years, and the semimajor axes in units of the Astronomical Unit (AU), the semimajor axis of the Earth's orbit. We then have

$$P^2 = a^3$$

For example, the semimajor axis of Venus' orbit is 0.7233 AU. We then have

$$P^2 = (0.7233)^3 = 0.3784$$

$$P = 0.6151 \text{ yr} = 224.7 \text{ days}$$

Kepler's achievements were no less revolutionary than those of Copernicus, and scientifically they were better grounded. With Kepler's advances and the simultaneous discoveries of Galileo, astronomy was at last ready to make the transition to the correct application of physical ideas soon to be introduced by Newton.

## CHAPTER REVIEW

**Babylonian astronomy:** planetary positions and eclipses noted; arithmetical methods used but no geometry.

**Greek astronomy:** developed geometrical models for operation of the solar system.

**Ptolemy:** developed the solar system model that had the Earth at the center and used nesting of circular motions, circles, and epicycles. Was able to produce very accurate predictions; tables based on his model continued to be used until sixteenth century.

**Arab astronomy:** preserved Ptolemy's work but added nothing conceptually innovative. Transmitted Ptolemy's astronomy to Western Europe via Spain.

**Copernicus:** recognized errors in Ptolemaic predictions (Alfonsine Tables)—developed solar system model with Sun at the center, planets in orbit around the Sun, and Earth rotating on its own axis. Model departed conceptually from Ptolemaic but was computationally no better. Did, however, produce a simple explanation for retrograde motions and also for a simple ordering of the planets from Venus out to Saturn. Heliocentric model also required distinction between synodic and sidereal periods.

**Tycho Brahe:** major improvement in observational precision, largely through design and use of better instruments. Observed 1572 supernova explosion and gave description that has been of use for present-day studies. Developed model for solar system that was intermediate between Ptolemaic and Copernican but gained little acceptance. Extensive observational data passed to Kepler on his death.

**Kepler:** after some attempts at developing a geometrical model of the solar system was influenced by recent discoveries in magnetism. Thought that magnetic force from Sun drove the planets. Analysis of Brahe's data together with idea of magnetic force led Kepler to three laws of solar system motions. Laws are correct; the Third Law, in particular, is very widely used today when it includes some additions developed by Newton.

## NEW TERMS

**astrolabe**
**conjunction**
**eccentricity**
**ellipse**
**epicycle**
**focus**
**heliocentric**
**parabola**
**semimajor axis**
**semiminor axis**
**uniform motion**

## PROBLEMS

**1.** What was the essential difference between Babylonian and Greek astronomy?

**2.** The geometry of ellipses was well known long before Kepler found that planetary orbits were ellipses. Why did not Copernicus, for example, try to use ellipses?

**3.** How does Copernicus' explanation for retrograde motion differ from that of Ptolemy?

**4.** Why are there two ways of defining the orbital period of a planet, and how are they related?

**5.** The sidereal period of Mars is 687 days or 1.881 years. Calculate its synodic period.

**6.** The synodic period of Venus is 584 days. Calculate its sidereal period.

**7.** Why is retrograde motion seen only with the planets and not with the stars or Moon?

**8.** Tycho Brahe devised a method of measuring angles to the nearest arc minute. Suppose you wanted to extend this method, to measure to the nearest arc second. Make an estimate of how large a wall quadrant you might need using the same general design as Brahe's.

**9.** Compare the solar system models of Copernicus and Tycho Brahe with the older model of Ptolemy. What astronomical evidence was then available to allow a clear choice to be made between these models?

**10.** When Tycho Brahe proposed his model for the solar system Galileo had not yet invented the telescope, and the phases of Venus had not been discovered. Can this discovery be explained by Tycho's model?

**11.** The eccentricity of the Earth's orbit around the Sun is 0.0167, and the semimajor axis is 1 AU. Calculate (in AU) the closest and farthest distances of the Earth from the Sun.

**12.** Mercury's distance from the Sun varies between 46 million km and 70 million km. Calculate the eccentricity of its orbit.

**13.** Sketch a planetary orbit for a planet that goes around the Earth and has an epicycle whose radius is one half of the orbit radius of 3 AU. What are maximum and minimum distances from Earth of this planet?

**14.** Why does Venus' brightness change markedly during the year, but the brightness of the stars does not seem to change?

**15.** What is Kepler's third law? Does it hold for circular orbits?

**16.** Calculate the period for a planet that moves around the Sun in an orbit with a semimajor axis of 10 AU.

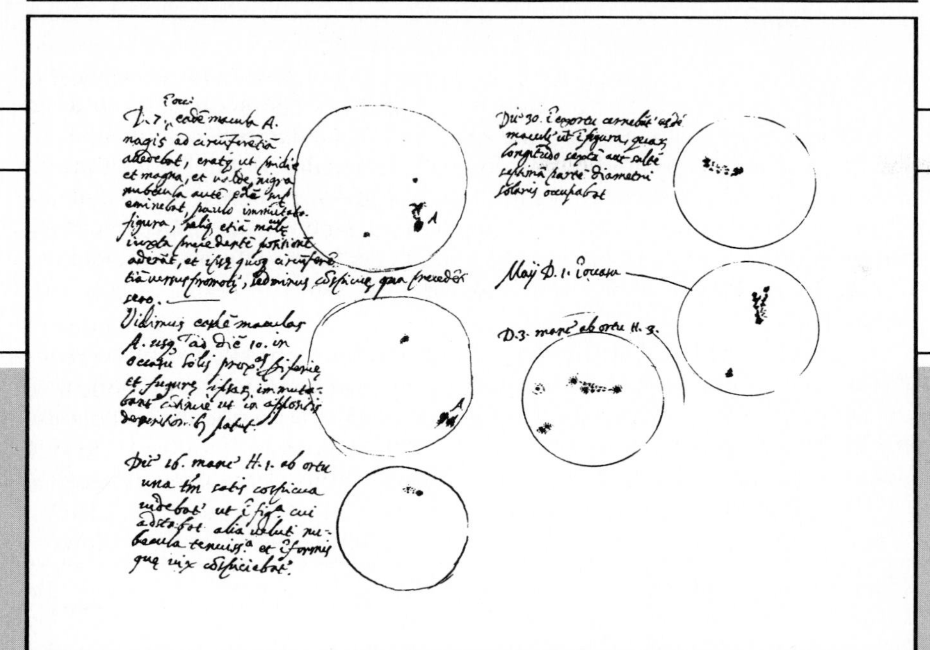

CHAPTER 6

# Galileo

Scientific discoveries come in all shapes and sizes. Many can be understood by only a small number of experts; this is particularly true in the fragmented and exotic science of today. Other discoveries can be appreciated in varying degrees by anyone. When this happens and when a discovery runs counter to some strongly held view that is important to a political or clerical power, repercussions can be expected. In this way a few ferocious rear-guard actions have marked attempts to divert science or halt its progress. Perhaps the most celebrated cases have been those of Galileo and Darwin.

There were many factors that finally combined to lead to Galileo's censure by the Church. One factor was the bitterness of the struggle during the Reformation and the Counter Reformation, another was the challenge to conservative scriptural interpretation. One factor was the internal rivalry between the Jesuits and the Dominicans, and another was Galileo's own argumentive nature.

How could astronomical views have produced such an intense reaction, especially after those views had been around for almost a century? The difference lay in a radical transformation of astronomy. As long as astronomical theories rested on complex geometrical arguments, they were not directly accessible to most people, and the central ideas—geocentric or heliocentric—could be debated in a sort of distant antiseptic way. The introduction of the telescope changed that permanently. With that tool, anyone, regardless of education or astronomical training could *see* the objects of the discussions.

## 6.1 GALILEO'S CAREER

Galileo Galilei was born in Pisa in 1564, son of a musician, Vincenzio Galilei. Lack of funds forced him to leave the University of Pisa before completing his formal education, but he continued with his own studies and started some experiments in physics. He became known for his exceptional mathematical abilities and was appointed by Ferdinand I de'Medici to the Chair of Mathematics at the University of Pisa in 1589. His lifelong interest in the science of the movements of objects (mechanics) appeared during his stay in Pisa in his lectures; his first book was *De Motu* ("On Motion"). His move to the University of Padua in northern Italy in 1592 was the result of financial pressures.

It was in Padua that Galileo made his great astronomical discoveries and achieved fame with the telescope, but he preferred to live in Florence. On the basis of his enhanced reputation, he was able to obtain the position in Florence as Chief Mathematician and Philosopher to Cosimo II, Grand Duke of Tuscany. There he remained through his various difficulties with the Church until his death in 1642.

## 6.2 THE TELESCOPE

The general optical behavior of glass was long known. By the end of the thirteenth century spectacles were in use in Italy, possibly helped by the excellence of the glass industry based in Venice and its nearby islands. The first clearly recorded use of a telescope comes from the Netherlands in 1608, but there seems to have been no use, certainly no systematic use, for astronomical purposes until Galileo's famous investigations in 1609 (Fig. 6.1). It is best described in his own words:

> First I prepared a tube of lead, at the ends of which I fitted two glass lenses, both plane on one side while on the other side one was spherically convex and the other concave. Then placing my eye near the concave lens I perceived objects satisfactorily large and near, for they appeared three times closer and nine times larger than when seen with the naked eye alone[1].

> It would be superfluous to enumerate the number and importance of the advantages of such an instrument at sea as well as on land. But foresaking terrestrial observations, I turned to celestial ones, and first I saw the moon from as near at hand as if I were scarcely two terrestrial radii away. After that I observed often with wondering delight both the planets and the fixed stars[2].

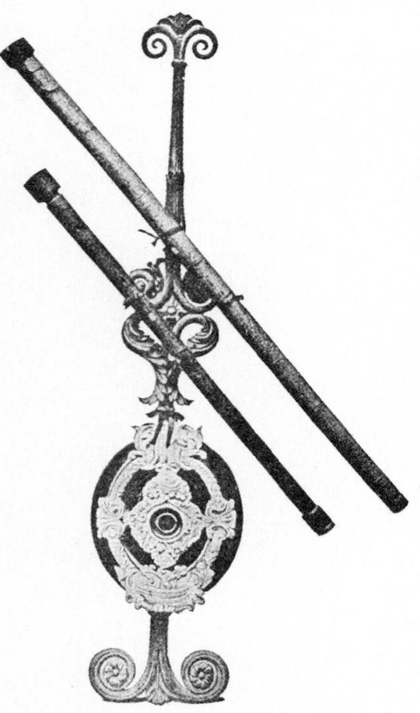

**Figure 6.1** Galileo's telescopes. The stand that holds them in the Museum for the History of Science in Florence is for display only; it is not a stand used by Galileo. In the circular opening at the center of the stand are the pieces of the lens used by Galileo in his first telescope (Yerkes Observatory).

Galileo's first telescope magnified three times, his next eight times, and then came the thirty magnification instrument. The lens quality would not allow the use of the full aperture without seriously degrading the image quality. Three of Galileo's lenses have survived; their diameters are 4 to 5 cm. One of these lenses is from the telescope with which he made his major discoveries; when used with one of Galileo's eyepieces in a modern examination, it showed detail down to 10 arc sec on the face of the Moon. Making the telescopes required more than simply fitting two lenses into a tube. For many years Galileo ground and polished his own lenses, and only about one lens in ten was good enough to use.

With his instruments Galileo revolutionized astronomy:

> Let us first speak of that surface of the Moon which faces us. For greater clarity I distinguished two parts of this surface, a lighter and a darker; the

[1] S. Drake, ed., *Discoveries and Opinions of Galileo* (Garden City, New York, Doubleday, 1957) p. 29.

[2] S. Drake, ed., *Discoveries and Opinions of Galileo* (Garden City, New York, Doubleday, 1957) p. 29.

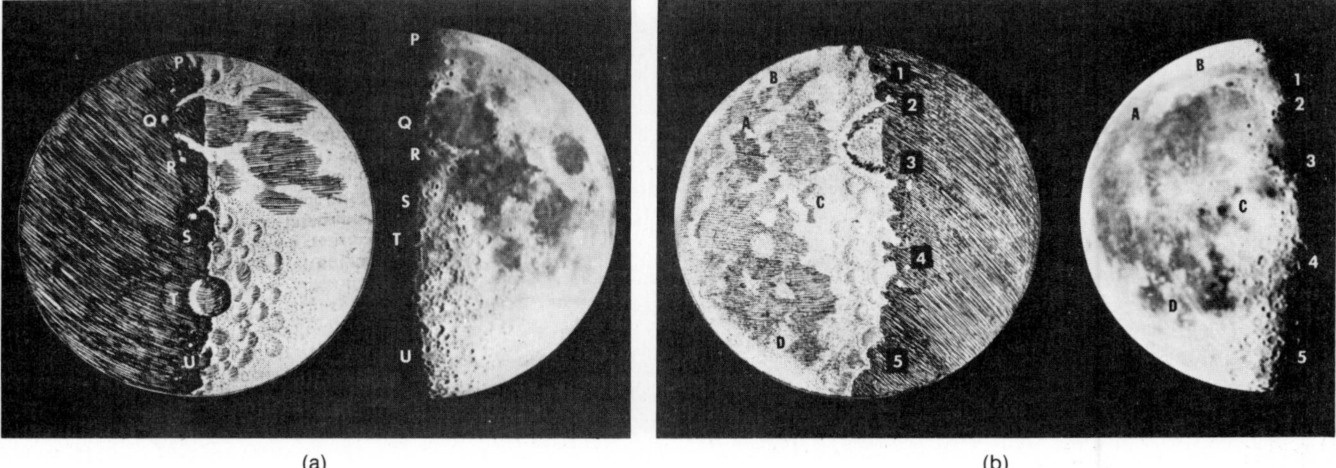


**Figure 6.2** Some of the earliest sketches made by Galileo observing the Moon through his newly built telescope. (a) December 3, 1609, (b) December 17, 1609. Modern photographs are also shown for the Moon at the same phases to show how well Galileo sketched prominent lunar features (Ewen A. Whitaker, Lunar and Planetary Laboratory, University of Arizona).

> lighter part seems to surround and to pervade the whole hemisphere, while the darker kind discolors the Moon's surface like a kind of cloud and makes it appear covered with spots. . . . From observations of these spots repeated many times I have been led to the opinion and conviction that the surface of the Moon is not smooth, uniform and precisely spherical as a great number of philosophers believe it to be, but is uneven, rough and full of cavities and prominences, being not unlike the face of the Earth, relieved by chains of mountains and deep valleys[3].

**Figure 6.3** Motions of the moons of Jupiter as recorded by Galileo in his notebook (Yerkes Observatory).

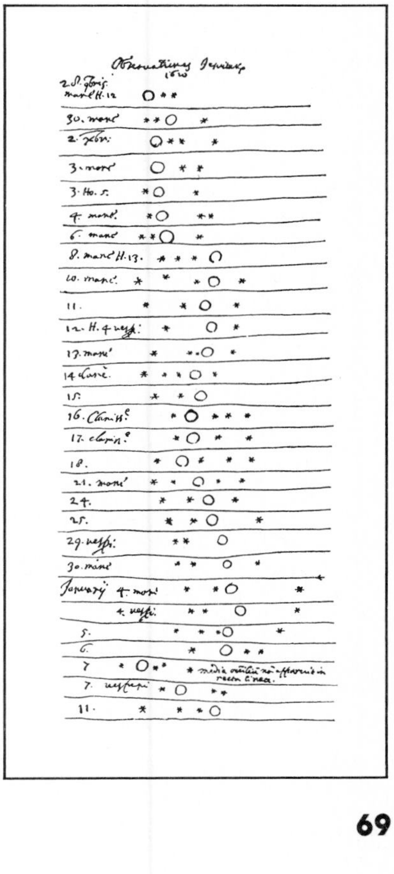

In 1610 Galileo described these discoveries in his *Siderius Nuncius* (The Starry Messenger), illustrated with his own sketches of his sightings. It has been incorrectly said that Galileo was a poor observer and that lunar features could not be recognized in these sketches. Ewen Whitaker, of the University of Arizona, has shown that many of the reproductions are indeed of low quality, but the original etchings are crisp with clearly identifiable features. By comparing Galileo's etchings with modern photographs and with knowledge of exactly what phase the Moon had on each day during the period of Galileo's observations, Whitaker has been able to date the observations precisely. Two of the sketches, made in December 1609, are shown in Fig. 6.2 along with photographs. From the lengths of shadows, Galileo was able to deduce the heights of some lunar mountains.

After December 1609 Galileo turned his telescope to the stars and the Milky Way, then towards Jupiter:

> On the seventh day of January in this present year 1610, at the first hour of night, when I was viewing the heavenly bodies with a telescope Jupiter presented itself to me; and because I had prepared a very excellent instrument for myself, I perceived (as I had not before, on account of the weakness of my previous instrument) that beside the planet there were three starlets, small indeed, but very bright. Though I believed them to be among the host of fixed stars, they aroused my curiousity somewhat by appearing to lie in an exact straight line parallel to the ecliptic, and by their being more splendid than the

[3] S. Drake, ed., *Discoveries and Opinions of Galileo* (Garden City, New York, Doubleday, 1957) p. 31.

others of their size. Their arrangement with respect to Jupiter and each other was the following:

East * * ○ * West

that is, there were two stars on the eastern side and one to the west. The most easterly star and the western one appeared larger than the other. I paid no attention to the distances between them and Jupiter, for at the outset I thought them to be fixed stars, as I have said[4].

Galileo watched these little stars every night and after about a week (by midJanuary), he noted

> I had now decided beyond all question that there existed in the heavens three stars wandering about Jupiter as do Venus and Mercury about the Sun, and this became plainer than daylight from observations on similar occasions which followed. Nor were there just three such stars; four wanderers complete their revolutions about Jupiter, and of their alterations as observed more precisely later on we shall give a description here. Also I measured the distances between them by means of the telescope, using the method explained before. Moreover I recorded the times of the observations, especially when more than one was made during the same night—for the revolutions of these planets are so speedily completed that it is usually possible to take even their hourly variations[5].

Galileo's phenomenal observational skill is shown by these records. His telescope was suddenly revealing thousands of new stars, stars too faint to be seen by the unaided eye. Although the constellations and patterns of the brighter stars were well known, the great number of new stars could have provided little in the way of recognizable patterns yet he was able to note night after night that four of them were moving and thus quite different from all others.

More observations were made and sketched, and by midMarch 1610 the *Siderius Nuncius* was published, dedicated to Cosimo II de'Medici with the satellites of Jupiter being named the Medicean Stars. (Today we refer to them as the Galilean satellites.) The importance of this particular discovery was clearly set out.

> Here we have a fine and elegant argument for quieting the doubts of those who, while accepting with tranquil mind the revolutions of the planets about the Sun in the Copernican system are mightily disturbed to have the Moon alone revolve about the Earth. . . . But now we have not just one planet rotating about another while both run through a great orbit about the Sun; our own eyes show us that four stars which wander around Jupiter as does the Moon around the Earth while all together trace a grand revolution about the Sun in the space of twelve years[6].

Moreover, this radical discovery was there for everyone to see (through the telescope) and did not depend on obscure mathematical computations.

Galileo's observations continued. He noted that the planets showed clearly circular discs, but the stars remained as bright points of light; Saturn appeared to have elongations like ears, but his telescope did not give clear enough images for him to see the true nature of the rings. He thought the ears were moons, as with Jupiter. (The discovery of the rings was made by

[4] S. Drake, ed., *Discoveries and Opinions of Galileo* (Garden City, New York, Doubleday, 1957) p. 51.

[5] S. Drake, ed., *Discoveries and Opinions of Galileo* (Garden City, New York, Doubleday, 1957) p. 53.

[6] S. Drake, ed., *Discoveries and Opinions of Galileo* (Garden City, New York, Doubleday, 1957) p. 57.

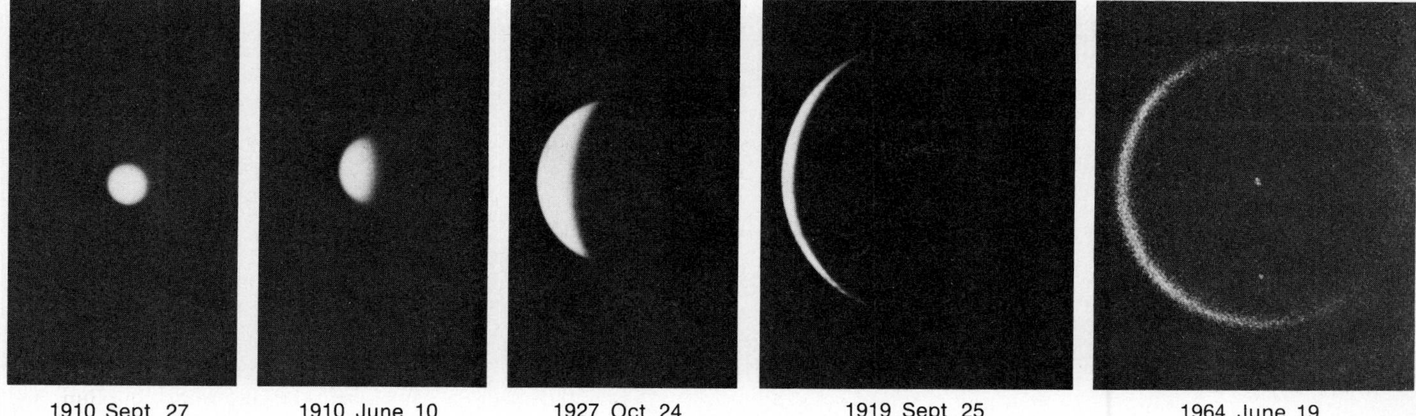


**Figure 6.4** Phases of Venus. Note also how the angular diameter changes markedly when Venus is closest and farthest from Earth (Lowell Observatory).

Christian Huygens, in the Netherlands about 45 years later.) He could see the full disc of Venus and observe its changing appearance over several months during which it displayed phases as does the Moon. This could be easily understood if one accepted the Copernican model with the orbit of Venus between the Earth and the Sun.

When Galileo wrote to Kepler in 1610, he wanted to tell him of his latest discoveries, but at the same time protect himself by not revealing exactly what it was he had found until he could publish a full account. He therefore used the following anagram:

SMAJSMRMJLMEPOETALEVMJBVNENVGTTAVJRAS

Kepler deciphered this as

> Salve umbistineum geminatum Martia proles (Hail, twin companionship, children of Mars)

and thought that it claimed the discovery of two satellites for Mars. Galileo was actually claiming (for Saturn)

> Altissimum planetam tergeminum observavi
> (I have observed the most distant planet to be three-bodied.)

There is a lesson that we might learn from this episode: If today we transmit coded messages away from the solar system, how will they be interpreted by extraterrestrial intelligent beings?

Finally, Galileo noted the spots on the Sun and by watching them move across the solar disc, he reached the correct conclusion that these were spots

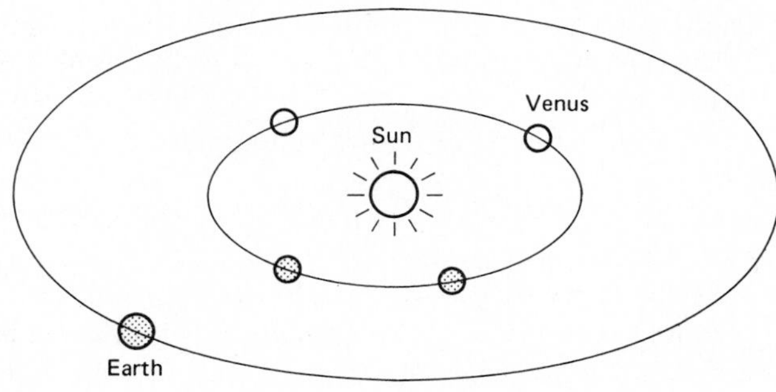


**Figure 6.5** Observation of the phases of Venus by Galileo could be understood if Venus had an orbit around the Sun that was smaller than that of Earth's orbit.

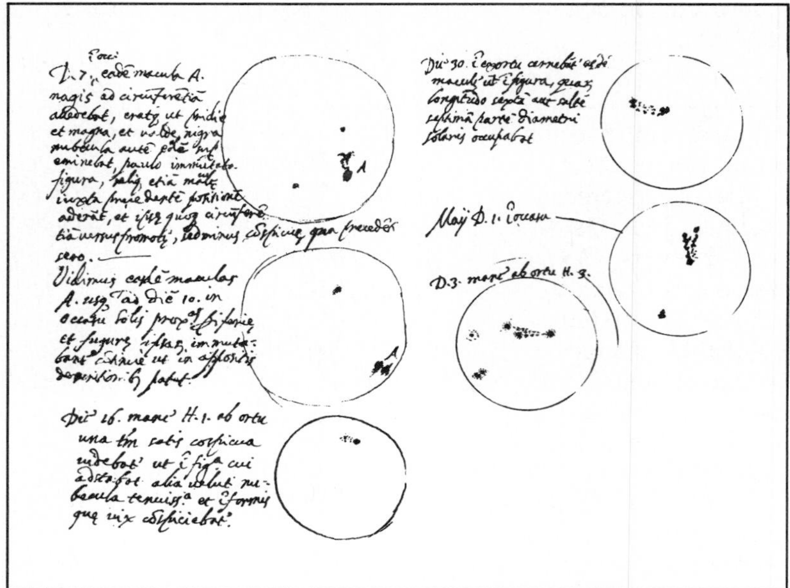

**Figure 6.6** Drawings of sunspots as recorded by Galileo in his notebook (Yerkes Observatory).

on the Sun (itself rotating) and were not simply nearby objects (possibly in the Earth's atmosphere) as had been thought (Fig. 6.6).

*The Starry Messenger* brought fame to Galileo. That fame, suitably augmented by the dedication of Jupiter's moons to Cosimo de'Medici, brought Galileo to Florence. Not only did he want a better position free from teaching (this sounds remarkably modern), but he probably made the move to try to put himself under the protection of a stronger patron before the outburst that he must have expected would ultimately follow his discoveries and his vigorous polemics.

## 6.3 CONTROVERSY

The inferences that could be drawn from Galileo's discoveries were in conflict with the Ptolemaic view of the universe, as supported by the Church, and a confrontation was probably inevitable. After an inquiry prompted by some complaints, the Inquisition, in 1616, instructed Galileo to abstain from any further teaching or defending the Copernican theory. Galileo's obedience was marginal. In 1632, he published his *Dialog about the Two Chief Systems of the World, Ptolemaic and Copernican,* which was a thinly-disguised attack on the Ptolemaic theory.

Again the Inquisition took action. He was sentenced to life imprisonment and his books were placed on the Index. They remained so listed until 1835. In the 1980s, the Church undertook a review of the whole case, and it seemed that some statement of exoneration might still be issued.

Galileo's imprisonment was in his home in Florence, where he continued to work. He returned to mechanics, his first interest, and in 1636 he completed another masterpiece. *Discorsi a due nuove scienze* (Two New Sciences) had to be published in a nonCatholic country; it appeared in 1639 in Amsterdam. The two sciences were mechanics and the strength of materials.

Despite his never-ending willingness to challenge accepted views and despite his many discoveries, Galileo did *not* do some things one might have expected from him. He did not follow his discoveries with any quantitative astronomical calculations. He did not even take note of Kepler's discovery of the elliptical shapes of the planetary orbits, and his *Two Chief Systems* presents an essentially Copernican view with circular orbits. On the other hand, the *Two New Sciences* were just as revolutionary in their introduction of a new approach and new ideas in mechanics, and these set the stage for the work of Newton.

## 6.4 MECHANICS

By this time astronomy had gone just about as far as it could with the methods at its disposal. Better observations had forced the abandonment of the use of the epicycles, and Kepler had produced a model that described the solar system extremely well. Galileo's telescopic discoveries supported the Copernican view in several ways. What was now needed in order to make any progress was a major change in the conceptual attack on the problem. What was needed was a new kind of mathematics that would handle the positions and speed and their changes in a convenient manner, and that was to be one of Newton's great contributions.

In grappling with his developing ideas on motion, Galileo introduced a new approach: the planned experiment. Galileo asked how moving bodies would behave and answered by conducting actual tests. The story of the dropping of two weights from the Tower of Pisa is a good story but almost surely untrue. The *point* of the story, however, is important. Until that time, Aristotle's view was accepted—that a heavy object would fall faster than a light object, and even the idea of a test was revolutionary.

From experiments Galileo was able to show that the distance fallen by an object increased with the square of the time taken, that is, an object falls nine times as far in 3 sec as it does in 1 sec. (Previously, it was thought that distance increased simply in proportion to time.) It followed that falling was an accelerated motion, and the idea of acceleration began to emerge. Another innovation was the concept of **inertia**, the property of a body that represents its resistance to changes in its motion—the effort to start a body from rest and the persistence of its motion once started. (We shall examine these ideas in more detail in the next chapter.)

Galileo is usually remembered for his championing of Copernicanism and the resulting clash with the Church. In the never-ending struggle for intellectual freedom that episode deserves to be recalled for there are comparisons that can be drawn with events today, and there are lessons to be learned. His contributions to science were more important. We cannot imagine astronomy without the telescope. As we shall see, we could not have modern science without his innovations in mechanics. Had he not made these advances others would have done so, but in no way does that detract from his achievements.

## CHAPTER REVIEW

**Galileo:** invention of the telescope.

**Discoveries:**

**Moon:** irregular surface with craters and mountains. Certainly not smooth as had been thought until that time.

**Stars:** discovered hundreds in Orion, far more than can be seen by eye alone.

**Jupiter:** discovered four moons. Observations over several nights showed beyond doubt that these moons were in orbit around Jupiter.

**Saturn:** telescope could not show rings clearly. Galileo thought he had seen two moons.

**Sunspots:** could show these were on the Sun and not in the Earth's atmosphere, and could also follow the Sun's rotation by noting movement of the sunspots across the solar disc.

**Controversy:** over his findings and the support they gave to the Copernican theory led to banning of his books and to Galileo's house arrest.

**Mechanics:** Galileo maintained a lifelong interest in mechanics. Pioneered the use of experiments to show how bodies moved. Showed that Aristotle was wrong: Falling objects fall with the same speed regardless of size. Introduced the concept of inertia. Studied the motion of projectiles.

## PROBLEMS

1. What did Galileo discover about the Moon?

2. Galileo discovered that Venus went through phases in much the same way as the Moon. Draw diagrams to see how this observation is easily explained by the Copernican theory. Can Tycho Brahe's theory explain these observations?

3. What did Galileo discover about Jupiter, and how did this contradict the Ptolemaic ideas?

4. The prohibitions placed on Galileo had their immediate desired effect, but in the long run the Copernican system has been accepted. Might it sometimes be best (or politic or prudent) to suppress a scientific discovery? Who should make such a decision, the scientists or some other group or person?

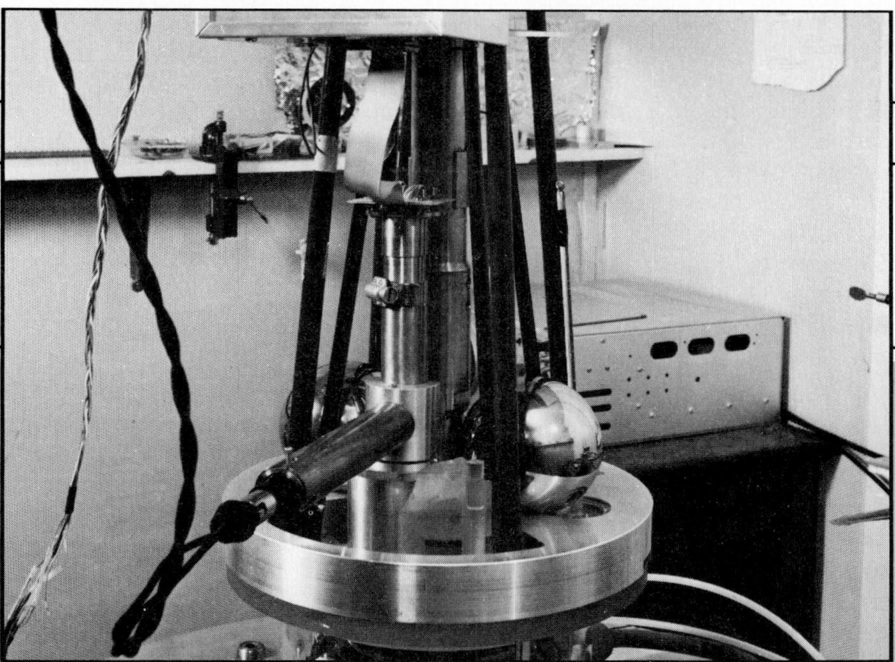

CHAPTER 7

# Newton: Mechanics and Gravity

Seventeenth-century England was very different from the Europe of religious clashes, where astronomical theories required protective language. England had its ancient universities, Oxford and Cambridge. The Royal Society, founded in 1662 after the restoration of the monarchy, had Charles II as its Patron. Science prospered.

## 7.1 NEWTON'S CAREER

Isaac Newton was born in December 1642 in the village of Woolsthorpe, about 60 miles northwest of Cambridge. He entered Trinity College in Cambridge in 1661 and studied under Isaac Barrow, a distinguished professor of mathematics, until 1665 when the rapid spread of the bubonic plague caused the university to close forcing Newton to return home.

Returning to the university in 1667, Newton resumed his studies. By then his remarkable ability and originality must have been clear to Barrow who the next year resigned his chair and nominated Newton to suceed him. Newton stayed at Cambridge until 1696 when he moved to London to work in the Royal Mint, becoming Master of the Mint in 1699.

Newton's monumental scientific advances were set out in Latin in his **Mathematical Principles of Natural Philosophy** (usually known as *Principia*) published in 1687, and his **Optics**, (in Eńglish) published in 1704. He was elected President of the Royal Society in 1703 and retained that position until his death in 1727.

## 7.2 THE PRINCIPIA

The *Principia* has three major sections. Book I is titled *The Motion of Bodies*; Book II: *The Motion of Bodies* (in Resisting Media); and Book III: *The System of the World.* In Book I, The Laws of Motion are set out after eight definitions; then the rest of that Book and all of Book II consists of an extraordinary web of derivations and proofs of new geometrical theorems and propositions.

The first two books give no encouragement to the casual reader, but Book III, starting with *Rules of Reasoning in Philosophy,* is more accessible. Here, set out in a way that we can immediately recognize, is the foundation of the *scientific method.* This method today is usually considered to be broader, extending to the endless sequence of experimenting and theorizing, but Newton's four rules still provide us with excellent guidance.

Rule I is directed to the economy of explanation; we try to get by with the simplest theory possible. Rule II is especially important in astronomy; we *assume* that the same scientific laws that we find on Earth will also hold at all other locations in the universe. Rule III is not immediately obvious to the reader today; it refers to older ideas now discarded. Rule IV asserts that *propositions* (*theories* or *models* we would call them today) derived from observations and experiments should be held to be "accurate or very nearly true," until additional results allow them to be improved or seen to be in error. In other words theories should not be viewed as absolute and final truths but at best as good working approximations.

Book III continues with its first nine propositions setting out the role of gravity and the formula that describes it. Thereafter, through the remainder of the Book, Newton pursues the consequences of his new mechanics in gravitational systems discussing the planets and their moons, comets, and the tides on Earth.

## 7.3 KINEMATICS

At many places in the preceding pages, we have referred to the speeds of the planets and the motions of the various heavenly bodies, without needing to supply precise definitions, for we could draw on our everyday experience of movement. We now need to define our terms with care.

The basic relations between distances, times, speeds, and accelerations comprise one part of the subject of mechanics, termed kinematics.

If an object moves a distance ($d$) in a specified interval of time ($t$), then speed is calculated as (distance/time) or ($d/t$). This can be expressed in miles per hour (mph), feet per second (ft/sec), kilometers per sec (km/sec), or whatever.

When the total distance can be divided into many small parts and the speed in all parts is the same, then we say that the speed is uniform or constant. Constant speed was assumed in all the classic models with circles and epicycles. The circumference of a circle is $2\pi R$ (where $R$ is the radius), and if $P$ is the time to travel exactly once round, then the average speed is given by $2\pi R/P$. $P$ is called the periodic time or simply the period.

Most movement, however, does not proceed with constant speed. Changes in speed are termed **accelerations** and are measured in (cm/sec) per sec, usually written as cm/sec$^2$. (If the speeds are measured in m/sec then the acceleration units are m/sec$^2$. In English units, an acceleration will be in units of ft/sec$^2$.)

Our most familiar example of acceleration is the motion of a falling object. Near the surface of the Earth, falling objects have an acceleration of 980 cm/sec$^2$ (or 9.8 m/sec$^2$ or 32 ft/sec$^2$). This means that a falling object gains an additional speed of 980 cm/sec during each second that it falls. This number is called the acceleration due to gravity and is denoted by 'g'.

The value of $g$ is not constant. It has slightly different values at different places on the Earth's surface, and it decreases if we go far above the Earth's surface. But at any given location, $g$ provides an excellent example of a constant acceleration in which equal increases in speed go with equal increases in time.

We still need to introduce one more important concept. Until now we have assumed that the speed changes are along a single straight-line path (such as for a falling object). Sometimes the motion is along a curved path, a circular or elliptical orbit for example. The speed along the path can be constant, but the **direction** of motion is changing, and this change in direction can be described as an acceleration.

In the case of uniform motion in a circle, we have the apparently contradictory situation of the speed being constant but there still being an acceleration. This concept of acceleration is often puzzling, but an example may help to clarify the idea. Suppose that you tie a string to a stone and swing it around in a horizontal circle. At all times, the stone is being accelerated toward the center of its circular path (Fig. 7.1). The force that produces this acceleration is toward the center along the radius, and there is no doubt about the reality of that force—you can feel it. If the string is cut, the stone will fly off—*not* out along a radius from the center but along a tangent. The acceleration while the stone is moving around the circle is toward the center, even though the stone's movement is always along the circumference of the circle. The stone is being accelerated, but its speed does not change.

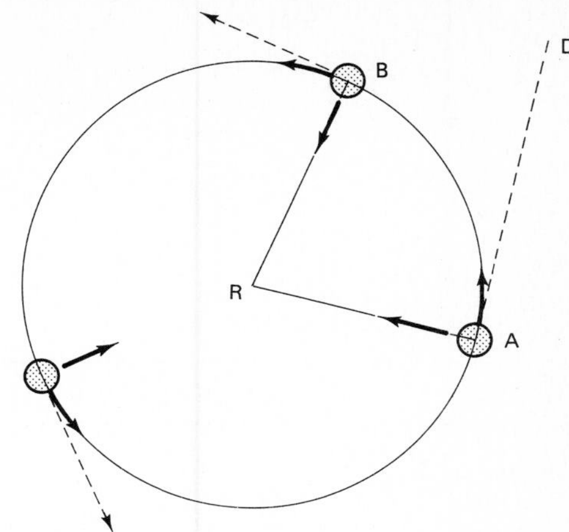


**Figure 7.1** Motion around a circle at constant speed. The instantaneous direction of motion is constantly changing, but the speed remains constant. At all points there must be an instantaneous acceleration towards the center of the circle in order to maintain the circular motion. If, when the object is at *A*, the force producing the acceleration is switched off, the body would move off along the tangent (towards *D*).

**BOX 7.1**
*Average Speed*

The Earth's orbit around the Sun is almost circular, with a radius $1.5 \times 10^{13}$ cm. The time taken for an orbit is 1 year, $3.16 \times 10^7$ sec. The average speed of the Earth along this circular path is thus:

$$v = \frac{2\pi R}{T} = \frac{2\pi 1.5\ 10^{13}}{3.16 \times 10^7}$$

$$= 2.98 \times 10^6 \text{ cm/sec}$$

$$= 29.8 \text{ km/sec}$$

**BOX 7.2**
*Acceleration Around a Circle*

If the body is moving with a constant speed, $V$, around a circle which has a radius, $R$, (Fig. 7.2), then the acceleration toward the center is given by

$$a = V^2/R$$

For example, the Earth is moving in its orbit around the Sun with an average speed of $2.98 \times 10^6$ cm/sec. If we take its orbit as circular with a radius of $1.5 \times 10^{13}$ cm, then the acceleration is

$$a = (2.98 \times 10^6)^2/(1.5 \times 10^{13})$$

$$= 0.59 \text{ cm/sec}^2$$

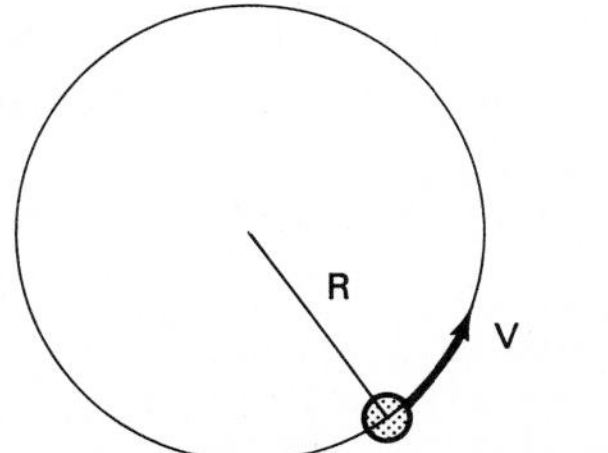


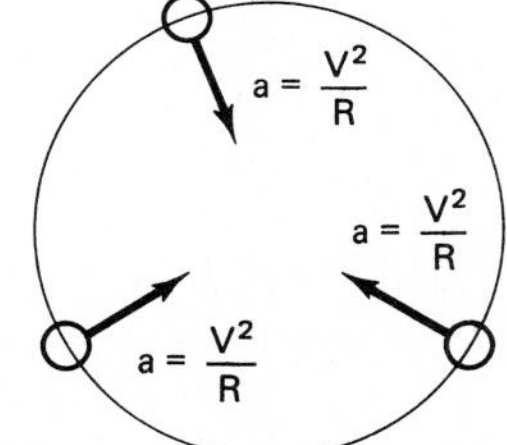


**Figure 7.2** Acceleration around a circle. Even though the object is moving with a constant speed, there must always be an acceleration $V^2/R$, always directed towards the center of the circle.

If we are told the speed and acceleration of an object, we are in the position to calculate the distance traveled in any time interval and thus predict future positions. We are not yet, however, able to deal with the question: *Why* does this object move in this way? To answer this question, we have to take account of the mass of the object and the forces that push and pull on it. This leads to the second part of mechanics: **dynamics**.

## 7.4 DYNAMICS

We are all familiar with pushes and pulls—the efforts we exert to move things: tables, chairs, luggage. We know that it takes a continued effort to keep something moving. Without this continued effort, friction causes motion to stop. We also know that it takes more effort to move a large and massive object than to move something small. It is then natural to wonder what is needed to keep the Moon in its orbit around the Earth, and the planets in their orbits around the Sun. It was Newton's genius to recognize that three brief laws of motion plus another for gravity could put the answers to these questions (and many others) into quantitative form.

The laws were first given in Latin, in the *Principia*, and the English versions given here have been slightly simplified (Fig. 7.3). These laws are:

> Law 1: Every body continues in its state of rest, or of uniform motion in a straight line, unless it is compelled to change that state by forces impressed on it.
>
> Law 2: A force applied to a body produces an acceleration. The acceleration is in the same direction as the force, and the magnitude of the acceleration is proportional to the size of the force and inversely proportional to the mass of the body.
>
> Law 3: For every applied force, there is an equal but oppositely directed reaction (force).

The first law provides a qualitative answer to the question: What happens when a force is exerted? The second law responds to the related question: How much happens? The second law is usually written in the form

$$F = m\,a$$

where the force, $F$, applied to the body having mass, $m$, produces an acceleration, $a$. (If $m$ is in grams and $a$ in cm/sec$^2$, then the units of force are termed dynes; if $m$ is in kilograms and $a$ in m/sec$^2$, then the units of force are Newtons; 1 Newton $= 10^5$ dynes.)

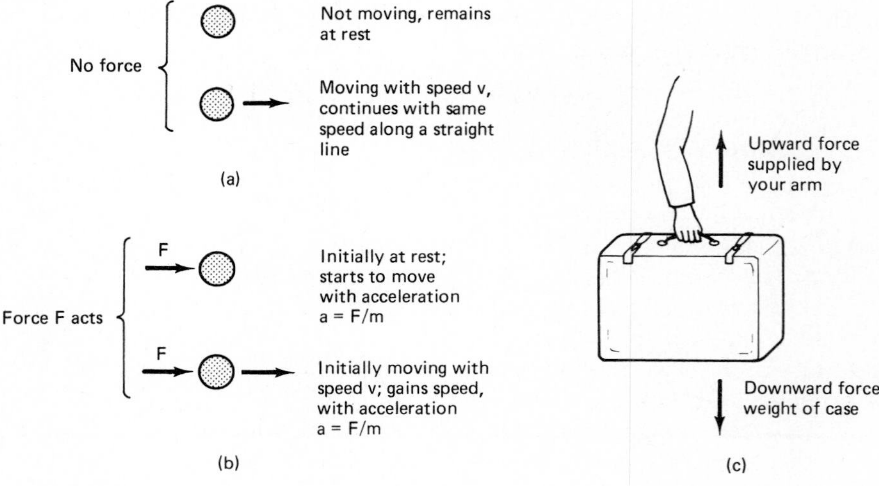


**Figure 7.3** Newton's Laws of Motion.
(a) In the absence of a force, an object remains at rest or (if already moving) continues with constant speed along a straight line.
(b) If a force acts, it produces an acceleration proportional to the size of the force and inversely proportional to the mass of the object. (c) For every applied force there is an equal and oppositely directed reaction.

It is common experience to see a moving object gradually slow down, but the heavenly bodies keep moving. It had long been accepted that a force was needed in order to maintain motion. Galileo and Newton both recognized that the slowing down was not due to the absence of a driving force but rather to the unseen presence of some other force, such as friction or air resistance. They went on to postulate that the motion would persist if the unseen force could be removed. Left to itself, a body would then remain at rest or keep moving in a straight line with constant speed. This property of **inertia**, of maintaining the state of motion, is asserted in Newton's first law, which specifies that a force is needed to change the motion of a body.

Acceleration lasts only as long as a force acts. When the force is removed, the body will continue with whatever speed it had attained in accordance with the first law and along the direction it was moving when the force was switched off.

Forces can be transmitted without a solid link to the object affected. A magnet can hold a piece of metal in midair; the gravitational force holding you to the Earth works just as well while you jump. On a grander scale, the attractive force due to the Earth that holds the Moon in its orbit has no solid link connecting them.

## 7.5 GRAVITY

The three laws of motion are quite general. They refer to the responses of bodies to forces without specifying the nature of the forces. For the laws to be of use in astronomy, what was still needed was a quantitative description of the force that controlled the celestial motions. We have seen how Kepler thought that magnetism provided the answer, but nowhere did he specify the strength of this force. Once again, it was Newton who formulated the solution. In the course of several propositions near the beginning of Book III of the *Principia*, Newton developed the basic formula written today as

$$F = \mathrm{G}\frac{M_1 M_2}{R^2}$$

where $F$ is the force of gravitational attraction between two masses, $M_1$ and $M_2$, that are a distance $R$ apart. G is the first of a number of **physical constants** that we shall encounter. These constants are quantities that enter into many physical laws. Their values must be found by experiment, for there is no way in which they can be derived from theory. The speed of light is another such constant. Today that speed can be measured to an accuracy of better than one part in a trillion ($10^{12}$), but the measurement of G is extraordinarily difficult, and the present value

$$\mathrm{G} = 6.6726 \times 10^{-8} \text{ dyne cm}^2/\text{gm}^2$$

is known to only about one part in $10^5$ (Fig. 7.4).

Newton's formula specifies the magnitude of the gravitational force. The direction was also specified, along the line joining the two masses tending to bring them closer together. The gravitational force is always attractive, never repulsive, that is, it cannot push objects apart.

There is no clear record that Galileo dropped weights from the Leaning Tower in Pisa, nor whether Newton was really stimulated by seeing an apple fall from the tree in Woolsthorpe. In the *Principia*, Newton does compare the acceleration of objects near the Earth with the acceleration of the Moon in its orbit. We can follow his reasoning. Newton used the value 27 days 7 hr 43 min ($27^d.32$) for the time taken by the Moon to make one trip around its orbit, and he took the mean orbital radius to be 60 times the radius of the Earth. Accordingly, since acceleration is proportional to force, and as the force decreases inversely as the (distance)$^2$, the acceleration of the Moon

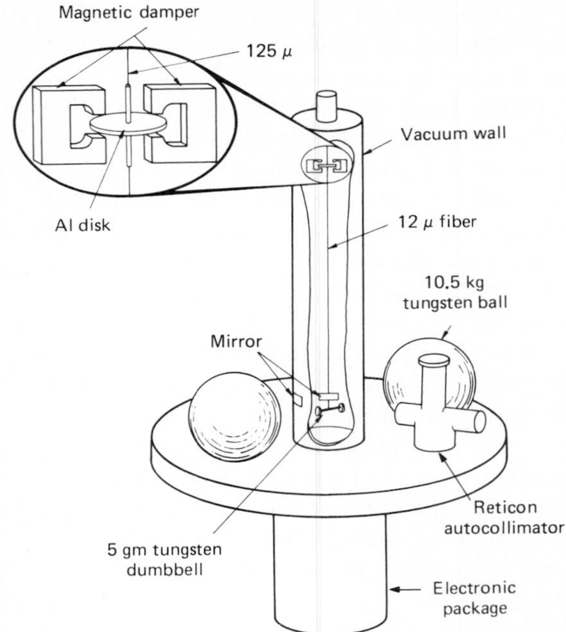


**Figure 7.4** Apparatus used at the National Bureau of Standards for the precision determination of the gravitational constant, G. The gravitational force between the large tungsten balls and the small tungsten dumbell is measured by noting the period of oscillation of the dumbell on its thin fiber support (Gabriel G. Luther, National Bureau of Standards).

due to the Earth's gravity should be $(\frac{1}{60})^2$ times the acceleration at the surface of the Earth (Fig. 7.5). Hence

$$\text{Accel}_{\text{Moon}} = 980/(60)^2 = 0.27 \text{ cm/sec}^2$$

The acceleration can also be calculated from the formula $v^2/R$ (see Box 8.2), where $R = 60$ Earth radii or $3.84 \times 10^{10}$ cm and

$$v = 2\pi R/T = 2\pi(3.84 \times 10^{10})/27.32 \times 24 \times 60 \times 60 \text{ cm/sec}$$

from which

$$\text{Accel}_{\text{Moon}} = v^2/R =$$

$$= 0.27 \text{ cm/sec}^2$$

This agreement between the values for acceleration obtained in two different ways confirmed the inverse square dependence on distance for the gravitational force, and as Newton expressed it

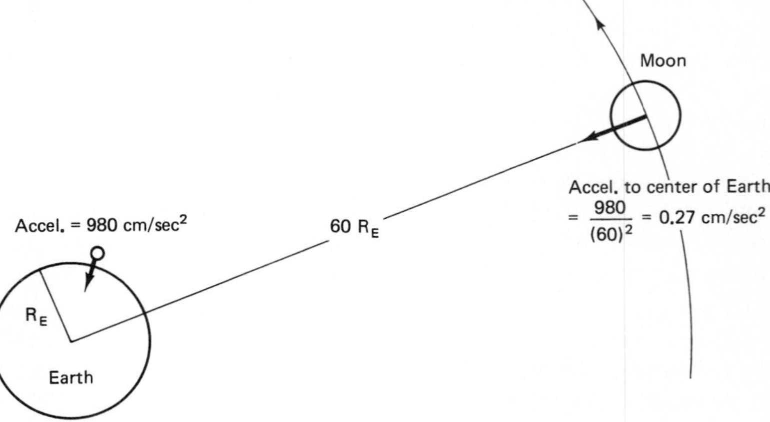


**Figure 7.5** The gravitational force decreases inversely with the square of the distance. Thus the Moon, 60 Earth radii from the center of the Earth, is accelerating towards the Earth with $(1/60)^2$ of the acceleration of a body falling at the Earth's surface.

The force by which the Moon is retained in its orbit is the very same force which we commonly call gravity.

## 7.6 THE APPLICATION OF NEWTON'S LAWS IN ASTRONOMY

Newton's laws have served us well. They form the basis of much of physics and they are still used for calculating the positions of the planets and for the trajectories of space probes.

Their first successes lay in astronomy and the understanding they provided for Kepler's laws. Kepler had produced laws that were accepted as accurate geometrical descriptions of planetary behavior: descriptions but not explanations. Now from Newton's laws Kepler's could actually be deduced.

Taking the laws of motion and the gravitational inverse square law, Newton could now *show* that the orbits had to be elliptical. In special cases the ellipses could become circular, and in other special cases an ellipse would open up to form a parabola. In all cases the focus was occupied by the controlling mass. For the planets, this mass was the Sun; for the moons orbiting Jupiter, it was Jupiter; and for the Moon, it was the Earth.

Kepler's second law could now also be derived. The essential requirement is that the force (such as between the Sun and the planet) be always directed along the line joining them. Finally, Kepler's third law could be derived from the inverse square nature of the gravitational force. Newton's clarification of Kepler's third law gave it a far greater importance than it had previously.

### BOX 7.3
*Kepler's Third Law*

Kepler had found a simple law that related the periodic time ($P$) for each planet to its average distance from the Sun ($a$):

$$P^2 = k\,a^3$$

The constant $k$ takes on the value 1 if $P$ is measured in years and $a$ in units of the semimajor axis of the Earth's orbit.

Newton was able to derive this law in a form where it could be applied with great generality. His version of Kepler's law is

$$P^2 = \frac{4\pi^2}{G(M + m)}\,a^3$$

In this, $P$ is the sidereal periodic time of any body of mass $m$ in an elliptical orbit (of major axis $2a$) around a second body of mass $M$. G is the universal gravitational constant ($6.67 \times 10^{-8}$ dyne cm$^2$/g$^2$). ($P$ is measured in seconds; $m$, $M$ in grams.)

Measurement of the period and orbital size permit a direct calculation of the sum of the two masses. If the central mass $M$ is much greater than the orbiting mass $m$, then the formula simplifies to

$$P^2 = \frac{4\pi^2}{GM}\,a^3$$

In this case, $M$ can be calculated alone. (In the more general case it is the sum of the two masses that can be calculated.) This type of calculation is not confined to planets around the Sun. It can just as well be applied to the Moon around the Earth (to yield a value for the Earth's mass), or moons around Jupiter (to give Jupiter's mass) or one star around another (to give stellar masses). The important application to pairs of stars will be discussed in chapter 20.

As formulated by Kepler, the third law amounted only to a rule relating the periodic times of different satellites to the sizes of their orbits around a common central object. Now via Newton, we can gain much more information. Once we have measured the $P$'s and the $R$'s, we can calculate the value of the composite quantity $\frac{4\pi^2}{G(M + m)}$, and since G is known from experiments in the laboratory, all that remains is to calculate $(M + m)$.

**BOX 7.4**
*An Application of Kepler's Third Law*

For the Earth's motion around the Sun, the period $P$ is 1 year or $3.16 \times 10^7$ sec, and $a$ is 1 AU or $1.5 \times 10^{13}$ cm. Then from Newton's version of Kepler's third law

$$P^2 = \frac{4\pi^2 a^3}{G(M + m)}$$

Solving for $(M + m)$,

$$M + m = \frac{4\pi^2 a^3}{GP^2} = \frac{4\pi^2(1.5 \times 10^{13})^3}{(6.67 \times 10^{-8})(3.16 \times 10^7)^2}$$

$$= 2.00 \times 10^{33} \text{ g}$$

This is the combined mass of Sun + Earth. A similar calculation for the Earth–Moon system will show that the combined mass of Earth + Moon is very much less that the combination of Sun + Earth. We thus deduce that the Earth is so much less massive than the Sun that the value $2.00 \times 10^{33}$ g must be effectively the solar mass. Further (with hindsight) we could have used the simpler form of Kepler's law,

$$P^2 = \left(\frac{4\pi^2}{GM}\right) a^3$$

where $M$ is the mass of the Sun alone.

The success of Newton's laws in describing the observed working of the solar system indicated that the inverse square law of force extended over vast distances. The force does get weaker, but it is still there to exert its influence. With large enough masses, gravitational effects should be observable over very great distances. We shall come across this again when we consider collections of stars that make up galaxies and collections of galaxies that make up our universe. The power of Newton's theories was clear to his scientific contemporaries, but the public admiration for his theories was increased after his death largely through Halley's computations on the orbits of comets.

## 7.7 PREDICTIONS

In 1680 a great comet appeared. Newton devised a method for determining its true orbit, and he was able to show that the comet's path was that of a very elongated ellipse, almost a parabola, with the Sun at a focus. His friend, Edmund Halley, used this new method to compute the orbits of twenty-four comets for which he felt the observations were reliable. The earliest comet on this list had been seen in 1337. Halley noted that three comets that had their closest approaches to the Sun in August 1531, October 1607, and September 1682 seemed to have followed the same path, whereas the other comets had paths that were scattered around the sky. With the intervals between the three comets being so similar (75 and 76 years between returns), Halley suggested that these were not three separate comets but rather the reappearances of a single comet, probably that which had also been seen in 1378 and 1456. He predicted that it would return in 1758.

We can use Kepler's third law to calculate the size of the orbit of Halley's comet if we assume that it is governed by the same law as the planets since its orbit, too, has the Sun at its focus. With $P = 76$ years, we have

$$76^2 = a^3$$

and $a$, the semi-major axis turns out to be close to 18 AU. This tells us that Halley's comet follows an orbit that stretches far beyond the orbit of Saturn, which is on average 9.5 AU from the Sun.

Alexis Clairault extended Halley's calculations allowing for the gravitational forces that the comet would experience when passing close to Jupiter and Saturn, and he predicted a **perihelion passage** (closest approach to the Sun) for April 1759. With much excitement in Europe the comet was sighted again in December 1758 when it was far from the Sun, and it passed perihelion on March 12, 1759.

Because Halley's comet does not approach Jupiter and Saturn within the same distance on each orbit, their gravitational effects on the comet vary. As a result the period between its returns is not constant but ranges between about 74 and 79 years. From ancient records, the returns have been noted as far back as 239 B. C. (Figs. 7.6 and 7.7.)

Within the solar system the dominant gravitational force on each comet and planet is that due to the Sun, but the smaller forces between planets also need to be taken into account for all accurate orbit calculations. The effects of these forces are termed **perturbations**. After the accidental discovery of the planet Uranus by William Herschel in 1781, its movements were followed and its orbit computed including the perturbations expected from Jupiter and Saturn, the two largest planets. It did not take long before it became clear that Uranus' orbit deviated from the calculated positions. Might there be yet another planet further out, whose presence had not yet been detected and accordingly whose perturbations on Uranus had been neglected? Turning the problem around, could one take the irregularities in Uranus' orbit and the known gravitational law and then deduce where this undiscovered planet might be?

**Figure 7.6** Edmund Halley (1656–1742; (Yerkes Observatory).

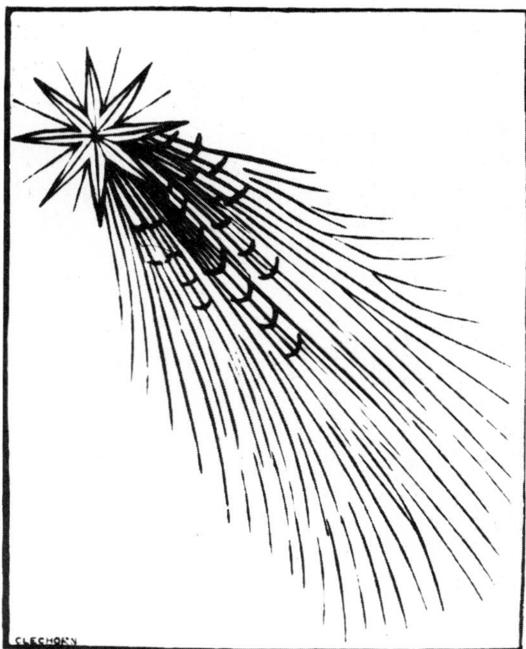

**Figure 7.7** Halley's comet, as it appeared in A.D. 684. This woodblock print is from the Nuremberg Chronicles (1493), where it was drawn in accordance with old descriptions of the great appearance (Yerkes Observatory).

John Couch Adams in England and Urbain LeVerrier in France independently undertook the calculations and derived remarkably similar positions for the suggested planet. While Adams in 1845 was unable to arouse the interest of the Astronomer Royal to undertake a search, LeVerrier was more successful. On receipt of LeVerrier's paper, J. G. Galle at Berlin Observatory promptly looked for the planet. He found it within 1° of LeVerrier's predicted position (and $2\frac{1}{2}$° from Adams') in his first search in September 1846. Another triumph for Newton's theories.

## 7.8 SUMMARY

The successes of Newton's theories has a profound effect, not only on science but also on the view of science held by the rest of educated society. Quantitative predictions could be made and verified, and it seemed as though the future course of celestial events was simply waiting to be calculated. All was already ordained. With this situation, why should not a similar state of affairs exist on Earth leaving what room for the free exercise of our own will? The view of science as cold, inhuman, mechanical, and deterministic dates from this period. More recent discoveries that show the incompleteness of this view have done little to change the popular misconceptions. But at this time, we are more concerned with the clear successes of the theories, and we shall return to use them on several occasions.

## CHAPTER REVIEW

**Newton:** Rules of Reasoning—try to get by with the simplest explanation possible; the same scientific laws that are found on Earth are assumed to hold at all other places in the universe; theories or models are considered as provisional but not final truths liable to be improved or disproved in the light of new evidence.

**Mechanics:** average speed = (total distance/total time)

**Acceleration:** change of speed or change of direction of motion even if the speed is unchanged.

- — a falling object has a constant acceleration; its speed increases steadily; the distance fallen increases with the square of the time.
- — acceleration due to gravity = $g$ = 980 cm/sec$^2$ = 32 ft/sec$^2$
- — acceleration in a circular path is always towards the center of the circle, and the magnitude of the acceleration is $V^2/R$.

**Newton's laws of motion:**

1. The law of inertia: a body remains at rest or continues to move in a straight line with constant speed if no force is acting,
2. When a force acts, the resulting acceleration depends on the size of the force and the mass of the body, $F = ma$
3. For every force, there must be an equal and opposite force (reaction).

**Law of Gravity:** $F = \dfrac{GM_1M_2}{R^2}$,

$$G = 6.67 \times 10^{-8} \text{ dyne cm}^2/\text{gm}^2$$

where G is the gravitational constant.

**Application of Newton's Law to astronomy:**

- — Kepler's laws could be deduced from Newton's laws.
- — Kepler's third law becomes: $P^2 = \left[\dfrac{4\pi^2}{G(M_1 + M_2)}\right] a^3$
- — Halley predicted the return of a comet using Newton's laws and older observations of several comets.
- — application of Newton's laws led to the successful prediction of the location of new planet (Neptune).

## NEW TERMS

**acceleration**
**acceleration due to gravity**
**average speed**
**calculus**
**g**
**G**
**inertia**
**perihelion passage**
**perturbation**
**speed**

## PROBLEMS

**1.** Newton's Rule II suggests that we assume that the same laws we find on Earth will also hold elsewhere. Suppose this assumption were wrong; how would we know? What evidence was available to Newton that his assumption was probably correct?

**2.** Which average speed is greater:

> *A* travels for 30 min at 30 mph, then for 1 hr at 55 mph, then for 15 min at 30 mph.
> *B* travels for 15 min at 20 mph, then for 1 hr at 60 mph, then for 30 min at 20 mph.

**3.** Which speed is larger, the Earth in its orbit around the Sun or the Moon around the Earth?

**4.** A falling object increases its speed by 32 ft/sec each second. Calculate its speed after 2, 5 and 10 seconds.

**5.** A falling object hits the ground with a speed 4900 cm/sec. How long had this object been falling?

**6.** A high jumper clears the bar at 7 ft (210 cm). It takes him 0.65 sec to fall to the ground. What is his speed at that time.

**7.** On the surface of the Earth the acceleration due to gravity is 980 cm/sec$^2$. What will it be at a distance of one Earth radius above the surface (i.e., two Earth radii from the center of the Earth)?

**8.** Calculate the gravitational force between a 100 kg astronaut and his/her 10 ton spacecraft (1 ton = $10^6$ g) when the astronaut is 5 m from the craft. If no other forces were present, what would be the initial acceleration of the astronaut toward the craft?

**9.** Kepler's laws provided geometrical rules that described the orbits of the planets, but Newton's laws provided an understanding. Discuss the differences between these two sets of laws.

**10.** A satellite is in orbit around the Earth at an altitude of 150 km above the surface. Calculate the acceleration of the spacecraft towards the Earth's center: (a) by computing the gravitational acceleration at that distance, and (b) from the speed in the orbit taking the orbital time to be 87.46 min.

**11.** The Earth's orbit around the Sun is almost circular with a radius of closely $1.5 \times 10^{13}$ cm. It takes $365\frac{1}{4}$ days for the Earth to complete one orbit.

a. Calculate the speed of the Earth in its orbit.
b. Calculate the acceleration of the Earth towards the Sun.
c. Calculate the gravitational force between the Sun and the Earth, and check that this is in agreement with $F = ma$.

**12.** Suppose a new planet is discovered in circular orbit around the Sun at a distance of 4 AU. Calculate the period of its revolution about the Sun.

**13.** Halley's comet has an average period of 76 years in its orbit around the Sun. At its next perihelion in 1986 it will be 0.3 AU from the Sun.

a. What is the length of the major axis of its orbit?
b. When will it next be at its greatest distance from the Sun?
c. What is the eccentricity of its orbit?

**14.** Suppose that Ptolemy was right: The Earth is at the center of the universe, and everything rotates around us once every day. At what distance from the Earth would the objects be moving with the speed of light?

CHAPTER

# 8

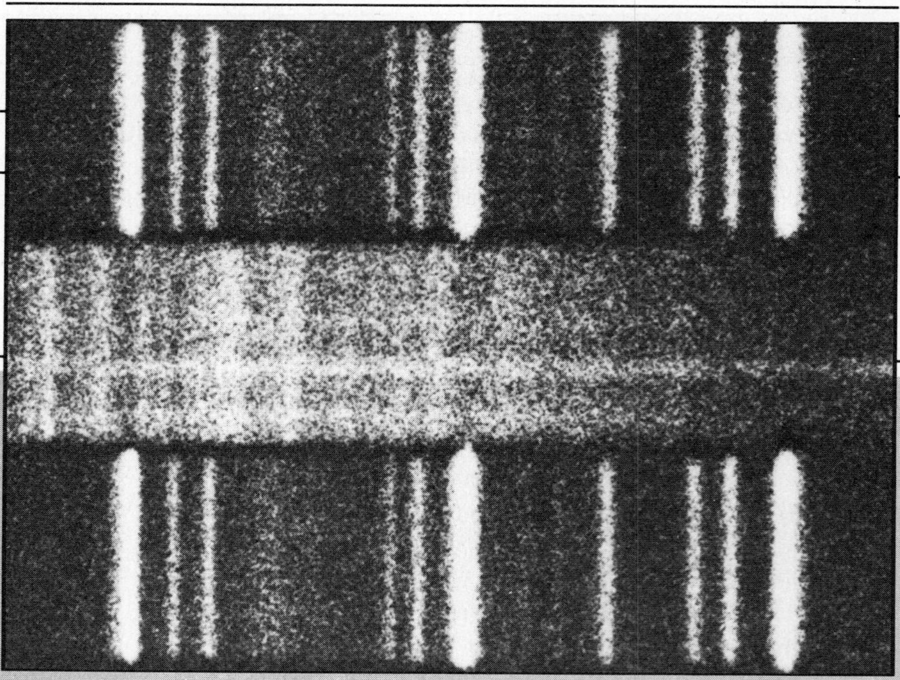

# The Signals of Astronomy

How do we find out about the planets, stars, and other bodies? We analyze the signals we receive. Galileo's *Starry Messenger* was a most aptly chosen title, but we have since found that our starry messenger carries far more information than Galileo ever could have imagined. In this chapter and the next, we shall look at the two complementary sides of this very important topic: the emission and detection of astronomical signals. First we shall see how the signals depend on factors such as the types of atoms present and on the temperature. Then we shall examine the ways in which we detect those signals and see how our methods of observing can influence our vision of the universe. These two chapters are vital to any further understanding of astronomy and its current progress. Without this material, we can read of the discoveries of new moons of Jupiter or new galaxies, but these will amount to no more than a catalog of facts however amazing. On the other hand, with this material we can provide a basis for understanding, so that long after the detailed facts of this course have been forgotten or become out of date, you may be able to follow newer developments, for the basic methods will remain in use even as they are refined or augmented.

At this point, we shall also change the form of presentation. Until now, our approach has been historical as we have traced our evolving picture of the universe. After the relatively simple science of pretelescopic days, astronomy became increasingly complex. New methods and discoveries came at an increasingly rapid rate, and it becomes more difficult to follow the historical line. Instead, we shall switch to decriptions of the methods that we now use and the knowledge that we now accept as our most reliable.

Before immersing ourselves in further detail, we should pause to take note of some milestones in astronomy, where there were major changes in the entire fabric of astronomical knowledge, and in the kinds of theories that became necessary. The first milestone was the introduction of the telescope. Until 1609 the properties of the human eye set the limits on the brightness and fine detail that could be seen. Even Galileo's first crude instruments transformed astronomy, as he was quick to realize. But despite this innovation, for about 200 years astronomy remained the same in one very important respect: It still consisted only in noting the positions of celestial objects and their apparent brightness. The next big change came with the analysis of the **quality** of the light, that is, spectroscopy, telling us of the processes occurring in distant parts of our universe. A third revolution was triggered by the introduction of photography giving us the ability to make simultaneous and permanent records of many objects, instead of having to keep notes of those objects observed one at a time.

All three of these revolutions were still within a single area: They deal only with radiation that the human eye (and photography) can detect. The explosion of astronomy into nonvisible radiations brought another revolution. Some objects emit most of their radiation in the infrared or radio or X-ray parts of the spectrum, but their importance was overlooked as long as attention was confined to their output in the visible. For example, the Crab nebula's bizarre nature has been recognized only since its radio and X-ray emissions have been examined, and Eta Carinae, one of the most luminous objects in our galaxy, radiates 95% of its energy in the infrared.

## 8.1 TYPES OF SIGNALS

To progress beyond the basic level of noting celestial positions and brightness, we need to be aware of the many ways in which astronomy now gathers and sets about making order of its information. Important developments have come through the laboratory examination of materials such as meteorites and lunar samples. Our astronomical signals thus come in very different forms, whose messages overlap at some times and supplement each other at others. We shall look at these methods and some of their limitations. As we do this, we shall be assuming some background knowledge of the structure of atoms and also some knowledge of energy and temperature. Brief surveys of those topics are provided in appendices at the end of this chapter. What now are the signals?

1. **Radiation:** light, radio waves, and X-rays are examples of **electromagnetic radiation**, the major topic of this chapter. Although we now have new sources of astronomical information (2–8 below), radiation has been and remains by far the most important means for sensing our astronomical environment.
2. **Lunar samples**, brought back to Earth by the astronauts during the Apollo missions from 1969 to 1972 and by the Russian Luna vehicles, have provided unique examples of extraterrestrial material (Fig. 8.1). The results of the lunar science program will be incorporated in other chapters—here we are concerned with techniques. Lunar samples can be subjected to a wide range of laboratory techniques. They can be analyzed chemically, bombarded at cyclotrons, and examined with microscopes and with X-rays. These methods are not confined to astronomy but draw on all of the laboratory sciences. It would take us too far from our main interest to describe all of the methods, and we shall content ourselves with brief mention when appropriate.
3. **Meteorites** (Fig. 8.2) are rocks that have come from distant parts of the solar system and landed on the Earth's surface. They provide the only

**Figure 8.1** Lunar sample brought back to Earth during the Apollo program (NASA).

**Figure 8.2** Meteorites: (a) stony meteorite Richardton observed to fall in North Dakota in 1918, (b) polished section of iron meteorite El Sampal found in Argentina (Center for Meteorite Studies, Arizona State University).

(a) (b)

extraterrestrial materials in additions to the lunar samples, and they can be subject to the same analytic techniques in our laboratories.

4. **Cosmic rays** are atomic nuclei and electrons, traveling at very high speeds, many near the speed of light. Most come from outside the solar system, some from the Sun (Fig. 8.3). They can be detected directly above the Earth's atmosphere, or indirectly within the atmosphere through the secondary effects they produce.
5. **Neutrinos** are subatomic particles emitted in certain nuclear processes, but they are extraordinarily difficult to detect. Neutrino experiments are few, are amongst the technically most difficult, and at present provide the most enigmatic results. They are discussed in Chapter 23.
6. **The solar wind** is a persistent but variable stream of electrons and atomic nuclei from the Sun spreading out through the solar system. We detect this wind by experiments on satellites and space probes. In many ways the solar wind is similar to the cosmic radiation, but the particle speeds are generally lower, and the effects that can be produced are therefore different.

7. **Gravitational radiation** is expected to be emitted in processes involving very massive objects, and as with neutrinos the detection is technically difficult. Gravitational radiation experiments will be reviewed in Chapter 25, but no gravitational radiation has yet been detected.
8. **Radar** (**ra**dio **d**etection **a**nd **r**anging) was developed for military purposes but is now much more widely used. With this technique radio signals are transmitted and their echos detected. The time taken between transmission and detection, together with the known speed of radio waves, allows the distance to the reflecting object to be calculated. Comparisons of the transmitted and reflected signals allows the speed of a moving object to be measured. Radar astronomy is confined to solar system studies, but the applications are diverse: mapping of lunar and planetary surfaces, measurement of planetary rotation speeds, tracking space probes so that their trajectories can be accurately mapped, observations of planetary atmospheres. Radar measurements can be carried out from space probes as well as from the Earth's surface.

The models we can construct for the universe must be based on the information gained through these channels. This requires an understanding of how these signals can be emitted, starting with a review of the properties of radiation.

**Figure 8.3** Tracks of cosmic-ray particles recorded in nuclear photographic emulsions. Nuclei with greater electric charge ($Z$) produce heavier tracks. These nuclei can eject electrons from atoms through which they pass, and the electrons themselves produce short tracks. This contributes to the track thickness and is especially marked for the heavier particle tracks. Particles can be identified from track structure. In these photographs a typical track thickness is about $2 \times 10^{-4}$cm. (P. H. Fowler, University of Bristol).

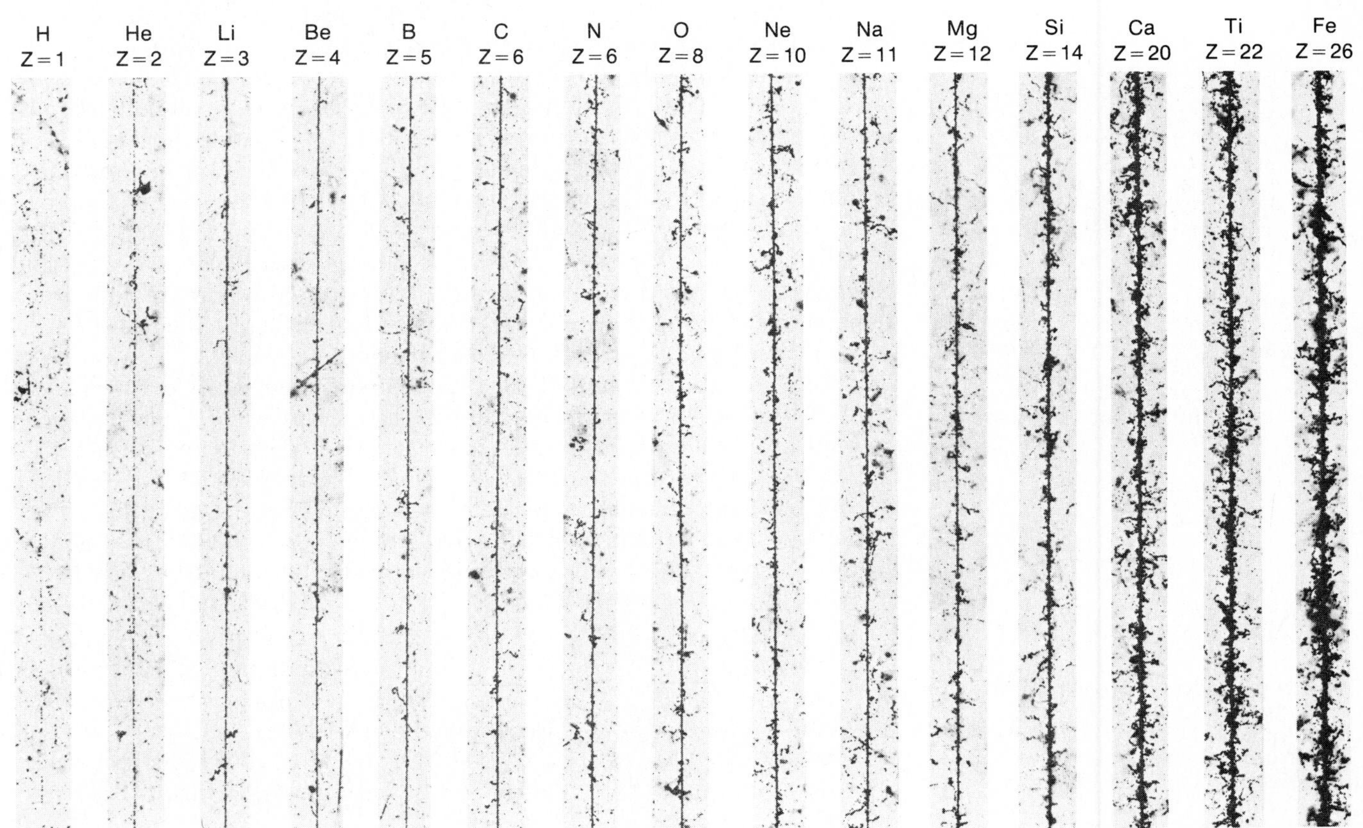

## 8.2 WHAT IS LIGHT?

This is a question that has given no end of trouble over the centuries. We can describe the behavior of light (and radio waves and X-rays) by trying to draw comparisons with other phenomena that are familiar, and we find that we need to use different (and apparently contradictory) comparisons for different properties. This is not a new problem, for we have seen that Ptolemy used different epicycles to describe different motions of a single planet. We are constructing a *model* of light (as we did earlier for the solar system), and a test of this model is its ability to lead to correct predictions. With light, there are two models. In describing the production and absorption of light, it is convenient to picture light as consisting of bundles of energy, **photons**, that seem to behave like little particles. In contrast, in describing the passage of light from one place to another and through lenses and optical instruments, it is more convenient to use a wave model and attribute wave-like properties to the light. Each model works well within its restricted region of validity, and a more comprehensive theory can be used when needed. We start with the wave picture.

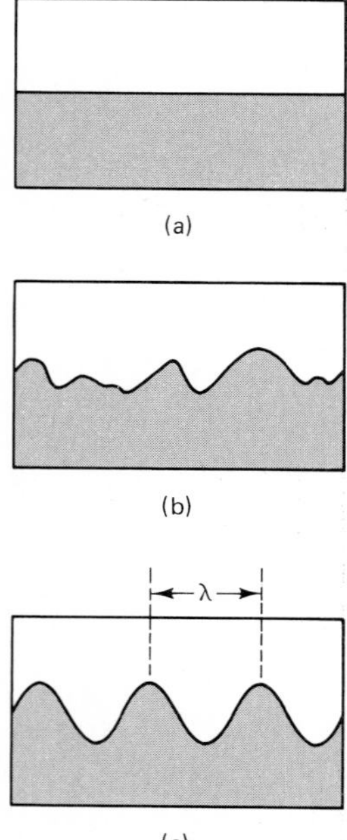


**Figure 8.4** Waves on the surface of water. An initially undisturbed surface (a) takes on an irregular profile, (b) under most conditions of disturbance, but careful paddling or the dropping of a stone into the water, can produce (c) a moving pattern of waves with regularly spaced crests and troughs.

We are familiar with waves. We hear sound waves, see water waves, and receive radio waves. But we do not usually analyze the properties of these waves, which is what we now need to do. Imagine a swimming pool with an observation window at the water level (Fig. 8.4). We can see the water surface, with crests and troughs as the water moves. When the water is completely calm, the appearance will be as in Fig. 8.4a, but when the water is disturbed, the surface may appear like that shown in Fig. 8.4b. If, however, the water is carefully paddled, a very regular pattern can be obtained (Fig. 8.4c). The crests and troughs will be evenly spaced horizontally and will be evenly displaced above and below the average (calm) level. In a large pool this pattern can be made to move steadily across our field of view. We then say that there is a wave traveling along the surface.

The distance between crests of this traveling wave is termed the **wavelength**, usually denoted by the Greek letter lambda, λ. This will also be the distance between troughs or between any two corresponding points. As the wave moves horizontally along, closer inspection shows a subtle feature. If you concentrate on the motion of a leaf or a cork floating on the surface, you will find that it is not moving horizontally with the wave but vertically. The appearance of the wave traveling horizontally is the result of each part of the surface bobbing up and down vertically, with adjacent parts slightly out of step so that the net result is the passage of the wave horizontally. This kind of wave is termed a **transverse** wave. Sound waves are different—the movement of the air is backwards and forwards along the direction the sound is traveling, and this is termed a **longitudinal** wave.

The vertical motion of the leaf can be specified by its **frequency**, the number of ups and downs (oscillations) it makes every second. This is usually denoted by the Greek nu, $\nu$. We can watch the whole wave pattern as it moves horizontally past some reference mark (such as $AA'$ in Fig. 8.5), and we will find that the number of crests passing in one second is just $\nu$. The frequency is measured in cycles (or oscillations) per second, abbreviated as c/s or cps. Older books will use that designation but the newer unit is Hz,

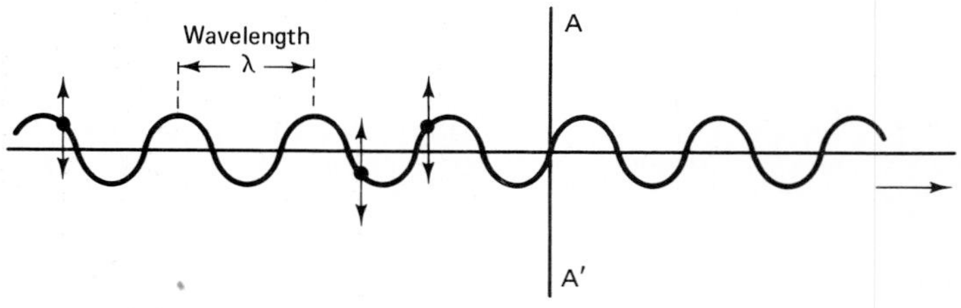


**Figure 8.5** In an harmonic wave, the crests and troughs are regularly spaced and define the wavelength. In a transverse wave, the direction of the vibration is at right angles to the direction in which the wave is moving.

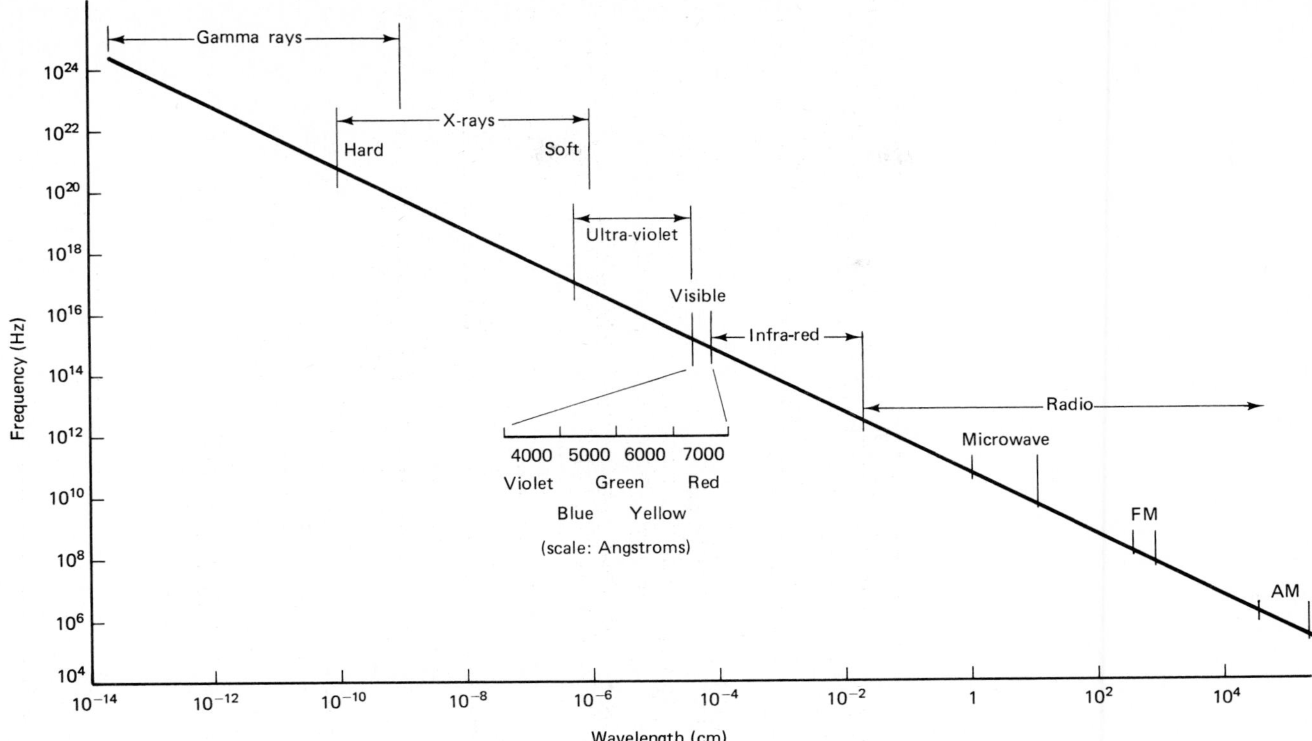


**Figure 8.6** The electromagnetic spectrum. All of these waves have the same velocity in vacuum. Wavelengths range from the atomic scale or less (the gamma rays) up through the visible to the long-wavelength radio waves.

named for Heinrich Hertz who did pioneer research on radio waves in the nineteenth century. High frequencies are designated by multiples of Hz or cps. Thus, 1000 oscillations per second will be designated as 1 kilocycle/sec or 1 kc or 1 kHz; 1 million cycles per second will be designated as 1 Mc or 1 Mc/s or 1 MHz. (The markings on the AM dial of a radio will probably be in kHz and the FM dial in MHz.)

The total length of the wave that passes the reference marker ($AA'$) in 1 sec is thus $\nu$ waves times $\lambda$ (the length of each wave). Because the distance moved by a wave in 1 sec is just its speed, a very important relation exists between the qualities describing the wave

$$\mathrm{v} = \nu\lambda$$

$$(\text{cm/sec}) = (\text{cycles/sec}) \times (\text{cm})$$

All electromagnetic waves share a common speed, usually designated as $c$ and referred to as the speed of light. Its value is another of the fundamental constants in science, known by now to better than one part in a billion: $2.99792458 \times 10^{10}$ cm/sec. We can round this to $3.0 \times 10^{10}$ cm/sec, quite accurate enough for all of our uses. The waves that we call radio, light, ultraviolet, X-rays and others differ in their $\nu$ and $\lambda$ values, but the product of those two quantities for each wave must always equal $c$ (Fig. 8.6).

**BOX 8.1**
*Wavelengths and Frequencies*

Various units can be used for wavelengths. If the wavelength is given in centimeters, the velocity of the wave will be in cm/sec. The wavelengths of visible light are much smaller than 1 cm; a typical value is $5 \times 10^{-5}$ cm. This will often be listed as 5000 Å, where the Angstrom unit (1 Å $= 10^{-8}$ cm) is named for the nineteenth-century Swedish spectroscopist.

In recent years a slightly different system has come into use, with the meter as the basic unit. The speed of a wave would then be given in m/sec, and the wavelength of our typical visible light would be listed as 500 nm, where 1 nm = 1 nanometer = $10^{-9}$ m.

In the infrared part of the spectrum, it is customary to list wavelengths in microns, where 1 micron = $10^{-4}$ cm.

Examples:

1. Calculate the frequency of the light vibration corresponding to a wavelength of 6328 Å:

$$\nu = \frac{c}{\lambda} = \frac{3 \times 10^{10} \text{ cm/sec}}{6328 \times 10^{-8} \text{ cm}}$$
$$= 4.74 \times 10^{14} \text{ Hz}$$

2. Calculate the wavelength corresponding to a frequency of $8 \times 10^{13}$ Hz:

$$\lambda = \frac{c}{\nu} = \frac{3 \times 10^{10} \text{ cm/sec}}{8 \times 10^{13}\text{/sec}}$$
$$= 3.75 \times 10^{-4} \text{ cm}$$
$$= 3.75 \text{ microns}$$
$$= 37500 \text{ A}$$
$$= 3750 \text{ nm}$$

We can see the water and its waves, and we can feel the wind even if we cannot see the air as it carries sound waves. The obvious question is then: for light, what is doing the waving? The answer seems unsatisfactory. Nothing material or substantial is waving. When electromagnetic waves travel, there are rapidly alternating electric and magnetic forces present. If we could put a single electron in the path of a radio wave, for example, the electron would move from side to side with the same frequency as we ascribe to the radio wave. This is, in fact, how we detect radio waves, except that we use very many electrons, not loose but in the metal of an antenna. The radio wave forces the electrons to rush back and forth in the antenna, and the moving electrons produce electrical signals that we can amplify in our radio and television sets. The frequencies of light or X-rays are far too high to be detectable in this way, but the underlying nature of the wave is the same: rapidly varying electric and magnetic forces that produce physical effects that we can detect and measure. No material substance is needed for these electromagnetic waves—they can travel in a vacuum. In transparent materials, such as glass, some waves can still travel, but their speed is reduced. Light travels through glass with about two-thirds of the speed it has in a vacuum. (The difference between the speeds of light in vacuum and air is sufficiently small that for many purposes we can ignore this and we use $c$ for the speed in air.)

BOX 8.2
*Polarization*

The electric field vibrations that constitute light and other radiation must be at right angles to the direction that the radiation is traveling, but otherwise there are usually no restrictions—the vibrations can be in any direction. In some special circumstances, however, the vibrations may be partly or totally in one direction, and the wave is said to be polarized (Fig. 8.7). Polaroid sunglasses have special crystals that allow only one direction of vibration

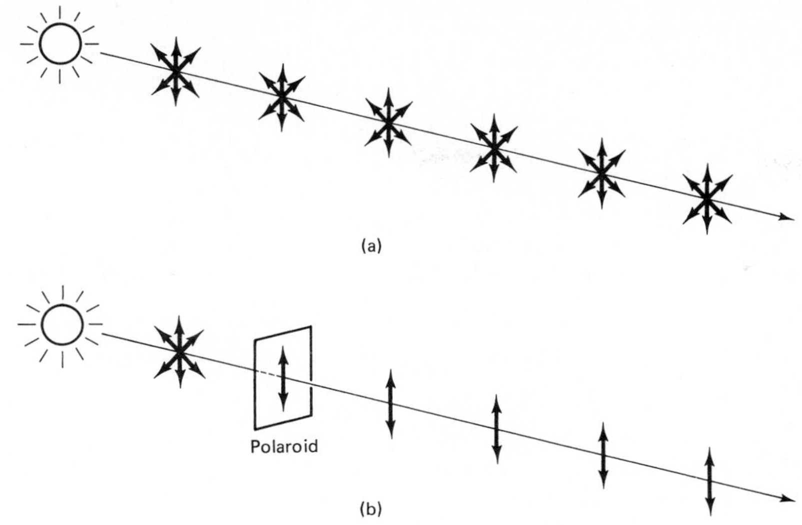


**Figure 8.7** Polarization (a) Electric vibrations must be at right angles to the direction of travel of an electromagnetic wave but otherwise can be in any direction. Thermal radiation such as from incandescent lamps is unpolarized. (b) Polarized radiation has its electric vibrations in only one direction. This can be produced by special materials such as Polariod.

to pass through. If you look through two polaroids together and rotate one slowly, you will see that there are positions where no light at all gets through. The light that passes through the first has been polarized, and whether or not it gets through the second polaroid depends on the rotation of the polaroid. Light from the sky and light reflected from a road surface are partly polarized—look at them through a polaroid and rotate the polaroid, and you will see the intensity change.

## 8.3 EMISSION OF THERMAL RADIATION

Hot surfaces give off radiation. We see because the Sun is hot and its light reaches the Earth where it bounces off trees, dogs, houses, and books to reach our eyes. We feel the warmth of the sunshine or the heat of a fire.

Thermal radiation was the subject of intense research during the nineteenth century. It was found that the total amount of energy (adding up the energy at all wavelengths) radiated by a hot surface depended very sensitively on the temperature of that surface and was accurately described by Stefan's Law:

$$\begin{bmatrix}\text{Total energy radiated from} \\ 1\text{ cm}^2\text{ in one sec}\end{bmatrix} = \sigma T^4 \text{ ergs/(cm}^2\text{sec)}$$

Here $\sigma$ is known as Stefan's constant, and it has the value

$$\sigma = 5.67 \times 10^{-5} \text{ ergs/cm}^2(\text{K})^4 \text{ sec.}$$

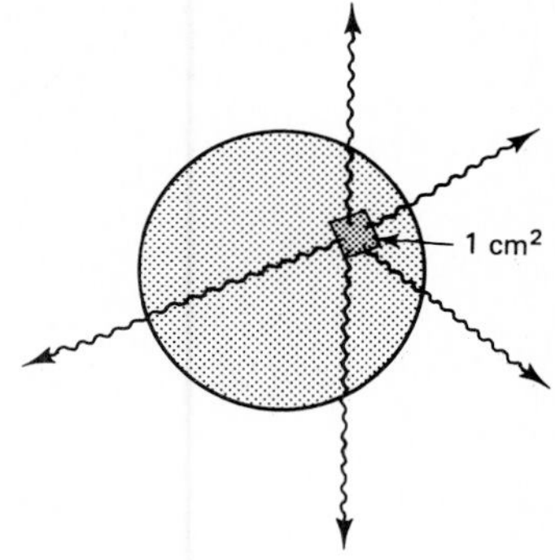


**Figure 8.8** Stefan's Law: Each square centimeter of a hot surface radiates energy at a rate that depends on the fourth power of the surface temperature.

The temperature, $T$, that is used in this formula is measured on the **absolute** or Kelvin scale closely related to the more familiar Celsius scale:

$$\begin{pmatrix}\text{temperature in absolute} \\ \text{or Kelvin degrees} \\ T_\text{K}\end{pmatrix} = \begin{pmatrix}\text{temperature in} \\ \text{Celsius degrees} \\ T_\text{C}\end{pmatrix} + (273)$$

$$T_\text{K} = T_\text{C} + 273$$

The divisions on the Kelvin and Celsius scales have the same size, but the scales have different zero points. Water boils at 0° on the Celsius scale and at 273° on the Kelvin scale. The reason for using the Kelvin scale is that many scientific formulas take on a simpler appearance than they would if the Celsius scale were used. For example, if we used Celsius degrees, Stefan's Law would read $\sigma(T_\text{C} + 273)^4$.

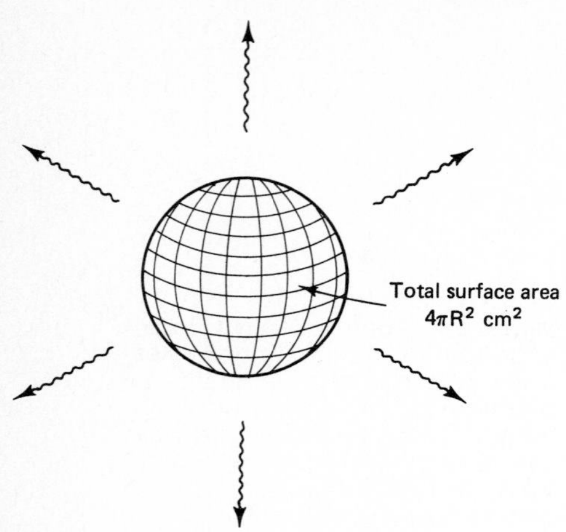


**Figure 8.9** For a spherical surface with an area $4\pi R^2$, the rate of radiating energy will be $(4\pi R^2)(\sigma T^4)$ ergs/sec, where $\sigma$ is Stefan's constant.

Strictly speaking, Stefan's Law and others we have yet to introduce, refer to an idealized surface that is known as a **black body**. A black body is defined as one that absorbs all radiation falling on it and reflects none. The absorbed energy heats the body so that it then emits its own radiation (see below). Most real surfaces will radiate somewhat less energy than a black-body at the same temperature, but it turns out that radiation from stars is well described by the laws for black body surfaces.

We now consider a star that has a radius $R$ cm and a surface temperature $T$ K (Fig. 8.9). It will have a surface area $4\pi R^2$ cm² and by Stefan's Law, in each second it will radiate a total amount of energy given by

$$L = (\text{surface area}) \times (\text{energy radiated per cm}^2 \text{ per sec})$$
$$= (4\pi R^2)(\sigma T^4) \text{ ergs/sec.}$$

We call $L$ the **luminosity** of the star. In many cases, we can measure $L$ and $T$, and this leaves $R$ as the only unknown quantity in the equation. We have thus found a way of measuring the size of the star.

We can now deduce a very important rule. Suppose that we have a star emitting $L$ ergs/sec and situated at a distance $D$ cm from us (Fig. 8.9). Because the energy radiated spreads out in all directions, at our distance $D$ it must cover a spherically shaped surface area of $4\pi D^2$ cm². Each square centimeter, at this distance from the star therefore receives $L/4\pi D^2$ ergs in every second. This is the quantity that we can actually measure—the amount of energy that each square centimeter of our telescope catches in every second. We call this the **flux** of the radiation:

$$f = \left(\frac{L}{4\pi D^2}\right) \text{ ergs/cm}^2\text{sec}$$

We note that although we can measure $f$ directly, we need to measure $D$ in order to determine $L$. Conversely, if we already know $L$ (which, in some cases we do, from other measurements we shall describe later), then we can use our measurements to find $D$, the distance of the star. These relations are used extensively in astronomy and we shall be referring to them on many occasions.

**Figure 8.10** Blackbody spectrum. At a given temperature, the intensity of the radiated energy has a maximum at a wavelength $\lambda_{max}$ that is related to the temperature $T$ by *Wien's Law:* $\lambda_{max} T = 0.2898$ *cm deg. The spectrum is steeper on the short-wavelength side of the peak than it is at the longer wavelengths.*

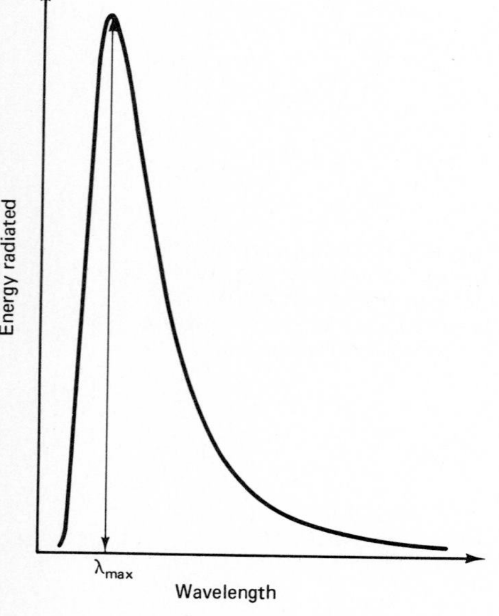


The luminosity includes the energy at *all* wavelengths, from the longest radio waves to the shortest X-rays, and so the next topic of importance is the way in which this energy is distributed over different parts of the spectrum. When we measure the amount of energy radiated by a black body at different wavelengths, we obtain results shown in Fig. 8.10. This is termed a **continuous spectrum**—there is a smooth variation in the amount of energy radiated from shorter to longer wavelengths. We note that there is a wavelength ($\lambda_{max}$) at which most energy is radiated. The radiated energy falls sharply at shorter wavelengths, whereas there is a much slower decrease at longer wavelengths. The value of $\lambda_{max}$ depends on the temperature of the radiator. If we increase the temperature, more energy is radiated (remember Stefan's Law) and the whole curve moves up the graph. The peak then shifts to a shorter wavelength. Figure 8.11 shows a family of curves corresponding to different temperatures.

There is a simple law, Wien's Law, relating temperature to $\lambda_{max}$:

$$\lambda_{max} T = 0.2898 \text{ cm degrees}$$

where the temperature is in degrees Kelvin and the wavelength, $\lambda$, is in centimeters. If we measure the energy radiated at several different wavelengths, we can in principle determine $\lambda_{max}$ and hence $T$ from Wien's Law. So, for example, for the Sun we find most energy being radiated close to $\lambda_{max} = 5.0 \times 10^{-5}$ cm, from which we deduce that $T_{sun} = 0.2898/5.0 \times 10^{-5}$ or $\sim$6000 K. The relative amounts of energy radiated at different wave-

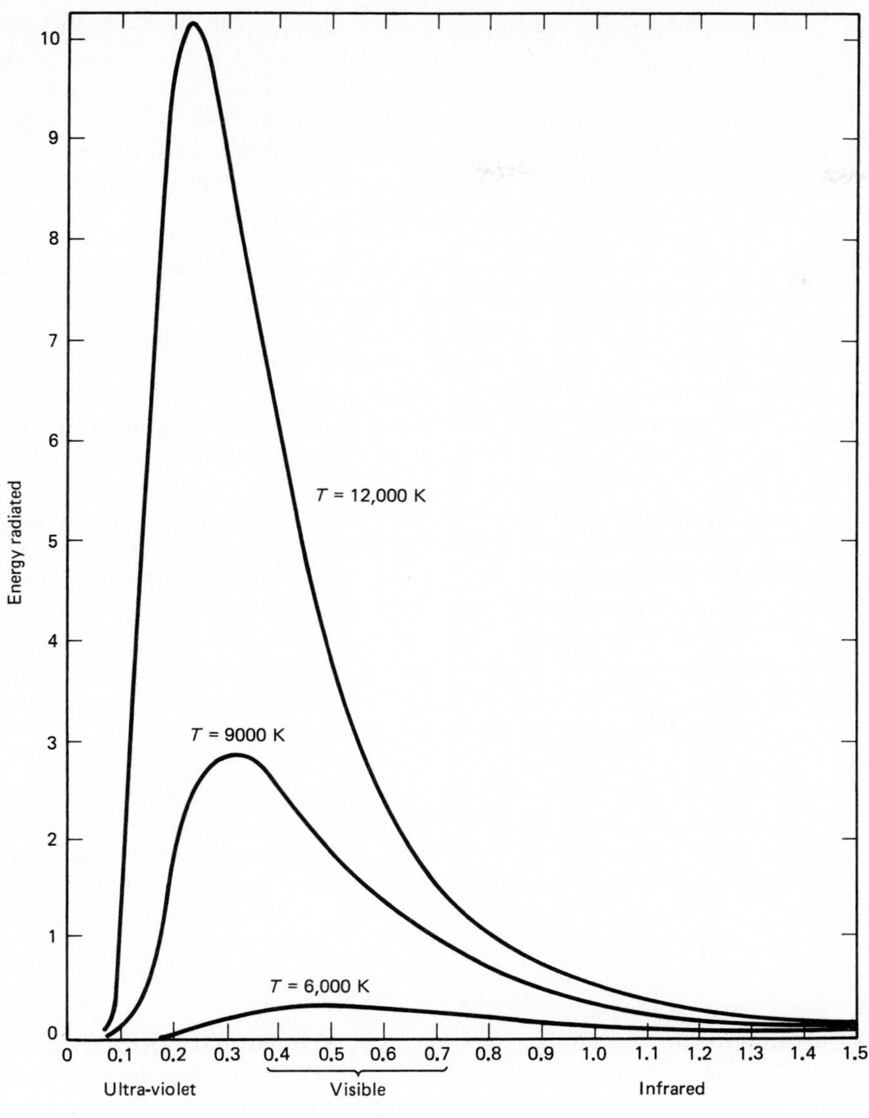


**Figure 8.11** Blackbody spectra for three temperatures. The curves are similar but with the positions of their peaks defined by Wien's Law, so that the radiation at a lower temperature has its maximum intensity at a longer wavelength. The peak will lie in the visible range for some temperatures but not for others.

lengths for different temperatures show up quite clearly: we can see the difference between red-hot and white-hot fires. The hotter the fire, the more short wavelength radiation is emitted and the more the peak emission shifts toward the shorter wavelength end of the visible spectrum.

The precise mathematical description for these black-body radiation curves was first derived by Max Planck in 1900. Earlier theories could explain the short and long wavelength regions separately but not both together, and Planck's success rested on his surprising assumption that the energy was being radiated and absorbed in packets that he termed **quanta** (today called **photons**). This **quantum theory** of Planck has been considerably expanded, and it is used with great success in many ways. Planck assumed that the energy carried by each photon was given by

$$E = h\nu$$

where $\nu$ is the frequency (associated with wavelength, $\lambda$) and $h$ is a new constant that we now call Planck's constant. The present best value is

$$h = 6.626 \times 10^{-27} \text{ ergs. sec}$$

Where Stefan and Wien had to determine their constants by experiment, with Planck's law it could now be shown that the values of those constants could be derived as combinations of other constants such as the speed of light and the new constant $h$. (Planck's constant must be measured, for it is another fundamental constant that cannot be derived from anything more basic.) We shall return to this analysis of thermal spectra in Chapter 21 when we come to discuss the classification of stars.

## 8.4 SPECTRAL LINES

Continuous thermal spectra give us no clue to the nature of the radiating surface, for their seamless shapes testify mainly to the temperature. All that is required is that the emitting material be opaque, as is the case for a solid or a very dense gas. For less dense regions of gas, the form of the spectrum is very different. In order to understand these effects, we need to review some properties of atoms and molecules. (See also appendix 8A at the end of this chapter.)

The electrons around an atomic nucleus can occupy various orbits, each orbit corresponding to a specific energy. Each electron has one unit of (negative) electric charge, whereas the nucleus has an opposite (positive) charge. The number of electrons in a normal atom is determined by the size of the charge on the nucleus so that the totals of positive and negative charges are equal. For hydrogen, the smallest and simplest atom, the nucleus has one positive charge and the atom has one electron. For carbon, with six units of positive charge in the nucleus, there are six electrons.

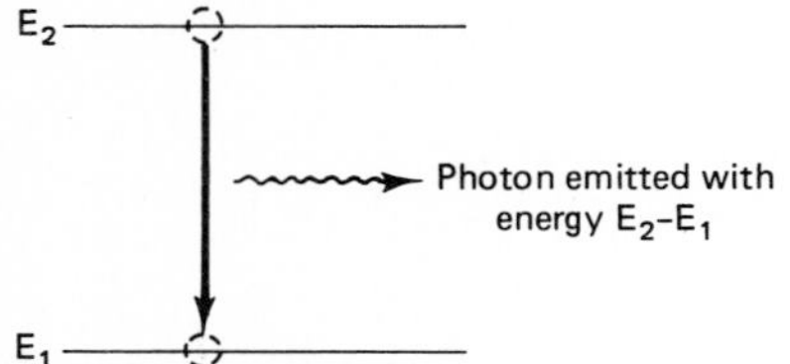


**Figure 8.12** An atom in a higher energy level $E_2$ can change to a lower energy $E_1$ and emit the energy difference ($E_2$-$E_1$) in the form of a photon. The energy levels correspond to electrons in different allowed orbits, and the transition to the lower energy involves an electron going from one orbit to another.

Only certain electron orbits and energies are possible. Thus an electron will occupy one or another of the permitted energy levels but no intermediate energy (Fig. 8.12). If an electron is in some energy level $E_2$ and moves to another level with a lower energy, $E_1$, the energy difference $E_2 - E_1$ is radiated as a photon whose energy and frequency are specified by the formula:

$$E_2 - E_1 = \text{energy of photon} = h\nu = hc/\lambda$$

$$(\text{remembering that } \nu = c/\lambda.)$$

Thus light of this particular wavelength is characteristic of the two energy levels, $E_1$ and $E_2$. Each different kind of atom (hydrogen, helium, carbon, etc.) has its own distinctive energy levels; the differences between energy levels are therefore as characteristic of each kind of atom as fingerprints are of each person. Measurement of wavelengths allows us to identify the energy levels responsible and thus the kinds of atoms present. In contrast to the thermal radiation that followed a continuous spectrum over a wide range of wavelengths, these well-defined energy levels result in energy being radiated only at some sharply defined wavelengths, and the spectrum is termed a **line spectrum**. (This name derives from the appearance of the spectrum when viewed through an instrument used to analyze spectra.)

The number identifying an energy level is known as the **quantum number**, $n$. The lowest energy level for an atom is designated as $n = 1$, the next higher level has $n = 2$, and so on. In an assembly of a large number of atoms of the same kind, some electrons will be in the lowest energy level and some will be in higher energy levels. The numbers of electrons with different energies depends on the temperature. A series of related spectral lines is obtained when electrons start from different higher energy levels but all end on the same lower level. In the case of hydrogen, spectral lines that are emitted when electrons end up in the lowest ($n = 1$) level are termed the Lyman series and all lie in the ultraviolet. In the Balmer series, the common point is the $n = 2$ level, and these lines are in the visible part of the spectrum. This series plays a major role in identifying hydrogen in stellar atmospheres.

Other series end at other levels and lie in the infrared, microwave, and radio regions of the spectrum. There is also an important hydrogen line at a wavelength of 21 cm. This is produced by a process somewhat different from that described, and it has played a critical role in the exploration of our galaxy by radio astronomers (see Box 8.3 and Sec. 26.5).

## BOX 8.3
### *The Hydrogen Spectrum*

The lines of the hydrogen spectrum were identified and their wavelengths measured long before Niels Bohr had proposed his atom model in 1913. This model still provides a good way of visualizing the behavior of atoms. Bohr imagined each atom to be something like the solar system: At the center was the atomic nucleus, and around it in orbits were electrons. In the solar system it is the gravitational force that holds the planets to the Sun: in an atom the force is electrical between the positively charged nucleus and the negatively charged electrons.

A relatively simple formula was found that relates the wavelengths in each series:

$$\frac{1}{\lambda} = R\left(\frac{1}{m^2} - \frac{1}{n^2}\right)$$

$R$ is known as the Rydberg constant, 109678 if $\lambda$ is in cm. For the Lyman series, $m = 1$ and $n$ takes in turn the values 2, 3, 4, . . . . For the Balmer series $m = 2$ and $n = 3, 4, 5, \ldots$ . There are additional series of spectral lines that correspond to other values of $m$. The significance of the $m$ and $n$ and the value of $R$ were not realized when this formula was developed in the 1880s, but all of these numbers (including a value for $R$) followed naturally from Bohr's model and from the quantum theory. This is yet one more example of a numerical regularity being recognized far ahead of any actual understanding. The situation is much more complicated for atoms with more than a single electron.

There is another characteristic line in the hydrogen spectrum that deserves special mention. At low temperatures, each electron is in the smallest allowed orbit. The electron and its parent nucleus (proton) each have **spin**. These two spins can be either in the same direction or opposed, and there is a very small difference in energy between these two possible states (Fig. 8.14). A hydrogen atom that changes from the higher energy state to the lower will emit a photon whose wavelength is 21 cm. This radiation has been

***Table 8.1*** WAVELENGTH OF LINES IN THE HYDROGEN SPECTRUM[1]

| Line | Wavelength |
|---|---|
| *Lyman Series (ultraviolet) (m = 1; n = 2, 3, 4 . . .)* | |
| Lα | 1215 |
| Lβ | 1025 |
| Lγ | 972 |
| Lδ | 949 |
| Lε | 937 |
| *Balmer Series (visible) (m = 2; n = 3, 4, 5 . . .)* | |
| Hα | 6562 |
| Hβ | 4861 |
| Hγ | 4340 |
| Hδ | 4101 |
| Hε | 3970 |
| *Paschen Series (infrared) (m = 3; n = 4, 5, 6 . . .)* | |
| Pα | 18751 |
| Pβ | 12818 |
| Pγ | 10938 |
| Pδ | 10049 |
| Pε | 9545 |

[1] Wavelengths are listed in units of Angstroms ($10^{-8}$ cm). Recent usage is in terms of nanometers (nm): 1 nm = $10^{-9}$ m = $10^{-7}$ cm. Thus the first Balmer line would be listed as 656.2 nm.

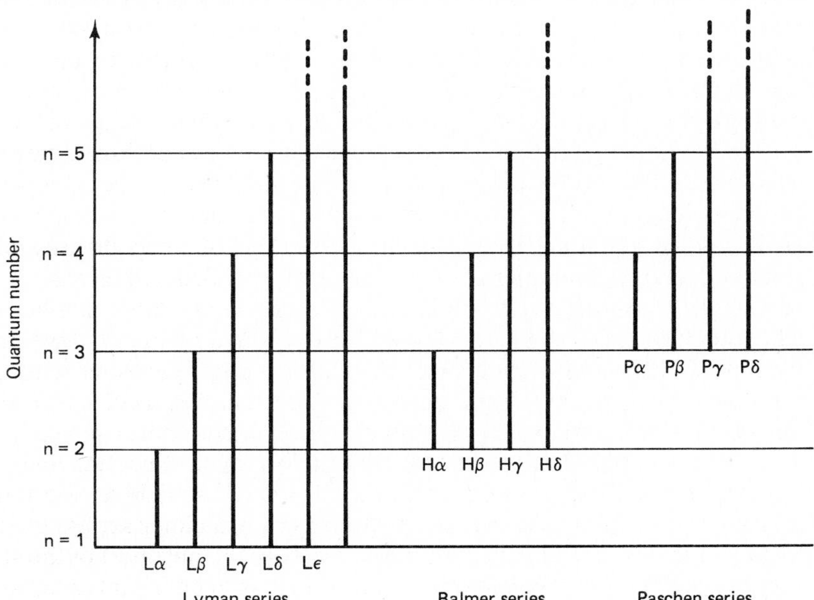


**Figure 8.13** The principal lines in the spectrum of hydrogen. Energy levels of the hydrogen atoms are designated by the quantum number $n$, corresponding to the orbit in which the atom's electron finds itself. Transitions that start or end on the lowest level ($n = 1$) constitute the Lyman series, and all the lines are in the UV. Transitions that start or end on the $n = 2$ level constitute the Balmer series, in the visible part of the spectrum. The Paschen series of lines is in the IR, and there are other series (not shown) that start or end on higher levels.

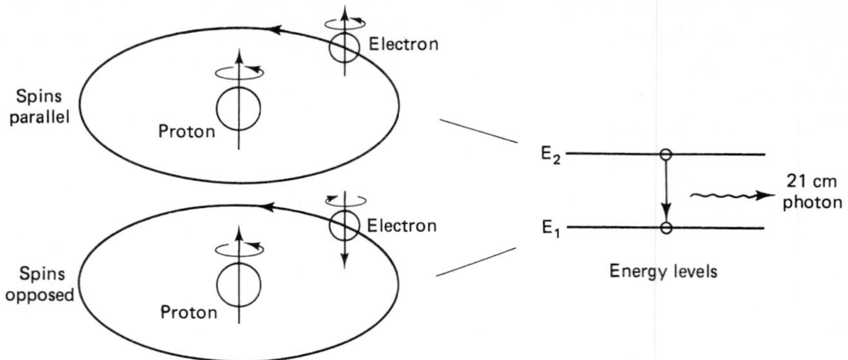


**Figure 8.14** Hydrogen atoms with their electrons in the smallest (lowest energy) orbits can still have slightly different energies depending on the relative directions of the proton and electron spins. Transition between these two energy levels leads to the emission of a photon that has a wavelength of 21 cm in the radio region of the spectrum.

of great importance to radio astronomers, allowing them to map the distribution of hydrogen in cold regions of our galaxy (see Sec. 26.5). At high temperatures, this spectral line is not seen.

What we have been describing is an **emission spectrum**. If glowing hydrogen gas is viewed through the appropriate instrument (a spectroscope), we will see a series of differently colored images, each image corresponding to a particular spectral line. At low temperatures most hydrogen atoms will be in the lowest quantum energy level ($n = 1$), but as the temperature is raised, collisions between the atoms result in some energy being given to the electrons, which become **excited**, that is, they move to higher energy levels. After a very short time (typically $10^{-8}$ to $10^{-9}$ sec), an excited electron will return to a lower energy level and radiate its excess energy. Calculating just what happens at any given temperature—how many electrons are in each of the many available levels and how do they return to their lower level—is a complicated but solved problem. For example, an electron that finds itself in the $n = 8$ level might return to $n = 1$ in a single step, or (less probably) pass through some intermediate steps to reach $n = 1$. The theory for all such possible jumps (and corresponding spectral lines) is well understood, and it is applied routinely in the analysis of line spectra. As you might expect, the spectra get steadily more complex for higher temperatures and especially for heavier atoms with their many electrons.

At high temperatures the colliding atoms have higher speeds and their collisions are more violent. Sometimes enough energy is transferred to an electron to remove it completely from its parent atom, which then has a net positive charge and is said to be **ionized**. At higher temperatures more than one electron can be removed from an atom. The spectrum of the atom with its remaining electrons is affected, but calculations and laboratory calibrations have given us tables of wavelengths so that, as with the spectral lines from nonionized atoms, many atoms can be identified by their characteristic lines. The degree of ionization is indicated in astronomy by Roman numerals: HI is neutral (nonionized) hydrogen, while HII is the second possible state, singly ionized. (The numeral is always one more than the number of electrons removed.) Nonionized carbon is CI, but since carbon has six electrons, we can find CII, CIII . . . , (corresponding to one, two, . . . , electrons removed). For iron with twenty-six electrons, many degrees of ionization are possible. Spectral lines from ionization states such as FeXII or FeXIV can be seen in the solar spectrum.

The spectra of some stars and galaxies show very great complexity indicating that the light is being emitted from several regions with different temperatures. Disentangling the spectra can be difficult, but spectral analysis is still a remarkably powerful tool and the only means of investigating many aspects of the structure of remote and inaccessible regions.

## 8.5 ABSORPTION SPECTRA

The emission spectrum of a hot, dense region will be continuous. If now a thin and cooler gas lies between that hot region and ourselves (Fig. 8.15), the spectrum that we detect carries some characteristic evidence of the cooler gas. The original continuous spectrum contains all wavelengths. Some

Hot dense gas (continuous spectrum)
Thin cool gas
Emerging radiation — spectrum has dark (absorption) lines
Bright continuous spectrum
Emission line spectrum

**Figure 8.15** Absorption and emission lines. A hot, dense gas will emit a continuous spectrum. If this radiation enters a cool gas cloud that is not very dense, some of the radiation will be absorbed at wavelengths that just match the energy level differences in the gas atoms. The emerging spectrum will then show a series of dark absorption lines whose wavelengths are characteristic of the gas cloud but not of the original hot source. Observing the cool cloud from a different direction, one will see a spectrum that consists of the same wavelengths as the absorption lines but now seen as bright emission lines.

**Figure 8.16** Solar Spectrum with dark absorption lines (Palomar Observatory).

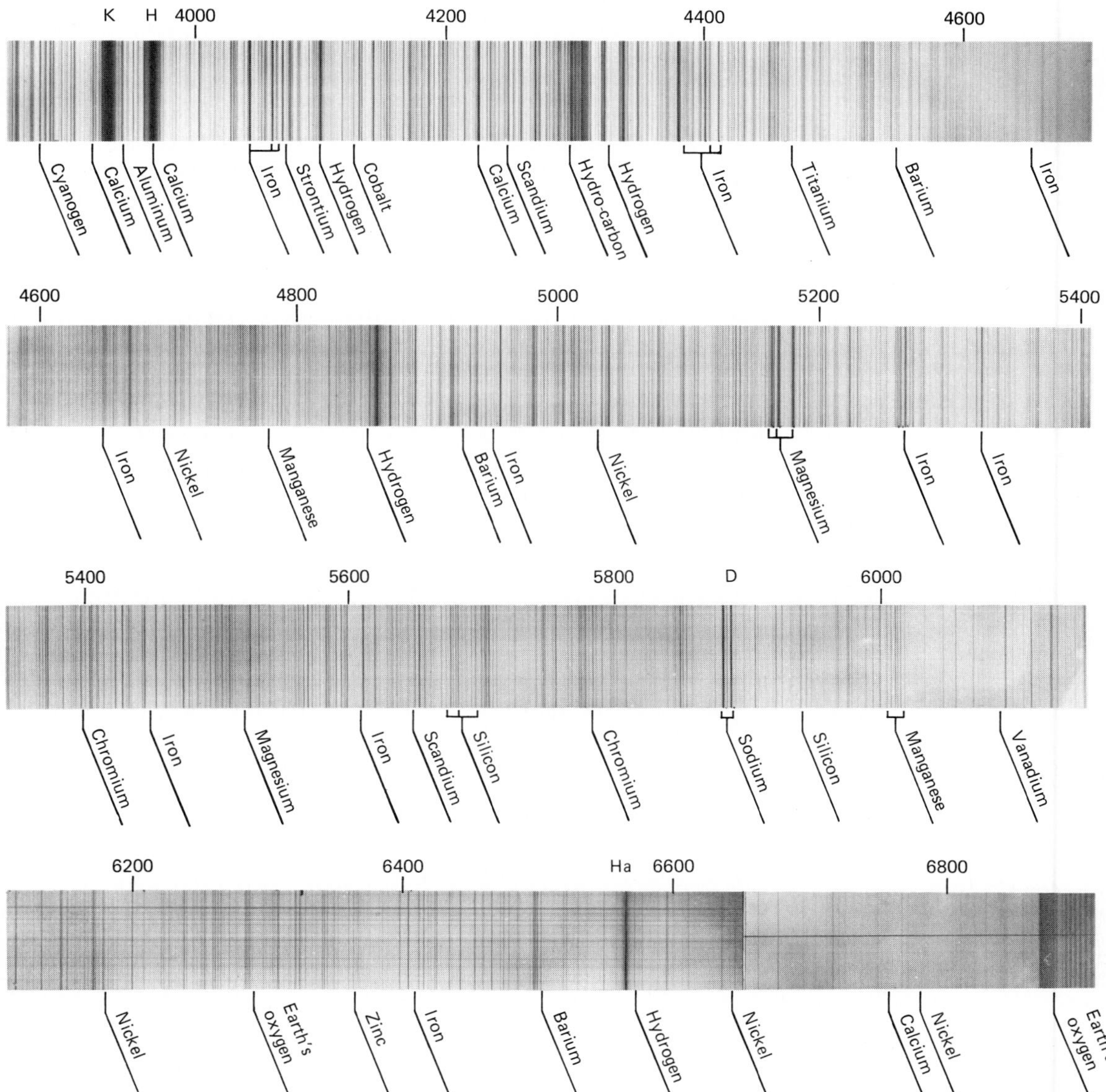

of its photons will have energies that match **exactly** with energy level differences in the atoms of the cool gas. As a result these photons can be absorbed. When the radiation emerges from the gas, there is less energy at the corresponding wavelengths, and the continuous spectrum shows a series of dark **absorption lines** (Fig. 8.15).

The absorbed energy **excites** the absorbing atoms to higher energy levels, and soon thereafter the electrons have to shed their extra energy. It would seem that this would restore the light to the radiation coming toward us, but this is only partly the case. The radiation that was absorbed had been traveling directly from the source to us; when the excited electrons return to their lower energy levels, the emitted radiation will go in all directions. A small amount will be emitted along the original direction, but the net effect is a reduction in the energy reaching us and the appearance of a dark line in the spectrum. This is not an effect that can be seen by the naked eye. On spectroscopic examination the solar spectrum shows many very pronounced absorption lines (Fig. 8.16). The bright light of the Sun is produced in a dense lower region of its atmosphere (the photosphere) that emits a continuous spectrum; the absorption occurs in a far thinner and cooler overlying atmospheric layer. Fraunhofer in 1814 discovered these absorption lines, whose wavelengths were later matched with emission lines observed in the laboratory for different types of atoms. This clearly showed that the Sun is made of the same materials (types of atoms) as the Earth, although the proportions and conditions are very different.

## 8.6 MOLECULES

Some spectral lines can be attributed to molecules, groups of atoms bound together. Molecular energy levels determine the wavelengths of their spectral lines, which have been seen in both emission and absorption. Where the energy levels in atoms depend on the electrons, there are additional factors to be considered with molecules. The distances between the atoms making up a molecule can change as the molecule vibrates. Molecules can also rotate. Vibrations and rotations require energy. The result is that molecular spectra tend to be far more intricate than those of atoms.

High temperatures lead to the dissociation of molecules into their constituent atoms, and only a few molecules such as titanium oxide (TiO) survive at the several thousand degree temperatures of the cooler stars. Spectra of molecules such as carbon monoxide (CO) are prominent when we observe interstellar clouds where the temperatures are far below freezing. Most of these molecular studies are carried out at millimeter wavelengths.

## 8.7 SYNCHROTRON RADIATION

Thermal spectra whether continuous or with emission or absorption lines were the only identified spectra in astronomy for a very long time. During the 1930s, radio signals coming from beyond the solar system had been discovered, but there was no understanding of their origin. When radio as-

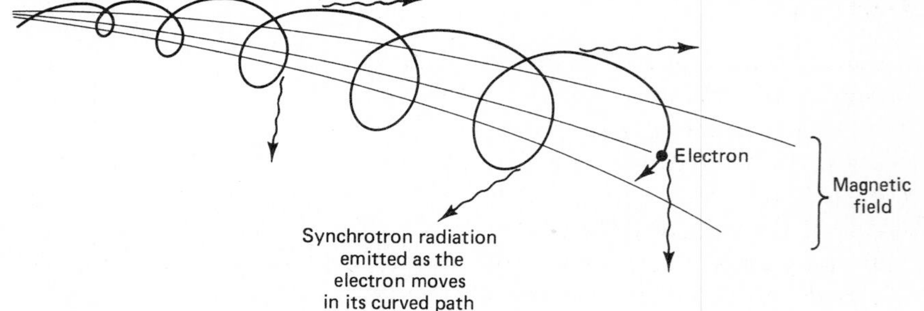


**Figure 8.17** Very-high-speed electrons follow curved paths and emit synchrotron radiation when traveling in a magnetic field. This radiation is emitted in a narrow cone along the instantaneous direction of motion.

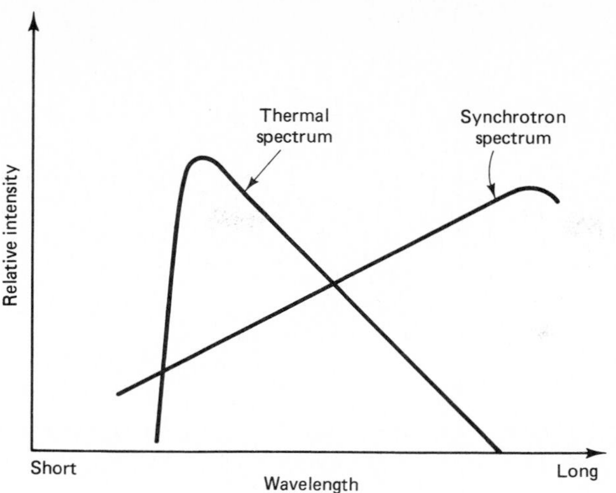


**Figure 8.18** Synchrotron radiation spectrum. The spectrum emitted by a large number of electrons is significantly different from the thermal spectrum of hot objects. Thermal radiation intensity drops sharply for wavelengths shorter than peak emission and decreases slowly for long wavelengths. In a synchrotron spectrum, there is a steady increase in intensity on the short-wavelength side towards the peak and then a rapid decrease. Depending on the electrons' energies and the strength of the magnetic field, the synchrotron radiation may be in the visible or radio parts of the spectrum.

tronomy was developing rapidly, after World War II, it was found that the radio spectra were quite unlike those of any thermal source. The term *non-thermal* came into use but without any physical understanding of the origin. The correct astronomical explanation came from Iosef Shklovsky, who was already mentioned when discussing ancient supernovae. The explanation involves some physical effects we have not yet encountered.

Magnetism is a familiar phenomenon. Close to a magnet, you can feel its pull on a piece of iron in your hand. The surrounding region contains the magnetic field. Magnetic fields are found around the Earth, in some stars and over vast distances between stars. In a magnetic field, a high-speed electron is forced to follow a curved path and in doing this it radiates energy, usually as radio waves but sometimes as light (Fig. 8.17). This mechanism is also encountered under laboratory conditions. Atomic nuclei can be accelerated to very high speeds by devices such as cyclotrons. For electrons a related device is the synchrotron, where the circular motion of the electrons in a magnetic field is accompanied by the emission of radiation. This kind of radiation is now generally termed **synchrotron radiation**, regardless of whether its origin is in the laboratory or an astronomical object or region (Fig. 8.18).

Synchrotron radiation has distinctive properties. It has a continuous spectrum, but the shape is quite different from the thermal spectra previously mentioned. In a thermal spectrum the intensity decreases very rapidly on the short wavelength side of the spectrum peak. In contrast, a synchrotron spectrum decreases much more slowly, and the shape of that part of the synchrotron spectrum provides information on the speeds of the electrons and on the strength of the magnetic field. In addition a synchrotron spectrum will be partly polarized.

Synchrotron radiation is of great importance in radio astronomy, and its appearance is a characteristic feature of supernova remnants. The detection of synchrotron radiation is not only a clear indication of the presence of very-high-speed electrons but also provides a means of measuring distant magnetic fields.

## 8.8 CHANGING THE WAVELENGTH

Once emitted, the radiation has to fend for itself. It may travel a relatively short distance (astronomically speaking) such as from the Moon to the Earth, or it may travel from a distant galaxy and reach us after millions of years. How do we know that the radiation we receive has not been changing during its travel? If there have been changes that we have failed to identify, we will end up by drawing erroneous conclusions about the origin of the radiation. It turns out that the radiation can indeed undergo some changes, but

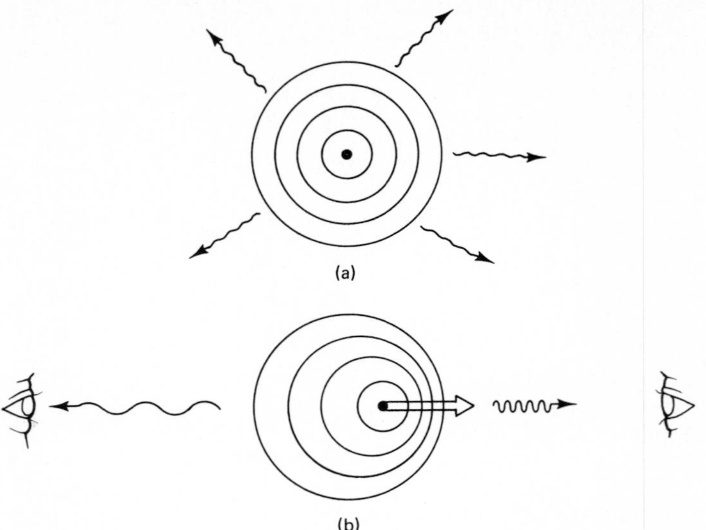


**Figure 8.19** The Doppler Effect. Radiation from a stationary source spreads out evenly in all directions with crests and troughs evenly spaced. When the source of radiation is moving (*b*), the crests appear to crowd together when detected in the direction in which the source is moving. The wavelength then seems shorter, the frequency greater, and the spectrum is blueshifted. At the same time, the radiation detected in the rear is redshifted—the crests and troughs appear farther apart and the frequency is lower.

we are often able to identify these and turn them to our advantage, using them to investigate the regions traversed. There are, however, two important effects that may take place at the source, that we should review here.

When the source of the radiation is not moving relative to us, the crests and troughs of the emitted radiation are evenly spaced, and the same wavelengths will be measured in the spectrum in all directions around the source. Appearances change if the source is moving toward or away from us while radiating. It makes no difference whether the source or ourselves are considered to be moving, all that matters is the relative motion.

When the source is receding from us (Fig. 8.19), the crests appear to be spread further apart. We can understand this by considering successive crests. In the time between emission of one crest and the next, the source has moved away from us. As a result, the two crests are farther apart than they would have been if the source had not moved. A greater distance between crests means a longer wavelength. Since longer wavelengths are toward the red end of the spectrum, this effect is termed the **redshift**.

Exactly the inverse occurs when the source is approaching us. Here, the crests pile up, for the source is running after crests already emitted. The wavelength is therefore shortened and this is termed as a **blueshift**. The size of the effect depends only on the relative speed of the source and observer, that is, how fast they are approaching or receding. When a spectrum is examined, a red- or blueshift will show up as a systematic shift of all the wavelengths by the same fraction. Measurement of the red- or blueshift provides a means for measuring the relative speed of source and observer.

A similar effect occurs with sound waves, and we are all familiar with the change in tone of a siren that is moving toward or away from us. The acoustic version was investigated by Christian Doppler in 1842 and checked experimentally for the first time in Holland in 1845. Some trumpeters were placed on a flatbed railroad car drawn behind an engine and musically trained observers both on the train and on the platform tried to estimate the pitch of the sound from the moving trumpets. More than 20 years later, the optical wavelength shift was discovered and today the effect is known as the Doppler effect. (No changes in frequency or wavelength are noted by an observer moving in the same direction as the source and with the same speed.)

**BOX 8.4**
*The Doppler Effect*

When source and observer are at rest, as in a laboratory experiment, the measured wavelength of a spectral line is denoted by $\lambda_0$. When the source and observer are moving (either apart or together) with relative speed $v$ cm/

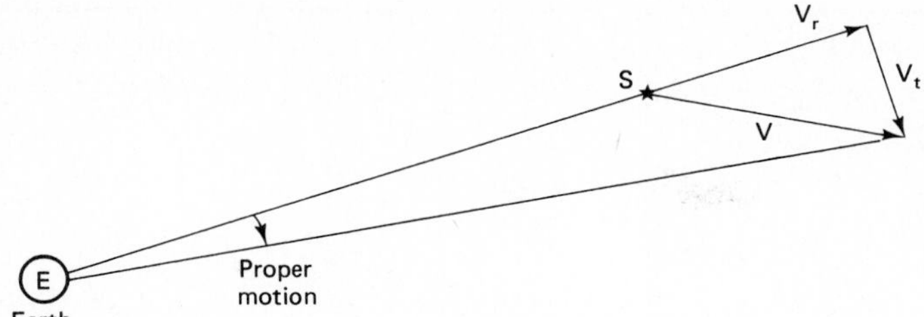


**Figure 8.20** The true (space) velocity of a star consists of two components. The radial component ($V_r$) is along the line toward or away from the Earth. This component can be measured from the Doppler shift in the spectral lines (if the shift is large enough). The transverse component ($V_t$) appears as a change in the direction of the line of sight from the Earth to the star. This is termed the proper motion of the star.

sec, the wavelength will be found to have a different value, $\lambda$. The change in wavelength ($\lambda - \lambda_0$) is denoted as $\Delta\lambda$ and the phenomenon of wavelength change is termed the Doppler effect. From the theory, we can derive a formula for the Doppler effect.

$$\text{Redshift} = z = \frac{\lambda}{\lambda_0} = \sqrt{\frac{1 + v/c}{1 - v/c}}$$

where $c$ is the speed of light. In many cases $v$ is very much smaller than $c$, and this formula can be simplified with quite sufficient accuracy to

$$\frac{\lambda - \lambda_0}{\lambda_0} = \frac{\Delta\lambda}{\lambda_0} = \frac{v}{c}$$

Following the discovery of quasars in 1963 very large redshifts have been detected. For many of these objects, $v$ is not small compared to $c$, and the simplified formula for the Doppler effect can no longer be used. The high-speed version will be needed when we discuss quasars, in Chapter 29. If $\lambda$ is found to be less than $\lambda_0$ (the line is blueshifted), then $\Delta\lambda$ is negative, and we interpret the negative value of $v$ as indicating that the source is approaching rather than receding.

What happens if the source and observer are not moving precisely along the line joining them? The Doppler effect then measures only that part of the speed that is along this line (Fig. 8.20). We term this $v_r$, the **radial** velocity. The velocity component perpendicular to the line of sight is the **transverse** velocity ($v_t$) which can be measured in other ways that do not involve spectral lines. (There is a transverse Doppler effect, but it is far too small to be measured in astronomical motions.)

Radiation also has its wavelength changed near very massive objects by an effect first described in relativity theory. Einstein's *Theory of Relativity* was published in two major parts. The first, the *Special Theory*, appeared in 1905, and the second, the *General Theory* in 1916. These papers are as important as Newton's *Principia* as turning points in physics. The 1905 papers provided among other items the relation $E = mc^2$ and the first full explanation of the Doppler effect. In the 1916 paper Einstein showed that light should be affected by gravity and that light leaving a massive object such as a star should have its wavelength shifted toward the red. This **gravitational redshift** has since been measured in the solar spectrum, where it is a very small effect, and in the spectra of some other stars. The effect arises through the loss of energy of the photons as they travel away from the massive source. The size of the effect depends on both the mass and the size of the source (see Box 8.5). Some masses can be sufficiently large and their dimensions so small that the photons that are emitted at their surfaces lose *all* of their energy and never emerge. We then have a **black hole**, *black* because its surface can never be observed because it immediately swallows its own radiation.

## BOX 8.5
*Gravitational Redshift and Black Holes*

If radiation of wavelength $\lambda_0$ is emitted from a point at a distance $R$ from the center of a mass $M$, then the gravitational redshift observed at large distances is given by

$$\frac{\Delta\lambda}{\lambda_0} = \frac{GM}{Rc^2}$$

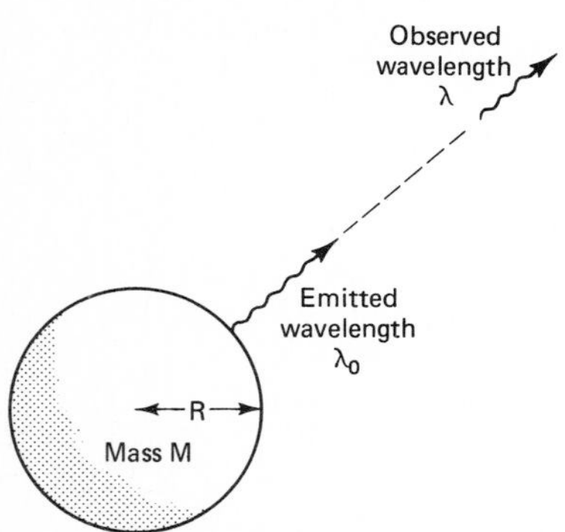


**Figure 8.21** Gravitational redshift. Photons that travel away from a massive object lose some of their energy because of the object's gravitational field. When those photons are detected some distance away, their wavelengths will be found to be longer, that is, they will be shifted towards the red end of the spectrum. The magnitude of this effect depends directly on the mass of the object and inversely on the distance from the object's center at which the photon is emitted.

where G, as before, is the universal gravitational constant and $\Delta\lambda$ is the change in the wavelength (Fig. 8.21).

Because $c^2$ is a large quantity in the denominator, the gravitational redshift is usually very small.

We can calculate the size of this effect for radiation leaving the surface of the Sun. It will be sufficiently accurate to take the solar mass as $2.0 \times 10^{33}$ g and its radius as $7 \times 10^{10}$ cm. We then have

$$\frac{\Delta\lambda}{\lambda} \sim \frac{(6.7 \times 10^{-8})(2 \times 10^{33})}{(7 \times 10^{10})(3 \times 10^{10})^2} \sim 2 \times 10^{-6}$$

and we see that the fractional wavelength change is about two parts in a million.

An even smaller wavelength change was detected in a demonstration of the gravitational redshift, carried out by Rebka and Pound at Harvard in the early 1960s. They used high-energy X-rays and noted the change in wavelength when these X-rays traveled a vertical distance of only 74 ft (about 22 m). Their measured wavelength change was 5 parts in $10^{15}$.

If $M$ is large and/or $R$ is small, then the redshift can become large. You will recall that, in the case of the Doppler redshift, the simple formula could no longer be used when the redshifts were large. A similar situation exists for gravitational redshifts. Large gravitational redshifts occur very close to large objects, where general relativity theory is needed for the full calculation. It turns out that photons will lose *all* of their energy if they are emitted at a distance $R_s$ from an object of mass $M$, given by

$$R_s = 2GM/c^2$$

$R_s$ is known as the Schwarzschild radius. Radiation emitted from positions closer than $R_s$ cannot emerge.

## 8.9 SUMMARY

The simplest use of light is to tell us where an object is. Noting a planet's or a star's position at different times can permit us to measure its speed from the changes in position, but that would be the end of our investigation if we did not examine the quality of the light. Spectral analysis provides a wealth of information. From a continuous thermal spectrum, we can infer the temperature; if the distance is known, we can also deduce the size of the object. From nonthermal spectra we can recognize the presence of high-speed electrons and measure the magnetic fields in which they travel. From the wavelengths and strengths of spectral lines, we can identify different types of atoms and their temperatures. This material represents the transition from astronomy, largely positional, to astrophysics, in which the structure and evolution of distant objects is studied. It is somewhat like a language that must necessarily be understood before we can make any sense of our *Starry Messenger's* reports. We must also have some understanding of the observational methods, for these often limit the quantities that we can measure and thus shape the pictures that we can construct. The next chapter will therefore complement the present one and shows how the astronomical signals can be detected and analyzed.

## APPENDIX 8A
## The Structure of Matter

So much of our understanding of the behavior of stars and radiation involves a knowledge of the structure of matter that this short note is being included, for those of you who have not yet encountered this topic.

Substances in the world around us can be broken down into basic material called **chemical elements**. To do this, we can combine different substances, heat them or torture them in other ways, but we find that what we call the elements cannot be further subdivided by any of these means. No matter how finely we divide an element, we always have material that has the same properties. Close to 100 chemical elements are now known. Most can be found naturally on Earth, but a few are the creations of nuclear physics experiments. The most abundant elements on Earth are hydrogen (H), carbon (C), nitrogen (N), and oxygen (O) which comprise water and most of the living animals and plants on Earth. Different elements can combine to form **compounds** whose properties can be dramatically different from the constituent atoms. Thus two atoms of hydrogen combine with one atom of oxygen to form a molecule of water ($H_2O$), which is a liquid at room temperature although hydrogen and oxygen are gases at the same temperature. Hydrogen peroxide ($H_2O_2$) is made of the same kinds of atoms, but its properties are very different from those of water. Most of our everyday world consists of compounds. Chemical methods are used on an industrial scale to separate compounds, extract elements, and recombine them into other compounds.

Each element is composed of atoms, the smallest particles of an element that still retains the chemical properties of the element. Atoms are minute. A typical atomic dimension is $10^{-8}$ cm, and there are about $10^{22}$ atoms in 1 $cm^3$ of water.

The internal structure of atoms is revealed by physics experiments, which show that the atom is largely empty. At the center is a minute **nucleus** (about $10^{-13}$ cm radius). Relatively far away by atomic dimensions (that is, around $10^{-8}$ cm) are one or more **electrons**, forming a system that in some ways seems like the solar system. The electrons surround the nucleus in different orbits. Whereas gravity is the force that holds the solar system together, it is the electrical force that holds the electrons to the nucleus.

Each nucleus has a positive electric charge, each electron has a negative electric charge, and they attract one another. The positive and negative terms are historical relics. The essential feature is that all electrons have the same kind of charge and they repel each other; all nuclei have the same kind of charge and they too repel each other, but the electrons and nuclei have different kinds of charge and are mutually attractive.

Hydrogen is the simplest atom: Its nucleus is the **proton** and it has one orbital electron. Together, they bring the net charge of the atom to zero. Heavy hydrogen (deuterium) has one electron, one proton, and one neutral particle in the nucleus joined to the proton. The extra particle is the **neutron**. The proton and neutron have very similar masses, each about 1836 times the mass of the electron so that most of the mass of an atom is in the nucleus. Because hydrogen and deuterium have the same number of electrons per atom (one), they have almost the same chemical properties; having the same number of protons in the nucleus but different atomic weights, they are termed **isotopes**. The neutron has no electric charge but is held to the proton through the **nuclear force**.

Heavier atoms have more protons, more neutrons, and more electrons. A well-known heavy atom is uranium, with its isotopes $U^{235}$ and $U^{238}$. Uranium has 92 protons in its nucleus, 92 electrons around it, and 143 and 146 neutrons in the nucleus (for the two isotopes mentioned).

## APPENDIX 8B
## Energy and Temperature

The mechanical universe of Newton's construction was driven by forces, but gradually it came to be realized that the successful and complete description of many phenomena required the use of additional concepts. Energy is one of these. We saw that a force would cause a body to move and that after removal of the force the motion would continue. If a force $F$ acts on a body of mass $m$ and pushes it for a distance $d$, then we define the product $Fd$ as the **energy** imparted by the force to the body. We can define a property of the moving body as **kinetic energy**, and we can derive an expression for it.

Recall from Newton's laws that a constant force produces a constant acceleration: $a = F/m$. A constant acceleration leads to a speed $v$ after a distance $d$, where $v^2 = 2\,ad$. Inserting $a = F/m$ into the last equation gives

$$Fd = \tfrac{1}{2}mv^2$$

(energy imparted by force) = (kinetic energy of $m$ moving with speed $v$).

In the metric system, the energy unit is the erg if $m$ is in g and $v$ in cm/sec. With $m$ in kg and $v$ in m/sec, the unit of energy is the joule, and 1 joule is $10^7$ ergs. (The joule is named after James Joule, who, around the middle of the nineteenth century, undertook important experiments that clarified our ideas of energy.)

Energy turns up in many guises. We have been considering the mechanical energy of a moving object. Thermal energy is another common form. Energy is needed in order to heat a body, and a refrigerator works by removing energy from its contents. What we call thermal energy, however, is none other than the kinetic energy of the atoms or molecules making up the substance. The basic unit of thermal energy is the calorie; it is the amount of energy involved in changing the temperature of 1 g of water by 1° Celsius. Other substances require different amounts of energy for the same temperature change.

The erg and the calorie are very small amounts of energy that are well suited for many scientific uses. For everyday purposes, it is more convenient to use larger basic units. Thus the "calorie" that is of so much concern in dieting is 1000 thermal calories. Electric energy consumption is measured in units of the kilowatt-hour, equal to $3.6 \times 10^{13}$ ergs. The explosive energy of a 1 megaton (1 MT) nuclear bomb is $4.2 \times 10^{22}$ ergs.

Energy is often changed between thermal and kinetic forms. Some of the thermal energy obtained from combustion of gasoline ultimately emerges as the kinetic energy of your automobile. In the other direction, when you want to slow down, the kinetic energy must be removed from the automobile, and this is done by converting it into thermal energy (heat) in the brakes.

The conversion between thermal and mechanical energy was first established by Joule, and the conversion factor is

$$J = 4.18 \times 10^7 \text{ ergs per calorie}$$

As is so often the case, social needs shape scientific interests. Joule, the son of a brewer, worked in Manchester in England, immersed in the full flow of the industrial revolution. It was quite natural, therefore, for him to take an intense interest in the relation between thermal and mechanical energy, for the efficiency of the new machines was of great current concern.

## CHAPTER REVIEW

**Astronomical signals:** radiation, lunar samples, meteorites, cosmic rays, neutrinos, solar wind, gravitational radiation, radar.

**Electromagnetic radiation:** speed ($c$) = frequency × wavelength = $3.0 \times 10^{10}$ cm/sec.

- — polarization: electric vibrations all or partly in one direction
- — thermal radiation: $\sigma T^4$ ergs/cm$^2$ sec from a hot surface at temperature $T$.
- — $4\pi R^2 \sigma T^4$ ergs/sec from a sphere of radius $R$ cm.
- — Wien's law: most intense thermal radiation at $\lambda_{max}$ given by $\lambda_{max} T = 0.2898$ cm deg.

**Spectral lines:** emitted/absorbed when atom changes energy.

- — hydrogen spectrum: Lyman series in ultraviolet; Balmer series in the visible; Paschen series in the infrared.

**Absorption spectra:** dark lines on a bright, continuous background, produced when dense hot region is viewed through a cooler less dense region.

**Emission spectrum:** bright lines with no continuous background.

**Molecules:** can produce lines or bands (many lines very close together).

**Synchrotron radiation:** produced by highspeed electrons moving in a magnetic field.

- — Shape of the spectrum provides information on the magnetic field and on the speeds of the electrons. Radiation is partly polarized.

**Doppler shift:** change in wavelength produced by movement between source and observer. Redshift (longer wavelengths) produced if source and observer are moving apart, blueshift if they are approaching. Size of the effect depends on mutual speed.

**Gravitational redshift:** redshift produced when radiation leaves a massive object. For massive and very compact bodies, radiation may not be able to leave: hence a black hole.

## NEW TERMS

**absorption lines**
**black body**
**black hole**
**blue shift**
**calorie**
**Celsius**
**compounds**
**continuous spectrum**
**cosmic rays**
**Doppler effect**
**electromagnetic radiation**
**electron**
**elements**
**emission spectrum**
**erg**
**flux**
**Fraunhofer lines**
**frequency**
**gravitational radiation**
**gravitational redshift**
**ionization**
**isotopes**
**Kelvin**
**line spectrum**
**longitudinal wave**
**luminosity**
**lunar sample**
**meteorites**
**neutrino**
**neutron**
**nucleus**
**photons**
**Planck's constant**
**proton**
**quantum number**
**radar**
**radiation**
**radial velocity**
**redshift**
**Rydberg constant**
**Schwarzschild radius**
**solar wind**
**spin**
**Stefan's constant**
**synchrotron radiation**
**transverse velocity**
**transverse wave**

## PROBLEMS

**1.** What physical processes produce continuous spectra? How are line spectra produced?

**2.** Most optical astronomy is carried out at night, but radio astronomers can observe by day and by night. Why?

**3.** A sound wave has a speed of 330 m/sec in air. Calculate the wavelength for a sound wave that has a frequency of 1024 Hz (two octaves above middle C).

**4.** Thermal X-rays have wavelengths around $2 \times 10^{-8}$ cm. What temperature is needed in order to radiate strongly in this wavelength region?

**5.** Wien's law relates the temperature of a surface to the wavelength at which it radiates most intensely.

a. The human body has a temperature of about 300 K. At what wavelength does its radiant spectrum peak?

b. A star is observed to radiate most intensely at a wavelength of $2 \times 10^{-5}$ cm. What is the temperature of its surface.

**6.** An AM radio station advertises itself as being at

1120 on the dial. Is this number a wavelength or a frequency? What are the units? If this is a wavelength, what is the corresponding frequency? If this is a frequency, calculate the wavelength.

**7.** Calculate the energy of a photon of visible light that has a wavelength of 5000 Å. The luminosity of the Sun is roughly $4 \times 10^{33}$ ergs/sec. How many photons per second are emitted by the Sun (assuming for simplicity that all the photons have the same 5000 Å wavelength).

**8.** Different kinds of waves travel at different speeds. Thus sound waves travel in air at about 300 m/sec and light at $3 \times 10^8$ m/sec. Calculate the length of the time interval between seeing a lightning flash and hearing the crackle if the flash is 3 km away from you.

**9.** Suppose that you had a container of hydrogen and could observe all wavelengths of radiation that might be emitted. What changes in the radiation could you observe as the temperature of the hydrogen is increased from room temperature to 10,000 K?

**10.** How are the following spectra produced: a continuous spectrum with absorption lines, and an emission line spectrum with no continuous spectrum?

**11.** Calculate the gravitational redshift for light emitted by a star that has a radius 80 times larger than the Sun and a mass 1.5 times the Sun's mass.

**12.** A star explodes. Some of the debris moves towards the solar system at a speed of 100 km/sec and emits light. The wavelength of that light is 6000 Å when emitted. At what wavelength will we detect that light?

**13.** There will be a gravitational blueshift for light approaching the Earth even if the Earth and the source are not moving. Calculate how large this effect is for light of wavelength 5000 Å.

**14.** The Voyager spacecraft used two radio transmitters to send signals back to Earth. One transmitter operated at a wavelength of 3.60000 cm and the other at 13.00000 cm. When the spacecraft was near Jupiter, the 3.60000 cm signal was observed at Earth at a wavelength of 3.60060 cm.

a. What was the speed of Voyager relative to Earth at that time?
b. At what wavelength would the 13.00000 cm signals be received?
c. How long did it take the signals to reach the Earth? (Assume Jupiter and Earth were then as far apart as they could be.)

**15.** A star is located in the ecliptic plane but is not moving relative to the Sun. Because of the Earth's orbital motion, lines in that star's spectrum will sometimes appear redshifted and sometimes blueshifted.

a. For a line emitted at a wavelength of 6000 Å, what are the longest and shortest wavelengths that will be observed?
b. How many months will separate those maximum red- and blueshifted observations?
c. What wavelength will be seen midway between those observations?
d. Another star is observed, but its spectrum always shows the same wavelengths. In what direction must that star lie relative to the ecliptic?

**16.** Use the formula in the text to calculate the wavelengths of the first three lines in the Balmer series of hydrogen.

**17.** Use the formula given in the text for the lines in the hydrogen spectrum. Calculate the wavelengths of the first lines in the Lyman series ($m = 1, n = 2$), and Bracket series ($m = 4, n = 5$). Where do these lines fall in the electromagnetic spectrum: ultraviolet, infrared, radio, or where?

**18.** A low-calorie diet allows consumption of 1000 calories per day. Each of these calories is 1000 of the calories used in thermal energy measurements. Calculate the equivalent energy in ergs for the 1000 diet calories.

**19.** The Sun radiates about $4 \times 10^{33}$ ergs per second. How many thermal calories is this equivalent to? How many diet calories?

**20.** An electric lamp radiates 100 watts when operating at a certain temperature. What change in temperature is needed if it is to radiate 1600 watts? (1 watt = 1 joule/sec)

**21.** Make an estimate of how much energy you radiate per second. Assume your temperature is 300 K. To calculate your surface area, consider yourself a cylinder with your known height and radius 30 cm. (Adjust these as needed.) If you calculate your area in sq cm, the energy radiated per second will be in ergs. Convert to watts (1 w = $10^7$ ergs/sec) and compare yourself to a lamp. You can now understand why a room heats up if there are many people in it.

**22.** Thermal radiation and synchrotron radiation are produced in different processes. Describe briefly the observable differences that allow astronomers to identify which type of radiation is being detected.

**23.** Investigate some properties of polarized light. Borrow a pair of Polaroid sunglasses. Looking through them, rotate the glasses while observing the blue sky (not looking close to the Sun) and also looking at reflections off a window and glare from a road surface. The intensity of the light should change noticeably as you rotate the glasses. If you can get two pieces of polaroid, hold one steady and rotate the other while looking through both, and note how the light intensity changes.

**24.** Hydrogen is the most abundant chemical element. Astronomers detect hydrogen in different forms and refer to HI and HII. What do these terms mean?

CHAPTER

# 9

# Telescopes and Radiation Detectors

All over the universe things are happening: stars evolve, planets and comets move, complicated molecules are being formed in interstellar space. From each of these, characteristic radiation is emitted, to spread out and in doing so diminish in intensity. Our telescopes will be collecting very dilute radiation, and each combination of telescope and detector has its own limit of sensitivity; fainter objects will not be seen. As we assemble pictures of these distant events by detecting their radiation, we are using our instruments in many distinct but complementary ways. Some systems are designed for surveys to note the position and brightness of objects radiating within some broad wavelength range. Other systems can focus on very small regions, and from their data we can draw maps, showing contours of equal brightness. Quite different systems are needed for separating the close but different wavelengths in a line spectrum. In the radio, infrared, and visible, each technique has it special strengths and limitations and must be selected for the particular problem at hand. Many of the methods are relatively new and are based on technical advances outside astronomy: electronics, computers, new photographic materials, cryogenic (low temperature) methods. Before entering into a survey of these modern methods, it should be noted, even briefly, that the human eye is still one of the most versatile and valuable instruments and a good example of an optical system.

The human eye collects radiation and focuses it onto the retina where chemical effects produce electrical signals that are transmitted to the brain for processing. A detector cannot absorb radiation with a wavelength much larger than its physical dimensions. In the case of the eye, the sensitive

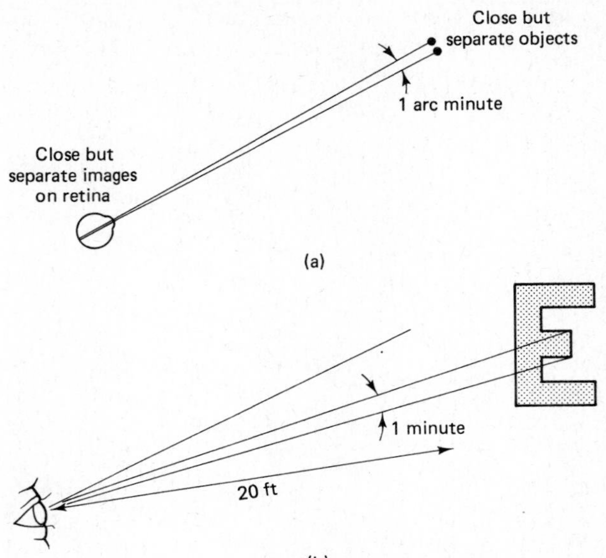


**Figure 9.1** (a) The angular resolution of an optical system is the smallest angular distance between two objects that can be seen to be separate. For the human eye, this limit is about 1 arc min. (b) The standard letters on an optical test chart have an overall size of 5 arc min and a thickness of 1 arc min for 20/20 vision.

element—the retina—is made up of rods and cones; their smallest diameters are about $10^{-4}$ cm or about $1\frac{1}{2}$ to $2\frac{1}{2}$ times the wavelength of visible light. We cannot, therefore, 'see' the long radio or infrared wavelengths. Even within the visible, we do not see all wavelengths with the same sensitivity—there is a clear maximum in the yellow/green region, and the sensitivity falls rapidly at longer and shorter wavelengths. The spacing of rods and cones sets the limit to the smallest detail we can see in an object, for if we wish to see two points as separate, then their images must fall on separate rods or cones. The closest separation at which we can distinguish clearly between two objects is the **angular resolution**, which depends on the objects' distance from us. It is the angle between them that we really note. The angular resolution of our eyes is about 1 arc min (= 1/60°) (Fig. 9.1).

The limit of sensitivity is set by the amount of light collected and thus by the maximum size of the pupil (about 8 mm diameter) and by the process occurring in the retina when the light is absorbed. Under exceptional conditions, the eye can detect a signal that is so weak that only 200 photons enter the eye each second. (A 100-watt light bulb emits more than $10^{18}$ photons per second in the visible part of the spectrum). This visual limit corresponds to a star of about 8.5 magnitudes, but a more usual limit is considered to be sixth magnitude, and indeed, Ptolomey's catalog of stars listed none fainter than the sixth magnitude. Newer instruments have improved on the eye's performance in many respects, but the combination of the optics of the eye and the information processing in the brain continues to provide the basis for important research. The combination of optical scanner plus computer can search for and compare images without tiring, but the human observer is still the best for spotting unexpected patterns or events.

## 9.1 THE ATMOSPHERE

The evolution of life on Earth has been strongly influenced by the optical properties of the atmosphere, which protects us from much of the Sun's ultraviolet radiation. The astronomical aspect of this circumstance is that some astronomically important information never reaches ground level, being absorbed in the atmosphere at various levels depending on the wavelength. Figure 9.2 shows how the transmission of different wavelengths varies markedly.

Traditional astronomy has been carried out in the narrow *optical window* where transmission is generally high, but at some wavelengths there is

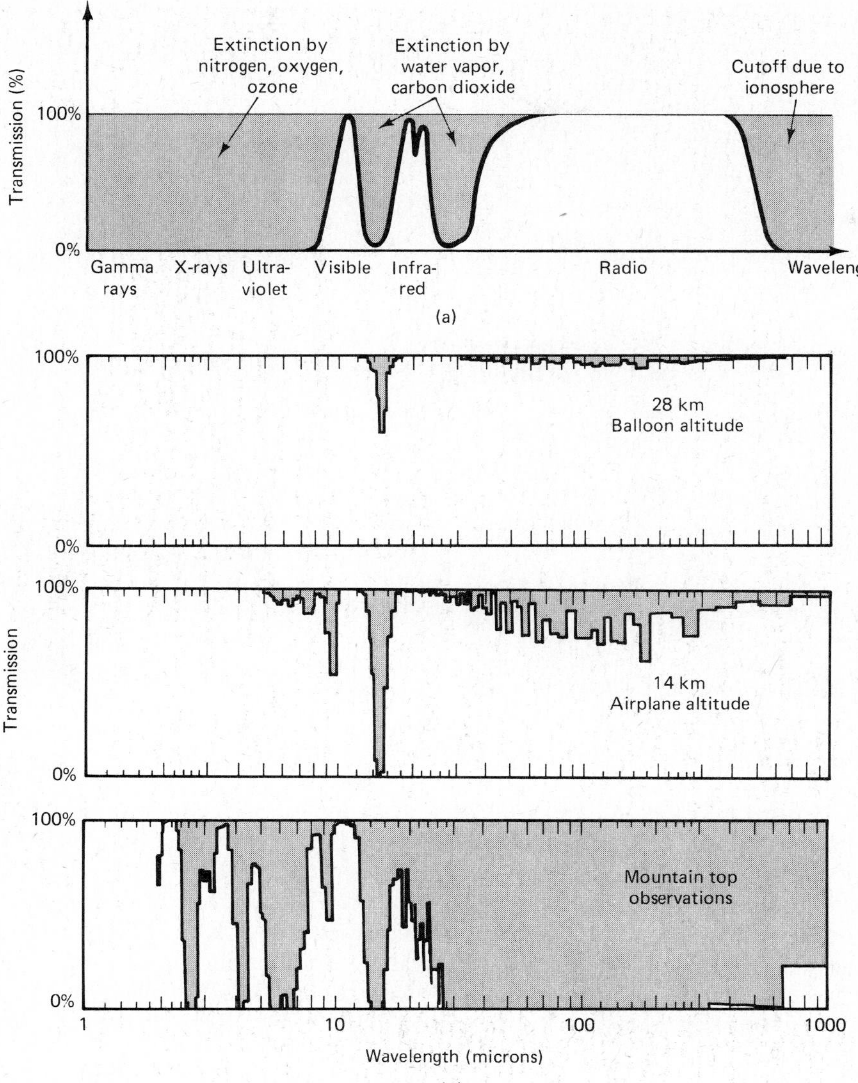


**Figure 9.2** (a) The Earth's atmosphere does not transmit all wavelengths in the electromagnetic spectrum with equal efficiency. The shortest wavelengths (gamma rays, X-rays, and most UV) are heavily absorbed, whereas most visible radiation is transmitted. A band of wavelengths in the IR is transmitted, but the short and medium-to-long IR are heavily absorbed. Most radio waves are transmitted. (b) Atmospheric absorption of IR wavelengths is particularly sensitive to altitude. This diagram shows the transmission for altitudes typical of mountain-top observatories, airplanes, and balloons. (Based on data by W. A. Traub and M. T. Stier).

very selective absorption by air molecules and by pollutants. This absorption shows up as dark **telluric** lines in spectra of the Sun and stars. (As an aside, we note that measurement of these lines in solar spectra observed over many years has provided us with a means of monitoring the growth of various pollutants in the atmosphere.)

The radio window is far larger than the optical, with the ratio of longest to shortest wavelength being well over 1000. In this range of wavelengths there is a serious problem of interference from man-made signals. Radio waves can bounce in a region of upper atmosphere (ionosphere) and be detected by receivers well over the horizon—indeed, this is the basis of some radio transmission. There is the well-known aphorism "one man's signal is another's noise," and this is especially true in radio astronomy where a strong transmitter on the ground could swamp weak cosmic signals. Different parts of the radiofrequency spectrum are now allocated for scientific use through the International Telecommunications Union. Thus, the 21 cm line of the hydrogen atom is protected by having frequencies between 1400 MHz and 1427 MHz (21.0 cm to 21.4 cm wavelength) reserved exclusively for radio astronomy, but other bands such as that from 37.37 MHz to 38.25 MHz (7.84 m to 8.05 m), are shared with nonscientific users. In all,

Figure 9.3 Photographs of the globular cluster M.13 using the Lick Observatory 36 in. refractor. The effect of atmospheric "seeing" on the limiting magnitude and resolution can be observed. (Lick Observatory).

there are ten frequency bands assigned solely for radioastronomy with another seven shared.

Most ultraviolet, X, and gamma radiation can be detected only at altitudes accessible to rockets and satellites. Carbon dioxide and water vapor are the main atmospheric absorbers of infrared radiation. This absorption blankets a very wide range of wavelengths, but there are some important infrared (IR) windows—wavelength intervals where the absorption is small. Observations can be made effectively through these IR windows from mountain-top observatories in places such as Arizona and Hawaii where the humidity can be extremely low. Infrared measurements can also be made from airplanes and balloons, and satellites.

The atmosphere imposes other restrictions on astronomy. Atmospheric movement, and fluctuations in its density and temperature at different heights cause images to defocus and shimmer, smearing the finest detail that the telescopes would otherwise be capable of recording (Fig. 9.3). Under good conditions, this **seeing** is as small as 1 arc sec, and occasionally a factor of two to four better than that. This is one of the factors that must be considered during extensive site surveys before deciding where to locate a new observatory.

The atmosphere introduces yet another distortion of astronomical signals. Light is scattered in the atmosphere, by molecules and dust. The blue (short) wavelengths are scattered more than the red, so that when we look at the setting sun it appears redder than it does at midday. At sunset the Sun's rays have to travel a longer distance through the atmosphere, and the reddening is immediately noticeable. We also notice the blueness of the sky during the day—this blue light comes from the Sun and is scattered so that it appears to be coming from all around (Fig. 9.4).

During the day this scattered light is brighter than the stars, but even at night scattered moonlight can interfere with the observation of very faint stars and galaxies. Unfortunately the same scattering effect combines with high-intensity street lights and urban smog to create an increasing problem for some observatories near large cities, such as Mount Wilson, where the

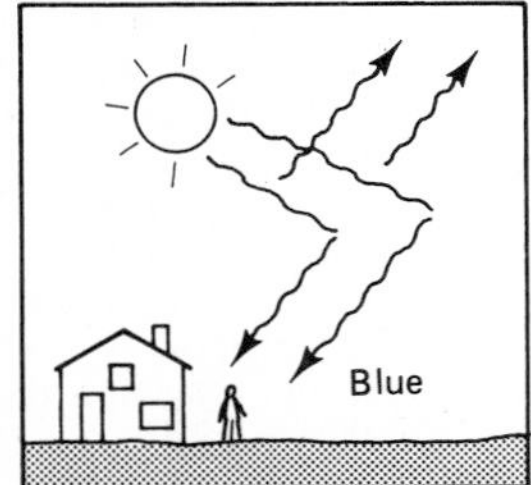


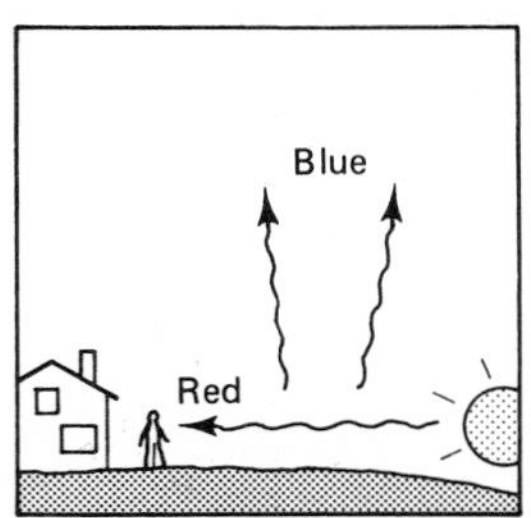


Figure 9.4 The shorter (blue) wavelengths in sunlight are scattered more than the longer (red) wavelengths. During the day, the sky therefore appears blue from the scattered sunlight. When the Sun is close to the horizon enough of the blue light is scattered away from the line of sight that the light that reaches us is distinctly reddened, and the Sun takes on its well-known flaming appearance.

light from Los Angeles already poses a severe problem. The combination of high altitudes and good seeing led to the decision to place major telescopes on Kitt Peak in Arizona, but some years later it became necessary to persuade the city of Tucson to place restrictions on the types of street and commercial lighting.

The long-wavelength radiations are far less affected by atmospheric scattering. At IR wavelengths, the sky is nearly as dark by day as it is at night, and daytime IR observations can be carried out as long as one in not looking close to the Sun. At even longer wavelengths, radio astronomy is free of this problem. None of these atmospheric difficulties would be encountered if telescopes could be carried to the Moon or when placed in orbit. Astronomy has already been transformed by the satellite observations of the past 20 years.

## 9.2 FORMATION OF IMAGES

We can imagine the energy spreading out from a luminous object being carried by **light rays**, traveling along straight lines. If we can persuade some of these rays to change their directions so as to converge we can produce an *image* of the original object; we refer to this process as **focusing**. We can achieve this in two ways.

Light rays suffer a sharp deviation in traveling from one transparent medium into another, such as in going from air into glass. This bending is known as **refraction**. The amount of bending depends (slightly but importantly) on the wavelength of the light and on the type of medium (glass, water, lucite, etc.). Snell's Law describes this process accurately. It was discovered by Willebrod Snell in the Netherlands in 1621, although approximate versions had been known for centuries. Each transparent medium has its own **index of refraction** that can be measured and then used to predict the behavior of light in that medium (Fig. 9.5).

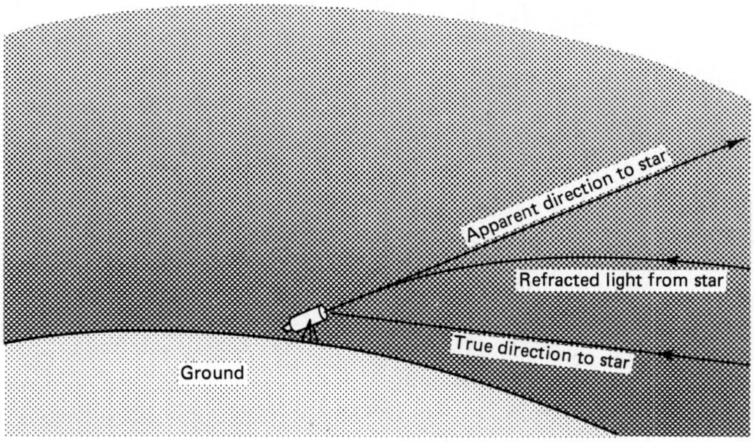

**Figure 9.5** Atmospheric refraction. Light from a distant object that enters the Earth's atmosphere will be refracted, the deviation depending on the direction of travel of the light ray and the amount of atmosphere that must be penetrated. The result is a change in the apparent position of the distant star or in the case of a nearby and large object such as the Sun or Moon, a change in the apparent size and shape.

### BOX 9.1
*Snell's Law*

A ray of light arriving in air at a surface at some **angle of incidence** $i$, will travel in the second medium along a new direction specified by the **angle of refraction** $r$, Snell's Law states that

$$\frac{\sin i}{\sin r} = \text{a constant}$$

This constant is the **index of refraction** usually denoted by $n$. Each value of $i$ will have a related value of $r$, given by Snell's Law (Fig. 9.6). Thus for light passing from air to water, the refractive index is 1.33; for light passing from air to glass, the index is in the range 1.5 to 1.7 (depending on the type

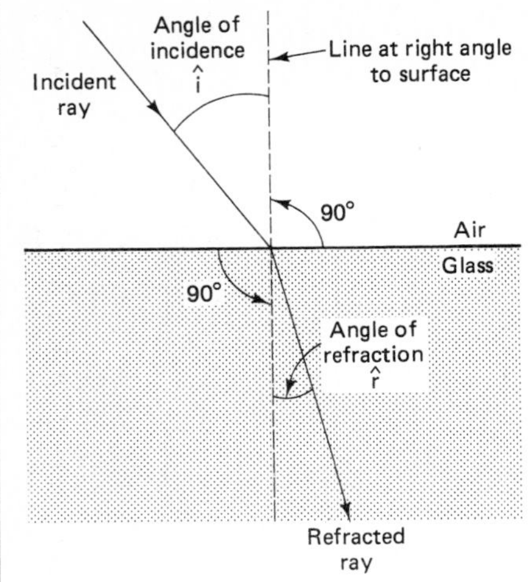


**Figure 9.6** Refraction. Snell's Law relates the angle of incidence, the angle of refraction, and the index of refraction of the material. Light is refracted (bent) away from the surface when passing from a less dense medium (such as air) to a more dense medium (such as glass).

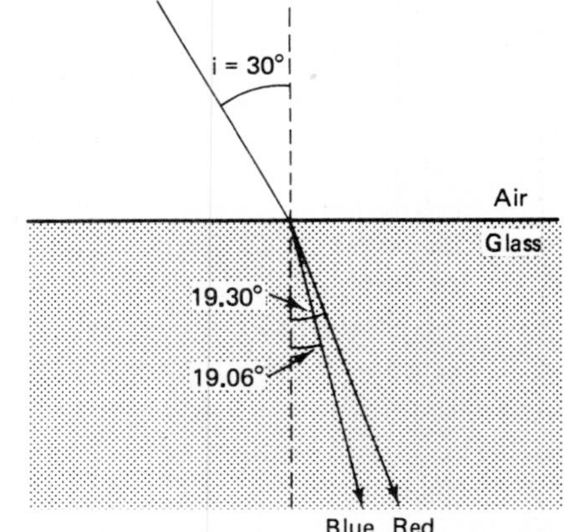


**Figure 9.7** The degree of refraction depends on the wavelength of the light. The shorter (blue) wavelengths are refracted more than the longer (red) wavelengths, and light that originally appears white will be split into its constituent colors.

of glass). The refractive index varies slightly with wavelength. For example, for one kind of glass $n = 1.531$ for blue light and $n = 1.513$ for red. Snell's Law can now be derived from electromagnetic theory

**Example**: Suppose a beam of light containing red and blue wavelengths strikes the surface of a glass block at an angle of incidence of 30°. The directions of the red and blue components inside the glass can be calculated (Fig. 9.7). For the blue light, the angle of refraction is given by

$$\sin r = \sin i/1.531 = 0.5/1.531 = 0.3266$$

from which $r = 19.06°$. For the red light

$$\sin r = \sin i/1.513 = 0.5/1.513 = 0.3305$$

and thus $r = 19.30°$. The blue and red components are refracted through angles that differ (in this case) by 0.24°.

Although the refractive index for a given wavelength is a constant for a given medium, the change in direction (deviation) depends on the angle of incidence of each ray. With a lens, the light is deviated both on entering and leaving the glass. Because of the curvature of the surfaces, the angles of incidence and thus deviation will differ for light striking the lens at different places. The overall result is that the lens can focus the light into an image even if the original rays are parallel or diverging. In the special case where the incident rays are parallel, as will very nearly be the case for a distant object, the distance of the image from the lens is termed the **focal length**, $F$ (Fig. 9.8). The actual value of $F$ is determined by the curvature of the two lens surfaces and the type of glass used so that specific lenses can be designed as needed. For ease in manufacture most lenses have spherical surfaces. However, non-spherical lenses are used for special purposes, most notably in eyeglasses where cylindrical lenses are used to correct for astigmatism, a common eye defect.

Each point of the object radiates its own rays. From each point these rays will be brought to a separate focus, and the combination of all these

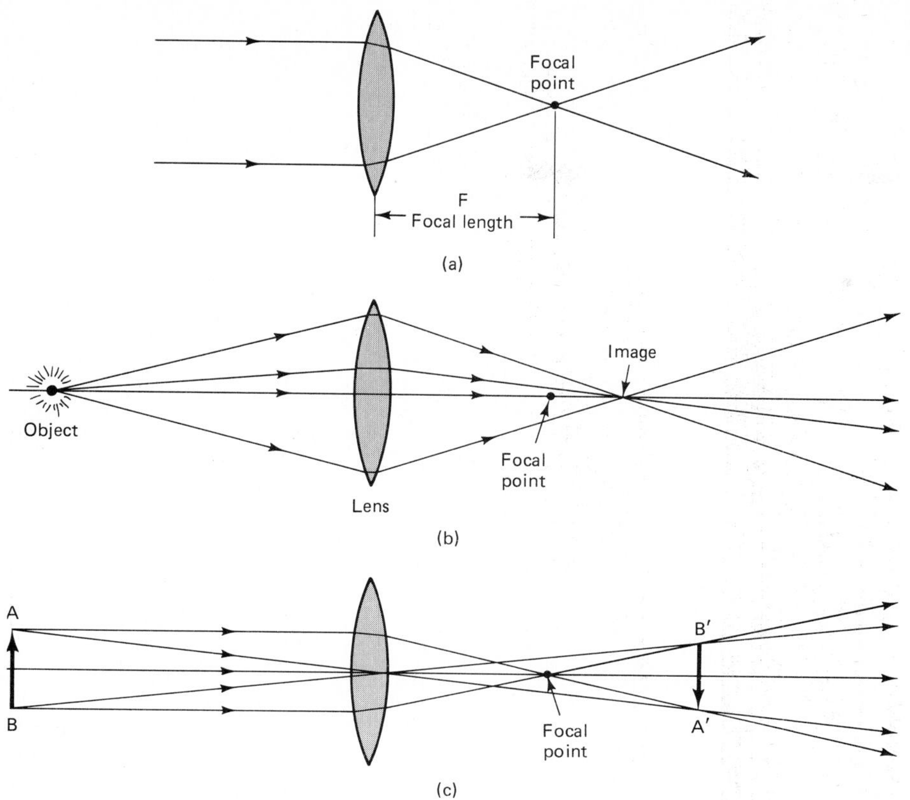


**Figure 9.8** Operation of a converging (convex) lens. (a) Light rays from a very distant point are effectively parallel and after passing through the lens will be brought to a focus at the focal point. The focal length of the lens is the distance between this focus and the lens. (b) Light rays from a nearby object will also be brought to a focus, but this focus is farther from the lens than the focal length. (c) Light from a distant object will be focused to produce an image. With a single lens (as shown), the image will be inverted and smaller than the object.

focused points is the image. The image of a distant object viewed through a single converging lens will be inverted. This happens all the time in our eyes, but our brains correct for this. For astronomical purposes it does not usually matter that the image is inverted as long as we keep track of which way is up. For telescopes and binoculars intended for nonastronomical use, inversion would be a nuisance, so an additional lens or prism is inserted to supply the needed second inversion.

With lenses, we encounter an effect similar to that of scattering: the shorter (blue) wavelengths are refracted more than the longer (red). The result is that a lens has slightly different focal lengths for different colors, and the images show colored edges (Fig. 9.9). This **chromatic aberration** can be corrected by using two or more lenses together with the lenses being made of different types of glass to produce compensating effects. The re-

**Figure 9.9** (a) Chromatic aberration. Because blue wavelengths are refracted more than are the red, multi-colored light from an object will be brought to a series of focused images with the blue image closest to the lens and the red farthest away. This effect occurs in all simple lenses. (b) An achromatic lens. Chromatic aberration can be corrected by using in combination two lenses made from different types of glass. The types of glass (usually crown and flint) and the curvatures of the surfaces are chosen so as to eliminate the aberration at two wavelengths and reduce it for intermediate wavelengths. Combinations of as many as six to ten lenses are used in cameras and microscopes to further reduce chromatic and other lens aberrations.

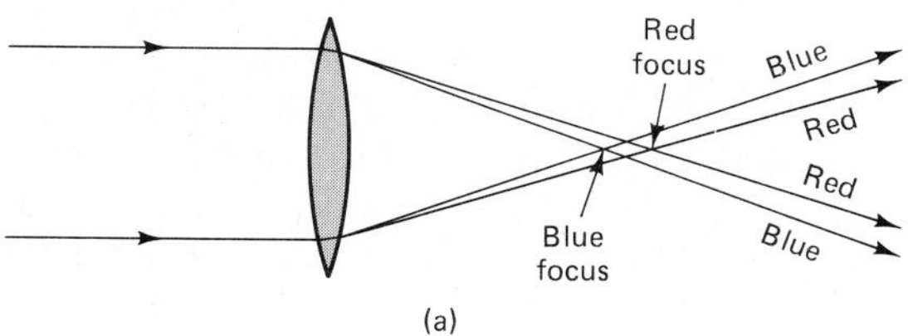


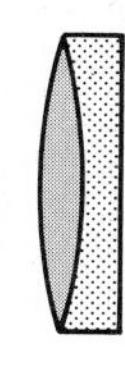

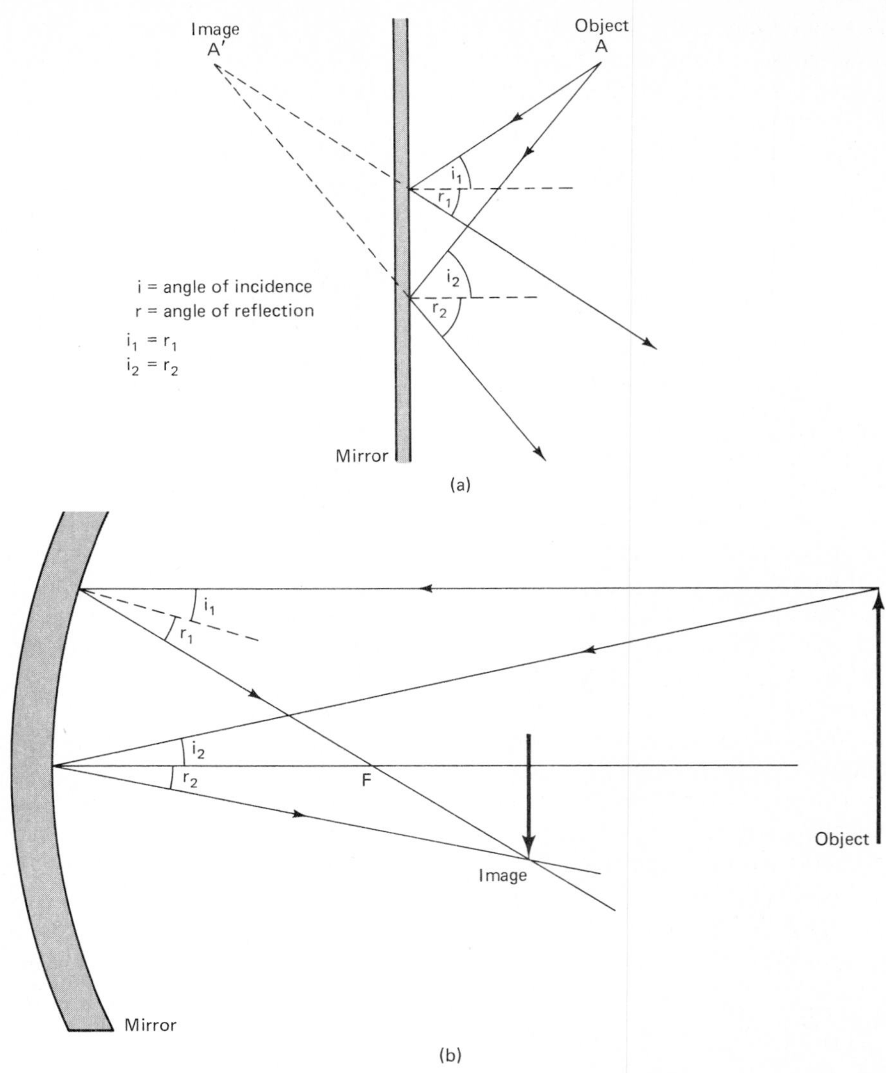

**Figure 9.10** (a) Reflection of light from a plane mirror produces an image as far behind the mirror as the object is in front. The law of reflection does not depend on the wavelength of the light: For all wavelengths, the angle of incidence ($i$) is equal to the angle of reflection ($r$). (b) Formation of an image by reflection of light from a concave mirror. For each light ray at each reflection, the angle of incidence is equal to the angle of reflection. Because the mirror surface is curved, the reflected rays can be brought to a focus. (c) The mirror for the 200 in. (5 m) telescope on Mt. Palomar during polishing. To reduce the weight of the mirror without sacrifice of strength, it was cast with a honeycomb structure beneath the reflecting surface. This honeycomb can be seen through the still transparent top surface, which had not yet been silvered.

sulting combination (an **achromat**) will produce a precise focus at two wavelengths and a compromise at others. Using additional lenses can extend the color correction to more wavelengths, but this adds greatly to the complexity of the lens systems and is usually restricted to microscope and camera lenses, where the lens dimensions are small.

There is another way in which radiation can be focused to produce an image. We are all familiar with mirrors, but probably few of you have analyzed their working. Light rays from some point $A$ (Fig. 9.10) are reflected by a flat mirror. Each ray strikes the surface at a slightly different angle of incidence, and the direction of the reflected ray is governed by a very simple law:

$$\text{angle of incidence} = \text{angle of reflection}$$

This law has been known since antiquity and can now be deduced from electromagnetic theory.

The rays diverge after reflection from a flat mirror, but they now appear to have come from some point $A'$ instead of their true origin $A$. This image of $A$ is exactly as far behind the mirror as $A$ is in front. If the mirror is curved, the reflected rays can be focused to an image that can be captured on photographic film to give a permanent record.

As with lenses, it is easy to make the mirror surface spherical, but better images can be obtained if the mirror has a parabolic shape. Although the use of lenses is effectively restricted to visible light and to the shorter-wavelength infrared, mirrors can be used for focusing all wavelengths from radio to the longest X-rays. What is required in all cases is that the refracting and reflecting surfaces be smooth on the scale of the wavelength to be focused. Thus lenses and mirrors for visible light have to be polished to a smoothness of better than $10^{-5}$ cm. For radio astronomy, the reflecting surface can be a metal mesh, provided the smoothness of the surface and the spacing between wires of the mesh are less than the wavelengths, which are typically of centimeter or meter dimensions.

## 9.3 PROPERTIES OF IMAGES

The usefulness of an image is determined by its size, brightness, and resolution (detail). The size of an image is usually described by its **scale**. In astronomy, dimensions of objects in the sky and apparent distances between objects are usually measured and denoted by angles. It is thus convenient to describe the scale of an image as (centimeters of image) per (degree of object). The scale is defined by the focal length ($F$) of the lens or mirror used in the telescope, and is given numerically by

$$s = (1.75 \times 10^{-2})\ F \text{ cm/deg}$$

The Hale telescope on Mount Palomar in California is probably better known than any other. Its mirror has a diameter of 200 in. (now usually listed as 5.08 m) and a focal length of 660 in. (16.764 m). Its images therefore have a scale of

$$s = 29.3 \text{ cm/deg}$$

but this is often expressed as $3.41 \times 10^{-2}$ deg/cm, which converts to 12.3 arc sec/mm in the units most widely used by astronomers.

The intrinsic wave nature of light sets a limit to the detail that can ever be obtained in an image even with a perfect lens or mirror. From the full theory of the behavior of electromagnetic waves, it can be shown that the angular resolution ($\theta$) of a lens or mirror of diameter $D$ is given by

$$\theta = 2.5 \times 10^{5} \frac{\lambda}{D} \text{ arc sec}$$

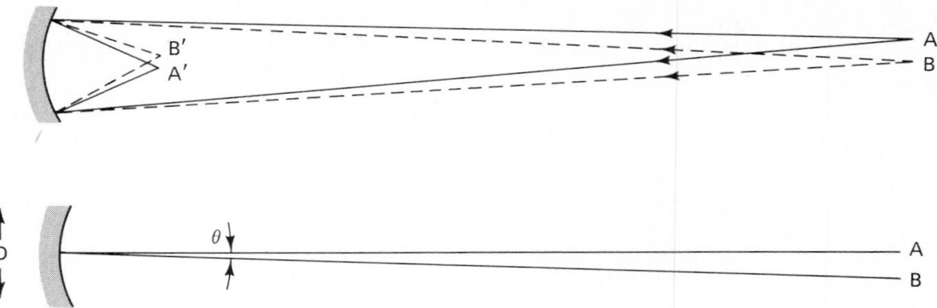


**Figure 9.11** Angular resolution. Two objects ($A$, $B$) are said to be resolved if their images ($A'$, $B'$) can be seen to be separate. The limiting resolution is specified by the angle between $A$ and $B$, and theory shows that this angle depends on the ratio of the wavelength of the radiation to the size of the lens or mirror of the optical instrument.

where the wavelength, λ, and diameter, D, are measured in the same units (Fig. 9.11).

Objects that are closer together (in angle) produce overlapping images whose separation cannot be detected. So, for example, the eye has a typical pupil diameter of about 0.4 cm. For $\lambda = 5 \times 10^{-5}$ cm (a typical wavelength) the eye's angular resolution is $2.5 \times 10^5 \times 5 \times 10^{-5}/0.4 \approx 30''$. This is quite close to the usually quoted value of 1′. For the Hale telescope with $D = 508$ cm, $\theta = 0.02''$, but in practice other factors such as the behavior of the atmosphere prevent this limit from being reached. Atmospheric "seeing" usually limits the resolution to around 1″. The Space Telescope which will operate above the atmosphere has a mirror of 240 cm diameter and is expected to resolve at about 0.15″. For comparison, the largest single radio telescope is at Arecibo, in Puerto Rico; with a diameter of 305 m and operating at a wavelength of 25 cm, its resolution is $\theta = 208''$ or just over 3′. This is typical of radio telescopes where the angular resolving power is vastly worse than those of optical telescopes. There are other means of improving the resolution, using more than one radio telescope at a time, and these will be described in Sec. 9.5.

The difference in resolution between optical and radio telescopes is shown dramatically in Fig. 9.12. This shows how the galaxy M51 appears through a large optical telescope and how it would appear optically if the resolution were only as good as the best resolution attainable with a large radio telescope.

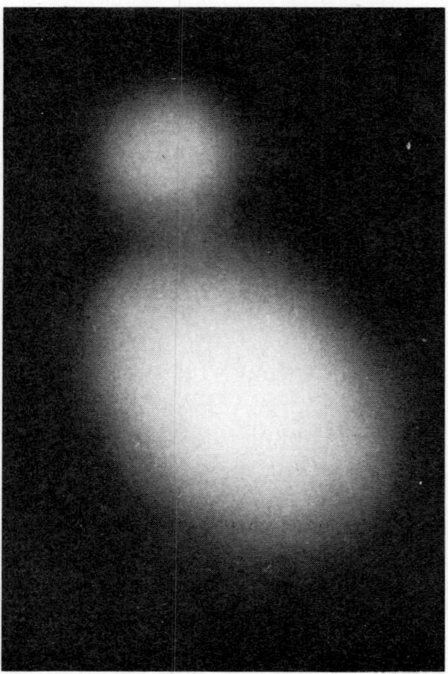

**Figure 9.12** The effect of resolving power. With a typical large telescope and good seeing, an angular resolution of 1 arc sec is attainable as shown in the photograph of the Whirlpool galaxy (M.51). Because of their use of much longer wavelengths, single radio telescopes typically achieve a resolving power of around an arc min. The right-hand picture demonstrates this by showing how the Whirlpool galaxy would appear if the optical resolution were only 2 arc min (NRAO).

## 9.4 OPTICAL TELESCOPES

As many of the design features of optical telescopes are also used in radio telescopes, and as optical telescopes are so widely used, they will be described first.

Within a surprisingly short time after its introduction by Galileo, the telescope had evolved into the two basic types that are still in use today: **refractors**, which use lenses for gathering light, and **reflectors**, which use mirrors. For many years, first one type was favored and then the other depending on technological innovation, but today all large telescopes are reflectors.

In a telescope, the front (**objective**) lens or mirror produces an image at the prime focus, where it can be photographed directly or examined through an eyepiece (consisting of several small lenses) (see Fig. 9.13).

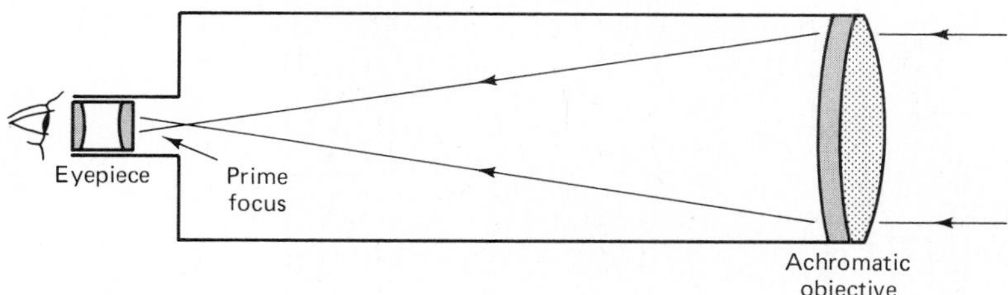


**Figure 9.13** In the refracting telescope, the achromatic objective lens produces an image at the prime focus, where it is examined and magnified by viewing through the eyepiece.

### BOX 9.2
*Angular Magnification*

We judge the size of an object by comparing it to other objects of known size or by seeing how large it appears to be, that is, by noting its angular size.

By using a suitable combination of lenses or mirrors, we can increase the apparent size of an object (Fig. 9.14). The angular magnification in defined as the ratio (angular size of image)/(angular size of object).

The angular magnification produced by the telescope is given by the ratio

$$M = \frac{\text{focal length of objective}}{\text{focal length of eyepiece}}$$

**Example**: A telescope has an objective lens with a focal length of 240 cm, and is used with an eyepiece that has a focal length of 2 cm. The overall magnification is thus

$$M = 240/2 = 120.$$

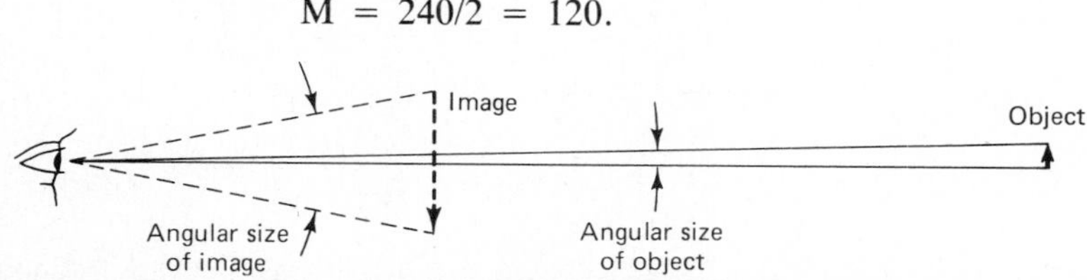


**Figure 9.14** Angular magnification. The angular size of the image is larger than the angular size of the original object, and the ratio of these angles defines the angular magnification.

It is a simple matter to change eyepieces and thus adjust the magnification. It might be thought that the magnification could be made as large as desired simply by selecting eyepieces of shorter focal length. One might, in this way, hope to see finer and finer detail in distant objects. Up to a point this is possible, but the limit to the detail that can be seen is set by the seeing and by the angular resolution of our telescope. That, in turn, is set by the size of the objective lens or mirror and the wave nature of light as pointed out in the previous section. We can indeed magnify further, but we will not see greater detail. Instead, we will see a larger but fuzzier image. We are then using "empty magnification."

**Figure 9.15** Albert Einstein at the Yerkes Observatory, 1921. This telescope is still the largest refractor in the world, with an objective diameter of 40 in. Einstein visited it during his U.S. tour, two years after the dramatic confirmation of his predictions of gravitational bending of starlight. Others in this group include E. B. Frost director of the Observatory (on Einstein's right) and Harriet Parsons (in plaid skirt), among the many astronomers (Yerkes Observatory).

**Figure 9.16** The 36 in. refractor at the Lick Observatory, one of the world's largest refractors (Lick Observatory).

Galileo's telescopes were refractors, and as such, they suffered from the chromatic behavior of lenses (Sec. 9.2). Newton's discovery of the reason for this effect and the consequent limitations of refracting telescopes led him to the invention of the reflector in 1668. Quite independently, James Gregory in 1663 (in Scotland) and Cassegrain (in France) in 1672 developed different forms of the reflector. The essential elements of the various designs are shown in Fig. 9.17. In each reflector the objective element is a concave mirror producing an image at its prime focus. A second mirror is needed to get the image out to a more convenient location.

Newton used a flat secondary mirror to direct the converging rays to a tube at the side where an eyepiece was used as with refractors. This form of telescope is still the simplest to construct and is popular with amateurs as a first telescope. Gregory used a concave mirror just beyond the prime focus to direct the light back down the tube and out through a hole cut in the center of the primary. Cassegrain had a very similar design but with a convex secondary placed ahead of the prime focus.

All of these reflectors eliminated the chromatic aberration, but they were scarcely practical instruments as the curved mirrors proved extremely difficult to grind to the correct shapes. The first mirrors were made of speculum. This alloy of copper and tin could be polished to a bright finish, but it tarnished in the air and gradually lost its reflecting efficiency. Glass was

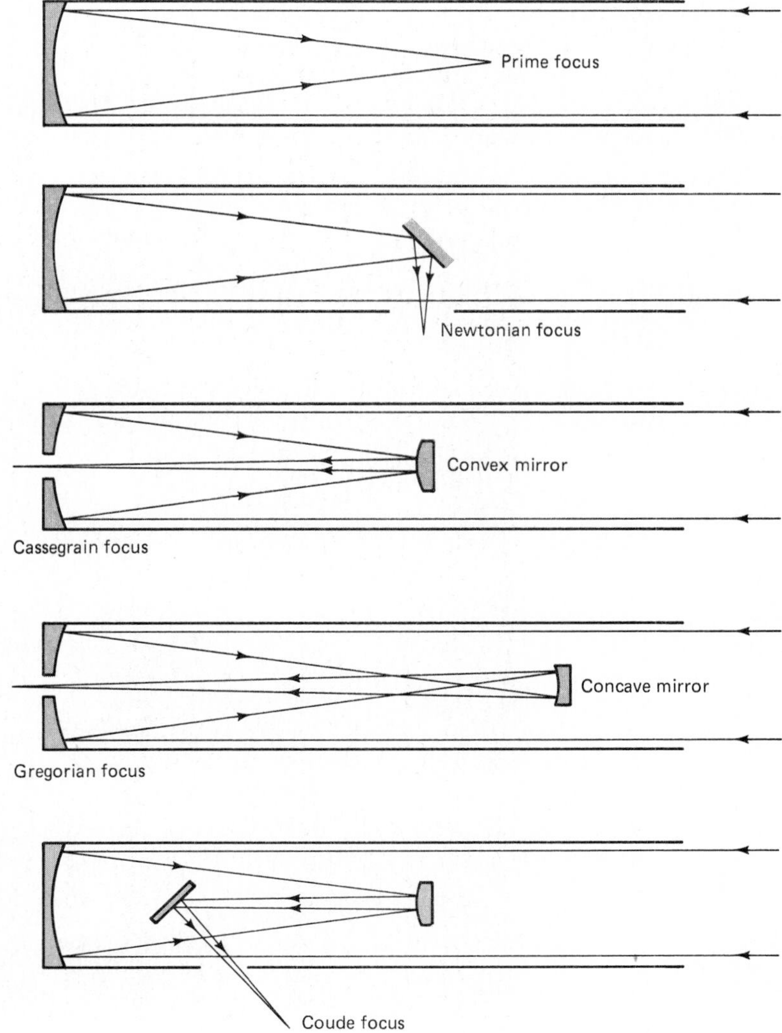


**Figure 9.17** Different types of reflecting telescopes.

**Figure 9.18** A modern reflecting telescope: the 4 m Mayall telescope at the Kitt Peak National Observatory (KPNO).

1. Building is 56.7 meters (186 feet) or approximately 18 stories high
2. Dome cranes—50-ton and 5-ton capacity
3. Prime focus cage
4. Ritchey-Chrétien focus
5. Coudé focus laboratory
6. Telescope control room
7. Visitors' gallery
8. Visitors' scenic walkway—26.8 meters (88 feet) above ground
9. Telescope pier
10. Second floor—dormitory
11. Ground floor—visitors' entrance at 2082 meters (6830 feet) above sea level

**Figure 9.19** The Shane 120 in. (3.05 m) telescope at the Lick Observatory, California. Note the black tube at the top end supporting the secondary mirror (Lick Observatory).

not used for telescope mirrors until much later. The instrumental difficulties delayed the widespread use of reflecting telescopes for a long time. The invention of the achromatic lens by John Dollond in London in the mid-eighteenth century, and improvements in glass manufacture were responsible for the dominant position of the refractors until the twentieth century.

As the need for larger telescopes developed, the problems of the refractors increased, for an objective lens can be supported only around its edge. The largest refractor at the Yerkes Observatory in southern Wisconsin has an objective of 40″ (102 cm) diameter (Fig. 9.15). Lenses larger than this will distort under their own weight and degrade the image. In addition too much light is absorbed in the considerable thickness of the glass of the large objective, and with an achromat consisting of two component lenses, there is the task of precision finishing for four surfaces. It seems unlikely that any larger objective lens will be made.

In contrast, a reflecting telescope has a mirror with only one large surface requiring optical-quality finishing. The lower side of the mirror can be supported at many places, and those supports can even be automatically adjusted to compensate for changes in the mirror shape as the telescope is rotated to point in different directions. Where a lens requires clear glass, free of major bubbles or distortions, it is only the top surface of the mirror that is important, and even that surface is coated with an aluminum layer that nowadays is protected against the atmosphere by special coatings. This *silvering* of mirrors has been a relatively modern development, and large reflectors were not developed until this century.

At modern research telescopes where sophisticated and heavy auxiliary equipment is to be used, accessibility of the image is of great importance. The prime focus of a reflector lies in the path of the incoming light. A photographic plate can be held at that focus without obscuring much of the light. In the largest reflectors, an observer can actually sit near the prime focus, in a small *cage* (Fig. 9.20). For many purposes this arrangement is not suit-

**Figure 9.20** (a) An observer in the observer's cage at the prime focus of the 5 m Hale telescope on Mt. Palomar (Palomar Observatory photograph). (b) Using an elevator to reach the cage at the prime focus (Palomar Observatory photograph).

(a)

(b)

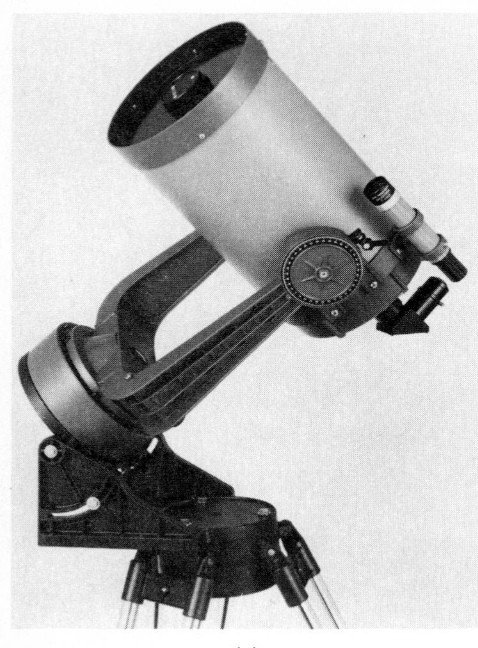

(a)

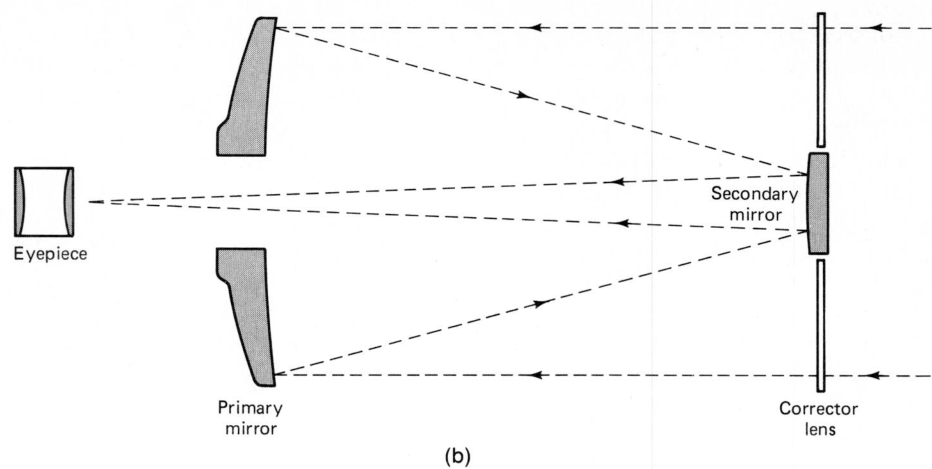


(b)

**Figure 9.21** (a) Celestron C8 telescope, with 8 in. primary mirror and folded optical design for compactness. At the center of the correcting lens is the support for the secondary mirror. The telescope mount can be adjusted for viewing at different latitudes, so that the polar axis can be aligned and the telescope then driven to compensate for the Earth's rotation. (b) Optical structure of the Celestron telescope. The convex secondary mirror reflects the light back through the aperature in the primary and also increases the effective focal length of the system over that of the primary alone. The corrector lens is not flat but is shaped to compensate for the image defects that the spherical mirror would introduce.

able, and a second mirror will be used to deflect the light to more convenient locations. The Cassegrain and coude versions are convenient for this, and many telescopes are designed to permit switching between different secondary mirrors to change the mode of operation.

Telescopes of different designs have been devised for specific purposes. For example, the Schmidt system that was used for the National Geographic Sky Survey-Palomar Sky Survey, uses a mirror that is spherical rather than the usual parabolic shape. This provides a large field of view, and to correct for image distortions that would otherwise be seen, a special lens covers the entrance aperture. This telescope design has been adapted to form the basis of the very compact and popular telescopes used by so many amateurs (Fig. 9.21). In this version, at the center of the correcting lens there is a convex secondary mirror that directs the light back to the Cassegrain focus where the images can be photographed or examined as usual with an eyepiece.

Another special-purpose instrument is the McMath Solar Telescope, one of several telescopes of the Kitt Peak National Observatory in Arizona (Fig. 9.22). A single flat mirror 2 m in diameter follows the Sun and reflects

(a)

**Figure 9.22** (a) McMath Solar Telescope at the Kitt Peak National Observatory (KPNO).

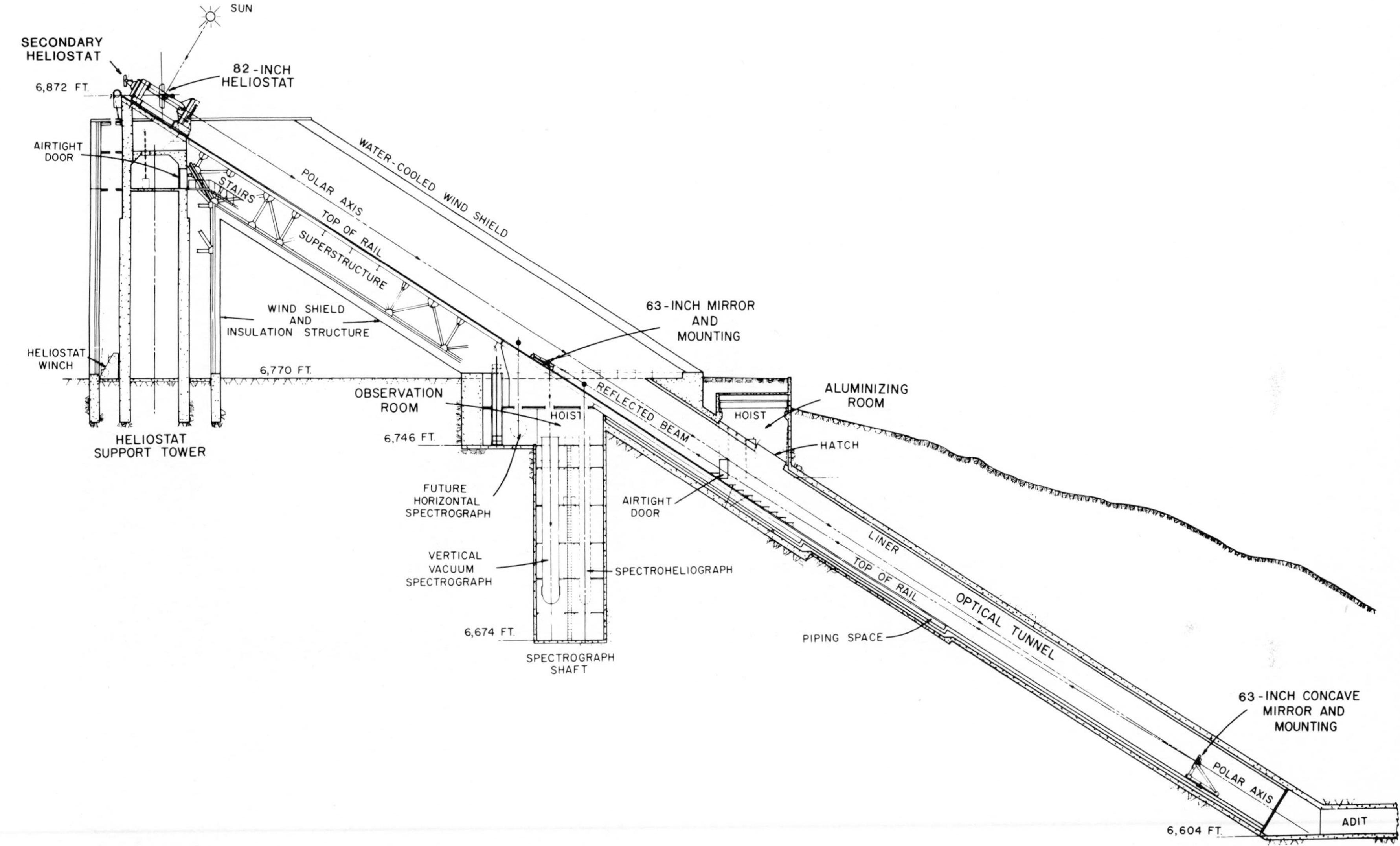


(b)

**Figure 9.22** (b) Light from the Sun is reflected from a flat mirror down the inclined structure to the observatory room that can be maintained at a steady temperature. Movement of the Sun is followed by rotation of the 82 in. heliostat mirror at the top (KPNO).

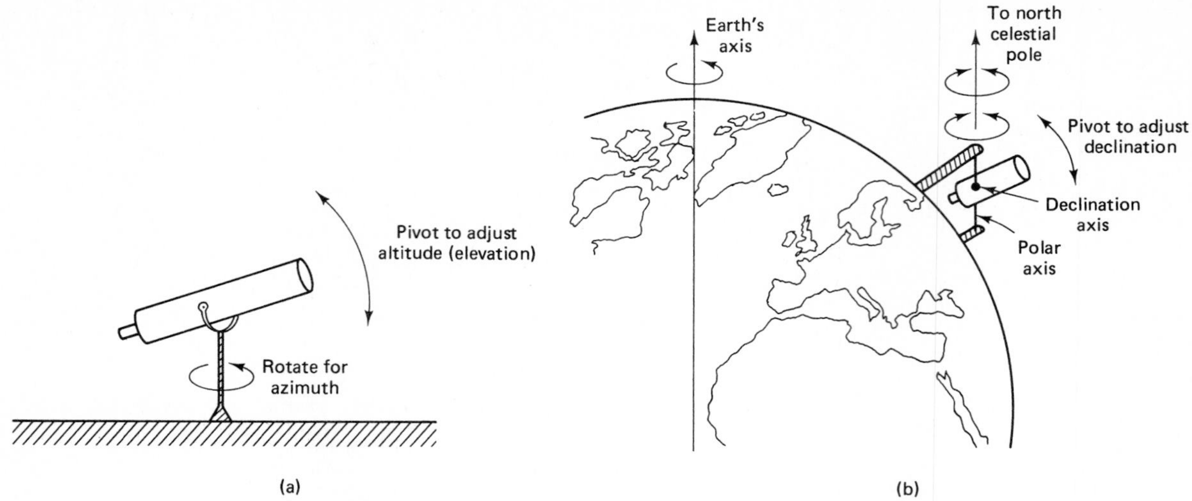


**Figure 9.23** (a) With the alt-azimuth system a telescope is pivoted to permit adjustment in altitude (elevation) and supported on a base that can rotate in azimuth. (b) With the equatorial system a telescope is mounted so that one axis of rotation is parallel with the Earth's rotation axis and thus aimed at the celestial pole. Rotation about this axis compensates for the Earth's rotation. At right angles to the polar axis is the declination axis.

the solar beam to a fixed 1.6 m concave mirror that focuses the beam to produce a solar image 76 cm in diameter.

To allow a telescope to be pointed to any part of the sky, it must be pivoted on two axes. The simplest type of mounting has one vertical axis, allowing rotation around the horizontal (azimuth) circle through north, east,

**Figure 9.24** (a) The Multiple Mirror Telescope on Mt. Hopkins, Arizona. Six separate mirrors each have a diameter of 1.8 m. The entire structure rotates in azimuth on its base, and the telescopes can be simultaneously adjusted in elevation (MMT). (b) Schematic diagram showing how the light from separate mirrors is combined to produce a single image (MMT).

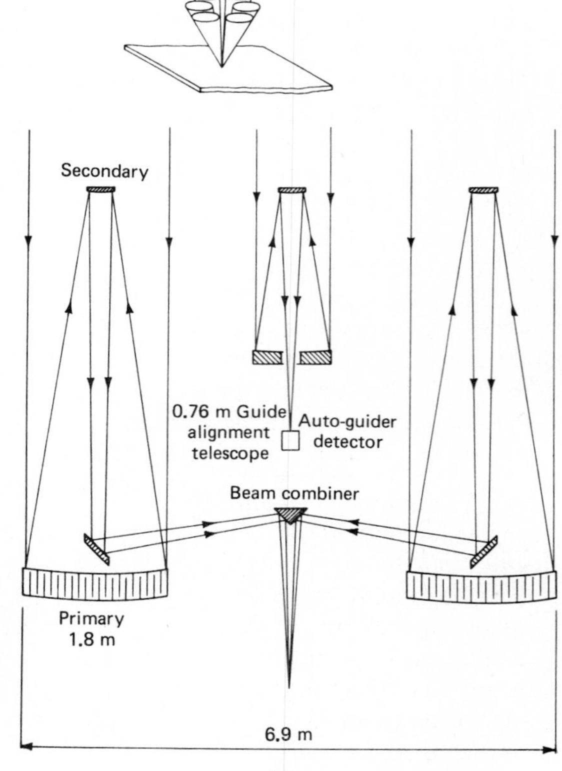

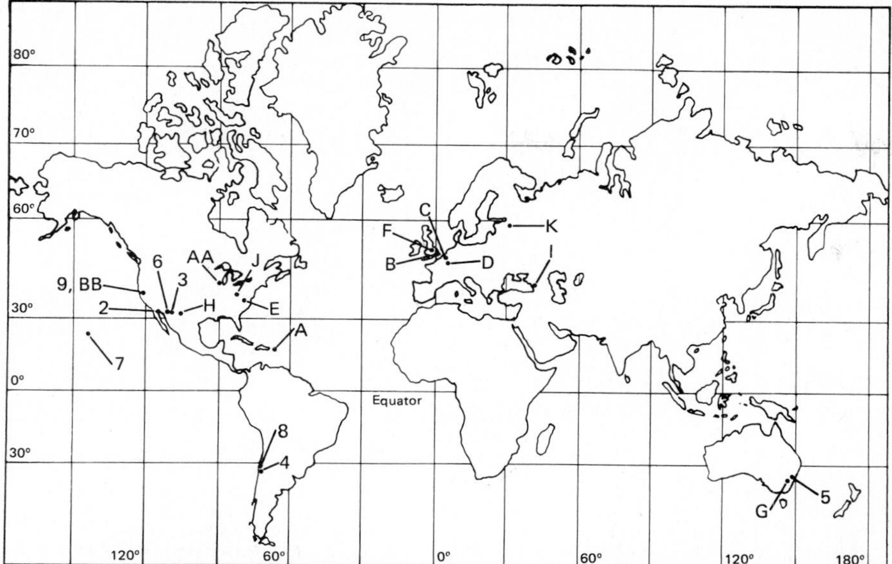


**Figure 9.25** Locations of some major optical and radio telescopes.

Reflectors:
1. Special Astrophysical Observatory, U.S.S.R., 6 m
2. Palomar Observatory, California, 5 m
3. MMT, Arizona, 4.6 m
4. Cerro Tololo Inter-American University, Chile, 4 m
5. Anglo-Australian Observatory, 3.9 m
6. Kitt Peak National Observatory, Arizona, 3.8 m
7. United Kingdom IR Telescope (3.8 m) and Canada-France-Hawaii Observatory, (3.6 m), Hawaii
8. European Southern Observatory, Chile, 3.6 m
9. Lick Observatory, California, 3 m
10. Mauna Kea Observatory (IR), Hawaii, 3 m

Refractors:
AA. Yerkes Observatory, Wisconsin, 1.01 m
BB. Lick Observatory, California, 0.9 m

Radio Telescopes:
A. Arecibo Observatory, Arecibo, Puerto Rico
B. Mullard Radio Astronomy Observatory, Cambridge, England
C. Radio Observatory, Dwingeloo, Netherlands Radio Observatory, Westerbork, Netherlands
D. Max Planck Institute for Radio Astronomy, Effelsberg, West Germany
E. National Radio Astronomy Observatory, Green Bank, West Virginia
F. Nuffield Radio Astronomy Laboratories, Jodrell Bank, England
G. Australian National Radio Astronomy Observatory, Parkes, Australia
H. National Radio Astronomy Observatory, (VLA), Socorro, New Mexico
J. Ohio State Radio Observatory, Delaware, Ohio
K. Central Observatory, Pulkovo, U.S.S.R

south, and west (Fig. 9.23a). The other axis, set horizontally, allows changes in elevation or altitude. This system is easy to construct but not convenient to operate, for the altitude and azimuth of an object in the sky depend on the location of the telescope, and adjustments must continually be made in both directions to compensate for the rotation of the Earth if it is desired to follow some object for a long time. Instead, the **equatorial** system is most widely used (Fig. 9.23b). The **polar axis** is aligned parallel to the Earth's own axis of rotation. The second (or declination) axis is perpendicular to the polar axis. Once the telescope has been set to the desired declination, only a single motor is needed to maintain a synchronous drive about the polar axis to compensate for the Earth's rotation.

For many years the equatorial system has been almost universally adopted. With very large telescopes, however, the increasing engineering complexity and the cost of the equatorial system and the protective dome led to a major experiment. With computers it is now possible to have precise and simultaneous control of the two drive motors needed for an alt-azimuth system. The overall mechanical design is thereby simplified and becomes more economical. This system has been used since 1979 with the Multiple Mirror Telescope (MMT) in Arizona.

The MMT has another design innovation. Instead of using a single large mirror, it has six mirrors, each 1.8 m diameter and held in an open but rigid structure so that the six beams can be combined to produce a single image (Fig. 9.24). The effective collecting area is equivalent to that of a single 4.5 m telescope, but it is simpler and less expensive to make six small mirrors.

## 9.5 RADIO ASTRONOMY

Radio astronomy dates only from 1931 and the observations of Karl Jansky, an engineer working at the Bell Telephone Laboratories at Holmdel, New Jersey. Jansky was investigating radio static that accompanies thunderstorms and found that his radio antenna, operating at a wavelength of 14.6 m, detected signals that at first he could not identify. He proceeded to show that they came not from the Earth but from the center of the galaxy. Jansky was not able to get funds for the construction of a large reflector, but Grote Reber, a radio engineer who lived at Wheaton, Illinois, had become interested in Jansky's work, and in 1937 constructed the first radio astronomy reflector in the backyard of his home. Reber's first observations at wave-

**Figure 9.26** Karl Jansky's radio antenna. Designed in 1929, the antenna was 100 ft in diameter and tuned to a wavelength of 14.6 m (20.5 MHz frequency). Using this antenna at the Bell Telephone Laboratories in Holmdel, New Jersey, Jansky made the first observations of radio waves coming from outside the Earth. The antenna was mounted on four Ford Model T wheels and rotated in azimuth once every 20 min (NRAO).

lengths of 9 cm and 33 cm were unsuccessful. In 1939, after rebuilding his receiver to operate at 1.87 m, he obtained positive results. He continued to work for several years surveying large parts of the sky and identifying several regions from which strong signals came.

At this time, World War II intervened and radio astronomy research effectively ceased. During the war there were major advances in electronics, so that research blossomed when the war ended. Electronics was not the only military contribution to radio astronomy. When the giant radio telescope was built at Jodrell Bank for the University of Manchester in England, the director of the project, Bernard Lovell, was able to get gun turret mounts and bearings from a surplus battleship.

Jansky's first antenna (Fig. 9.26) was a more elaborate version of those that are widely used domestically for television reception: an array of wires or rods on a frame that could be rotated. Except for a few special projects, this type of antenna has been superseded for research by large reflectors operating essentially in the same way as optical reflectors. The incoming radiation is focused by a large curved mirror (metal rather than glass) onto a relatively small antenna at the focus to convert the radiation into electric signals. These signals must then be amplified as in an ordinary radio. The amplifier can be tuned to different frequencies, over a range that is set by the design of the telescope and antenna.

For two reasons, radio telescopes have to be much larger than their optical counterparts (Fig. 9.27). Cosmic radio signals are weak so that large antenna areas are necessary to collect enough energy to be detectable. Beyond this minimal requirement, a large size is needed to improve the angular resolution (Sec. 9.2), which depends on the ratio $\lambda/D$. As mentioned earlier, even the largest radio reflectors cannot match the optical resolution of arc second detail, and an ingenious technique has been developed to get around this limitation.

If a row of similar radio reflectors or "dishes" is set out along an east-west line, radio waves from a distant source will arrive at successive dishes at slightly different times. For another row of dishes along a north-south line, the pattern of arrival times will be different. The difference in arrival times of the signals at the various dishes depends on the direction of the source. By recording the signal at each dish along with very accurate time

(a) (b)

**Figure 9.27** (a) 140 ft radio telescope at the National Radio Astronomy Observatory in Greenbank, West Virginia. Note the similarity in design to the optical Cassegrain reflecting telescope. (b) The radio waves are reflected to the secondary radio mirror and then down to the receiver antenna (NRAO).

**Figure 9.28** (a) The Arecibo radio telescope in Puerto Rico consists of a fixed metal mesh reflector with a diameter of 1000 ft suspended in a natural bowl between hills. Because the reflecting surface is spherical in shape, there is not a single well-defined focal point, but signals from different directions can be received by moving the system that carries the receiver antenna. (b) View of the movable antenna looking up through the mesh and supporting structure. A band of the sky 40° wide and centered on the local zenith can be scanned (Arecibo).

(a) (b)

markers, the signals from all of the dishes obtained over many days can be combined by computer, and a radio map can then be constructed. This map will show the angular detail that would have been obtained by a single large dish having the same aperture (area) as the product of the lengths of the two rows of dishes.

Many variations have been made in this basic design, involving the number of dishes, their distances apart and the angle between the lines of dishes. The first design of this type was known as the Mills' Cross, after its Australian inventor, Bernard Mills. Other designs are termed interferometers. In some the dishes along one line are kept fixed, whereas some or all of the dishes are shifted along the other line.

The largest system of this kind in the U.S. is the Very Large Array (VLA) in New Mexico (Fig. 9.29). This is one of several radio telescopes operated by the National Radio Astronomy Observatory. The VLA consists of twenty-seven dishes set out along the 13-mi arms of an enormous Y pattern. Each dish is 25 m across, mounted on a track so that its position can be changed depending on the observations needed. When fully extended, the VLA can resolve 0.2″, an angle equivalent to a dime at a distance of 20 km (13 mi).

The method is practicable because of the great accuracy with which time can now be measured. Signals from different dishes taken over many nights can be recorded with precise time signals and later be synchronized so that all signals from the same part of the sky can be combined. The optical equivalent is impossible, for atmospheric seeing varies too much from night to night. Successive and overlapping photographic exposures on the same plate would soon become blurred.

Radio interferometry has been pushed to its Earth-based limit. Radio telescopes at sites across the U.S. have been used in combination with others in Hawaii and Europe to provide an interferometer with dimensions comparable to the size of the Earth. This Very Long Base-Line Interferometry (VLBI) has produced *pictures* of radio sources with detail down to one thousandth of an arc second, far better than that of any optical telescope. (One milli-arcsecond is the angular diameter of a quarter at about 3000 miles.)

Radio telescopes have limitations that do not affect optical telescopes, which are always housed in protective domes. The size of radio dishes means that most must be exposed to the weather, and part of the mechanical design must be concerned with the stability of the structure even under strong winds. In some climates there may be a problem with snow—depending on the size of the mesh of the reflecting surface, snow can accumulate and cause distortions. These problems have limited the maximum size to about 300 ft for radio telescopes that need to be pointed in all directions, but some telescopes with restricted steering capabilities have been constructed with 2 or 3 times larger size.

In astronomy, necessity is the mother of ingenuity as well as invention. In 1963 another Australian radio astronomer, Cyril Hazard, devised a completely different method of measuring the size and position of compact radio sources. As the Moon moves across the sky it eclipses some radio sources, so that an observer will see their signals diminish and then reappear. Unlike the better-known solar and lunar eclipses with their fuzzy penumbras, eclipses of radio sources display a pattern of brightness variations that is well known in optics. This succession of increases and decreases in the intensity of the signal comes from the wave nature of radiation and the theory is fully understood. Comparison of an observed pattern (Fig. 29.2) with the theory, then allows the size and position of the source to be defined to arc second accuracy. Clearly this technique is restricted to those parts of the sky that are covered by the Moon's movement, but its uses can be dramatic.

(a)

(b)

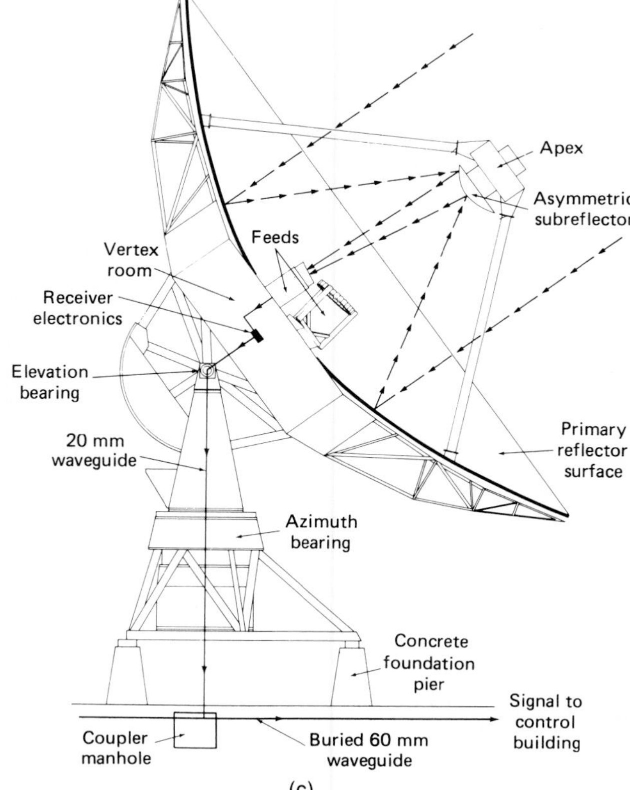


(c)

**Figure 9.29** The Very Large Array (VLA) radio telescope. (a) View of the central 2 km of the VLA showing the three arms of the Y-shaped array. Two of the arms extend for 21 km and the other for 19 km. Individual dishes can be moved to different configurations for particular observing requirements. (b) There are twenty-seven individual dishes each with its own pointing system. (c) Each dish has a diameter of 25 m (82 ft) and operates similarly to an optical reflecting telescope (NRAO).

Indeed some X-ray astronomers observed the occultation of the Crab nebula from a rocket that was timed precisely to reach maximum altitude during the Crab's brief eclipse by the Moon.

## 9.6 DETECTING THE RADIATION

All that telescopes do is collect and focus the incoming radiation. What is still needed is a detector, a device to absorb and record radiation. Since the classic exploratory stages, the eye has been supplemented, first by photography and then by electronic devices, which are both more sensitive and can be automated so as to run without fatigue or attention for long periods.

Photography still finds widespread use. Special photographic plates are manufactured mostly by the Eastman Kodak Company. Long exposures can reveal the presence of very faint objects or wispy features in galaxies and in interstellar space, and the photographs can be examined by densitometers to measure the blackness of the images and thence the brightness of the sources. Many of these tasks can now be performed more quickly and with greater precision by newer electronic techniques, but the record-keeping quality of photography is unique. Old photographs can be reexamined to look for changes in appearance recorded over many years but whose importance has only now been realized. When the first quasars were discovered in 1963, it was possible to use the collection of plates in the Harvard College Observatory to check positions and brightness as far back as the 1880s. In no other part of the electromagnetic spectrum is this possible; photography will continue to play this vital, archival role.

Unfortunately photographic plates are inefficient in that only a few percent of the incoming photons get converted to grains in the image. Long exposure times (several hours) are needed for the faintest objects, but a limit is set by the accumulation of sky **background**, very faint light produced and scattered in the atmosphere. This accumulates slowly on the photographic plate during a long exposure, causing an overall darkening against which the faint images can no longer be seen.

For many purposes, electronic techniques have replaced photography at optical and near-infrared wavelengths. The image produced by the telescope can be focused onto a photosensitive surface, which is less wasteful of its photons than are the photographic plates. Electrons are emitted from the specially coated surface to be accelerated by a high voltage and focused through two or more stages. The final image can then be recorded by a photographic plate or a television-type camera. Different techniques have different names such as image intensifier, electronographic camera, Digicon, but they share the same basic features.

These devices allow very faint objects to be observed in much shorter times than are needed for photography, and thus allow more objects to be examined in the time available. In addition, very rapid variations in light can be followed. For example, the Crab nebula has a central star that switches on and off thirty times each second. This effect is easily detected photoelectrically but cannot be seen with conventional photography.

Radio astronomy requires quite different types of detectors. The power in the focused waves is very weak; $\sim 10^{-15}$ to $10^{-20}$ watt is typical. In the antenna, these signals will produce minute electrical voltages that must be amplified by enormous factors ($10^{10}$ is not uncommon) to reach more easily handled voltages. Electronic components such as resistors and transistors produce *noise*; even with no incoming signal, there will always be a small voltage produced by the electrons in the circuitry. The smallest cosmic signals that can be detected will depend on the size of this noise. One way to reduce the noise is to cool many of the components. This reduces the movement of the electrons and thus the noise voltage.

Cooling is essential to many infrared detectors. After the radiation has been focused by the reflecting telescope, it is absorbed by a special crystal, typically a few millimeters or less in size. The type of detecting element is determined by the wavelength of interest. The detector might, for example, be a specially prepared crystal of germanium. Most of these detectors work best when cooled to temperatures below about 4 K, and this requires the use of liquid helium. These **cryogenic** detectors were developed in the early 1960s by Frank Low, then working at Texas Instruments Company. They are now standard items used at many observatories and on high-altitude research balloons and on satellites (Fig. 9.30). The operating principle of all IR detection is the heating of the detector when the radiation is absorbed. This produces a change in the electrical properties of the detector or a small change in its voltage. But the heating is minute—less than 0.01 K—and the voltage is around one billionth of a volt. These effects are quite unmeasurable unless everything is kept ultracold.

At radio and IR wavelengths, the energy per photon is so low that all detectors must work by absorbing large numbers of photons. For the traditional optical techniques this is still the case, but some newer electronic techniques can see single optical photons. Detection of X- and gamma rays requires completely different techniques. With each X-ray and gamma ray photon far more energy is carried, so that single photons are capable of producing many effects by which they can be detected. Articles reporting discoveries in X- and gamma ray astronomy will often list the intensity of a source as so many counts per second—literally reporting how many photons have been detected.

The units of energy that we have encountered thus far have been ergs, and we could specify a photon's energy as $h\nu$ ergs, where $\nu$ is the frequency and $h$ is Planck's constant, $6.63 \times 10^{-27}$ erg·sec. The erg is too small a unit for everyday use but far too large for describing single photons (or atomic particles), and the **electron volt** is generally used. The conversion is

$$1 \text{ eV} = 1.6 \times 10^{-12} \text{ ergs}$$

**Figure 9.30** The 1 m balloon-borne IR telescope of the Smithsonian Astrophysical Observatory and the University of Arizona. This Cassegrain system is flown in a gyro-stabilized supporting structure that allows the telescope to be pointed and controlled from the ground (Giovanni G. Fazio, Smithsonian Astrophysical Observatory).

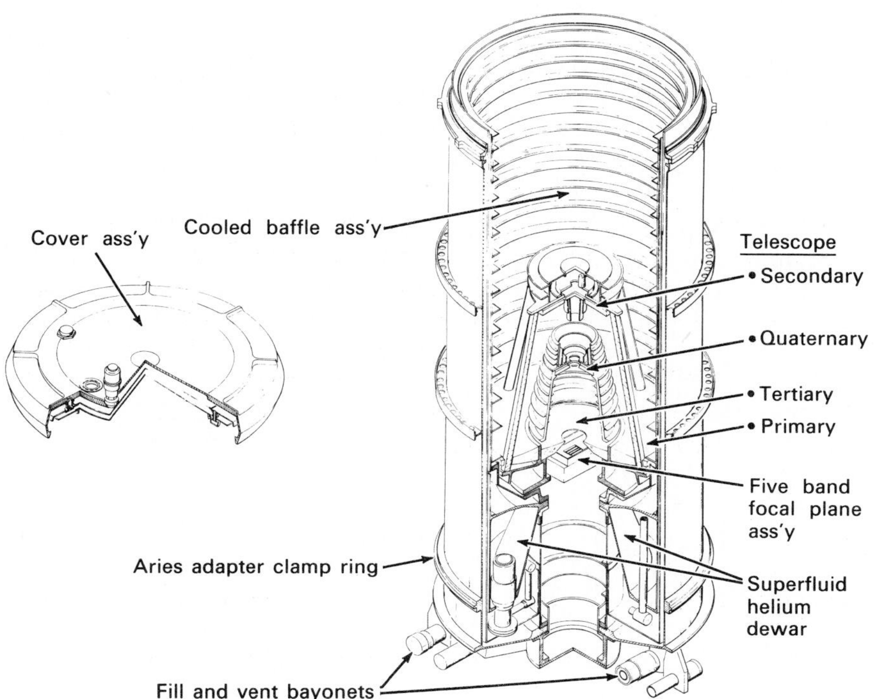


**Figure 9.31** Far Infrared Sky Survey Experiment (FIRSSE) for rocket-borne observing in a joint experiment between the Air Force Geophysical Laboratory and the Naval Research Laboratory. The liquid helium cooled system is carried to a peak altitude of 380 km providing 450 sec of observing time (Stephan D. Price, Air Force Geophysical Laboratory).

*Table 9.1* PHOTON ENERGIES

| Spectral region | Radio | IR | Optical | X-Ray | Gamma |
|---|---|---|---|---|---|
| Typical wavelength (cm) | 10 | $10^{-3}$ | $5 \times 10^{-5}$ | $10^{-7}$ | $10^{-11}$ |
| Frequency (Hz) | $3 \times 10^{9}$ | $3 \times 10^{12}$ | $6 \times 10^{14}$ | $3 \times 10^{17}$ | $3 \times 10^{21}$ |
| Energy/Photon (ergs) | $2.0 \times 10^{-17}$ | $2.0 \times 10^{-14}$ | $4.0 \times 10^{-12}$ | $2.0 \times 10^{-9}$ | $2.0 \times 10^{-5}$ |
| Energy/Photon (eV) | $1.25 \times 10^{-5}$ | $1.25 \times 10^{-2}$ | 2.5 | $1.25 \times 10^{3}$ | $1.25 \times 10^{7}$ |

**Figure 9.32** X-ray telescope. X-rays can be reflected off specially shaped metal surfaces and focused to produce an image. Shown here is a nested system of cylindrical mirrors used in the ATM (Apollo Telescope Mount) solar X-ray experiment on Skylab. The mirrors were made of specially coated beryllium, (G. Vaiana, Harvard/Smithsonian Center for Astrophysics).

Larger units will often be used: 1 keV = 1000 ev and 1 MeV = 1 million eV. Table 9.1 lists the energies of some photons.

At very high photon energies, the usual optical techniques will not work. No lens will focus X-rays or gamma rays, and they will penetrate a mirror rather than be reflected in the usual way. The only focusing technique that works with long-wavelength X-rays but not with gamma rays is **glancing angle reflection**. X-rays that arrive at very small angles to the surface (angle of incidence very close to 90°) will be reflected. Cylindrical surfaces that are elaborately curved will provide focusing. Figure 9.32 shows the shape and actual hardware of one of these X-ray telescopes.

To be detected, the X- and gamma rays have to be absorbed in crystals such as sodium iodide (NaI) or in a gas such as argon, where electrical signals are produced by each photon. The size of the signal depends on the photon energy. The detector must be surrounded by screening materials so that the photons will be counted only if they arrive within a specific range of directions, and the result is often very coarse angular resolution of several degrees. Better resolution can be obtained with some X-rays by a **modulation detector** that has two grids of wires (Fig. 9.33). Photons will pass between

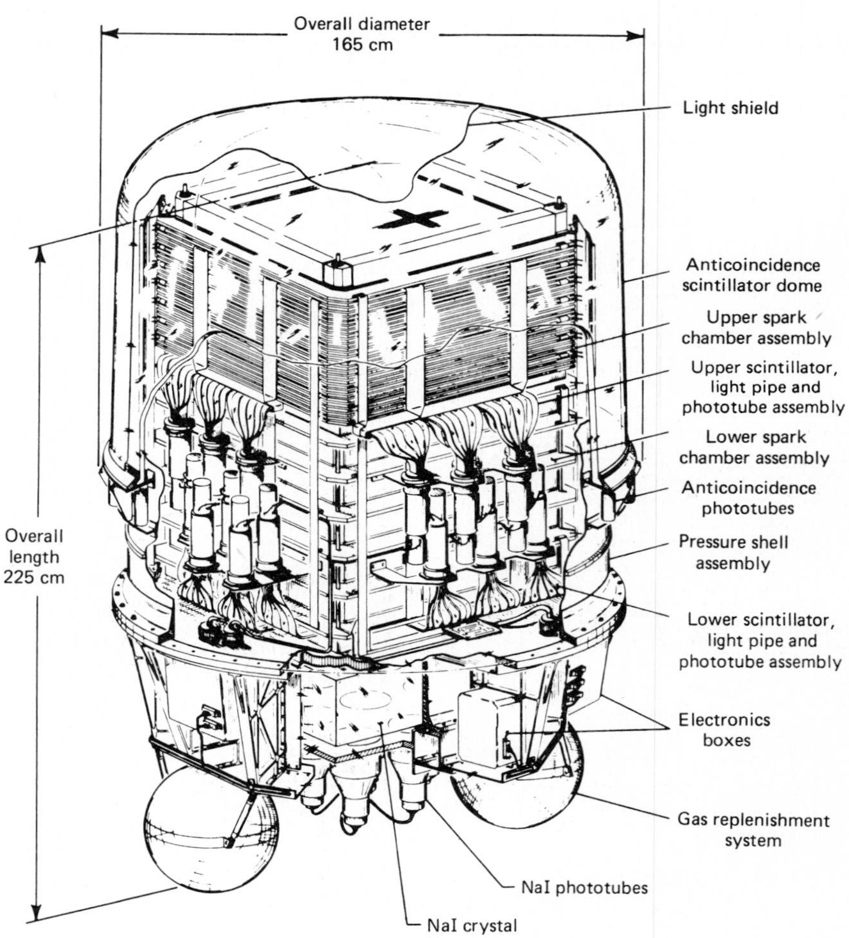


(a)

**Figure 9.33** (a) Gamma-ray telescope. This system was flown on the second Small Astronomical Satellite (SAS-2). It consists of two spark chambers.

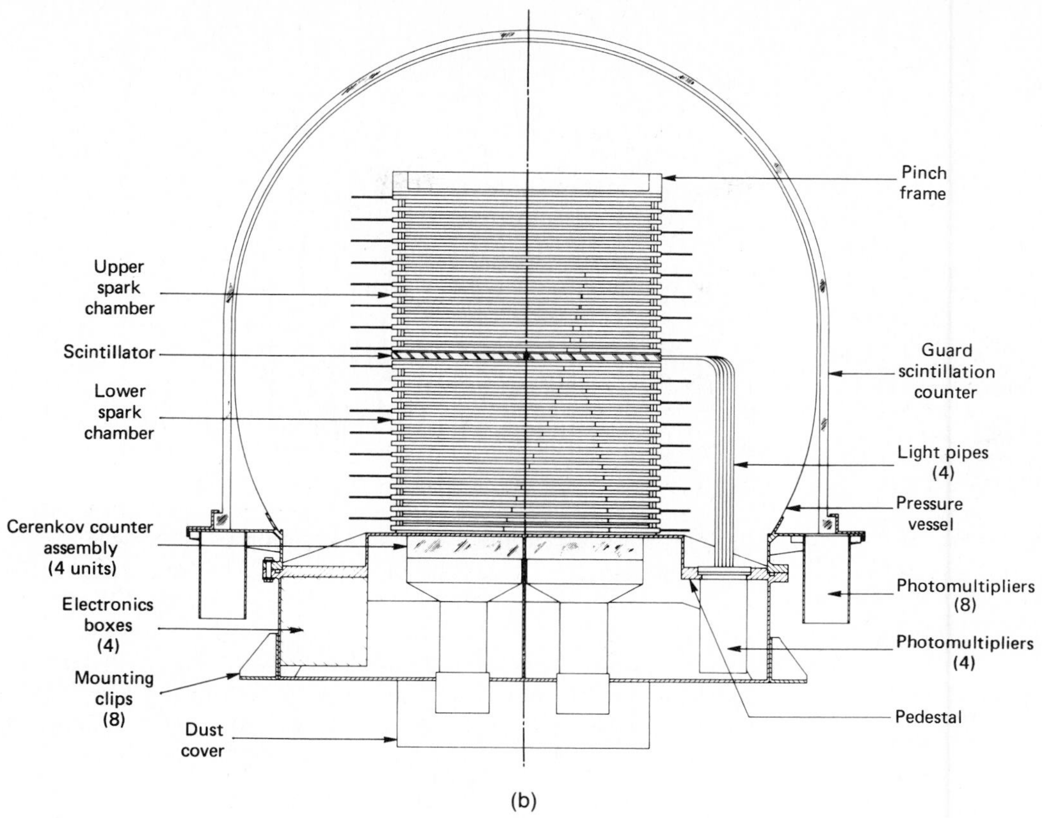


(b)

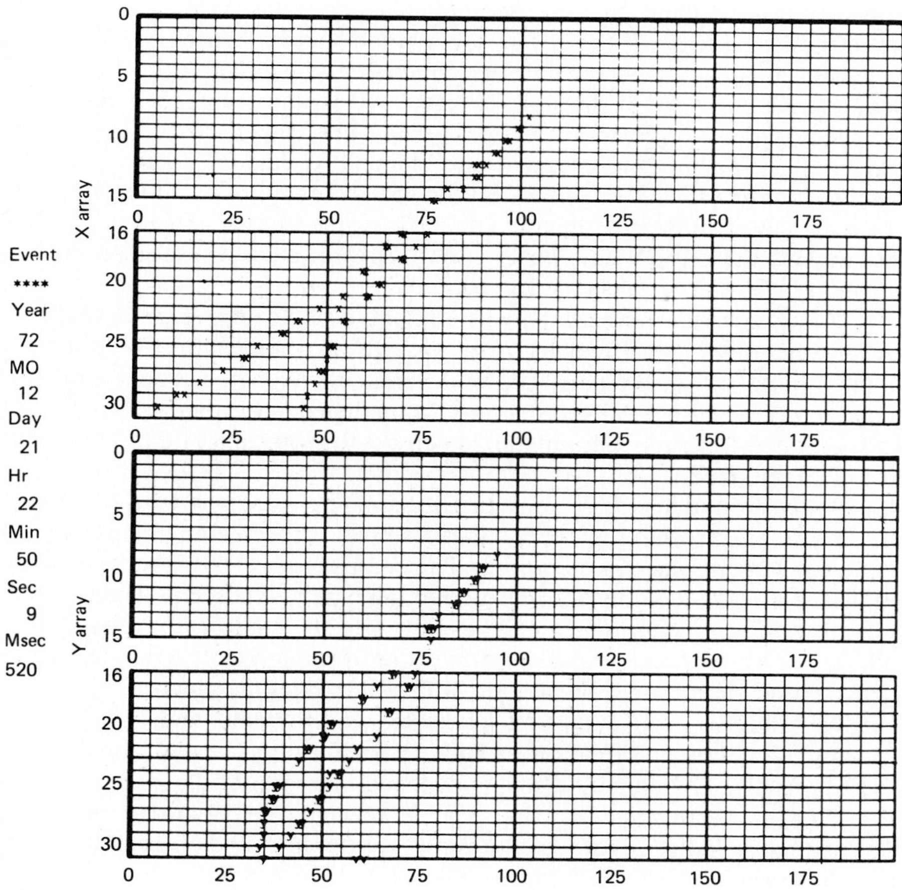


(c)

**Figure 9.33** (b) High-energy gamma-ray photons convert into electron-positron pairs, and these electrically charged particles produce tracks in the spark chambers. The original gamma-ray energy and its arrival direction can be calculated from measurements of the spark tracks. (c) Stereo views of an electron-positron pair produced by a gamma ray (Carl E. Fichtel, NASA Goddard Space Flight Center).

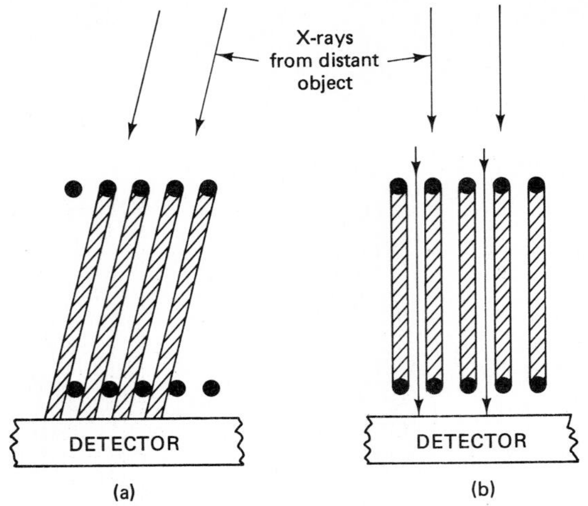

X-rays
from distant
object

DETECTOR

(a)

DETECTOR

(b)

**Figure 9.34** X-ray modulation collimator. With two wire grids, X-rays can pass through the detector or will be intercepted by the grid depending on the arrival direction. Direction of the source and its angular size can be computed from the way in which the X-ray signal varies as the source moves across the system aperture.

both grids only in some directions; from other directions those photons that get through the first grid will be intercepted by the second grid. The angular resolution depends on the diameter of the wires and the separation between grids, and about 1 arc min resolution has been obtained.

A different technique can be used for the detection of high-energy gamma rays (Fig. 9.34). Each of these photons has enough energy to produce an electron-positron pair. The energy of the gamma ray and its direction of arrival can be deduced from the trails of sparks produced by the electron and positron in a device known as a spark chamber.

## 9.7 MEASURING THE WAVELENGTH

The radiation reaching us from some distant star consists of a mixture of wavelengths with more energy at some wavelengths than at others. This spectrum must be dispersed into its components in order to begin to understand the structure of the star. Different types of instruments will spread the spectrum out to greater or lesser extents. When we discussed fine detail that could be seen in the image of an extended object, we defined angular resolution. Similarly, we can now talk of **spectral resolution** and the **spectral resolving power** of an instrument. The more a spectrum is spread out by an instrument, the more detail we can see resolved, such as emission or absorption lines that are very close together.

At radio wavelengths, the measurement is electronic. A given antenna will absorb radiation efficiently over a range of wavelengths. Within that range, tuning of the amplifiers allows different wavelengths (frequencies) to be selected. Although the spectral resolution depends to some extent on the strength of the signal, radio methods can detect wavelength differences of less than one part in $10^5$.

In the IR and visible regions, broad-band filters are widely used. These are sheets of absorbing material placed across the detector or inserted at some convenient place in the optics. The composition of the filter is chosen so that some wavelengths are transmitted, all others absorbed. (The colored cellophane used for wrapping candies falls into the category of broad-band filters, except that its spectral properties are probably not carefully specified.) These filters have been standardized covering a selection of wavelengths from the near ultraviolet to the IR, and they are widely used, for example, for the classification of stars (hot stars, cool stars, giants, etc., see Chap. 21). The standard filters are listed in Table 9.2.

For greater resolution in the visible spectrum, a prism can be used. The refractive index of glass is different for different wavelengths: the short wavelength (blue) light is refracted more than the long wavelength (red). A

*Table 9.2* STANDARD FILTERS FOR THE ULTRAVIOLET, VISIBLE, AND NEAR IR[a]

| *Ultraviolet To Near IR* | | | | | |
|---|---|---|---|---|---|
| Filter | U<br>ultraviolet | B<br>blue | V<br>visual | R<br>red | I<br>infrared |
| Central Wavelength (Angstroms) | 3650 | 4400 | 5500 | 7000 | 9000 |
| ***Middle IR*** | | | | | |
| Filter | J | K | L | M | N | Q |
| Central Wavelength (microns) | 1.25 | 2.2 | 3.4 | 5.0 | 10.2 | 22.0 |

[a] Ref: Allen, C. W.: *Astrophysical Quantities*, 3rd ed. The Athlone Press, University of London, 1973.

beam of light that contains a mixture of wavelengths will emerge dispersed after passing through a prism, and intensities can then be measured in different wavelength intervals. With some telescopes a small-angle prism is set across the whole objective, so that each star's image is a small spectrum. Only the strongest features will be seen in this way, but that is still sufficient to permit the bright stars to be examined.

For very high resolution, a **diffraction grating** can be used. A grating consists of a reflecting surface on which thousands of very fine parallel lines have been ruled. The wave nature of light results in each wavelength coming off most strongly in a direction that depends on the wavelength and on the separation of grating lines. The result is a wide dispersion of the spectrum in which great spectral detail can be seen (Fig. 8.16).

The brevity of this section should not be mistaken to imply simplicity of spectroscopic techniques. Many of the important astronomical discoveries have come through the painstaking and precise analysis of spectra measuring the wavelengths of emission and absorption lines and then trying to identify the kinds of atoms responsible. Sometimes the wavelengths are red- or blue-shifted, and the detective work is complicated. The instruments themselves are the products of great skill, and long experience plays a role in the extraction of new results. With electronic aids and with computers the techniques are continually under development, but the basic principles that we have covered will allow us to understand the material we shall encounter in the later chapters.

## 9.8 NEW DEVELOPMENTS

Instrumental innovation is continuous. Astronomers want to see fainter objects at greater distances and in greater detail. The resulting need for larger telescopes soon collides with mechanical problems that limit the size of conventional designs. In addition, larger telescopes require larger and much more expensive domes. Considerable ingenuity is therefore being invested in radically new designs to improve the performance while holding the cost to manageable proportions.

In an attempt to open the way to much larger telescopes, the Multiple Mirror Telescope (MMT) was designed, and it started operation in 1979 (Fig. 9.24). The MMT consists of six reflecting telescopes (each 1.8 m in diameter) mounted together and having their separate beams combined at a central location. The immediate advantage is that each mirror is of manageable size to make, polish and align, yet together the combination has the collecting power of a single mirror of 4.5 m diameter. Each mirror is separately supported and mounted, and a defect in one can be corrected at far less cost than would be the case if a single large mirror had to be resurfaced.

In the U.S. several groups are working on the designs of telescopes with effective diameters in the range of 8 to 15 m. Some have single mirrors of new lightweight construction, others are extensions of the MMT design.

**Figure 9.35** Large telescope designs. Designs for very large optical telescopes are being studied; these scale models show two of the many possible configurations. On the left, the large primary consists of sixty hexagonal segments, each with its individual control for alignment. On the right is a larger version of the Multiple Mirror Telescope, using four 7.5 m primaries. In the center for comparison of scale is the existing 4 m Mayall telescope (KPNO. Photo: © 1983 AURA, Inc., Kitt Peak National Observatory).

A consortium of European observatories is looking into a 16 m design, whereas the Russians are considering a 25 m telescope. It takes a long time to design, finance, and construct a major telescope.

Some new designs are shown in Fig. 9.35. A mosaic of flat mirrors nested together makes up a large surface with each component separately mounted and continuously aligned under computer control. In this way, the large cost associated with the manufacture and finishing of a very large mirror can be avoided. The focusing of the many flat mirrors can still give a good image with large collecting power and lower cost despite the complexity of the control system needed. In addition, computer control will permit very rapid adjustment for variation of atmospheric seeing across the anticipated large diameter. A prototype of this active control has already been tested, with impressive results.

A factor in the new designs is the mechanical support and drive of the telescope. The equatorial systems that are so widely used do most of their tracking in right ascension, with a single motor keeping the declination fixed. This compensates for the Earth's rotation. Now that these drives can be accurately controlled by computers, alt-azimuth systems (requiring two continuous and synchronized drives) are very attractive because of their greater mechanical simplicity and resulting lower cost.

A completely different approach is represented by the Space Telescope. This will be placed in Earth orbit by the Space Shuttle. Above the atmosphere its 2.4 m diameter mirror will collect enough light to detect objects fifty times fainter than the present limit. Its angular resolution of 0.15″ will be about ten times better than is now generally obtainable by ground-based telescopes. Part of this improvement will come from the absence of the overlying atmosphere, part from the reduction in distortion of the mirror in its weightless orbit. The design lifetime of the Space Telescope is 15 years in orbit.

Other instrumental research involves low-noise amplifiers for radio (especially for the shorter wavelengths), interferometers (for very high resolution in the optical and IR), and detectors themselves. The new electronic charge-coupled device (CCD) can detect 70 percent of the photons falling on it—far less wasteful of photons than conventional photographic materials with their 1 percent efficiency. Our ability to see detail in faint and distant objects depends just as much on these less-well-publicized parts of astronomy as they do on the telescopes themselves.

## 9.9 SUMMARY

Astronomy is more than just pointing a telescope. It has expanded to cover the entire electromagnetic spectrum and shows no sign of slackening in its rate of discoveries. Astronomy is currently one of the most vigorous and exciting of the sciences, and with this survey of methods and hardware complete we are now in the position to understand the pictures that we receive and the theories we invent to explain the spectra.

## CHAPTER REVIEW

**The eye:** angular resolution about 1 arc min.

**Atmosphere:** radio and visible light are transmitted, ultraviolet, X- and gamma rays are absorbed. Infrared is partly transmitted at some wavelengths. Scattering: more for short (blue) wavelengths.

**Refraction:** bending of light when passing from one transparent medium to another.

**Lens:** brings light to a focus, but different colors have different focal points.

— Achromat: lens that consists of two or more components to correct for chromatic aberration.

**Concave mirrors:** can also produce images, and position of image does not depend on wavelength. Widely used for telescopes.

**Refractors and Reflectors:** two basic types of telescope, apparent size of image depends on angular magnification of telescope.

**Telescope Mounts:** equatorial mount has one axis parallel to the Earth's axis of rotation, other axis at right angles to this. Alt-azimuth mount requires two continuous drives, to compensate for the Earth's rotation.

**Radio Telescope:** similar in principle to optical reflectors with spherical or parabolic reflecting surface.

**Angular Resolution of Telescope:** depends on ratio (wavelength of radiation)/(size of telescope objective lens or mirror). Good resolution for optical telescopes but poor resolution for single radio dishes.

**Arrays of radio dishes:** constitute radio interferometer.

**Use of lunar occultations:** to obtain precise information on position and size of sources.

**Detection of Radiation:** photographic or electronic (depends on wavelength). Infrared detectors require liquid nitrogen or helium to cool them to required temperatures.

**X- and gamma rays:** individual photons detected.

**Optical wavelengths:** measured by use of filters, prisms, or spectrometers.

## NEW TERMS

**achromat**
**angle of incidence**
**angle of reflection**
**angular magnification**
**angular resolution**
**chromatic aberration**
**electron volt**
**equatorial mount**
**focal length**
**focus**
**image**
**index of refraction**
**objective**
**polar axis**
**prime focus**
**radio interferometer**
**reflector**
**reflection**
**refraction**
**refractor**
**scale of image**
**seeing**
**spectral resolution**
**telluric lines**
**VLA**
**VLBI**

## PROBLEMS

**1.** The atmosphere transmits visible light. Why is it still desirable to fly a telescope in the Space Shuttle or other satellites?

**2.** Why must astronomical observations at some wavelengths be made above the atmosphere? For which wavelengths is this essential?

**3.** Why are the largest modern telescopes reflectors rather than refractors?

**4.** Why is the alt-azimuth mount in inexpensive telescopes inconvenient for tracking a star or a planet?

**5.** A satellite with a diameter of 10 m is in orbit 150 km above the Earth's surface. The satellite is tracked through a ground-based telescope with a focal length of 3 m. How large an image is formed by the telescope objective?

**6.** The primary mirror of the Palomar telescope has a focal length of 16.7 m. How large an image will it produce of Jupiter?

**7.** The human eye has a resolving power of 1 arc min.

a. How small an object has this apparent size if viewed at a distance of 30 cm (1 ft)?
b. How small a surface feature can be seen on the surface of the Moon at this limit of resolution?

**8.** A pair of binoculars is listed as $7 \times 50$. This means that it magnifies seven times and that the aperture of its objective lenses is 50 mm diameter. For a wavelength of 5000 Å, what is the resolving power of these binoculars expressed as an angle? At a distance of 400 ft, what is the smallest detail that can be resolved?

**9.** The Earth's atmosphere degrades the images that telescopes produce, and 1 arc sec resolution is often the limit attainable.

a. How large a telescope objective is needed to match this resolution for a wavelength of 5000 Å?
b. What is the advantage gained by using telescopes much larger than the size you calculated in (a)?

**10.** The largest optical telescope has a diameter of 6 m. Why are radio telescopes very much larger?

**11.** What angular resolution can be attained by a radio telescope array that has antennae 1000 km apart and operates at a wavelength of 10 cm? What distance does this resolution correspond to if Jupiter is being observed when it is closest to the Earth?

**12.** The VLA often operates at a wavelength of 6 cm. What is the angular resolution attainable when the antennae are located over a distance of 30 km?

**13.** A satellite in orbit 150 km above the Earth's surface is used for a photographic survey. If it is planned to see detail as small as 30 cm on the ground, what must be the minimum diameter of the camera lens? Assume of wavelength of 5000 Å.

**14.** A small telescope has a primary mirror of 15 cm diameter. Will this be able to resolve a double star where the stars are 2 arc sec apart? Assume a wavelength of 5000 Å.

CHAPTER

# 10

# Space Exploration: New Opportunities for Astronomy

Mention of the *space age* will usually produce the responses of manned flight, interstellar travel, and space colonies. Indeed, the space age has often been greeted as the materialization of the visions of Jules Verne and Arthur C. Clarke, a natural stage in the progress of our liberation from confinement to the Earth. Although manned flights have circled the Earth and gone to the Moon and back, more ambitious manned travel still remains very far in the future. If this were all that the space age comprised, it would represent technical achievement involving impressive new technologies and unsurpassed quality in engineering design, construction, and operation—analogous to the Pyramids and Stonehenge as a summation of a culture but not necessarily leading to new scientific knowledge or insights.

However, the space age is literally reshaping our view of the universe. From our manned and unmanned space vehicles, we are obtaining not simply more or better quality astronomical information—we are now able to undertake measurements that are totally different from those that we could make with even the largest Earth-bound telescopes. New experimental techniques can now be used—some new only to astronomy, others developed primarily for the space sciences. As a result, our knowledge of the Moon, the planets, our solar system, and stars and galaxies beyond has expanded in unexpected directions. We must now devise theories and cosmologies that include physical processes very different from those that we thought sufficient as recently as midcentury. In this chapter, we shall survey these methods and the new kinds of information that have now become available, and in later chapters we shall see how they are contributing to our understanding.

**Figure 10.1** Lunar panorama, Apollo 16. Astronaut Charles Duke (lunar module pilot) is standing near the rim of Plum Crater (left) and is also seen walking to the right. The time lapse between the pictures can be gauged from the change in direction of the astronaut's two shadows (NASA).

## 10.1 LUNAR SAMPLES

No science-fiction writer managed to invent anything remotely similar to the Lunar Curatorial Facility (LCF). Located in Houston at the Johnson Space Center and operated by the National Aeronautics and Space Administration, the LCF has received and processed all samples brought back on the Apollo flights.

In addition to the Apollo samples, 130 g of lunar material were returned by the Russian Luna 16 and Luna 20 vehicles (Fig. 10.3).

Each year the Lunar and Planetary Institute (also operated by NASA) has been host to a conference at which scientific results obtained during the preceding twelve months are described. The first conference was a tour-de-force. Although held less than 6 months after the first landing (Apollo 11) and barely 3 months after the first samples were distributed, 143 papers were presented based on a most impressive range of experimental techniques.

Since then about 200 teams of scientists from the U.S. and fifteen foreign countries have obtained samples, of average size 100 g, but the bulk of the lunar material is being kept for future experiments, because theories change and suggest new experiments and experimental techniques improve.

**Figure 10.2** Apollo 17 Lunar Module and Lunar Rover with astronaut Eugene Cernan. In the background is the South Massif, 5 km away and 2500 m above the valley floor (NASA).

*Table 10.1* LUNAR SAMPLES FROM APOLLO MISSIONS

| | | *Total Sample Weight (kg)* | *Percent Allocated for Study* | *Weight Allocated (kg)* |
|---|---|---|---|---|
| Apollo 11 | July 1969 | 21.7 | 19.0 | 4.1 |
| Apollo 12 | November 1969 | 34.4 | 7.8 | 2.7 |
| Apollo 14 | January 1971 | 42.9 | 6.6 | 2.8 |
| Apollo 15 | July 1971 | 76.8 | 3.9 | 3.0 |
| Apollo 16 | April 1972 | 94.7 | 3.1 | 2.9 |
| Apollo 17 | December 1972 | 110.5 | 3.1 | 3.4 |
| Totals | | 381.0 | 5.0 | 18.9 |

Until 1959 our observations of the Moon were carried out at a respectful distance, but since then we have had lunar orbiters and landers and now we have samples in our laboratories, and under our microscopes and in our test tubes. What are the experimental methods that can now be applied to these lunar samples? *Wet chemistry* as it is so aptly labeled leads to the identification of the proportions of the 100-odd known chemical elements and their more numerous compounds through the use of well-known tests based on chemical reactions. Modern chemical analysis can work with micrograms; the difference between having no sample and a sample of even one-millionth of a gram is immense.

Physical techniques can, of course, also be used. In contrast to the wet chemistry, the combinations of the atoms are often left unchanged, but the physical properties are examined. Crystal structures can be identified through X-ray and optical examinations, and the spacing of the various atoms in their crystals can be deduced. When some chemical compounds form under very hot conditions and then cool and solidify, the resulting crystalline form—the arrangements of atoms in the crystal—is strongly influenced by the prevailing temperature and pressure. Accordingly, if we identify a particular crystal type, we can at once specify the range of temperatures and pressures during formation.

Measurements of radioactivity provide information not only on the well-known elements such as radium and uranium and thorium, but also on the equally important lighter atoms, such as potassium. Their isotopic abundances are used along with abundance of their nonradioactive products, in calculations that provide us with values for the ages of the lunar materials.

**Figure 10.3** Apollo 15 mission: collecting samples from a lunar boulder (NASA).

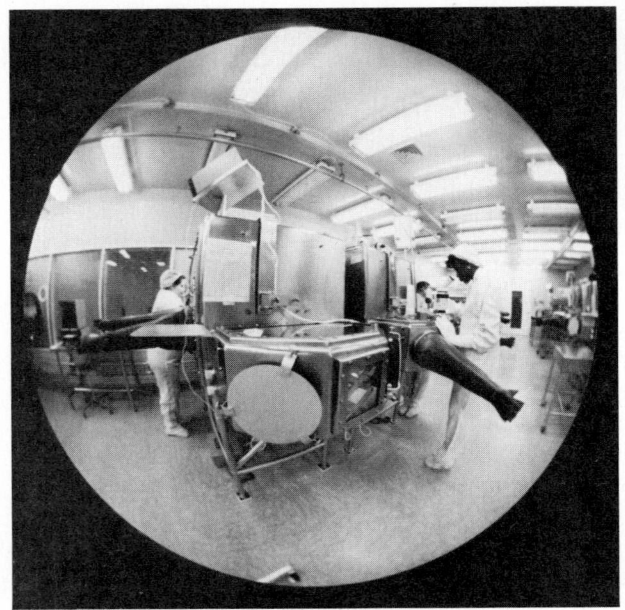

**Figure 10.4** Processing lunar samples at the Lunar Science Institute (NASA/Johnson Space Center).

We can examine some of the sample in mass spectrometers in which a minute quantity of the sample is vaporized into hot, high-speed atoms that can be sent through electric and magnetic focusing fields to bring each isotope to a separate focus; isotopes can be identified and their relative proportions measured.

We can take individual crystals, often far smaller than 1 mm across, polish them and etch them in caustic soda to render visible the tracks left by bombarding cosmic ray particles that have come from the Sun and from farther away in the galaxy. The tracks can be accumulated in crystals for hundreds of millions of years, until revealed by etching. From many measurements made near the Earth, we know how many of these particles arrive each day. Thus when we have measured the total number of tracks in a crystal, we can calculate its *exposure age* that is, how long it must have taken to accumulate the observed number of tracks.

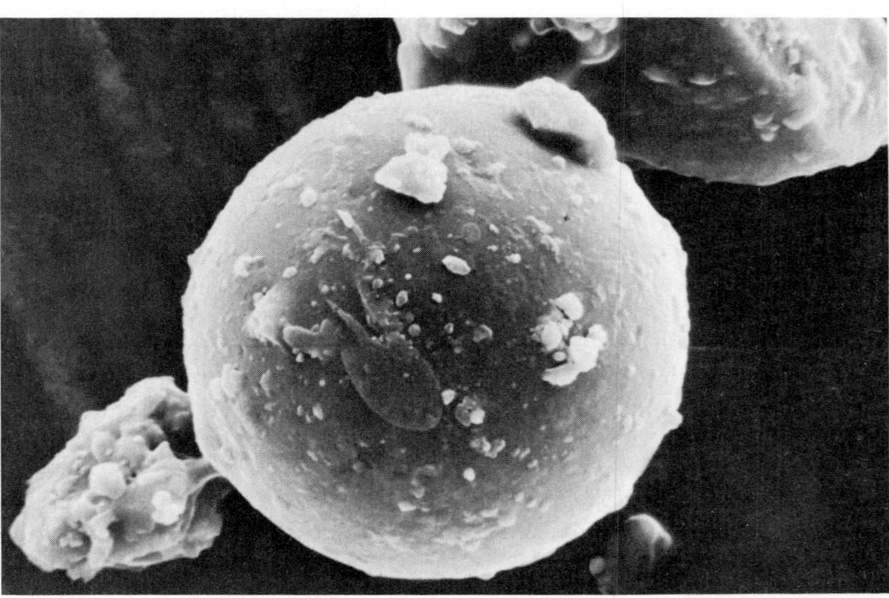

**Figure 10.5** Glass spherule from the lunar surface. Not all lunar samples were large rocks. Samples of the lunar soil were found to contain many minute constituents probably formed during the bombardment of the lunar surface. This electron microscope photograph shows a spherule that has a diameter of 1/100 mm (Robert M. Walker, Washington University).

In certain types of crystal, some energy lost by cosmic ray particles can be stored by electrons. Carefully controlled heating releases this stored energy through the emission of light, a phenomenon known as **thermoluminescence** (TL). Here again is an indicator of the total number of cosmic ray particles that have hit the crystal. The track and TL techniques have been applied to core samples obtained by driving a hollow tube into the lunar surface and thus preserving the profile of changes below the surface when the tube is removed with its sample intact inside. The variation of the TL and track numbers with depth allows us to deduce how rapidly (or slowly) the upper layers of the crust were turned over (*gardened*) by the vigorous but intermittent impacts of meteorites over millenia.

From all of this, we build up our increasingly complex picture of the lunar crust, those layers closest to the surface. At this stage, the separation between astronomy and geology becomes blurred, for modern planetary astronomy uses a range of geologic techniques and, in return, adds to our knowledge of the Earth's evolution.

## 10.2 MEASUREMENTS ON THE LUNAR SURFACE AND FROM LUNAR ORBITERS

Lunar samples are obviously important, but we should not overlook the complementary information that we can obtain from experiments actually carried out on the surface of the Moon and from observations made from lunar orbiters. For several years before the first Apollo landing, a series of flights, first unmanned and later manned, provided increasing detail of the appearance of the lunar surface. Starting with Luna 3, which in October 1959 transmitted the first pictures of the far side of the Moon, the mapping has been extended over the entire surface. The Ranger series of vehicles (1964 and 1965) was designed to crash into the lunar surface, transmitting the pictures with increasing detail virtually until the moments of impact.

The later vehicles, in the Luna and Surveyor series, all landed softly and continued to operate and transmit data from the surface. Scientific payloads became increasingly complex, graduating from television monitors to ingenious experiments for chemical and physical analysis, performed by remote control under radio command from the Earth. Although these latter experiments were important at the time, their results have been quite overshadowed by the measurements on the returned samples. The visual surveys, however, have no substitute and are vital to our attempts at understanding the early period of the Moon's existence when it was being so heavily bombarded, presumably by meteorites. The detail seen in these local surveys can never be seen from Earth.

The lunar orbiters have, in addition, provided us with very different kinds of information. Precision tracking of the vehicles in their paths revealed completely unexpected orbital irregularities which have been attributed to variations in the density of the lunar material beneath the surface. The locations of these concentrations of mass (*Mascons*) and some of their properties are now known, and we have some theories of their origin.

The absence of a lunar atmosphere is well known, and we are all familiar with the specially designed pressure suits worn by the astronauts. Not as well known is the effect of an atmosphere on radiation. On the Earth, the radiation emitted by naturally occurring uranium and radium and other radioactive atoms in layers close to the surface is quickly absorbed in the air, but on the Moon this radiation can be detected from quite some distance above the surface. As a result, lunar orbiters could survey the radioactivity of the lunar surface even from the altitude of their orbits. These measurements go far beyond the very limited surveys that could be carried out in the vicinity of the few landings.

**Figure 10.6** Lunar Ranging Experiment: laser beam from the 107 in. telescope of the McDonald Observatory, University of Texas (McDonald Observatory, University of Texas).

Some experiments have been left on the lunar surface to continue operating even after the departure of the astronauts, with the data being telemetered (transmitted) back to Earth. So, for example, it has been known for some time that the Moon has no magnetic field, and it was something of a surprise to find in samples brought back with Apollo 11 evidence for the presence of a magnetic field some billions of years ago. A magnetometer was therefore included among the instruments in later missions. The Apollo 15 magnetometer telemetered data for about 6 months, well after the departure of the visitors. Seismic observations with instruments on the surface allow us to understand some of the properties of the interior of the Moon that are otherwise completely inaccessible from the surface. Seismic data have been telemetered back, and Apollo 17 data have included the seismic signals produced by the impacts of the Saturn rocket booster and the lunar module ascent stage, as well as moonquakes and meteorite impacts.

Less dramatic but equally important have been the other on-site measurements: Temperature variations through the lunar day and night provide a measure of the heat flow from the interior and the thermal conductivity of the crust. The solar wind, a variable stream of electrically charged particles continually blowing away from the Sun, induces effects on the lunar surface that tell about the electrical properties of the surface.

Several landers carried **retroreflectors**, special mirrors that return light beams in the direction from which they came (Figs. 10.6 and 10.7). Using large telescopes, laser beams can be directed from Earth to these reflectors and back. It is then possible to send coded signals via a laser beam and to time this signal's round trip with great precision. Since the speed of light is accurately known, we are in a position to measure the distance of the Moon from the Earth and detect small changes in this distance. These measurements can be used to monitor the slowly changing Earth-Moon distance. Changes as small as a few centimeters can be detected.

**Figure 10.7** Retroreflector placed on the lunar surface during the Apollo 11 mission. A laser beam, sent from Earth and reflected back, can be used to measure the Earth-Moon distance with precision (NASA).

## 10.3 ABOVE THE ATMOSPHERE: AIRPLANES, BALLOONS, AND ROCKETS

Missions to the Moon and closer planets have attracted so much attention that we tend to overlook the scientific contributions that have come from other vehicles.

Some astronomical measurements have a modest need: simply to get above most of the atmosphere to detect the heavily absorbed IR or UV or X-rays. For over 30 years, high-altitude balloons have been used, and these can now routinely lift payloads of 2 tons to altitudes of 140,000 ft (Fig. 10.8). The balloons are enormous—30 million cu. ft ($8 \times 10^5$ $m^3$) volume, made of polyethylene less than $\frac{1}{1000}$ in. thick and filled with helium. Scientific ballooning in the U.S. is now coordinated with the National Scientific Balloon Facility in Palestine, Texas. Flights have been made all over the world. Typical flight durations are 10 to 20 hr, although much longer flights can be obtained by taking advantage of favorable winds in the stratosphere. Telescopes on these flights operate under radio control from the ground, and the telemetered data are recorded at the ground base. Telescopic pointing of arc minute accuracy can be maintained. One of the earliest systems (in the late 1950s) was Princeton University's Project Stratoscope, that yielded major solar discoveries. More recently the Smithsonian Astrophysical Observatory and the University of Arizona have collaborated in an IR project. Both of these systems have used reflecting telescopes of about 1 m diameter, and there have been many flights with much smaller systems.

**Figure 10.8** Launch of the balloon with the Arizona/Smithsonian 1-m IR telescope. As the balloon rises, the helium will expand to fill out the entire balloon volume. This photograph was taken shortly before the release of the telescope gondola from the launch vehicle (Giovanni G. Fazio, Smithsonian Astrophysical Observatory).

**Figure 10.9** Balloon inflation and launch preparation for flight of cosmic-ray experiment. The gondola weighed over 2 tons, and was lifted to over 35 km altitude (Herb Weitman, Washington University).

(a)

(b)

**Figure 10.10** (a) Kuiper Airborne Observatory. Designed primarily for IR observations, the 91-cm telescope observes through the square aperture visible in the plane just ahead of the wings (NASA/Ames). (b) Close-up of the telescope aperture showing the mount for the secondary mirror. Turbulence of the airflow across the telescope aperture is minimized by the row of deflectors (at left) (NASA/Ames).

Airplanes with scientists on board are also used. For IR studies, NASA operates a Lear Jet with a 30 cm telescope and a modified C141 transport with a 91 cm telescope (Fig. 10.10). Both of these heavily instrumented planes operate at about 40,000 ft altitude, but their flights are limited to about 7 hr. In recent years, U2 planes which can fly at 90,000 ft have found astronomical use, collecting microscopic dust particles (probably from comets or micrometeorites) and carrying monitors for cosmic radio signals.

Far greater heights can be reached by rockets but only during very short flights. With a peak altitude of around 300 km, a typical rocket can collect about 5 min of useful data. To achieve stability in flight, the rocket is given a high initial spin, but for astronomical observing it must be slowed down to permit star-trackers to provide accurate pointing information. Their extreme altitudes make the rockets useful for UV work or for some X-ray and cosmic-ray observations when brief phenomena need to be observed. For example, during some periods of high solar activity, rockets have been kept ready for launch should a large but unpredictable flare erupt on the solar surface. Neither balloons nor airplanes can respond so quickly, and a satellite could well be on the wrong side of the Earth just when needed.

It might seem that we have stretched the term *space* when we include these quite local flights—balloons, airplanes, rockets—and so we should point to another important role that they play. Many of the highly successful systems carried on satellites have been developed from earlier balloon prototypes, where the systems can be tested and modified at a relatively modest cost. The satellites and space probes have attracted most of the publicity, but the pioneering role of the less glamorous vehicles was essential.

## 10.4 SATELLITES AND SPACE PROBES

A detailed study of the vehicles used for space research would be a fascinating survey of high technology—fuels, computers, guidance systems, telemetry—but it would also be too wide a digression from our main interest. A quick survey is presented here to show what kinds of astronomical observations have become possible.

The first satellite placed in orbit around the Earth was launched by the U.S.S.R. in 1957 as part of the program of the collaborative International Geophysical Year. Sputnik was typical of near-Earth satellites—flying at an altitude of about 100 km and orbiting the Earth in about $1\frac{1}{2}$ hr. Scientific payloads were small, a few pounds. By now scientific satellites have become massive observatories. For example, the High Energy Astronomical Observatory (HEAO) series was designed to detect X-rays, gamma rays, and cosmic rays and carried scientific payloads weighing well over a ton.

## BOX 10.1

*Orbits of Near-Earth Satellites*

We can use Kepler's Third law to calculate the time taken for satellites to orbit the Earth (Fig. 10.11). As a satellite's mass is so much smaller than that of the Earth, the following formula can be used:

$$P^2 = \frac{4\pi^2}{GM} a^3$$

In this, the mass of the Earth is $M$, $5.98 \times 10^{27}$ g, and G is the gravitational constant $6.67 \times 10^{-8}$ dyne cm$^2$/g$^2$. For a circular orbit, the radius will be the sum of the Earth's radius ($6.378 \times 10^8$ cm) and the altitude of the orbit above the Earth's surface. We shall take 100 km ($10^7$ cm) as a typical orbit altitude, so that the orbit radius is the $6.478 \times 10^8$ cm. Then

$$P^2 = \frac{4\pi^2(6.478 \times 10^8)^3}{(6.67 \times 10^{-8})(5.98 \times 10^{27})}$$

from which

$$P = 5187 \text{ sec}$$
$$= 86.45 \text{ min}$$

The same formula can be used for calculations for other kinds of orbits. Some satellites are **geostationary**, or **synchronous**; that is, they stay directly above the same place on Earth. But *staying directly above* is only in appearance; the satellite must actually be in orbit with an exact 24-hr period. Other satellites are in very elliptical orbits that take them from altitudes of 100 km at closest approach (**perigee**) to maximum distances as large as 30,000 km (**apogee**). Kepler's law can be used for all of these orbits, provided that the appropriate values of $a$ are used.

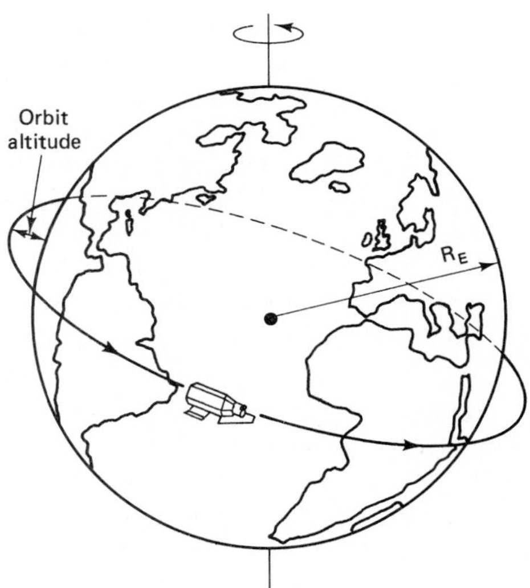


**Figure 10.11** Earth-orbiting satellite. The radius of the satellite orbit is the sum of the Earth's radius and the orbit altitude. Kepler's Law relates the orbit radius, the periodic time, and the mass of the Earth. The inclination of the plane of the orbit to the equatorial plane depends on the launch conditions and can be designed to suit the purposes of a particular mission.

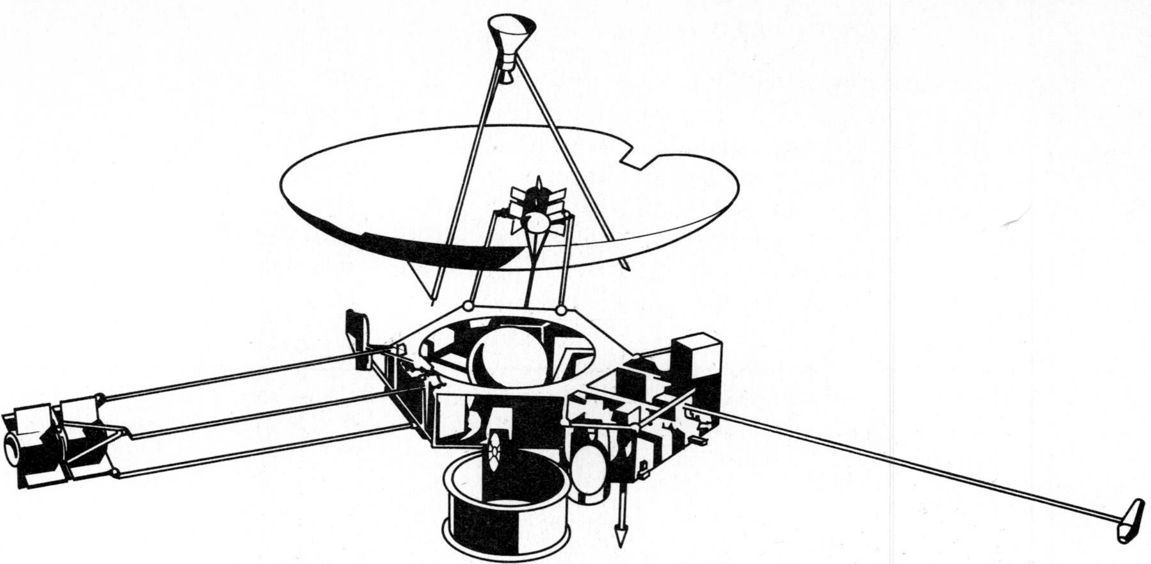

**Figure 10.12** Pioneer Jupiter Spacecraft is typical of many probes that have been used to explore the farther reaches of the solar system (TRW Defense and Electronics).

Probes have visited every planet from Mercury to Saturn, and Voyager 2 is now on its way towards Uranus. What is so different about these missions? They can carry out observations during their flights and not only at their destinations. As a result, we have been able to explore conditions in interplanetary space—the flow of the solar wind and cosmic rays and the strength of magnetic fields, and the changes in all of these both with distance from the Sun and with time.

Some probes are designed to land on planets. Planetary atmospheres and surface features can be examined in detail quite impossible from the Earth. Some probes fly past one planet on the way to another, using the gravity of the first to deflect them into just the right path towards the second. From these varied missions we can measure wind speeds and observe lightning in an atmosphere, note seasonal changes such as affect the Martian pole caps, discover volcanic activity, see surface craters (or the lack of them), and measure planetary magnetic fields (if any). We can also observe the natural satellites of these planets and even discover new rings. Television cameras can scan the sides of planets and satellites not observable from Earth. Finally, no less important or costly, we can search for signs of life.

Among the many planetary probes, a few deserve special mention. The Pioneer 10 and 11 vehicles were launched in 1972 and 1973, respectively. After about a year's travel, they reached Jupiter, and for the first time, we obtained local sampling of the regions of trapped radiation, similar to the Van Allen belts around the Earth but containing many more particles. Saturn was reached in 1979, and television viewers were treated to the remarkable views of that planet, its rings and major moon, Titan.

In 1975 Viking I and II were launched toward Mars. Arriving a year later, they first went into orbit around the planet. Pictures transmitted back to Earth were used for the selection of landing sites. The vehicles then separated, the landers settling gently into the surface, (Fig. 10.13) and the remaining orbiters staying aloft to relay information back to Earth and accept commands from Earth.

The two Voyagers set off in 1975, reaching Jupiter in 1979, and Saturn in 1980 and 1981. Voyager 2 is now on its way toward an approach to Uranus in 1986. Their pictures sent back to Earth were more than simply spectacular. They revealed previously unsuspected occurrences such as volcanic activity on Jupiter's satellite, Io, and a bizarre complexity in Saturn's ring structure.

**Figure 10.13** Viking lander (during testing on Earth). Two of these units landed on the surface of Mars. Surface samples were collected with this scoop and then analyzed (TRW Defense and Electronics).

Some satellites have been designed primarily for solar studies. During the 1960s and early 1970s, the Orbiting Solar Observatories (OSO) made observations at wavelengths shorter than the visible, in the UV, X-ray, and gamma-ray parts of the spectrum. More recently, the Solar Maximum Mission (SMM) was launched to cover the period of maximum solar activity during its 11-year cycle.

Completely different types of spacecraft have been used for observations of much more distant objects beyond the solar system. These spacecraft have been placed in Earth orbit above the atmosphere and have produced a remarkable yield of scientific data. For example, the second Orbiting Astronomical Observatory (OAO 2) operated in the UV (1200 to 4000 Å), and OAO 3 ("Copernicus") in the UV and X-ray regions. The International Ultraviolet Explorer (IUE) viewed radiation in the wavelength range between 1150 and 3300 Å.

In 1971 the Uhuru satellite produced the first major sky survey at X-ray wavelengths (photon energies between 2 and 20 keV), and this was followed by the Einstein satellite (HEAO 2) in 1978 (Fig. 10.14), with its greater

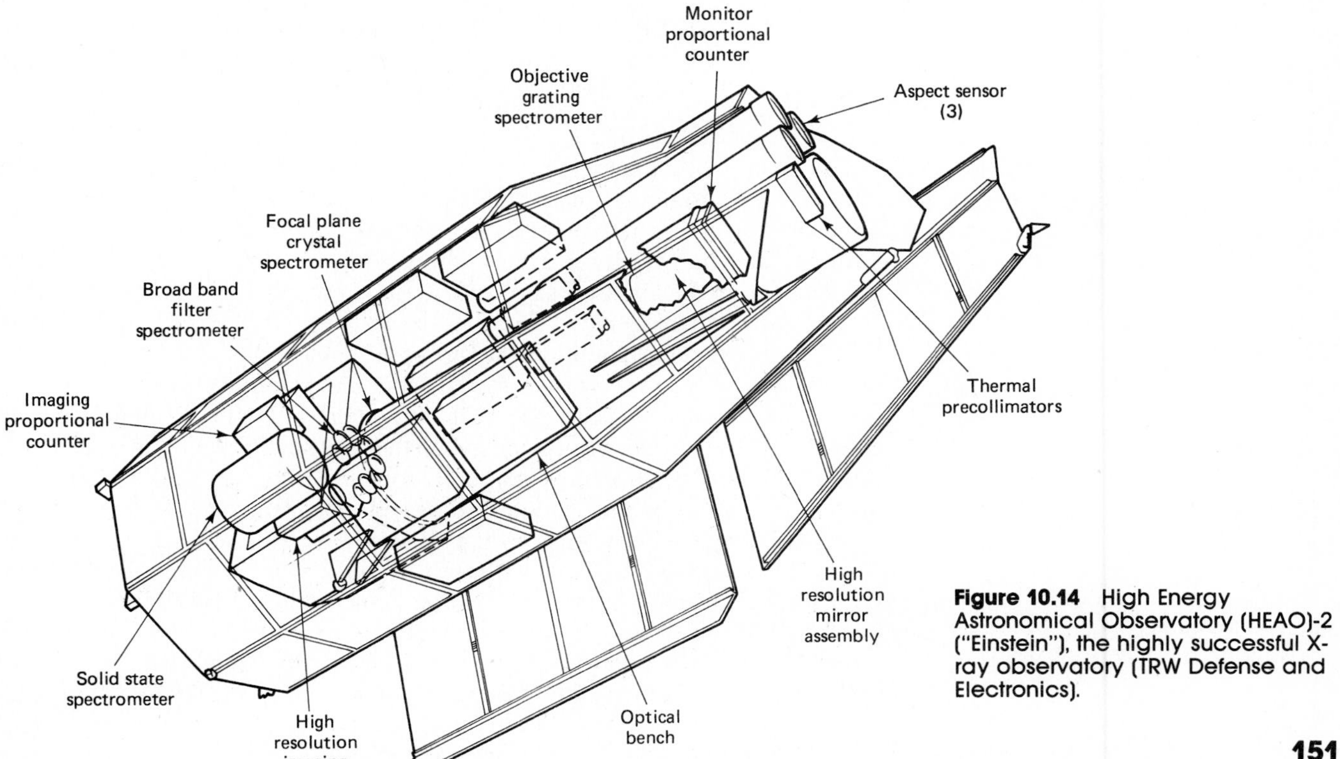


**Figure 10.14** High Energy Astronomical Observatory (HEAO)-2 ("Einstein"), the highly successful X-ray observatory (TRW Defense and Electronics).

**Figure 10.15** Installing a cosmic-ray experiment into the HEAO-3 satellite (TRW Defense and Electronics).

sensitivity and considerably improved angular resolution (2 arc sec). At the highest energies and shortest wavelengths, gamma-ray photons have been detected by the U.S. satellite SAS-2 and the European COS-B.

In addition to the many unmanned scientific satellites and space probes that have been described, there have been relatively few that have carried astronauts who have made scientific observations. The Apollo series, ending in 1972, focused on lunar science. Astronauts in Skylab (1973–1974) made lunar, solar, cometary, stellar, and Earth observations.

The later chapters of this book are very different from what they would have been had this book been written in 1955 before the first space flights, or even as recently as 1975.

## 10.5 FURTHER PROJECTS

The cost of space exploration is high. The annual NASA budget is in the region of 3 to 4 billion dollars. This is large, but many ambitious proposals do not develop into actual hardware, and many are reduced substantially in scale. Despite these fiscal uncertainties, long-range planning is absolutely necessary: It takes many years to develop a complex system and build the components. Launch dates have to be carefully chosen, to minimize the amount of fuel needed and take advantage of the changing positions of Earth and planets. A delayed launch might make a mission useless.

Recognizing the existence of many uncertainties between an idea and an actual launch, what are the missions that the scientific community would like to have in the next few years? European astronomers are working on an X-ray satellite, while the infrared astronomical satellite (IRAS), is a collaborative venture between scientists in the U.S., Holland, and Britain. The Space Telescope is now well advanced. It will contain a reflecting telescope with a 2.4 m primary mirror and will be launched from the Shuttle, perhaps in 1986 (Fig. 10.16). A research institute has been established, near Baltimore, for the coordination of its observations and the analysis of its data. The director of the institute is Riccardo Giacconi, who played a major role in the pioneering and consolidation of the X-ray astronomical field, from the early rockets through the later Uhuru and Einstein satellites.

The European Space Agency is now working on the Hipparcos satellite to be launched (if all goes well) in 1986. The function of this satellite is **astrometry**, the very precise determination of stellar positions. Hipparcos is designed to yield positions to 1 or 2 thousandths of an arc second, a hundred to a thousand times better than can now be obtained for most stars.

Halley's comet will be returning to its perihelion passage in 1986. Hopes for a U.S. probe that would rendezvous with the comet have already evaporated, but fly-by missions are being planned by Japanese and European astronomers. Hoped for but still uncertain are other missions, such as a

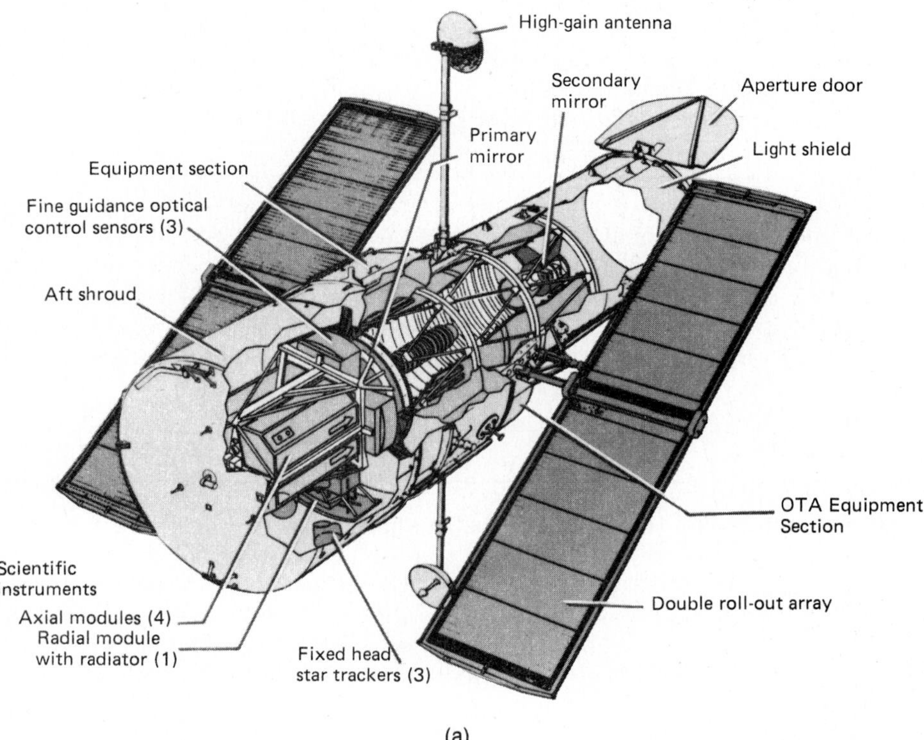


(a)

(b)

**Figure 10.16** The Space Telescope. (a) schematic overall design, and (b) the primary mirror. This telescope is due to be launched into orbit in 1986 (NASA/Marshall).

Mars lander that would scoop up material from the planet's surface and return to Earth, and the Galileo mission to be launched from the Shuttle to circle Jupiter studying its atmosphere. Another ambitious project is the out-of-the-ecliptic mission. Until now, all probes have stayed close to the ecliptic for this is where the planets are. In addition, it is easiest to fly near the ecliptic by taking advantage of the Earth's own orbital speed when launching a probe. Far less is known about the conditions well away from the ecliptic.

Other wavelength regions also figure in NASA's planning of scientific space vehicles. The gamma-ray observatory (GRO) may be launched in the late 1980's, and an Advanced X-ray Astronomical Facility (AXAF) is being designed for 100 times greater sensitivity than Einstein. A radio antenna is being considered for Earth orbit, to be used in conjunction with ground-based radio telescopes to provide very long base line (VLB) surveys. An advanced solar observatory is also at the planning stage.

Although not all of these ambitious projects will progress beyond the planning stage to actual flight, the prospects are still exciting.

## 10.6 SUMMARY

Space flight—close to the Earth and over vast distances—has revolutionized astronomy. Traditional methods simply could not have produced many of the discoveries that now seem so ordinary that they are included in introductory textbooks. The price of this progress is the permanently increased complexity and cost of future observations as the new areas are consolidated. This is an exhilarating and fascinating time to be involved, even to follow.

## CHAPTER REVIEW

**Lunar samples from the Apollo missions:** about 5 percent distributed, the rest retained for further experimenting.

— laboratory techniques can be applied to lunar samples: wet chemistry; physical techniques such as X-ray crystal analysis, mass spectrometry, track etching, thermoluminescence.

**Measurements on the lunar surface and from orbiters:** photographic surveys, magnetic and gravity surveys (*mascons* discovered), radioactivity survey.

— retroreflector for ranging measurements; seismometer to detect moonquakes.

**Airplanes, ballons, and rockets:** complement satellite observations. Can get high enough for cosmic-ray, IR, UV, and X-ray observations. Very large experiments can be carried by balloons. Airplanes can carry telescopes to about 40,000 ft. Rockets can get to great heights but have very short observing times.

**Satellites: different types for different purposes:** Earth orbiters, in close orbits and very distant orbits. Planetary probes, orbiters, and landers can also investigate interplanetary space while en route. Orbiting observatories for stellar and galactic studies.

## NEW TERMS

**apogee**
**astrometry**
**geostationary**
**mass spectrometer**
**perigee**
**retroreflector**
**thermoluminescence**

## PROBLEMS

**1.** What kinds of astronomical measurements can be carried out from airplanes or balloons but not from the Earth's surface?

**2.** There have been photographic surveys of the far side of the Moon, but no manned landings and no samples brought back from that side. Why?

**3.** Apart from the collection of lunar samples, describe two experiments that have been carried out on the lunar surface that cannot be performed without landing on the Moon.

**4.** List four types of measurements that can be made on samples returned from the Moon that cannot be carried out by telescopic observation from Earth.

**5.** A satellite in near-Earth orbit takes about $1\frac{1}{2}$ hr for an orbit. The plane of the orbit is almost fixed in space, but the Earth continues to rotate on its axis within that orbit. As a result, the satellite does not go over the same parts of the Earth's surface on each satellite orbit. How much (in degree) does the Earth rotate per satellite orbit, and how many orbits are needed before the satellite is again over the starting point?

**6.** Communications satellites are placed in orbit above the Earth so that the orbital period is exactly 24 hr. These satellites are called synchronous satellites. Calculate the altitude of such an orbit above the Earth's surface.

**7.** A spacecraft is in orbit with radius 3500 km around Mars. Calculate the orbital time for this spacecraft.

**8.** How long does it take the laser beam to make the round trip from the University of Texas' telescope to the retroreflector on the lunar surface and back again? If this time interval can be measured to an accuracy of 1 part in $10^9$, how accurately (in cm) can we measure the distance to the Moon?

# CHAPTER 11 The Solar System—A Survey

Before entering into our detailed review of the various bodies that comprise our solar system, it will be useful to conduct a quick survey—to identify those bodies, set the scale and, in general, provide the outlines that we shall fill in through the next few chapters.

At the start, our task will be largely descriptive, but after we have reviewed our present state of knowledge of the solar system, we shall turn our attention to those great questions that have tantalized mankind for so long: How old is the Earth? How did the solar system start? Are we alone, or are there other inhabited places either in the solar system or further away? In Chapters 14 to 19 we shall be able to make at least a start on some of the answers, but others will have to wait until later chapters.

## 11.1 THE SUN

If we had to select a typical star from among the hundred billion in our galaxy, the Sun would be a good example. It is a large sphere of very hot and luminous gas composed of the same chemical elements that we find on Earth. With a mass of almost $2 \times 10^{33}$ g (about 300,000 times that of the Earth), it contains 99.86% of the mass of the entire solar system. The Sun's large mass is held together by gravitational attraction, which produces an immense pressure at its center. This pressure, in turn, produces a central temperature of over 10 million degrees—sufficient to sustain the thermonuclear reactions whose released energy both prevents the collapse of the Sun and also keeps it in gaseous form. That energy percolates to the surface and is then radiated as light that we see. The surface that we see is not a rigid, solid boundary. From the outside going in, the gas gets denser and denser and correspondingly more and more opaque. Less radiation can emerge directly from the progressively deeper regions. We have much the same effect in a dense fog: we can see a few feet ahead but not to any great distance. Automobile lights cannot be seen in the fog until they get close to us.

The solar surface that we see in visible light has a temperature of about 5800 K and a diameter of $1.39 \times 10^6$ km (about 109 times the diameter of the Earth). From the Earth the Sun appears to have an angular diameter of about $\frac{1}{2}°$, just about the same apparent size as the Moon.

The total energy radiated in each second by the Sun is $3.83 \times 10^{33}$ ergs ($3.83 \times 10^{23}$ kilowatts power). Each square centimeter of the solar surface radiates $6.3 \times 10^{10}$ ergs per second. By the time that this has spread out and reached the Earth, only 1370 watts ($1.37 \times 10^{10}$ ergs/sec) are actually received on each square meter at the top of the atmosphere. (Practical applications of solar power, therefore, require very large collector areas.)

Above the visible solar surface, very hot gases extend and expand for great distances and some (the solar wind) drifts out well beyond the Earth.

## 11.2 THE PLANETS

The complicated way in which the planets appear to move provided the focus for classical astronomy, but our interest in those objects now extends far beyond the simple catalog of their orbits. Nine planets are now known. In addition to the Earth, five have been known for millennia: Mercury, Venus, Mars, Jupiter, and Saturn. There seems to have been no curiosity regarding their small number; no one raised the question whether there were other planets in addition. Even after the invention of the telescope, no search for new planets was conducted, and it was only well after the accidental discovery of Uranus by William Herschel in 1781 that this line of research gradually emerged.

Neptune was discovered in 1846 as a result of a search for the cause of the irregularities in Uranus' orbit (chap. 8). The obvious next question was whether there were any more planets farther out. Perhaps such an undiscovered planet might first reveal its presence by perturbations it would impose on Neptune's orbit. Several calculations and unsuccessful searches followed before Pluto was discovered in 1930. This story has been well documented, especially in the moving account given by Pluto's discoverer, Clyde Tombaugh.

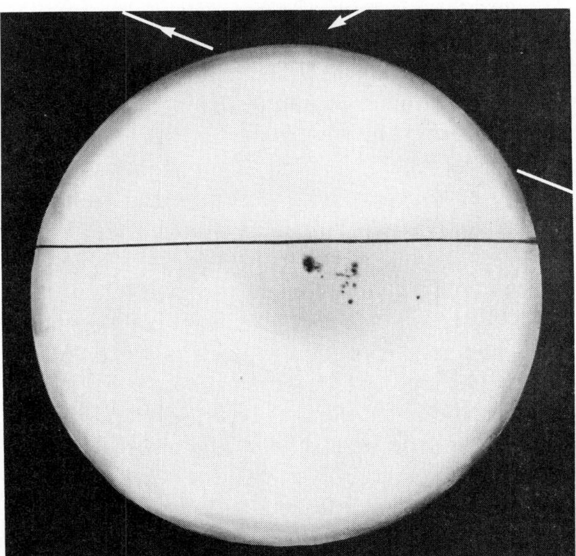

**Figure 11.1** Mercury, photographed during its transit across the disc of the Sun on November 14, 1907. (One arrow points to Mercury, the other indicates the path of the planet) (Yerkes Observatory.)

Venus

Mars

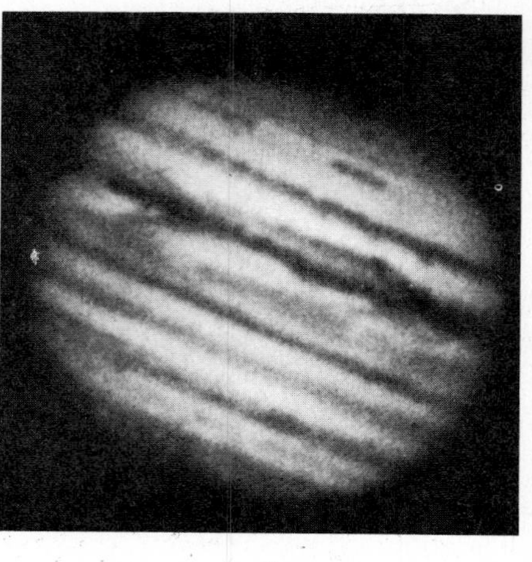

Jupiter

Saturn

Pluto

**Figure 11.2** A selection of planets: the crescent of Venus (Palomar Observatory), Mars, Jupiter, and Saturn (Lick Observatory), and Pluto (Palomar Observatory photograph).

**Figure 11.3** Sir William Herschel (1738–1822), discoverer of Uranus, which he suggested be named the Georgian planet after his patron George III (Yerkes Observatory).

Percival Lowell, a wealthy Bostonian with an amateur interest in astronomy, had been stimulated by reports of the sightings of canals on Mars. He seems to have been among the first to recognize the need for major observatories to be located at sites chosen for their seeing rather than simply for convenience of access, and he selected the then-isolated Flagstaff in Arizona for his new observatory.

Although Mars dominated his initial efforts (see chap. 15), from around 1902 Lowell became increasingly concerned with the possibility of finding a trans-Neptunian Planet X through a systematic search. After his death in 1916 no progress was made for several years. A major search finally got under way in 1929. Clyde Tombaugh was hired because it was thought that he had sufficiently little astronomical training that he would be content to undertake the labor of search without wanting to go off in other directions on his own lines of research. Additionally, since he had no college education, his pay could be less than that of a professional astronomer. Tombaugh very quickly became more than simply an assistant.

The search required taking repeated photographs of each region of the sky in a broad band centered on the ecliptic plane, where new planets might be expected to have their orbits. Movements of distant objects in the interval between the photographs would show up when seen against the images of the stars whose relative positions changed far more slowly. Tombaugh soon found that he was locating so many asteroids that there was a good chance that a faint planet might be missed. It was his inspiration to devise a schedule for the photographing that greatly magnified the differences between the size

**Figure 11.4** Thirteen-inch telescope with which Clyde Tombaugh discovered Pluto (Lowell Observatory).

of movements of asteroids and planets. With the senior astronomers at the observatory fully occupied with other tasks, Tombaugh resumed his routine comparing plates taken several days apart and looking for traces of movement. On February 18, 1930 he found just that in plates taken in late January (Fig. 11.5).

Pluto's orbit and its other properties make it a strange member of the solar system. More space is devoted to it in chapter 16. Its discovery did, however, revive the question: Are there any more planets farther out than Pluto or perhaps closer to the Sun than Mercury? We cannot categorically exclude this possibility. Calculations and unsuccessful searches suggest that any further planets must be very small or very far and, accordingly, very difficult to detect.

Controlled in their orbits by gravitational attraction, mostly to the Sun but slightly influenced by one another, the planets obey Kepler's third law and thus allow us to calculate the mass of the Sun (chap. 7). Most planets also have their own satellites, so that we can again apply Kepler's Law to deduce the planets' masses. Two planets (Mercury and Venus) do not have satellites, but their masses are now accurately known from observations of

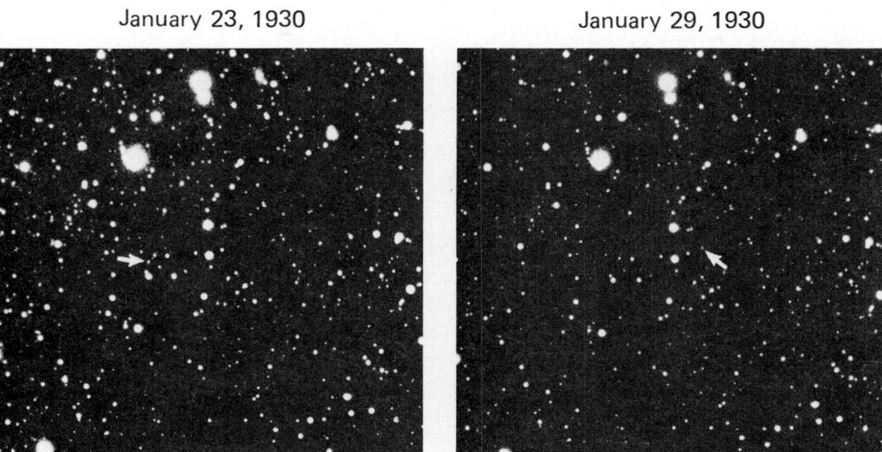


**Figure 11.5** Discovery plates for Pluto. The plates were taken on January 23 and 29, 1930 and examined on February 18. The comparison showed the movement of Pluto (marked by arrows) against the background of stars (Lowell Observatory).

the perturbations they have produced in the paths of space probes that have passed close by.

The Earth's orbit around the Sun provides a natural unit for describing the dimensions of the solar system. The average of the Earth's maximum and minimum distances from the Sun (the semimajor axis of the elliptical orbit) is defined as the astronomical unit.

$$1 \text{ astronomical unit (1 AU)} = 149{,}597{,}892 \text{ km}$$
$$= 1.49597892 \times 10^{13} \text{ cm}$$

For our purposes, the approximate value $1.5 \times 10^{13}$ cm will be sufficiently accurate.

With the masses and dimensions known, we can calculate the densities: mass per unit volume (g/cm$^2$) (Table 11.1), and here we have a surprise: the planets fall into two well-defined groups. The **terrestrial planets** have densities in the range 3.9 to 5.5 g/cm$^3$; the **Jovian planets** (Jupiter, Saturn, Uranus, and Neptune) and Pluto, all have much lower densities. Any theory of the origin of the solar system must be able to explain this, as well as other properties we shall review later.

In their orbits the planets all travel in the same direction. Viewed from far above the North Pole, the planets travel in counterclockwise orbits. At the same time, each planet spins on its own axis, and with two exceptions all of these rotations are also counterclockwise or **prograde**. One exception is Venus, which rotates very slowly in the **retrograde** (clockwise) direction on its axis. The other exception is Uranus, which has its rotation axis only a few degrees from the plane of the ecliptic. As with the densities, so with the spins: A division exists between the terrestrial and the Jovian planets, the former having rotation periods of a day or longer, the latter much less. Pluto is deviant here too: Its rotation period is nearly a week. The planetary orbits define planes that are all close to the ecliptic except for Pluto.

There is another regularity in the scheme of planetary orbits that has been noted but not understood. The orbits lie close to a single plane, but what about the spacing of the orbits? Kepler's Laws do not tell us **where** a planet must have its orbit. What those laws **do** tell us is that once the size of the orbit is specified so is the periodic time (and thus the speed). Is there a formula that relates the sizes of the orbits to one another? Why does Venus have an orbit with an average radius of 0.72 AU, and the Earth an orbit with 1 AU, and nothing between?

A curious numerical sequence was discovered by the astronomer J. D. Titius (in Wittenberg) in 1766 but, it was not widely known until Johan Bode, director of the Berlin Observatory, publicized it in 1772. The sequence is constructed as follows: Start with 0, add 3, then double for each successive member of the series. We thus obtain the numbers 0, 3, 6, 12, 24, 48, . . . .

*Table 11.1* SOME PLANETARY DATA

| | *Semimajor Axis of Orbit (AU)* | *Sidereal Period* | | *Mean Density of Planet (g/cm$^3$)* |
|---|---|---|---|---|
| | | *(days)* | *(years)* | |
| Mercury | 0.39 | 87.97 | 0.24085 | 5.43 |
| Venus | 0.72 | 224.70 | 0.6512 | 5.24 |
| Earth | 1.00 | 365.26 | 1.0000 | 5.52 |
| Mars | 1.52 | 686.98 | 1.8809 | 3.9 |
| Jupiter | 5.20 | — | 1.86 | 1.3 |
| Saturn | 9.54 | — | 29.46 | 0.7 |
| Uranus | 19.19 | — | 84.01 | 1.3 |
| Neptune | 30.06 | — | 164.8 | 1.5 |
| Pluto | 39.53 | — | 248.5 | ≈0.5 |

*Table 11.2* BODE'S LAW

| *Planet* | *Bode's Law Calculation* | *Average Radius of Actual Orbit (AU)* |
|---|---|---|
| Mercury | (0 + 4)/10 = 0.4 | 0.39 |
| Venus | (3 + 4)/10 = 0.7 | 0.72 |
| Earth | (6 + 4)/10 = 1.0 | 1.00 |
| Mars | (12 + 4)/10 = 1.6 | 1.52 |
| — | (24 + 4)/10 = 2.8 | — |
| Jupiter | (48 + 4)/10 = 5.2 | 5.20 |
| Saturn | (96 + 4)/10 = 10.0 | 9.54 |
| Uranus | (192 + 4)/10 = 19.6 | 19.19 |
| Neptune | (384 + 4)/10 = 38.8 | 30.06 |
| Pluto | (768 + 4)/10 = 77.2 | 39.53 |

Now add 4 to each, and then divide the result by 10. The final results are in remarkably good agreement with the sizes of the planetary orbits (in AU) as shown in Table 11.2. The discovery of Uranus in 1781 seemed to confirm the law. Then came the 1801 discovery of the minor planet, Ceres, at 2.77 AU filling the gap in the sequence and giving yet more support. Although Bode's Law gave some guidance to Adams and Leverrier in their calculations that led to the discovery of Neptune, the actual orbit deviates significantly from that expected. Pluto's orbit is even further from the next term in the series. We are left with a puzzle, even if we ignore Pluto, whose orbit poses other problems too. We know of no physical basis for the law. The generally held view is that it represents nothing more than a numerical coincidence, rather than some new and general rule.

We usually use the term law to denote a physical relation, but we know of no law here. Further, in fairness, Titius' role should be recognized. A correct designation would be the *Titius–Bode sequence,* but *Bode's Law* is so firmly entrenched that we shall simply take note of this historical misusage.

**Figure 11.6** Comet West on March 7, 1976 (Lick Observatory).

## 11.3 COMETS, MINOR PLANETS, METEORITES, AND DUST

Whereas the planets contribute about 0.14 per cent of the mass of the solar system, the comets and dust amount to less than one-half of one-millionth of the total. Next to eclipses, comets provide probably the most spectacular sights in the skies, far out of proportion to their masses, which range from less than one-billionth that of the Earth down to not much more than the combined mass of all the people on Earth.

After much dispute, it is now generally accepted that comets are mostly frozen gases with some solid particles included. Most comets have orbits around the Sun that are highly elongated ellipses, almost parabolic, with most of their time being spent at such vast distances from the Sun that they remain cold. They warm up only when coming close in, thus vaporizing some of their constituents that are then seen as their spectacular tails pushed away from the Sun by the solar wind and the pressure of the solar radiation. When close to the Sun and being warmed, the gaseous head of the comet may be from $10^4$ to $10^5$ km across with the tail streaming away from millions of kilometers.

Most comets have such elongated orbits that the time between successive returns may be thousands of years or more, but about twenty periodic comets have been recognized in their returns. Their periodic times range from 3.3 years for Comet Encke to up to 76 years for Halley and 151 years for Comet Rigollet. About five to ten new comets are discovered each year. Most do not develop to the appearance usually conjured up by the term *comet* and so do not attract much attention outside the astronomical circles.

Their orbits may be strongly perturbed when passing close to the giant planets, Jupiter or Saturn, and may even be so altered as to cause them to leave the solar system. The fragile nature of comets has been shown by some that have passed too close to the Sun or Jupiter, where tidal forces have torn them apart. Biela's comet was discovered in 1772 and found to have a period of about 6 years. After several returns it did not reappear when expected in 1865. Brooks' comet of 1889 had an orbit that brought it close to Jupiter and was found to have three small companions. In 1976 Comet West broke up in its perihelion passage (closest approach to the Sun), and several fragments could easily be seen.

The origin of the comets is not known with certainty. One widely accepted theory is due to the contemporary Dutch astronomer, Jan Oort. He postulated that there are billions of comets, left over from the formation of the solar system and in orbits hundreds of times further out than Pluto. A passing star may disturb some of them sending them into new orbits that take them into our well-patrolled and well-scanned inner solar system.

The **minor planets** or **asteroids** differ from the major planets mainly in size. The largest is Ceres whose diameter is about 1000 km. A few hundred have diameters larger than 40 km, and there are probably thousands large enough to be seen, larger than a few kilometers. Most have orbital radii between 2.5 and 3.0 AU in the Asteroid belt between the orbits of Mars and Jupiter, but a few such as Eros have orbits that cross that of the Earth.

**Figure 11.7** Meteorite recovered from the Barringer Crater, Arizona (Yerkes Observatory).

Far below the limits of individual visibility, there are in orbit billions of small solid objects, the **meteoroids**. Some are so small that we 'see' their trails only when their orbits cross that of the Earth and they burn up after entering the Earth's atmosphere. It is estimated that hundreds of millions of these **meteors** (shooting stars) collide with the Earth every day. Some meteors arrive in groups producing showers, and the Earth moves through some groups regularly each year.

From the orbits of the groups of meteors that are seen annually, it has been deduced that most, perhaps all, of these meteors are debris of comets following the original cometary orbits. Thus the disruption of Biela's comet

**Figure 11.8** Zodiacal light (Yerkes Observatory).

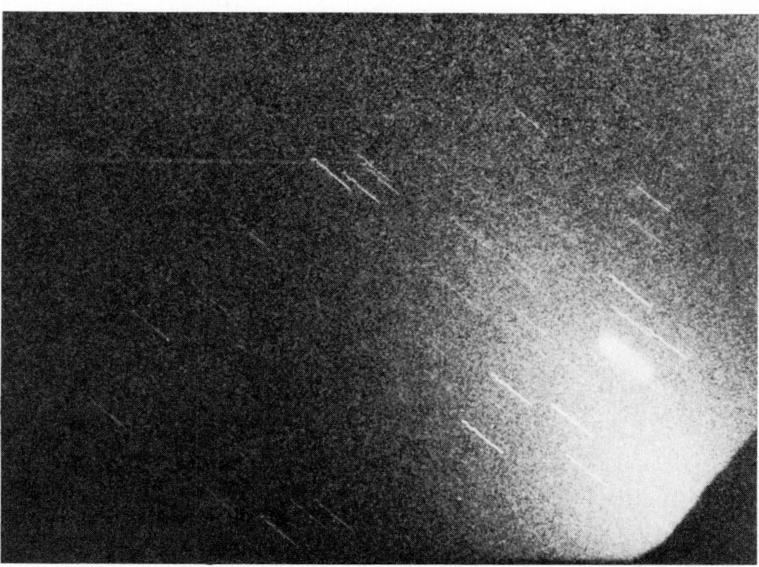

**Figure 11.9** Zodiacal light photographed from the lunar surface during the Apollo 17 mission. A Nikon camera with 55 mm f/1.2 lens and Kodak type 2485 film were used with a red filter. This 40-sec exposure was taken on December 12, 1972 facing east along the ecliptic plane, with the Sun 15° below the lunar limb (lower right corner) (NASA).

was followed by meteor showers that were seen annually for a few years when the Earth crossed Biela's orbit. That shower has now dispersed, but others have persisted for much longer times. Other meteoroids have more nearly planet-like orbits around the Sun.

On occasion a meteoroid may be large enough to reach the Earth's surface despite the intense heating in the atmosphere (Fig. 11.7). Such **meteorites** are rare (only a few thousand have been found), but they provide us with otherwise unobtainable solid extraterrestrial samples whose composition and structure are important clues to the conditions during the very early years of the solar system. Very recently, the remains of the smallest particles, the **micrometeorites**, have also been collected, using high-altitude research planes. These micrometeorites can also be detected by the small impact pits they produce when there is no atmosphere to slow them before they collide with rocks on the lunar surface or with special collectors flown outside spacecraft.

Even smaller than the micrometeorites are the interplanetary dust particles whose presence we observe by the sunlight they reflect. On very clear nights, the faint **zodiacal light** from these particles can be seen along the ecliptic (Figs. 11.8 and 11.9); it can be studied photometrically and spectroscopically, from which we can deduce the sizes of the particles and their abundance.

## 11.4 TEMPERATURES IN THE SOLAR SYSTEM

Every object radiates energy in amounts that depend on its temperature. Wien's Law (chap. 8) relates that temperature to the peak wavelength in the emitted spectrum. So, for example, an object as hot as the Sun radiates strongly in the visible part of the spectrum. The planets in our solar system, however, are at temperatures where most of their own radiation is not in the visible part of the spectrum, but their IR and radio emissions can readily be detected. When we see them with the naked eye, we are actually seeing reflected sunlight.

A relatively simple calculation will allow us to derive an important relation connecting a planet's temperature ($T$) with its distance ($r$) from the Sun:

$$T^4 = \frac{(1 - A)(1.37 \times 10^6)}{4\sigma D^2} K$$

Here $A$ is the albedo, the fraction of light that a planet reflects, $D$ is the distance of the planet from the Sun in AU, and $\sigma$ is Stefan's constant. The derivation is given in Box 11.1.

## BOX 11.1
*Planetary Temperature and Distance from the Sun*

Measurements of the solar radiation above the Earth's atmosphere show that $1.37 \times 10^6$ ergs fall on each square centimeter in each second. This energy flux is termed the **solar constant**. At sea level, somewhat less is received because of absorption in the atmosphere and, especially at higher latitudes, the oblique illumination. For another planet at a distance $D$ (in AU) from the Sun, the corresponding flow of energy will be

$$1.37 \times 10^6 \times (1/D)^2 \text{ ergs/cm}^2 \text{ sec}$$

(see Chap. 8 for the 1/(distance)$^2$ dependence of energy flow).

If a planet has a radius $R$, then it will present an area $\pi R^2$ (cm$^2$) to the beam of solar radiation and thus intercept an amount of energy (Fig. 11.11).

$$(\pi R^2)(1.37 \times 10^6)(1/D^2)\text{erg/sec}$$

Not all of this intercepted energy will be absorbed by the planet. Depending on the nature of its atmosphere and surface, some fraction $A$ (the albedo) will be reflected and only the fraction $(1 - A)$ absorbed. The energy absorbed is thus

$$(1 - A)(\pi R^2)(1.37 \times 10^6)(1/D^2)\text{ergs/sec}$$

Absorbing this radiation causes the planet to warm up; when it is at a temperature $T$, it will be radiating an amount of energy each second given by

$$4\pi R^2 \sigma T^4 \text{ ergs/sec}$$

This assumes that the energy is radiated and absorbed uniformly from the entire surface, not unreasonable for the rapidly rotating planets.

A steady condition will be reached when the energy absorbed is exactly balanced by that radiated

$$\text{energy radiated} = \text{energy absorbed}$$

$$4\pi R^2 \sigma T^4 = (1 - A)(\pi R^2)(1.37 \times 10^6)(1/D^2)$$

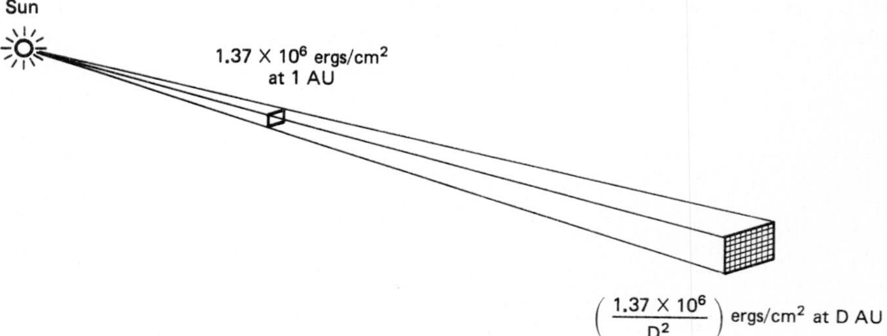


**Figure 11.10** Intensity of sunlight decreases inversely with the square of the distance from the Sun. At the Earth's orbit (1 AU), the intensity is $1.37 \times 10^6$ ergs/cm$^2$ sec. When the sunlight reaches a distance $D$ (AU), the intensity will be down to $1.37 \times 10^6/D^2$ ergs/cm$^2$ sec.

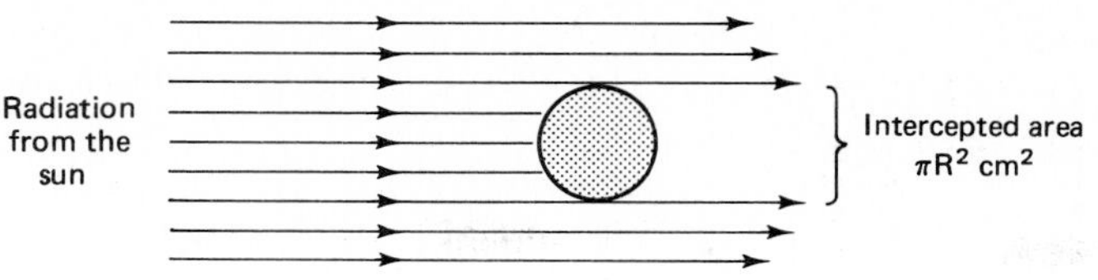


**Figure 11.11** A round object such as a planet presents an area $\pi R^2$ to the beam of solar radiation, and all of the radiant energy passing through this area will be either absorbed or reflected by the planet.

Note that the planet's size, $R$, enters on both sides of this equation and thus cancels, so that the final result will depend only on the distance from the Sun, $D$, and the albedo, $A$, but not on the size of the planet. Solving for $T$, the temperature

$$T = \left[\frac{(1 - A)(1.37 \times 10^6)}{4\sigma D^2}\right]^{1/4}$$

This relation tells us what the temperature of each planet would be if the only source of heat were the Sun's radiation. We have made some simplifying assumptions in our derivation: For example, we have ignored possible variations of albedo with wavelength, the circulation of the planetary atmosphere, and different planetary rotation rates, but the resulting relation is still a very useful guide.

Although there is fair agreement in the general trends of the calculated and observed temperatures, there are enough differences to suggest that our model was too simple. Jupiter in particular is much warmer than expected and radiates about twice as much energy as it receives from the Sun. This strongly suggests the presence of an internal source of heat. We will review this in chapter 16.

## 11.5 PLANETARY ATMOSPHERES

We have just seen how a planet's temperature is largely dependent on its distance from the Sun. An additional calculation will show how this temperature and the size of the planet combine to control the gases that the planet can retain in its atmosphere.

Consider first the effect of temperature on a planet's atmosphere made of a mixture of gases. At any given temperature, the molecules in a gas can have a range of speeds, well known from the studies of gases that date back to Boyle in the seventeenth century and put into the present theoretical form by Maxwell and Boltzman about 100 years ago (Fig. 11.12). As the molecules collide with each other, their individual speeds may be changed, but there are so many molecules that on average there are as many that gain as lose energy in their collisions, and the distribution remains unchanged if the temperature doesn't change. There is some speed that is most popular, but a significant number of molecules will have higher speeds and some lower. The higher the temperature, the faster will be the movement of the molecules and the average speed will be higher, but the general appearance of the curve that displays the distribution will be the same. The molecules will be most frequently affected by their collisions on each other, but the gravitational attraction of their parent planet plays an important role.

For a planet of a given size and mass, there is a well defined **escape velocity**. This is the minimum velocity needed for an object to escape from the planet's gravitational attraction. An object (rocket or molecule) that is aimed vertically up with the escape speed ($v_{esc}$) may get away, but one with a lower speed will first go up and then back down. For a rocket leaving the Earth, an initial speed of 11.2 km/sec (about 25,000 mph) is needed to escape. For a molecule, having this speed will not guarantee escape, because it might have a collision with another molecule and either lose much of its speed or

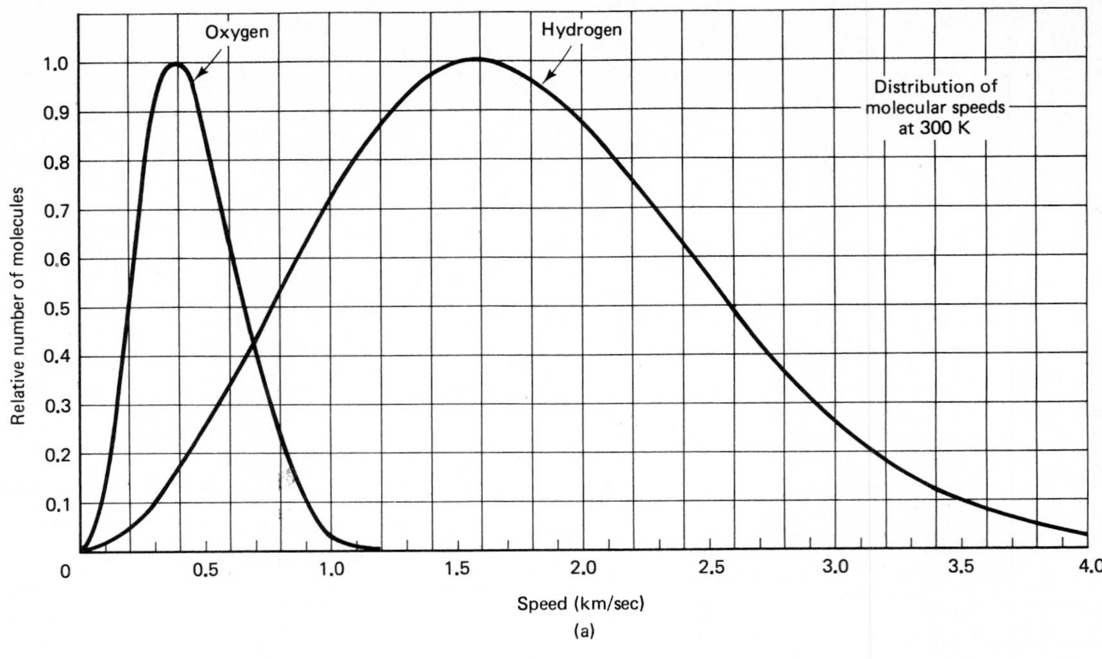


**Figure 11.12** The Maxwell distribution of speeds for gas molecules. (a) At any temperature, hydrogen molecules have on the average greater speeds than the heavier oxygen molecules. (b) Distributions of speeds of oxygen molecules at 300 K and 600 K. At the higher temperature the average speed is higher, and there are many more high-speed molecules. In this example, although the temperature has only doubled, there are about ten times as many molecules at 1 km/sec at 600 K as at 300 K.

be deflected into a downward direction. As a rough guide, a planet will retain a particular gas in its atmosphere if the average speed of those gas molecules is less than about one-tenth of $v_{esc}$.

In a mixture of gases, the average kinetic energy per molecule will be the same for all constituents, but we must remember that since kinetic energy is $\frac{1}{2} mv^2$, the lighter molecules will have on average higher speeds. Thus in a mixture of oxygen ($O_2$) and hydrogen ($H_2$), the hydrogen molecules will move much faster than the oxygen on average. Accordingly a larger fraction of the hydrogen than oxygen will have speeds above $v_{esc}$. The ability of a planet to retain its atmosphere will depend on its size (which influences $v_{esc}$) and its temperature (which depends most strongly on distance from the Sun). The data for the major bodies of the solar system are shown in Fig. 11.13.

From Fig. 11.13 we can see that the Moon's escape velocity is so low that it cannot retain an atmosphere of any of the common gases. Mercury is too hot to retain anything lighter than carbon dioxide ($CO_2$). The Earth

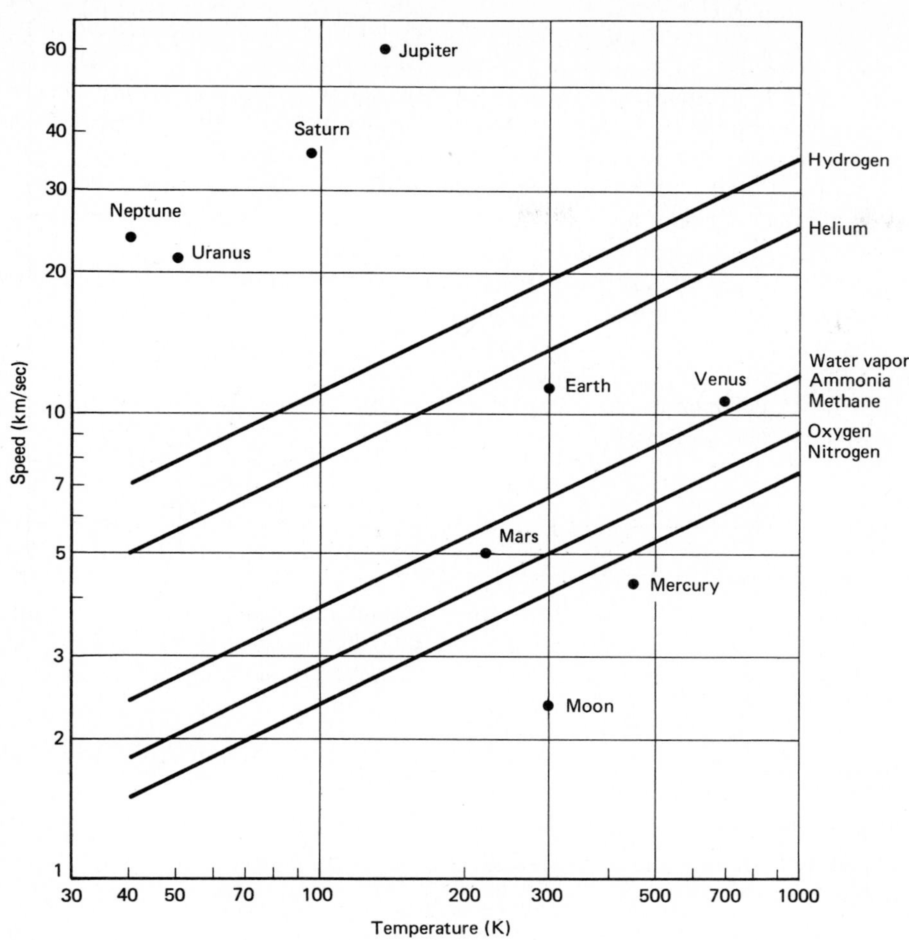


**Figure 11.13** Molecular speeds compared to escape speeds from the Moon and planets. The lines represent ten times the mean molecular speed for each designated type of molecule for the indicated temperatures. The point for each planet represents the escape speed for that planet and the surface temperature. The Earth is located in this diagram below the hydrogen and helium lines and will tend to lose those gases, but it will retain the heavier water vapor, oxygen, etc. The outer planets are sufficiently cold and/or massive that they can retain hydrogen atmospheres, whereas the Moon is both warm and insufficiently massive to retain any of these common gases. Venus, with its heavy cloud cover and resulting greenhouse effect, is not well represented on this diagram.

is continually losing hydrogen. The Jovian planets are sufficiently large and cold to retain extensive hydrogen atmospheres, as is actually observed.

**BOX 11.2**
*Gases*

The average kinetic energy of a gas molecule at a temperature $T$ is given by $\frac{1}{2} mv^2 = \frac{3}{2}$ k$T$, where $T$ is in Kelvin, $m$ is molecular weight in g, and k, Boltzmann's constant, is $1.38 \times 10^{-16}$ erg/degree, a value obtained from extensive experiments with gases. From this we can deduce that $v = \sqrt{3kT/m}$ cm/sec. Because of the spread of speeds, this value of $v$ is not the true average, but it provides a useful guide to the behavior of the gases. At a typical room temperature (about 300 K), $v$ = 525 m/sec. This is comparable to the speed of sound (330 m/sec).

For a planet having a mass $M$ and a radius $R$, the escape speed from near the surface is given by $v_{esc} = \sqrt{2GM/R}$, where G is the gravitational constant that we first encountered in chapter 7.

To retain a gas species that has a molecule of mass $m$, it is required that the typical speed of a molecule be less than one-tenth of the escape speed:

$$v \lesssim \frac{1}{10} v_{esc}$$

$$\sqrt{3kT/m} \lesssim \frac{1}{10} \sqrt{gM/R}$$

The planetary dimensions ($M$ and $R$) and temperature ($T$) thus determine whether a particular gas (with molecules of mass $m$) can resist escape. This relation is shown in Fig. 11.13.

## 11.6 SUMMARY

The general information on the solar system that was reviewed in this chapter has been obtained by traditional observations from the Earth, together with the application of some very basic principles of physics. Now, in proceeding to more detailed reviews, we shall be using the results of new types of Earth-based measurements and add the truly revolutionary information that we are still obtaining from lunar landers and space probes. With our present knowledge, we can now pretty well rule out certain theories for the formation and evolution of the solar system, but we do not yet know enough to be able to settle on *the* correct theory. We have made it harder to construct a theory, for, in a sense, we know too much. Solar system studies have come a long way from the old study of orbits, and, as we shall see, many new problems have emerged.

## CHAPTER REVIEW

A quick survey of the solar system:

**Sun:** dimensions, luminosity

**Planets:** discovery of Pluto; are there any more planets waiting to be discovered?

- — terrestrial planets and Jovian planets have very different densities.
- — all orbit in the same direction. All except two spin in the same direction. Venus rotates in reverse direction, and Uranus' spin axis is almost in ecliptic plane.
- — Titius-Bode law: a curious numerical coincidence.

**Comets:** Some have orbits that bring them back repeatedly, others probably pass only once through the inner solar system. May be broken up by passing too close to the Sun or Jupiter.

- — origin: possibly a vast reservoir far beyond Pluto's orbit.

**Asteroids (minor planets):** in orbit around the Sun, sizes range from 1000 km down. Some have Earth-crossing orbits, but most are in asteroid belt around 3 AU from the Sun.

**Meteoroids:** smaller than asteroids, in orbit, probably debris from comets.

**Meteor:** trail seen when a meteoroid burns up in the upper atmosphere.

**Meteorite:** meteoroid that is large enough to survive the heating in the Earth's atmosphere and reach the surface.

**Zodiacal Light:** faint glow of sunlight scattered from interplanetary dust.

**Atmospheres:** composition of a planetary atmosphere depends on albedo, distance from Sun, and the escape velocity.

## NEW TERMS

**albedo**
**asteroid**
**Bode's Law**
**Boltzmann's constant**
**escape velocity**
**Jovian planets**
**meteor**
**meteorite**
**meteoroid**
**micrometeorite**
**minor planet**
**prograde**
**solar constant**
**terrestrial planets**
**zodiacal light**

## PROBLEMS

**1.** At the top of the Earth's atmosphere, the intensity of sunlight is $1.37 \times 10^6$ ergs/cm$^2$ sec. How many photons per second does this represent assuming all the photons have the same wavelength of 5000 Å.

**2.** At the top of the Earth's atmosphere, the intensity of the solar radiation is $1.37 \times 10^6$ ergs/cm$^2$ sec. What will the intensity of the solar radiation be at the orbit of Neptune?

**3.** What fraction of the energy radiated by the Sun falls on the Earth?

4. Which planets have been known for as long as we have records, and which planets have been discovered during historic times?

5. How large is one astronomic unit (AU)? How long does it take light to travel 1 AU? How many Earth diameters would fit into 1 AU?

6. Draw a diagram to represent the orbits of Venus and Earth, as they would be seen from far above the north pole. Indicate the directions in which they move in their orbits, and also indicate the directions they rotate on their own axes.

7. Densities of the planets are listed in Table 11.1. Why could Kepler and earlier astronomers not have compiled those data?

8. Using data from Appendix E, calculate the greatest angle that an observer on Mars would see between the Earth and the Sun.

9. If the Earth's mass and orbit remained unchanged, how much larger diameter would the Earth need to have in order not to be able to retain water vapor in its atmosphere?

10. If the Earth were located at the orbit of Jupiter, what other gases might be found in our atmosphere?

11. Suppose that we could invent a form of life that required a methane atmosphere. Which planets could possibly support that form of life?

12. Suppose that there was a planet at 4 AU from the Sun with a mass, albedo, and radius equal to those of Jupiter. What atmospheric gases would that planet be able to retain?

13. Suppose the intensity of the solar radiation were to increase to sixteen times its present value. Which planets would no longer be able to retain water vapor in their atmosphere?

14. In a comic strip some years ago, one character claims to have discovered a new planet that will make him famous. His friend puts him down with the question "Suppose it burns out early and turns into a real dud?" What is wrong with this exchange?

15. A white dwarf has a mass equal to that of the Sun, but a size that is the same as the Earth. Calculate the escape velocity from the surface of the dwarf.

16. Suppose that you were aboard a spacecraft 1 million km from Jupiter.

   a. What angular diameter would Jupiter have?
   b. What would be the angular diameter of the Sun?

17. Suppose that the Earth's atmosphere reflected visible sunlight but was transparent to IR radiation from the Sun. How would the temperature on the Earth's surface change from its present value?

CHAPTER 12

# The Earth—Our Home

It has not been easy to demote the Earth from its long-accepted position as the center of the universe to its present status as simply the best-known planet among several. Kepler and Galileo may have taken the first scientific steps that confirmed Copernicus' geometrical ideas, but to many, perhaps most people, the tangible and comforting solidity of the Earth must have still remained as a token of silent dissent from these fantasies of the astronomers. It has taken the Apollo program with its dramatic photographs

**Figure 12.1** The Earth as seen from Apollo 17.

of the Earth—isolated against the darkness or rising over the lunar horizon (Fig. 12.1)—to show so vividly our home's true insignificance on the astronomical scale.

The purpose of the present chapter is to describe the Earth as an astronomical body almost as though we were observing it from Venus or Mars or some other planet. Of course, by living on it and after so much exploration, we know far more about the Earth than any other planet, but much of this knowledge is not relevant to an astronomical survey. We shall therefore describe those features which we can use later in comparison with those of other planets, to try to develop a theory of the origin and evolution of the solar system. We shall omit most of geology and the life sciences. We shall, however, include a brief discussion in Chapter 30 of conditions on Earth that have supported life.

## 12.1 THE EARTH AS A PLANET

The Earth is the third planet from the Sun and at just the right distance to keep us comfortably warm. From this base, we have observed the other planets, whose brightnesses depend on both the Sun-planet and planet-Earth distances as well as the planet's albedo. The Earth's albedo is now most accurately measured from spacecraft, but before the space era it had to be calculated from measurements of the Moon's faint glow at new moon when it is illuminated by earthshine. With a value of 0.35, the Earth's albedo is less than half that of Venus because of the great difference in the extent of cloud cover on these two planets. Observers on the more distant planets would see Venus much brighter than the Earth.

## 12.2 SIZE, SHAPE, AND MASS

For astronomical purposes the Earth's shape can be taken as spherical to a first approximation. There is a slight flattening because of rotation. The equatorial radius is 6378 km, the polar radius 6357 km. The mass of the Earth could be deduced once Newton had shown how gravity operated and the gravitational constant, G, had been measured (see chap. 7). The present best value for the mass is $M_E = 5.98 \times 10^{27}$ g. We can compute the average density (Fig. 12.2) of the Earth. Using an average radius of 6368 km, the

**Figure 12.2** Density is a measure of compactness of a body or of a gas, and is defined as mass/volume. For the Earth, the average density is 5.5 g/cm$^3$. For air at atmospheric pressure, each cubic centimeter contains only 0.0013 g of gas.

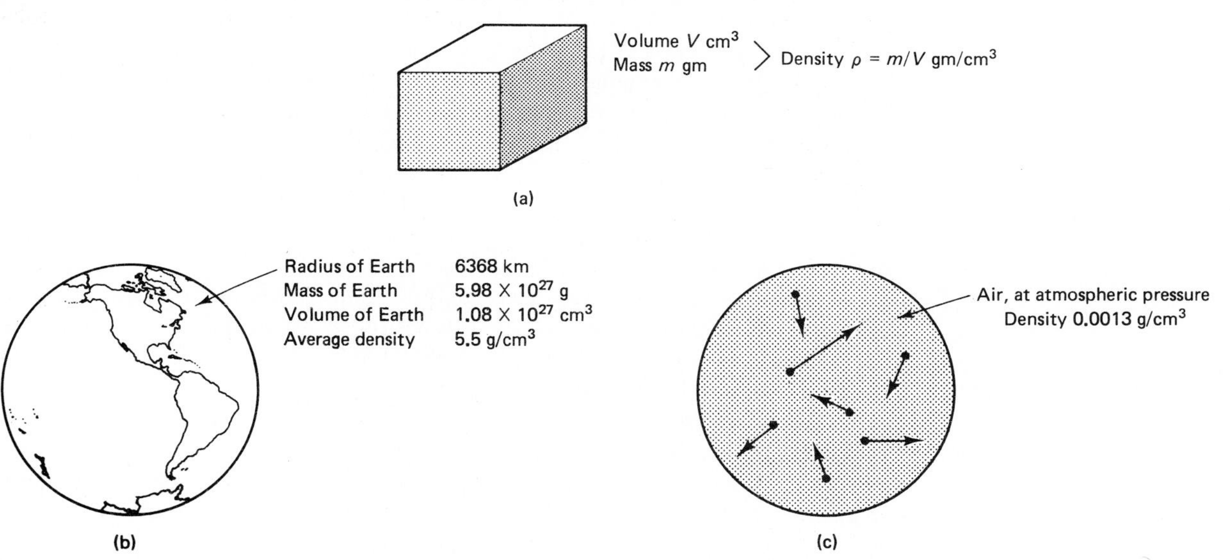

volume is

$$\frac{4}{3}\pi R^3 = 1.08 \times 10^{27}\ \text{cm}^3$$

leading to

$$\text{density} = \text{mass/volume} = \frac{5.98 \times 10^{27}}{1.08 \times 10^{27}} = 5.5\ \text{g/cm}^3$$

This is an average value for the whole Earth, but we at once take note of a problem. Much of the Earth's surface is covered by water of density 1 g/cm$^3$. Further, when we measure the densities of rocks that are easily available to us at the surface, we obtain values of density around 2.7 to 3.5 g/cm$^3$. Our immediate inference is that the deeper and inaccessible regions of the Earth must have much higher densities in order for the average to be as high as 5.5 g/cm$^3$.

**BOX 12.1**
*Density*

The volume and mass of an object can be combined into a single quantity that is a very useful and descriptive measure of compactness (Fig. 12.2). **Density**, denoted by the Greek letter rho ($\rho$), is defined as (total mass)/(total volume), or

$$\rho = m/v$$

With masses in grams and volumes in cubic centimeters, $\rho$ will be in g/cm$^3$. Water has a density of 1 g/cm$^3$, lead has 11.3 g/cm$^3$. Gases have very low densities. Under normal conditions air has a density of 0.0013 g/cm$^3$, but out in interstellar space the gas densities may be as low as $10^{-24}$ g/cm$^3$. We can tell quite a lot about the properties of a body simply from knowing its density, as we shall see when we start to compare the planets.

## 12.3 THE COMPOSITION AND STRUCTURE OF THE EARTH

Only the most superficial layer of the Earth, the **crust**, is within reach for direct sampling. We can estimate the volume of the oceans and the quantity of the dissolved matter they hold, and we can analyze the composition of the crust. With volcanic eruptions, we can obtain material from somewhat deeper down, but for the most part we have to rely on indirect measures in arriving at a picture of the interior of our planet. Seismology provides an effective probe.

The crust of the Earth is estimated to vary in thickness between 16 and 40 km. Within this crust, there is a slow creeping of some adjacent rock layers that is sometimes accelerated leading to the sudden release of energy in an earthquake. Earthquakes trigger off waves in the solid Earth much as an explosion in the atmosphere sends out a pressure wave. Seismic waves spread rapidly and are monitored by an international network of seismometers. This continual earthquake-watch is especially important for countries in and around the Pacific Ocean, since some undersea earthquakes can produce damaging **tsunamis** or tidal waves.

The often quoted magnitudes on the Richter scale relate to the amounts of energy released in earthquakes. For example, an eighth-magnitude quake corresponds to an energy release of about $10^{25}$ ergs or as much energy as the Sun radiates in ten billionths of a second.

One kind of seismic wave, the L wave, travels only along the Earth's surface, but two other kinds of wave can penetrate to greater depths (Fig. 12.3). One kind, the P wave, comprises back and forth (longitudinal) pressure variations and ground motion, whereas the other (S) wave transmits its energy by transverse vibrations (we encountered this sort of transverse vi-

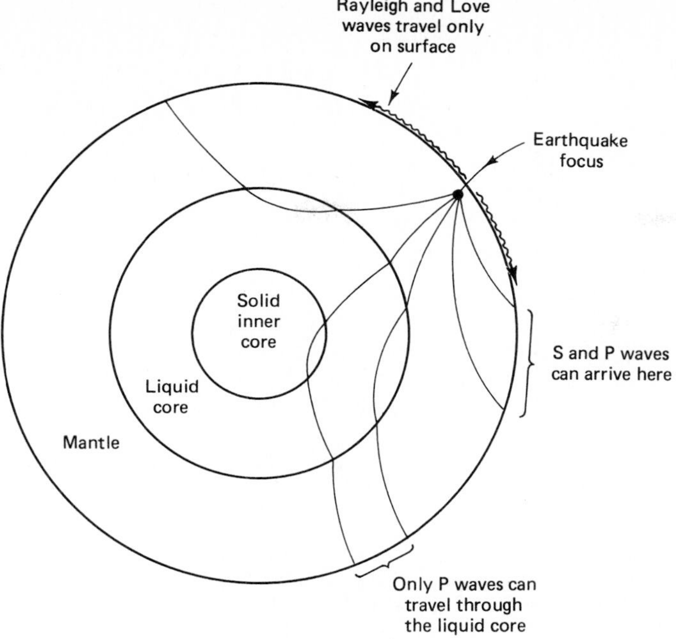


**Figure 12.3** Several different types of waves are generated by earthquakes. Rayleigh and Love waves travel only along the surface. Other types of wave (P and S) can travel through the interior, but they do so at different speeds, and only the P waves can travel through the liquid core. As a result, the seismic wave pattern observed at any detecting station will depend on a complex way on the depth of the earthquake focus and the distance of the station from the focus.

bration when discussing light waves in chap. 8.) The speeds of these different waves depend on the composition and crystal structure of the regions through which they travel. There is another important difference between the S and P waves: Only the P waves can travel through a liquid.

Seismometers can distinguish between S and P waves, measure their size, and note the times of arrival at different recording stations (Fig. 12.4).

**Figure 12.4** Seismograms from the earthquake of April 26, 1981. The focus was in southern California, at a depth of 6 km, and the body-wave magnitude was 5.3. These seismograms were recorded near St. Louis at a distance of 2350 km from the focus. Note the different arrival times of the P, S, and Rayleigh waves. Actual ground movement peaked at about $\frac{1}{10}$ mm. (Otto W. Nuttli, St. Louis University).

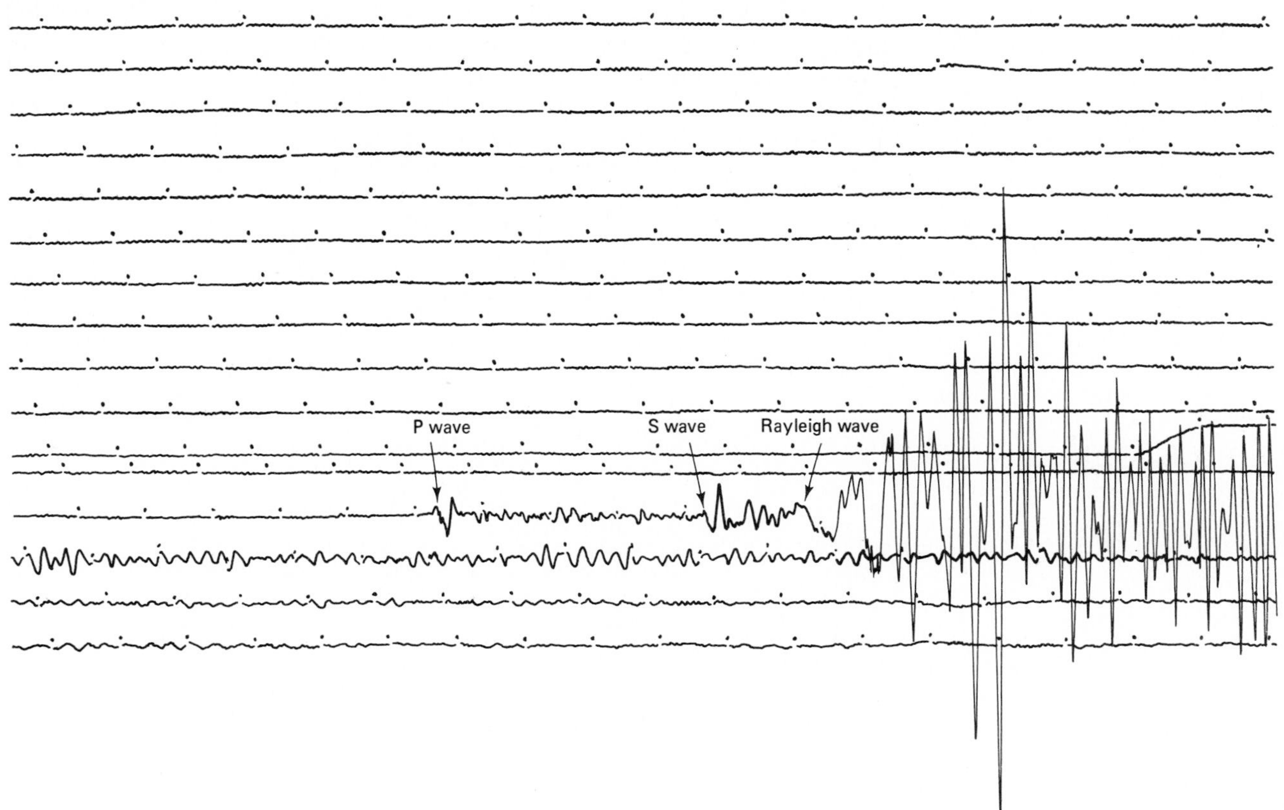

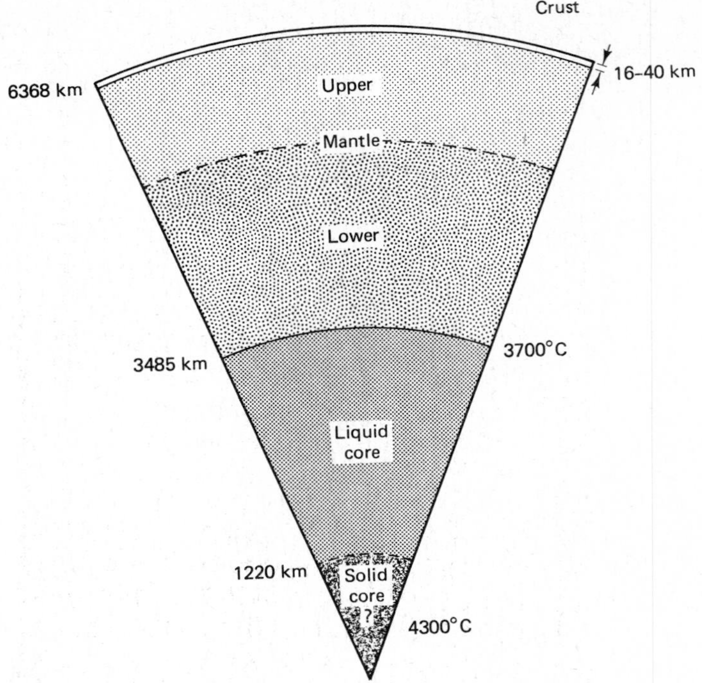


**Figure 12.5** The internal structure of the Earth.

From these records it is possible to locate the point of origin (the **focus**) of the earthquake and calculate the transit times of the waves. These can then be compared to theoretical models that are based on assumptions regarding the variation of density and composition with depth below the surface. We also need laboratory studies of the behavior of materials at very high pressures. Since laboratory studies cannot yet duplicate the extreme pressures that exist deep in the interior of the Earth, we add some inspired "guesstimates" and arrive at a model such as that displayed in Fig. 12.5.

Below the crust there is a thick solid **mantle** of rock nearly 2900 km thick overlying a liquid core of radius 3485 km. The existence of this liquid core is clearly indicated by the failure of S waves to be detected at some stations (see Fig. 12.3). There is evidence, not yet conclusive, that there may be a solid inner core of radius 1220 km.

*Table 12.1* BULK COMPOSITION OF THE EARTH[a,b]

| Element | Percent by Weight |
|---|---|
| Iron | 34.6 |
| Oxygen | 29.5 |
| Silicon | 15.2 |
| Magnesium | 12.7 |
| Nickel | 2.4 |
| Sulfur | 1.9 |
| Calcium | 1.1 |
| Aluminum | 1.1 |
| | 98.5 |
| All other elements (none more than 0.6 percent) | 1.5 |

[a] There have been several compilations of the abundances of the chemical elements that comprise the Earth. Because of the variations between experimental samples drawn from different locations on the Earth's surface and the different importance attached to these by different authors, compilations will differ in detail, but the general features should be consistent.

[b] B. Mason and C. B. Moore, *Principles of Geochemistry* (4th ed.), New York: John Wiley and Sons, 1982.

Calculations of the variations of pressure and density with depth lead to the depth profiles of temperature. After a rapid rise in temperature in going down through the crust and outer mantle, there is little change all the way through to the center (Fig. 12.6).

Although we can make direct measurements of the chemical abundances at the surface, we have to infer (or juggle) the corresponding numbers for the interior in order to match the densities that we have deduced from the seismic observations and from the total mass and size. The results are listed in Table 12.1.

In arriving at the abundances calculated for the interior, additional information is drawn from the observed composition of meteorites and from an understanding of the formation of the elements in stellar processes.

The combination of the elements in different proportions and under different physical conditions leads to rocks of different types. **Basaltic** rocks—largely oxygen, silicon, aluminum, magnesium, and iron—are found in the oceanic crust. The upper continental crust is more **granitic** (oxygen, silicon, aluminum, sodium, and potassium) and is not as dense as the basaltic material. The mantle in turn is denser than both the basalts and the granites. This layering of densities is termed **differentiation** and results from partial melting and recrystallization.

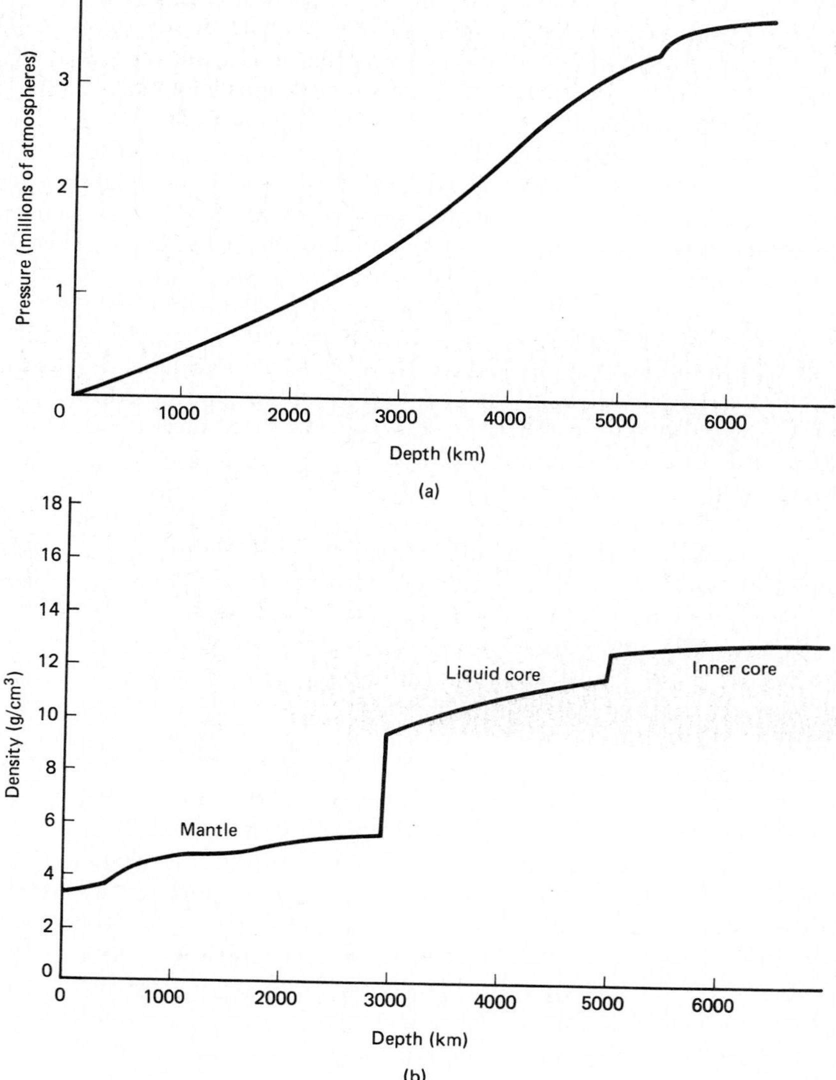


**Figure 12.6** The interior of the Earth. (a) The pressure increases steadily towards the center. (b) The density also increases towards the center, with sharp changes at the boundaries between the mantle, liquid core, and solid inner core.

## 12.4 THE AGE OF THE EARTH

How can we measure the age of the Earth? Even to pose this question represents an advance from an earlier time when neither a natural origin nor any changes were even considered. The old legends abound with gods, stars, and planets, but do not lend themselves to the construction of a precise chronology. The scriptures, however, provided an apparently firmer basis for examining this problem even though a scientific foundation was still lacking. Theophilus of Antioch (around A. D. 150) seems to have been the first to try working back through the Bible's chronicle of generations, to arrive at 5529 B. C. as the time of creation. Many others tried to refine this approach: Luther's value was 4000 B. C., and best known of all Bishop Ussher in 1650 placed the creation on Sunday, October 23, 4004 B. C. For comparison, it is interesting to note that the Mayan calendar has as its starting date 3313 B. C.

The discovery of fossils, some clearly of marine life but located on high mountains far from any ocean, presented early difficulties. Later the steady accumulation of geological knowledge (such as the recognition of the process of sedimentation—the slow accumulation of successive layers of deposits at river mouths) might be thought to have presented even greater problems for chronologies that covered at most only a few thousand years. But we must remember how difficult it is to deal with time intervals much longer than our own experience. The slowness of the geological processes was not yet known when the natural philosophers of the seventeenth century were trying to answer the question: *Given* that the Earth was created in 4004 B. C., how can the geological observations be made to fit into this very restricted time scale?

Today, our approach is quite the reverse. We can identify physical and geological processes and study the rates at which they work their changes in the continents and mountains, oceans and ice sheets, and we can measure the abundances of certain isotopes. We now ask: how long has the Earth existed while these changes have occurred?

Many different methods have been used for estimating the Earth's age. For example, the Earth's age has been calculated from the measured concentration of salt in the oceans and the rate at which rivers add to that salt concentration by bringing dissolved salt from the land. A requirement for any reliable calculation is that we identify all of the contributory factors and know their rates, and this is especially difficult when dealing with the river-ocean method, although it is appealingly simple at first sight. Difficulties are immediately encountered when one tries to obtain reliable estimates for the total rainfall, the rate at which chemicals dissolve in the rivers, the evaporation rates from the rivers and oceans, and how these various quantities might have changed over the centuries.

The only method that appears to avoid most problems is radioactive dating, the basis for our present calculations (see Box 12.2). Radioactive decays are unaffected by temperature, pressure, or weather conditions and are even unaffected by chemical combination. One of the preferred isotopes is potassium 40 ($K^{40}$), which decays to argon 40 ($Ar^{40}$) with a half life of 1.3 billion years. Measurement of the $K^{40}/Ar^{40}$ ratio have shown that the oldest rocks known on Earth (in Greenland) solidified about 3.8 billion years ago.

Other radioactive isotopes have also been used. For example, uranium 238 ($U^{238}$) decays through a sequence of stages until lead 206 ($Pb^{206}$) is reached, and that isotope is not radioactive. Uranium 235 ultimately leads to lead 207, and thorium 232 has lead 208 as its stable end product.

Radioactive dating techniques have been applied to lunar samples (Chap. 13) and meteorites (Chap. 17), and ages up to 4.6 billion years have been obtained. Estimates of the relative abundances of the lead isotopes

206, 207, and 208 on Earth appear to agree with values measured in many meteorites and are thus consistent with an age of 4.6 billion years.

The age of the Earth is thus at least 3.8 billion years and quite probably 4.6 billion years, but the gap in the record (between 3.8 and 4.6 billion years) is still not understood.

## BOX 12.2

### *Using Radioactivity to Measure Ages*

Atomic nuclei consist of combinations of protons and neutrons. Some combinations are **stable**; if left to themselves, these nuclei will last permanently. Thus six protons and six neutrons make up the stable carbon nucleus $C_6^{12}$ essential to all living organisms. Some mixtures of protons and neutrons are **unstable** or **radioactive**; even if left to themselves, they will ultimately fall apart. Thus $U_{92}^{238}$ has 92 protons and 146 neutrons in each nucleus, and it is radioactive.

A measure of the instability is the **half-life**, the length of time in which one half of an original group of nuclei will **decay** (Fig. 12.7). At the end of that time, one half of the atoms will remain. During the passage of another half-life, one half of the remainder will decay leaving 25 percent of the original number. After yet another half-life, the fraction surviving will be 12.5 percent, and so the process continues.

The decay of an unstable nucleus can have several results depending on the type of nucleus. In the case of $U^{238}$, each decay involves the ejection of a high-speed **alpha particle** (the nucleus of a helium atom); the remaining nucleus is that of thorium ($Th^{234}$), another chemical element. The alpha particles from uranium can be detected. The number of alpha particles produced per gram of material per second provides a measure of the quantity

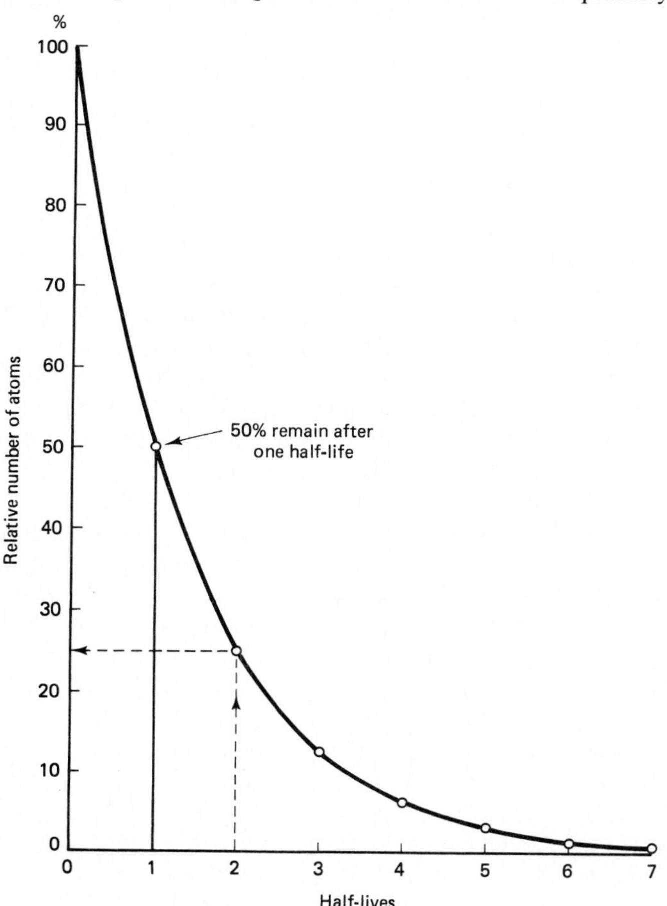


**Figure 12.7** Radioactive decay. During the first half-life, 50 percent of the original atoms decay, leaving 50 percent unchanged. During the next half-life, half of the these survivors decay. After six half-lives, only about 1.5 percent of the original atoms remain.

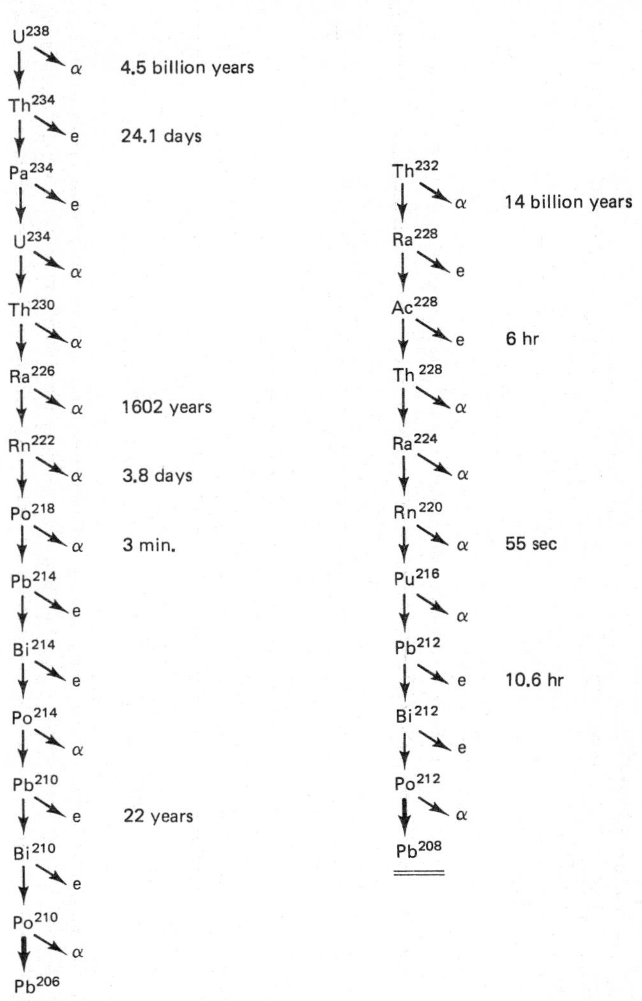


**Figure 12.8** The Uranium and Thorium radioactive decay chains. Each chain starts with a nucleus that has a half-life comparable to the age of the Earth. After many successive decays, each ends at a stable isotope of lead. Decays in which an alpha-particle is emitted are indicated by α, those in which an electron is emitted by *e*. A few representative half-lives are indicated.

of uranium present. The thorium atoms are also radioactive, decaying to protoactinium (Pa), which in turn also decays. This process continues through many stages terminating in the stable isotope of lead, $Pb^{206}$ (Fig. 12.8). An initially pure sample of $U^{238}$ will thus slowly transform into a complex mixture of lighter atoms along the chain, each one having a different half-life. Another radioactive series starts with thorium 232 ($Th^{232}$) and ends at a different lead isotope, $Pb^{208}$.

The half-lives of the various isotopes have been measured so that the rate of transformation of uranium into lead is now well known. For example, the half-life of $U^{238}$ is 4.5 billion years, the half-life of $Th^{234}$ is 24.1 days, and the half-life of $Ra^{226}$ is 1602 years. If, then, we measure the relative amounts of the different isotopes in a rock, we can calculate the age of that rock, defining "age" as the time since that rock solidified.

In recent years it has been found that other radioactive isotopes provided a better base for the measurement of some types of very old rocks. Rubidium ($Rb^{87}$) decays to strontium ($Sr^{87}$) with a half-life of 47 billion years; potassium ($K^{40}$) decays to argon ($Ar^{40}$) with a half-life of 1.3 billion years. These isotopes have also been used for the dating of lunar samples and meteorites.

For measurement of much shorter time scales, other radioisotopes must be used. Thus radiocarbon ($C^{14}$) is widely used for the dating of archaeological finds. All living material contains stable carbon atoms ($C^{12}$) and a minute trace of the radioactive isotope $C^{14}$, which has a half-life of 5730 years. After death, the relative proportions of $C^{12}$ and $C^{14}$ slowly change as the $C^{14}$ decays. Measurement of the proportion of $C^{14}$ allows the age to be calculated. This method is limited to organic remains and will not work on metals or ceramics for which other methods must be used.

## 12.5 THE ATMOSPHERE

Our previous encounter with the Earth's atmosphere was as an astronomical nuisance: It is selective in the wavelengths it transmits, and it degrades the images formed by our telescopes. On the other hand, we would not be here without it and a brief survey of its properties can add to our understanding of the Earth and the other planets.

Figure 12.9 shows how the principal properties of the atmosphere vary with altitude above the Earth's surface, and Table 12.2 lists the major constituents. As we might expect from the discussion of atmospheric temperatures (chap. 11), there is little of the two lightest gases, hydrogen ($H_2$) and helium (He); 99 percent of the atmosphere consists of nitrogen ($N_2$) and oxygen ($O_2$). Carbon dioxide, vital for plant growth, constitutes less than $\frac{1}{10}$ of 1 percent, and the noble gases helium (He), neon (Ne), krypton (Kr), and xenon (Xe) together amount to about 0.002 percent. The surprisingly high argon (Ar) concentration probably comes from the radioactive decay of potassium 40 ($K^{40}$) in rocks. Water vapor is also found, mostly, at lower altitudes.

Almost all of the abundant atmospheric gases are present as molecules, but at altitudes above about 100 km solar UV and X-radiation is intense enough to eject electrons from many atoms and disrupt the molecules. This produces electrically charged **ions**: atoms or molecules that have lost one or more of their electrons. These ions and their electrons populate the region known as the **ionosphere** and constitute a reflecting layer for short-wave-

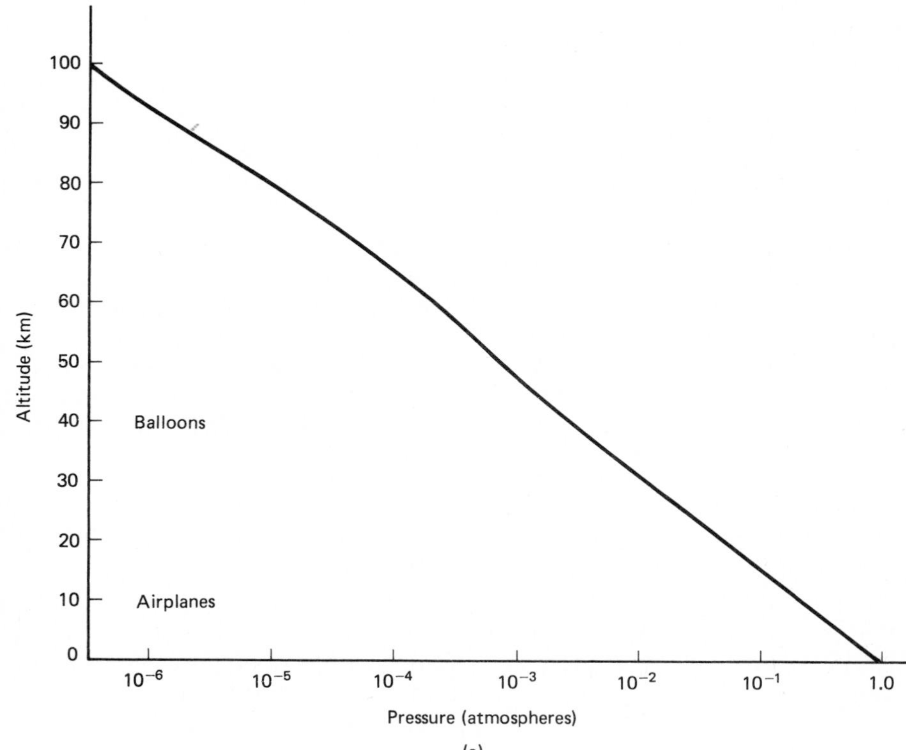


**Figure 12.9** The Earth's atmosphere. (a) decrease of pressure with altitude. At a typical airplane altitude of 10 km (30,000 ft), the air pressure is about one-quarter atmosphere, and at the altitude of research balloons it is a few thousandths of an atmosphere.

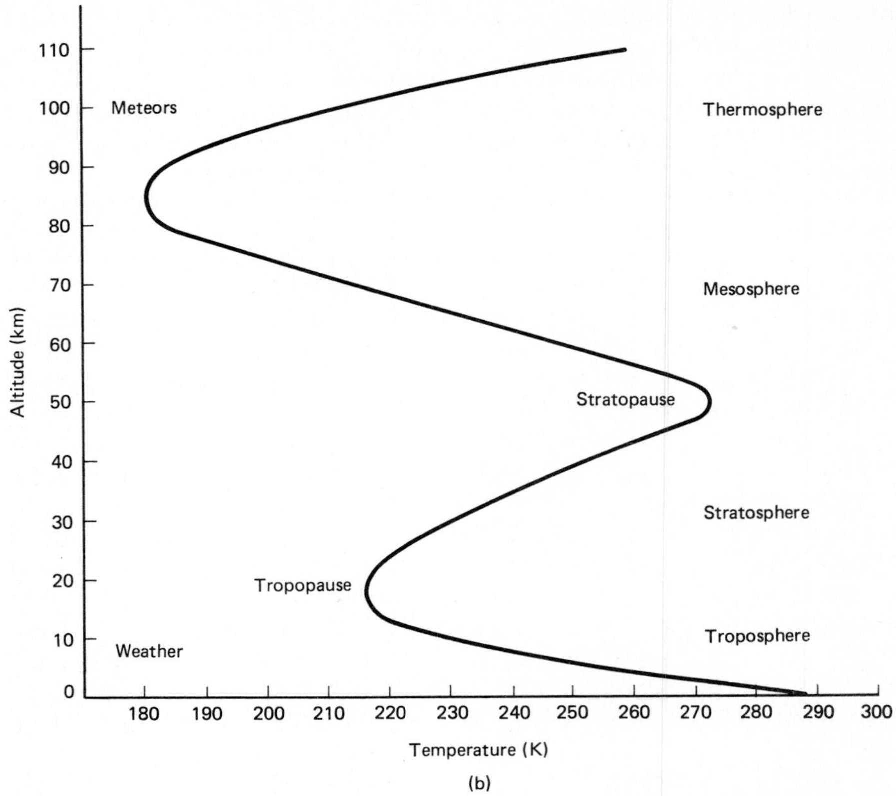

Figure 12.9 (Cont.) (b) Variation of temperature with altitude. Major features in the temperature profile mark major changes in the atmospheric behavior. Most of our weather is contained within the troposphere, the lowest 12 to 15 km. Not shown in these figures is the ionosphere that is above about 350 km. Most of the atmospheric mass is contained within the lowest 30 km, and it is in this region that most absorption of radiation occurs. Meteor trails are generally produced in a band around 100 km.

*Table 12.2* COMPOSITION OF THE EARTH'S ATMOSPHERE[a] (FOR DRY AIR)

| | *Percent by Volume* | *Percent by Weight* |
|---|---|---|
| Nitrogen | 78.0 | 75.5 |
| Oxygen | 20.9 | 23.1 |
| Argon | 0.9 | 1.3 |
| Carbon dioxide | 0.03 | 0.05 |
| Neon | 0.0018 | 0.0013 |
| Helium | 0.0005 | 0.00007 |
| Krypton | 0.0001 | 0.00033 |
| Xenon | 0.0001 | 0.00004 |

[a] C. W. Allen, *Astrophysical Quantities*, 3rd ed., The Athlone Press, University of London, 1973.

length radio waves. Without this radio mirror, direct radio communication would be restricted to line-of-sight and could not extend over the horizon. The altitude and properties of the ionosphere are sensitive indicators of solar activity, and major sunspot activity is often accompanied by interference with radio communications because of disturbances produced in the ionosphere.

## 12.6 THE EARTH'S MAGNETIC FIELD

Magnetism is such a familiar phenomenon that we have transferred the concept to the picturesque description of "personal magnetism." The idea of a strong attraction is clearly conveyed. (We might pause to note that magnetism also involves an equally strong repulsive force, as you can show very quickly with a pair of magnets, but this aspect has not been taken into the language.) The earliest demonstration of magnetism involved the strange properties of some iron-rich rocks, the lodestones, which could align themselves north-south when loosely suspended. We still use this property for navigation with the compass, a small magnetized iron pointer in a protective glass-topped container.

If you were to follow the directions that the compass needle points on the Earth's surface, you would find yourself following paths that end at the north and south magnetic poles. The region where the Earth's magnetism extends is the **magnetic field**, and it is not confined to the Earth's surface (Fig. 12.10). The magnetic field extends below the Earth's surface to the core where it originates, and it extends out to vast distances beyond the Earth's atmosphere where it is termed the **magnetosphere**.

The magnetic poles do not coincide with the geographical poles but are at north and south latitudes of about 78° with the north magnetic pole in northern Canada. In many ways, the Earth behaves as though it had a strong bar magnet near but not exactly at its center. The alignment of this magnetic field relative to the rotation axis and equator is drifting slowly but persistently, and the locations of the magnetic poles have shifted markedly even within the past few hundred years. Magnetic compass directions require correction for reliable navigation, and this correction must be updated every few years. Modern navigation now needs a more accurate method and is based on the use of gyrocompasses.

Not only do the magnetic poles wander, but the strength of the magnetic field changes so that at times the field vanishes then reappears but in the opposite direction, that is, the pole to which the north compass needle points will switch to the southern geographic hemisphere. Many old rocks show traces of magnetism, recording the conditions when they solidified. Through this **paleomagnetic** record we have been able to trace the history of the Earth's magnetic field back over millions of years. The evidence shows that the magnetic field takes about 10,000 years to reverse itself with intervals between reversals ranging from hundreds of thousands to tens of millions of years.

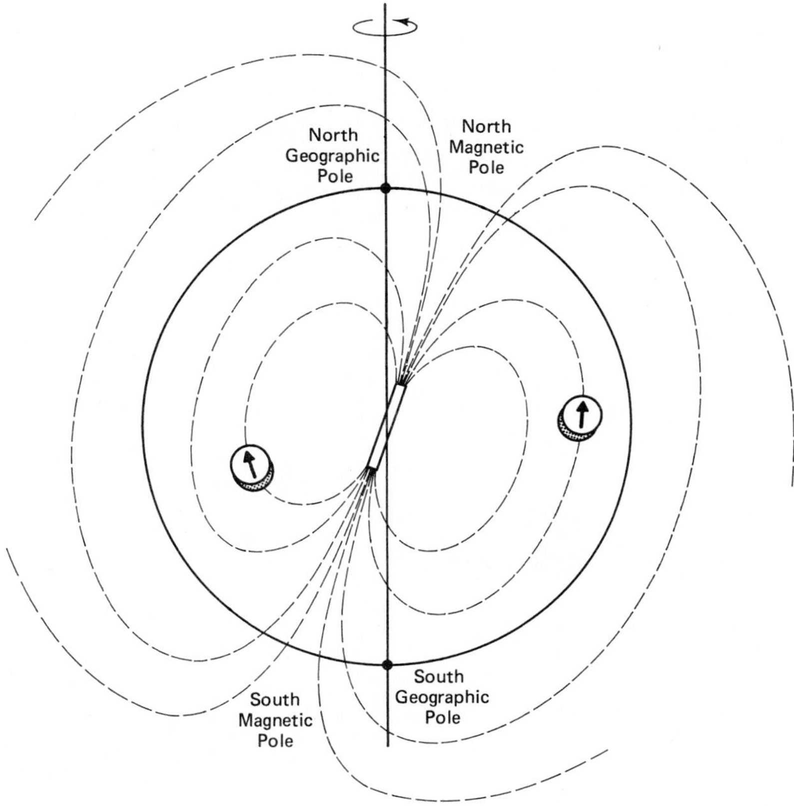


**Figure 12.10** The Earth's magnetic field. Close to the Earth, the field is similar to that of a bar magnet located close to the Earth's center but not quite aligned with the Earth's rotation axis. Lines of force can be traced by following the directions in which compass needles point. At greater distances the magnetic field is distorted by the solar wind, a stream of electrically charged particles from the outer parts of the solar atmosphere.

It is now generally agreed that the source of the Earth's field is in the liquid core where circulating electric currents act like a dynamo. The details of this theory are by no means established for we do not yet know, for example, the electrical properties of the core. These depend on the composition (the type and quantity of impurities in the mostly iron core) and on probably complex current patterns. As for the reservoir of energy that drives the dynamo, we are even less certain. Our present best guess is that the energy comes from the slow cooling of the core.

Although the strength of the magnetic field decreases with distance from the Earth's surface, its effects show up in several ways. The pattern of the Earth's field allows some fast electrically charged particles to be trapped in fairly well-defined regions (belts) around the Earth. The presence of these regions of trapped particles was quite unsuspected before James Van Allen of the University of Iowa discovered them with his geiger counters aboard the satellite Explorer I in January 1958. The inner belt with its protons extends from about 500 km above the Earth's surface out to about 5500 km, and the outer belt (mostly electrons) is located between about 15,000 and 20,000 km. Some of the particles in the belts probably come from the collision of cosmic rays with atoms in the Earth's atmosphere, others directly from the Sun. The location, extent, and population of the belt are strongly influenced by solar activity, and at times many of the trapped particles are dumped into the atmosphere near the poles, producing auroras. In addition, as it flows by the Earth, the solar wind distorts the belts. Extensive mapping from satellites and deep space probes has produced a typical pattern such as that shown in Fig. 12.11. In our daily existence, we are quite unaware of these radiation belts, but instruments on spacecraft that fly through them are affected by the intense radiation. Astronauts in near-Earth orbits are below these regions, but those who fly to the Moon or farther receive a small radiation dose as they go through the belts.

The most dramatic demonstration of the Earth's magnetic field comes from the **auroras** (northern hemisphere auroras are termed aurora borealis or "northern lights," and those in the southern hemisphere, aurora australis) (Fig. 12.12). These are brilliant displays of lights in the sky, generally at latitudes about 20° from the magnetic poles. Auroras are produced in the atmosphere by high-speed electrons and protons that have been released from the trapped radiation belts in the magnetosphere. The collisions of these particles with air molecules produce the auroras. The dumping of particles

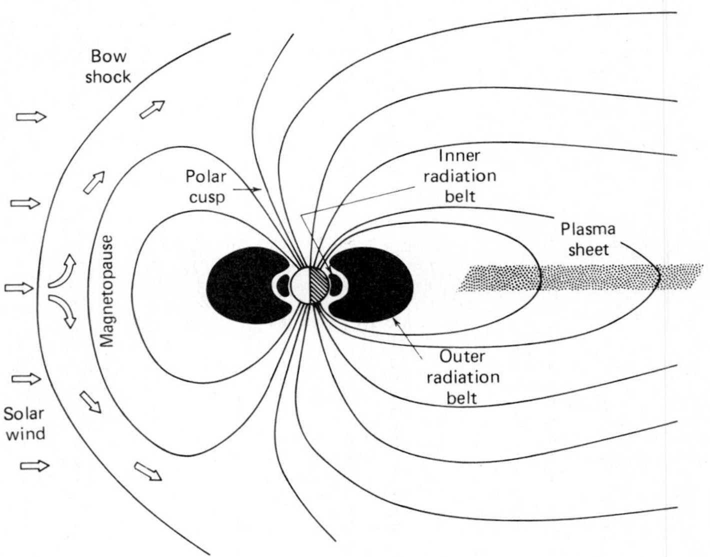


**Figure 12.11** Van Allen belts and the Earth's magnetic field at large distances. Note how the solar wind distorts the magnetic field. Some cosmic-ray particles can be trapped to populate the Van Allen radiation belts (James A. Van Allen, University of Iowa, and Addison-Wesley Publishing Company).

**Figure 12.12** Aurora. High-speed electrically charged particles can follow magnetic field lines and approach into the relatively denser upper atmosphere. Collisions between these particles and atmospheric atoms and molecules produce extensive auroral displays, typically at altitudes around 180 km (Vic Hessler, Institute of Geophysics, University of Alaska).

from the Van Allen belts is strongly influenced by solar activity. The shape and extent of the auroras are affected by both the Earth's magnetic field and another magnetic field produced by electric currents in the ionosphere.

## 12.7 THE EVOLUTION OF THE EARTH

We shall defer until chapter 18 our discussion of the theories of the origin and formation of the Earth, for these require a comprehensive view of all the bodies of the solar system. In this section, we shall confine ourselves to the later evolution of the Earth, setting aside the critical problems of the earlier stages.

First, we must note that the Earth has changed markedly over its 4.6 billion years of existence. The atmosphere with its winds and water is a major force in the erosion and alteration of surface features, sweeping much material into the oceans. The atmosphere itself has changed from its primordial composition with the addition of water vapor, carbon dioxide, and nitrogen from the crust, probably released under the influence of heating. The earliest atmosphere probably contained hydrogen and helium, but those would have been lost quite quickly. Outgassing from volcanoes and the heating of the crust would then have produced carbon dioxide ($CO_2$), methane ($CH_4$), ammonia ($NH_3$), and water vapor ($H_2O$). The methane and ammonia were then dissociated quickly by sunlight, and the water vapor could condense (as the Earth cooled) to produce the oceans. Hydrogen from the dissociation of the methane and ammonia would soon be lost, but the nitrogen would accumulate. Dissociation of some of the water vapor would lead to an increase in the oxygen and then to an ozone ($O_3$) layer which would shield the lower layers from the solar UV. About 2 billion years ago, biological (plant) activity probably initiated the major change to our present oxygen-rich atmosphere. Another indicator of change in the atmosphere is the argon 40, most of which comes from radioactive potassium 40 in the crust.

The balance between $O_2$ and $CO_2$ depends on several processes such as photosynthesis by plants, the burning of wood and organic (fossil) fuels, and the weathering of rock. The absorption of solar radiation can be influenced by small changes in atmospheric composition, and the concentration of $CO_2$ is a matter of some concern, lest it increase and lead to a change in the average surface temperature of the Earth. (This effect is the subject of considerable research and equally considerable disagreement as to the environmental consequences.) Just as important in influencing the absorption of solar radiation, at least on a short time scale, is the presence of large amounts of dust in the atmosphere. Major volcanic episodes such as the historic Krakatoa (1883) and the Mt. St. Helen's eruptions can produce very marked but temporary weather changes.

Volcanoes also indicate that considerable activity is taking place beneath the surface. The present view is that the upper layer (**lithosphere**) consists of about a dozen major plates supported by and moving slowly over the next lower layer, the **asthenosphere**. At one particular time the continents were contiguous, but they have drifted to their present positions over hundreds of millions of years. In earlier times, the continental crust was probably broken in different ways and a different set of continents existed. Precise measurements show that, for example, Africa and South America are still drifting apart by a few centimeters per year. At the boundaries between plates the crust can be pushed up forming mountains, and plates can be pushed down, one beneath another, producing fault lines and regions prone to earthquakes and volcanic activity. In other places where plates are pulling apart such as in the mid-Atlantic, new material slowly flows up from the mantle to the ocean floor. All these changes require energy. Radioactivity and the slow loss of heat remaining from the Earth's early years are the two sources that now seem most reasonable for this energy. Radioactive heating is thought to provide about 50 to 80 percent of the heat lost through the Earth's surface. There are also external agents of change but on a globally much smaller scale. The surface of the Earth is marked with craters from

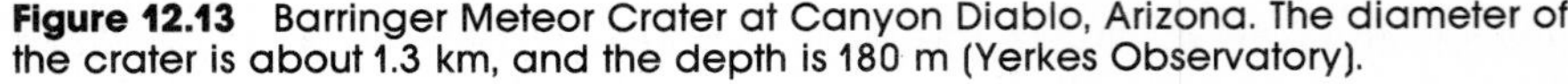

**Figure 12.13** Barringer Meteor Crater at Canyon Diablo, Arizona. The diameter of the crater is about 1.3 km, and the depth is 180 m (Yerkes Observatory).

the bombardment by meteorites. Some craters, such as Meteor Crater in Arizona, are prominent and well known (Fig. 12.13). Others have been discovered only in the past few years, as air surveys revealed the almost completely eroded features of very large craters that had escaped detection on the ground.

In summary, the face of the Earth is changing slowly but continuously. Because of our proximity we have been able to document and develop an understanding of many of these changes. We can use this knowledge to guide us as we explore the other planets and their satellites, but we must remember how spotty our information is about those more distant bodies. We must start by documenting conditions as they are today. Trying to work our way back to the formation of the solar system requires some inspired guesses to fill the gaps and help us to bridge the 4.6 billion years. We should not be offended or surprised if later discoveries show that some of our present theories are wrong, but that should not deter us from a modest amount of theorizing, for in that way we can be guided in our further observing.

## CHAPTER REVIEW

The Earth's albedo is 0.35. To a distant observer, it would appear much fainter than Venus.

The **Earth's shape** is not quite spherical with an average radius of 6367 km. Its mass is $5.98 \times 10^{27}$ g, and its average density is 5.5 g/cm$^3$. The interior must have a much higher density than this value.

The **structure** of the interior has been studied with seismic methods. Beneath a relatively thin crust, there is a thick mantle and then a liquid core. There is probably a solid inner core. The continents are drifting slowly apart. Movement of the continental plates leads to the production of mountain ranges and the release of energy in earthquakes.

**Age** of the Earth has been determined by radioactive methods. The oldest rocks on Earth are 3.8 billion years old. Comparison of uranium, thorium, and lead isotopes on Earth and in lunar samples and meteorites suggests that the Earth is probably 4.6 billion years old. The gap between 3.8 and 4.6 billion years is not understood.

**Atmosphere:** almost surely the result of many changes. The original atmosphere was probably hydrogen and helium, but these would soon be lost. Heating of the surface probably produced carbon dioxide, methane, ammonia, and water vapor. Hydrogen from the methane and ammonia was probably lost, but the nitrogen accumulated. The start of plant life probably led to the production of oxygen. Argon is probably from radioactive decay of potassium in rocks.

A **magnetic field** extends from inside the Earth out to great distances where high speed protons and electrons can be trapped in radiation belts. There is evidence for many reversals of the Earth's field over geologic times.

## NEW TERMS

**asthenosphere**
**alpha particle**
**aurora**
**basalt**
**crust**
**decay**
**density**
**differentiation**
**granite**
**half-life**
**ions**
**ionosphere**
**lithosphere**
**magnetic field**
**magnetosphere**
**mantle**
**paleomagnetism**
**radioactivity**
**radiation belts**
**seismology**
**stable nuclei**

## PROBLEMS

**1.** What fraction of the solar energy that falls on the top of the Earth's atmosphere is absorbed?

**2.** Make a rough calculation of the total mass of the Earth's atmosphere, and compare it to the mass of the Earth. Assume that the air has the same density from sea level up to 10 km and zero beyond that. Use $10^{-3}$ g/cm$^3$ for the density of the air.

**3.** How can the Earth's mass be determined from astronomical observations?

**4.** Why does examination of samples from the Earth's crust not give us a good guide to conditions in the very early stages of the solar system?

**5.** The surface area of the Earth is 71 percent water and 29 percent land.The density of water is 1 g/cm$^3$ and that of average land about 3.0 g/cm$^3$. What fraction of the weight of the topmost kilometer of the Earth's surface is water, and what fraction is land?

**6.** Calculate the fraction of the Earth's volume occupied by the core, the mantle, and the crust.

**7.** How can the age of the Earth be determined?

**8.** Uranium 238 has a half-life of $4.5 \times 10^9$ years, and uranium 235 has a half-life of $7 \times 10^8$ years. Suppose that they are initially present in equal amounts in a rock on Earth. Explain in words (without any calculations) how the relative numbers change with the passage of time.

**9.** We are accustomed to thinking of the atmosphere consisting mostly of nitrogen and oxygen with some carbon dioxide. Why is there so much argon?

**10.** A tree is cut down today for use in a building. What fraction of the $C^{14}$ in that tree will still remain, in A. D. 4000? (Use Fig. 12.7.)

**11.** Radium has a half-life of 1602 years. Why is there any radium left on Earth considering that the Earth is over 4 billion years old?

**12.** What is the important difference between the S and P seismic waves, and how do we use this to probe the Earth's interior? Assume a value of 0.001 gm/cm$^3$ for the density of air.

**13.** Why do we think that the Earth has a liquid core?

**14.** What is the probable source of the Earth's magnetic field, and why do we think this is so?

**15.** Why does the Moon show more evidence of cratering than the Earth?

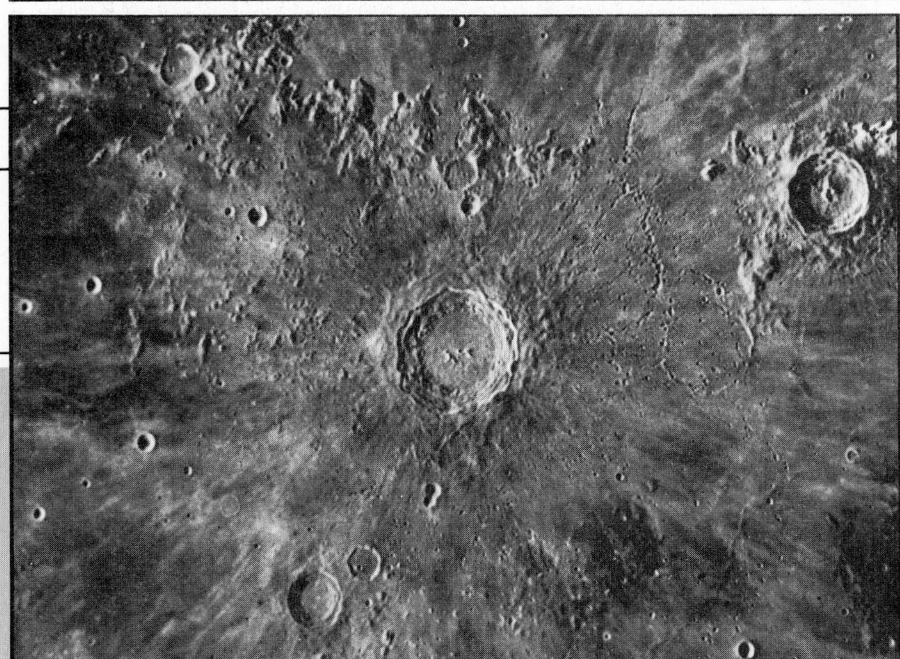

CHAPTER 13

# The Moon

By far the brightest object in the night sky, the Moon has always attracted attention and speculation, inspired poets, musicians, and artists, and at a more utilitarian level provided the basis for the earliest calendars. Even Galileo's introduction of the astronomical telescope in 1609 did not remove the sense of mystery and awe that we can still feel when looking at the Moon.

After years of space spectaculars, visitors to an observatory are still thrilled by their first look at the bleak and shadowed face of the Moon. The space program has brought a revolution in our knowledge of the Moon and our theories of its origin. We shall certainly include recent information as we proceed, but we should start this chapter with a brief description of the Moon as we knew it from a distance.

## 13.1 THE MOON—AS SEEN FROM THE EARTH

**General features** We always see the same face on the Moon, even as it goes through its phases and the different features take their turns in being illuminated. In this case, familiarity breeds oversight, for the simple observation of always seeing the same face presents us with the illusion that the Moon is not rotating. In fact, it must rotate exactly once on its own axis in the same time that it takes to go precisely once round the Earth (Fig. 13.1). An external observer would see this easily—at the center of the motion, we do not notice it. More careful observation shows that the Moon has a small wobble in its rotation (libration), so that we see very slightly more than half of its surface, 59 percent to be precise (Fig. 13.2). The other 41 percent is permanently hidden to Earth—not dark but visible only to lunar orbiters.

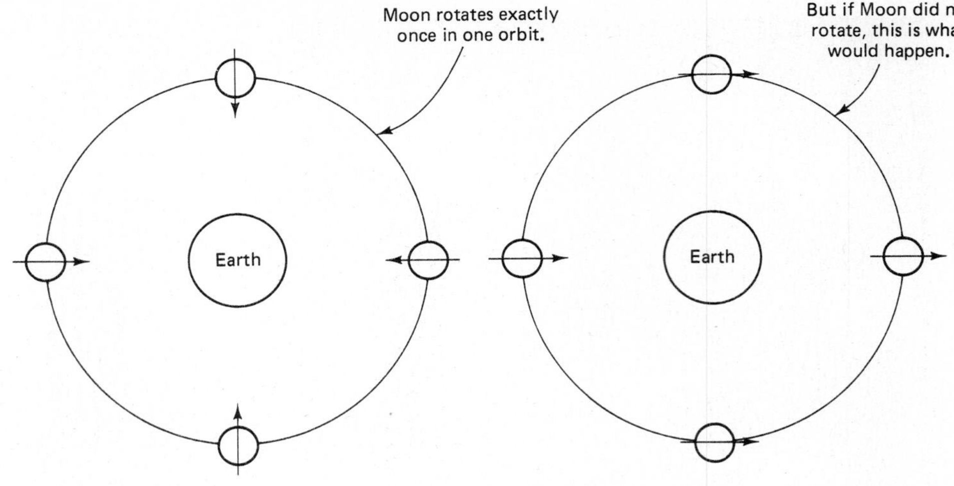


**Figure 13.1** Rotation of the Moon. Because the Moon rotates on its own axis in exactly the same time that it takes to make an orbit around the Earth, it always presents the same face to us.

**Figure 13.2** Lunar libration. Small irregularities in the lunar rotation and some geometrical effects allow us to see slightly more than half of the lunar surface. In these two views, the small dark area (Mare Crisium, lower left) can be seen to have moved distinctly from the lunar limb (Lick Observatory).

With the naked eye, we see bright areas (**highlands**) and dark areas (the seas or **maria**) (Figs. 13.3 and 13.4). Through the telescope we find more detail: prominent mountain ranges (Figs. 13.5 and 13.6) and valleys in the highlands; narrow gulleys or **rilles** about a kilometer wide, some stretching for hundreds of kilometers. There are ridges. There are craters, some overlapping, from which we can deduce their relative ages. Some craters show patterns of ejected material (Fig. 13.7). From the lengths of shadows, it had been deduced that the maria are on the average about 3 to 4 km lower than the brighter highlands. The battle-scarred nature of the highlands suggested a history in which the craters were formed at a time when the Moon was solid enough to retain those features. The flatness of the maria pointed to their later filling, and after those "seas" had solidified, some cratering still persisted. Until the Apollo landings there was no way to settle conclusively the dispute as to which if any of the features arose from impacts and which arose from volcanism. Nor was there any way of calibrating the time scale of the sequence of events that had sculpted the surface.

By watching closely as stars were occulted behind the Moon, it was shown that the Moon had no atmosphere—just as we would expect from the discussion in Chapter 11.

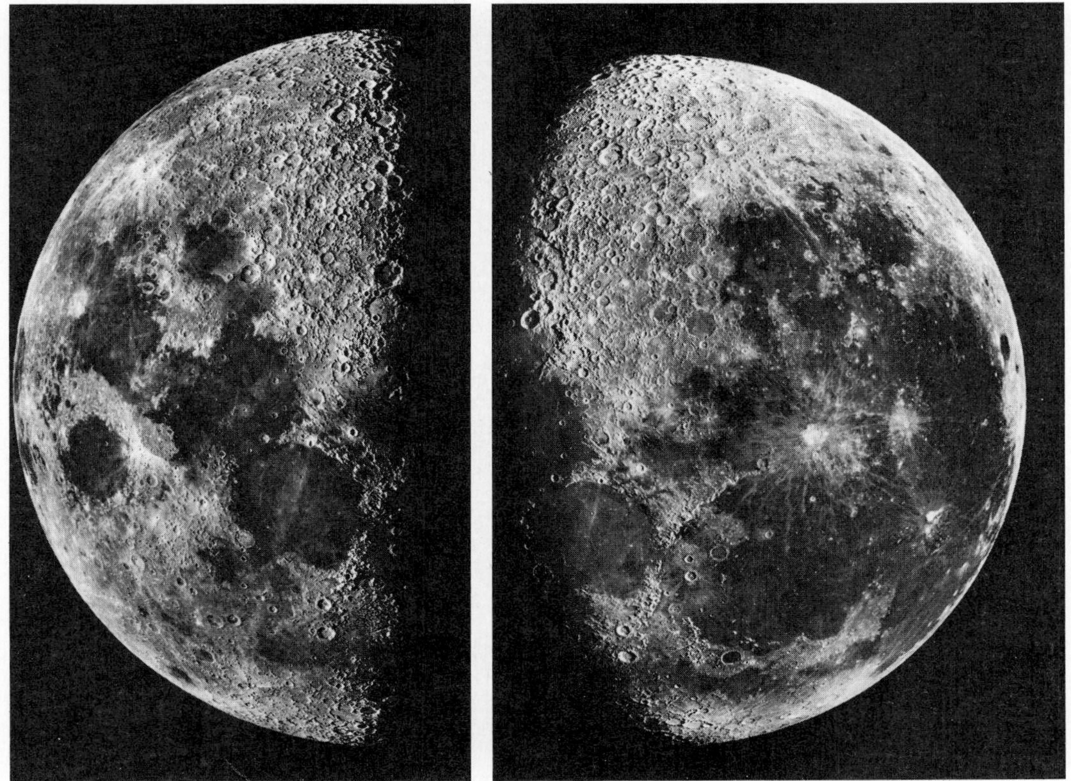

**Figure 13.3** Photographs of the moon taken at first and last quarters (Lick Observatory).

**Figure 13.4** Key to some of the major features on the first and last quarter Moon.

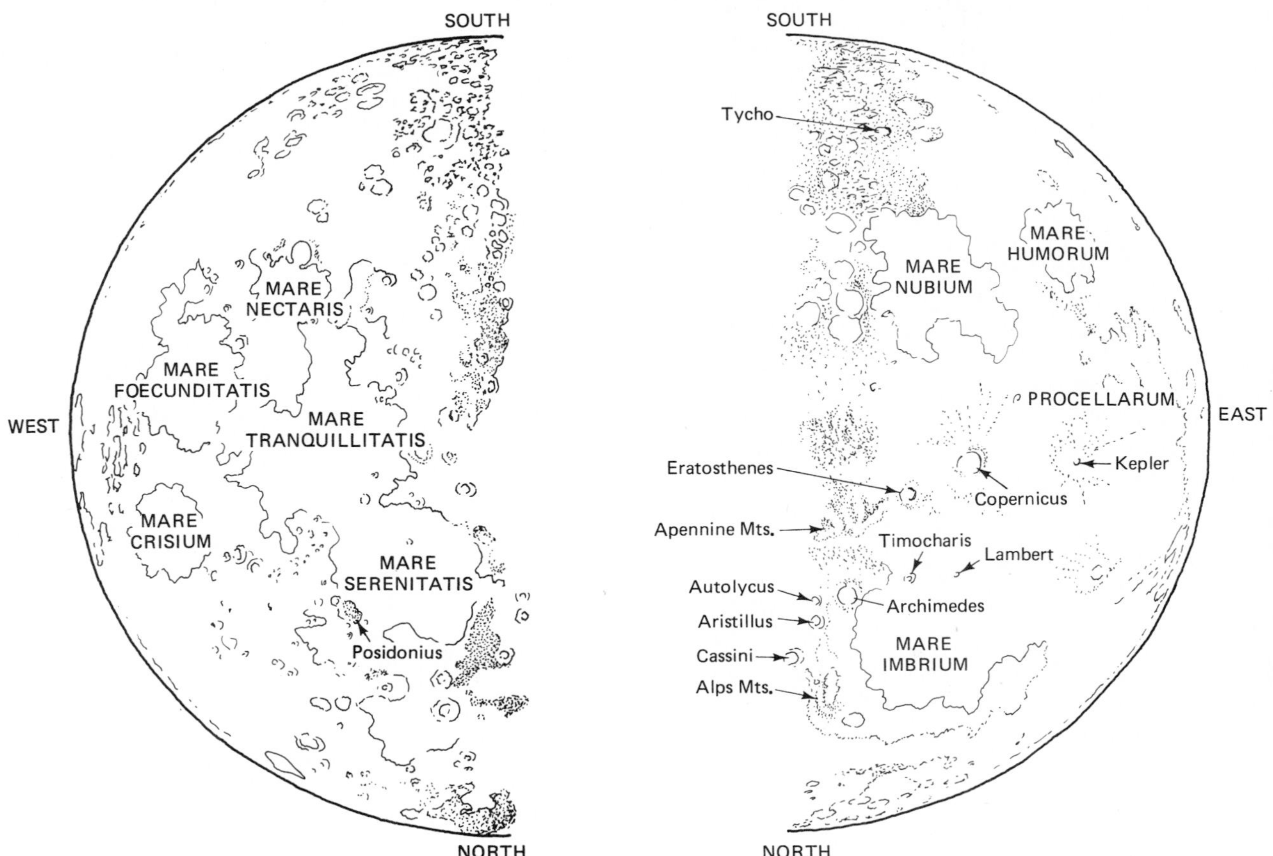

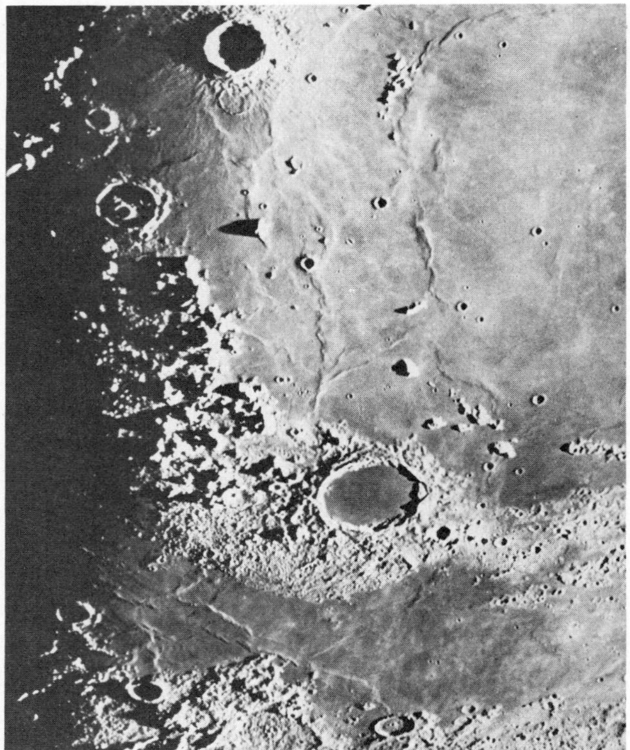

**Figure 13.5** Lunar mountains: the Alps (Lick Observatory).

**Figure 13.6** Lunar mountains, the Appenines, with crater Archimedes (largest crater in this view), Aristillus and Autolycus (to right) and Timocharus (left, with central peak) (Lick Observatory).

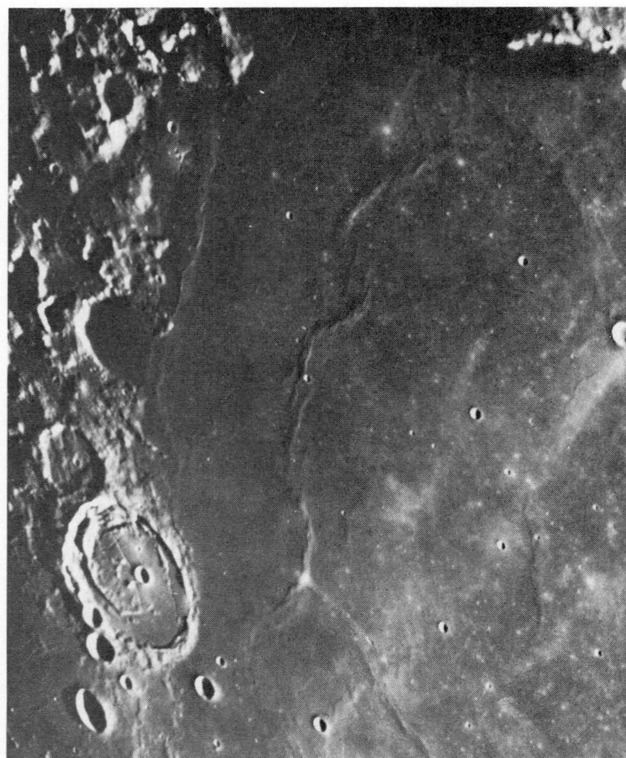

**Figure 13.7** Crater Posidonius at the edge of Mare Serenetatis (Lick Observatory).

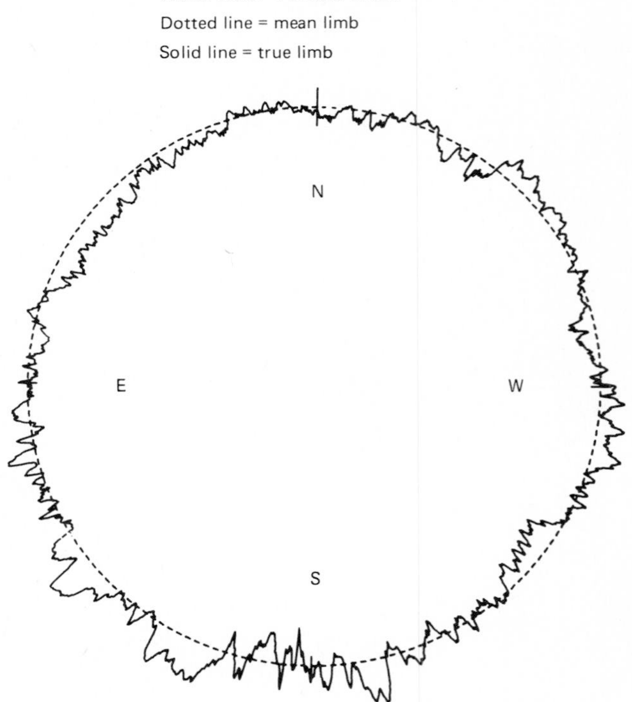


**Figure 13.8** Lunar limb profile. The scale of the limb variations is enlarged sixty times relative to the overall lunar radius (U.S. Naval Observatory).

Once its distance was known, the size of the Moon could be calculated. Not quite spherical, its mean radius is 1738 km, a little over one quarter the size of the Earth (Fig. 13.8). From its orbit around the Earth, the Moon's mass was reasonably well known (see Box 13.1).

## BOX 13.1
*Center of Mass*

Along the line between any two masses ($M_1$ and $M_2$) a distance $D$ apart, there is an important point known as the **center of mass** (CM) (Fig. 13.9). Its position is defined by

$$M_1 L_1 = M_2 L_2$$

where $L_1$ and $L_2$ are the distances from CM and $D = L_1 + L_2$.

A related but more familiar term is the **center of gravity**. If the two masses, $M_1$ and $M_2$, are suspended from a horizontal beam near the Earth's surface, the beam will balance if it is pivoted at the center of gravity (CG) between $M_1$ and $M_2$. The location of the CG is calculated in the same way as is the CM; the "gravity" in the name derives from the balancing of the gravitational forces exerted by the Earth on the two masses.

If one mass is much greater than the other, then the CM will be very close to the center of the larger mass. A good example is provided by the motion of the Moon around the Earth. Although it seems as though only the Moon is moving, in fact both Moon and Earth are in orbit about their mutual CM. Careful measurement of the movements of the Earth and Moon have shown that the CM is 4645 km from the center of the Earth. Since the average radius of the Earth is 6368 km, this places the CM of the Earth-Moon system inside the Earth.

The average Earth-Moon distance is 384,000 km, so the Moon must be on average 379,755 km from the CM. The ratio of the two masses is then given by

$$\frac{M_E}{M_M} = \frac{379755}{4645} \equiv 81$$

The Moon's mass is thus $\frac{1}{81}$ of that of the Earth, or $7.35 \times 10^{25}$ g.

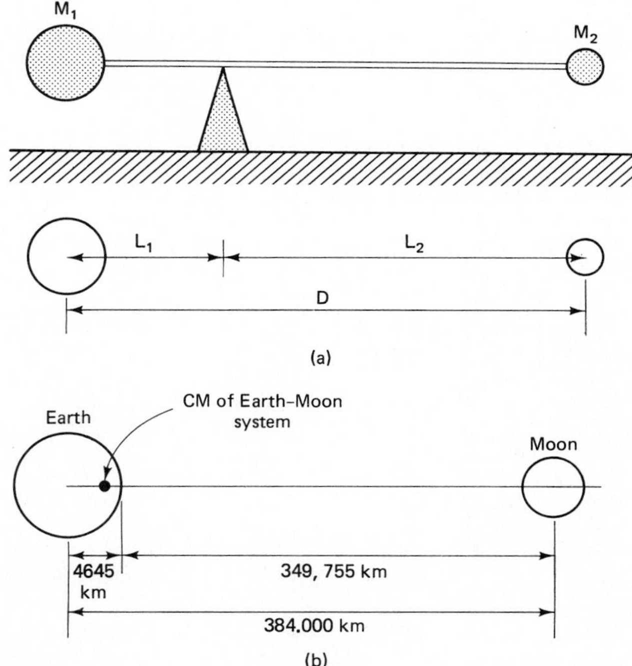


**Figure 13.9** Centers of gravity and mass. (a) Two masses will balance near the Earth's surface when they are supported at their center of gravity. (b) The center of mass of the Earth-Moon system is inside the Earth.

**Figure 13.10** View across the central highlands towards the Oceanus Procellarum and crater Landsberg (39 km diameter) from Apollo 16. Mountains in the foreground (Montes Riphaeus) are the unflooded remnants of the rim of a large basin (NASA).

Combining the mass and diameter for the Moon, we find its mean density to be 3.3 g/cm$^3$, similar to the Earth's crust but significantly less than the average for the whole Earth. We must conclude that the Moon cannot have a high-density core, and therefore it is unlikely to have a magnetic field which a dense (probably iron) core would have implied.

**Lunar dynamics** A perfectly uniform sphere will respond to external forces in a fairly simple fashion, and a billiard-ball-type Moon would follow a simple orbit around the Earth. The ball used for bowling in the U.S. is smooth and uniform, but in England a different form of bowling uses a ball with an off-center weight so that the ball rolls along a curved path that depends on the precise position of the weight at release. In a somewhat similar way, we can see from the Moon's motion whether there is any sign of nonuniform distribution of its internal mass. The fact that the same side always faces us then tells us that there must be some internal handle so that the Earth's gravitational attraction can over a long time pull more strongly on one side than the other, not enough to pull the Moon out of its orbit but enough to slow its initial speed of rotation into its present synchronous rate. (The complexity of the Moon's motion indicates that other processes probably also play important roles.)

**Figure 13.11** Sinuous rille across lava that shows very little cratering. Length of rille shown is about 100 km and width about half a kilometer. The crater diameter is about 2.5 km (NASA).

**Figure 13.12** Ridge of a buried crater can be seen with later small craters (NASA).

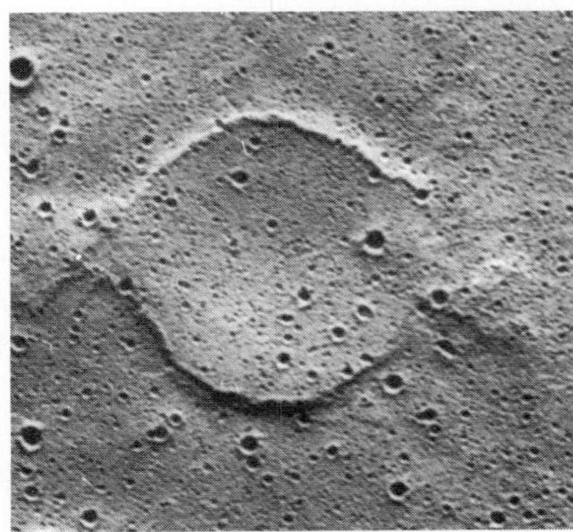

The Moon's motion around the Earth produces the ocean tides. The Moon's gravitational attraction on the Earth is strongest on the closest side and weakest on the farthest side. The continents are not free to respond to this difference in force, but the oceans are, and they move in the daily pattern we call tides. The tidal patterns, which are obviously of great importance on Earth, are hard to calculate with precision being influenced by the shapes and depths of the oceans, but there is one astronomical aspect we should note. The regular sloshing back and forth of the oceans uses up some of the rotational energy of the Earth. As a result the rotation of the Earth is decreasing at present by about 2 msec per century.

Calculations show that this effect is accompanied by a slow drifting of the Moon away from the Earth at a rate now actually measured to be about 3 cm per year. Billions of years ago the Moon and Earth must have been very much closer, with the Earth's ''day'' only 5 or 6 hr. Some corals and shellfish show daily and monthly growth patterns that correspond to the tides. Careful counts of these growth rings on fossils give support to the estimate of different lengths of the day and month far back in the past. (A word of caution is again needed: These fossil measurements have generated considerable professional controversy, and their reliability does not yet match the ingenuity of their concept.)

**Heating and cooling** Most moonlight is reflected sunlight, but we can observe the Moon's own IR radiation at new moon and during an eclipse. With a measured albedo of 7 percent, the Moon absorbs 93 percent of the incident sunlight. Because of the slow rotation of the Moon this leads to a very high surface temperature of about 370 K (nearly the boiling point for water) at high lunar noon but around 120 K at lunar midnight. During a lunar eclipse, the surface of the Moon cools rapidly as it enters the Earth's shadow. In an hour, the temperature may drop by around 150 K. Measurements of the rate of this cooling tell us something about the thermal properties of the lunar surface. A solid rock surface would cool very slowly—just check how brick and stone buildings on Earth retain their heat at night. The rapid lunar cooling indicated a surface of finely divided dust, as was later confirmed during the Apollo landings. The effect is not uniform, however, and lunar hot-spots have been noted in some of the craters.

## 13.2 THE LANDING SITES ON THE MOON

Three types of probes have landed on the lunar surface. The early Rangers were designed to crash into the surface transmitting pictures almost until the moment of impact, but the following Surveyors landed softly and made measurements on samples of the surface that they scooped up. The period between 1969 and 1972 was one of great excitement, as the Apollo missions returned with their lunar samples. In addition to collecting carefully selected samples, the astronauts made measurements and set up automated monitoring devices that could continue to operate after the astronauts departed. The Soviet probes were all unmanned. Their Luna series retrieved samples of the soil and returned them to Earth. Two missions carried automated Lunakhod vehicles for traversing the surface and telemetering their data back to Earth. The locations of the soft landings are shown in Fig. 13.13.

The Apollo landing sites were

Apollo 11: in the Mare Tranquillitatis, a generally smooth and level area but marked with craters ranging from less than 1 cm diameter to tens of meters. Samples included

1. **basaltic** rocks (Fig. 13.14), that is, similar to the rocks associated with volcanoes on Earth.

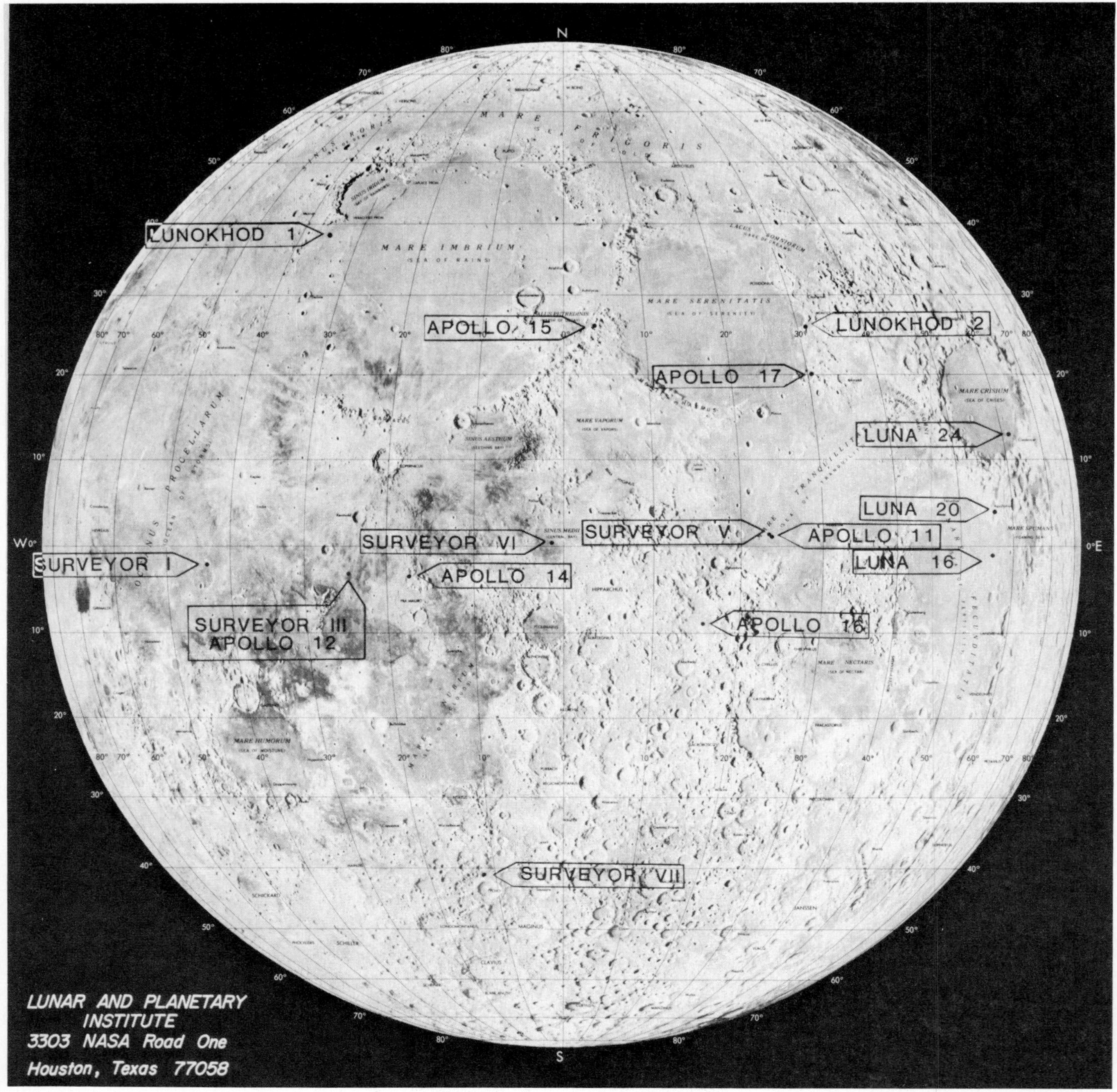


**Figure 13.13** Landing sites for the Apollo, Surveyor, Luna, and Lunokhod missions that had soft landings (Lunar and Planetary Institute, NASA).

2. **breccias** (Fig. 13.15), a name given to rocks formed by compaction of a mixture of soil and smaller rocks.
3. a soil of dustlike, fine grains including many glassy fragments (Figs. 13.16, 13.17 and 13.18).

Apollo 12: in Oceanus Procellarum not far from Surveyor 3 in a ray that seems to have come from the prominent crater Copernicus. This region has fewer craters than the Apollo 11 site. As with Apollo 11, samples included

**Figure 13.14** Vesicular basalt rock collected by Apollo 15. Note the many bubbles produced by trapped gas as the rock cooled and solidified (NASA).

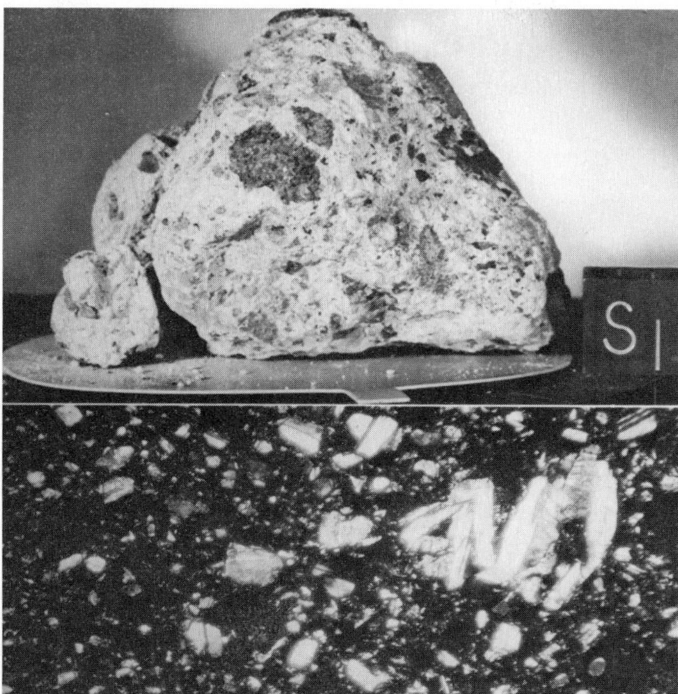

**Figure 13.15** Breccia collected by Apollo 15. In the upper photograph, different types of small rocks can be seen compressed into the single large rock. The lower photograph shows an enlarged view of a thin section in which the different components stand out clearly (NASA).

**Figure 13.16** Fine texture of the lunar soil is shown by the great detail in this footprint. A coarse material would not have shown such fine detail (NASA).

**Figure 13.17** Particle from the lunar surface. Included in the soils are fine particles with a great variety of shapes. Glassy spherule has 0.175 mm diameter with green color and surface fractures probably due to impacts and also with irregular additions (NASA).

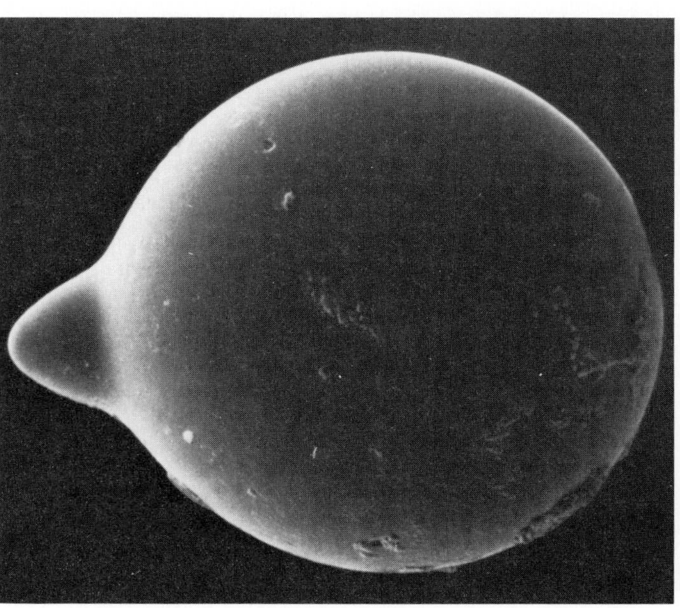

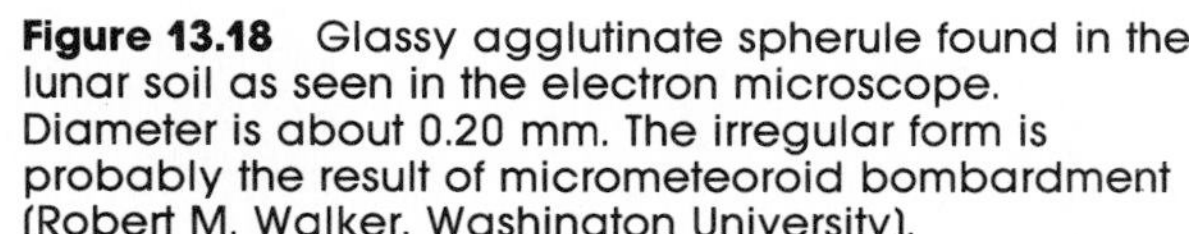

**Figure 13.18** Glassy agglutinate spherule found in the lunar soil as seen in the electron microscope. Diameter is about 0.20 mm. The irregular form is probably the result of micrometeoroid bombardment (Robert M. Walker, Washington University).

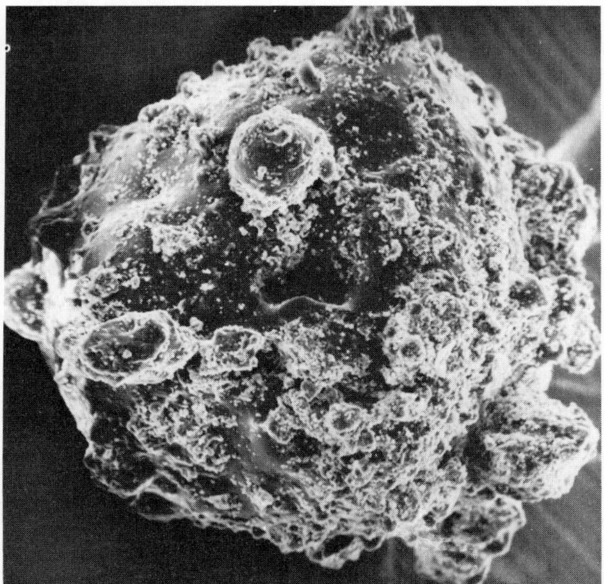

basaltic rocks, breccias, soil, and also parts of Surveyor 3 brought back to be examined for effects of solar wind and the cosmic radiation. In many features the Apollo 12 samples resembled those of Apollo 11, but a new type of rock fragment was found with considerable enrichment of trace elements potassium (K), the rare-earth elements (REE, those with atomic numbers between 58 and 71), and phosphorus (P)—hence designated KREEP. This new material is sufficiently different from that of the rest of the site to suggest that it came from the crater Copernicus. The age of Copernicus can then be estimated to be about 1 billion years.

(Apollo 13 did not land. A malfunction en route led to the decision to return to Earth.)

Apollo 14: in the Fra Mauro formation, surrounding the Mare Imbrium and believed to be ejecta from a giant impact that produced the Imbrium basin. Very large blocks of breccia were observed. The returned samples were almost all breccia that themselves consisted of breccia of breccia. The picture that emerged is one of continued impacts each leading to the compaction of surface material into the next generation of breccia.

Apollo 15: on a mare surface of Palus Putrendis on the eastern rim of Mare Imbrium and at the base of the Appenine Mountains near Hadley Rille (Fig. 13.21). The samples were similar to those from Apollo 12, and the cratering in the two regions was also similar. High-resolution photographs showed very clearly that strata were formed by successive lava flows, often meters in thickness. Samples from the highland massif, Hadley 6, were of a quite different type of rock, mostly **anorthosite** (basaltic rock, composed mostly of oxides of silicon, aluminum, and calcium). Some were crystalline, indicating far less brecciation in this region. Another feature of the anorthosite was the low concentration of radioactive elements.

**Figure 13.19** Crater Copernicus, with Eratosthnes to the right, located at the start of the Appenines. Note the rays of ejecta from Copernicus. (Lick Observatory).

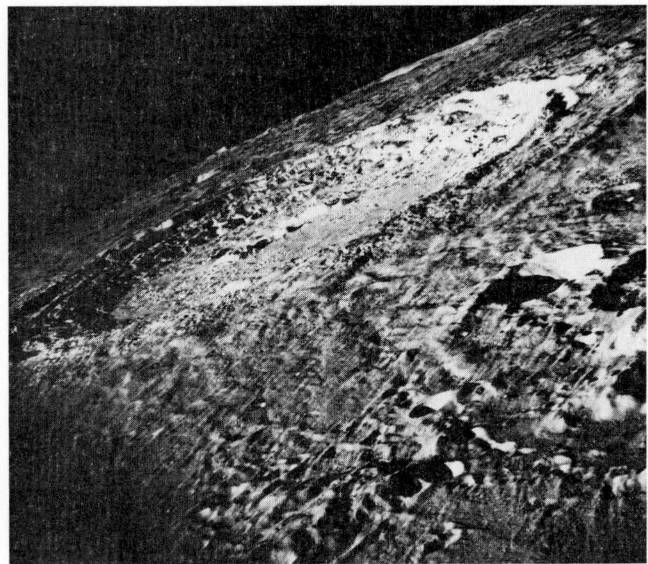

**Figure 13.20** Copernicus as seen from Apollo 17 (NASA).

**Figure 13.21** James Irwin with the Apollo 15 rover at the rim of Hadley Rille (NASA).

Apollo 16: in the central highlands north of the crater Descartes, representative of a region thought to have been subjected to volcanic activity. It was quite a surprise, therefore, for the astronauts to find that most of the surface rocks were breccias rather than igneous (formed from solidification of molten lava). Analysis of returned samples confirmed this showing a high abundance of aluminum and calcium (Fig. 13.22). Some samples were found to have a rustlike coating consisting of an iron hydroxide now thought to have been possibly formed on the lunar surface after a cometary impact that was followed by water vapor in the comet's tail.

Apollo 17: on the southeast of Mare Serenetatis in a valley between two large massifs, one of the oldest lunar features (Fig. 13.23). Surface obser-

**Figure 13.22** Collection of a sample during Apollo 16 mission (NASA).

**Figure 13.23** Apollo 17 landing site in the Taurus-Littrow valley (NASA).

vations suggested relatively recent cratering with many blocks at the base of the massifs as though they had slid down the slopes. Samples were largely basaltic, with high titanium content being correlated with the dark appearance. Some samples from the Shorty crater had an orange-colored layer that was found to consist of very fine grained glass spheres probably from volcanic eruptions.

## 13.3 FINDINGS OF THE LUNAR MISSIONS

A measure of the richness of the scientific results of the lunar program is provided by the published papers from each year's lunar (now lunar and planetary) science conference, totaling 3500 pages each year. To these must be added the content of journals such as *Icarus* and *The Moon*. Trying to cover this torrent of information in only this one chapter requires some ruthless editing.

**The lunar craters and basins** The earliest lunar probes, the Soviet Luna series and the U.S. Rangers and Surveyors, produced two major results. They extended the surveys of the surface to that side always hidden from view and showed a moonscape apparently more cratered and scarred than the familiar face (Fig. 13.26). Maria were few and small, and the cratering record was thus well preserved. The range of known crater sizes has been extended up to the gigantic Orientale (Fig. 13.27) (with an outer ring of 930 km diameter) and, as was shown by the Ranger landers, down to smaller and smaller craters or impact pits at the limit of resolution of the cameras. Closer inspection of lunar samples shows that the spectrum of sizes of impact pits continues down to the microscopic: Micrometeorites arriving at high speeds produce pits less than $\frac{1}{1000}$ mm diameter, visible only after magnification by an electron microscope.

The largest craters are marked by extensive blankets of ejecta often with spectacular rays. There are in all 125 craters with diameters greater than 100 km. With no atmosphere to weather them, these features will remain for eons until altered by newer and overlapping craters and their debris (Figs. 13.28, 13.29, 13.30, 13.31, and 13.32). The near circular form of the giant maria look like filled-in craters, and this impression is supported by the second surprise of the early missions, the discovery of **mascons**. In 1968, spacecraft orbiting the Moon displayed small but quite unpredicted devia-

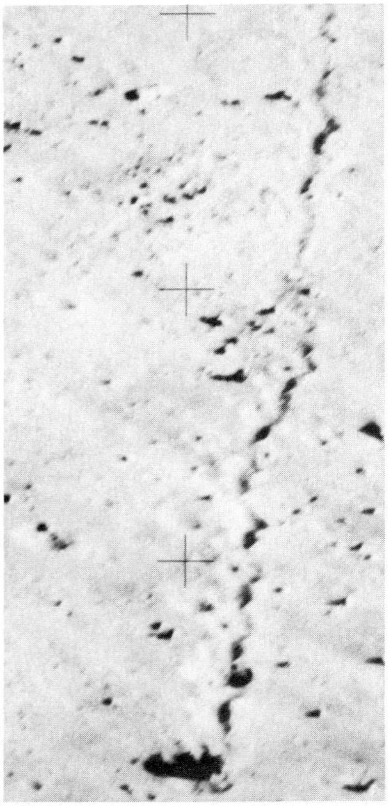

**Figure 13.24** Track left by a boulder that had rolled downhill, photographed from the Apollo 17 lunar orbiter. Boulder diameter is about 5 m (NASA).

**Figure 13.25** Far side of the Moon: This view, from Apollo 16, is centered on what is usually the eastern limb of the Moon that we see from Earth. Mare Crisium is the dark area shown at the limb, and the other two are Marginis (to the right of Crisium) and Smythii (below). Unlike the familiar face, the far side is almost devoid of mare and shows an almost entirely cratered surface (NASA).

**Figure 13.26** The large double crater is Van de Graaff, 250 km long filled with mare materials and also with some ejecta from crater Birkeland (largest crater, top left). This area showed the highest levels of natural radiation on the lunar far side (NASA).

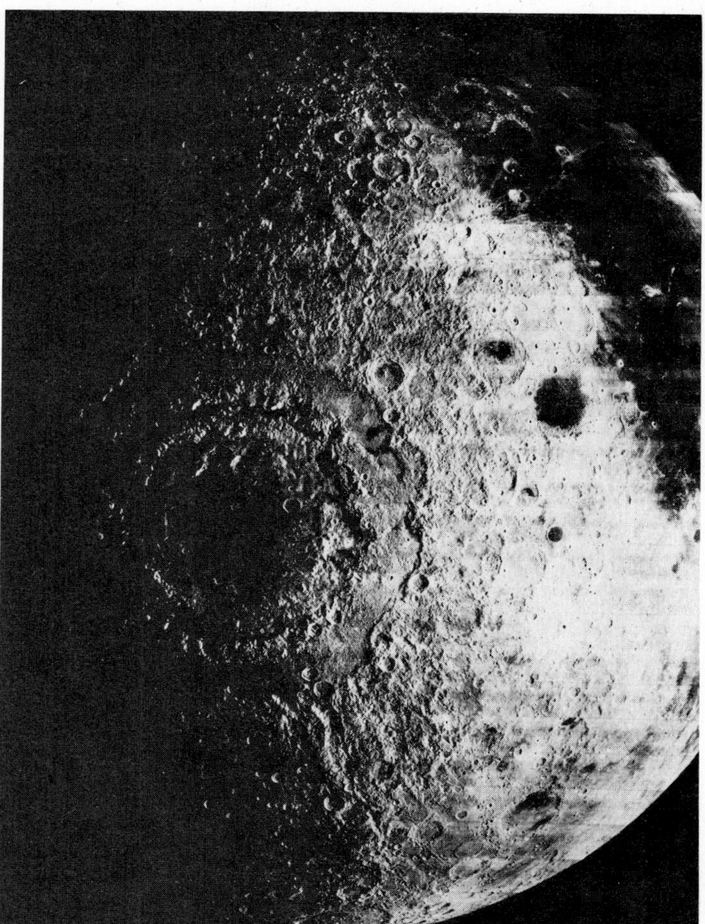

**Figure 13.27** Orientale crater on the far side, 930 km diameter. Unlike the largest craters that are seen on the near side, Orientale was not subsequently flooded with basalts. Its multiringed structure is still relatively intact indicating that this is one of the youngest basins on the Moon (NASA).

**Figure 13.28** Crater Lambert, 30 km in diameter. To the right of the crater there is a relatively young lava flow that has partly covered some of the ejecta from Lambert (NASA).

**Figure 13.29** View at liftoff of the Apollo 17 lunar module showing the craters Eratosthenes (60 km diameter) and Copernicus (on the horizon). The central spike of Eratosthenes is typical of many craters (NASA).

**Figure 13.30** Crater Euler (35 km diameter) with central spike. Later lava flow has partly covered some of the ejecta (NASA).

**Figure 13.31** Crater Kriger (center) is in a region that shows many of the features considered to indicate basaltic lava flow. Relative youth of Kriger is shown by its ejecta lying on top of the mare basalts. Kriger is considered to be volcanic, and some of its lava flows are indicated. Rilles wind from Kriger as well as from another crater just out of this field of view. A small impact crater (8 km diameter) is prominent near the top of this photograph (NASA).

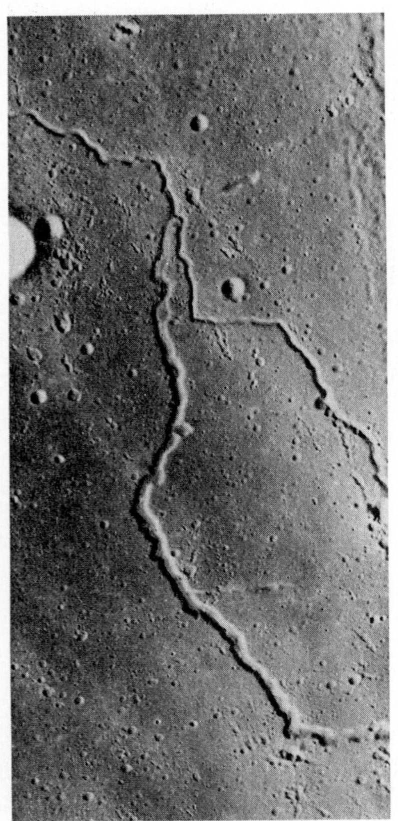

**Figure 13.32** Close-up of area shown in Fig. 13.31, showing sinuous rilles. Origin of rilles is not fully understood. They seem to be a product of lava flow or of fracturing of the surface (NASA).

tions in their flight paths, which can be explained by gravity anomalies due to concentrations of more massive material hidden close to the lunar surface. Mascons have been located at most of the circular maria but not at the irregularly shaped maria.

Two classes of theories have been considered to explain the mascons. The first suggested that the mascons were the remains of large, colliding objects that had produced the maria basins and subsequently been covered over. Later observations showed that the mascons were very shallow and could not conceal deeply hidden objects. The weight of opinion has now settled on the second model: The maria basins were indeed large craters, but the lava that later flooded them was substantially denser than the surrounding material thus producing a very local increase in the gravitational pull on an orbiting object.

Not yet fully understood is the frequent appearance in a crater of a central spike in an otherwise flat base. It is thought that this represents a complex rebound effect, but to date a precise theory has not been produced.

The first lunar samples settled one major question very quickly: the nature of the maria. Whereas earlier ideas included asphalt, dust, or sedimentary rocks, the samples showed strong similarities to basaltic lavas that we find on Earth near volcanoes, and the consensus now is that lava flows filled up the gigantic basins, not recently but at least by 3 billion years ago.

**Samples from the lunar surface** The bombardment that produced the craters spread an uneven blanket of debris across the lunar surface. Although many of the older features have survived for billions of years, the continuous rain of small meteorites, even to the present day, has churned this surface layer (a process termed **gardening**) and shattered many of the rocks so that

*Table 13.1* CHEMICAL COMPOSITION OF LUNAR MATERIAL MAJOR CONSTITUENTS[a]

| | *Lunar Highlands Surface (Percent)* | *Moon: Bulk (Percent)* | *Earth: Mantle (Percent)* |
|---|---|---|---|
| $SiO_2$ | 45 | 42 | 46 |
| $Al_2O_3$ | 25 | 8 | 4.3 |
| CaO | 14 | 6 | 3.1 |
| MgO | 8.6 | 31 | 38 |
| FeO | 6.6 | 12 | 8.2 |

[a] Ross Taylor, "Structure and Evolution of the Moon," *Nature: Vol. 281*, pp. 105–110, 1979.

much of the surface is now covered by a fine dust, as shown so clearly by the footprints of the astronauts (Fig. 13.16) and by the returned samples. This layer of pulverized debris that now covers the Moon is 4 to 5 m thick and is termed the **regolith**. This is the layer whose general features had been indicated by those measurements of cooling during eclipses. The fine lunar soil samples also included glassy spherules indicative of an origin in impact melting.

Many of the rock samples brought back by the astronauts resembled rock types familiar on Earth. Maria rocks were mostly **basalts** that form from the cooling of molten material such as lava, as can be seen by their crystalline structure. These tend to be dark in color. Rocks from the highlands were anorthosite, light in color and of a type relatively rare on Earth. Their structures show that like the basalts they formed from molten material but much more slowly. From highlands and maria, breccia samples were obtained, their forms showing their history of compaction from rock fragments and minerals. Finally, a less common type of rock was found in the KREEP (see Apollo 12) with its high concentrations of potassium, phosphorus, and rare-earth elements. The chemical composition of lunar samples showed broad similarities to terrestrial material but also some striking differences (see Table 13.1), and the same pattern persists when attention is focused on isotopic abundances of some elements. The more we know of the elemental and isotopic abundances, the harder it becomes to invent simple scenarios for the formation of the bodies of the solar system, and we shall return to this theme in Chapter 18, after we have accumulated the data on the planets and other bodies.

In pursuing comparisons and drawing conclusions, we must remember just how fragmentary our sampling has been—a few places on the Moon, many more on Earth, supplemented by inspired guessing as to just how representative these sites are. To emphasize this problem, just consider what your view would be of the composition of the Earth if you were confined to about a dozen sites chosen at random. Most would be in the oceans since they cover most of the Earth's surface. Would the remainder include all of deserts, the arctic or antarctic, mountains, glaciers, forest, plains, jungles, lakes, rivers, etc.?

The lunar materials generally show smaller proportions of the **volatile** elements, that is, those that have low melting points and can evaporate more easily. There are relatively high proportions of **refractory** elements, that is, those with high melting points such as calcium, aluminum, and titanium. Uranium, thorium, and the rare-earth elements (see Apollo 12 section) are all relatively more abundant on the Moon than on Earth. No water at all was found on the Moon.

**Lunar chronology** Before the Apollo landings, dating of the lunar surface was relative. Overlapping craters and the smoothness of the maria could only tell us the sequence of some events, not the length of time involved nor the delays between formation of different features. Radioactive dating of the samples has now given us an absolute scale of dates—again, with a caution. To date a rock as being 3.5 billion years old tells us only the age since it last solidified and retained the isotopes we are now measuring. Once solidified, the relative abundances of radioactive atoms and their daughter products are governed by the well-established laws of radioactive decay, but rock material that is still molten can be mixed, lose some of its volatile constituents, and change its composition. Lead isotope and rubidium (Rb)-strontium (Sr) ratios measured in the lunar rock samples all yielded ages in the range 3.1 to 4.46 billion years. The highland samples (3.9–4.46 billion

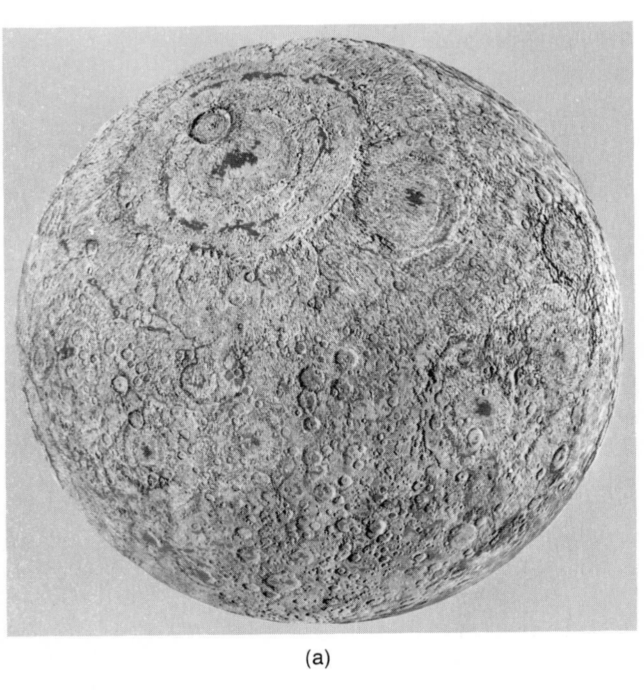

(a)

(b)

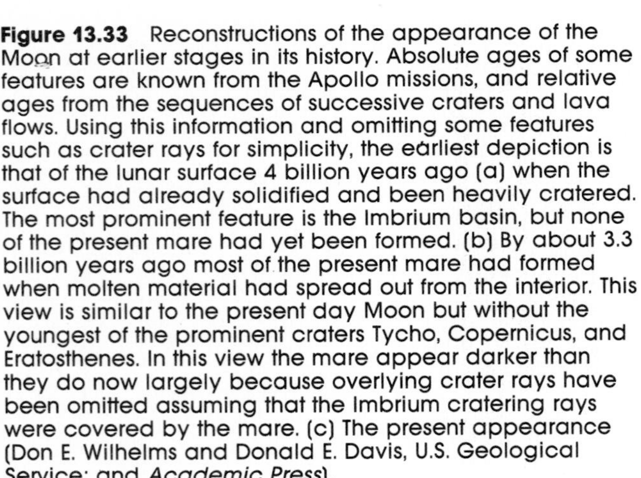

(c)

**Figure 13.33** Reconstructions of the appearance of the Moon at earlier stages in its history. Absolute ages of some features are known from the Apollo missions, and relative ages from the sequences of successive craters and lava flows. Using this information and omitting some features such as crater rays for simplicity, the earliest depiction is that of the lunar surface 4 billion years ago (a) when the surface had already solidified and been heavily cratered. The most prominent feature is the Imbrium basin, but none of the present mare had yet been formed. (b) By about 3.3 billion years ago most of the present mare had formed when molten material had spread out from the interior. This view is similar to the present day Moon but without the youngest of the prominent craters Tycho, Copernicus, and Eratosthenes. In this view the mare appear darker than they do now largely because overlying crater rays have been omitted assuming that the Imbrium cratering rays were covered by the mare. (c) The present appearance (Don E. Wilhelms and Donald E. Davis, U.S. Geological Service; and *Academic Press*).

years) were older than the maria basalts (3.2–3.9 billion years); KREEP basalts were in the range of 3.4 to 3.6 billion years.

The general picture that emerges is as follows: The Moon formed 4.6 billion years ago with the top surface still molten at about 4.4 billion years ago. Heat for this melting came (in a presently unknown proportion) from internal radioactivity and from intense bombardment (converting kinetic energy to heat). The major period of cratering was between 3.9 and 4.2 billion years ago, followed 3.9 billion years ago by internal heating that produced the volcanism and lavas that flooded the largest of the bombardment basins. The lava flow stopped about 3.1 billion years ago, and since then there have been occasional large impacts with very steady gardening of the surface but no large-scale melting (Fig. 13.33).

The gardening can be investigated by analysis of core tube samples. On several Apollo missions, the astronauts hammered hollow tubes 2.4 m down into the surface collecting samples with the relative positions of their strata preserved. The microscopic effects in crystals caused by the cosmic rays and the solar wind decreases with depth below the surface. It is affected by the gardening that slowly brings fresh material to the surface and moves surface material down. From the core measurements, it has been inferred that the upper 0.5 mm of material is turned over almost 100 times in a million

**Figure 13.34** (a) Seismograms recorded on the lunar surface July 17, 1972. The signal is believed to be from a meteoroid impact on the far side of the Moon (time indicated as I). Arrival of the P and S waves are also indicated (NASA).

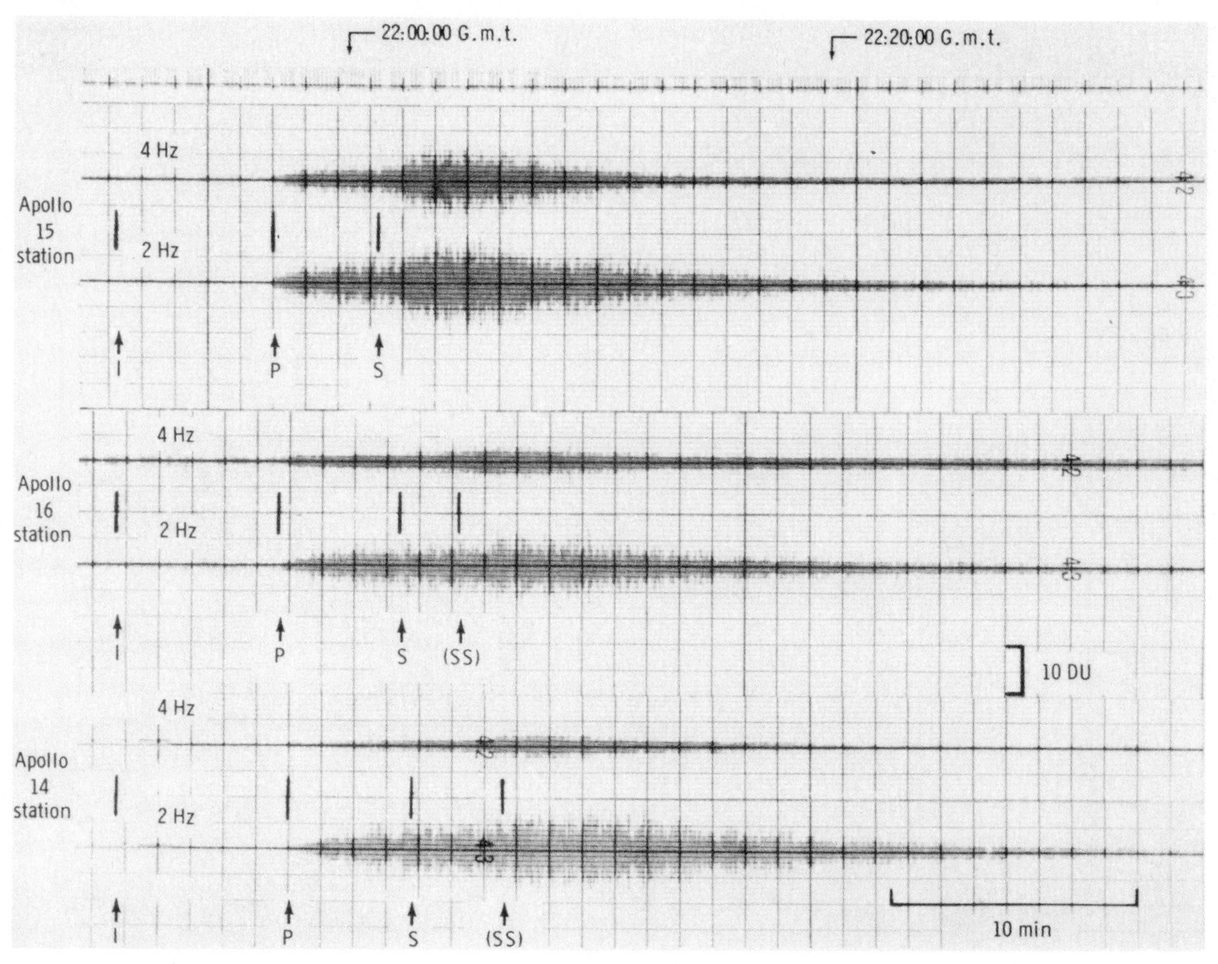


(a)

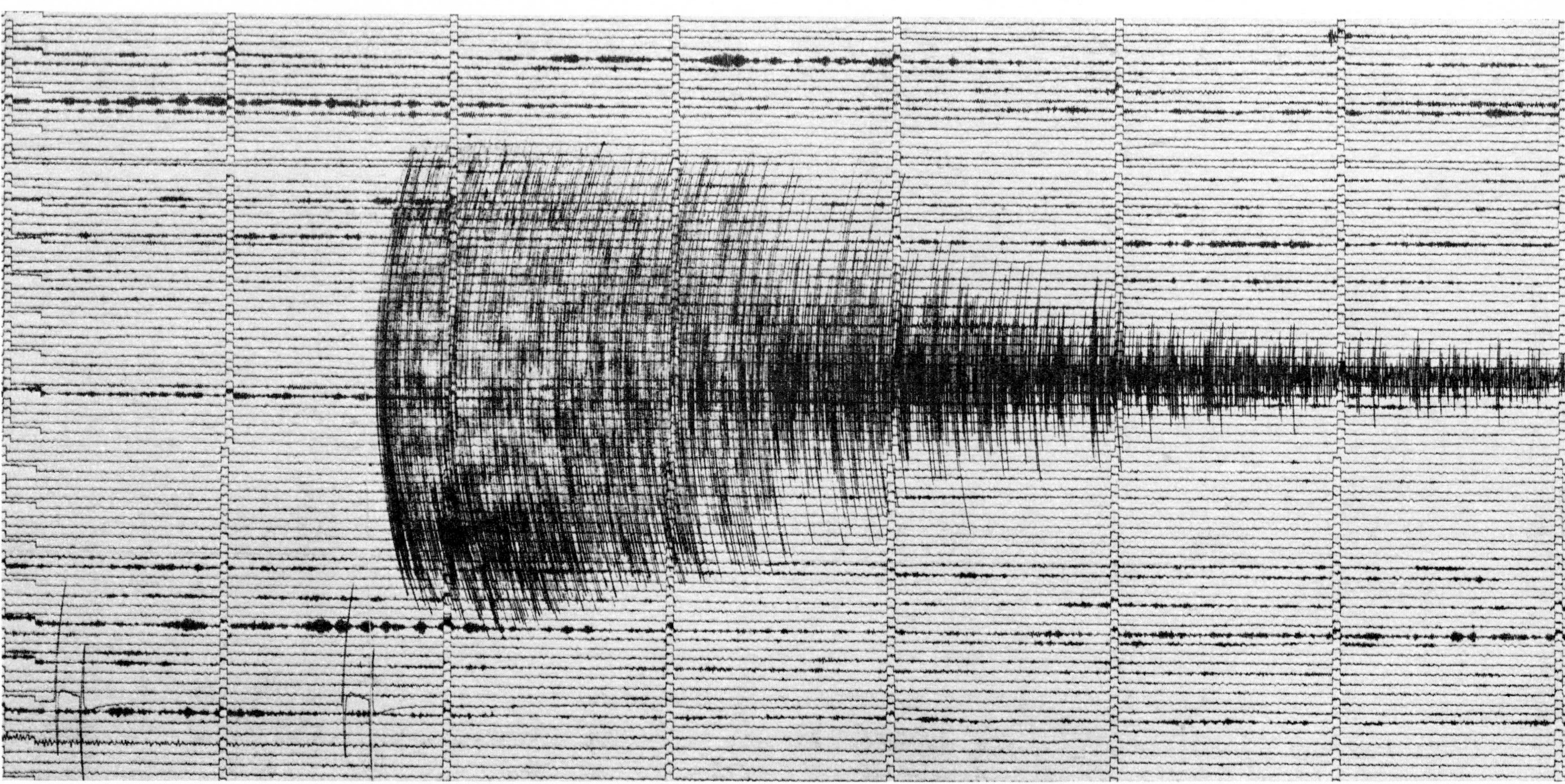

**Figure 13.34** (b) Seismogram recorded near St. Louis on May 14, 1983. The earthquake that produced these signals had a magnitude of 4.4 and was located in Illinois 113 km from the recording station and at a depth of 12 km. Relatively long duration and high wave frequencies (typical of eastern United States earthquakes) makes this seismogram somewhat resemble the lunar seismogram (Otto W. Nuttli, St. Louis University).

years, whereas at 1 cm depth perhaps 50 percent of the material has been untouched. Only after 10 million years has the top 1 cm been turned over once, and at depths between 10 and 100 cm, the time scale is around a billion years. The Apollo 15 core showed no turnover at 2 m for 500 million years. Within the top 0.5 mm, the activity was intense and complex. Electron microscope and X-ray studies of individual mineral grains show many to have a damaged surface layer less than $10^{-6}$ cm thick, the result of the arrival of many solar wind particles.

**The interior** As on the Earth, the Moon's physically inaccessible interior can be probed by seismometers. Four seismic stations were set up by the astronauts and continue to telemeter their data back to Earth. Meteorite impacts and moonquakes have been detected. The Moon is far quieter seismically than the Earth. Most moonquakes are less than 2 on the Richter scale, and about 3000 per year are recorded. The total annual seismic energy thus released on the Moon is $2 \times 10^{13}$ erg, about $10^{12}$ times less than on Earth or about the equivalent of 1 lb of TNT. Analysis of the seismic wave patterns suggests the presence of a small, molten core surrounded by a solid mantle and then the outer solid crust that we see and sample.

The magnetic field on the lunar surface is generally far smaller than that on Earth, but some of the Apollo 11 rock samples showed evidence of there having been a slightly stronger field present when they solidified. The origin of this field and the reasons for its subsequent decrease are not yet understood.

## 13.4 THE ORIGIN OF THE MOON

The Apollo program has led us to a very broad understanding of the major stages in the evolution of the Moon over the 4.6 billion years since it solidified. There are, of course, many details that are not well understood, but there is little dissent from the scenario outlined briefly in the preceding section. When we turn to the next questions consensus vanishes. What was the history of the Moon before 4.6 billions years? What, indeed, is the origin of the Moon?

Before the Apollo landings, there were three models that were seriously considered.

1. Condensation or accretion: formation of the Moon near the Earth as a double planet and at about the same time as the Earth was itself forming. In this model the proto-Moon and proto-Earth condensed while the solar nebula, the extended and hot region surrounding the evolving Sun, was cooling.
2. Fission: The Moon was a part of an already formed Earth and for reasons not known split off.
3. Capture: The Moon was formed in some other part of the solar system and was captured from its passing path by the Earth's gravity.

The double-planet theory faces the problem of detailed differences in chemical abundances. Direct condensation of Earth and Moon within the same part of the solar nebula seems unlikely and some intermediate stages are needed to permit separation of heavier and lighter chemical elements and explain the difference observed between the terrestrial and lunar elemental abundances.

Fission seems an attractive alternative because of the similarity of the densities of the Moon and crust of the Earth, but here again the chemical differences are probably more important than the similarities.

The capture model which seems so attractive to elementary speculation soon runs into severe difficulties. Studies have shown that such tight limits must be placed on the precapture path and speed of the Moon as to make capture seem a most unlikely occurrence. But we cannot argue probabilities on the basis of a single example. There is a more serious objection to the capture model: Calculations show that the Moon would have to be captured into an orbit that would take it so close to the Earth that the difference in the gravitational attraction on the two sides of the Moon would have torn it apart. There is the added problem of having to assume that the captured Moon was formed somewhere else in the solar system with properties that are quite similar to the Earth.

## 13.5 CONCLUSION

There is an important question that needs to be posed. Could unmanned missions have obtained as much information giving us such critical insights, if not yet full understanding, into the Moon's structure, composition, and age? Unmanned missions, such as the Luna probes, can collect lunar soils, but the rock samples selected by the perceptive astronauts could have been obtained in no other way. Now, with our present knowledge, further automated missions to the Moon and the planets can perhaps be designed, but the contribution of the astronauts should be emphasized.

One of the results of the Apollo program has been to provide us with a more sophisticated basis for being critical of every hypothesis of lunar origin. There is no general agreement; some scientists are strongly attached to one theory, others equally strongly to another, and none can yet be cleanly excluded. As Peter Goldreich (California Institute of Technology) put it, "It is more likely that we shall still be speculating about the origin of the Moon a century hence."

## CHAPTER REVIEW

**Moon:** always keeps the same face to the Earth because the rotation period is the same as the orbital period. Small wobble in lunar rotation actually allows 59 percent of the surface to be seen.

- — surface has bright areas (highlands) and dark areas (maria).
- — mass of Moon is $\frac{1}{81}$ of mass of Earth and average density of 3.3 g/cm$^3$.
- — present Moon-Earth distance is 384,400 km and increasing by about 3 cm/year.
- — surface temperature of the Moon is 370 K at lunar noon, 150 K at lunar midnight.
- — Apollo landing sites were chosen for their diversity, and different types of samples were retrieved.
- — lunar craters are mostly the result of impacts, with lava later filling some to form the maria. Dense lava produces local gravitational effects, "mascons."
- — lunar surface is regularly turned over ("gardened") by meteorite impacts, and upper region ("regolith") has pulverized debris from these impacts.
- — radioactive dating shows that the highlands formed about 4.4 billion years ago, with maria forming until about 3.1 billion years ago when lava flows stopped.
- — interior probably a liquid core, with solid mantle and crust on top.

**Origin of Moon:** no clear answers yet.

## NEW TERMS

| | | | |
|---|---|---|---|
| **anorthosite** | **center of mass** | **libration** | **regolith** |
| **basalt** | **gardening** | **mare (plural; maria)** | **rille** |
| **breccia** | **igneous** | **mascon** | **volatile** |
| **center of gravity** | **KREEP** | **refractory** | |

**Plate 17** Frost on the Martian surface, photographed by Viking Lander 2. Frost is probably less than .01 mm in thickness (NASA).

**Plate 18** Deimos, satellite of Mars, photographed from Viking 1 and showing detail down to 200 m (NASA).

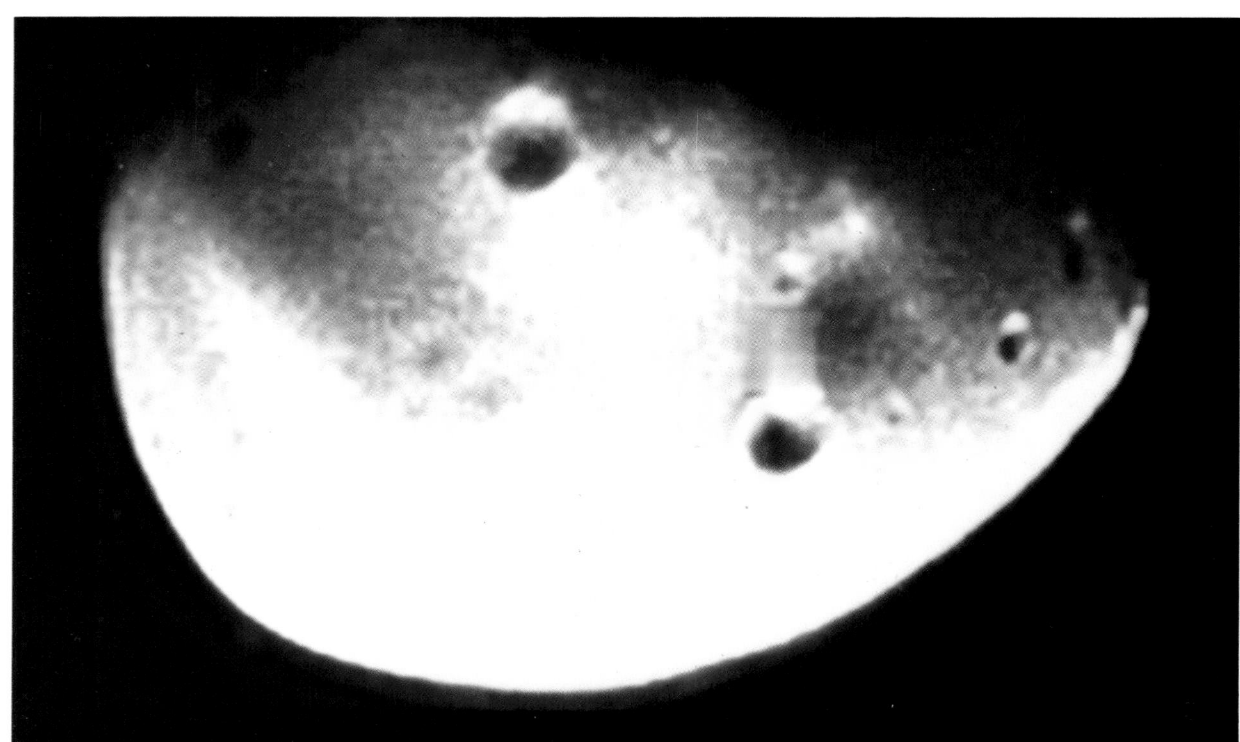

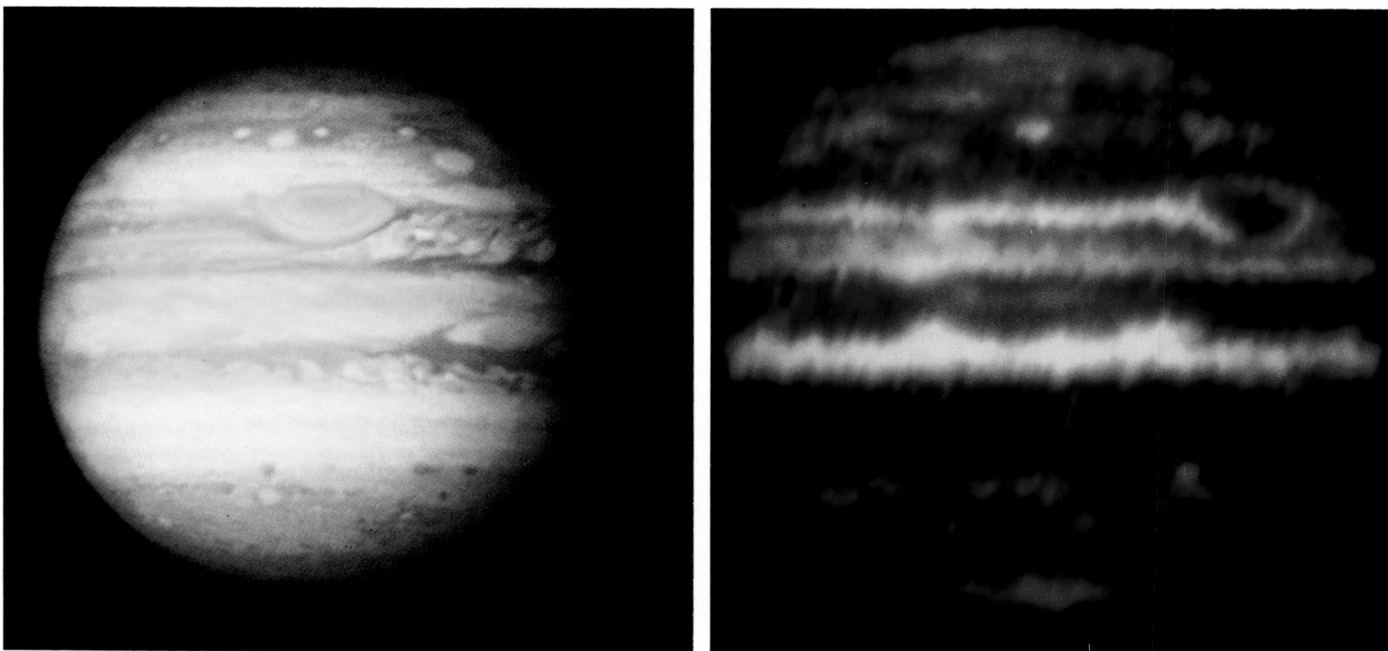

**Plate 19** Jupiter, in visible light (right) and IR light (left), at 5 microns wavelength (NASA).

**Plate 20** Great Red Spot on Jupiter seen from Voyager. Color enhancement for greater contrast. Note the rotating vortices around the edges of the spot (NASA).

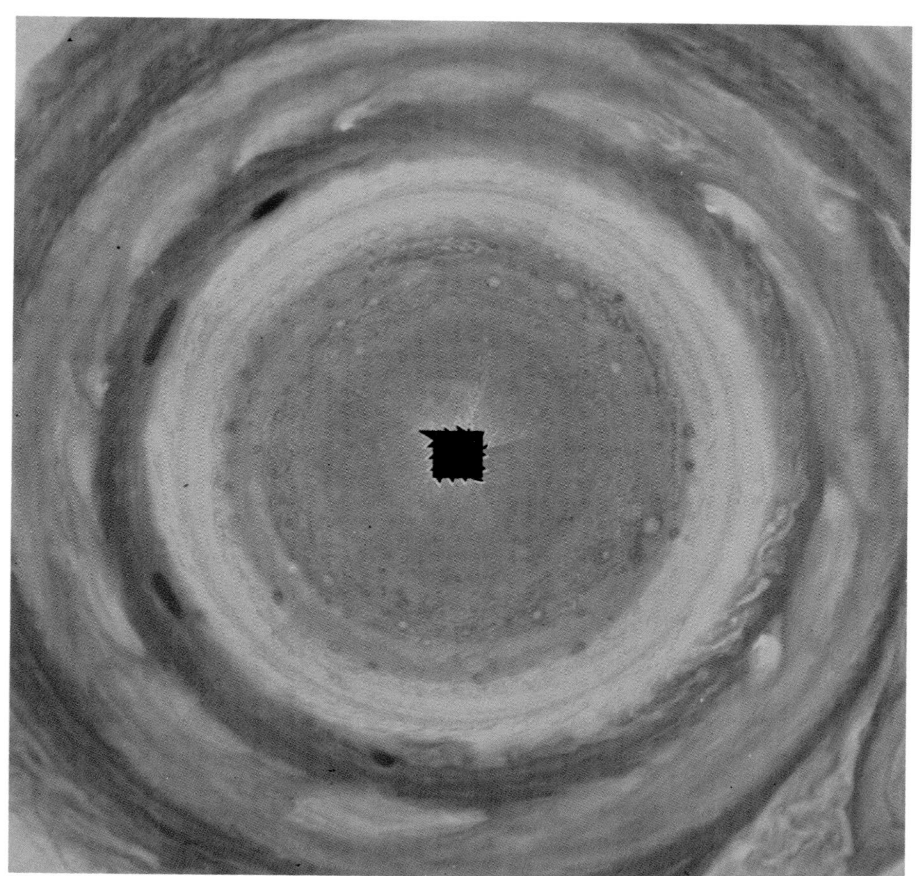

**Plates 21 and 22** North (above) and south (below) poles of Jupiter showing the strongly banded structure. The Great Red Spot can be seen along with patterns showing the turbulence. Dark areas were not covered by Voyager data (NASA).

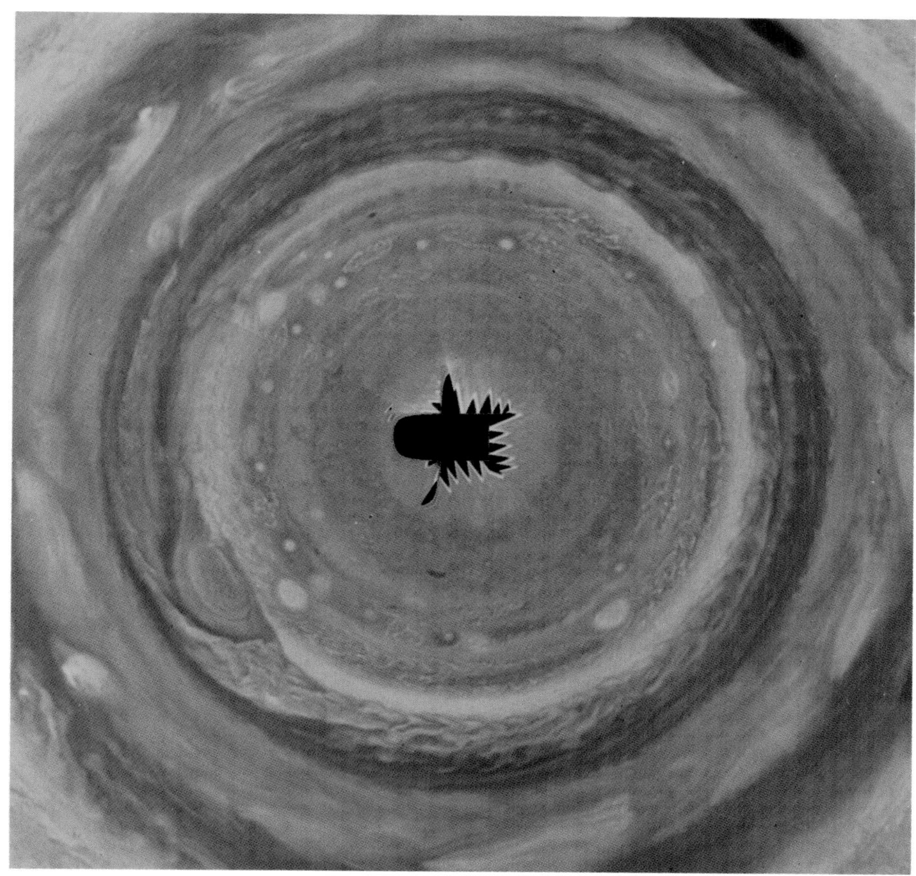

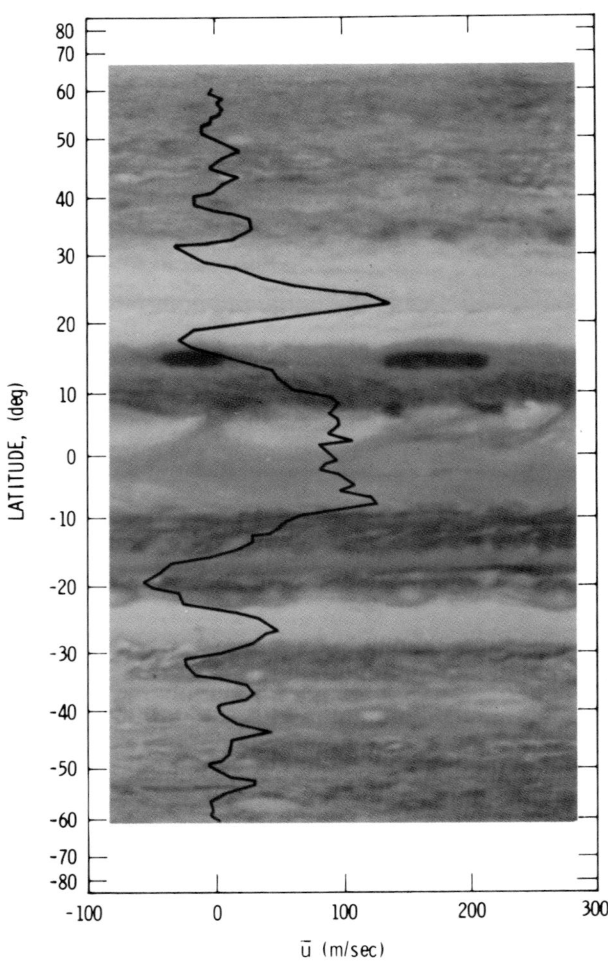


**Plate 23** Profile of wind speeds in Jupiter's atmosphere. The variations in the profile correspond well with the visual boundaries of the bands and zones (NASA).

(a)

(b)

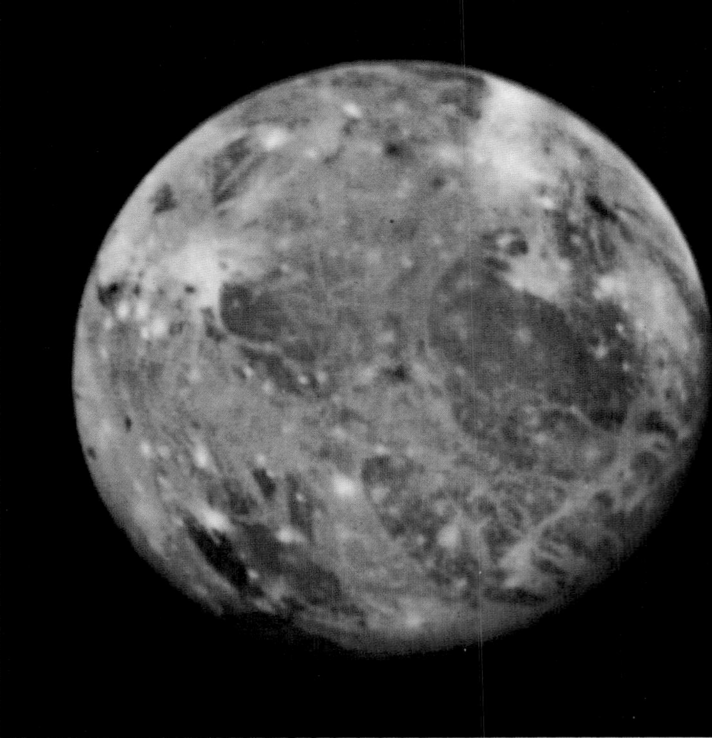

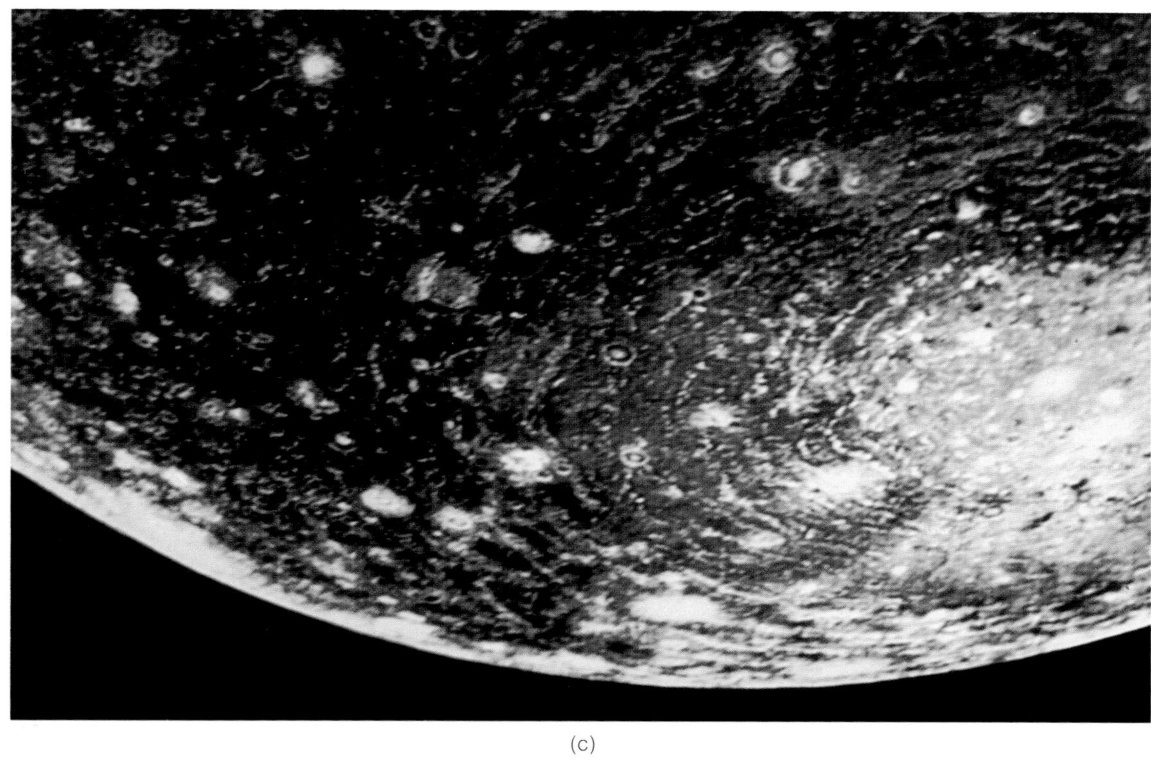

(c)

**Plate 24** Gallilean satellites of Jupiter. (a) Io, showing craters and volcanoes, (b) Ganymede, largest of Jupiter's satellites, (c) Callisto, and (d) Europa.

(d)

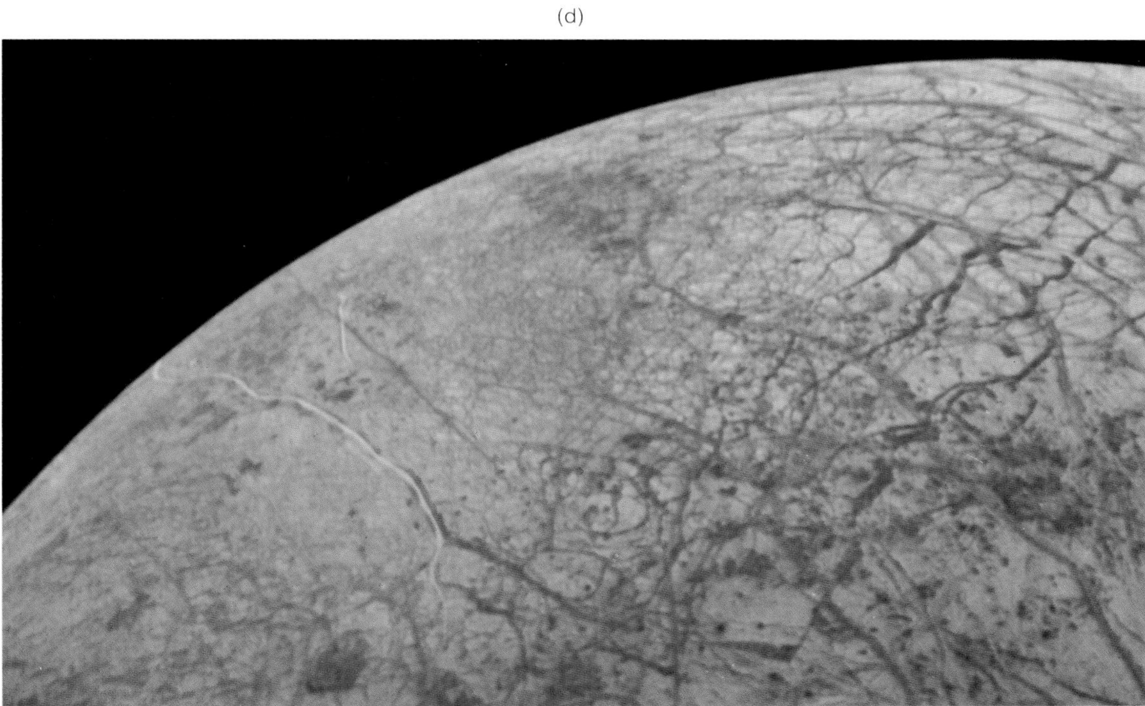

**Plate 25** Voyager 1 photograph of Jupiter with Io (on left) and Europa. Io was about 350,000 km above the Great Red Spot. Smallest detail on Jupiter is 400 km (NASA).

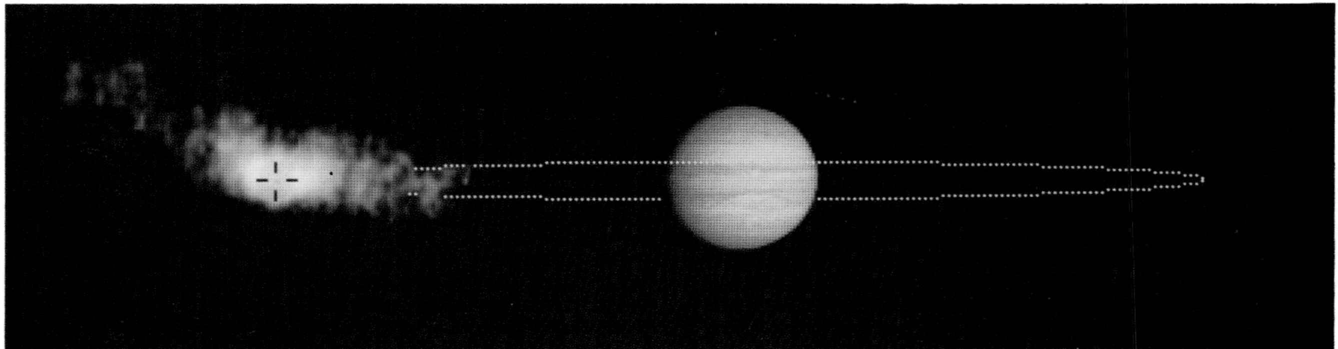

**Plate 26** Sodium cloud surrounding Io as seen from Voyager in May 1981. Io's orbit and Jupiter are also shown (Bruce A. Goldberg, J.P.L.).

**Plate 27** Io photographed from Voyager 1, showing detail as small as 10 km. The coloring of the surface is thought to come from sulfur and various salts, mixed with volcanic deposits. Lack of impact craters suggests that surface features are probably volcanic in origin (NASA).

**Plate 28** Volcanic eruption plumes on the limb of Io. Photographed from Voyager 2, these plumes are 100 km high and were erupting for at least 4 months, since their earlier discovery by Voyager 1 (NASA).

**Plate 29** Saturn, from Voyager 1, four days after passing the planet. Saturn's shadow falls across the rings, and the planet's surface can be seen through the rings (NASA).

**Plate 30** Saturn from Voyager 2, seen from a distance of 34 million km (NASA).

**Plate 31** False-color image of Saturn's rings, constructed from Voyager 2 photographs. This processing emphasizes the difference between the *B* ring (orange) and the *C* ring (blue), indicating different surface compositions of the particles making up the rings. Note also three thin rings with the yellow color within the *C* ring (NASA).

## PROBLEMS

1. The semimajor axis of the Moon's orbit around the Earth is 384,000 km, and the eccentricity of the orbit is 0.055. Calculate the closest and furthest distances of the Moon from the Earth.

2. The phases of the Moon, as seen from the Earth are well known. How will the phases of the Earth appear as seen from the Moon? Draw a diagram to illustrate the changing appearance of the Earth during a month.

3. Suppose that the Moon had been hit by a giant comet and disintegrated several hundred years ago. Apart from any debris that might have reached the Earth, what other consequences can you think of that would affect the Earth or the science of astronomy?

4. A manned spacecraft is in orbit 50 m above the lunar surface. Calculate the time it takes for a complete orbit around the Moon.

5. The Moon is slowly receding from the Earth. When it is twice its present distance from Earth, how many days will it take for one orbit?

6. The Moon's surface is very much colder than the Sun, yet the visible spectrum of the Moon resembles the solar spectrum. Why?

7. Define the following terms, which have been used to describe features on the Moon: breccias, regolith, KREEP, mare.

8. What evidence points to the lunar maria having formed after the highlands?

9. Describe two scientific findings that have come from analysis of lunar samples that cannot be made by remote observation from the Earth.

10. The Lunar Rovers were electrically driven vehicles. Why could gasoline-powered vehicles not be used (apart from problems with flammability of the gasoline)?

11. The Moon and the terrestrial planets show many scars from bombardment mostly during their early histories. Why does debris from these impacts go further on the Moon than it does on Earth for collisions involving the same impact speed?

12. What evidence is used for

a. deciding on relative ages of some lunar features?

b. determining the absolute ages of some lunar features?

13. Not all craters formed on the Lunar surface are still visible. What processes take place that tend to hide or destroy older features? (There is no atmosphere as on Earth to help with this.)

14. What are "mascons," and what is thought to be their origin?

15. Before the lunar landings, there was good reason to believe that the lunar surface was covered with a layer of fine dust. What was the basis for this belief?

16. Because the Earth is so much more massive than the Moon, their center of mass is close to the center of the Earth. Suppose that the Earth were only four times as massive as the Moon. Where would the center of mass be, assuming that their distance apart were unchanged?

17. What are the models that have been suggested for the origin of the Moon?

18. Refer to the drawings of the reconstructions of the Moon's earlier appearance. Identify three prominent craters and three of the maria on the drawing for 3.3 billion years ago.

CHAPTER

14

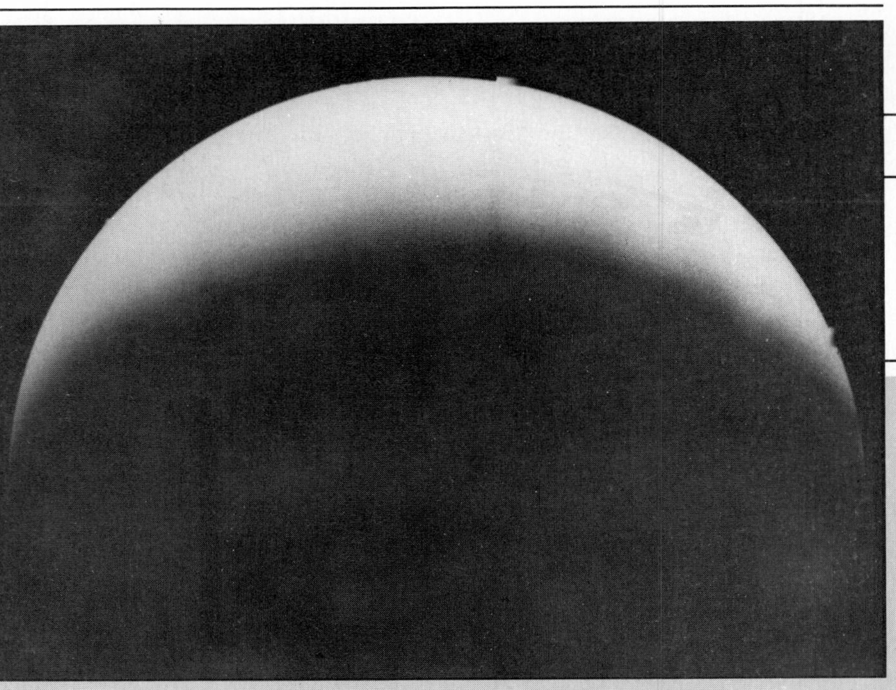

# The Inner Planets: Mercury and Venus

Our nearest neighbors, the terrestrial planets Mercury, Venus, and Mars, all approach to within about half an astronomical unit of the Earth, whereas Jupiter, the next closest, is about eight times farther away. Despite their occasional closeness, Mercury and Venus have not been easy to observe. Having orbits between the Earth and Sun, they can never appear far from the Sun (Fig. 14.1). As the space age developed, their closeness to the Earth

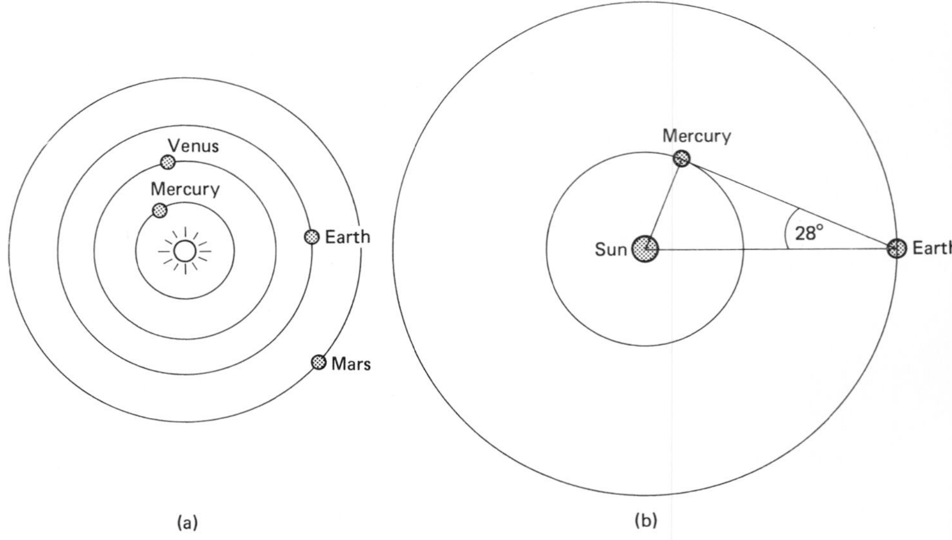


**Figure 14.1** Orbits of the terrestrial planets: Mercury has semimajor axis 0.39 AU, Venus 0.72 AU, Earth 1 AU, and Mars 1.4 AU. As a result, the closest approach between Earth and the others is about 0.5 AU, far less than the distance to the next planet, Jupiter (4.2 AU at the closest approach). (b) Mercury never appears to be far from the Sun when viewed from Earth. Maximum elongation is 28°.

made it easier for space probes to reach these three planets, and all have been the subjects of intense investigation, often with surprising results.

## 14.1 MERCURY

**Telescopic appearance** Long known for its alternating appearance as a morning or evening "star," Mercury is never more than 28° away from the Sun, and so it is visible to the naked eye only in the twilight. Observing or taking photographs at those times requires sighting through a long atmospheric path, close to the ground with its air currents, and this is not conducive to clear images. With modern telescopes much observing is, therefore, done during the day through a shorter atmospheric path, but one must then cope with a bright sky. As a result of these viewing conditions which are so much worse than those for the outer planets, we have never been able to see much detail from Earth, and the best that could be achieved was to note the phases and record some fuzzy variations in shading.

At closest approach, the planet shows a disc with an angular diameter of about 13″; many of the best photographs were taken when Mercury was farther from Earth at about 20° elongation and presented a disc of only 6″. The pictures obtained from spacecraft have been strikingly better.

**Orbit** Mercury's orbit, with an eccentricity of 0.206, is the most elliptical of any planet except Pluto. Its distance from the Sun varies between 46 and 70 million kilometers, with an average of 58 million km or 0.39 AU. The orbit is inclined at 7° to the ecliptic—a greater angle than for any other planet again apart from Pluto. Each orbit around the Sun takes 88 days (the sidereal period), but as seen from the moving Earth, the periodic time appears to be 116 days (synodic period).

As seen from the Earth, the major axis of Mercury's orbit appears to drift (precess) slowly around the Sun about 1.56° per century (Fig. 14.2). Most of this apparent change is due to the Earth's own movement; when that effect is subtracted, the remaining precession (in space) is found to be 593″ per century (about 0.16°).

The effect of the other planets' gravitational attraction on Mercury accounts for 550″ per century, leaving a residue of 43″ per century still to be explained. In 1859, thirteen years after his dramatic and successful prediction of the position of the new planet Neptune, Leverrier drew attention to this anomalous precession of Mercury's orbit. (Leverrier's value for this was 38″ per century.) There appeared to be two possible causes. Perhaps the inverse-square nature of the gravitational force might not hold at such close distances to the Sun—but the successful application of that force law,

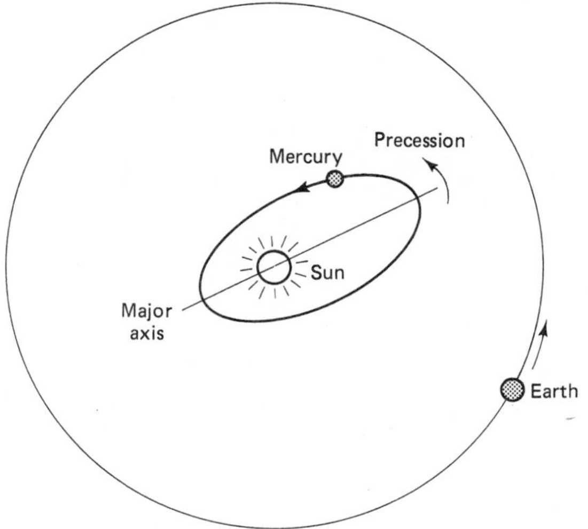


**Figure 14.2** Mercury's orbit is more elliptical than that of any other planet except for Pluto. The major axis of Mercury's orbit rotates (precesses) about 1.56° per century as viewed from Earth. Most of this is due to the Earth's own movement and to gravitational attraction of the other planets, leaving 43 arc sec per century that could only be explained after the development of the General Theory of Relativity.

by Halley with comets and by Leverrier himself and by Adams argued against this explanation. Leverrier, therefore, took the same line of attack as he had with the perturbations of Neptune. He tried to show that the observed effect could be due to a new planet between Mercury and the Sun (and thus very hard to observe directly) or perhaps a ring of particles (also hard to detect). Soon thereafter, a French physician, Lescarbault, announced his discovery of just such an inner planet. It was named Vulcan, and its sighting was reported by several observers allowing Leverrier to calculate its orbit and predict its future positions. Leverrier died in 1877 before further observations failed to confirm his predictions, and Vulcan now remains in the books only as an interesting example of the powers of suggestion.

Simon Newcomb and Asaph Hall, American astronomers, reopened the question of deviations from the $1/R^2$ law. Hall in 1894 suggested that the observations could be explained by using the value 2.00000016 instead of precisely 2 in the law. This explanation does not seem to have attracted much support, and the correct explanation for Mercury's anomaly had to await Einstein's General Theory of Relativity in 1916, which correctly yielded 43″ per century for Mercury, and later 8.5″ per century for Venus, and 5.0″ per century for the Earth, which were also observed.

**General properties** With no satellites of its own, Mercury's mass must be deduced from its perturbing effects on the paths of transient visitors. Icarus, a minor planet, came within 15 million km in 1968. From its perturbed orbit, a value equal to 0.053 times the mass of the Earth was obtained with an accuracy of about 2 percent. The trajectory followed by the Mariner 10 vehicle (1976) provided many more measurements, and Mercury's mass is now known to about one tenth of one percent.

$$M_M = 0.05527\ M_E$$
$$= 3.303 \times 10^{26}\ g$$

The planet's size has been well known for rather longer not from telescopic observations (although that was the older method) but from radar echoes later confirmed by Mariner 10. The diameter is 4878 km, slightly larger than our Moon and about $\frac{1}{3}$ the size of the Earth. Gravitational acceleration at the surface of Mercury is 362 cm/sec$^2$, about 0.37 of $g$ on Earth. The average density of the planet is 5.43 g/cm$^3$, very similar to the Earth.

**Rotation** Repeated observations of the hazy markings on Mercury's surface were used to determine the planet's rotation period, and a value of 88 days was widely listed. Although the absence of clear surface markings did give rise to some doubts about the reliability of the 88 day period, its coincidence with Mercury's orbital period was a strong factor in the acceptance of the 88d value (Fig. 14.3). The equality of orbital and rotation periods was a phenomenon long known for the Moon, and Mercury was considered as another example with the underlying cause attributed to an oblateness.

It therefore came as a considerable surprise when in 1965 Pettengill and Dyce, using the 300 m diameter radio dish at Arecibo, Puerto Rico, for their radar observations, found that the rotation period was close to 59d. An Italian astrophysicist, G. Colombo, then drew attention to this being $\frac{2}{3}$ of the orbital period—the planet rotates three times about its axis while it makes two orbits round the Sun. The orbital time is 87.97d, so that $\frac{2}{3}$ of this is 58.65, exactly as found by radar astronomers.

Since that time, the subject of planetary rotation has received far more theoretical attention and is now much better understood. Synchronism of the rotation with the orbital circuit in the simple 3:2 ratio is thought to result from slow internal dissipation of energy through tides. The acceptance of

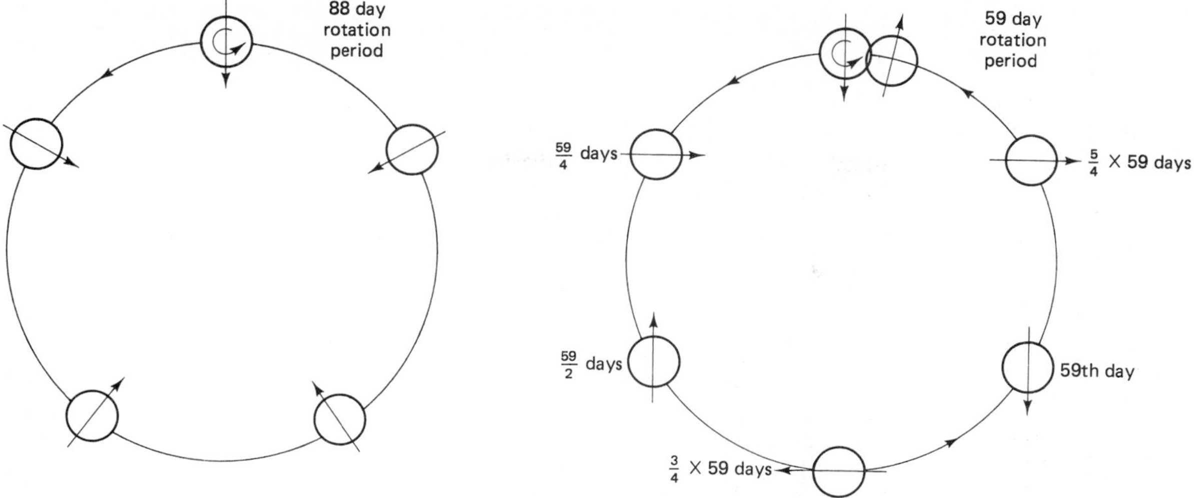


**Figure 14.3** Rotation of Mercury on its own axis. (a) Until 1965 it was thought that Mercury rotated once on its axis during each orbit around the Sun having an 88-day periodic time for each motion. (b) Radar observations showed that the correct period for Mercury's rotation on its axis is 59 days, so that it actually completes $1\frac{1}{2}$ rotations in each orbit.

the erroneous 88d rotation period raises some interesting questions similar to those we saw with the planet Vulcan and often termed the bandwagon effect. A reexamination of older photographs has since shown that they are consistent with both the erroneous 88d and correct 59d periods, the ambiguity arising from the accident of just which days the photographs were taken and the difficulty in obtaining good observations of Mercury through some parts of its orbit.

**Observations from Mariner 10** Mariner 10 left the Earth in November 1973 and made observations of Venus in February 1974 as it flew close enough to have its trajectory altered, so that it was deflected into an orbit around the Sun in which it approached Mercury every 6 months. This is shown in Fig. 14.8.

From the first flyby and two subsequent approaches, we have received streams of data. Soon thereafter Mariner used up the fuel that would have allowed it to be stabilized against tumbling, and no further data have been received. Without stabilization the on-board antenna cannot be aimed at Earth to send signals or receive commands.

Mercury's surface temperature was measured more accurately than is possible from Earth: It ranges from a peak of 775 K in full sunshine to below 100 K on the dark side.

No true atmosphere was expected or found. Traces (about a trillionth of the Earth's atmospheric pressure) of atomic hydrogen and helium were noted. Perhaps the helium was trapped temporarily from the solar wind, but the origin of the hydrogen is not known. More surprising was the discovery of a planetary magnetic field, very weak at about 1 percent of the Earth's field but still strong enough to affect the trajectories of protons and electrons in the solar wind. Mariner 10 did detect electrons, but the survey was not detailed enough to determine whether anything like the Van Allen belts exist for Mercury. The presence of the magnetic fields suggests that the planet's core may be iron, and that would be consistent with the high average density.

Finally and most immediately impressive was the survey of the surface (Fig. 14.4). Craters were seen to be dotted all over. The Caloris Basin, at 1400 km diameter one of the largest impact signatures and similar to Orientale

**Figure 14.4** View of Mercury from Mariner 10 (NASA).

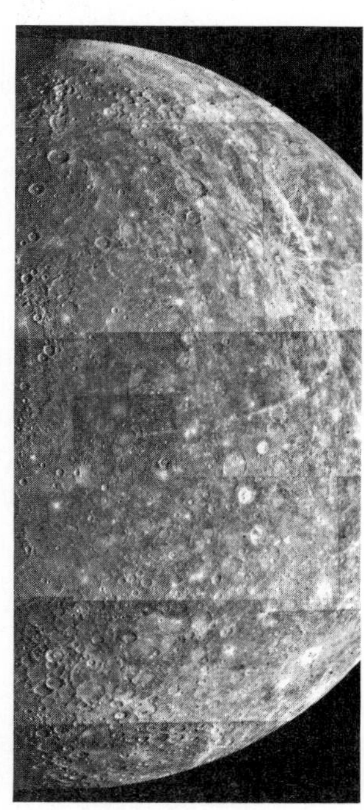

**Figure 14.5** Mercury: craters with 300 km cliff curving across some craters (NASA).

on the Moon, was easily detected, but as Mariner 10 came to within 327 km of the surface on its third pass, craters as small as 50 m could be seen. At first sight, the surface seemed remarkably similar to the Moon, but closer inspection showed some important differences (Fig. 14.5). There were fewer craters per unit area than on the Moon, and even those areas most heavily cratered had clear plains. This might be explained by Mercury's surface gravity, which is about twice that of the Moon. Debris from impacts will, therefore, not rise as high and thus will not be able to travel as far as on the Moon (Fig. 14.6), and secondary craters thus will also be closer to the main impact locations.

Mercury also differs from the Moon in having shallow cliffs (**lobate scarps**) that extend for hundreds of kilometers. One suggestion for their origin is a wrinkling that resulted from cooling and shrinking of the planet several billion years ago. This mechanism was once thought to explain the mountain ranges on Earth, before the theory of continental drift was so widely accepted. On Earth, however, continental drift is accompanied by features that are not seen on Mercury: folded mountains, spreading ridges, and rifts (Fig. 14.7).

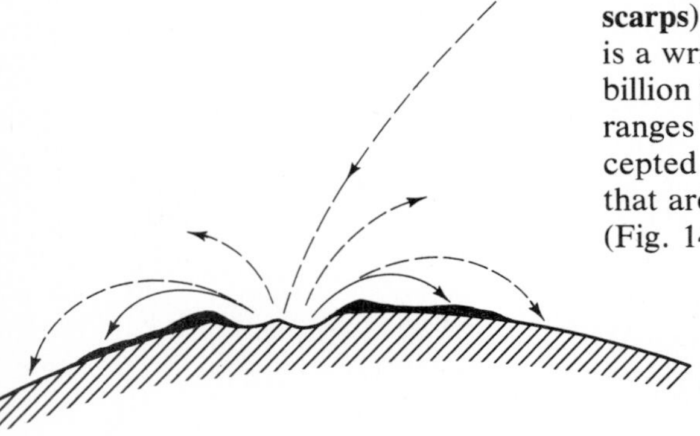

**Figure 14.6** Local acceleration due to gravity has a large effect on distance that debris can travel from an impact that produces a crater. Mercury has a surface gravity about twice as large as that of the Moon; debris will fall much closer to the crater.

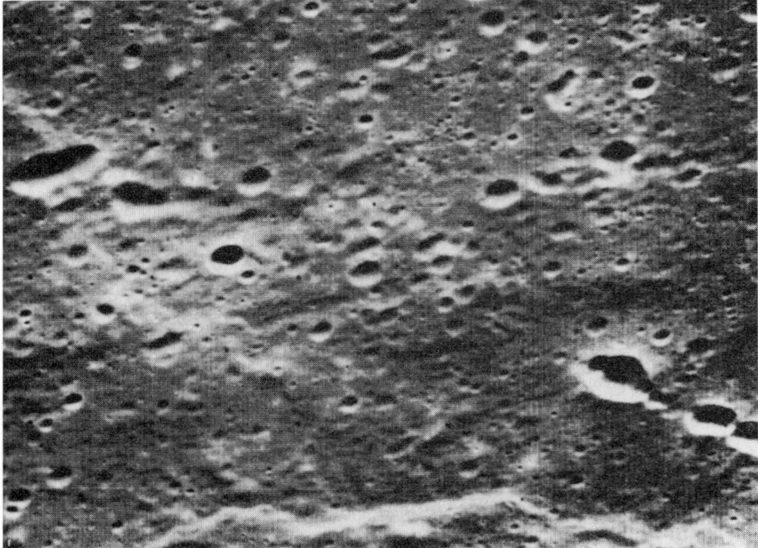

**Figure 14.7** Closest approach of Mariner 10 to Mercury, on March 29, 1974: 5900 km (3700 miles). Craters as small as 150 m (500 ft) can be seen (NASA).

*Table 14.1* A POSSIBLE GEOLOGICAL HISTORY OF MERCURY

| *Epoch* | *Description* |
|---|---|
| 1. | Accretion and differentiation: accumulation of most of the solids for the planet; presumed settling of the heavier iron and nickel to the core leaving the lighter silicates for the crustal region with basalts between. Final crustal stage of accretion leaves the surface with no major features, a relatively flat surface now seen as intercrater plains. Cooling and shrinking then produce lobate scarps mostly visible in the plains. |
| 2. | Heavy bombardment onto a generally featureless surface but with less cratering than on the Moon. |
| 3. | Formation of Caloris Basin. Major modification to one hemisphere followed by— |
| 4. | Volcanism with lava flows—larger areas of smooth plains formed, some older basins filled. Low density of craters on smooth plains indicates little subsequent bombardment. |
| 5. | From end of lava flooding until now: little cratering, probably similar to Moon, with gardening by meteorites and exposure to the solar wind. |

**Geological history** Drawing for comparison on the (far greater) body of experimental evidence from the Moon and the noted similarities and differences, we can reconstruct a plausible geological history for Mercury (Table 14.1). A critical omission from this chronology is an absolute time scale. Without samples whose radioactivity can be measured, we cannot determine the ages of the various features nor the lengths of the different stages of Mercury's evolution.

## 14.2 VENUS

**Telescopic appearance** The blue-white brilliance of the evening "star" can be seen in the western sky for as long as three hours after sunset, and the equally bright morning "star" can be seen before sunrise. Around the sixth century B. C. the identity of these two "stars" was recognized. Apart from the Sun and Moon, Venus is the brightest object in the sky, bright enough to cast shadows on a dark night, but its brightness and visibility vary considerably depending on the distance from the Earth and the angular distance from the Sun (never more than 47°) (Fig. 14.9).

With its orbit having an average radius of 0.72 AU, its distance from the Earth varies greatly. As a result its angular diameter also varies, between 10″ and 62″. With its orbit between the Earth and Sun, Venus has its maximum brightness at neither its closest approach (inferior conjunction) when its disc appears largest, nor at greatest elongation but about midway between

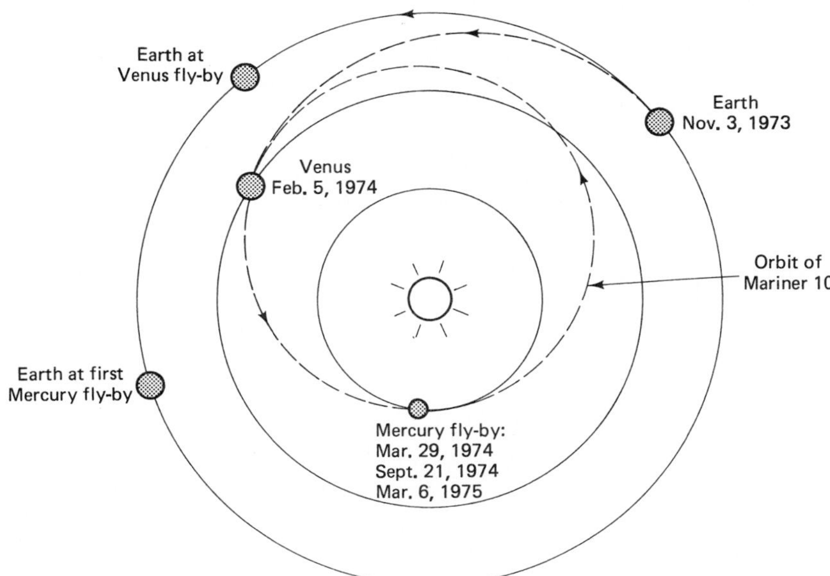


**Figure 14.8** Trajectory of Mariner 10 passing Venus and then approaching Mercury, thereafter going into orbit around the Sun with repeated encounters with Mercury.

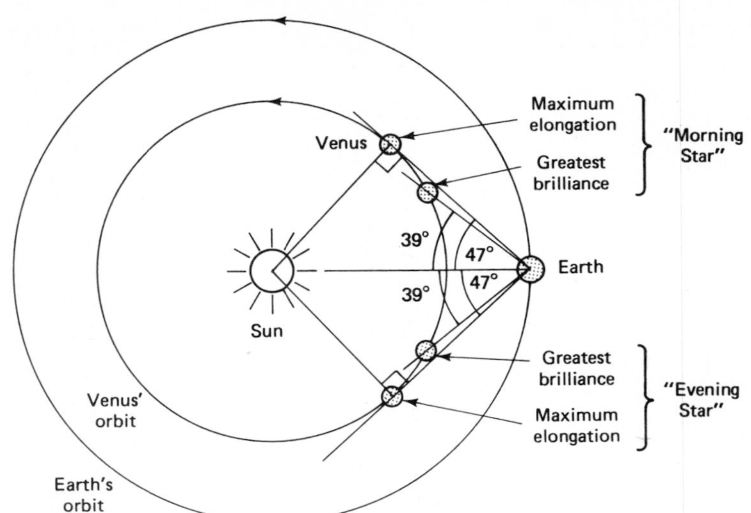


**Figure 14.9** Like Mercury, Venus is never seen far from the Sun. Its maximum elongation is 47°. Maximum brightness is seen when the elongation is 39°, the result of competing factors of closeness to Earth and illuminated area visible.

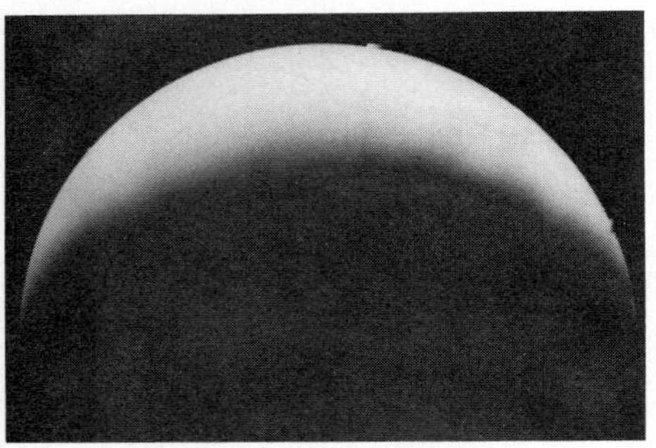

**Figure 14.10** Crescent Venus as photographed from Pioneer Venus December 1978 (NASA).

when its elongation is 39°, and it is thus not at its highest altitude in the predawn or sunset skies. Galileo's telescopic discovery of the phases of Venus (like the Moon, Fig. 14.10) came as a surprise, since all that had previously been noted was the variation of brightness.

Even with the most modern telescopes Venus has shown us only a bright disc with its surface completely covered by clouds that had no markings or distinguishing features. In 1761 the Russian scientist Lomonosov had noted at inferior conjunction the appearance of a thin halo around the planet suggesting the presence of an atmosphere that would scatter the sunlight, but the first Earth-based observation of any markings came only in 1928 with UV observations. Reported observations of those clouds showed a retrograde direction of rotation with speeds up to 100 m/sec. All the time and until the use of the radar signals and space probes, the surface conditions remained hidden.

**General properties** The radius of the cloud disc has long been known to be 6110 km, but the size of the underlying solid surface was first measured in the 1960s when radar (radio waves) was used to penetrate the cloud cover. The best current estimate for the size of the solid surface now comes from the altimeter on the Pioneer Venus probe and the radar tracking of the probe (Figs. 14.11 and 14.12). The 1980 value is 6051.4 km with an uncertainty of only 100 m. The radar mapping shows a planet that is very nearly spherical to within a few hundred meters (Fig. 14.13). Twenty percent of the surface lies within 125 m of the average level, 60 percent within 500 m, and the maximum excursions are only about 3 km down and 12 km up (the Maxwell Montes). The center of mass is within 500 m of the geometric center, indicating a more uniform distribution of the internal mass than is the case for the Earth, where the center of mass is offset by more than 1 km. Gravity anomalies are very small, and nothing like the lunar mascons has been detected.

The uplift areas tend to be rougher than average, with some 70 percent of the area rolling plains. Some circular features have been "seen." The number of these features per square kilometer is comparable to that on the Moon, Mercury, and Mars, suggesting that they are of impact rather than volcanic origin. The lowlands tend to be uncratered.

Like Mercury, Venus has no satellites and its mass must be inferred from its effect on the paths of minor planets or space probes. The most accurate value comes from the radar tracking of Mariner 10 and can be

**Figure 14.11** Venus, full disc, photographed from Pioneer Venus: More detail can now be seen showing details of clouds and circulation pattern (NASA).

**Figure 14.12** View from Mariner 10 in 1974, using a UV filter. THe surface cannot be seen beneath these dense clouds. Strong circulation patterns can be seen in upper atmosphere (NASA).

**Figure 14.13** Radar relief map of Venus based on Pioneer Venus data. Higher altitudes are shown light-colored. Maxwell is the peak at 10° longitude, and the Lakshmi Plateau centered on 330°, both around 65° north latitude (Gordon H. Pettingill, Massachusetts Institute of Technology).

240° 270° 300° 330° 0° 30° 60° 90° 120° 150° 180° 210° 240°

80° 60° 30° 0° -30° -60° -70°

±65° ±30° 0° 0 1000 5000 KILOMETERS

NASA, Ames Research Center
U.S. Geological Survey
Massachusetts Institute of Technology
NOVEMBER 1981

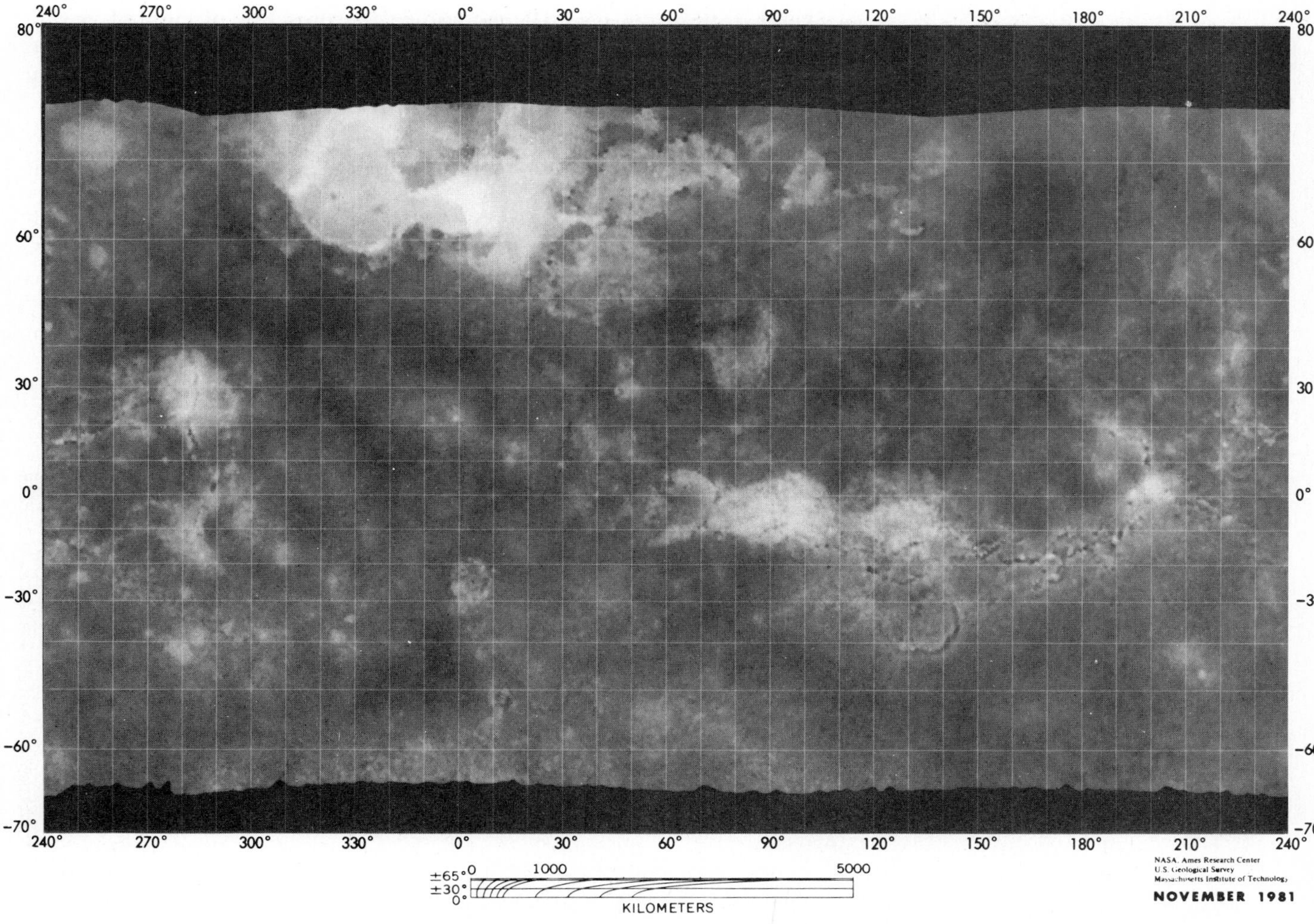

expressed as a ratio to the Earth's mass

$$M_V/M_E = 0.8150$$

which translates to $4.87 \times 10^{27}$ g with a standard error of 3 parts in a million.

The density of the solid sphere then turns out to be 5.25 g/cm$^3$. In its dimensions, Venus is very similar to the Earth: Its diameter is only 5 percent less, its mass 18 percent less, its density almost the same, and the escape velocity 10.3 km/sec compared to Earth's 11.2 km/sec. Surface gravity is 0.88 g or 862 cm/sec$^2$.

**Rotation** With the surface completely hidden by clouds no measure of Venus' surface rotation was available before the radar observations in 1962. The cloud tops had been seen to be moving at high speed, taking about 4 days to circle the planet (Fig. 14.14), but there was no hint of the surface's retrograde rotation, unique among the planets. Because the planet is rotating, one edge is moving toward us at the same time that the other edge is moving away. The rotational speed then produces a Doppler shift in the reflected radar wavelength; the approaching side is "blueshifted" and the receding side "redshifted." The speed of rotation can then be deduced from the observed Doppler shifts. Venus takes 243d for one complete rotation. Its rotation axis is within a degree of perpendicular to its orbital plane around the Sun. We have no understanding of the reasons for either the direction of rotation or its leisurely pace.

**Atmosphere** From the Mount Wilson Observatory in 1932, spectroscopic measurements in the near IR had shown the presence of carbon dioxide ($CO_2$), and the temperature of the cloud tops had been measured at 230 to 240 K, but, as is so often the case, space probes have provided completely unexpected results.

Venus has been the prime target for Soviet planetary exploration starting with Venera 1 in 1961, and it has been visited on the average by about one probe a year.

The intense surface heat (750 K) and high atmospheric pressure (about ninety times that at the Earth's surface) are clearly too hostile for life as we know it and also too hostile for many space probes. Venera 4 used parachutes to reduce its speed through the Venusian atmosphere, and it transmitted data for 94 min of descent before failing at an altitude of 25 km. The tem-

**Figure 14.14** Rotation of upper clouds on Venus as photographed at 7-hr intervals from Mariner 10 in February 1974 (Wind speed is about 100 m/s or about 200 mph.) (NASA).

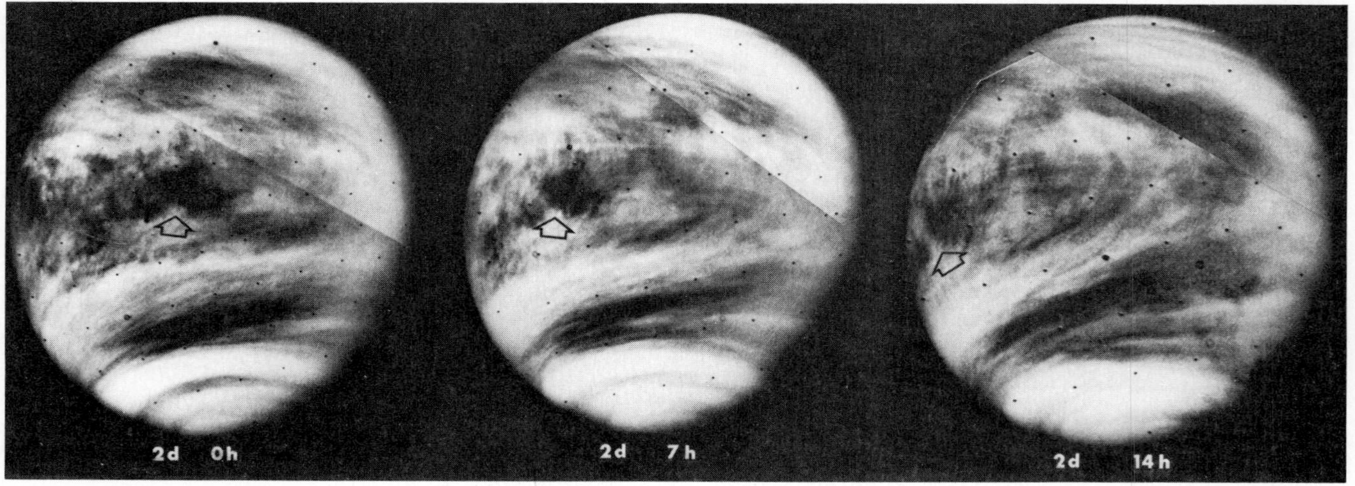

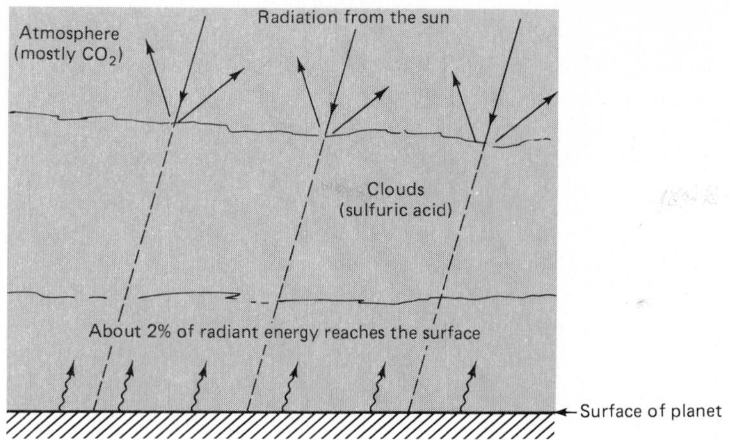


**Figure 14.15** The Greenhouse Effect. Most of the solar radiation that falls on Venus is reflected from the uppermost clouds, but about 2 percent of the energy does manage to penetrate to the surface. This energy warms the surface which then radiates mostly at IR wavelengths. These relatively long wavelengths cannot penetrate the clouds, so that the radiated energy is absorbed in the lower regions of the atmosphere producing the very high temperatures observed.

perature at the cloud tops was 240 K, with a steady increase to the surface. Venera 7 transmitted 2 min of data from the surface. The survival time increased to nearly 2 hr for Venera 12.

The atmosphere is predominantly carbon dioxide (97 percent), with less than 2 percent nitrogen ($N_2$). For water vapor ($H_2O$), carbon monoxide (CO), hydrochloric acid (HCl), hydrofluoric acid (HF), and ammonia ($NH_3$), only traces were detected. A host of other interesting gases could not be detected at the sensitivity limit of about 1 part per million. These include oxygen ($O_2$), oxides of nitrogen (NO and $NO_2$), methane ($CH_4$), and sulfur dioxide ($SO_2$).

The detection of krypton and argon has posed a puzzle. These are rare gases that do not combine chemically. The mass spectrometer on Pioneer Venus detected krypton at a level of about 50 parts per billion. The Soviet probes Venera 11 and 12 found much more, between 700 and 1400 parts per billion. The discrepancy between these two sets of values is not understood, and the Pioneer value is well below the abundance of 1 part per million in the Earth's atmosphere. There was, however, agreement on the argon abundance at more than fifty times the abundance in the Earth's atmosphere. A possible source of the krypton is the solar wind, but the argon abundance is not understood.

The clouds that are responsible for Venus' brilliance (by reflecting so much sunlight) seem not to be water droplets (as are clouds on Earth) but sulfuric acid. These corrosive clouds extend from about 35 km above the surface to a maximum altitude of around 80 km. Most of the clouds lie between 49 and 62 km. Liquid droplet sizes in the clouds are in the range $0.5 \times 10^{-4}$ to $2.5 \times 10^{-4}$ cm. The wind speed varies from a gentle 1 m/sec at the surface (too low to erode surface features) to 50 m/sec at 45 km and to higher speeds at great heights as already known from the UV pictures.

Only about 2 percent of the incident sunlight reaches the surface, the rest being absorbed or reflected by the clouds. The extreme heat on the surface can now be understood. Being closer to the Sun than the Earth, Venus receives about three times as much solar radiant energy. Some of this radiation manages to penetrate through the clouds to the surface where it is absorbed. This heats the surface, which radiates in the IR, but the atmospheric carbon dioxide is opaque to that long-wavelength radiation. The net result is a trapping of solar energy until a steady (and high) temperature is reached at which the energy losses just balance the absorption of yet more solar energy. This balance is reached for Venus only when the surface temperature is as high as 750 K. This heating effect is often termed the **greenhouse effect** (Fig. 14.15). Greenhouse glass windows allow the visible solar radiation to enter and they trap the IR that is emitted by the heated interior, but a more important factor in the greenhouse heating is that the windows prevent the heated air from escaping. Greenhouses work as much by trapping the air as the radiation. You can observe the same effect when you open the doors and windows of an automobile that has been parked in the sun on a hot day. The internal temperature drops rapidly as the hot air escapes.

**The surface** General features of the terrain have been revealed by the Earth-based radar and by the Mariner and Pioneer orbiters. The first direct visible-light photographs of the surface came from Venera 9 and 10, which landed in October 1975. Venera 9 showed a rocky, desertlike view, with many sharp-edged stones up to about 60 cm long by 20 cm wide. The view from Venera 10 was smoother with what appeared to be a stony base covered by a dark, fine-grained soil.

The density of the surface soil has been measured to be 2.3 g/cm$^3$, much less than the planets' overall density of 5.3 g/cm$^3$. Gamma-ray spectrometers on the Venera 8 probe found high concentrations of the radioactive uranium (U), thorium (Th), and potassium (K) similar to terrestrial granites. The instruments on Veneras 9 and 10 found very low radioactivity as is the case with terrestrial basalts. No seismic observations have been made.

**Commentary** We are now in a position of being able to unravel some of the complexities in the puzzling chemistry of Venus. At the very high surface temperatures observed, volatile chlorine (Cl) and fluorine (F) evaporate into the atmosphere where traces of their combinations HCl and HF have been detected. On Earth the rocks and oceans contain large amounts of $CO_2$ in the form of carbonates (in the rocks) and in solution (in the oceans), but on Venus the high temperature bakes the gas out into the atmosphere. Water vapor is probably rare on Venus because the needed hydrogen is not available. Hydrogen is present but already chemically combined in the acids, and any free hydrogen would rapidly escape because of the high temperature.

## CHAPTER REVIEW

**Mercury:** maximum elongation from Sun is 28°; very difficult to photograph clearly from Earth because long atmospheric sightlines or daytime viewing are needed.

- — precession of major axis of orbit is faster than could be explained by known gravitational effects—General Theory of Relativity provides accurate explanation.
- — no satellites, so mass is known only from effects of Mercury on minor planets or spacecraft.
- — rotation: was thought to be 88 days (same as orbital period), but radar observations showed it to be 59 days ($\frac{2}{3}$ of orbital period).
- — surface is heavily cratered, but fewer craters per unit area than on Moon, and some areas clear of craters. Long, shallow cliffs (lobate scarps) observed, perhaps a result of contraction on cooling.
- — no atmosphere.

**Venus:** brilliant appearance due to heavy cloud cover.

- — radius of solid surface known only since use of radar (to penetrate clouds).
- — relatively flat surface.
- — rotation is retrograde: no explanation for this.
- — atmosphere: 97 percent carbon dioxide, small quantities of nitrogen. Clouds are sulfuric acid droplets.
- — greenhouse effect involved in production of intense heat of surface. Little sunlight can penetrate the clouds, but even that little solar energy heats up the surface, which radiates in IR. Infrared cannot penetrate the clouds, so energy is trapped, and surface heats up.

## NEW TERMS

**greenhouse effect**
**lobate scarp**

## PROBLEMS

1. Why are Venus and Mercury hard to observe from the Earth?

2. The Moon has an angular diameter of almost half a degree. How much smaller are the angular diameters of Mercury and Venus at their largest?

3. Calculate the escape velocity from Mercury.

4. The peculiar precession of Mercury's orbit was known long before the correct explanation was produced by Einstein. What explanations had been previously considered to explain this effect?

5. How were the masses of Mercury and Venus estimated before spacecraft approached them?

6. In what way does the cratering of Mercury appear different from that on the Moon?

7. What is the composition of the clouds that make Venus appear to be bright?

8. Why is the surface of Venus so hot?

9. The rotational speed of Venus can be determined by bouncing radar signals off the two edges and noting the change in frequency. Use data from the tables to calculate the fractional difference in frequencies between signals reflected off the edges, one approaching us and the other moving away.

10. Mercury and Venus do not have moons. How were their masses determined?

11. Why has the atmosphere of Venus evolved in a different way from the atmosphere on Earth?

12. Venus has been studied from Earth using optical telescopes and radar. What different information do these two spectral regions provide?

13. Suppose that the Earth always kept the same side toward the Sun (as Mercury was thought to do). If you lived on the dark side of the Earth, would you then always see the same stars? Illustrate your answer with a sketch.

14. The diameter of Venus is 12,220 km. What will be its greatest and least angular diameters as viewed from Earth?

15. The surface temperature of Venus is about 750 K. Calculate the wavelength at which the radiation from the surface is most intense.

CHAPTER

15

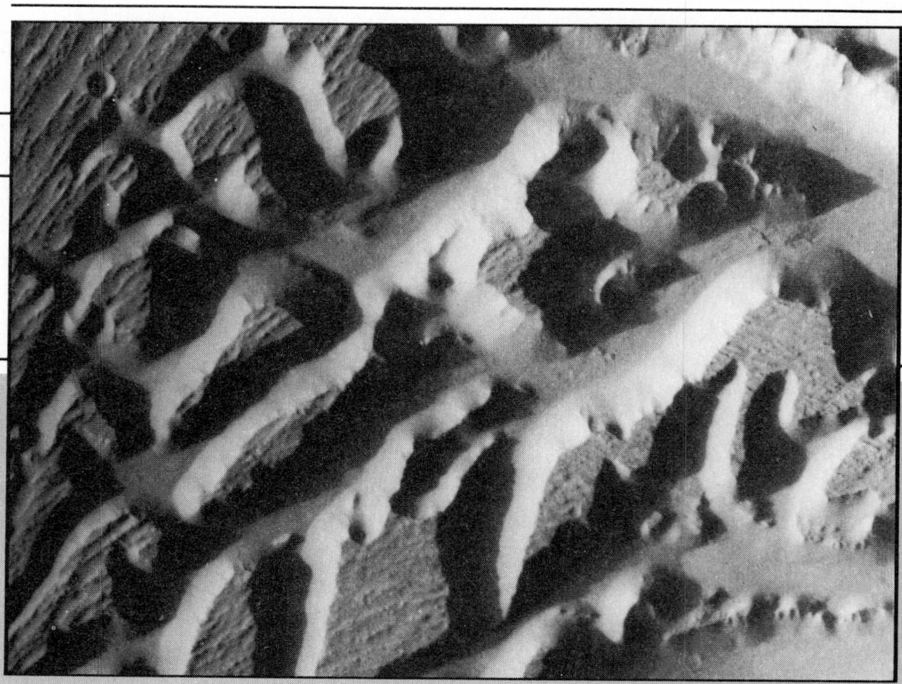

# Mars

## 15.1 APPEARANCE

At its brightest, Mars is second only to Venus among the planets and more than twice as bright as any star. Its dull-red appearance suggested to the ancients an association with the god of war. With its orbit further from the Sun than is the Earth, Mars can often be observed throughout the night, but its brightness and angular size vary greatly through the year. Because of the eccentricities of the orbits of Earth and Mars, at opposition (closest approach) the distance between the planets can be as little as 56 million km or as much as 101 million km. The corresponding angular sizes of Mars' disc are 25″ and 14″, respectively. The most favorable oppositions occur at intervals of 15 or 17 years, the last being in 1971, and the next is due in September 1988 (but the July 1986 opposition will find Mars almost as close) (Fig. 15.1).

At 1.52 AU average distance from the Sun, Mars completes each orbit in 686 days (its sideral period), but as seen from the moving Earth it appears to take 780d (the synodic period). Mars' orbit has an eccentricity of 0.093, almost twice as large as that for any other planet except for Mercury and the always deviant Pluto.

As the quality of telescopes improved steadily over the years, detail could be seen on the Martian disc (Fig. 15.2). What looked like polar icecaps changed their sizes with the seasons, appearing to contract to the poles (but never vanishing) with the Martian summer, then expanding with the approach of winter. Other markings were noted, coming and going, and occasionally, near perihelion, what seemed to be dust storms could be seen.

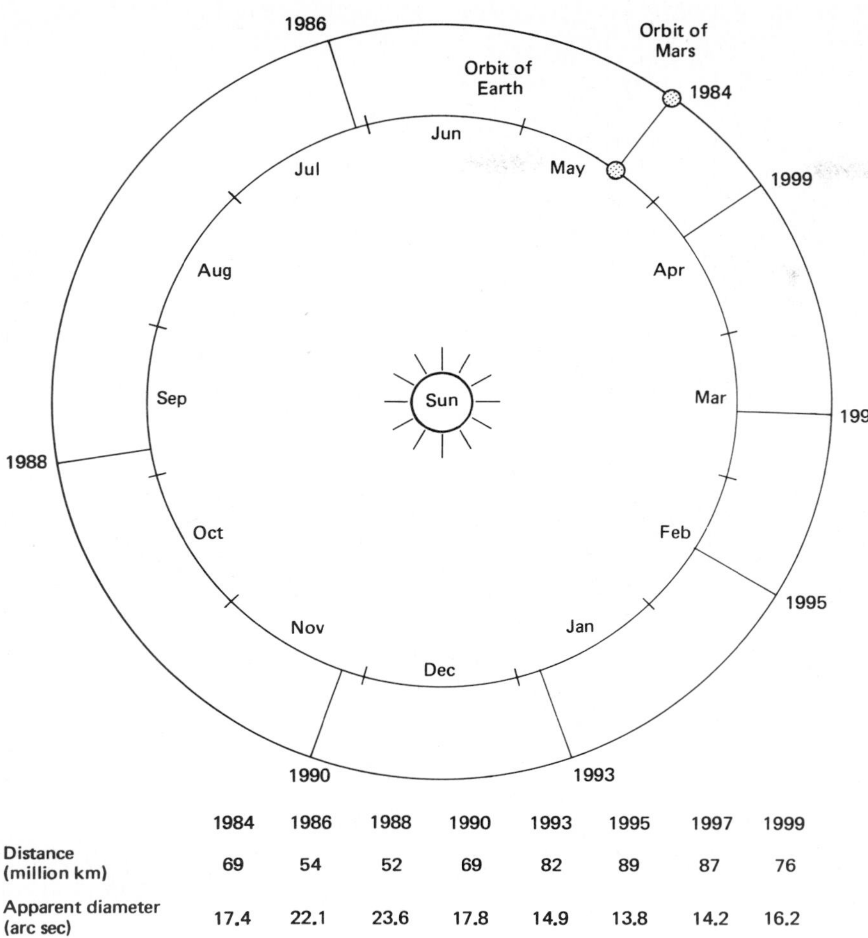


| | 1984 | 1986 | 1988 | 1990 | 1993 | 1995 | 1997 | 1999 |
|---|---|---|---|---|---|---|---|---|
| Distance (million km) | 69 | 54 | 52 | 69 | 82 | 89 | 87 | 76 |
| Apparent diameter (arc sec) | 17.4 | 22.1 | 23.6 | 17.8 | 14.9 | 13.8 | 14.2 | 16.2 |

**Figure 15.1** Oppositions of Mars. Mars and Earth have close approaches roughly every two years providing the best opportunities to observe this planet. Mars' distance and disc size vary markedly from one approach to the next because of the elliptical shape of the two orbits.

**Figure 15.2** Mars as seen from Earth: The polar ice cap is clearly visible in these views at the 1971 opposition (Lick Observatory).

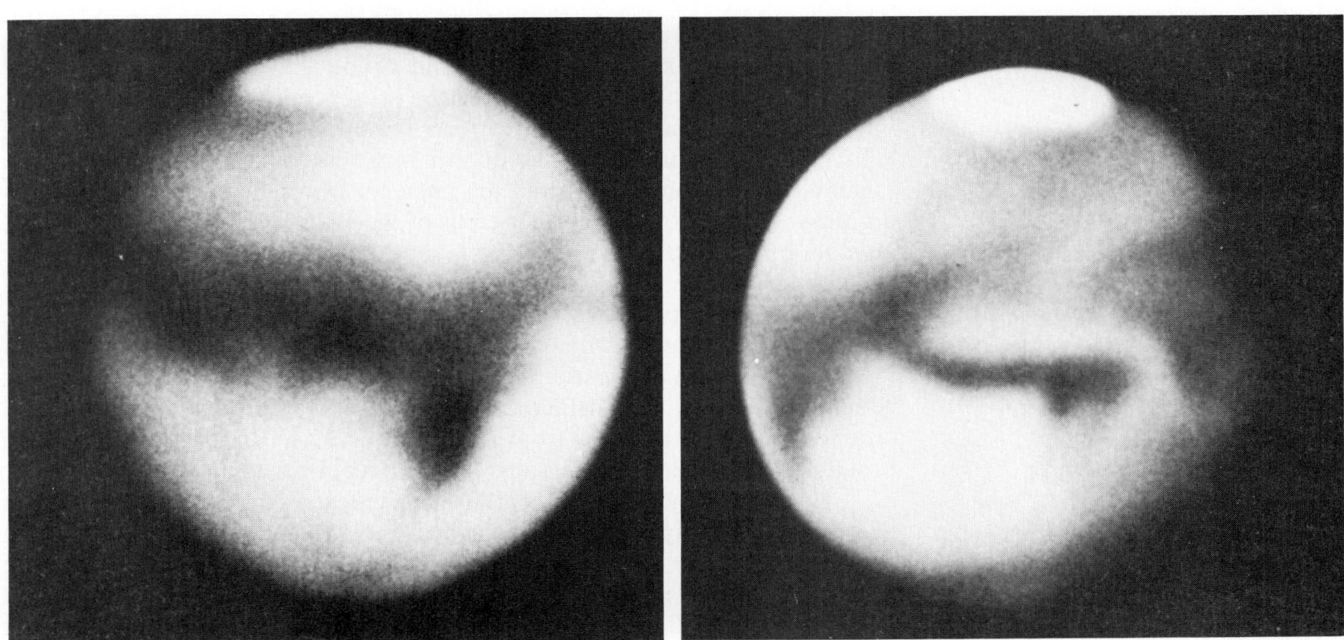

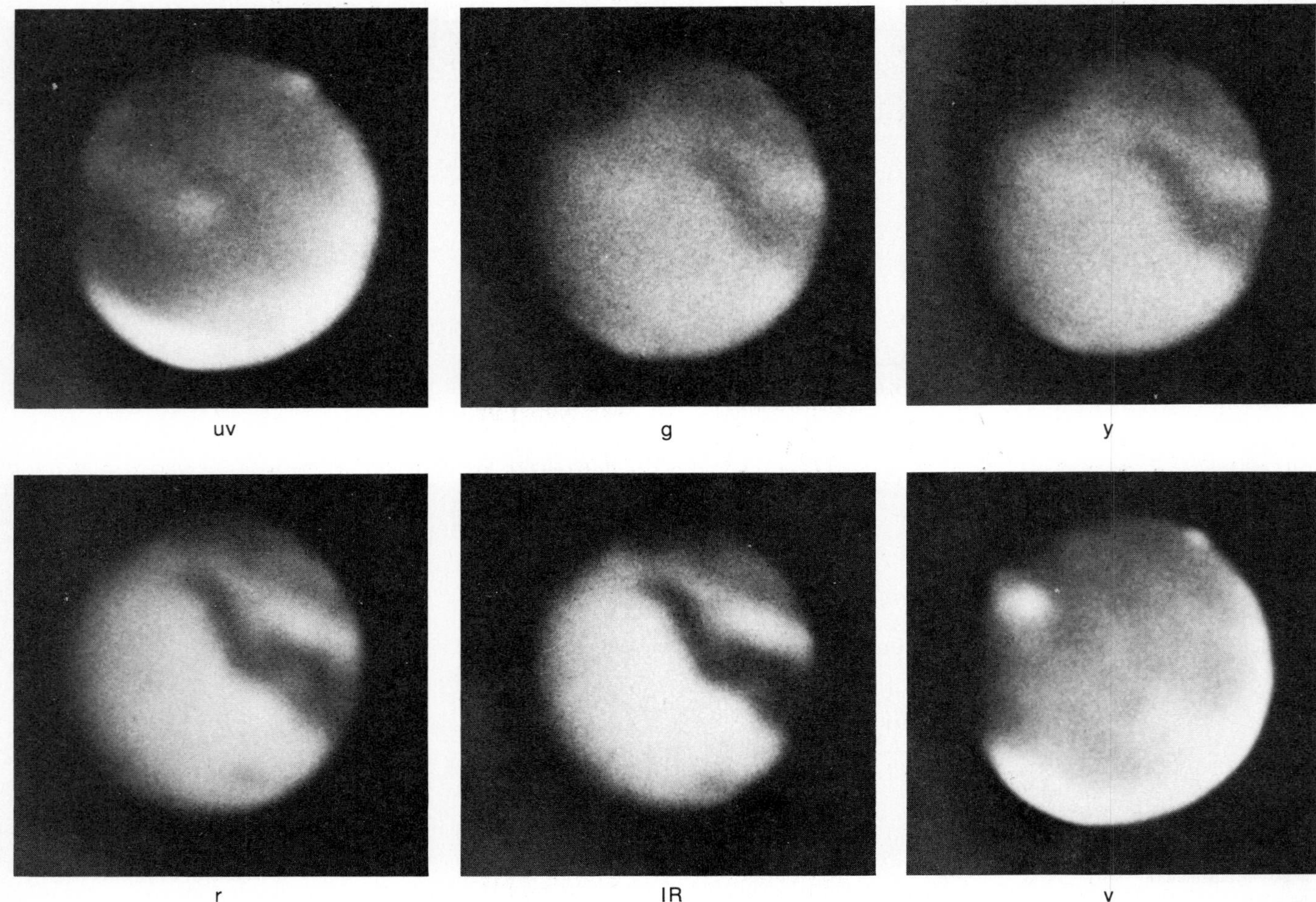


**Figure 15.3** Mars as seen from Earth at different wavelengths: UV, violet, green, yellow, red, and IR. Note how the polar caps are more clearly defined at the shortest and longest wavelengths (Lick Observatory).

In 1863 Father Angelo Secchi in Rome produced the first colored map of Mars, showing shadings of green and yellow areas. Along with several other observers at that time, he noted the presence of some streaky features not very well-defined, which he termed "canali." In the translated form of *canals*, these features gave rise to an intense interest in Mars built on the fantasy that the canals indicated the presence of intelligent living creatures on Mars. Father Schiaparelli, Director of the Rome Observatory, produced an elaborate map of Mars after the favorable opposition of 1877. After the 1879 opposition he noted that some of the canali had doubled.

The reports of canals were seized by others and embroidered upon. They caught the imagination of Percival Lowell, who was already mentioned in connection with the search for Pluto. Lowell's new observatory was excellently equipped. He installed first an 18 in. refractor, later a 24 in. instrument, both with lenses from the shops of Alvan Clark and Sons, then probably the best lens makers for astronomical telescopes.

Unfortunately, Lowell's enthusiasm was unrestrained, and he wrote extensively not only of the increasing detail he thought he saw on the Martian surface (Fig. 15.5), but also of the canal builders and the society that he imagined, even to its political organization. Other astronomers, notably at the University of California's Lick Observatory, contested his findings, for they saw no canali even with their 36 in. telescope, and they were highly critical of Lowell for extrapolating far beyond his data. The general public, however, seems to have had little difficulty in deciding that Lowell's exuberance was preferable to the sober approach of the professionals. The prospect of encountering visitors from other planets was even more widely encouraged through H.G. Wells' *The War of the Worlds* (1898) and later by Edgar Rice Burroughs' series of eleven novels, starting with *A Princess on Mars* in 1912.

However, scientific criticism increased. The use of newer and larger telescopes during the 1909 opposition failed to confirm Lowell's observations, and the result was a rapid decline in interest in the canals as indicators of intelligent life on Mars. Interest in the markings as such did persist, however, but the arrival of the space probes has finally shown what the Martian surface is really like.

Here is indeed a fine example of the creation of a myth barely a hundred years ago. The Oxford Dictionary sets 1892 as the year when ''Martian'' was first used in the current sense of inhabitants of that planet. We must refrain from the easy temptation of writing off Fathers Secchi and Schiaparelli. Both were respected astronomers of their time with important observations to their credit. They both used refractors of 9 in. aperture, large instruments for those days. But we should note that with a Martian disc of 25″, a resolution of 1 sec corresponds to a distance of around 250 km on the Martian surface, impressively wide for any canal. (Lowell recognized this and considered that he was observing not the canals but the adjacent irrigated areas.) ''Martian,'' however, remains in the language as a monument to Lowell. More recent attempts to find life on Mars (described later in this chapter)

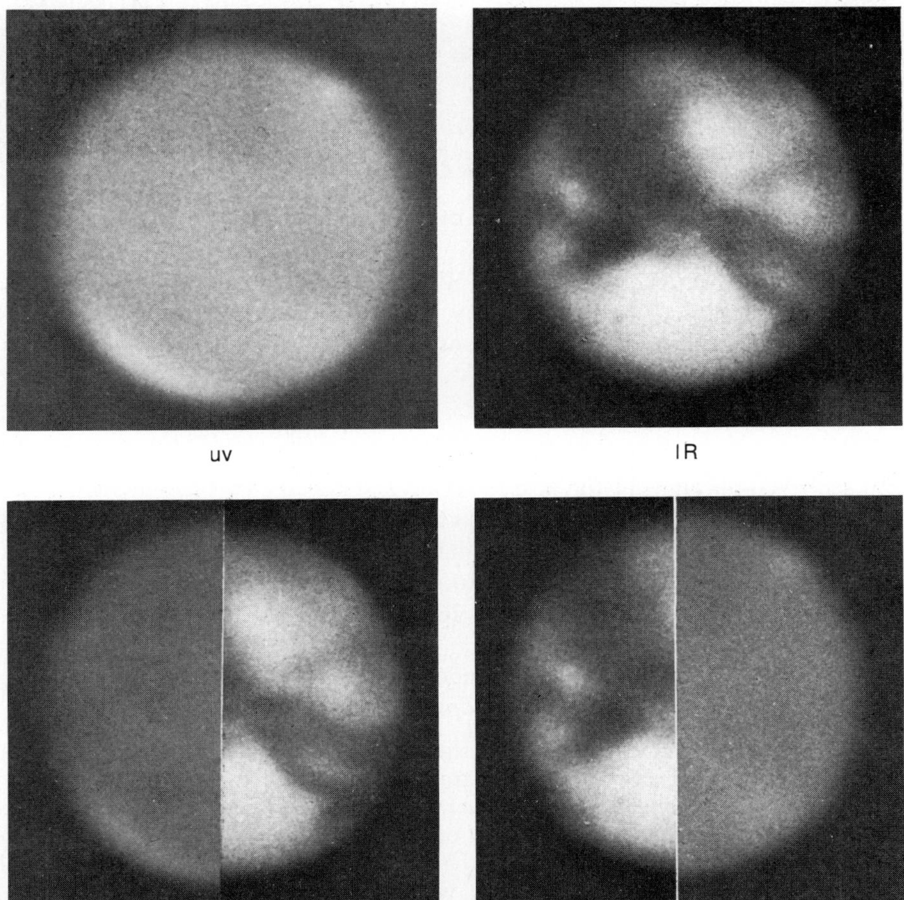


**Figure 15.4** Mars as seen from Earth: Mars appears slightly larger at UV wavelengths than in the IR. Note also how the canals show up more strongly at the longer wavelengths (Lick Observatory).

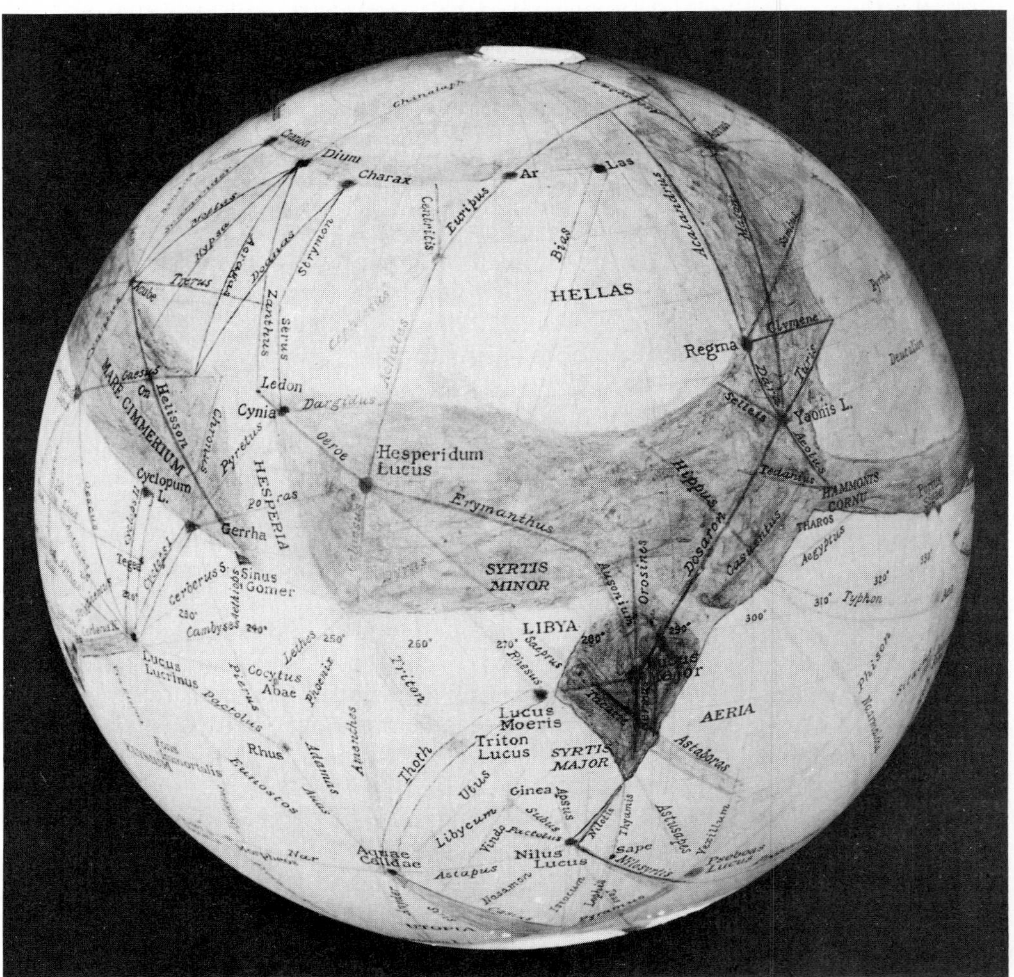


**Figure 15.5** Canals on Mars as drawn by Lowell in 1907 (Lowell Observatory).

have been directed towards detecting microorganisms not humanoids, and searches for intelligent beings away from the solar system have become sophisticated exercises in communication. It is clearly a subject that continues to attract popular interest, and one to which we shall return in Chapter 31.

## 15.2 SATELLITES

In 1879 Asaph Hall, at the U.S. Naval Observatory in Washington, D.C., discovered the two satellites of Mars and named them Phobos (''fear'') and Deimos (''panic'') after the horses that drew the chariot of Mars, the Greeks' god of war. These satellites were too faint to be seen until the Naval Observatory obtained its Alvan Clark 26 in. refractor, at that time the largest in the world.

This discovery had, in a strange way, been foreshadowed in 1726 by Jonathan Swift in the *Voyage to Laputa* in *Gulliver's Travels*

> They have discovered two lesser stars, or satellites, which revolve about Mars, whereof the innermost is distant from the center of the primary planet exactly three of his diameters, and the outermost five; the former revolves in the space of ten hours, the later in twenty-one and a half; so that the squares of their periodical times are very near the same proportion with the cubes of their distance from the center of Mars, which evidently shows them to be governed by the same law of gravitation that influences the other heavenly bodies.

Swift's values for the orbital radii and periodic times are indeed in the correct ratio to satisfy Kepler's third law (Fig. 15.6). On the other hand, if we use Swift's figures to calculate a value for the mass of Mars, we find a value that is about six times too high leading to a density of about 23 g/cm$^3$, far higher than anything on Earth. Swift's choice of the orbital radii remains something of a puzzle. He clearly knew enough to ensure that Kepler's Law was obeyed but was apparently satisfied to leave matters there even though the implications for Mars' mass were out of line and could have been calculated at that time. There is no question, though, that Swift could not possibly have known of the existence of the moons, for no telescope in Swift's day could have gathered enough light to reveal them.

The Mariner and Viking probes have given us close-up views of Phobos and Deimos showing them as barren pieces of rock, decidedly nonspherical and with many craters visible on otherwise featureless surfaces (Figs. 15.7, 15.8, 15.9 and 15.10). Their dimensions are small: Deimos is only 15 km at its longest and Phobos 27 km. The Viking Orbiter 1 made seventeen close passes by Phobos in 1977 coming to within 300 km. The surface was scanned at visual and IR wavelengths. The reflectance was found to be similar to those of asteroids Ceres and Pallas, and to some stony meteorites. From perturbations of the Orbiter's path, the mass and mean density could be calculated. The density was found to be 2.0 g/cm$^3$, about half that of Mars. This suggests that Phobos was not formed at the same time and in the same place as Mars but might have been captured long after its formation much further out in the solar system.

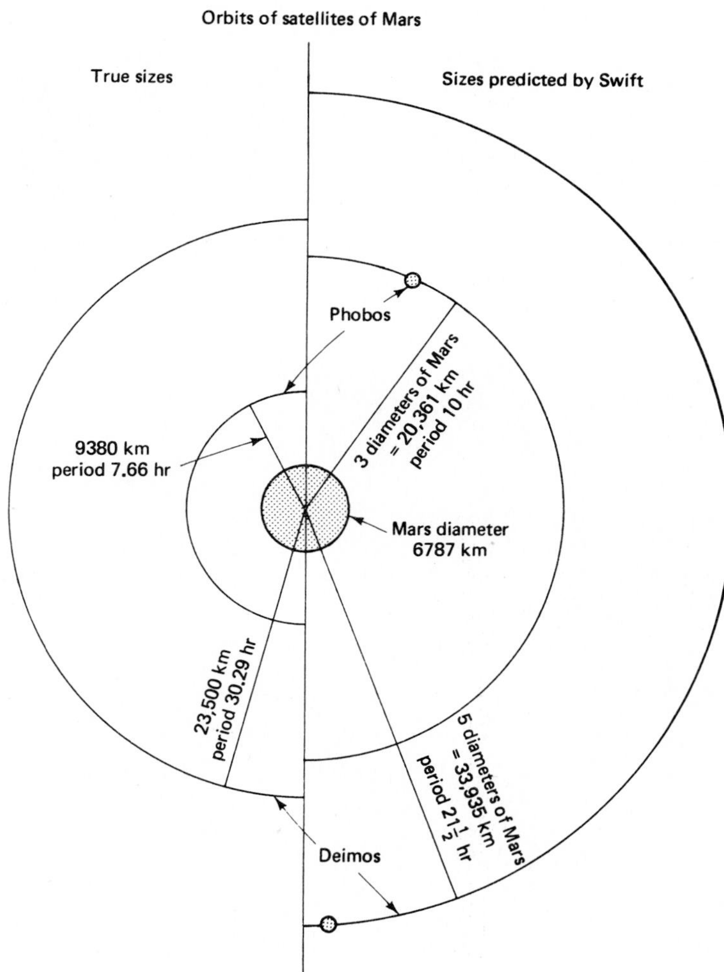


**Figure 15.6** Comparison of orbits of Mars' satellites as described by Swift, and modern values.

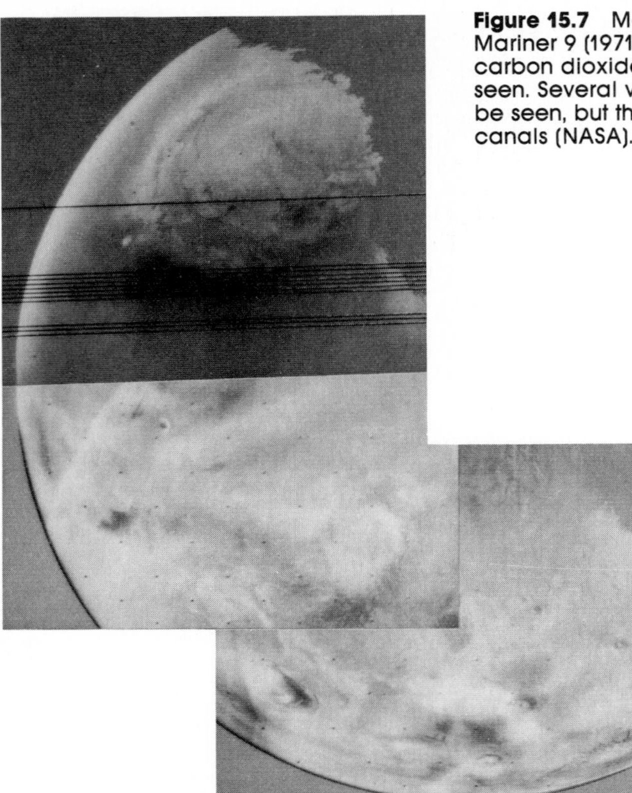

**Figure 15.7** Mars as seen from Mariner 9 (1971). North polar cap of carbon dioxide frost is clearly seen. Several volcanoes can also be seen, but there is no sign of canals (NASA).

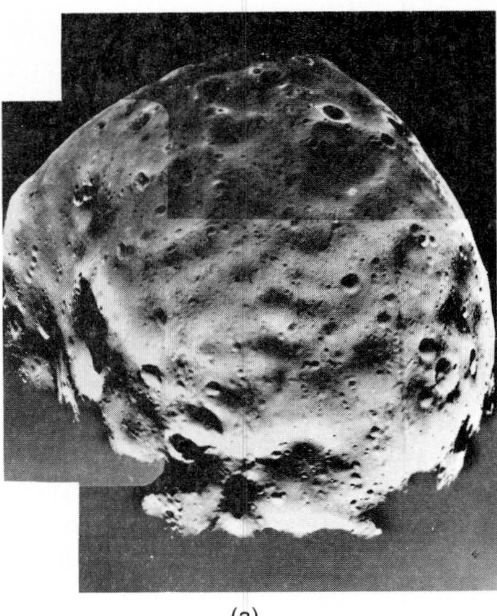

(a)

(b)

**Figure 15.8** Satellites of Mars as seen from Viking orbiter: (a) Phobos from a distance of 480 km, and (b) Deimos from a distance of 330 km (NASA).

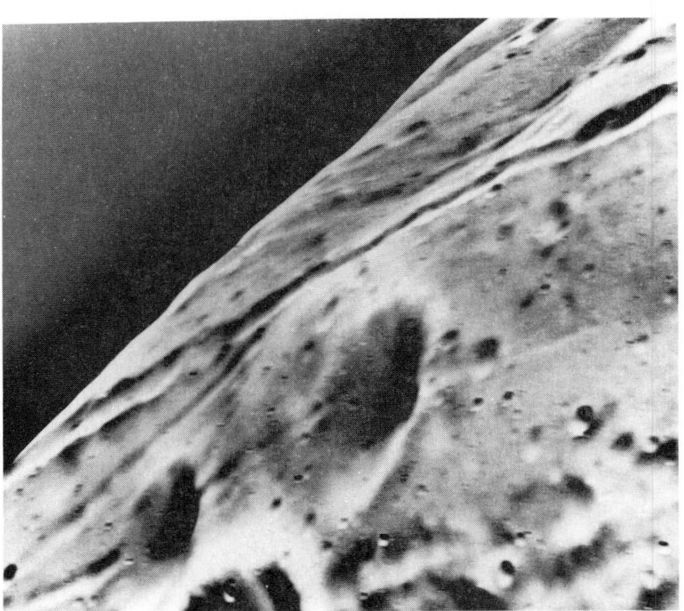

**Figure 15.9** View of Phobos' surface from Viking Orbiter 1 at a range of 120 km. Smallest surface feature identifiable is about 15 m across. Craters have diameters from 10 m to 1.2 km, and the groove running across this field of view is 100 to 200 m wide and extends for some tens of kilometers (NASA).

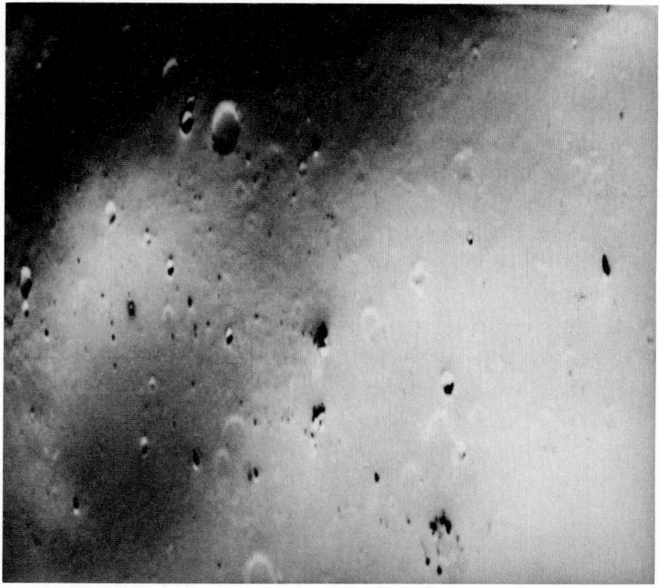

**Figure 15.10** Surface of Deimos from the Viking Orbiter 2 at a range of 30 km. From greater distances, the surface seems to be smooth, but at this close approach it can be seen that the surface has actually been heavily cratered and the craters later partially filled or covered (NASA).

## 15.3 GENERAL PROPERTIES

With a diameter of 6787 km, slightly over half the Earth's size, Mars is the second smallest planet (after Mercury). By applying Kepler's law to the motion of its satellites, we find its mass to be 0.108 times that of the Earth, or $6.40 \times 10^{26}$ g. The average density then turns out to be 3.9 g/cm$^3$, much less than the Earth and slightly higher than the Moon. At the surface, the gravitational acceleration is 372 cm/sec$^2$. Although being appreciably further from the Sun and therefore cooler, its atmosphere is not expected to retain the lightest gases, since with Mars' small mass the escape velocity is low.

Although the misinterpretation of the surface markings produced the canali episode, the presence of markings was well enough established for the planet's rotation rate to be measured. Huygens in 1659 had already drawn attention to some markings, and Cassini in 1666 had obtained the value $24^h$ $40^m$ for the rotation period. The best value today is $24^h$ $37^m$ $23^s$, with the axis of rotation tipped at 24° to the orbital plane, an almost precise replication of the Earth's inclination. The consequence, as on Earth, is a succession of seasons, so clearly shown in the cyclic changes in the size of the polar icecaps.

## 15.4 SPACE PROBE EXPLORATION: ATMOSPHERE AND SURFACE

A succession of Mariner probes, in the years 1965 to 1972, yielded increasing detail on the surface, atmosphere, and moons, and then came the Viking missions with orbiters and landers in 1976.

Mars has an atomosphere, but with a surface pressure of only about 0.006 of that on Earth. This is about the pressure at an altitude of about 40 km above the Earth's surface. As with Venus, carbon dioxide is the major constituent of the atmosphere, comprising 95 percent at low altitudes, decreasing slightly above 100 km. Nitrogen makes up 2 to 3 percent, oxygen about 1/4 percent, and argon about 1 1/2 percent. Water vapor, hunted by the IR spectrometer on the orbiters, is generally less plentiful than 1/10 percent but increases markedly in northern latitudes as one approaches the polar icecap. Even this increase amounts to a tiny total of water vapor. If all the water vapor were condensed, it would form a layer less than 1/10 mm thick over the entire surface.

The polar icecaps are the most striking feature of the Martian surface. The Viking 2 orbiter has taken hundreds of high-resolution pictures of the northern icecap (Fig. 15.11). From these and other observations, we now have a picture of a permanent cap of frozen water ice (whose thickness is estimated to be between 1m and 1 km), augmented seasonally by dry ice (solid carbon dioxide) as the temperature drops to 150 K. During the Martian summer, the temperature rises to 205 K in light regions and to 235 K in dark regions. This is too warm for dry ice but still cold enough for water ice to remain frozen. It is not possible for the water ice to melt to water—at the very low atmospheric pressure on Mars, the ice will evaporate directly, and for the same reason water droplets cannot exist in the atmosphere, for they would promptly evaporate.

Very dark dunes extend all around the North Pole, and there are no fresh impact craters in the entire region, from which we infer that erosion or deposition of fresh material must be rapid and extensive.

The rest of the Martian surface shows a bewildering diversity of features. There are, of course, many craters (Fig. 15.12). Some are surrounded by patterns of explosively distributed debris with little indication of erosion. Other patterns have the appearance of semiliquid ejecta, whereas yet others

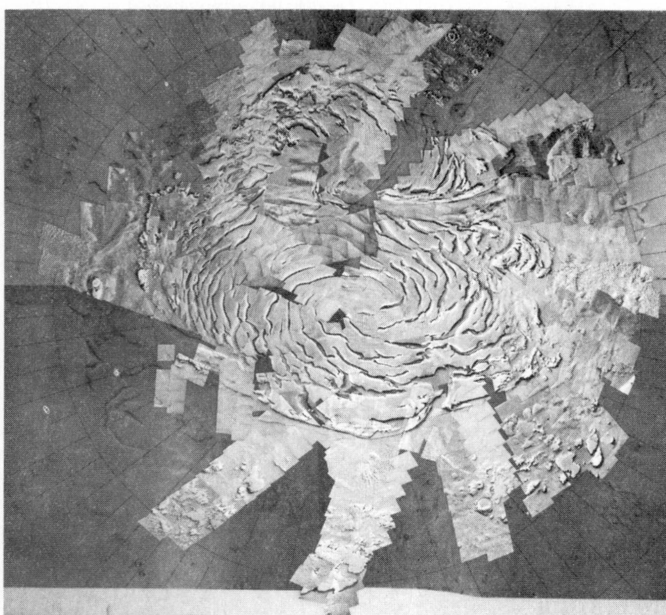

**Figure 15.11** Martian north pole. Mosaic of over 300 frames taken from Viking Orbiter 2. Darker bands indicate places where there is no frost. The polar cap is surrounded by a giant ring of sand dunes (NASA).

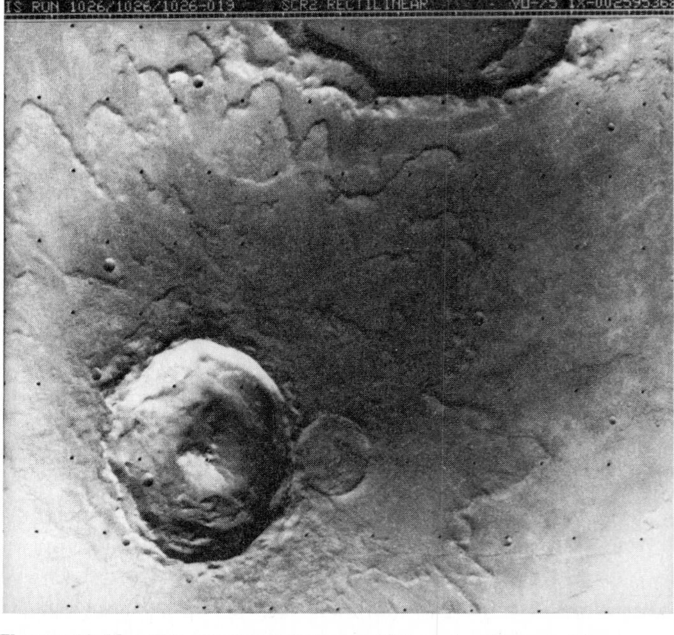


**Figure 15.12** Yuty Crater (diameter 25 km), with an older crater adjacent. Note the central spike also seen in many lunar craters (NASA).

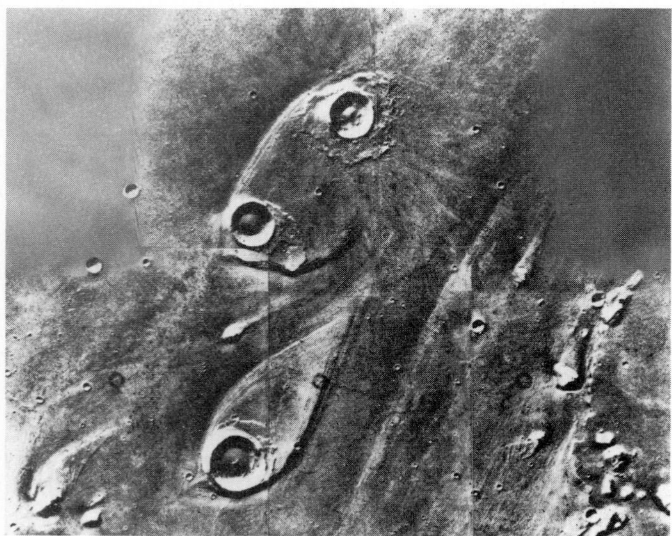

**Figure 15.13** Martian floods were diverted around older craters, to form these "islands." The largest of these craters photographed from the Viking Orbiter is 10 km (NASA).

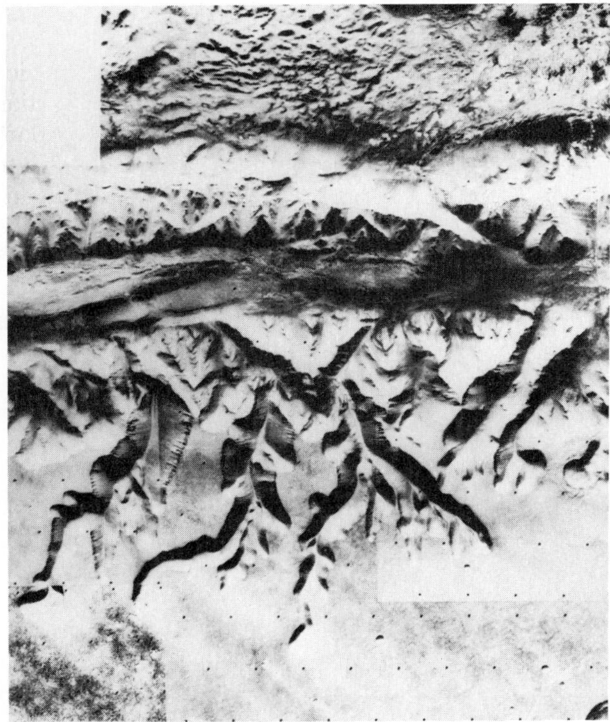

**Figure 15.14** Part of the Valle Marineris showing 120 km of its total 4000 km. Tributaries on the south side do not have a supporting network of smaller scale branches (NASA).

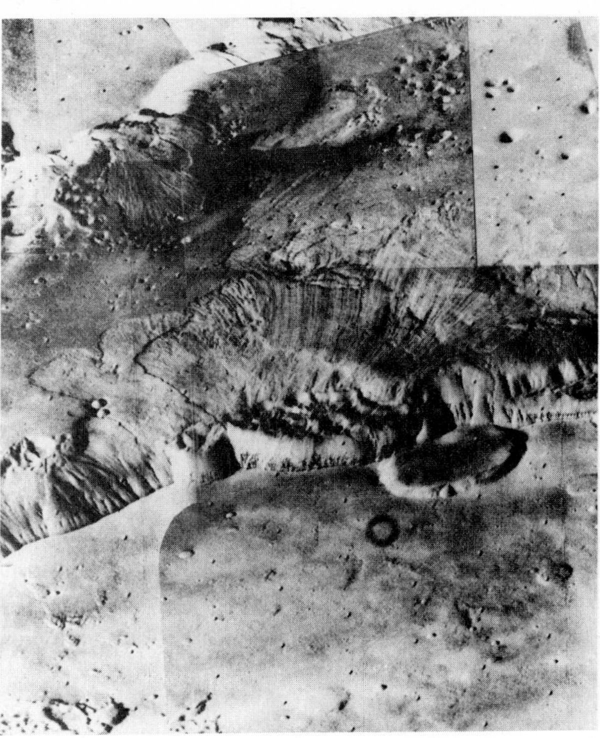

**Figure 15.15** Close-up of Valle Marineris (about 70 km shown) photographed from Viking 1. Canyon gradually enlarges itself by landslides that flow into the canyon floor (NASA).

look as though there had been flash-flood inundation (Fig. 15.13). Elsewhere, there are patterns that look like the eyes on peacock tails showing the characteristic flooding/erosion patterns of large amounts of surface water, and other patterns that show the effects of windblown dust with long shadows trailing from rocks, crater rims, and other surface protrusions. There are extensive cliffs with landslides. It is clear that there were very large quantities of water on the Martian surface at some time, far more than is now known to be in the icecaps.

There are craters with flat floors and low rims probably weathered by intense dust storms. There is an enormous chasm, the Valles Marineris (Figs. 15.14 and 15.15), 4000 km long, up to 200 km wide, and 6 km deep, that

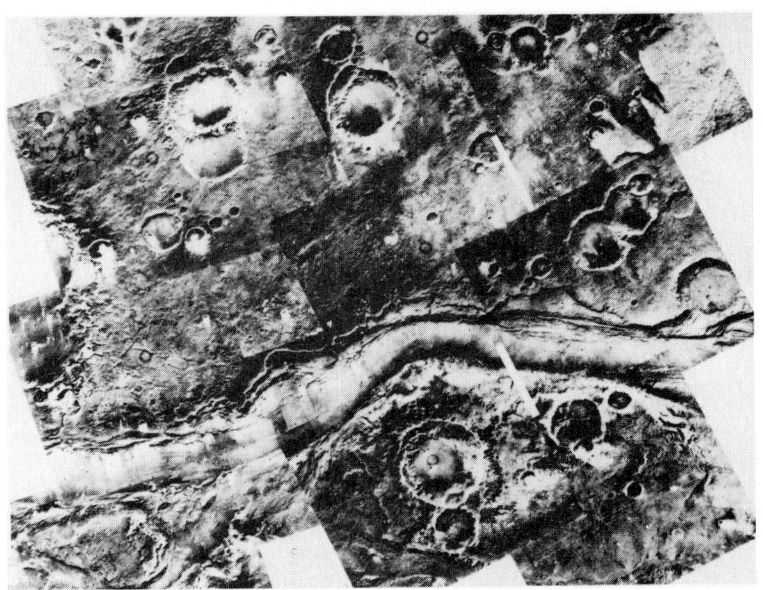

**Figure 15.16** Ares Vallis channel, 25 km wide and 1 km deep, across heavily cratered uplands. Note very few craters on channel floor (NASA).

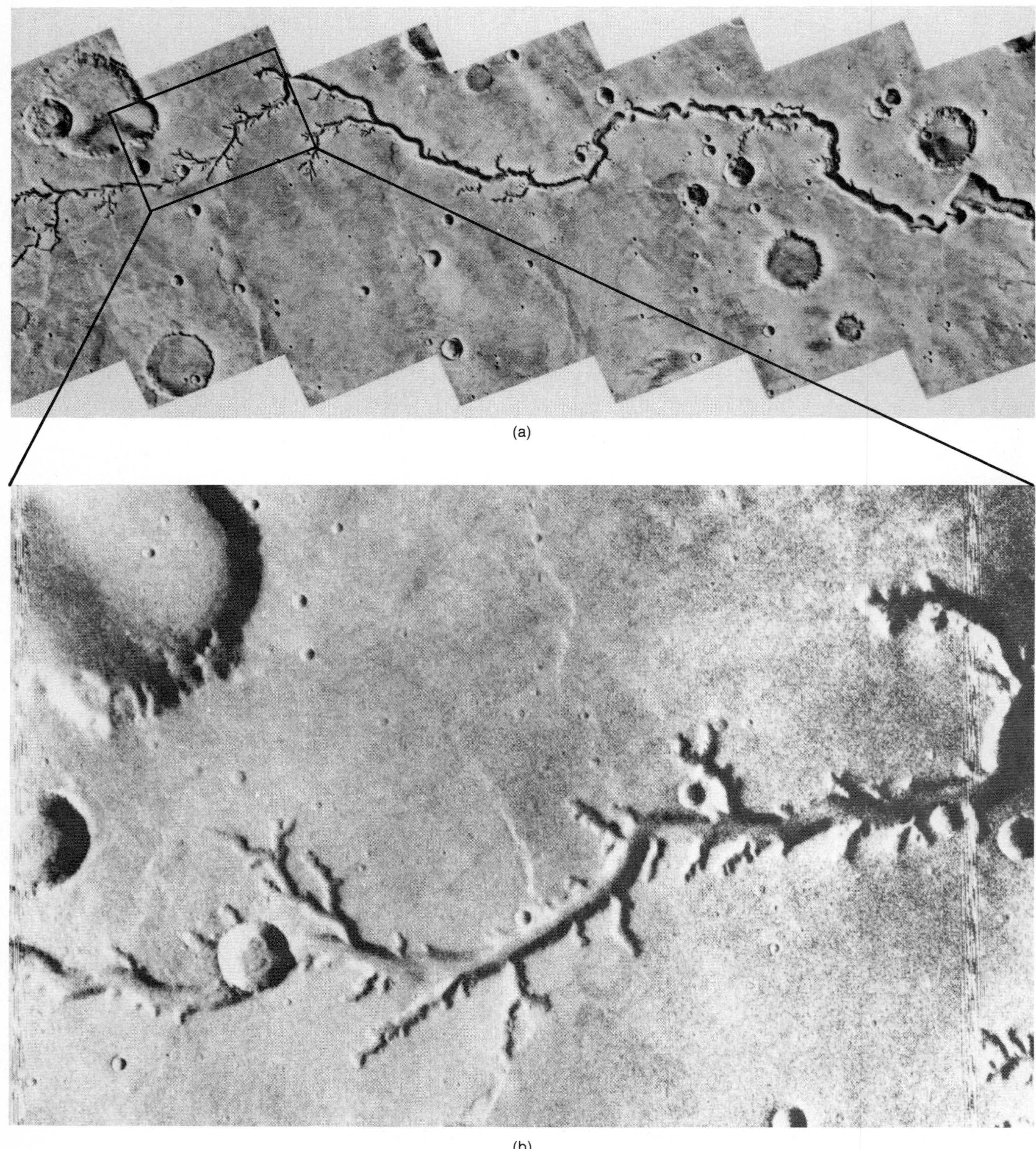

(a)

(b)

**Figure 15.17** (a) Viking view of Nirgal Vallis with (b) close-up of an 80 km section. Open network of tributaries with steep walls and well-separated branches, as with many other such systems on Mars. Surface runoff of water would not produce these features; sapping of ground water seems more likely, from ice on the ground or from water beneath a permafrost layer (NASA).

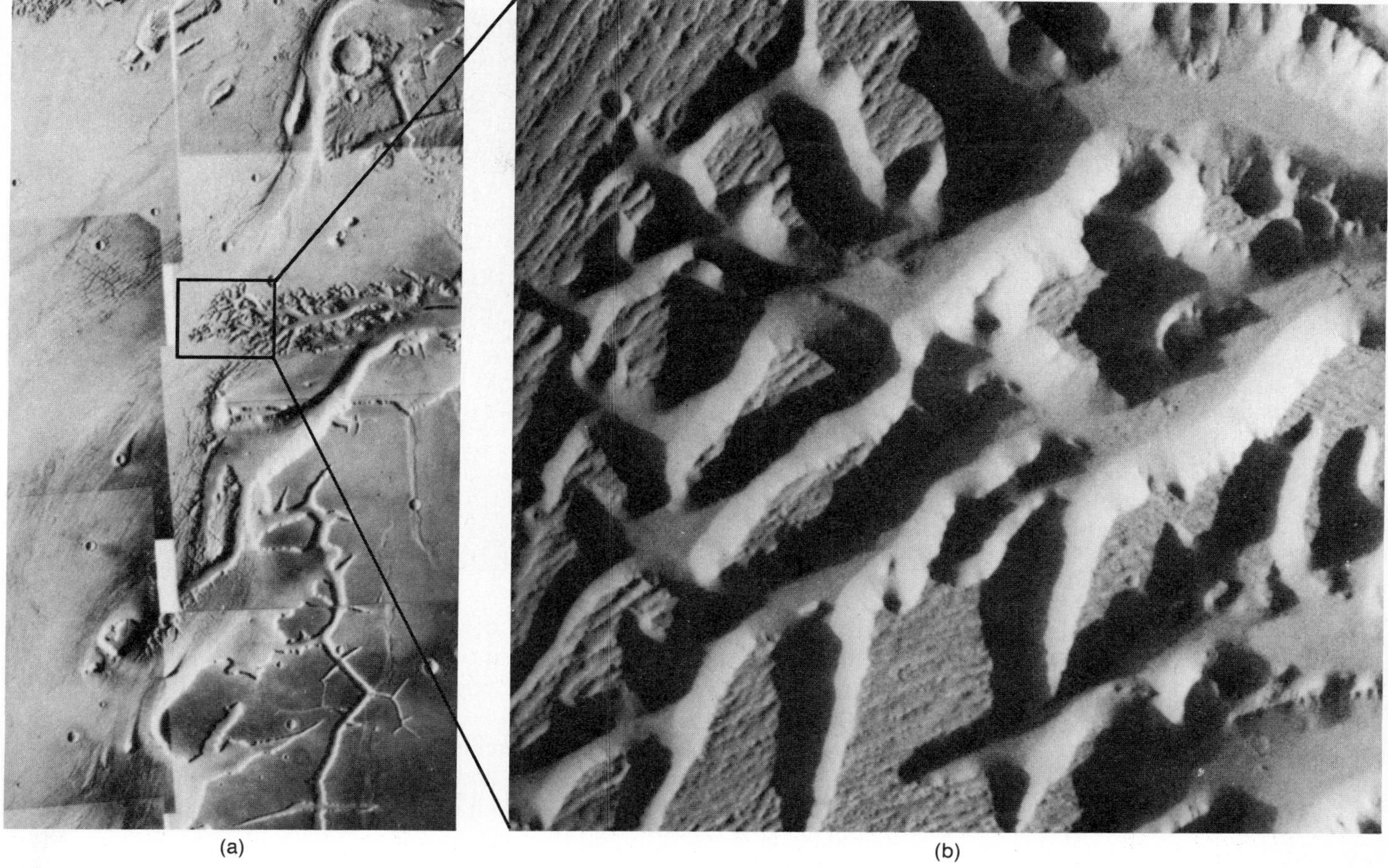


**Figure 15.18** (a) Kasei Vallis, another channel system that is more than 300 km across in places. (b) Close-up of a 30 km section shows islands left by the deep erosion. Note very few craters on top of the flow pattern (NASA).

could hold 2000 Grand Canyons. And finally there are volcanoes, with Olympus Mons the most spectacular (Fig. 15.19). Its base is 500 km across, its height is 25 km (twice Everest), and its lava-covered slopes are topped by a crater about 80 km diameter. An extensive cloud system mostly of ice crystals forms around Olympus during spring and summer in the Martian northern hemisphere.

Close-up surface detail was shown by the television cameras on the Viking 1 and 2 landers, which relayed dull-red scenes of Martian desolation (Fig. 15.20). Viking 1 showed a wide range of rock types and also a glimpse of bedrock indicating the work of erosion. The Viking 2 site had more homogeneous rocks, many being highly pitted. Atmospheric clarity was better in the morning than in the late afternoon. During dust storms there were many particles in the atmosphere, but the appearance of the Sun at different times suggests that there is almost continually a suspension of very fine dust particles.

Analysis of samples scooped up into the landers from depths of 5 to 10 cm showed little difference between the two lander sites, probably the result of the regular dust storms. Chemical composition analysis showed silicon (21 percent) and iron (13 percent) most abundant; magnesium, aluminum, sulfur, and calcium each 3 to 5 percent, all being present as oxides such as $SiO_2$ (44 percent), $Fe_2O_3$ (18 percent); there were traces of other

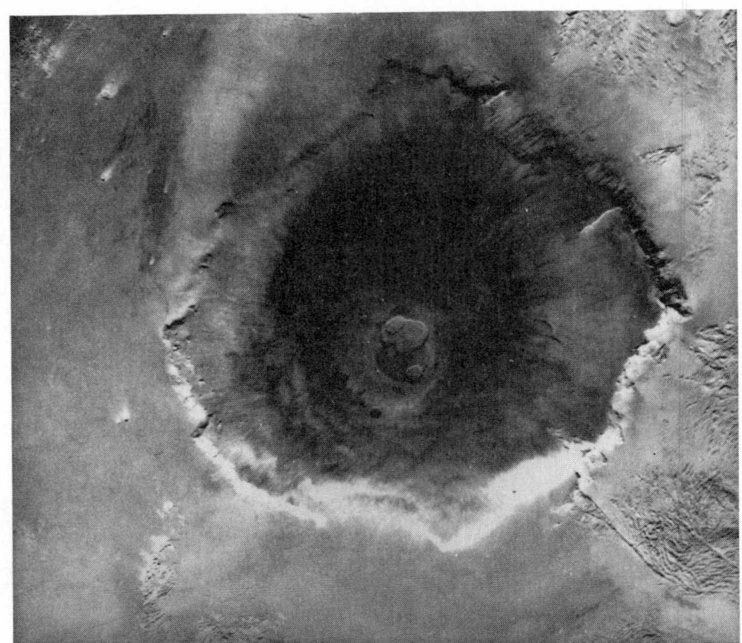

**Figure 15.19** Olympus Mons, 25 km high and 500 km in diameter. This and other large volcanoes on Mars are very similar to those in Hawaii. Relative youth of Mons is indicated by absence of impact craters. It is thought that the large size indicates continued activity for at least a billion years (NASA).

elements. The characteristic red color is probably due to highly oxidized iron, $Fe_2O_3$. This is not a magnetic compound, but a high (5 percent) abundance of some as yet unidentified magnetic particles was detected in the soil by scoops from the landers.

During the first year after landing, frost was noted at the Viking 2 site. This frost lasted more than one third of a Martian year and consisted of water ice and dry ice. Pictures taken over an extended time have been compared by computer, and in this way small changes in detail could be noted (Fig. 15.21). For example, some soil near a large rock was found to have slumped, probably indicating a slip between two different types of soil, although the cause of the slump remains unclear—perhaps a long-delayed echo of the shock of landing. In another area, dark particles grew on the surface of a previously light area, probably as the result of wind erosion. A thin dust deposit has also been noted, perhaps forming with the frost and then left behind when the ice evaporated. Some of the surface measurements will

**Figure 15.20** First photograph taken on the surface of Mars. The distance from the Viking 1 camera to the rock near the center of the photograph is about 1.5 m, and the rock size is about 10 cm (NASA).

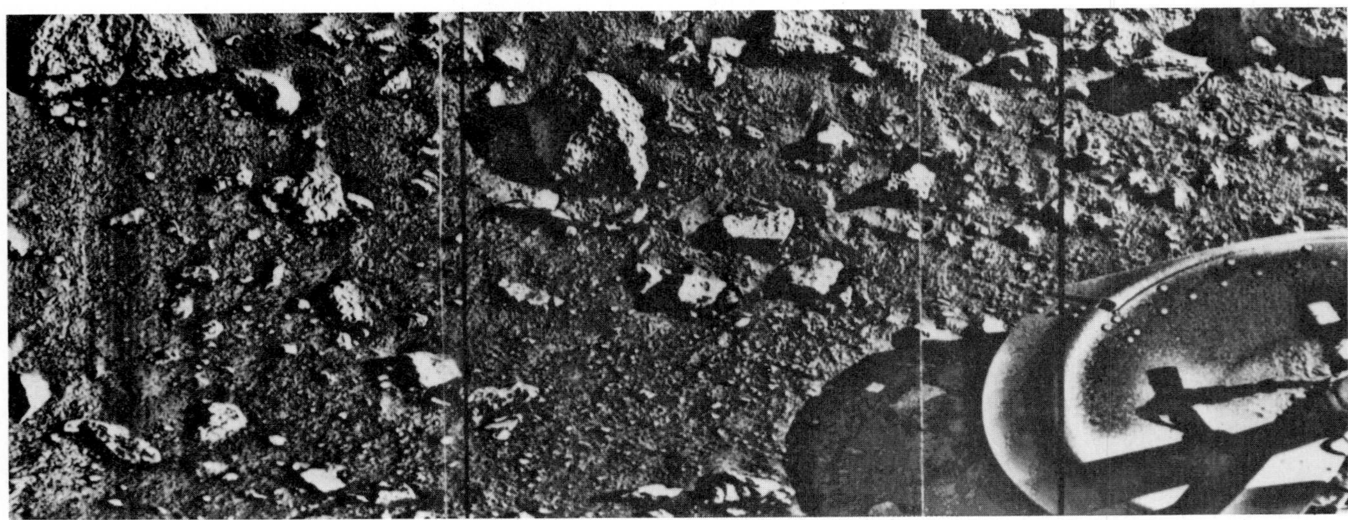

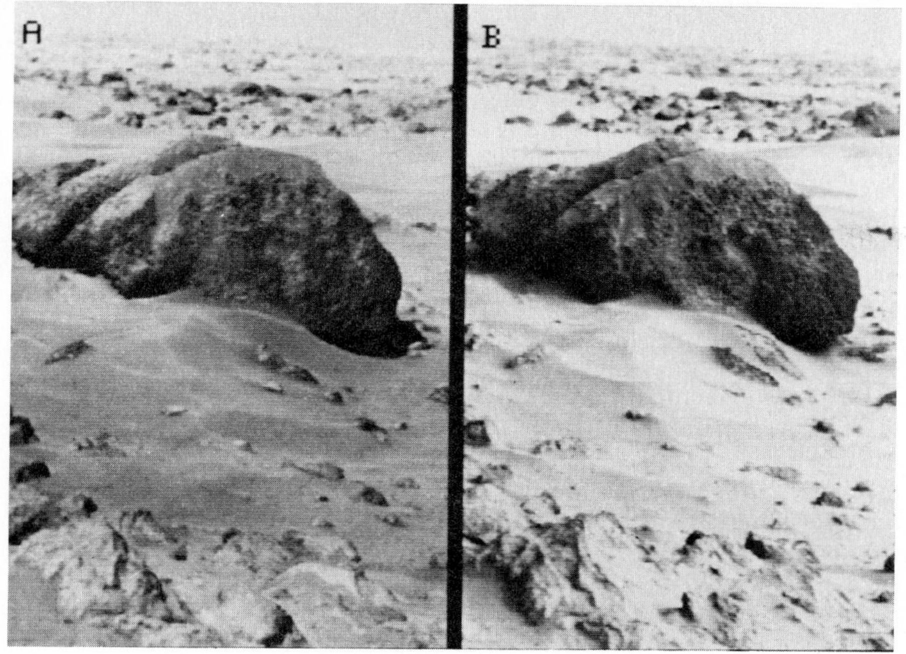


**Figure 15.21** Changes seen on Martian surface. Repeated photographs were taken of the same view. The left photo was taken in August 1976 and the right in September 1978. Dust movement is thought to come from high winds (NASA).

continue. Viking 2 ceased operating in 1979; but Viking 1 should continue to telemeter data at intervals until 1990.

The landers have also followed the local weather conditions. Two surprise findings have emerged. First, the temperature, pressure, and wind speed show extremely uniform variations through each day (Fig. 15.22), far less variable than on Earth where water and water vapor play an important

**Figure 15.22** Daily temperature variations as recorded by the Viking 1 lander. The daily pattern is similar to those observed at China Lake, California but generally shows less day-to-day variation than on Earth.

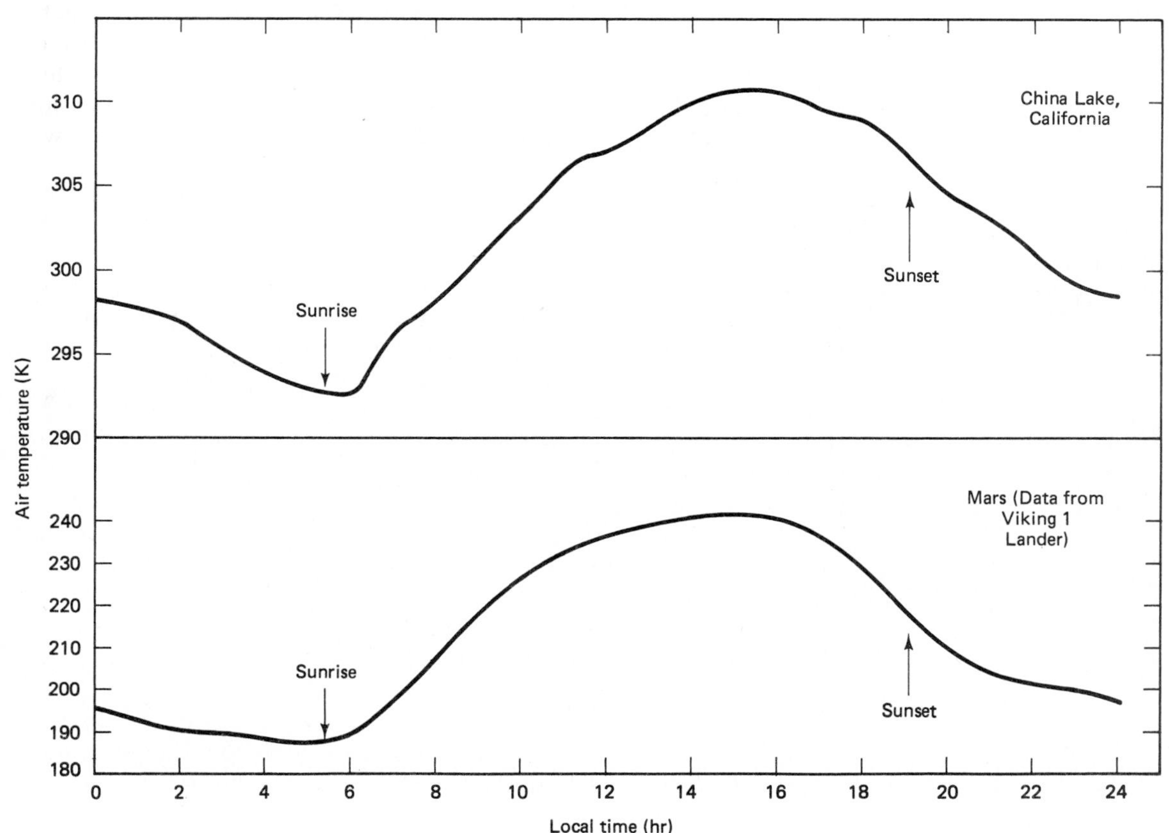

role. The day-night temperature range is typically from 186 K to 240 K. Second, and in contrast to the uniformity of the daily weather, the seasonal change in atmospheric pressure is large, probably due to the condensation of large quantities of carbon dioxide at the winter poles.

## 15.5 GEOLOGIC MODELS

Models can be constructed for the interior of Mars, although important information is still lacking. The Viking 1 seismological instrument did not work; so, even though the Viking 2 package has operated, the internal structure cannot be investigated in the same way that has been possible on the Earth and the Moon, by noting differences of arrival times of seismic signals at separated detectors. What the Viking 2 instrument does show is that down to the detectable limit of about 3 on the Richter scale, the quakes have patterns similar to those seen in southern California with an identical detector. The mean crustal thickness can be inferred to be about 16 km, half that of the Earth, but there must be parts of the crust that are considerably thicker (50–100 km) to be able to support volcanoes of the size observed, such as Olympus. Such a great crust thickness even in parts probably also means that there is no crustal movement similar to continental drift on Earth.

Knowing Mars' total mass, we can calculate the interior pressure and temperature at different depths, recognizing that we have to make some assumptions about the behavior of materials at high pressures. We then expect the central region to be molten, and as the planet's rotation rate is similar to the Earth's, we would expect conditions to be ripe for a magnetic field. There is no clear evidence for a field. If it exists, it is more than 5000 times weaker than the Earth's. We must therefore conclude that the central core of Mars cannot be of iron-nickel as is almost surely the case for the Earth. It probably contains iron, perhaps with iron sulfide (FeS), which will have very different electrical properties from iron-nickel. An iron sulfide mixture would also be consistent with the planet's low average density.

Building models in this way might seem a very arbitrary process, but it can be useful. In the present case, we can draw an important conclusion (the probable absence of nickel in the core), and we can thus confine further calculations to models with only certain properties. In some cases, we may be able to point to further measurements that could allow us to narrow our choice of models even more. We can then, depending on the budget, design space probes and their experiments to enable us to carry out those tests.

## 15.6 THE SEARCH FOR LIFE ON MARS

If there is life anywhere in the solar system in addition to that on Earth, then the terrestrial planets would seem to be the most likely sites. Mercury is too hot to sustain biological molecules of the types we know, and conditions on Venus are even more hostile. The Jovian planets are so far away that their low temperatures would not seem to be hospitable and thus, by elimination, we come to consider Mars.

The canali and Lowell's *Mars and Its Canals* (1906) and *Mars as the Abode of Life* (1908) can now be discounted, but the scientific question remains: Could there be any living organism on Mars, any organism that could be detected by carefully planned experiments carried to the Martian surface by a spacecraft? The Viking missions had among their prime experiments the search for indications of living organisms (Fig. 15.23).

No clear, unambiguous positive results were obtained. On the largest scale the photographic surveys of the surface were negative. They showed no track or footprints, no changes other than those that we have already described, which find natural explanations in physical causes: the slumping of some soil and the formation and evaporation of frost. The mass spectrometer detected no large molecules of the sort found on Earth. All of our

**Figure 15.23** The Viking landers contained experiments to search for biological activity. Samples were scooped from the Martian surface and reaction products examined at intervals after exposure to controlled quantities of added liquids or gas (TRW).

living organisms are based on the chemistry of carbon atoms joined together in chains to form molecules, with varying types and groups of other atoms attached along the chain. The mass spectrometer was sufficiently sensitive to detect organic molecules at the level of a few parts per billion, but all that were found turned out to be residual traces of the cleaning fluids that had been used in the preparation of the spacecraft and its experiments. It would even have been possible to detect organic matter in the small amounts expected to arrive with impacting meteorites, and the null result indicates that organic molecules cannot exist for long on the Martian surface before they are destroyed by physical or chemical reactions.

Searches for living organisms can be carried out in different ways. The mass spectrometer can search for molecules that we expect from what we know of life on Earth. Another method involves testing for metabolic activity, that is, the ability of living organisms to effect changes in the chemicals they ingest. Three experiments on the landers were designed to this end. In each of these a scoop was extended from the lander and a sample retrieved for examination within the experimental package.

In the first, the gas exchange experiment, a nutrient solution was introduced into the sample chamber so as to increase the humidity but not actually wet the soil (Fig. 15.24). Any metabolic activity would change the

**Figure 15.24** Results of two of the biological experiments. (a) Carbon dioxide was promptly released, but then tapered off. Oxygen was also promptly released but then decreased steadily. (b) In the second experiment, radioactive tracer atoms were used. Again, after an initial response, the effect levelled off. Both results are taken as an indication of chemical rather than metabolic activity.

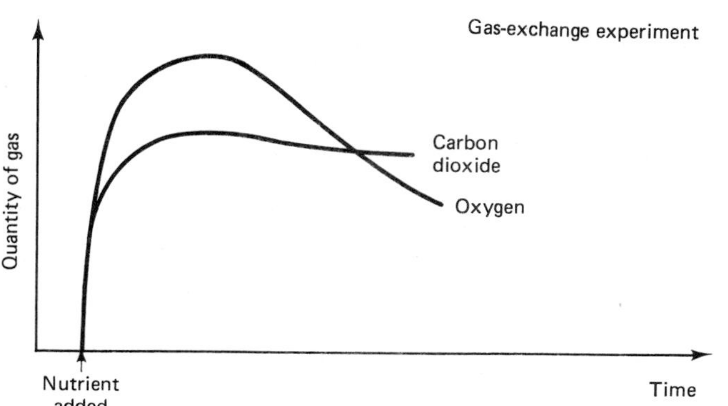


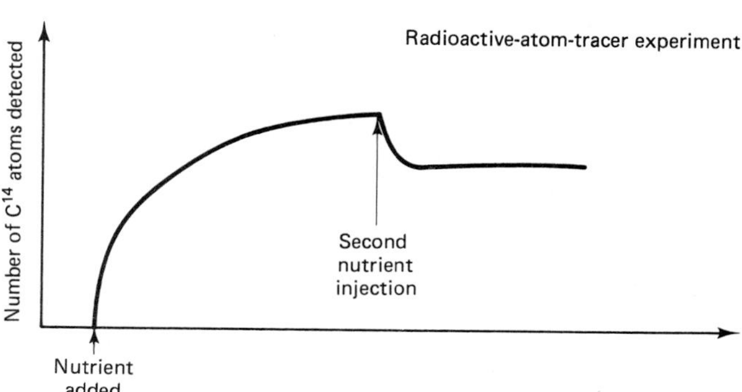

gas composition; the gases in the chamber were analyzed several times. Then some more nutrient was added to saturate the sample. Chamber gases were again analyzed at intervals for the next 6 months. Large amounts of carbon dioxide and oxygen were detected—very promptly after the first humidifying, but after the saturation the production of carbon dioxide gradually stopped, leaving a net amount of gas in the sample chamber, while the amount of oxygen after the initial release actually decreased steadily to zero (probably through combining with some of the nutrient).

The interpretation of these observations is that in the initial humid stage the nutrient water was present in amounts far larger than the Martian surface had seen for a long time, and there was a chemical reaction at the surface. Had any genuine metabolic activity been taking place, the production of the gases should have continued. Clearly this was not the case.

In the second experiment, radioactive $C^{14}$ atoms were incorporated in many of the molecules of the nutrient. These labeled atoms act as tracers. Their participation in any metabolic reactions will be signaled by their appearance in gases that are later released. Once again the addition of nutrient produced a prompt response, and as before, a second injection of nutrient was followed by a reduction in the effect, in this case a reduction in the radioactivity of the released gas (mostly $CO_2$).

The results of these two experiments can best be understood if the Martian surface contains peroxides such as $H_2O_2$, which act as a plentiful source of oxygen. In the labeled release experiment, it is thought that the formic acid (HCOOH) in the nutrient probably reacted with the hydrogen peroxide ($H_2O_2$) to produce a molecule of $CO_2$ and two molecules of $H_2O$, because the measured amount of $CO_2$ corresponded closely to what would be expected from this reaction. A small deficit of $CO_2$ was attributed to solution in the water.

The third experiment was designed to detect the synthesis of organic matter and not its decomposition. Some soil was scooped into the sample chamber where it was illuminated by a lamp that simulated the Martian sunlight. Radioactively labeled $CO_2$ and CO were introduced, and after 5 days the soil and atmosphere were analyzed several times with heating to different temperatures between stages. Seven of nine tests gave positive results, that is, they did show that some of the radioactive gases had been involved in some reactions, but a more detailed analysis including consideration of the temperature changes in the successive stages of the tests suggests that the reactions were most probably chemical and not biological.

In summary, the three Viking biological experiments gave results which in the prevailing view can be most simply understood in terms of chemical rather than biological reactions. There is a remote chance that some biological explanation might be found, especially for the third test, and it is impossible to *prove* that no biological activity exists. The ambiguities will not be removed until new experiments can be conducted, but there is no immediate prospect for this. At best the evidence for life on Mars is very weak indeed.

## 15.7 THE TERRESTRIAL PLANETS—A SUMMARY

When we grouped the inner planets and labeled them as terrestrial, it was on the basis of their densities, which separated them clearly from Jupiter and those beyond. Now, after a detailed survey of each planet, it will be useful to take note of their other major similarities and their differences, and we shall include the Moon at this time, for in many ways it seems to be a part of the group.

Venus and the Earth are the largest terrestrial planets and the only ones that retain substantial atmospheres. The Earth has an internal heat

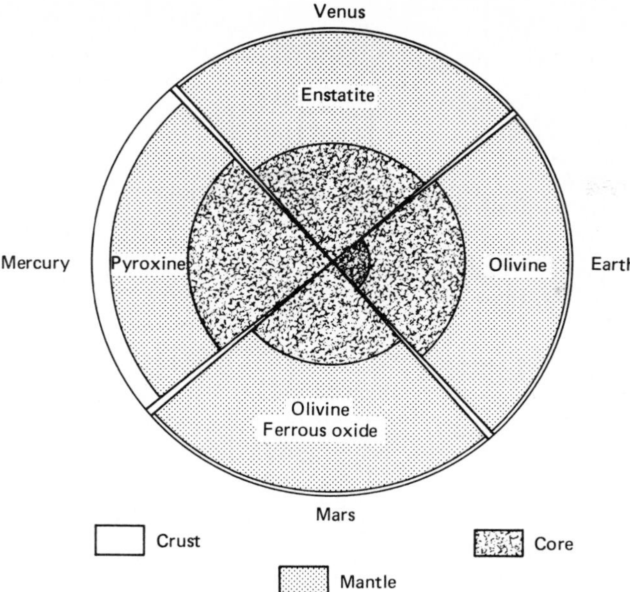


**Figure 15.25** Comparison of models of interiors of the terrestrial planets.

source that is still important, as demonstrated by volcanic activity, by the drifting of the continents, and by the presence of the geomagnetic field.

Mercury and the Moon are the smallest of the group. Their small sizes have resulted in an inability to retain any atmosphere and in rapid cooling so that geological activity ended relatively early in their histories. On both, the surface appearances are dominated by the impact craters that persist uneroded.

Mars is intermediate in size and properties. Its internal geological activity probably continued for longer than on the Moon but not as long or as strongly as on Earth, and it shows no evidence for crustal plates with their movement. Its atmosphere is thin, but still sufficient to be the agent of major surface changes.

The present atmospheres of Venus, Earth, and Mars are very different. To try to draw conclusions about their primordial atmospheres, allowance must be made for their different evolutionary histories, which include the effects such as the searing surface temperature on Venus and the condensation of the icecaps on Mars. Further, in comparing the amounts of carbon, hydrogen, and oxygen, allowance must be made for the exchange between atmosphere and oceans and solid surfaces. With considerable uncertainties, this analysis shows that Venus and the Earth probably had quite similar atmospheres at the start, but Mars continues to pose problems. Its low abundances of carbon, hydrogen, and nitrogen are exactly opposite of what would be expected for a planet that formed that far from the Sun, and the fate of the Martian water remains a mystery.

More or less plausible models can be devised for the planetary interiors, but here again we are left with puzzles (Fig. 15.25). All of these planets have solid surfaces that display a record of heavy bombardment from a very early stage in the evolution of the solar system. There remain many unanswered questions about the bombardment, not the least its abrupt cessation.

Progressing beyond the present state of knowledge is going to be slow. Major advances have been made with the space program, but no new probes will visit the terrestrial planets during the 1980s. Some information will continue to come from Apollo and Viking landers, and the lunar samples in Houston are still available, but new experiments built on the experience of the old are not in sight.

## CHAPTER REVIEW

**Mars:** great variations in distance from Earth, so that favorable conjunctions are of great interest for observing.

- — canals on Mars? a delusion, vigorously pursued by Lowell.
- — satellites: two, discovered in 1879, 153 years after Swift had mentioned satellites in "Gulliver's Travels."

Spacecraft observations showed these moons (Phobos and Deimos) to be barren, cratered rocks, less then 30 km (20 miles) across. Density of moons about half that of planet—probable origins elsewhere.

- — density of planet 3.9 g/cm$^3$, less than Earth and other terrestrial planets.
- — very thin atmosphere, mostly carbon dioxide, with some nitrogen and traces of other gases.
- — polar icecaps expand and contract seasonally. Probably dry ice plus water ice with water ice remaining frozen longer.
- — surface shows cratering, rocks, dust, and much evidence for past flooding (but no direct evidence for water still present in the quantities suggested by the flow patterns).
- — giant volcano (Olympus).
- — seismometer on Viking lander indicates crust thickness about half that of Earth. However, great size of Olympus volcano means that crust must be thicker is some places, so probably no movement of crust like continental drift on Earth.
- — central region should be molten but no evidence for magnetic field: this probably means that interior cannot have iron-nickel as on Earth.

**Search for life:** several experiments on Viking landers. No clear evidence for any living organisms or metabolic activity in samples scooped up by landers for analysis. Reaction products that were observed were probably the result of chemical (inert) reactions.

**Comparison of terrestrial planets:**

- — all are cratered, but with significant differences in cratering patterns.
- — Venus and Earth have dense atmospheres, others do not.
- — Mercury and Moon are smallest, would have cooled fastest. Without atmospheres, craters have not been eroded.
- — two planets closest to Sun have no moons.

## PROBLEMS

**1.** Mars' orbit is one of the most eccentric of any planet. Which planets have more eccentric orbits?

**2.** Using Swift's values for the orbital sizes and periods of Mars' two satellites calculate the mass of Mars. Compare this to the presently accepted value.

**3.** Should Swift receive any credit for correctly predicting the number of moons for Mars? Indeed, does the section in *Gulliver's Travels* constitute a prediction?

**4.** Mars is much closer to the Earth than Jupiter is, yet it does not appear as bright. Why?

**5.** The surface gravity on the Moon is about one sixth that on Earth. On Mars the surface gravity is 0.38 of that on Earth. How will this difference affect the distribution of debris from craters formed on these two bodies?

**6.** What evidence is there for the Martian crust being very thick in at least some places?

**7.** What is the evidence for there being large amounts of water on the Martian surface in the past?

**8.** What evidence exists for the composition of the polar icecaps on Mars?

**9.** Why is the sky over Mars pink and not blue?

**10.** If Mars and the Earth were to exchange orbits, how might their atmosphere differ from their present properties?

**11.** Describe briefly one of the experiments that was carried out during the Viking mission to search for life on Mars.

**12.** Why (apart from the possibility of canals) was Mars considered a likely place for life to exist?

CHAPTER 16

# The Jovian Planets

The contrast between the terrestrial and Jovian planets is very marked. Many of us have had the experience of looking at several children in a family and wondering just how brothers and sisters could develop so differently, and something of this feeling emerges as we catalog the properties of the planets. The Jovian planets are much larger than the terrestrial planets, they are far less dense, and their compositions must, therefore, be very different. They each have more moons than the terrestrial planets do, and several have systems of rings.

This puzzling diversity has been emphasized by the spectacular pictures of the moons of Jupiter and Saturn obtained by the Voyager probes. The appearances of these moons show a remarkable and totally unsuspected variety. The fly-by missions produced some dramatic pictures and whetted our appetites for further exploration, but that will be a long way in the future because of budgetary tightening.

## 16.1 JUPITER

**Appearance and orbit** At its brightest, Jupiter is brighter than any star and nearly as bright as Mars. Even Galileo's small telescope could collect enough light to reveal the presence of Jupiter's four main moons, but he made no mention of any detail on the Jovian disc, which varies in size between 32″ and 48″. At closest approach, 1″ corresponds to nearly 3000 km so that any features observed from Earth must be enormous.

Modern telescopes show a pattern of bands encircling the planet (Fig. 16.1 and 16.2). There are light-colored bands (zones) alternating with darker bands (belts), with contours that are slightly fuzzy. The overall pattern seems

**Figure 16.1** Jupiter as seen from the Earth in blue light (Palomar Observatory photograph.).

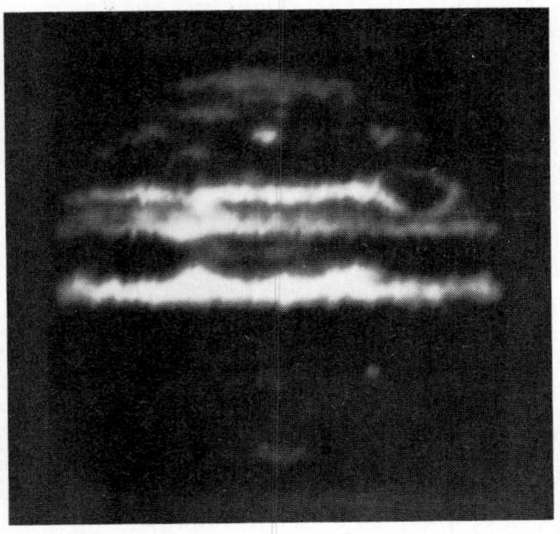

**Figure 16.2** Jupiter as seen from the Earth in the IR (wavelength 5 microns) (NASA).

to be permanent, and the bands have been given descriptive names, such as equatorial, tropical and temperate.

The single most prominent feature is the Great Red Spot, in Jupiter's midsouthern latitudes. We do not know for certain whether this was the feature recorded by Cassini in the 1660s; after it was discovered in 1878 older records of its sightings could be traced back only to 1831 and so we do not know whether this feature on Jupiter is permanent or only temporary but long-lasting. We do know that its color and size both change, though not in any regular fashion. Its width is fairly steady at 14,000 km, but its length has been seen to change significantly. Its present length of 21,000 km is about half the size of a century ago.

Jupiter's average distance from the Sun is 5.2 AU, in an orbit of eccentricity 0.048 inclined to the ecliptic by only 1°18′. Its sidereal period is 11.9 years; its synodic period is 399 days.

**General properties** With a diameter of 143,000 km Jupiter is the largest of the planets, being about eleven times larger than the Earth. From observations of its satellites, its mass is known to be 318 $M_E$ or $1.90 \times 10^{30}$ g. This is about $\frac{1}{1000}$ that of the Sun but still two and a half times the total mass of all others planets together. At its surface the acceleration due to gravity is two and a half times that on the Earth's surface.

The average density is 1.3 g/cm$^3$, far less than Earth and only slightly greater than the density of water. Thus, although we cannot see far into the clouds nor have we yet had any landing probes such as those on Venus or Mars, we infer that Jupiter must be largely composed of lighter elements and cannot have a large, solid core.

The Red Spot and the irregular markings in the zones and belts serve as markers for measuring the planet's rotation. We find that the rotation period varies slightly, from $9^h50^m30^s$ at its equator to $9^h55^m41^s$ at the poles. As with Venus, what we are seeing is the topmost part of its cloud system, but the rotation of deeper layers remains unexplored. The short rotation period, coupled with the large size, result in a very high surface rotational speed—12 km/sec at the equator. Jupiter is, therefore, appreciably flattened by this rotational speed: The difference between its equatorial and polar diameters is 6 percent, compared to only $\frac{1}{3}$ percent for the Earth. Because Jupiter's equator is inclined at only 3° to the plane of its orbit, the northern and southern hemispheres will not experience marked seasonal changes as we have on Earth.

Despite its distance from the Sun and Earth, Jupiter's large size and

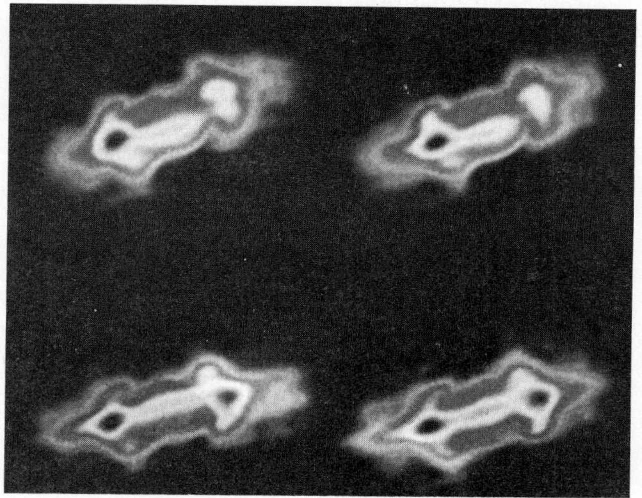

**Figure 16.3** Jupiter as observed at 20 cm radio wavelength by the VLA. Four successive pictures showing rotation of the planet have been reconstructed from the radio observations. The sequence of the radiographs is clockwise from the upper left, and during this sequence Jupiter rotated through 120°. The distance between the two brightest spots is about 50″, so that the radio emission extends appreciably beyond the visible disc (48″) of the planet (NRAO).

high albedo (0.51) make it appear very bright in reflected sunlight. Its own radiation can be measured in the IR. The temperature of its cloud tops has in this way been found to be 135 K (very cold indeed) but when we calculate the total energy being radiated from its large surface ($4\pi R^2\sigma T^4$—chap. 8), we find that Jupiter is radiating more than twice as much energy as it receives from the Sun. It must thus have some internal source of heat, and its temperature would be much lower if it depended only on the Sun for its warmth.

**Atmosphere** Knowledge of the properties of the Jovian atmosphere comes from Earth-based spectroscopy augmented by the Pioneer 10 and 11 and Voyager 1 and 2 probes. The atmosphere is about 1000 km thick; below this the pressure is sufficiently great that the gases must be liquefied. The atmosphere is mostly hydrogen (80 percent) with some helium (19 percent). All other gases total about 1 percent, and this includes methane ($CH_4$), ammonia ($NH_3$) and water vapor. Ethane ($C_2H_6$) and acetylene ($C_2H_2$) have also been detected, and significant changes in abundances were observed to take place in the 4 months between the arrivals of the first and second Voyagers.

Theoretical modeling suggests an atmosphere something like that shown in Fig. 16.5. Below the cloud tops at 135 K, there is first a region of

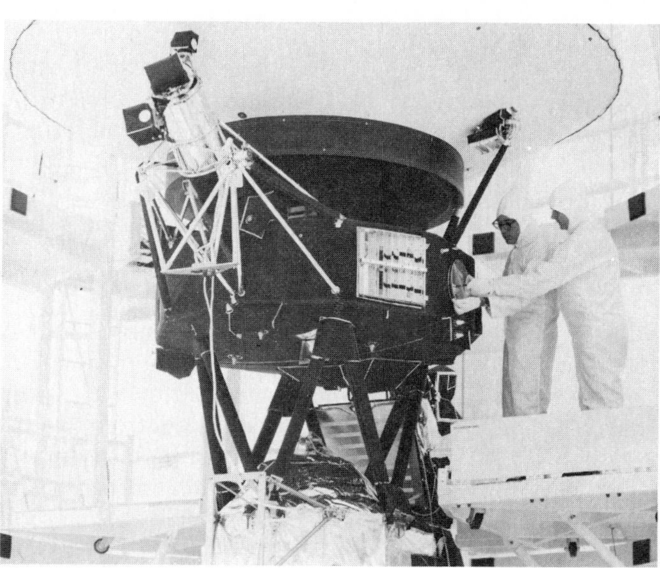

**Figure 16.4** Preparation of Voyager 2 (NASA).

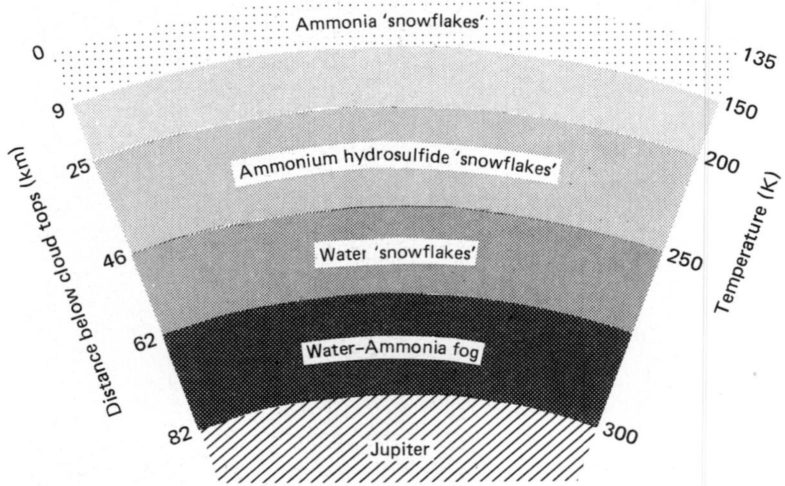


**Figure 16.5** Jupiter's atmosphere: composition.

ammonia snowflakes, then at somewhat higher temperatures and great depths, a layer of ammonia hydrosulfide ($NH_4SH$). Further down there is probably a layer of water snowflakes and deeper yet a region of water vapor and ammonia droplets. The various clouds are suspended in hydrogen and helium with smaller quantities of water vapor, methane, and ammonia.

Infrared observations from the Pioneers and Voyagers showed that the light-colored zones were warmer than the darker belts (Fig. 16.6). In the zones the gases rise, then cool in the upper atmosphere, and spill over to the adjacent belts and descend. The energy for this extensive circulation pattern is available internally, as we know from the measurements of total IR radiation. Although the band structure does have this reasonable explanation, the colors do not. The zones are mostly pale yellow or white and the belts reddish brown, but the known constituent gases are all colorless. It is thought that the coloring could be due to sulfur or phosphorus, and only minute traces would be needed, but this remains a puzzle.

Within this churning atmosphere the stability of the Red Spot is also not fully understood. It has a 2° lower temperature but a higher pressure than its immediate surroundings; it is thought to be similar to hurricanes on Earth and thus an atmospheric phenomenon not related to some underlying surface feature. Within the Spot there is a counterclockwise circulation of the winds, taking 12 days for one rotation. Similar but much smaller features have been seen at other places on Jupiter and typically persist for about 2 years.

Lightning storms occur all over the planet, perhaps the result of photochemical reactions that produce large numbers of electrons and ions. The Voyagers provided almost continuous monitoring of the surface for 8 months, with surface detail as small as 3 km. The rapid rotation of the planet shows that most of the weather is within each zone with little north-south flow between zones. The wind speeds are as high as 150 m/sec in the north temperate zone and still as high as 100 m/sec at the equator. There is a 3 K difference in temperature between equator and poles.

**Magnetosphere** In addition to the visible radiation (reflected sunlight) and the IR (from its internal energy), Jupiter radiates very strongly at radio wavelengths. This radiation, discovered by accident in 1955, is **nonthermal**, that is, its radio spectrum is quite different from that expected from an object at 135 K. Some of this nonthermal radiation is synchrotron (chap. 8), coming from very-high-speed electrons spiraling in the planet's magnetic field. Calculations show that to explain the intensity of the radio radiation we detect,

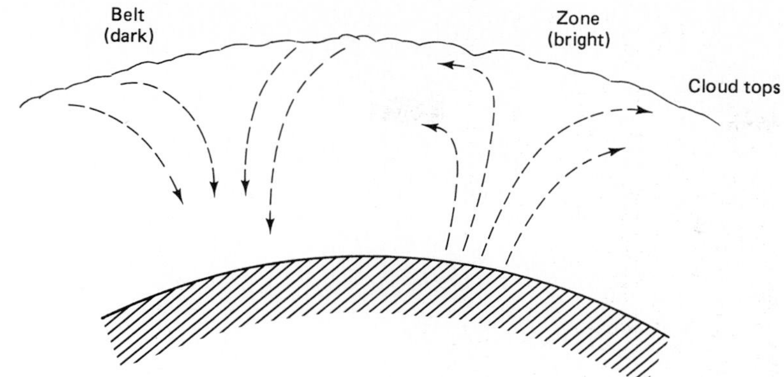


**Figure 16.6** Jupiter's atmosphere: structure.

Jupiter must be surrounded by a strong and extensive magnetic field that is populated by very large numbers of electrons. Perhaps those electrons were captured from the solar wind. The situation is similar to that at the Earth with its Van Allen belts. This was confirmed in 1973 and 1974 when the Pioneer 10 and 11 probes found the magnetic field to be about ten times that of Earth. There are more than 100,000 times as many electrons around Jupiter as around the Earth. At long wavelengths (tens of meters) there are additional radio signals that are highly sporadic, coming in bursts of short duration. The timing of these bursts has been found to be strongly associated with the position of Io, the innermost of the four Galilean moons. Neither the radiation burst mechanism nor Io's role are understood; a plausible guess is that the bursts are atmospheric in origin, something like lightning.

Exploration by the Voyagers has provided more detail. The magnetosphere is similar to the Earth's in general shape but extends to very great distances, about 100 Jupiter radii or 660,000 km. The extent of the magnetosphere is strongly influenced by the solar wind and can change in size by a factor of two. Whereas our Moon is far outside the Earth's mangetosphere, seven of Jupiter's moons have orbits inside its magnetosphere. An enormous electric current of 5 million amps flows between Jupiter and Io. The power involved is twenty times greater than the total electric power available on Earth.

**Structure** Theoretical models have been devised for the structure of Jupiter (Fig. 16.7). The outer 1000 km is gaseous with the topmost 100 km having the cloud layers discussed in the previous section. With increasing depth, the pressure is high enough to liquefy the hydrogen, and this liquid hydrogen layer is considered to extend for about 25,000 km. Below that, the next 35,000 km are probably liquid metallic hydrogen. This term needs clarification. We usually think of metals such as iron and silver in terms of their density and strength, but an important property is their ability to conduct electricity. This property is the result of having many electrons that are free to move in the metal and are not permanently attached to their parent atoms. At the very high pressures in the interior of Jupiter, the hydrogen atoms will be pushed so close together that their electrons will be able to behave much as electrons in metals. This unusual behavior might allow electric currents to circulate and thus be responsible for the extended magnetic field. It must be emphasized though that there are no direct laboratory observations of hydrogen at high enough pressure to confirm this conjecture, and so this feature of the model remains unproven. The central region, with radius around 12,000 km, is thought to be composed of iron, magnesium oxide (MgO), and silicon dioxide ($SiO_2$) at a temperature of roughly 30,000°. Because of the pressure, this core will be solid despite the temperature.

**Figure 16.7** Jupiter's internal structure: a model.

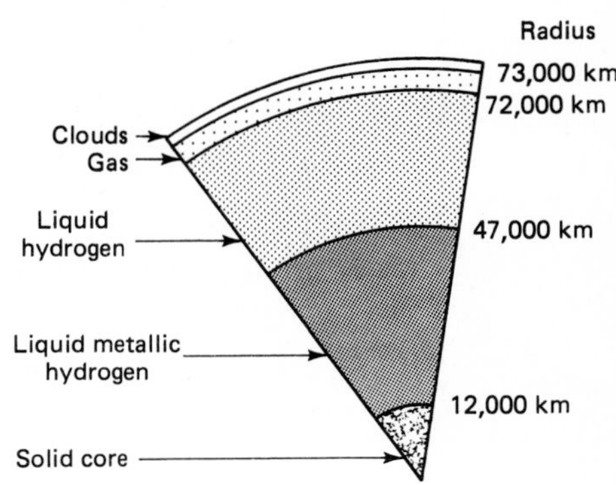

**Figure 16.8** The Galilean moons of Jupiter: from the top left: Europa, Io, Ganymede, and Callisto (Lick Observatory).

**Satellites** The four moons that Galileo discovered (Table 16.1) can be seen with a good pair of binoculars (Fig. 16.8). Their paths around Jupiter are very nearly circular, and their orbital periods are short enough that their relative positions can be seen to change markedly from night to night. The positions of these moons are listed in the *Astronomical Almanac* and in each month's *Sky and Telescope*. Their regular disappearance behind Jupiter and subsequent reappearances can be accurately predicted and are listed, for example, in each month's *Sky and Telescope*. When viewed from the Earth these occultations will appear at slightly different times that depend on the longitude of the place of observation. Together with our own Moon, one of Saturn's moons, and Neptune's Triton, these are the largest satellites in the solar system.

Io is unusual in several respects (Fig. 16.9). We have already noted its association with radio bursts and the large electric current flowing between it and Jupiter. Its orbit around Jupiter is enclosed in what appears to be a

**Figure 16.9** Io viewed from a distance of 8 million kilometers against Jupiter. Surface detail of 100 km is resolved (NASA).

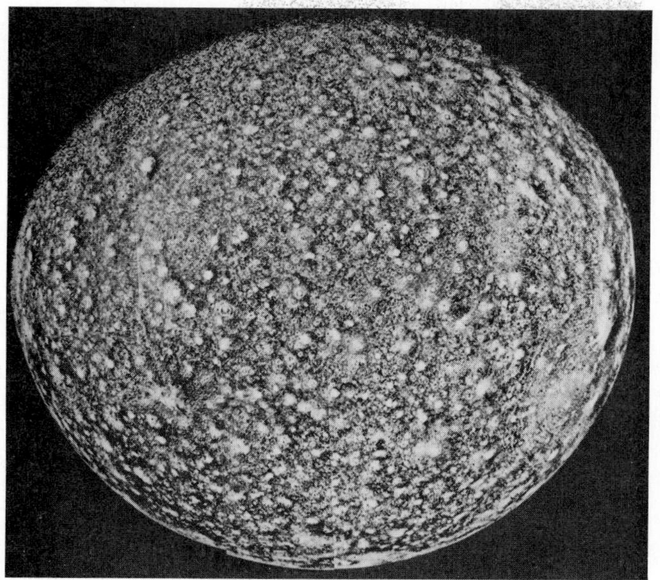

**Figure 16.10** Callisto: view from Voyager 2 from a distance of 400,000 km. The surface is uniformly covered with craters of about 100 km diameter. Because there are very few craters larger than 150 km, it is inferred that large craters cannot survive on this surface (NASA).

tubular cloud of sodium atoms, and it trails a cloud of hydrogen behind. It almost seems as though its orbit is enclosed in a doughnut shaped region of sodium and hydrogen. Most spectacular of all, as discovered by Voyager, are the volcanoes whose plumes extend 100 km up from the surface. Of the eight volcanoes discovered by Voyager 1, six were still erupting when viewed by Voyager 2.

Shortly before this discovery, volcanism on Io had been predicted in a paper written by Peale (University of California at Santa Barbara), and Cassen and Reynolds (NASA). Their calculations had shown that tidal forces in Io, due to the large mass of Jupiter and a complex pattern of orbital perturbations, could lead to a molten core and probably to volcanism. Io's surface has a red-orange color, probably due to sulfur-rich materials brought to the surface by the volcanic activity, and it is thought that collisions of trapped radiation particles are responsible for ejecting sodium atoms from the surface to form the observed cloud. No craters have been seen on Io's surface.

Europa, the next moon, seems to be covered by a thin layer of water ice with an intricate pattern of fracture lines and very few craters. In complete contrast Ganymede, the next moon, has a more heavily cratered surface (Fig. 16.11) and smooth areas showing long faults and grooves as though from contraction during cooling. Callisto, the outermost of Galileo's moons (Fig. 16.10), is probably more heavily cratered than any body in the solar system with two large, circular impact rings 600 km across, reminiscent of the Mare Orientale on the Moon.

***Table 16.1*** THE GALILEAN SATELLITES

| | *Io* | *Europa* | *Ganymede* | *Callisto* |
|---|---|---|---|---|
| Diameter (km) | 3638 | 3126 | 5276 | 4848 |
| Mass (Jupiter = 1) | 0.000047 | 0.000025 | 0.000078 | 0.000057 |
| Average density ($g/cm^3$) | 3.5 | 3.1 | 1.9 | 1.6 |
| Orbital radius (km) | 421600 | 670900 | 1070000 | 1880000 |
| Orbital period (days) | 1.769 | 3.551 | 7.115 | 16.689 |
| Albedo | 0.63 | 0.64 | 0.43 | 0.17 |
| Maximum surface temperatures (K) | 140 | 140 | 154 | 167 |

**Figure 16.11** Ganymede, showing grooved region of mountains. Impact craters on top of the ridges indicates that they were most probably formed several billion years ago (NASA).

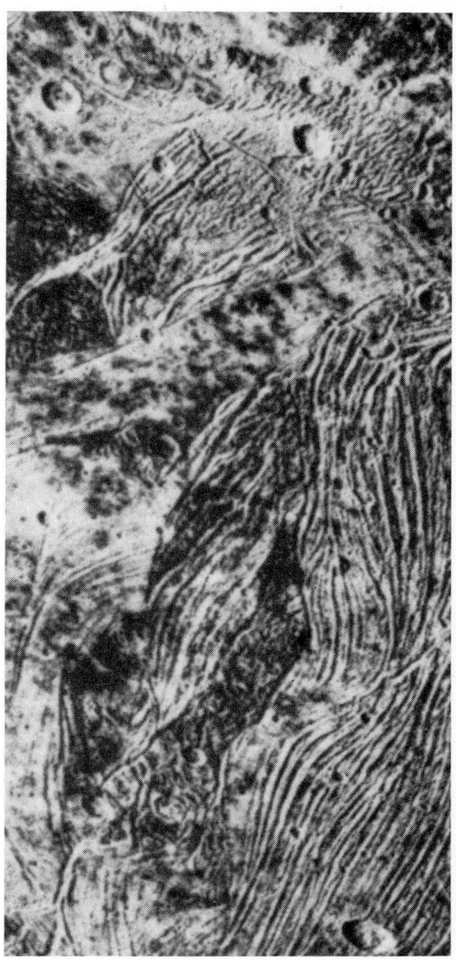

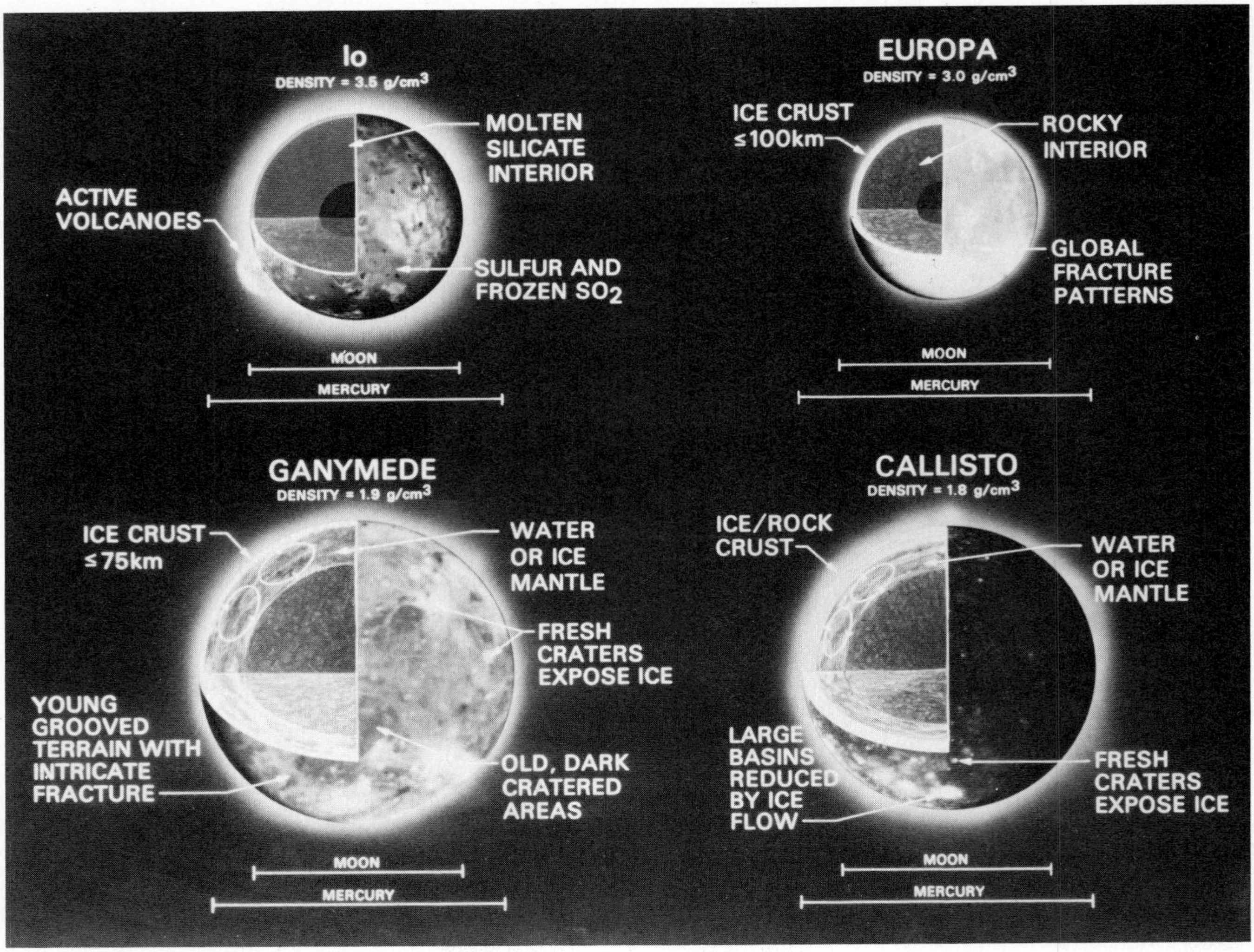

**Figure 16.12** Structure of the Galilean satellites. Their properties differ far more than was imagined from Earth-based observations. Colors range from bright white (Europa) to orange (Io); only one exhibits volcanism (Io); surface structures differ widely; even the average densities are different indicating significant differences in internal composition (JPL).

**Figure 16.13** Jupiter's rings seen from Voyager 2. The bright core has a thickness of about 800 km, and the ring material extends further in and out for some thousands of kilometers (NASA).

Precision tracking of the Voyagers' trajectories has led to refined values for the masses and densities of the larger moons through the perturbations they produce. Io and Europa have densities similar to Mars. Ganymede and Callisto have much lower densities, but all are more dense than Jupiter (Fig. 16.12). Amalthea, the innermost of all of Jupiter's moons, was also viewed by the Voyagers and found to have an irregular shape about 265 km by 150 km.

New moons are still being found around Jupiter, with two in April 1980 bringing the total to sixteen. It is increasingly difficult to detect new moons; they tend to be smaller (and therefore fainter) the further they are from Jupiter, and it is not easy to distinguish them from faint stars.

**Rings** Voyager also provided a surprise in the discovery of rings around Jupiter, which are less complex than those around Saturn. Jupiter's rings start roughly 55,000 km from the top of the clouds and extend out for about 6000 km. The thickness is estimated to be less than 1 km, and the rings are thought to be made up of particles with a range of sizes from a few meters down to the microscopic. No hint of this had ever been seen from the Earth (Fig. 16.13). Jupiter's fourteenth moon is 30 to 40 km in diameter, and its orbit is at the outer edge of the rings where it sweeps up particles that would otherwise stray outward.

## 16.2 SATURN

**Appearance and orbit** In the seventeenth century telescopic observations transformed the vision of Saturn from simply a planet brighter than most stars to a beautiful creation (Fig. 16.14). We can try to imagine the legends that the ancients might have invented had they been able to see those rings. Even in this space age Saturn is far enough away that it has received few visiting probes. Pioneer 11 came by in 1979 and Voyagers 1 and 2 in 1980 and 1981, respectively.

Galileo's simple telescope was not good enough to reveal the rings' structure. To him they appeared as round extensions of the planet, perhaps a pair of moons. Forty-five years later Huygens could describe the rings correctly for the first time, and after another 20 years Cassini could see some detail including the dark circumferential lane that separated the structure into inner and outer regions. We now refer to this dark band as the Cassini Division (Fig. 16.15). The pictures of Saturn's rings transmitted back to Earth from the Voyagers are among the most spectacular ever received from a space probe. They show structure in the rings exotic far beyond anything that had been imagined.

At an average distance of 9.5 AU from the Sun, Saturn was the farthest of the classic planets, taking 29.5 years for a circuit of its nearly circular

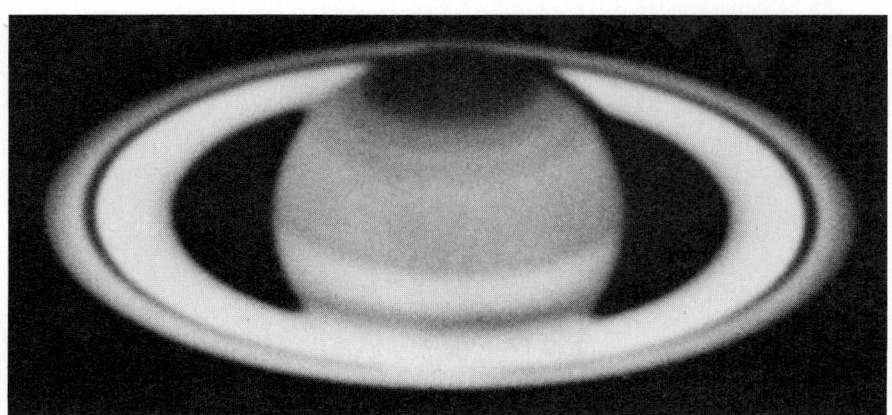

**Figure 16.14** Saturn and its rings as seen from Earth (Palomar Observatory photograph).

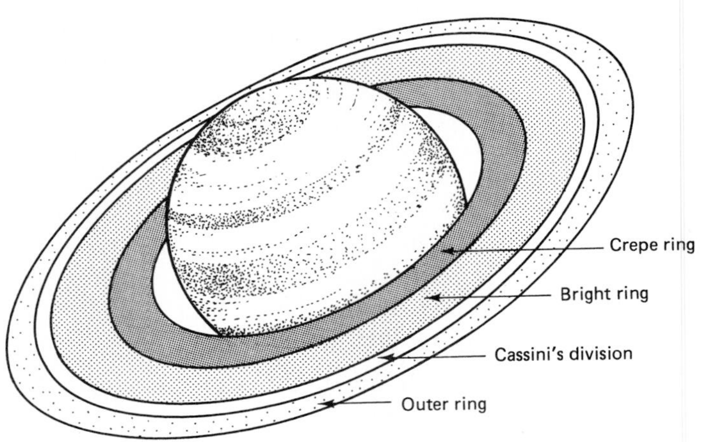


**Figure 16.15** Schematic diagram to show the main features of Saturn's rings as seen from Earth.

orbit. With an average angular diameter of about 7″ (as seen from Earth) the disc of Saturn is about one sixth that of Jupiter. One arc sec thus corresponds to nearly 17,000 km on Saturn's face, very coarse resolution indeed. The Voyagers, however, approached close enough to provide detail down to 1 km.

**General properties** Saturn's average diameter is 120,000 km, and its mass, as deduced from the motion of its satellites, is ninety-five times that of Earth or $5.7 \times 10^{29}$ g. The average density then turns out to be 0.7 g/cm$^3$, by far the least dense of the planets (except Pluto). As with Jupiter, we must conclude that any solid and dense core must be relatively small and that the bulk of the planet must be hydrogen in gaseous or liquid form.

We can see a band structure on its cloud tops, similar at first glance to Jupiter's but not as pronounced and with its bands extending closer to the poles. From markings on these bands, the rotation period has been found to vary from $10^h14^m$ near the equator to $10^h38^m$ at higher latitudes. This high rotation rate results in a nonspherical shape, with the difference between equatorial and polar diameters close to 10 percent. Saturn's equator is inclined to the ecliptic by $27\frac{1}{2}°$, and this should produce seasonal weather changes as we have on Earth.

With Jupiter as a guide, models for Saturn's structure have been constructed. It probably has a dense, rocky core but with the difference that this is probably surrounded by a thin icy and solid layer before the liquid metallic hydrogen layer and then the outer liquid hydrogen. Outermost there is the atmosphere whose cloud tops we can see (Fig. 16.16).

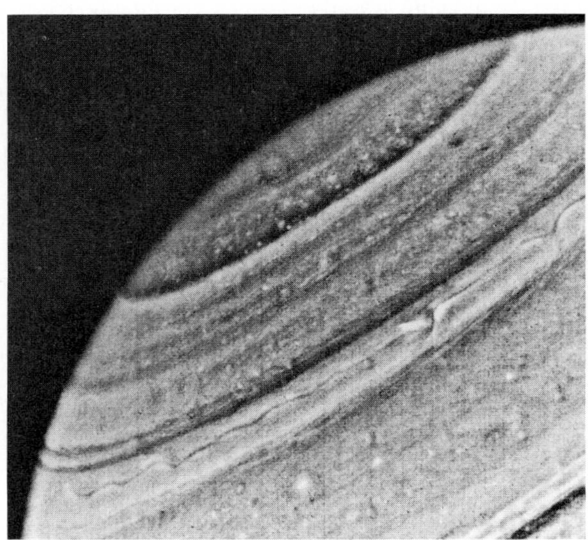

**Figure 16.16** Northern hemisphere of Saturn seen at 10.7 million kilometers from Voyager 2. Smallest features are about 100 km. The wavy band is a high-speed westerly jet, with two dark parallel cloud bands just to the north (NASA).

The total outflow of radiant energy from Saturn exceeds that which it receives from the Sun. Infrared measurements of the cloud tops yield an effective average temperature of 96 K with the southern region a few degrees warmer than the northern. Voyager observations have shown the albedo to be 0.36, rather lower than the value of 0.564 deduced from older observations.

**Atmosphere and magnetosphere** With a relatively high escape velocity (because of its size) Saturn would be expected to retain even such light gases as hydrogen and helium. This has been confirmed by spectroscopic observations that have also identified methane (0.08 percent) and ammonia (0.02 percent) as minor constituents. Hydrogen at 89 percent is the major constituent.

Voyager found that Saturn's atmosphere had 11 percent helium, considerably lower than Jupiter's 19 percent. The Voyagers' IR instruments also detected about one part per million of phosphine ($PH_3$), acetylene ($C_2H_2$), and ethane ($C_2H_6$). The abundance of ammonia ($NH_3$) was lower than on Jupiter, probably because at the lower temperature ammonia will be frozen. The wind velocities are the highest on any planet: a ferocious 500 m/sec (about 1000 mph) near the equator. A red spot, 11,000 km long and about 5000 km wide, similar to Jupiter's, was discovered.

A strong planetary magnetic field was discovered by Pioneer 11—not as strong as Jupiter's but still much stronger than the Earth's. Within this magnetic field there is a region of trapped high-speed particles similar to the Earth's Van Allen belts. Some of the particles in the belt may have come from cosmic-ray collisions with dust particles in the rings.

**Figure 16.17** Wind speeds on Jupiter and Saturn. On Jupiter the ups and downs in the wind-speed profile generally follow the structure of bands, but on Saturn the situation is quite different. There, an equatorial jet stream with a speed of 500 m/sec dominates the flow system and only at latitudes above about 40° does Saturn's wind profile resemble Jupiter's (NASA).

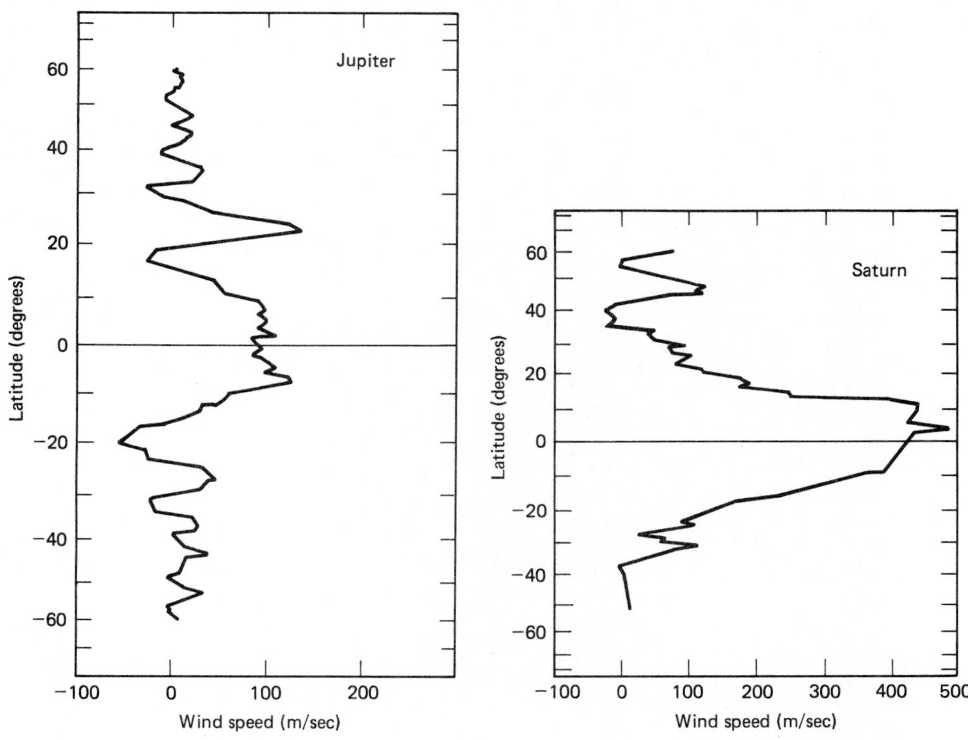

**Satellites** Saturn has an extensive system of moons, most of them outside the well-known rings (Figs. 16.18 and 16.19). Several moons are so small that they cannot be seen from the Earth, and their discovery was impossible until the approach of the Voyagers, which also revealed spectacular detail in the moons and rings. At the same time new puzzles have been posed.

The moons divide into three general groups (Table 16.2). The best-known moons are those that have diameters greater than about 400 km. A second group comprise the five that are closest to Saturn, all with diameters under about 200 km and all discovered by the Voyagers in 1980. The three members of the third group, also discovered in 1980, are little more than very large rocks littered between a few of their better-known companions.

Titan had been thought the largest satellite in the solar system, but close-up measurements have shown it to be 2575 km in radius and thus smaller than Jupiter's Ganymede and Neptune's Triton. This still leaves Titan large enough to be able to retain a surprisingly complex atmosphere.

In 1944 Gerard Kuiper had spectroscopically detected a methane ($CH_4$) atmosphere on Titan. Voyager now showed that this gas comprised only about 2 percent of the atmosphere with almost all of the other 98 percent being nitrogen. The surface atmospheric pressure was found to be 1.6 times that on Earth, and the total mass of atmosphere about nine times larger. Small amounts of hydrocarbons and carbon-nitrogen compounds were also detected in Titan's atmosphere. Many disintegration products of methane can be recognized among the hydrocarbons, for example ethane ($C_2H_6$) and acetylene ($C_2H_2$). The carbon-nitrogen compounds include hydrogen cyanide (HCN), cyanoacetylene ($HC_3N$), and cyanogen ($C_2N_2$). In 1982, Earth-based IR observations revealed the presence of carbon monoxide, about 10,000 times less abundant than nitrogen. Notably lacking is free oxygen;

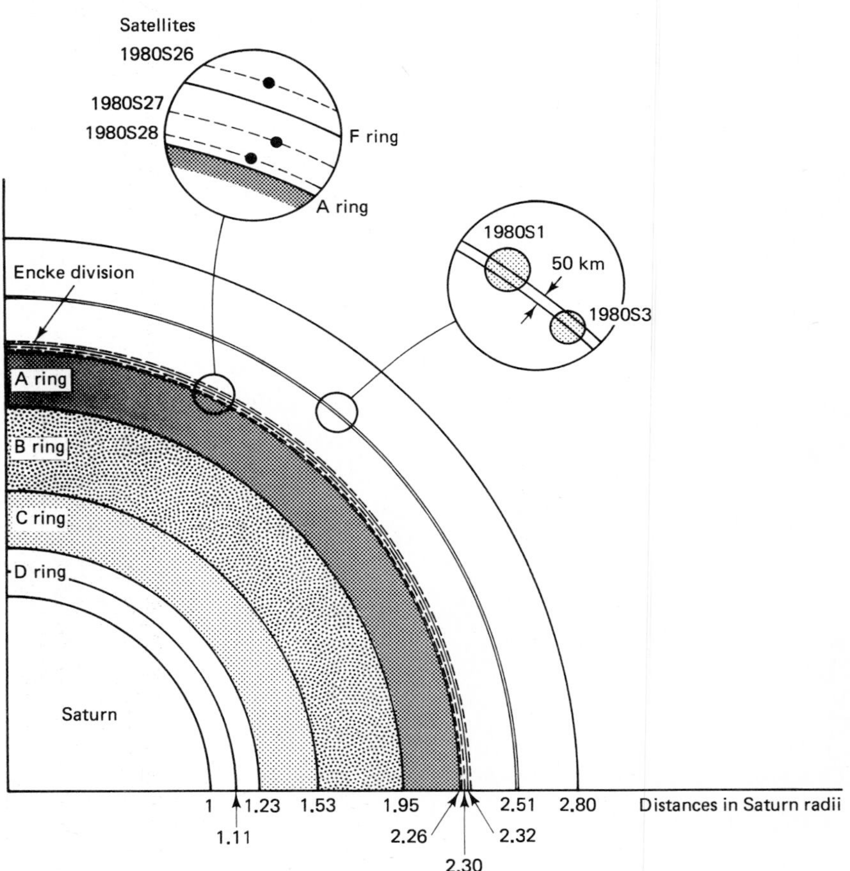


**Figure 16.18** The rings and inner satellites of Saturn.

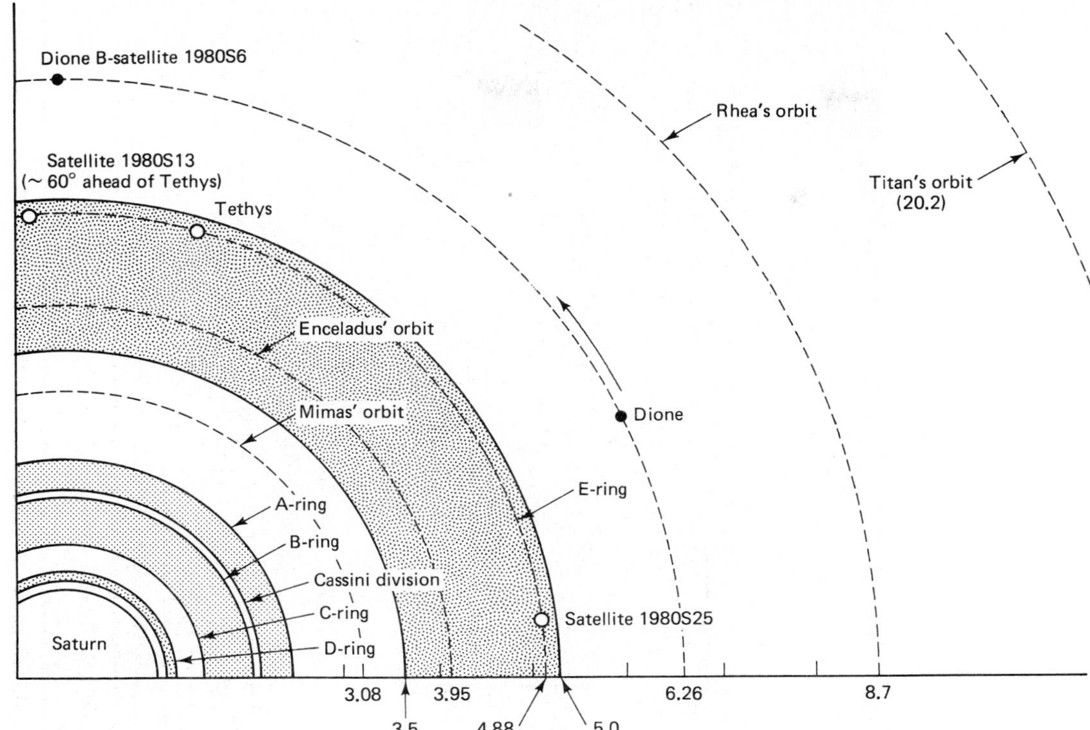


**Figure 16.19** The rings and outer satellites of Saturn.

apart from oxygen in the carbon monoxide, it is probably locked in the water ice, which is thought to constitute a layer 100 km thick beneath the surface of this frigid satellite.

The inhospitable atmosphere is so dense that we cannot see through to Titan's surface. The orange appearance is due to haze and opaque clouds. Very little sunlight can penetrate to the surface where the temperature is around 90 K as measured by Voyager's IR scanners. Under these conditions methane can be solid or liquid, so it is possible that the surface is covered by a liquid methane ocean with methane ice at the polarcaps.

*Table 16.2* SATELLITES OF SATURN

| *Name* | *Orbit radius (km)* | *Orbit radius (Saturn radii)* | *Orbit period (hrs)* | *Radius maximum dimension (km)* | *Density ($g/cm^3$)* |
|---|---|---|---|---|---|
| Atlas | 137,670 | 2.28 | 14.446 | 20 | — |
| 1980 S27 | 139,350 | 2.310 | 14.712 | 70 | — |
| 1980 S26 | 141,700 | 2.349 | 15.096 | 55 | — |
| Janus | 151,422 | 2.510 | 16.664 | 70 | — |
| Epimetheus | 151,472 | 2.511 | 16.672 | 110 | — |
| Mimas | 185,540 | 3.075 | 22.618 | 196 | 1.4 |
| Enceladus | 238,040 | 3.946 | 32.885 | 250 | 1.2 |
| Tethys | 294,670 | 4.884 | 45.310 | 530 | 1.21 |
| 1980 S13 | 294,670 | 4.884 | 45.310 | 17 | — |
| 1980 S25 | 294,670 | 4.884 | 45.310 | 17 | — |
| Dione | 377,420 | 6.256 | 65.686 | 560 | 1.43 |
| 1980 S6 | 378,110 | 6.267 | 65.686 | 18 | — |
| Rhea | 527,100 | 8.737 | 108.42 | 765 | 1.34 |
| Titan | 1,221,860 | 20.253 | 382.69 | 2575 | 1.88 |
| Hyperion | 1,481,000 | 24.550 | 510.64 | 205 | — |
| Iapetus | 3,560,800 | 59.022 | 1903.94 | 730 | 1.16 |
| Phoebe | 12,954,000 | 214.7 | 13210.8 | 110 | — |

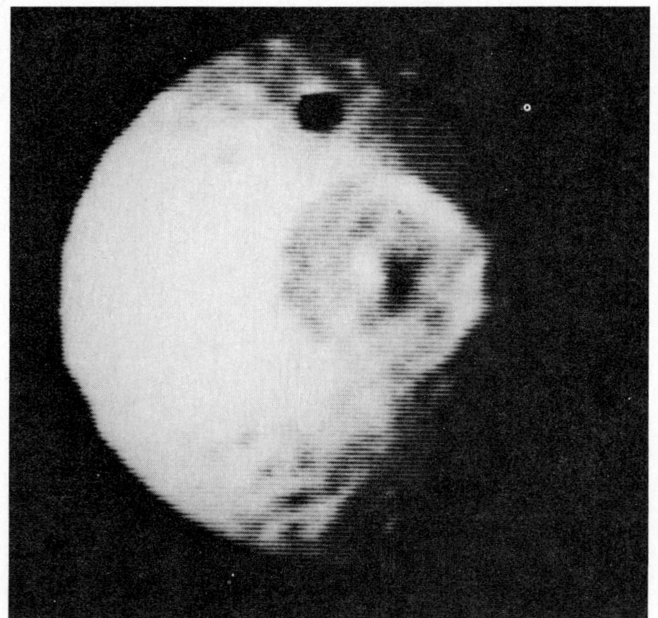

**Figure 16.20** Mimas, innermost of the larger satellites. The crater in the center is 130 km across, more than one quarter of Mimas' diameter. Other craters with sizes 15 to 45 km can also be seen. Note the raised rim and central peak in this large crater, as often seen in craters on the Moon and terrestrial planets (NASA).

The other large moons are also strange objects. The main feature on the surface of Mimas is a giant crater 130 km across with a central peak rising 6 km above the crater floor (Fig. 16.20). This crater, named for the legendary King Arthur, is about the largest that could have been produced without shattering Mimas. The entire surface is covered with impact scars. Enceladus has a very bright surface with an albedo of more than 0.9. It absorbs so little solar energy that its surface temperature is a very cold 70 K, about 20 K colder than the other moons or the parent planet. Its surface shows far less cratering than does Mimas, and no crater is larger than 35 km.

Tethys shows dense cratering and a system of valleys that extends three quarters of the way around its circumference (Fig. 16.21). Perhaps these valleys were formed in the freezing and expansion of its icy surface. Dione (Fig. 16.23) has an orbital period of 2.7 days, which appears to cor-

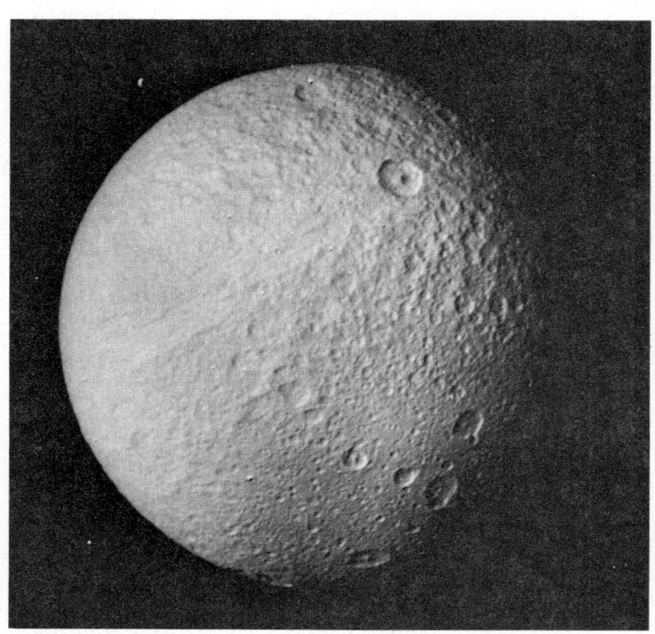

**Figure 16.21** Tethys, just over 1000 km diameter. View from Voyager 1 from 1.2 million kilometers showing features down to 25 km across. The surface is heavily cratered. On the left is a structure of valleys, several hundred kilometers in length (NASA).

**Figure 16.22** Rhea, seen from Voyager 1 at its closest approach, 73,000 km. Features down to 1 km can be seen, but the largest crater is only 75 km diameter (NASA).

relate with some periodic changes in Saturn's radio emission, but no connecting mechanism has been identified. Rhea shows an unevenly cratered surface (Fig. 16.22), and Iapetus has a generally bright surface with an inexplicably black hemisphere. Phoebe is the only satellite in retrograde orbit.

The densities of these satellites, though generally not well-determined, are all larger than that of Saturn. Titan is the densest at 1.88 g/cm$^3$. As water ice has a density of slightly less than 1 g/cm$^3$, we infer that these moons must have rocky cores whose sizes vary greatly one from the other.

Two of the tiny satellites, S13 and S25, share Dione's orbit, an interesting situation that is well understood on the basis of Newtonian dynamics but had not been seen previously. Three innermost satellites appear to play the role of sheepdogs, sweeping up particles that stray from the edges of the A and F rings.

**Figure 16.23** Dione, viewed from 240,000 km, shows heavy cratering. The bright lines are thought to be debris from some impacts. Irregular valleys with craters on top are probably old faults (NASA).

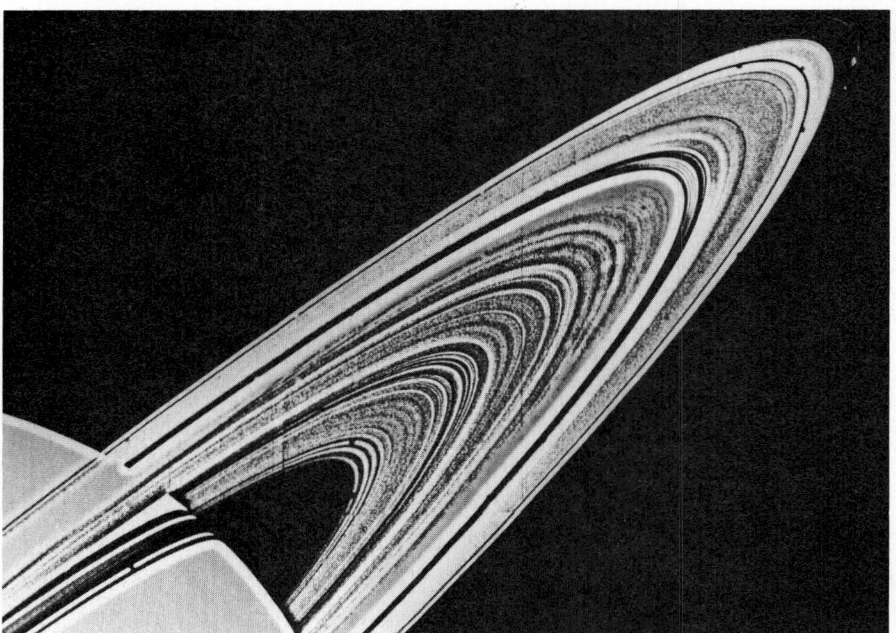

**Figure 16.24** Saturn's rings as seen from a distance of 8 million kilometers from Voyager 1. Nearly 100 concentric rings can be counted. Satellite S.14 (discovered by Voyager) can be seen just beyond the main rings and slightly inside the very faint F ring.

**Rings** For a long time the ring structure was thought to be unique to Saturn (Fig. 16.24). Now that rings have been found around Jupiter and Uranus and more is known of Saturn's rings, the problem of the origin and behavior of planetary rings is being reexamined.

Saturn's rings are tipped at $27\frac{1}{2}°$ to the ecliptic and retain their orientation as Saturn journeys around the Sun (Fig. 16.25). The result is that the appearance of the rings changes. Sometimes they are seen as an instantly recognizable brim around Saturn, but at other times, viewed edge on, they are so thin as to be almost invisible.

The outer diameter of the classic rings in 274,000 km, the inner edge of the innermost being 12,000 km from Saturn's clouds. Three inner rings

**Figure 16.25** Over several years the rings are seen from different angles because of Saturn's rotation relative to the Earth (Lick Observatory).

**Figure 16.26** Spokes of Saturn's rings. Viewed from 4 million kilometers, the spokes appear dark against the reflected light from the rings (NASA).

**Figure 16.27** Spokes viewed from a distance of 936,000 km, after Voyager had passed Saturn and the spokes were seen bright against the forward-scattered light from the B ring. Light scattering behavior of the spokes indicates that they are composed of particles no larger than the wavelength of light (NASA).

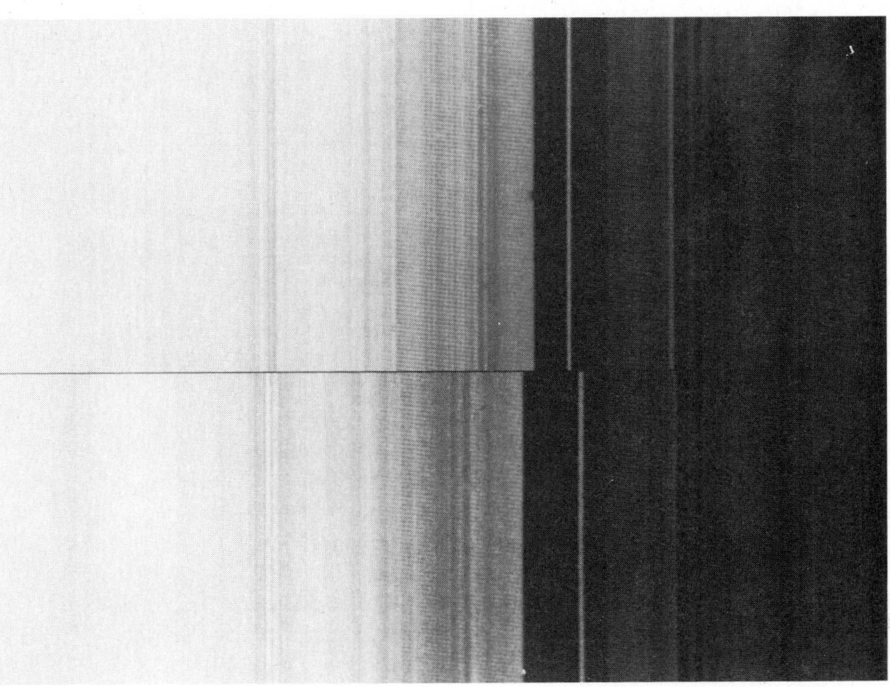

**Figure 16.28** Composite image showing two views of the outer edge of the B ring (bright, on the left) and the inner part of the Cassini division (at right). Images were taken on opposite sides of Saturn from a range of 610,000 km. Structure has different dimensions from one side of planet to the other. The mismatch of the thin ringlet on the inside of the Cassini division is about 50 km indicating an eccentric shape. The edge of the B rings is thought to be distorted by the satellite Mimas, and the fine structure variations within the B ring may indicate wave phenomena (NASA).

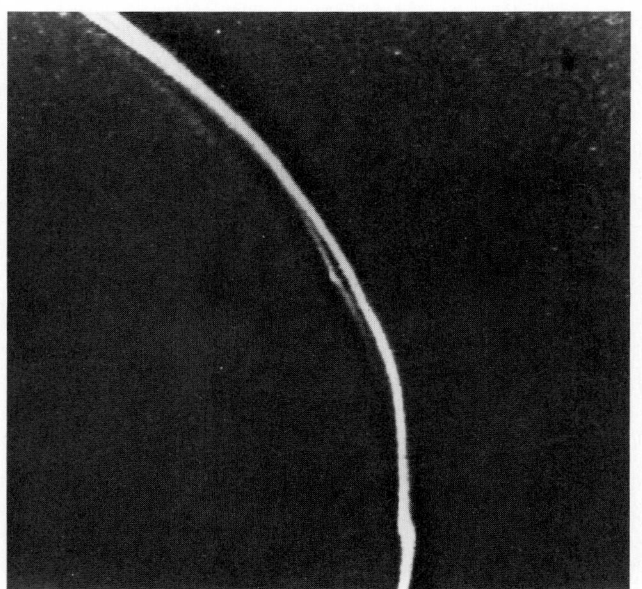

**Figure 16.29** Outermost (F) ring, photographed from 750,000 km. The two narrow bright rings follow braided orbits. Ring structure is about 35 km wide with localized clumps of material (NASA).

(A, B, C) have long been known. The D ring close to Saturn was the first discovered by space probes. An E ring had been suspected (outside A) some years ago. Space probes confirmed its presence and added more. The F ring (Fig. 16.29) is outside the A ring and separated from it by the Pioneer division, named for the probe from which the discovery was made in 1979. Still farther out is the G ring. Evidence for it first came from IR detection on Pioneer 11 with confirmation obtained by the Voyagers.

The thickness of the rings varies but is generally less that 2 km, and distant stars can be seen through them. The total mass of the rings is less than about one hundred millionth that of Saturn itself. The rings' temperature ranges from 70 K on the illuminated side to 55 K on the shadowed side, and their composition is of low-mass ices, probably lumps of water and ammonia ice.

The rings themselves rotate, which is nicely demonstrated by the Doppler shift of the lines in their spectra. Along one edge the approaching material shows lines that have been shifted to shorter (blue) wavelengths,

*Table 16.3* SATURN'S RINGS

| Feature | Distance from planet center (km) | (Saturn radii) | Rotation period (hr) |
|---|---|---|---|
| Cloud tops | 60,330 | 1.00 | 10.657 |
| D ring, inner edge | 67,000 | 1.11 | 4.91 |
| C ring, inner edge | 73,200 | 1.21 | 5.61 |
| B ring, inner edge | 92,200 | 1.53 | 7.93 |
| B ring, outer edge | 117,500 | 1.95 [a] | 11.41 |
| A ring, inner edge | 121,000 | 2.01 [a] | 11.93 |
| Encke division | 133,500 | 2.21 | 13.82 |
| A ring, outer edge | 136,200 | 2.26 | 14.24 |
| F ring | 140,600 | 2.33 | 14.94 |
| G ring | 170,000 | 2.8 | 19.9 |
| E ring, inner edge | 210,000 | 3.5 | 27.3 |
| E ring,[b] maximum | 230,000 | 3.8 | 31.3 |
| E ring, outer edge | 300,000 | 5.0 | 46.6 |

[a] The Cassini Division *is* the gap between the outer edge of the B ring and the inner edge of the A ring.
[b] The Encke division is 200 km wide.

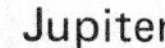

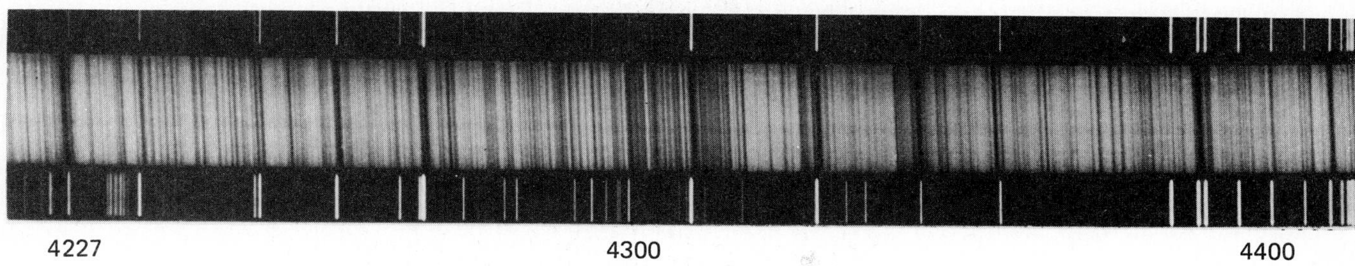


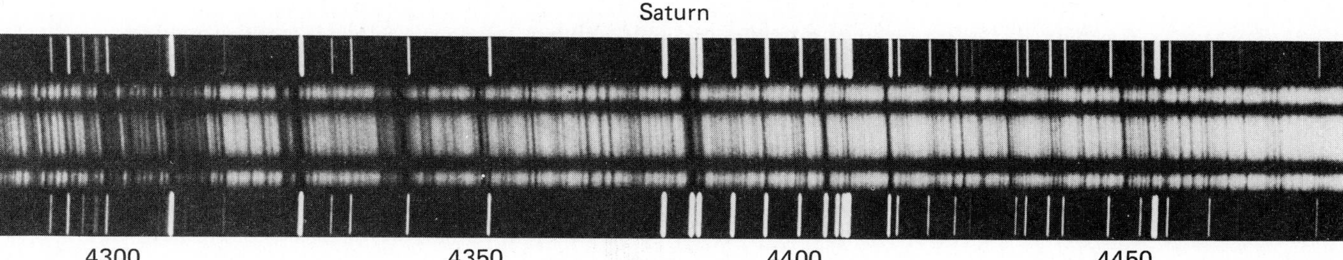


**Figure 16.30** Spectra of Jupiter and Saturn. At the top and bottom of each spectrum there is a laboratory calibration spectrum. The broad central band comes from the disc of each planet. For Saturn the two thin bands come from the rings, one side approaching Earth and the other receding. The Doppler shift of the lines produced by planetary rotation is well shown as is the rotation of Saturn's rings (Lowell Observatory).

whereas on the opposite side the receding material shows an equal shift to longer (red) wavelengths (Fig. 16.30). The rotation speeds and radii follow Kepler's third law as one would expect.

The pictures of the rings that the space probes have sent back to Earth show a system of hundreds of thin ringlets sometimes braided and sometimes looking like a phonograph record. Within the classic rings there are several large gaps. Most of the gaps are not totally empty but have some particles circulating in them. Particle sizes, as determined by a variety of visible, radio, and IR measurements, seem to range from less than a thousandth of a millimeter up to tens of meters.

**Figure 16.31** View from Voyager 1 from 1.5 million, just after closest approach. Saturn can be seen through the A, B, and C rings, and the shadow of the rings can be seen on the planet's disc (NASA).

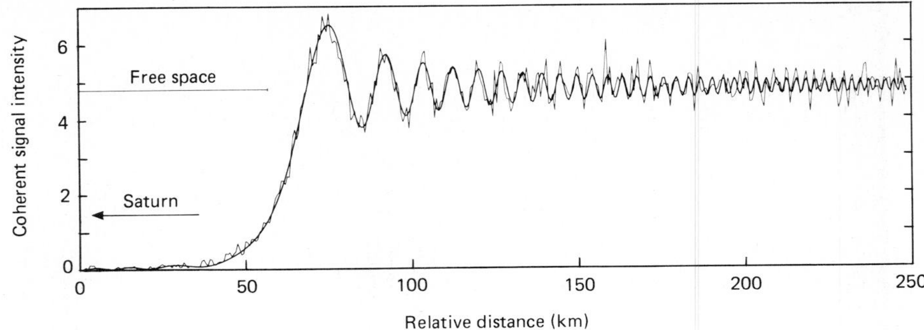


**Figure 16.32** Diffraction of radio signals from Voyager 1 at the edge of the Encke gap. Radio signals at a wavelength of 3.6 cm were transmitted when Voyager 1 was 215,000 km beyond the rings. Dark, smooth curve shows theoretical predictions for diffraction at the edge of the gap, and the jagged light line shows the actual signals received at Earth. Close agreement indicates that the particles density in the rings tapers to almost zero over a distance of less than 200 m (E. A. Marouf and G. Leonard Tyler, Stanford University, and *Science*).

Some of the rings and the divisions between them seem to be controlled by small satellites. Thus the F ring has a total width of around 100 km with strands less than 10 km wide. It lies between the tiny satellites 1980 S26 and S27. Just beyond the A ring there is a small sheepdog satellite 1980 S28. The gravitational effect of these small satellites will tend to sweep up small particles that would otherwise stray from their rings.

The extreme sharpness of the ring edges was shown when radio signals from Voyager 1 were received through the Encke gap, which is 270 km wide. As Voyager passed behind the gap edges, the signals showed the typical diffraction maxima and minima that are seen under laboratory conditions. The analysis of these Voyager signals shows that the transition from the open gap to the full ring thickness takes less than 200 m (Fig. 16.32).

The rings are thought to have originated from the breakup of a satellite that came too close to Saturn and was literally torn apart through tidal forces. If this had happened, the debris would drift close to the original orbit forming a ring. Over a long time the gravitational force of the nearest other moons (surviving safely at their larger distances) will work to clear a gap at some radii but not at others. Some of the ring features can be understood in this way, but others remain unexplained. It is thought that some of the narrow gaps might be due to sheepdog satellites that are so small that they have so far escaped detection.

## 16.3 OUTER PLANETS: URANUS, NEPTUNE, AND PLUTO

All of our information on the outer planets has been gathered from the Earth. The first space probe visit might occur in 1986 for Uranus and 1989 for Neptune. Since its encounter with Saturn in 1980, Voyager 2 has followed a path to those distant planets, but the continuing good health of its orienting telemetry and observing systems will determine the scientific outcome of that long journey.

Uranus presents us with a disc of only 3.6″ (Fig. 16.33) whose diameter has been best measured by Princeton University's Stratoscope II, which uses a 1 m reflecting telescope carried to 15 mi altitude by a balloon. With little atmosphere above, resolution of 0.15″ was obtained, far better than from the ground. No surface markings could be seen, even with this resolution. The diameter of the planet from these high-altitude measurements turns out to be 51,800 km.

Neptune has a disc of 2.5″ and can be seen (without detail) with a good 30 cm reflector. The most accurate measurement of its diameter comes from the 1968 timing of its occultation of a distant star: 49,500 km.

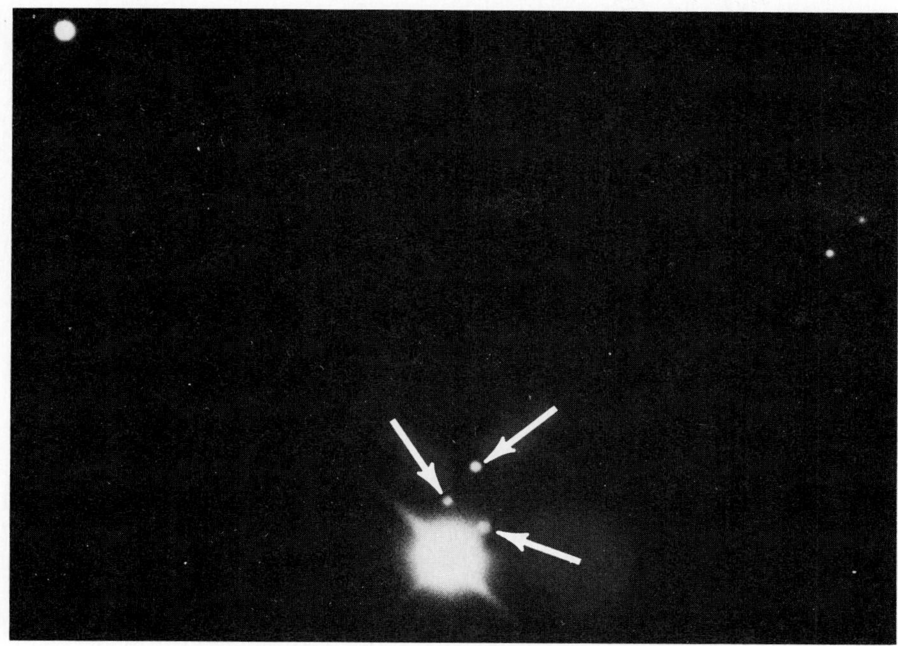

**Figure 16.33** Uranus and three of its satellites (Lick Observatory).

Pluto's image looks starlike, far below the normal resolution of telescopes that have to peer through our atmosphere. A reliable measurement of its size was reported only as recently as 1979 by Arnold Boksenberg (University College, London) and Wallace Sargent (California Institute of Technology) applying a new technique, **speckle interferometry**, and using the 5 m telescope on Mount Palomar. With this method, images of Pluto were recorded electronically using a relatively narrow wavelength interval centered on 5000 Å. The very short exposure time of 0.02 sec was used for each image. During such a short time the planet's image is actually an interference pattern that can be computer analyzed and added to the results from the other images; in all, 170,000 images were analyzed, Pluto's angular diameter was thus found to be $0.17 \pm 0.02''$, corresponding to a linear dimension of 3600 km. More recent measurements, some with speckle interferometry and some with measurements of Pluto's IR emission, point to a much smaller diameter, around 1500 km.

Planetary masses are known from observations of their moons; Pluto's single moon was discovered only in 1978. The masses and densities of the outer planets are listed in Table 16.4.

Earlier estimates of Pluto's mass, based on perturbations of Uranus' orbit after allowing for the effects of Neptune, had given a value of 0.11 $M_e$, about fifty times larger than is now accepted. From this we must now conclude that the perturbation values were probably inaccurate.

The outer planets are distant and cold, Uranus having a temperature of about 60 K, Neptune about 55 K, with Pluto around 40 K. Uranus and Neptune have orbits of small eccentricity inclined to the ecliptic at small angles, but Pluto is enigmatic. At an average distance of 39 AU, it will take 248 years to complete one orbit, and since its discovery we have seen less

*Table 16.4* MASSES AND DENSITIES OF THE OUTER PLANETS

| | *Mass (g)* | $M/M_{Earth}$ | *Diameter (km)* | *Density (g/cm$^3$)* |
|---|---|---|---|---|
| Uranus | 8.69 $10^{28}$ | 14.5 | 51800 | 1.31 |
| Neptune | 1.03 $10^{29}$ | 17.2 | 49500 | 1.77 |
| Pluto | 1.32 $10^{25}$ | 2.2 $10^{-3}$ | ~1800 | ~0.5 |

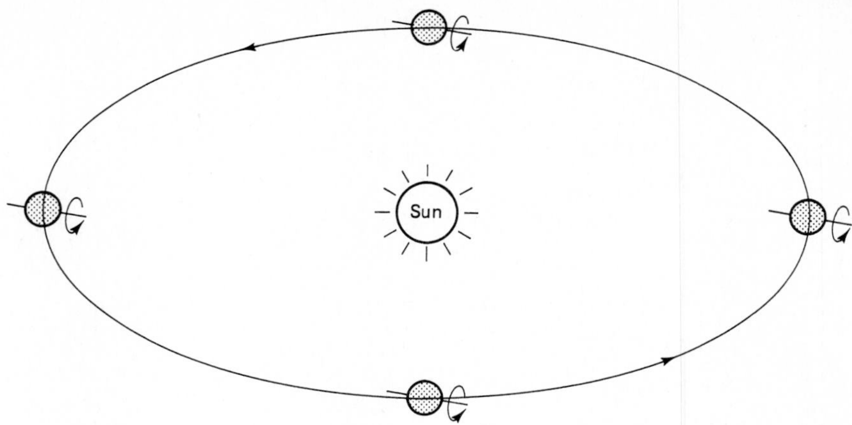


**Figure 16.34** Uranus rotates about an axis that is very nearly in the same plane as its orbit around the Sun.

than one quarter of this. The orbital eccentricity of 0.249 is very large—so large, in fact, that Pluto crossed Neptune's orbit in November 1978 and will remain closer to the Sun until May 2000 after which it will once again be the outermost of the planets. Pluto and Neptune will not collide, however. Pluto's orbit is inclined at 17° to the ecliptic and even at closest approach these two orbits are 385 million km apart.

In the absence of any well-defined surface markings, the only way of determining the rotational rates of these planets is by using the Doppler effect (as was done for Saturn's rings). The rotational period for Uranus has been found to be $10^h45^m$, that for Neptune $15^h50^m$, but the reliability is not better than about 15 percent. Neptune's rotation axis is inclined at 29° to its orbital plane, quite similar to those of the Earth and Mars. Uranus quite surprisingly was found to have its axis tipped by 98° to its orbit, that is, the axis is very nearly in the orbital plane (Fig. 16.34). As a result, we will sometimes find ourselves looking directly down the axis and at other times directly over the equator (but with its featureless surface this will not be at all obvious).

Pluto is too far away, small, and faint for surface markings to have been seen, but since 1956 it has been known that Pluto's brightness varies in a complex but regular pattern going through a cycle every 6.39 days. This has been attributed to rotation of Pluto on its axis with the surface having a patchy nature. Part of the surface is supposed to have an albedo of 0.5 (probably a covering of methane frost) and part an albedo of 0.13. Pluto's recently discovered moon, Charon (Figs. 16.35 and 16.36), has the same period for its orbital motion that Pluto shows for its cycle of brightness changes. It would seem that this moon hovers in its orbit over the same place on the planet's surface. In April 1980 an accidental observation of Charon's occultation of a faint star allowed a value of 1200 km to be obtained for that moon's diameter.

Hydrogen and methane have been discovered (spectroscopically) in the atmospheres of both Uranus and Neptune. There have been no observations of helium or ammonia, but these are expected to be present. Pluto is probably too small to retain hydrogen or helium and too cold for other gases not to feeeze out. The spectrum of frozen methane has been observed.

Plausible models can be constructed for Uranus and Neptune showing solid cores, icy regions, and an outer layer of liquid hydrogen, but it seems unlikely that any of these planets has the liquid metallic hydrogen that Jupiter probably has.

Uranus has five moons, Neptune four, but little is known about them. One, however, deserves note: Neptune's Triton (Fig. 16.37), which has often been listed as the largest satellite in the solar system. Newer measurements reported in 1982 place its diameter at 3200 ± 400 km, still large and com-

Circular No. 3241

Central Bureau for Astronomical Telegrams

INTERNATIONAL ASTRONOMICAL UNION

Postal Address: Central Bureau for Astronomical Telegrams
Smithsonian Astrophysical Observatory, Cambridge, MA 02138, U.S.A.
TWX: 710-320-6842 ASTROGRAM CAM Telephone: (617) 864-5758

1978 P 1

Capt. J. C. Smith, U.S. Navy, reports: "Elongation of the photographic image of Pluto has been detected by J. W. Christy on plates taken with the U.S. Naval Observatory's 155-cm astrometric reflector on 1978 Apr. 13, 20 and May 12; 1970 June 13, 15, 16, 17 and 19; and 1965 Apr. 29 and May 1. The maximum elongation is ∿ 0″.9 in p.a. 170°-350°. Observed position angles are consistent with a uniform revolution/rotation period equal to the lightcurve period of 6.3867 days. The data suggest that there is a satellite, 2-3 magnitudes fainter than Pluto, revolving around Pluto in this period at a mean distance of ∿ 20 000 km; the implied sun-Pluto mass ratio is 140 000 000:1. The other orbital elements are: e ∿ 0, Ω = 350°, i = 105° (with respect to the plane of the sky), T(Ω) = 1978 May 12.2 UT. The probable satellite was confirmed on exposures with the 155-cm reflector on 1978 July 2 and 5 and by J. A. Graham with the 400-cm reflector at Cerro Tololo Interamerican Observatory on July 6. Further observations are very desirable; a brief ephemeris follows."

| 1978 UT | p.a. | Sep. | 1978 UT | p.a. | Sep. |
|---|---|---|---|---|---|
| July 8.2 | 357° | 0″.8 | July 12.2 | 165° | 0″.9 |
| 9.2 | 342 | 0.8 | 13.2 | 129 | 0.3 |
| 10.2 | 277 | 0.2 | 14.2 | 5 | 0.6 |
| 11.2 | 181 | 0.7 | 15.2 | 348 | 0.9 |

The designation 1978 P 1 conforms with the temporary nomenclature system for announcing discoveries of new satellites. This system was approved at the meeting in June of the IAU Working Group for Planetary System Nomenclature and has been submitted for endorsement at the forthcoming meeting of the IAU Executive Committee.

**Figure 16.35** IAU Notification of the discovery of Charon (B. G. Marsden, Central Bureau for Astronomical Telegrams).

parable to Jupiter's Io and Europa but decidedly smaller than Saturn's Titan and Jupiter's Callisto and Ganymede. Its orbital direction is retrograde.

Science is too often perceived as the coldblooded pursuit of yet another decimal place, so it is good to find an example of a carefully planned experiment turning up a completely unanticipated result. NASA operates the Kuiper Airborne Observatory, a research airplane that carries a 91 cm telescope and can fly at about 40,000 ft, above about four-fifths of the atmosphere. In 1977 a group of astronomers from Cornell University had planned

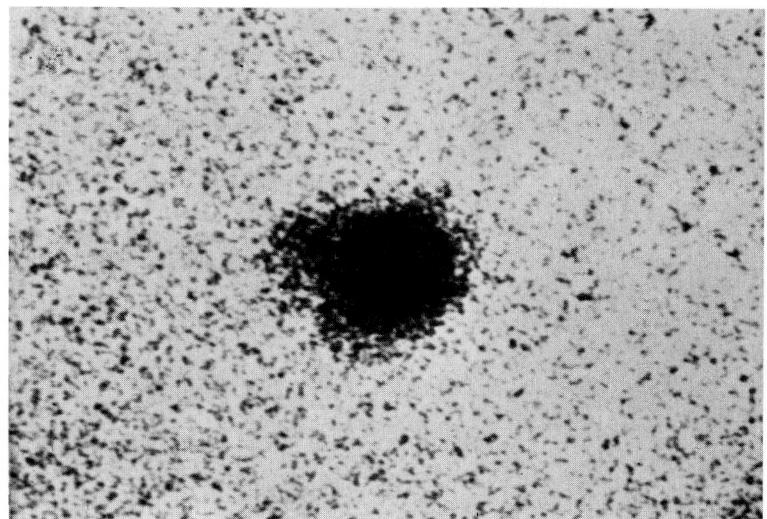

**Figure 16.36** Pluto and its Moon, Charon, discovered as a bulge on the image of the planet. The position of the bulge is seen to move, from one photograph to another, indicating an orbital period of 6.39 days (U.S. Naval Observatory).

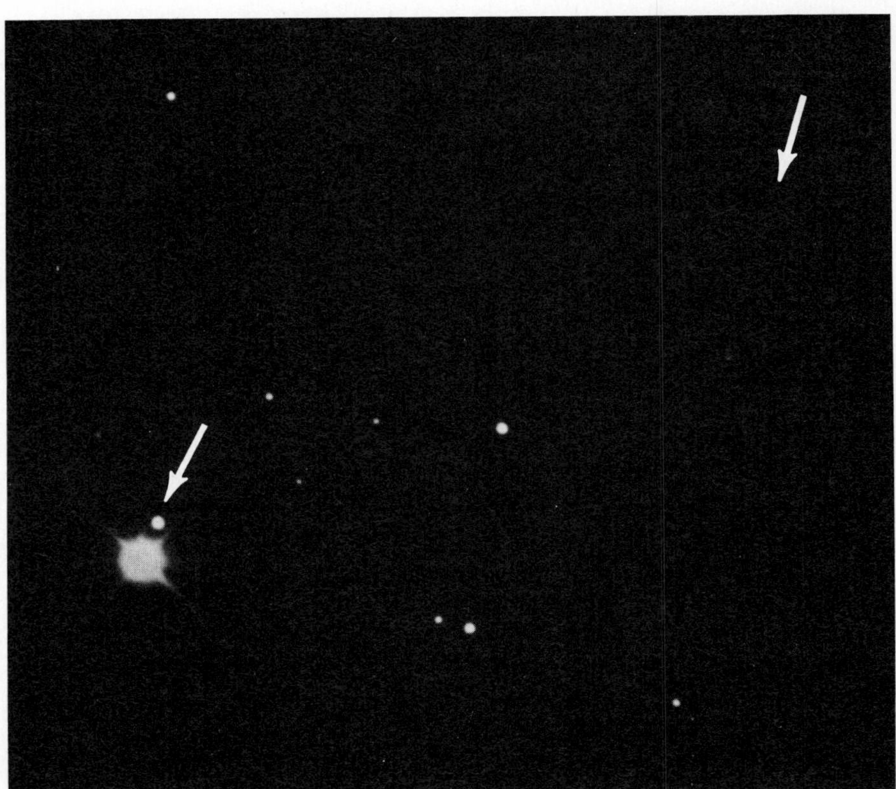

**Figure 16.37** Neptune with its satellites Triton (closest) and Nereid (barely visible) (Lick Observatory).

**Figure 16.38** Rings of Uranus. Diagram to show the positions and dimensions of the rings as deduced from occultations observed at different observatories. Rings have been labeled with numbers or Greek letters (James L. Elliot, M. I. T. and *Astronomical Journal*).

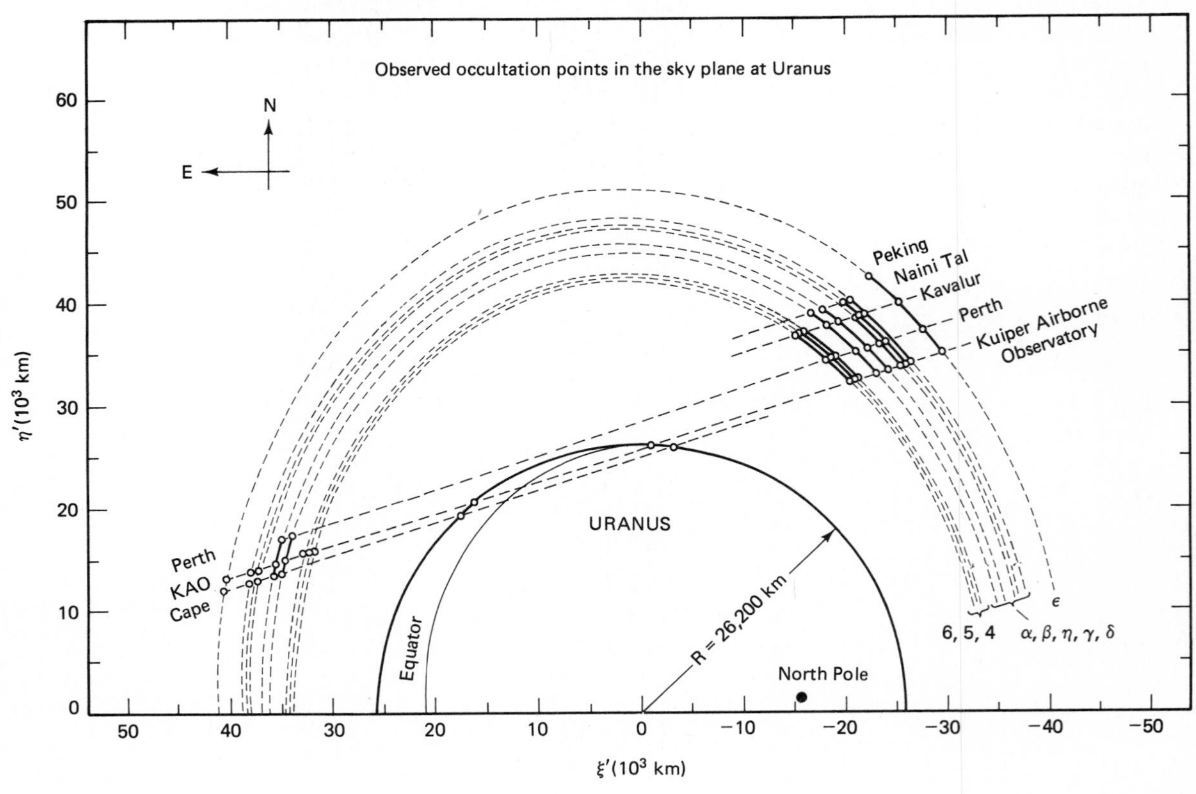

to observe an occultation of a star by Uranus to give an accurate diameter for the planet. To their surprise, the star was occulted 40 min before the expected time and then repeated this four more times in the next 9 min. Half an hour after the expected planetary occultation, another series of occultations was observed. The Cornell group interpreted their observations as the result of Uranus being surrounded by five rings with a system similar to Saturn. By coordinating their airplane observations with those of five ground-based observatories, a partial picture of the ring system could be reconstructed (Fig. 16.38). The rings lie between 42,000 and 52,000 km from Uranus and are too thin to be seen by normal photography. They might be visible from a passing space probe but otherwise can be detected only by their occultation of faint stars. More recent oservations have shown a total of nine rings.

## 16.4 JOVIAN PLANETS: A SUMMARY

Grouped together initially because of their great distances and low densities, the Jovian planets have displayed other common properties, all of which set them apart from their terrestrial counterparts.

All (except Pluto) are far larger than the Earth. Their great masses produce very high internal pressures and, probably, solid central cores. Jupiter and Saturn both radiate significantly more energy than they presently receive from the Sun but whether their internal heat comes from residual cooling following contraction or from radioactivity is not known. Theoretical models for Jupiter and Saturn include the unusual structural feature of liquid metallic hydrogen. The outer three planets are too cold to sustain such an internal layer. Jupiter and Saturn are also warm enough and large enough to support highly complex atmospheres (Fig. 16.39).

When structural models are assembled, a trend in the planets' compositions emerges, as shown in Table 16.5.

The models that satisfy the known properties of these planets show that gaseous hydrogen and helium decrease as the distance from the Sun increases. At the same time, there is a steady increase in "ices," whereas the solid cores show much less variation. It will be interesting to see how Pluto fits into this sequence when we come to know its properties.

Systems of multiple satellites and extended rings of debris seem to be a feature of the Jovian planets. At this time, we do not know whether the relatively small number of moons of Neptune is real or simply the temporary result of observational difficulties. In 1981 when Neptune occulted two stars, a careful search failed to produce any evidence for rings. If any ring system is present, it must be much less extensive than those of Uranus and Saturn. Most of the outer satellites of the Jovian planets are very small, and many of them move in retrograde orbits. It is thought that some may have been captured from trajectories that brought them sufficiently close to the planets.

Pluto departs so much from virtually every general feature and trend that it has been suggested that it may have formed elsewhere and was subsequently captured into its present orbit.

***Table 16.5*** COMPOSITION OF THE JOVIAN PLANETS

| | | *Overall composition (percent by weight) (oberservation + theory)* | | |
|---|---|---|---|---|
| | *Major atmospheric constituents* | *Gas ($H_2$, He)* | *Ices (water, methane, ammonia)* | *Core (minerals, iron)* |
| Jupiter | $H_2$, He, $CH_4$, $NH_3$ | 82 | 5 | 13 |
| Saturn | $H_2$, $CH_4$ | 67 | 12 | 21 |
| Uranus | $H_2$, $CH_4$ | 15 | 60 | 25 |
| Neptune | $H_2$, $CH_4$ | 10 | 70 | 20 |

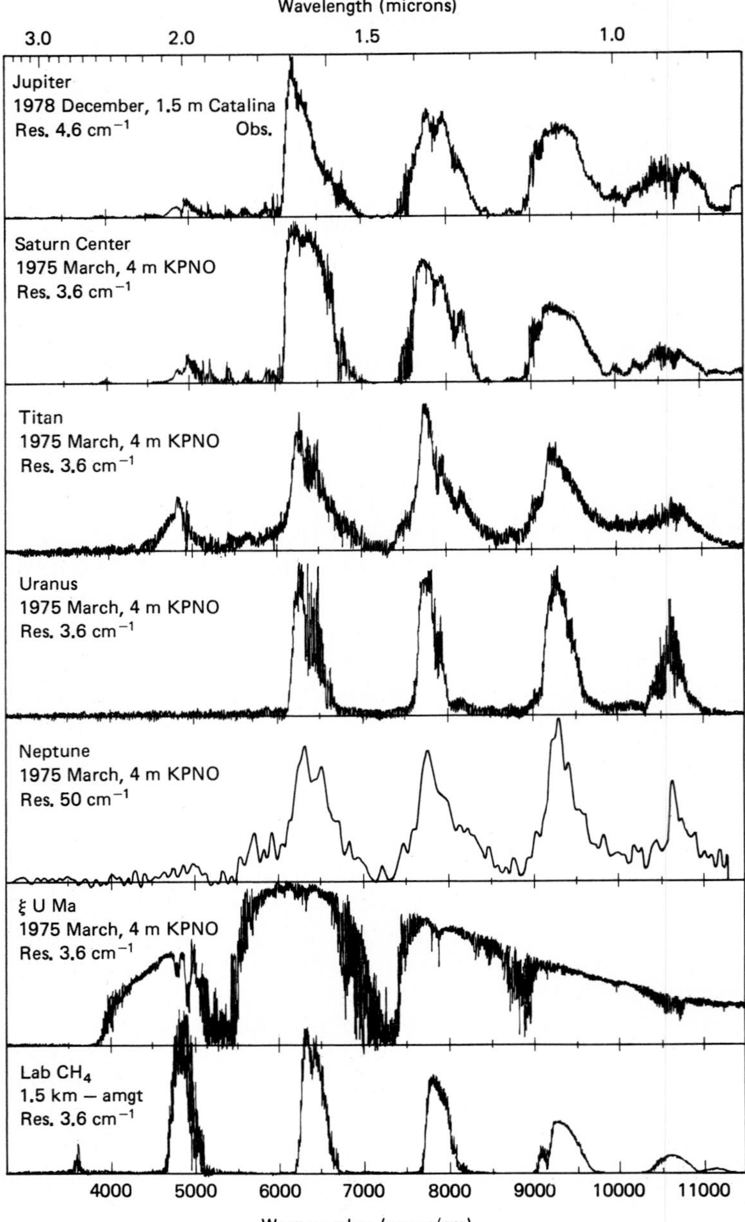


**Figure 16.39** Infrared spectra of Jupiter, Saturn, Titan, Uranus, and Neptune showing presence of methane ($CH_4$) in their atmospheres. Compare with laboratory spectrum of methane (bottom). Solar-type star for comparison shows good atmospheric transmission at the methane bands (Harold P. Larson and Uwe Fink, Lunar and Planetary Laboratory, University of Arizona, and *Astrophysical Journal*).

In Chapter 18, we shall show how modern theories for the origin of the solar system incorporate many of these observations, as well as the quite different properties of the terrestrial planets, to provide a framework that is plausible but still well short of being definitive.

## CHAPTER REVIEW

**Jupiter:** angular diameter varies between 32 and 48 arc sec; planet is brighter than any star. Earth-based observations show light- and dark-colored bands and Great Red Spot, but surface detail smaller than 3000 km cannot be seen.

— largest planet, about 318 times mass of Earth with low density of 1.3 g/cm$^3$.

— rotation period is just under 10 hr, and varies slightly with latitude.

— cloud top temperature 135 K is low, but total

energy radiated by Jupiter is about twice the energy it receives from the Sun. Source of this internal energy has not been definitely identified.

Atmosphere: hydrogen 80 percent, helium 19 percent; small amounts of other gases including methane.

— ammonia and ammonia hydrosulfide snowflakes in atmosphere constitute the clouds seen.

— high wind speeds with variations corresponding to bands and zones.

Magnetosphere: strong and very extensive with many trapped protons and electrons giving rise to intense radio noise.

Interior: probably liquid hydrogen with a region of liquid metallic hydrogen. Probable small, solid core.

Satellites: four Galilean moons show great diversity in properties including spectacular volcanism on Io, which orbits in a cloud of sodium and hydrogen. Many other moons discovered.

Rings: cannot be seen from Earth but discovered by Voyager.

**Saturn:** angular diameter is only 7 arc sec, so little surface detail can be seen from Earth; slightly smaller than Jupiter but considerably less dense (0.7 g/cm$^3$).

— rotation also about 10 hr.

— radiates more energy than receives from the Sun, so must have an internal energy source.

Atmosphere: hydrogen 89 percent, helium 11 percent, small quantities of other gases, including methane.

— extremely high wind velocity near equator but also high at other latitudes.

Magnetosphere: stronger than Earth's but not as strong as Jupiter's.

Satellites: extensive system, some very large, others little more than large rocks. One (Titan) is large enough to have its own atmosphere.

Rings: probably debris from a disintegrated satellites. Great complexity in ring structure.

**Outer planets:** sizes and densities now well-known, (except perhaps for Pluto), but as no space probe has approached close, data comes from Earth-based observations.

— all have satellites (Pluto has only one), and Uranus has rings.

— planets are very cold so ices must be abundant.

## NEW TERMS

**belts**
**Cassini division**
**liquid metallic hydrogen**
**"sheepdog" satellite**
**speckle interferometer**
**zones**

## PROBLEMS

**1.** Compute the rotational speed on Jupiter's equator and compare it to that on the Earth's equator.

**2.** The IR spectrum of Jupiter indicates a temperature of 135 K at the top of its atmosphere. Calculate the wavelength at which Jupiter's thermal radiation is most intense.

**3.** Jupiter and Saturn radiate more energy than they receive from the Sun. What does this tell us about those planets?

**4.** Jupiter and the Earth have significant radiation belts, regions of trapped, high-speed particles. Other planets have little or none. What does this tell us about the internal compositions of the planets?

**5.** Aurorae are observed on Earth in the polar regions. On which other planet might something similar be seen, and which planet would definitely not show aurorae?

**6.** The surface temperatures of Jupiter and its moons are very low. What evidence leads us to think that Io has a molten inferior?

**7.** In what ways do the surfaces of Jupiter's major moon resemble or differ from that of our Moon?

**8.** Which observations appear to provide a plausible explanation for Io's color?

**9.** Does any moon of Jupiter have an atmosphere? Calculate the escape velocity from the largest moon, and then refer to the diagram in chapter 11 relating escape velocities to temperature.

**10.** The radio map of Jupiter obtained with the VLA (Fig. 16.2) shows strong radio emission from well outside the planet. What is the explanation for this?

**11.** A satellite is to be placed into a synchronous orbit about Saturn. What must be the radius of this orbit, and where will it be in relation to the rings and moons?

**12.** How do we know that the rings of Saturn are not solid but are made up of a vast number of very small pieces of matter?

**13.** List two important discoveries made by space probes for each of Jupiter and Saturn.

**14.** Why does it seem probable that all of Saturn's moons do not have the same origin?

**15.** Which moon of Saturn has an atmosphere? What is known of that atmosphere's composition?

**16.** Refer to Fig. 16.30 showing spectra of Jupiter and Saturn. The spectrum of Saturn shows three bands, one for the planet itself and one for rings on each side. Which of the ring spectra in this illustration comes from the part of the rings which was approaching us and which from that part that was receding? Why are the lines for the planets slanting?

**17.** Check the compositions of the atmospheres of Jupiter and Saturn. How do these compare with the Earth's atmosphere?

**18.** What is the evidence that Saturn and Jupiter must have small, solid cores if they have any?

**19.** Pluto is about 40 AU from the Sun. What is the ratio of the intensity of sunlight at Pluto to that at Earth? Pluto has a surface temperature of about 50 K, and the Earth has a surface temperature of about 300 K. What is the ratio of radiant energies emitted per square centimeter from each of these planets? (The answers to these two parts are similar. Is this an accident?)

**20.** If another planet exists beyond Pluto at a distance of 50 AU, how long will it take to orbit the Sun? If photographs are taken of that part of the sky where it is expected to be, how far apart must photographs be in order to see a movement of 1 arc min?

**21.** Are there any planets beyond Pluto? How might one hope to find them? Review the ways in which Uranus, Neptune, and Pluto were discovered.

**22.** Use data from Appendix G for Saturn's moons Mimas and Titan to show that they are in agreement with Kepler's third law. (Use any units that are convenient.) (Caution: Distance from the planet is *not* the radius of the orbit.)

CHAPTER 17

# Minor Bodies of the Solar System: Comets, Asteroids, and Others

The solar system took shape about 4.6 billion years ago. From the primordial materials, the planets and their moons formed with the Sun at the center of this revolving parade. It seems that the condensation process also produced many bodies that never grew to planetary or even satellite size. Some (the **asteroids** or **minor planets**) are now in planetlike orbits around the Sun and a few have orbits that cross the Earth's. Others (the **meteoroids**) are much smaller and are in orbits that sweep across those of the planets, sometimes entering the Earth's atmosphere to burn up or, in a few cases, reach the Earth's surface. Another group, the comets, seem to inhabit a vast reservoir far beyond Pluto before they swoop through the central region of the solar system and pass by the planets.

The planets have evolved in numerous ways since their formation: Their atmospheres have lost some of their primordial gases and gained other gases from volcanic action or the heating of the planetary surface. Geological activity often assisted by bombardment has transformed some surfaces. In contrast, the comets and others are too small for most of these processes to have operated, and so their study may provide us with important clues to the properties of material that has survived with little or no change from the earliest years of the solar system.

## 17.1 MINOR PLANETS OR ASTEROIDS

**Asteroid** means starlike, but this descriptive term scarcely distinguishes these particular objects from so many others that the more correct term **minor planet** will be used (as is recommended by the International Astro-

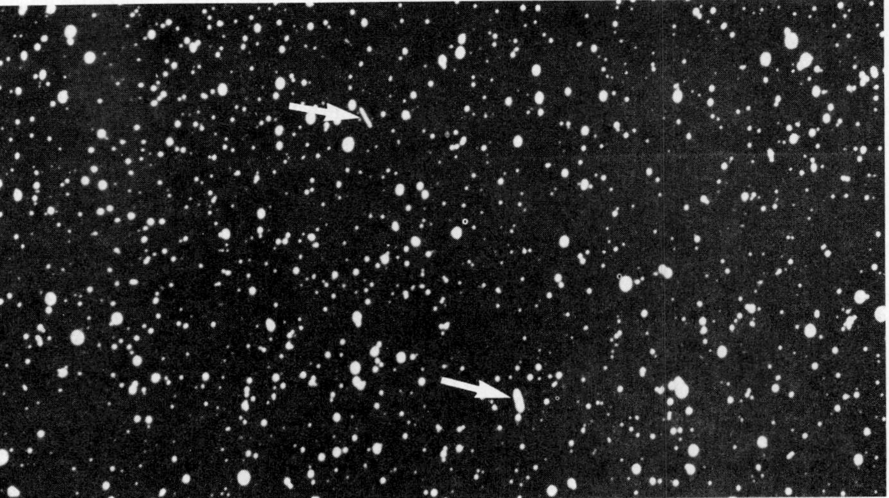

**Figure 17.1** Trails of two minor planets. During this 1-hr exposure, the telescope was driven to keep the stars fixed in the field of view, and the movements of the minor planets can be seen against this fixed background (Yerkes Observatory).

nomical Union). The first minor planet was discovered in 1801 by Giuseppe Piazzi who did indeed find an object starlike in appearance before it was lost to view in the daytime sky when it moved close to the Sun. Most unstarlike, Piazzi's object moved slightly from night to night (Fig. 17.1). The invention of new computational methods by the great mathematician Gauss allowed the orbit to be calculated from Piazzi's few observations. With this calculated orbit, the new planet was rediscovered later that year when it had moved some distance from the Sun. At Piazzi's request, it was named Ceres after the patron goddess of his native Sicily.

Ceres' average distance from the Sun turned out to be 2.77 AU and its periodic time 4.6 years (Remember $P^2 = R^3$). This orbit lay exactly at the distance that one would have expected from Bode's Law (Chap. 11). Indeed, Uranus' earlier discovery, also at a distance in good agreement with Bode's law, had spurred a search for the missing planet between Mars and Jupiter to fill the gap in Bode's sequence. In rapid succession Pallas (1802, also 2.77 AU), Juno (1804, 2.67 AU), and Vesta (1807, 2.36 AU) were found and it seemed as though the missing planet had somehow disintegrated with its fragments populating its former territory.

After these initial successes, no further minor planet was found until 1845 (Astraea, 2.56 AU), but thereafter a steady trickle of discoveries has accumulated so that the prime catalog, the Soviet *Ephemeris of Minor Planets,* now lists about 2000 of these objects.

All move in prograde orbits, mostly close to the ecliptic but with a few having steeply inclined orbits. Most have semimajor axes that lie between 2.3 and 3.3 AU, constituting the **asteroid belt**, but as in any crowd, there are some individualists (Fig. 17.2). Icarus, for example, has an elongated orbit with an eccentricity of 0.83, a semimajor axis of 1.078 AU, and it gets to within 0.19 AU (27 million km) of the Sun. Hidalgo, also with a distinctly noncircular orbit (eccentricity 0.66), has the largest known semimajor axis of 5.32 AU.

The minor planets' sizes are very hard to determine. Most are so small and so far away that like stars their image dimensions from the measured durations of their occultations are mainly the result of atmospheric seeing. In a very few cases, the sizes have been measured directly but not very accurately, and their albedos found to be about 0.1. For other minor planets, sizes can be calculated from the observed visible and IR brightness and the measured distances from Sun and Earth. This method uses a calculation similar to that explained in Chapter 11 to derive an approximate value for a planet's temperature.

Several other methods have been used for determining the sizes of the minor planets. Probably the most accurate of these makes use of occultations of stars by some of the minor planets whose sizes can be deduced from the measured durations of their occultations.

Only about 200 minor planets have reasonably well-known diameters that are larger than 100 km, 30 are larger than 200 km, and Ceres is the largest at 1000 km. Most are under 50 km diameter.

Masses have been calculated for only three minor planets, from perturbations they have produced in the trajectories of other minor planets. The masses have been found to lie in the range $2 \times 10^{23}$ to $10^{24}$ g, indicating densities in the range 2 to 3 g/cm$^3$. The total mass of all minor planets is estimated to be around $3 \times 10^{24}$ g, certainly far less than that of any of the major planets.

Most of the minor planets show regular variations in brightness that are attributed to rotation of nonspherical shapes, perhaps with albedos varying over different parts of their surfaces. Infrared spectrometry and radar observations suggest that their compositions may be similar to meteorites (later in this chapter) but with a dusty surface layer as on the Moon.

One group of about twenty minor planets, the Apollo objects, follow orbits that cross the Earth's path and some of them could possibly collide with the Earth. Assuming that there are another 1000 Apollo objects too small to have been seen but still larger than 1 km, it has been calculated that only about four of them might collide with the Earth in a million years. Such an impact, however rare, should be impressive. The release of about $4 \times 10^{27}$ ergs (equivalent to about 100,000 megaton nuclear bombs) will produce a crater of more than 20 km diameter. A careful search of the Earth's surface has revealed scars that seem to be eroded craters from long past collisions of this type.

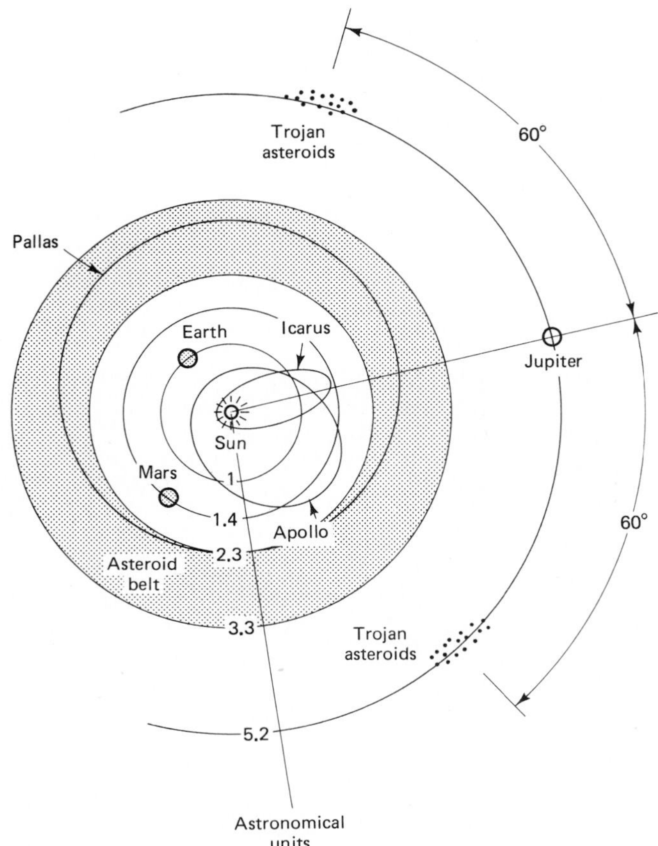


**Figure 17.2** Orbits of some minor planets. Most of these bodies have orbits in the asteroid belt between about 2.3 and 3.3 AU, but some have orbits that cross the Earth's orbit. There are two groups of Trojan asteroids that move in Jupiter's orbit, 60° ahead and behind the planet.

A recent discovery may possibly point to another group of minor planets but not in the usual region. In 1977 Charles Kowal at Mount Palomar discovered an object (Chiron) moving in an orbit with a semimajor axis of about 13.7 AU. The orbital eccentricity is very high at 0.38 so that its perihelion is inside Saturn's orbit, its aphelion close to the orbit of Uranus, and its sidereal period is 50 years. Its next perihelion passage will be in 1995. After its discovery, older observations as far back as 1895 were traced, and its orbit is now well documented. From its brightness, a size of 300 km has been calculated. At this time, it is not known whether this is an isolated object or a member of a family. Further discoveries of this type will be rare for Chiron's observation requires the largest telescopes, and there are many competing demands for their use.

## 17.2 COMETS

The movement of the planets had served to identify them, but those movements were slow and predictable. In contrast, the unforeseen appearance of a comet with its far swifter motion and sometimes with a tail that stretched far across the sky aroused fears and required reassurance. Chinese catalogs record the sighting of many guest stars, including Halley's comet (Fig. 17.3) as far back as 239 B. C. Comets were taken as omens, and the appearance of Halley's comet in A. D. 451 was considered to be associated with the defeat of Atilla the Hun at Chalon-sur-Marne; the same comet's appearance in 1066 at the time of the Norman conquest of England seemed more than coincidence. Of course at those times it was not known that these were reappearances of the same comet—had that been known the predictions of its reappearances would surely have led to panic.

Brahe in 1577 showed that the absence of any observable parallax placed that year's comet beyond the Moon and certainly far beyond the Earth's atmosphere where it had been thought the comets resided. Fears have persisted, however, that comet tails contain deadly vapors, and as recently as 1973 the appearance of Comet Kohoutek could find a public willing to be regaled with predictions of impending disaster.

The historic comets had to come close enough and be bright enough to be seen without a telescope, but the pattern of comet discoveries has changed in recent years. For some time, dedicated amateurs have found and given their names to new comets, but now large professional observatories have effectively taken over. With time-exposure plates, comets can be discovered long before they are bright enough to be seen with smaller telescopes. Photographs taken on consecutive nights are viewed in rapid alter-

**Figure 17.3** Halley's comet May 8, 1910 photographed with the 60-in. reflector (Mt. Wilson and Las Campanas Observatories).

nation (blinked) and the presence of a moving object can be readily detected against the motionless stars.

With the changing pattern of discovery, definitive numbers cannot yet be given for the total number of cometary objects, for the older searches had a strong bias in favor of those comets that had small perihelion distances. Typically, between five and ten new comets are found each year, and each is named for its discoverer (or up to the first three in cases of multiple discovery). Each comet is also listed and given a reference number for its year of discovery; thus 1973f was the sixth comet in 1973. This designation is later replaced by a permanent number to indicate the sequence of passing perihelion; and so 1976 III was the third comet to pass perihelion in 1976. (The sequence of discovery need not correspond to the sequence of perihelion passage.)

**Orbits** Until as recently as the seventeenth century all comets were unpredictable. Halley's calculations (Chap. 7) then showed how a cometary reappearance could be successfully predicted. By now out of over 600 comets discovered, about thirty have periods of less than 100 years and have been seen repeatedly. The shortest period is that of Encke's comet, 3.3 years, and more than fifty returns have been seen since its discovery (see Table 17.1).

The orbits of about 75 percent of all comets are such elongated ellipses that they are close to being parabolic (Fig. 17.4); the remainder have elliptical

***Table 17.1*** SELECTED COMETARY DATA

| *Comet* | *P (years)* | *Eccentricity of Orbit* | *Perihelion Distance (AU)* | *Semimajor Axis (AU)* |
|---|---|---|---|---|
| Encke | 3.3 | 0.847 | 0.34 | 2.2 |
| Temple 2 | 5.3 | 0.545 | 1.38 | 3.0 |
| Schwassmann-Wachmann 2 | 6.5 | 0.384 | 2.16 | 3.5 |
| Schwassmann-Wachmann 1 | 16.1 | 0.132 | 5.53 | 6.4 |
| Crommelin | 27.9 | 0.92 | 0.74 | 9.2 |
| Halley | 76.2 | 0.967 | 0.59 | 17.8 |

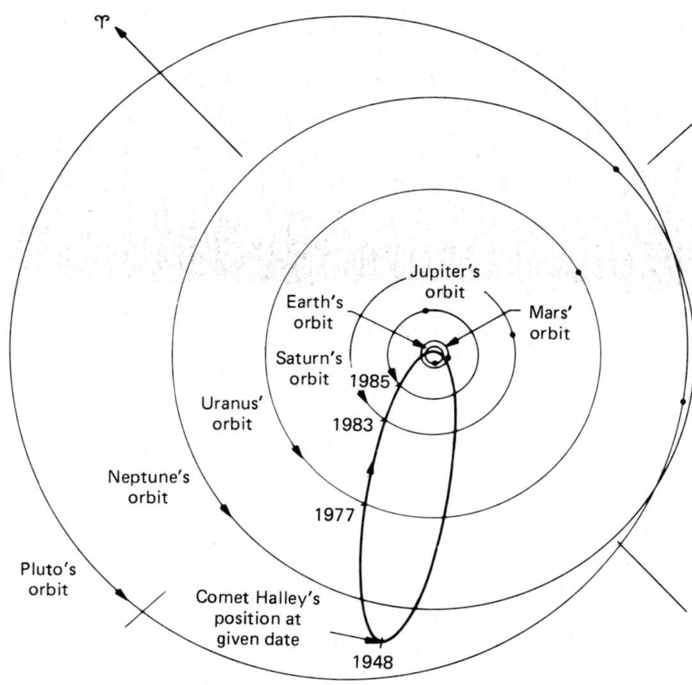


**Figure 17.4** Orbit of Halley's comet projected onto the ecliptic plane. The comet will have its next perihelion in 1986 and be back at its greatest distance beyond Neptune's orbit in 2022 (Jet Propulsion Laboratory, California Institute of Technology).

orbits with eccentricities in the range 0.3 to 0.7. There are noticeable differences between long-period comets (those with periods greater than 200 years) and those with shorter periods. Those with long periods tend to get closer to the Sun (median perihelion 0.85 AU), and their orbital planes come at all angles to the ecliptic. In contrast, the short period comets tend to have their orbital planes within about 30° to the ecliptic, and they do not get as close to the Sun (median perihelion 1.45 AU).

**Appearance** The bright head and flowing tail that we associate with a comet are short-lived phases in an otherwise frigid existence (Fig. 17.5). For most of their lives, comets are far from the Sun, frozen snowballs of ice and dust. Only as a comet approaches the Sun and gets within the orbit of Jupiter, does solar radiation provide enough energy to vaporize some of the ice and

**Figure 17.5** Comet Brooks, October 21, 1911 (Lick Observatory).

**Figure 17.6** Comet Mrkos, 1957. The tail of gaseous ions extends straight behind the comet's head, while the tail of dust particles is more diffuse and bends towards the right in these views (Palomar Observatory Photograph).

**Figure 17.7** Comet Arend-Roland, April 25, 1957. The antitail that seems to precede the comet is actually trailing but appears this way because of the perspective when viewed from Earth (Lick Observatory).

produce the first visible head. With closer approach more ice is vaporized and a tail appears. The head of the comet contains a solid nucleus surrounded by a glowing **coma** of gas. Measurements of the brightness contours in several comet heads show that the nucleus is typically 15 km across, sometimes as small as 2 km. For a density of 1 g/cm$^3$, a 10 km nucleus will have a mass of around $10^{18}$ g. Evaporation of some ice produces a coma of about 1 million km diameter, and recent UV observations have shown hydrogen envelopes that may extend as far as 40 million km.

Two types of comet tails are observed (Fig. 17.6). Type I tails are gaseous, composed of ions that emerge in an uneven stream from the comet and are blown about by the protons in the solar wind. Type I tails are clumpy because both their emission and the solar wind are quite variable. The appearance, therefore, is that of a patchy tail pointing straight outward along the Sun-comet line. Type II tails are made of dust particles from the comets, probably residue left from ices that have evaporated. Though small, the dust particles are still much larger than the gaseous particles in the Type I tails. They respond to the pressure of the solar radiation photons and form tails that curve gently behind the comets. The appearance of a comet as seen from the Earth will depend on the relative positions of Sun, Earth, and comet. Two separate tails might be seen, and one tail might even appear to precede the comet (Fig. 17.7).

The brightness of a comet depends on how much gas is being evaporated and how much dust is released. There is no formula that will allow the brightness of a comet to be predicted reliably. The solar heating will clearly depend inversely on the square of the Sun-comet distance ($R$), but it is the comet's albedo and structure that will determine just how much material will evaporate. Tracking of individual comets has shown that the brightness varies in proportion to $R^{-n}$, where $n$ generally lies between 3 and 6 but may even change with time for a given comet.

**Composition** A rich variety of molecules and ions has been revealed by spectroscopic observations of the comas and the Type I tails (Table 17.2).

*Table 17.2* ATOMS, MOLECULES, AND IONS IDENTIFIED IN COMETS

| *Coma* | *Tails* |
|---|---|
| H, OH, O, S | $CO^a$, $CO_2^a$, $OH^a$ |
| C, $C_2$, $C_3$, CH, CN, CO, CS | $CH^a$, $N_2^a$, $Ca^a$, $C^a$, $CN^a$ |
| HCH, $CH_3CN$, NH, $NH_2$, Na | |
| Fe, K, Ca, V, Cr, Mn, Co, Ni, Cu | |

[a] $H_2O$ has been reported in the coma but not confirmed.

Water vapor that evaporates from a comet is soon dissociated to provide the hydrogen envelope that was discovered only as recently as 1970 along with the OH radical in UV observations of Comet Bennett from the unmanned Orbiting Astronomical Observatory (OAO). Solar UV photons are probably responsible for the formation of the ions from the initially neutral particles.

A few comets that approach very close to the Sun (such as 1965 VIII, which had a perihelion distance of 0.14 AU) have shown spectral lines of metals such as Fe, Ni, and Cr.

The spectra of the Type II (dust) tails are continuous, rather redder than sunlight suggesting scattering from particles of about $10^{-4}$ cm in size.

**Halley's comet** Among the many spectacular comets that have been seen, Halley's comet has held an unique position since its 1759 return. The 1910 return (Fig 17.9) provoked widespread interest and concern, and already the 1986 return is beginning to attract attention. Every appearance of the comet since 239 B. C. can be identified in the old records.

Extensive calculations of the orbit have been carried out by Donald Yoemans of NASA's Jet Propulsion Laboratory, and the next perihelion

**Figure 17.8** Halley's comet, May 12 and 15, 1910, shortly before perihelion photographed with a telephoto lens (Mt. Wilson and Las Campanas Observatories).

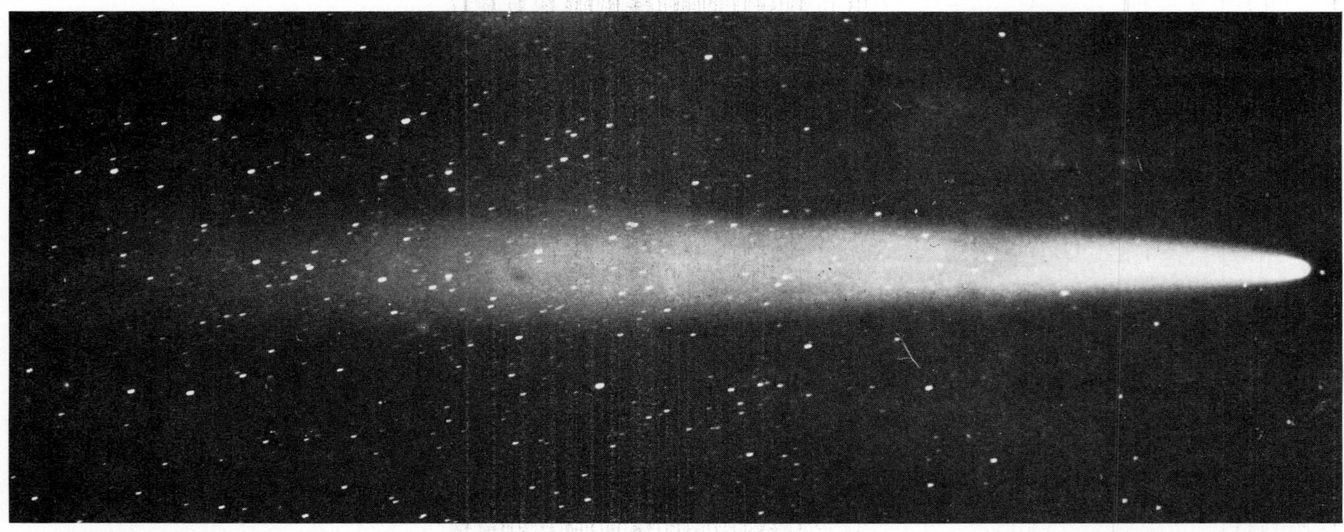

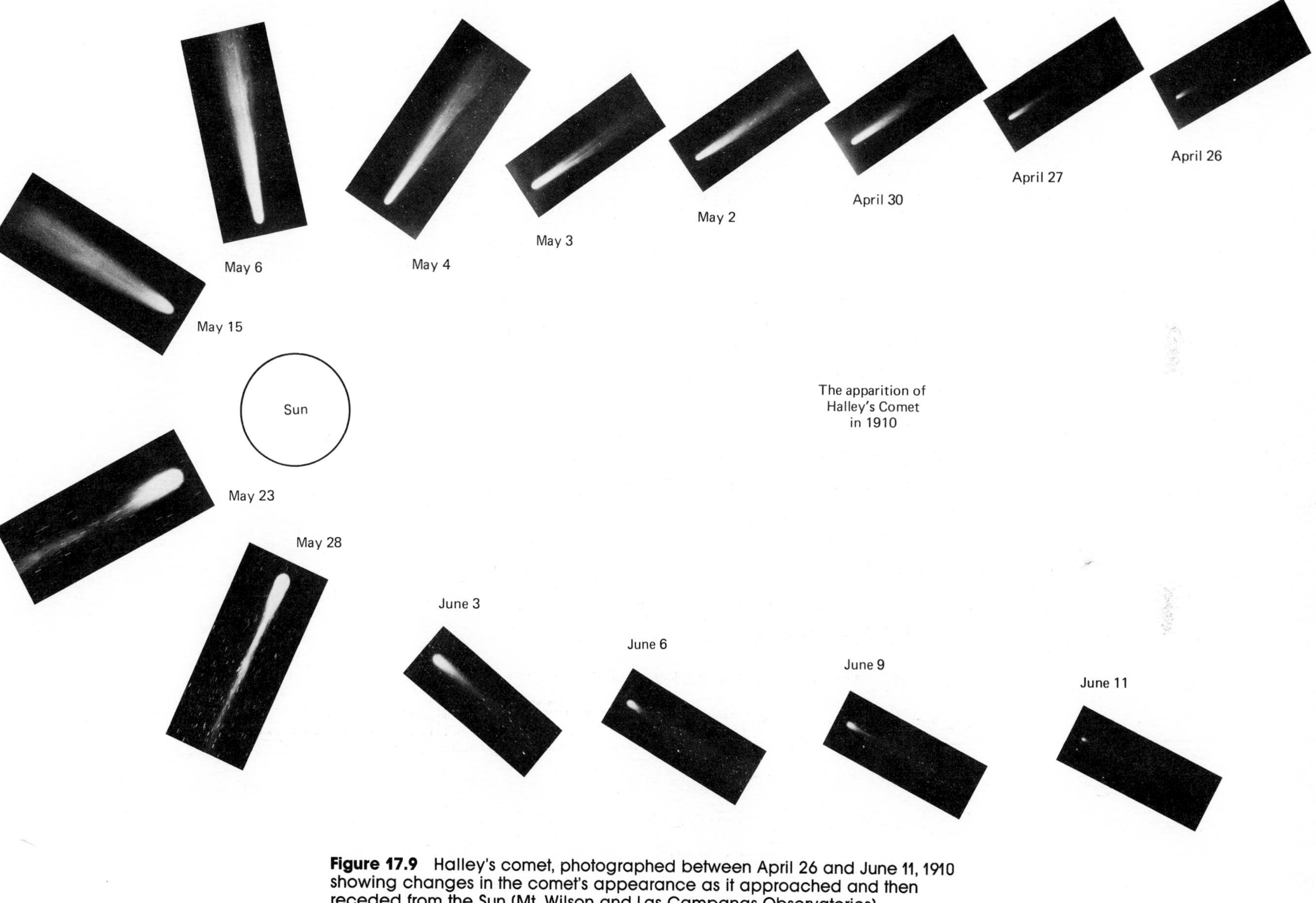


**Figure 17.9** Halley's comet, photographed between April 26 and June 11, 1910 showing changes in the comet's appearance as it approached and then receded from the Sun (Mt. Wilson and Las Campanas Observatories).

Circular No. 3737

**Central Bureau for Astronomical Telegrams**
**INTERNATIONAL ASTRONOMICAL UNION**

Postal Address: Central Bureau for Astronomical Telegrams
Smithsonian Astrophysical Observatory, Cambridge, MA 02138, U.S.A.

TWX 710-320-6842 ASTROGRAM CAM    Telephone 617-864-5758

PERIODIC COMET HALLEY (1982i)

D. C. Jewitt, G. E. Danielson, J. E. Gunn, J. A. Westphal, D. P. Schneider, A. Dressler, M. Schmidt and B. A. Zimmerman report that this comet has been recovered using the Space Telescope Wide-Field Planetary Camera Investigation Definition Team charge-coupled device placed at the prime focus of the 5.1-m telescope at Palomar Observatory. Five exposures of 480-s effective duration each (in seeing measured to be $1''.0$ fwhm) were taken on Oct. 16 through a broad-band filter centered at 500 nm. Definite images near the expected position and having the expected motion of P/Halley were noted. No coma was detected, and the object had a Thuan-Gunn magnitude of [g] = 24.3 $\pm$ 0.2 (corresponding to V ~ 24.2; and presumably B ~ 25). Two exposures were also made in the [r] band. Preliminary representative positions, which have an estimated external error of $\pm\ 0^s.35$ in $\alpha$ and $\pm\ 5''$ in $\delta$ but greater internal consistency, follow:

| 1982 UT | $\alpha_{1950}$ | $\delta_{1950}$ |
|---|---|---|
| Oct. 16.47569 | $7^h11^m01^s.9$ | $+ 9°33'03''$ |
| 16.49097 | 7 11 01.8 | + 9 33 02 |
| 16.52153 | 7 11 01.7 | + 9 33 00 |

The object is located some $0^s.6$ west of the position predicted by D. K. Yeomans (1981, The Comet Halley Handbook), suggesting that T = 1986 Feb. 9.3 UT. Confusion with a minor planet would be extremely unlikely. An attempt to confirm the recovery on Oct. 19 was successful in the sense that no objects were detected at the Oct. 16 locations and that the comet's image would then have been in the glare of a star; the dense stellar field has in fact thwarted other attempts to recover the comet during the past month. The recovery brightness indicates that the 1981 Dec. 18 attempt (cf. IAUC 3688) failed to record the comet by a very small margin and for an assumed geometric albedo of 0.5 leads to a radius of 1.4 $\pm$ 0.2 km. The comet's heliocentric and geocentric distances at recovery were 11.04 and 10.93 AU, respectively.

NOVA SAGITTARII 1982

Corrigendum. On IAUC 3736, line 17, the first astrometric position should be attributed to J. Hers, Sedgefield.

1982 October 21    Brian G. Marsden

**Figure 17.10** IAU Circular announcing the recovery of Halley's comet (Brian G. Marsden, Central Bureau for Astronomical Telegrams).

passage is predicted for February 1986. Charts have been produced for locating the comet in the several months of best viewing. Unfortunately the relative positions of Earth, comet, and Sun will not provide us with as good a view as was obtained in 1910. The best viewing period in the northern hemisphere will be before the perihelion passage, and after that passage for observers in the southern hemisphere (Fig. 17.10).

Cometary brightness is always hard to predict with reliability, but some calculations suggest that Halley will not be as impressive as it was in 1910.

Some attempts have already been made to detect the comet's nucleus while it is still far from the sun, before the tail has started to develop. Observations in December 1981 from the Kitt Peak National Observatory failed to find the comet. The next search, using the most sensitive electronic detectors at the prime focus of the Palomar 5 m telescope, was successful (Fig. 17.11). Halley's comet was recovered on October 16, 1982 when it had an apparent magnitude of 24.3 and was 11 AU from the Sun. Assuming that the comet's albedo is 0.05, the size of the nucleus is then calculated to be about 9 $\pm$ 1 km.

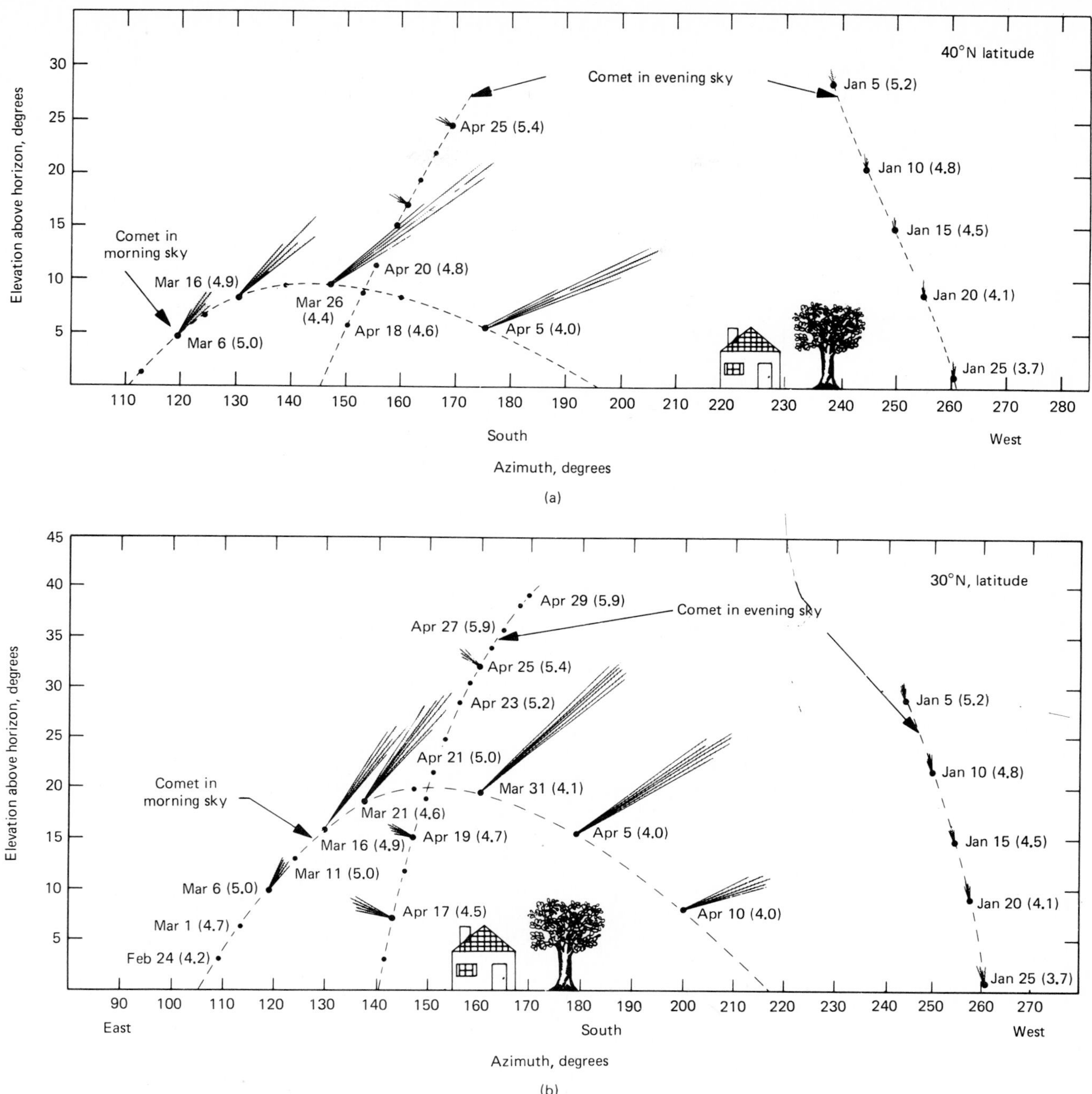


**Figure 17.11** Observing positions for Halley's comet for 1986. Comet positions are given for the beginning of morning astronomical twilight or end of evening astronomical twilight. Total visual magnitudes are given in parentheses for each date. (a) for 40° north latitude and (b) for 30° north latitude (Jet Propulsion Laboratory, California Institute of Technology).

**Nature and origin** From the changes in appearance as the comet approaches and then recedes from the Sun, and the identification of the volatile constituents, a model can be assembled. First prepared by Fred Whipple of the Harvard College Observatory in 1950, the "dirty iceberg" model is now widely accepted, not necessarily to describe every comet in complete detail but still as a very useful framework.

In a typical comet the nucleus is thought to consist of a solid core, perhaps similar in composition to some meteorites (Fig. 17.12). Around the

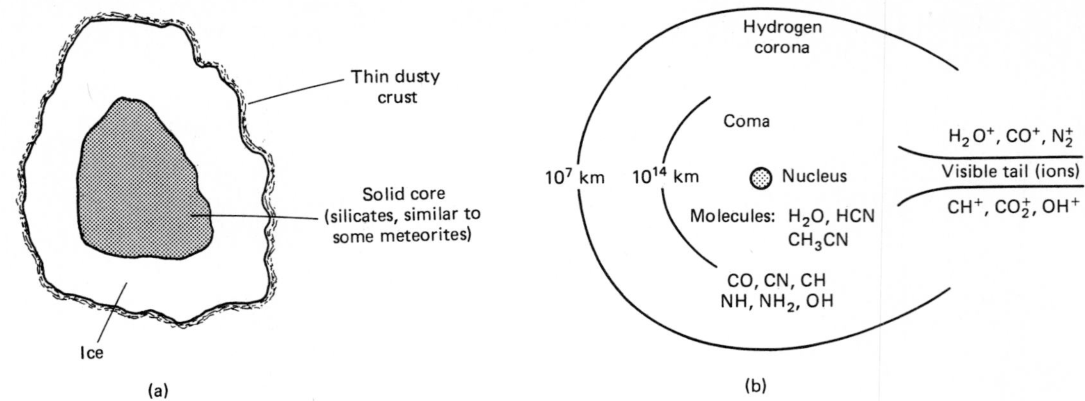


**Figure 17.12** (a) Schematic diagram of a comet nucleus. (b) Schematic diagram of main gaseous features. The hydrogen corona extends out for as far as 10 million km, much farther than the bright head that is usually seen. The bright coma of molecules and radicals is much closer to the nucleus, and the tail of ions can extend back as much as an astronomical unit.

core, there will be a thick region of dirty ice and then a thin (less than a meter thick?) surface layer of dust. On the side of the comet that faces the Sun, ice will steadily sublime (evaporate directly from solid to gas without going through a liquid stage). Meteorite impacts will knock some dust off and produce fractures through which gas from evaporation can escape.

Close to the nucleus, the inner coma will have an average density of a few million ions and molecules per cubic centimeters. The density of the coma decreases rapidly with distance. At greater distances, there will be ions in the tail and radicals in the outer coma. The dust tail is produced as the dust is ejected by impacts or is released as surrounding ice evaporates.

Where and how do the comets form, and where do they reside until their paths bring them within range of telescopes? In 1950 the Dutch astronomer Jan Oort suggested that a reservoir of comets exits in a giant cloud far beyond the planets. This Oort cloud (as it has come to be known) is now supposed to extend from about 10,000 to 100,000 AU from the Sun with its outer parts approaching the closest stars. Within the cloud, there is a population of around $10^{12}$ comets with a total mass of somewhat less than the Earth. The comets are not static but are in orbits around the Sun. Passing stars are thought to provide small perturbations that will deflect the comets, sometimes into orbits where we see them but equally often into orbits that will take them farther away. Some that do penetrate into the inner solar system will have their orbits further altered by passing close to the Sun and the giant planets. In this way, after many deflections, an initially long-period comet might have its orbit altered into one of short period. Some comets might be disrupted by tidal forces if they approach too close to the Sun or Jupiter or Saturn.

The place of formation of the comets is less well settled. One view is that they formed near the outer planets and slowly drifted to their present places in the Oort cloud. A different model suggests hat they were initially part of a large planet (about 100 times larger than the Earth) originally located at 2.8 AU from the Sun. Through some unspecified event this planet fragmented. Some of the debris remains as the asteroid belt, and some of the debris drifted out to form the Oort cloud. Some debris might have been captured by Jupiter to appear as satellites.

The speculative nature of these models is at once apparent, but this should not be the reason for their immediate rejection. If their predictions are not confirmed, then they will go the way of many other ideas, but a solution to the problem of the origin of comets may well be found in this way from imaginative models.

**Space probes** Of necessity, all of the comet observations so far have been made from the Earth or very close by from rockets and satellites. The scheduled returns of two comets have aroused considerable interest in the possibility of launching space probes to approach them. Comet Halley will be within reach in late 1985 and early 1986, and Comet Tempel-2 will approach in 1989. Budgetary tightening finally killed NASA's developing plans to send a probe to Halley, but the European Space Agency will be sending its Giotto probe, and Japan and the Soviet Union have also developed plans.

Giotto is to be aimed to approach within 1000 km of the comet in March 1986, about a month after perihelion passage. There is some concern that the orientation of the spacecraft might be disturbed by collisions with small grains in the dust surrounding the comet's nucleus. If this happens, the spacecraft's antenna might no longer aim directly back to Earth and contact could be lost. Protective measures are being designed.

It has turned out that an important view of Halley can be obtained from Pioneer Venus, in orbit around Venus. The comet will approach to only within about 40 million km of Pioneer, but the importance of this viewing is that it will occur in February 1986, 5 days before perihelion. The UV spectrometer on Pioneer will be able to look for atomic hydrogen, oxygen, carbon, nitrogen, and perhaps sulfur.

A close approach to a comet could provide a good estimate of its mass through the comet's effect on the probe's trajectory. Close-up images will certainly show far more detail than can be seen from Earth. Other instruments that can be carried include mass spectrometers, magnetometers, photometers for several wavelength regions, and a dust analyzer. This latter device would collect dust particles in the tail, vaporize them and examine the vapor in a mass spectrometer. The sizes and speeds of dust particles can be measured by their impacts: A different detector (a plasma probe) is needed to examine the ionized gases.

## 17.3 METEOROIDS, METEORS, AND METEORITES

The **meteoroids** are smaller than comets. In space, away from the Earth, they are not directly observable but those that enter the Earth's atmosphere heat up to produce spectacular **fireballs** or **meteors** (Fig. 17.13). Many meteors flash across the sky in random directions, but related showers are also

**Figure 17.13** Meteor trail seen near the Pleiades (Yerkes Observatory).

seen in which the meteors are moving along parallel trajectories in space but all *appear* to be coming from a single point in the sky, the **radiant**. Each shower is named for the constellation in which its radiant is located. Many meteoroids move along orbits that the Earth intersects annually so that showers of meteors can be predictably observed (on clear nights) with regularity. The scheduling of some of these showers is listed in Table 17.3.

The orbits followed by many meteoroid swarms are similar to those of periodic comets, and it is thought that all of the showers are related to comets. For example, debris from Halley's comet is strung out along its orbit, and the Earth plows through this collection twice a year producing the Eta Aquarids in May and the Orionids in October.

**Meteorites** are those few meteoroids large enough to survive the intense heating in their passage through the atmosphere and still reach the Earth's surface. They range in size from small rocks to ton-sized boulders. Some are found by accident, others by searching in the region where bright fireballs have been seen to burn out at low altitude. Until the arrival of lunar samples, meteorites provided the only solid samples of extraterrestrial materials available.

The exterior of a meteorite often gives at once a clue to a part of its history—a crust that looks as though it had been molten and then solidified more smoothly than ordinary Earth-bound rocks. Cutting the meteorite open will then reveal structures that allow the meteorite to be classified into three broad groups: stony meteorites or **chondrites** (92.8 percent), irons (5.7 percent), and stony irons (1.5 percent) (Fig. 17.14).

The characteristic feature of the chondrites is the generally rocklike appearance with small (millimeter-sized) spherical grains (**chondrules**) embedded in a matrix. Ordinary chondrites comprise about 80 percent of all meteorites. Some stones have very high carbon content (the carbonaceous chondrites, 5.7 percent), whereas a few more (the **achondrites**, 7.1 percent) have no chondrules but a rather coarse internal structure. Chemical analysis of the chondrites shows them to be made up of silicate materials such as **olivine** $(MgFe)SiO_4$, **pyroxine** $(MgFe)\ SiO_3$, and other minerals well-known to geologists. There is also a metal component (about 20 percent) and some lesser constituents. The overall abundances of the chemical elements are, however, much closer to solar abundances than to those in the Earth's crust. The nature of the crystalline structure indicates that the chondrites must have formed at a temperature around 500 K, in a region where the pressure was around $10^{-4}$ atmospheres. The apparently low pressure is actually much higher than typical interstellar space and is more suggestive of a dense region surrounding an evolving Sun.

***Table 17.3*** METEOR SHOWERS[a]

| Shower | Date of Maximum | Maximum Number (per hr) | Average (per hr) | Associated Comet |
|---|---|---|---|---|
| Quarantids | January 3 | 110 | 30 | — |
| Lyrids | April 21 | 12 | 8 | 1861 I |
| Eta Aquarids | May 4 | 20 | 10 | Halley |
| Delta Aquarids | July 30 | 35 | 15 | — |
| Perseids | August 12 | 68 | 40 | 1862 III |
| Orionids | October 21 | 30 | 15 | Halley |
| Taurids | November 4 | 12 | 8 | Encke |
| Leonids | November 16 | 10 | 6 | 1866 I, Tempel |
| Geminids | December 13 | 58 | 50 | — |

[a] Most showers can be seen for about 2 days on each side of maximum, whereas the Delta Aquarids and Taurids extend for about a month. Showers are best seen after midnight without telescope or binoculars and away from the glow of the city lights.

(a) (b) (c) (d) (e)

**Figure 17.14** Meteorites (a) Albin, stony-iron. (b) Bjurbole, stony. (c) Toluca, iron. (d) Whitman, stony. (e) El Sampal, iron (154 kg) (Center for Meteorite Studies, Arizona State University).

Next in abundance are the iron meteorites—largely an iron-nickel mixture whose complex crystalline patterns are revealed when a meteorite is sliced open and the newly-exposed face is polished and then etched with acid. These patterns are indicative of slow cooling under low pressure. The cooling must have been very slow—about one degree per million years—and this implies formation within a body of at least 200 km diameter, because smaller bodies would cool too rapidly. The meteorites that are found must, therefore, be fragments of an initially much larger body. Finally, the stony irons (1.5 percent) show a matrix of iron-nickel that holds mineral inclusions.

In some meteorites organic matter has been found, combinations of carbon and hydrogen in typically long-chain molecules. This is not taken as an indication of the presence of biological matter but rather as the result of a chemical, inanimate process.

It is very widely agreed that meteorites represent a scarcely-evolved sample of the very early solar system: What conclusions can we draw about those early conditions? How long ago did they form? We can use radioactive dating techniques (as already described for lunar samples and dating for the Earth). Analysis of the Sr-Rb and U-Pb ratios leads to ages close to 4.5 billion years in excellent agreement with the lunar sample results. In fact it was the meteorite data that first pinpointed this age for the solar system. A number of meteorites appear to have younger ages, and these can be understood if the meteoroids had been heated subsequent to formation. When heating occurs, some of the products of the radioactivity will escape, and the abundances of the critical isotopes will be distorted, no longer giving true ages since formation but rather ages since the last overall heating.

A very detailed analysis of the structure of the igneous meteorites, shows that they must have been subjected to an intense but short-duration heating before solidifying. The source of this heating has not yet been established. Heat from the radioactivity of U, Th, and K would not have been intense enough and would have lasted for too long. An ingenious suggestion has been only partially confirmed. This involves the postulated presence in the very early solar system of radioactive isotopes that have relatively short half-lives and are by now extinct. It was suggested that the heat might have come from their radioactive decays.

What is the evidence that such isotopes might have existed? The initial evidence came from meteorites and the confirmation came from nuclear physics. Small amounts of xenon, an inert gas, had been found in meteorites. Xenon has several isotopes, and the relative isotopic abundances of some of the xenon had been attributed to radioactive fission decay of now-extinct plutonium, $Pu^{244}$. In addition, in some meteorite crystals, etching had revealed large numbers of nuclear particle tracks that are also characteristic of fission. When $Pu^{244}$ was subsequently made in the laboratory and then observed to decay, analysis of the decay products provided confirmation for the extinct element hypothesis.

The early presence of other extinct isotopes such as iodine 129 ($I^{129}$), with a 17 million year half-life, has also been documented. Recently, attention has turned to aluminum 26 ($Al^{26}$), with a half-life of only 720,000 years, as a possible source of the heat necessary to melt objects in the early solar system. Although modern theories of the synthesis of isotopes in stars and stellar explosions lead us to believe that $Al^{26}$ can be produced, its short half-life made it seem unlikly that very much of the isotope would still be left by the time that the solar system formed. Surprisingly, evidence for the early presence of a relatively large amount of $Al^{26}$ has recently been found. Perhaps the stellar explosion that produced it also triggered the collapse of the solar nebula. A host of other isotopic anomalies has been found in meteorites and provide puzzles that must still be solved.

**Origin** The prevailing view is that most meteorites originally formed parts of minor planets and have come to the Earth on elliptical orbits from the asteroid belt typically at 2 to 4 AU from the Sun. The meteorites' properties are consistent with their formation in much larger bodies and not as objects that were only slightly larger than their present sizes.

Two recently discovered types of meteorites seem to have very different origins. The meteorite, ALHA 81005 (Plate 35), was discovered in 1981 by U.S. scientists in the Anarctic. This meteorite is generally similar to the Apollo 16 lunar breccias. Analysis of trapped gases and the mineral structure all point to a lunar origin. There are, however, some small differences from other lunar samples and it has been suggested that ALHA 81005 came from a part of the Moon not yet sampled, perhaps from the far side.

Another class of meteorites (the SNC) has recently been shown to have many similarities to the Martian soil that was analysed by the Viking landers.

The dynamical processes that permit these meteorites to be ejected from their parent bodies and reach the Earth intact are not yet understood.

The carbonaceous chondrite meteorites have played an exceptionally important role in studies of the overall composition of the solar system. The Sun of course contains most of the mass of the solar system, but our direct knowledge of its composition comes from spectroscopic analysis that is not as accurate as the elemental analysis of solid materials. In the solar spectrum, many lines overlap or are faint, and the spectral analysis is best used for the volatile elements. We do have analysis of the Earth's crust, but we know that the deeper interior is different and that the crust has evolved significantly, so that its elemental abundances are surely different from the solar system in general.

Where solar and meteoritic elemental abundances can be directly compared there is generally good agreement, lending strong support to the idea that meteorites are fossil remnants of the early solar system. The elemental abundances in meteorites can be measured with great accuracy especially for the nonvolatile elements, and these provide the basis for all modern tabulations.

## 17.4 DUST

The planets, comets, and meteoroids do not exhaust the list of the population of the solar system. The size spectrum extends down to micron-sized dust particles (1 micron equals $10^{-4}$ cm), and the evidence for their presence takes several forms. Along the ecliptic, a faint glow can be seen in the east a few hours before sunrise or in the west after sunset. The spectrum of this **zodiacal** light is the same as light that comes directly from the outer regions of the Sun's atmosphere. The zodiacal light appears to have been scattered by minute dust particles.

The zodiacal light is seen not too far from the Sun, but the **gegenschein** (counterglow) (Fig. 17.15) is seen as a faint glow about 8° wide also along the ecliptic but directly opposite the Sun. Its spectrum shows it also to be reflected sunlight.

Many of the dust particles can be detected as **micrometeorites**. With typical sizes of a few microns, some burn up in the upper atmosphere and their debris can be detected after it has settled slowly to the ground. Others collide at high speed with space vehicles, and their impacts have been detected with microphones and with collector plates on which the splashes of their destructive collisions leave measurable traces (Figs. 17.16 and 17.17).

A new type of related particle was discovered in the 1970s. Using a NASA high-altitude U2 research plane, Donald Brownlee (University of Washington) has collected dust particles on a plate covered with a very viscous oil. Particles with sizes between 10 and 50 microns (weighing only

**Figure 17.15** Gegenschein as photographed from Skylab (NASA/Marshall).

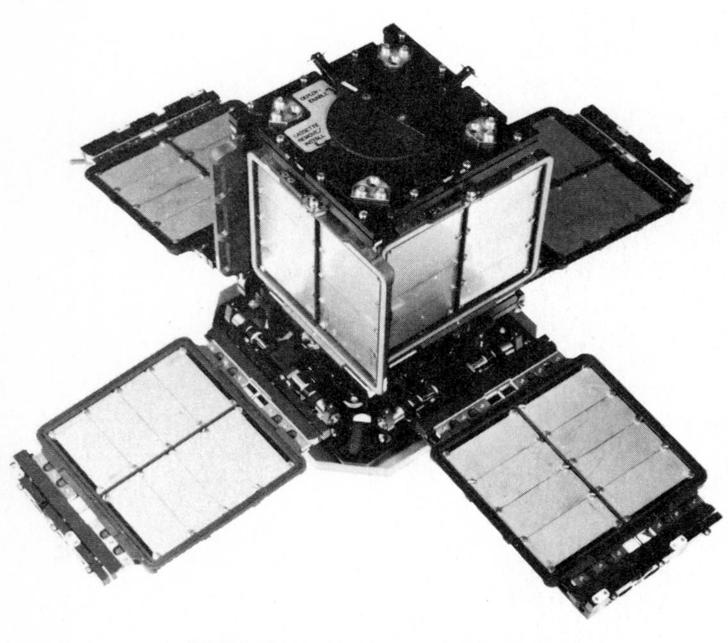

**Figure 17.16** Micrometeoroid collection experiment used on Skylab showing the exposed metal collector panels. Particles that collide with the panels produce impact craters (NASA/Marshall).

**Figure 17.17** Electron microscope photographs of impact craters produced by micrometeoroids. (a) crater in stainless steel, 2 micron diameter. (b) large crater in aluminum foil, 110 microns diameter probably produced by a particle with a diameter of 30 microns (NASA/Marshall).

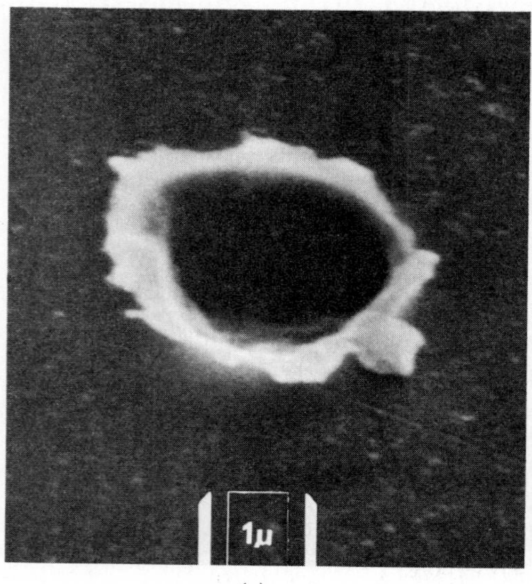


(a)

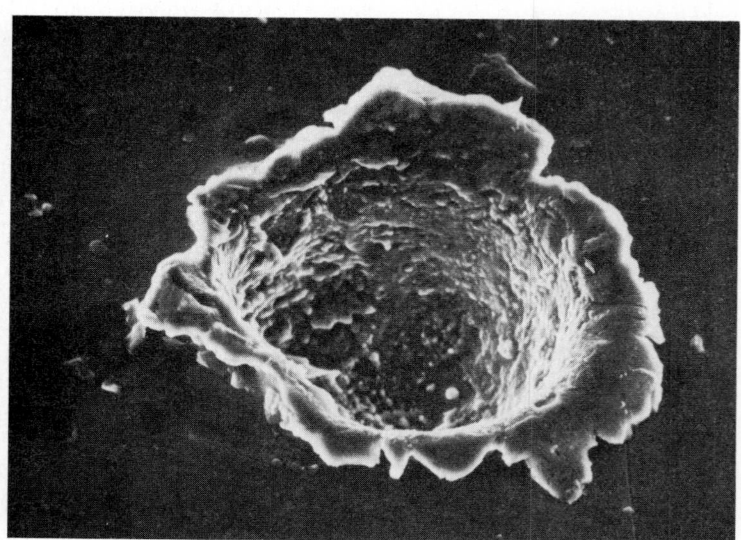

(b)

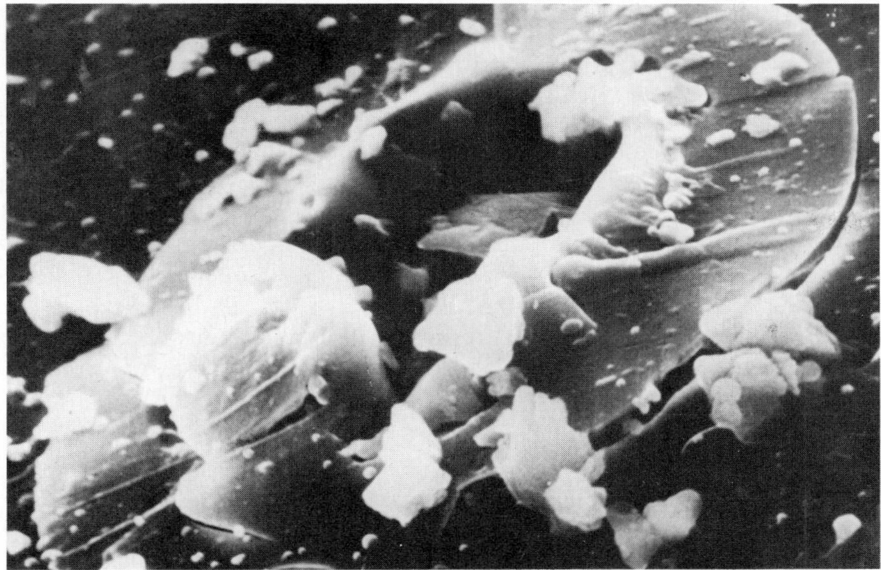

**Figure 17.18** Pit produced by impact of micrometeorite on a particle of the lunar soil; diameter about 0.1 mm (Robert M. Walker, Washington University).

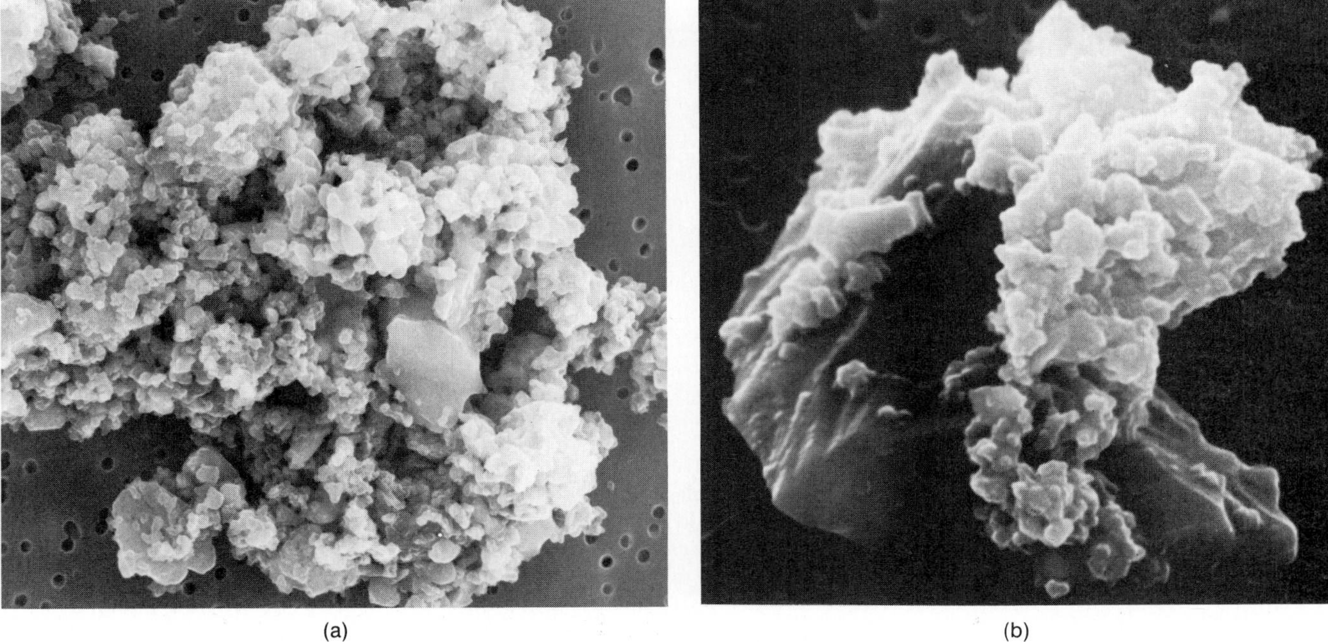

(a) (b)

**Figure 17.19** Electron microscope photographs of micrometeorite particles collected during U2 airplane flights in the stratosphere. (a) Porous chondritic particle, studded with mineral grains. Scale bar in lower right-hand corner is one micron ($10^{-4}$ cm) (Donald Brownlee, University of Washington). (b) chondritic particle, a rare single crystal of the mineral olivine embedded in fine-grained adhering material; diameter about 20 microns (Robert M. Walker, Washington University).

around a billionth of a gram) can survive their impacts and be removed intact from the collection plates (Fig. 17.19). Some very delicate analyses have shown them to be unlike terrestrial particles (such as pollutants or particles from volcanoes). In many ways they are similar to carbonaceous chondrites, but there are important differences in composition and structure. Electron microscope studies show some to have a spongy structure made up of tiny components. It is thought that these particles are micrometeorites and as such are another important clue to the structure and composition of the early solar system.

## 17.5 SUMMARY

The comets and minor planets of the solar system probably grew to their present sizes and never existed as parts of much larger bodies. How we detect them depends on their size and distance. The minor planets reflect sunlight as do the planets; the comets remain invisible to the naked eye until close enough to the Sun to be sufficiently warmed for some constituents to evaporate. The meteors and meteorites are detected because of their collisions with the Earth and its atmosphere. The interplanetary dust that produces the zodiacal light and gegenschein can so far be studied only at a distance. At the moment the meteorites are probably most important for astronomy for the insights they have given us on chemical composition and physical conditions in the early solar system, and we have this only because we have actual samples in the laboratory. The contrast between the vast amount of diverse data on the meteorites, and the far more restricted data on all of the other minor bodies, once again shows how important it is to have samples.

## CHAPTER REVIEW

**Minor planets (asteroids):** mostly in a region between 2.3 and 3.3 AU from the Sun but some have Earth-crossing orbits.

- — largest (Ceres) has 1000 km diameter, but most are much smaller.

**Comets:** probably come from a reservoir (Oort cloud) far beyond Pluto's orbit.

- — some have orbits that bring them repeatedly back through the solar system, others probably make only one passage.
- — two types of tails: Type I are gaseous—ions, radicals, molecules; Type II consist of dust particles. Type I tails have clumpy appearance, point straight along Sun-comet line, and are directed by particles in the solar wind. Type II tails curve and are composed of dust particles that are not as much affected by the solar wind but rather by solar radiation pressure.
- — composition of comets (deduced from spectra): atoms, molecules, and ions of the light elements: H, C, N, O, and S, with some heavier atoms shown in comets that have small perihelion distances.
- — Halley's comet will have next perihelion passage in March 1986 and will be observed by several spacecraft, some within less than 1000 km.
- — composition of comets: dirty iceberg model—solid core surrounded by dirty ice with a surface layer of dust.

**Meteoroids:** objects in orbit around the Sun—we detect them as meteors and meteorites.

**Meteors:** flashes seen when a meteoroid enters the upper atmosphere and burns up.

- — meteor showers are regular and occur when Earth crosses orbits of comets that have broken.

**Meteorites:** meteoroids that are large enough to reach the Earth's surface, and extremely valuable for the information they provide on conditions in the early solar system. Meteorites are thought to represent very primitive solar system material.

- — different types: stony, iron, and stony-iron. Internal structure provides information on temperature and pressure during formation and cooling.
- — origin: probably originally part of minor planets.

**Dust:** in solar system—detected by reflected sunlight (zodiacal light, gegenschein), also collected in high-altitude flights and with collectors on spacecraft.

## NEW TERMS

**achondrites**
**asteroid belt**
**carbonaceous chondrites**
**chondrites**
**chondrules**
**coma**
**dirty iceberg**
**gegenschein**
**irons**
**olivine**
**pyroxine**
**radiant**
**sublime**
**Type I, II tails**

## PROBLEMS

**1.** What is the difference in structure and behavior of the dust and ion tails of a comet?

**2.** Draw a diagram to show how a comet can sometimes appear to have one tail behind and a spike in front, yet both are truly on the same side of the comet relative to the Sun.

**3.** Under what conditions does a comet's tail travel in front of its head?

**4.** Describe the changing appearance of a typical comet from the time it crosses the orbit of Jupiter, through perihelion passage and until it is once again beyond 5 AU.

**5.** Why are some comets brighter after perihelion than they are before?

**6.** Why have some comets broken up when passing close to the Sun or to Jupiter?

**7.** Using data from Table 17.1, calculate the aphelion distances for Comets Encke and Halley.

**8.** An early estimate of the periodic time for Comet Kohoutek was 750,000 years. Assuming its orbit was an elongated ellipse, calculate the size of the semimajor axis. The comet had a perihelion distance of 0.14 AU. How far from the Sun was its aphelion?

**9.** The Oort cloud of comets may be at 50,000 AU from the Sun. How long does it take an object at that distance to make one orbit around the Sun if the perihelion distance is 0.5 AU?

**10.** What are the major categories of meteorites?

**11.** What are the differences between meteoroids, meteors, and meteorites? (Refer to the text as needed.)

**12.** In what way does the study of carbonaceous chondrites contribute to our knowledge of the early history of the solar system?

**13.** What is unusual about the Allan Hills meteorite?

**14.** What is the relation between meteor showers and comets?

**15.** The sizes of minor planets are estimated from their brightness knowing their distances from the Sun and Earth and assuming a value for the albedo. Suppose that the albedo for a minor planet is thought to be 0.25 but is really 0.125. How large an error does this make in the computed diameter for the planet?

**16.** A minor planet has an orbit with a semimajor axis of 15 AU and an eccentricity of 0.85.

a. calculate its closest and farthest distances from the Sun.

b. calculate its maximum and minimum temperatures assuming an albedo of 0.1.

**17.** In what way have some minor planets aided in the study of Mercury and Venus?

**18.** Which observations indicate the presence of dust within the solar system?

# CHAPTER 18

# The Origin of the Solar System

## 18.1 THE PROBLEM

How was the solar system formed? This is a very old question with a succession of answers that chart the progress of scientific knowledge. The question was considered sufficiently important that plausible answers were devised thousands of years ago within the context of beliefs and legends, such as the description in Genesis.

Modern science had its first great successes in mechanics, and dynamical theories of the solar system dominated the scene until around a hundred years ago. Then spectroscopic evidence revealed the presence in an astronomical setting of chemical elements well-known on Earth. A place had then to be found in the theories for this accumulating knowledge of the chemical composition. Another important factor that distinguishes today's theories from the ancient legends is the tacit acceptance of a body of physical laws that govern the processes of interest—laws that are generally assumed to have remained constant over the eons. (At present there is some important research into possible changes in these laws, but it seems that any changes have been very small, and we shall omit this complication for the present.)

The study of scientific models of the history and evolution of the solar system is sometimes termed **cosmogony**, a most unfortunate term that is often confused with **cosmology**, the study of models of the origin and evolution of the entire universe. Cosmogony is occasionally used in the wider sense, but increasingly it seems to be applied more narrowly, such as in Fred Hoyle's 1978 book, *The Cosmogony of the Solar System*.

As we have seen in the preceding six chapters, an impressive store of information on the planets and other bodies of the solar system has now been assembled. We know the general features of all of these objects and some details of the structure of a few. We have been able to construct plausible models and gain a general understanding of such aspects as the retention of a planetary atmosphere or the temperature at which a meteorite has solidified.

To construct a coherent picture of the formation and evolution of the solar system, which are the most important items that we must include? We would hope that once we have a model that incorporates the major features, we can pursue the fine details. The most important properties that we have discovered can be grouped in three broad categories as in Table 18.1

*Table 18.1* MAJOR FEATURES AND PROPERTIES OF THE SOLAR SYSTEM

*I. Dynamic Properties*

1. The orbital planes of the planets lie very close to one another (Fig. 18.1).
2. The Sun's equatorial plane is close to the median plane among the planets' orbits.
3. The planetary orbits are nearly circular.
4. The directions of the planets in their orbits are all prograde, as is the Sun's own rotation.
5. Apart from Venus and Uranus, all of the planets rotate in the same prograde direction on their own axes.
6. Most of the **angular momentum** (Box 18.1) in the solar system is carried by the planets (and most of that by Jupiter). The Sun, by far the most massive object in the system, has only ½ percent of the total angular momentum (Box 18.2).

   (Note: we do not list Kepler's Laws along with the regularities above since they are completely understandable through Newton's Laws of dynamics and gravity and provide us with no clues to the origin of the solar system.)

*II. Physical and Chemical Properties*

1. The chemical composition of the planets and their atmospheres and moons; trends in the composition with increasing distance from the Sun.
2. The age of the solar system (4.6 billion years) and the duration of

**Figure 18.1** Orbital planes of planetary orbits. The inclination of Pluto's orbit is by far the greatest; after that, only Mercury's exceeds 5°.

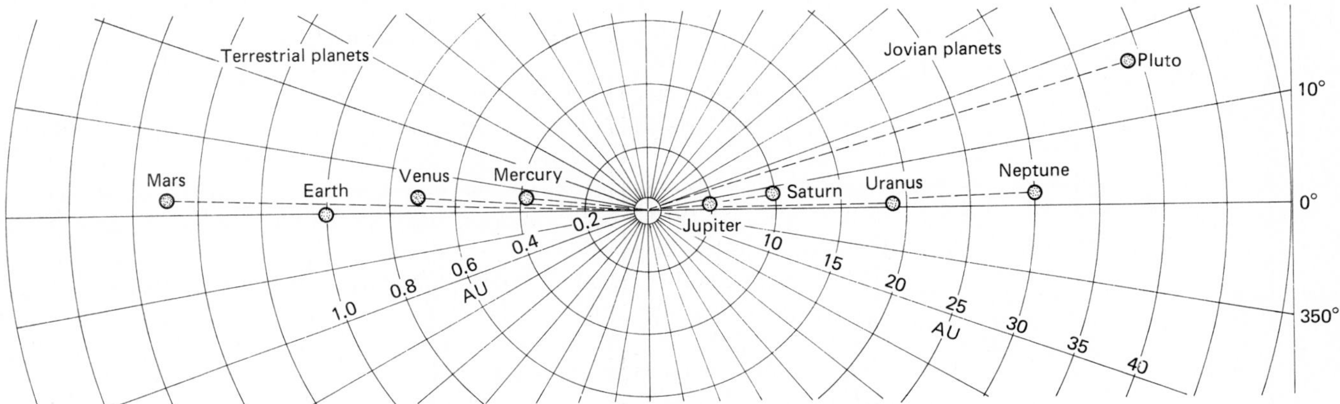

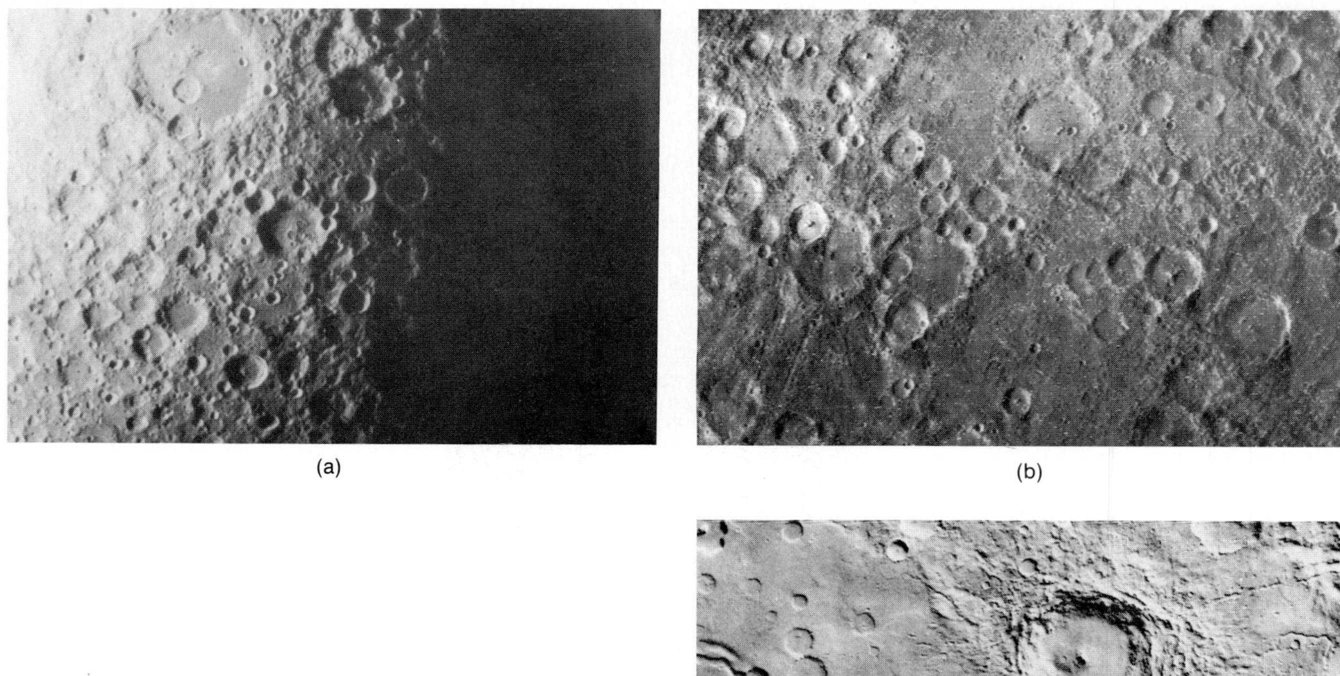

(a) (b)

(c)

**Figure 18.2** Comparison of surface craters for Mercury, Mars, and the Moon. These three bodies have insufficient atmosphere to slow the bombarding bodies, but their cratering appearances differ. Mars and Mercury have plains between craters and have few 10 to 15 km craters compared to the Moon. Craters on Mercury, however, appear more like those on the Moon than those on Mars, where surface flows have been extensive (NASA).

the formation stage (no longer than about 100 million years) as defined by the abundances of radioactive isotopes.

3. The temperatures, pressures, and densities at solidification as shown by the structure of meteorites and planets.
4. The isotopic abundances which show that certain process must have been operative, and that other processes can be excluded.

*III. Miscellaneous Features*

1. Systems of satellites and rings.
2. Comets, minor planets, and meteoroids.
3. The bombardment phase that produced the cratered surfaces (Fig. 18.2).

---

BOX 18.1
*Angular Momentum*

Until now, when we have been describing the movements of a body, we have specified its position, speed, and acceleration. Predicting its later behavior requires us to know its mass and the forces that push or pull on it. To deal with the motion of a planet around an orbit or the rotation of a planet on its own axis, we find that we often need to introduce another quantity, the **angular momentum**. This is defined as a product of mass, speed, and distance (or length) (Fig. 18.3).

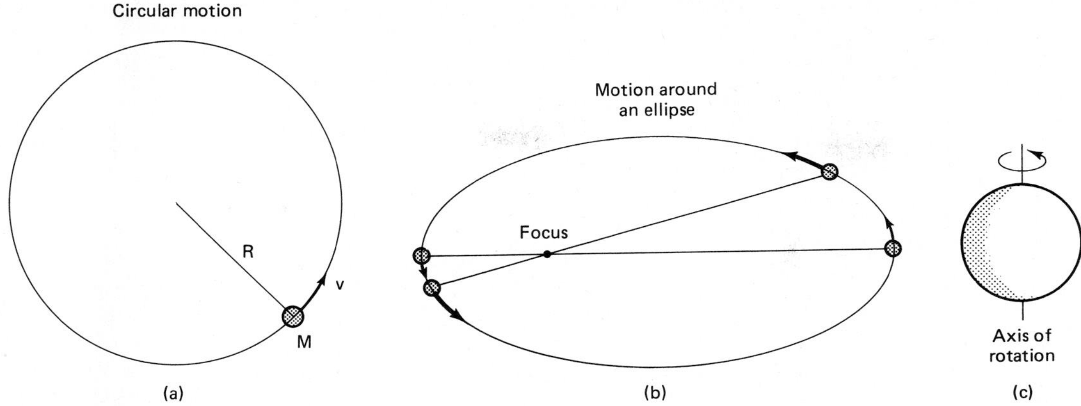


**Figure 18.3** Angular momentum. (a) A body with mass *M* moving with speed *v* around a circle of radius *R* has angular momentum *MvR*. (b) An object traveling in an elliptical orbit will go faster when close to the focus of the ellipse and slower when far from the focus, but the angular momentum remains constant. (c) An object rotating about an axis through its center also possesses angular momentum. For a sphere this is 2/5 *MvR*, where *v* is the speed of a point on the equator.

In the case of orbital motion in a circle, we define the angular momentum as $L = MvR$, where $R$ is the radius of the circle around which the mass $M$ is travelling at speed $v$. If there are no outside forces acting, then the angular momentum remains constant. This is required by theory and observed in practice. Thus for uniform motion around a circle, both $R$ and $v$ remain constant. Along an elliptical path, however, the distance from the focus is constantly changing, and so therefore must the speed, as was noted by Kepler for the planets. When $R$ is smallest, such as at perihelion, then $v$ will have its greatest value. When $R$ is greatest, at aphelion, $v$ reaches its minimum value.

To calculate the angular momentum of a planet rotating about its own axis, the product $MvR$ requires modification by a numerical factor because the different parts of the mass are at different distances from the axis of rotation and moving at different speeds. For a rotating spherical object of radius $R$, the factor is 2/5 and the angular momentum is given by $L = 2/5\ MvR$, where $v$ is the speed of a point on the planet's equator. (In some bodies the density is not uniform but varies with distance from the center. This is especially true for the Sun so that its angular momentum of rotation cannot be calculated by the simple use of the formula $2/5\ MvR$.)

The importance of all of this to the formation of the solar system and to the formation and condensation of stars, is that during the contraction of a gas and dust cloud, the angular momentum of the system cannot change. As the dimensions become smaller, the rotational speed must increase, although some of the angular momentum might be separated by throwing off some mass. In general, inclusion of angular momentum into a theory makes it much more complicated, and so the simplest models usually do not consider rotation. On the other hand, we observe rotation in many celestial systems, and we realize that it is something that we cannot permanently ignore.

The angular momenta of the Sun and the planets are listed in Table 18.2, and we are immediately struck by the uneven distribution of mass and angular momentum. The Sun, overwhelmingly the most massive object in the solar system, possesses a surprisingly small part of the total angular momentum. Any credible model for the formation of the solar system must be able to explain this.

*Table 18.2* ANGULAR MOMENTUM IN THE SOLAR SYSTEM

| | *Mass* | | *Angular momentum* | |
|---|---|---|---|---|
| *Object* | *(g)* | *(percent of mass)* | *(gcm/sec)* | *(percent of total)* |
| Sun | $1.99\ 10^{33}$ | 99.865 | $1.6\ \ 10^{48}$ | 0.51 |
| Mercury<br>Venus<br>Earth<br>Mars | $4.87\ 10^{28}$ | 0.0006 | $4.95\ 10^{47}$ | 0.157 |
| Jupiter<br>Saturn<br>Uranus<br>Neptune<br>Pluto | $2.66\ 10^{30}$ | 0.134 | $3.14\ 10^{50}$ | 99.3 |

## 18.2 OLDER MODELS

Cosmogonic models divide into three broad classes:

1. **catastrophe models,** in which the material for the planets comes from the Sun, forced or pulled out by some assumed unique event (Fig. 18.4).
2. **capture models,** in which the Sun captures the planets or the material to form them,
3. **evolutionary or nebular models,** in which the Sun, planets, and other bodies form in the same region and at about the same time out of a common pool of material.

The first scientific theory was evolutionary and came from René Descartes in 1644. He assumed that the universe had been filled with a swirling gas and that the Sun, planets, and other bodies formed through the condensation of these turbulent currents. It was an ingenious idea, with some features that have survived, but without Newton's dynamics there was no way in which it could be put into quantitative form.

Nearly 100 years later, in 1745, the first catastrophe theory came from the Comte de Buffon who suggested that a comet had collided with the Sun; the planets had then formed from the material thrown out in the impact. Many elaborate catastrophe theories have since been proposed, culminating in 1917 with the best-known of all, by the English mathematical physicist, J. H. Jeans. This model was based on the idea that a cigar-shaped streamer of gas, drawn from the Sun by a passing star, condensed to form the planets. Although this model has an immediate appeal, several detailed calculations in the 1930s have identified major objections, and catastrophe theories are no longer considered.

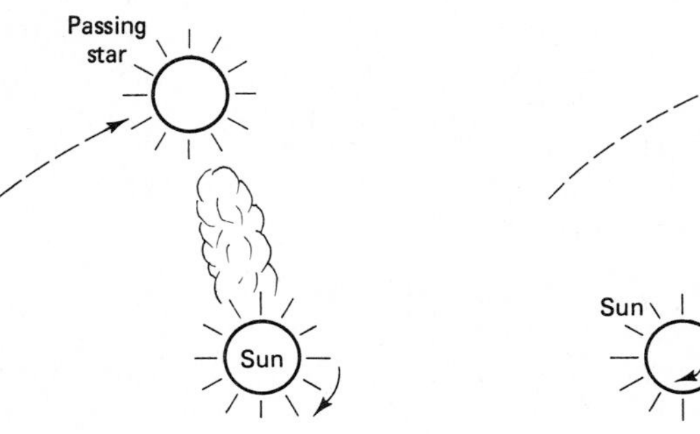


**Figure 18.4** Catastrophe model for the origin of the solar system: (a) Gravitational attraction between the Sun and a passing star draws out a cylinder of solar matter, which later (b) condenses to form the planets.

The first objection is the inherently low probability for the required close encounter. More substantial objections come from the calculations that show ejected material will be disrupted by solar tidal forces. In addition, the ejected material, because of its high temperature, will expand and disperse rather than condense. Further, the chemical composition of the terrestrial planets is now known to be significantly different from that of the Sun, contrary to what would be expected if the planetary material had been drawn from the Sun.

Capture models have found no serious support, for they would require not just one rare event but many, in order to assemble the solar system with its present dynamic and chemical properties by capturing the planets and moons one at a time.

Only nebular or evolutionary theories remain. Early evolutionary theories such as those proposed in 1755 by the philosopher Immanuel Kant, and in 1796 by the mathematician Laplace had only to cope with the dynamics, but as we have seen (Table 18.1), far more is now expected of a theory. There have been many evolutionary models proposed, and the framework that now has widest acceptance comes from ideas put forward independently by Fred Hoyle and W. H. McCrea in the late 1950s and early 1960s.

In the following section we will describe a plausible scenario to show how an evolutionary model is thought to have proceeded. Some of the quantities used in this description are by no means well known, and you may find different values or processes used in other books or articles. You should not be disturbed by this, since it is a normal occurrence in science when knowledgable and inventive people are proposing and testing different models to accommodate the same set of observations. Nevertheless, although there are still dissenting views on many points of detail, the model outlined here has gained a broad consensus. The general features are probably correct and the model does organize much of the present knowledge in a coherent way.

## 18.3 AN EVOLUTIONARY MODEL

Evolutionary models are built around the idea of the collapse of a large cloud of gas and dust, with the Sun being formed from the central condensation, and the planets and other bodies forming from the outlying material (Fig. 18.5). Our main interest at this time is in the formation of the planets. We shall not, therefore, treat the details of the Sun's development as a star, but shall defer that to chapter 24 when we deal more generally with star formation and evolution.

The starting point is a cloud of interstellar gas mixed with some dust. In interstellar space away from the stars, there is on the average about one atom in each cubic centimeter, but the cloud we are considering will probably have around 1000 times this density. Most of the gas (nearly 75 percent by mass) will be hydrogen, with an appreciable amount of helium (nearly 25 percent by mass) and traces of heavier elements. There may also be some dust particles but almost certainly less than 1 percent of the cloud mass. Internally, turbulent currents will probably cause the density to vary from place to place, and the whole cloud will be rotating slowly, perhaps once in 10 million years. The overall temperature will be no more than 50 K.

Left to itself, the cloud will collapse, for there is a net inward gravitational force. As the cloud collapses, the need to conserve angular momentum will cause the cloud to flatten and the rotation will speed up. Perhaps it is during this stage that the cloud breaks apart into smaller condensations, and it is from one of these that the Sun and planets will form. The original cloud might have had a total mass of 500 to 1000 solar masses. By the time that fragmentation occurs, the cloud that will produce the solar system

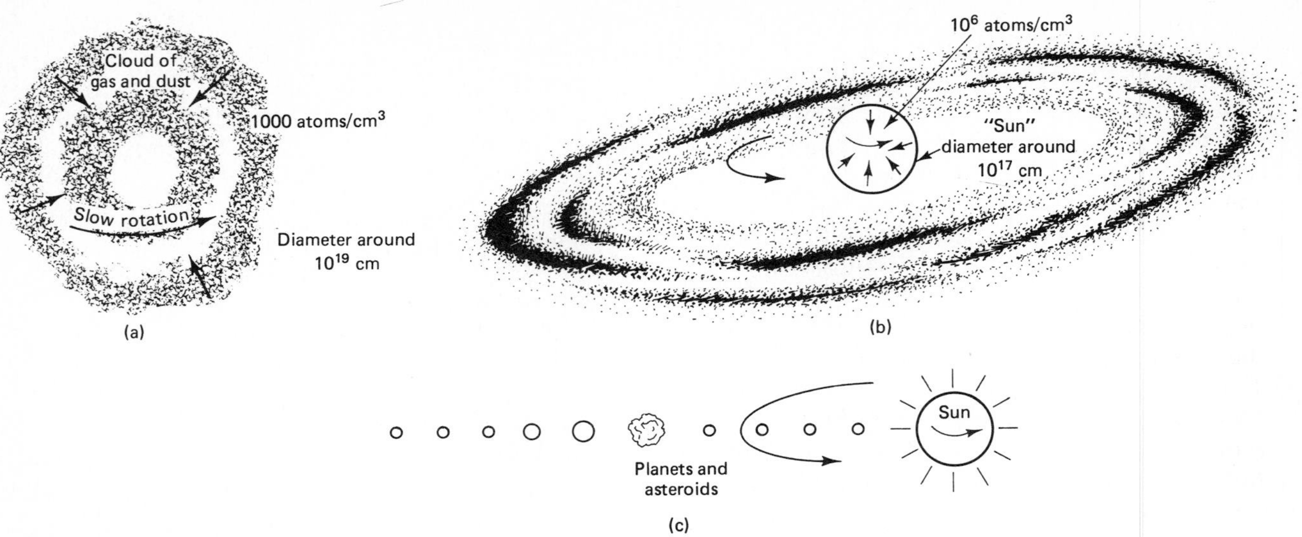


**Figure 18.5** Schematic representation of the modern theory for the origin of the solar system. (a) A large cloud of cold gas and dust rotating slowly contracts through gravitational attraction. (b) As the contraction proceeds, the angular momentum remains so that rotation speed increases. The condensing matter forms a star (the Sun) at the center with a rotating disc of matter around. (c) Through collisions, matter in the disc collects to form the planets.

might have a radius of about $10^{17}$ cm (about 7000 AU) and a density of $10^6$ atoms/cm$^3$. We will follow the evolution of this smaller cloud, which has a mass of about 3 M.

Contraction releases gravitational energy that heats up the nebula. Initially, much of this energy can be radiated away, but as the contraction continues, the density of the nebula increases and begins to obstruct the outward flow of energy. The interior of the nebula, therefore, heats up more, and the contraction is slowed. This first phase of contraction stops when the sun has a radius of about $2 \times 10^{12}$ cm (about thirty times its present size) and a surface temperature of 3500 K.

When this stage is reached, the Sun will be surrounded by a less dense nebula in which the temperature and density decrease steadily from the center out. The results of a typical set of calculations are shown in Fig. 18.6. The temperatures range from close to 1500 K at 0.4 AU, where Mercury will later be in its orbit, down to around 200 K near the future orbits of Jupiter and Saturn.

The Sun continues to contract but at a slower pace than before. Meanwhile, the surrounding material in the outer nebula begins to condense, and an important separation takes place. The original cloud would probably have been fairly uniform in its composition, and the contraction thus far would probably not have produced any separation of the different materials. However, at any given temperature, some materials will be able to condense, but others can only remain gaseous. Because of this and the marked temperature gradient in the nebula from the center out, we find that different materials condense at different distances.

Materials that can condense at high temperatures are termed **refractory**, and are mostly compounds of silicon and oxygen with calcium, aluminum, magnesium, iron, sodium, and sulfur. Apart from oxygen, these are not the most plentiful chemical elements. The most abundant atoms are those of the elements hydrogen, helium (which does not normally form chemical compounds), carbon, nitrogen, and oxygen. Compounds of these elements can condense only at low temperatures, and they are termed **volatile**. These compounds include ammonia ($NH_3$), water, and methane ($CH_4$). Their solid forms are often collectively termed **ices**.

Calculations by A. G. W. Cameron of Harvard College Observatory and by John Lewis of the University of Arizona have shown that the compositions of the planets can be understood from a knowledge of the condensation properties of the different materials under the different conditions of temperature and pressure encountered at various positions within the evolving nebula (Fig. 18.6).

For example, at Mercury's orbit, iron, nickel, and magnesium silicate ($MgSiO_3$) can condense, but the temperature would still be too high (1500 K) to allow iron sulfide (FeS) or iron oxide (FeO) to solidify. The temperature in the nebula would be around 600 K at the Earth's orbit, cool enough to allow FeS and the heavier Fe and Ni to condense. Much further out in the region of the Jovian planets, ices can condense.

This type of model thus provides an understanding of the very different compositions of the terrestrial and Jovian planets. The differences between the planetary masses can also be understood. Those materials that can condense in the high-temperature part of the nebula are the refractories, and they are made of the less abundant elements. There will be less of those materials than hydrogen, and planets made from them can only be small—as is observed for the terrestrial planets. On the other hand, the volatile compounds that can only condense at low temperatures are abundant and so the planets they produce can be large—as is observed with Jupiter, Saturn, Uranus and Neptune.

The formation of the planets may have proceeded in stages, with the first stage being governed by the condensation temperatures leading to small particles and then to larger particles (planetesimals) that in turn collide with or are gravitationally attracted to others to form protoplanets. By this time, those bodies closest to the evolving Sun will have lost their volatile con-

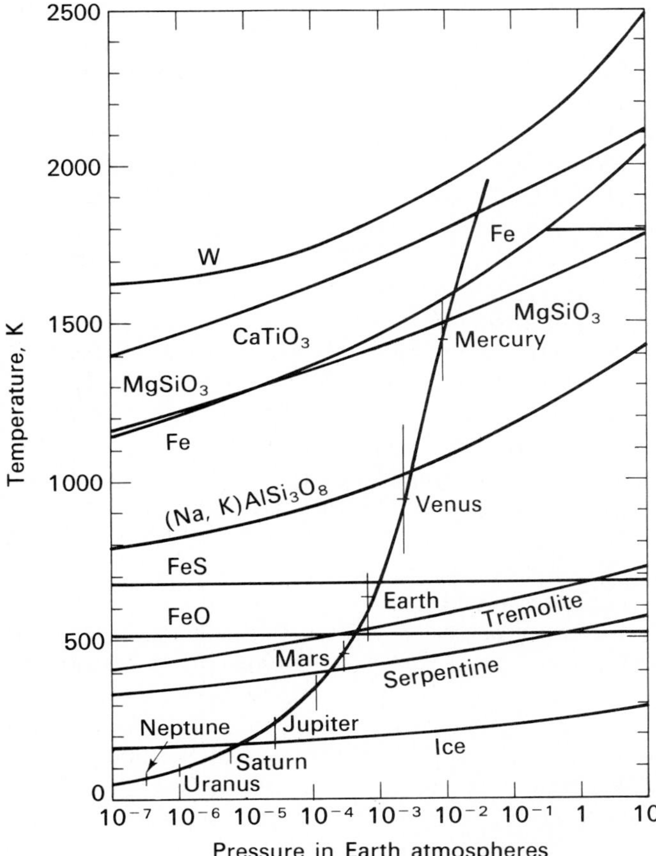


**Figure 18.6** Condensation temperatures for different substances. As the solar nebula cools and the planets begin to form, the temperature and pressure decrease steadily from the center of the nebula outwards. Since different materials can condense only under different conditions of temperature and pressure, the compositions of the newly forming planets will vary in a progressive way from the center out. This diagram shows the condensation temperatures and pressures for a variety of substances that are found in the solar system. Iron (Fe) is represented by a line that is lower than that for calcium titanate ($CaTiO_3$), which means that there will be places where the conditions will allow calcium titanate to condense but where it will still be so hot that iron can exit only as a gas. The iron sulfide (FeS) line is at much lower temperatures and ice even lower. Conditions for the evolving planets are shown along a curving line from the lower left. Thus at the orbit of the Earth, it is too hot for ice to form but cool enough for iron sulfide to condense. At the orbit of Mercury, it is so hot that only a few substances can solidify, but these do not include iron sulfide or oxide. On the other hand by the orbit of Jupiter, it is cool enough even for ices to solidify (Based on a diagram by John S. Lewis, University of Arizona).

stituents, and the giant planets Jupiter and Saturn may have developed to the stage where their own gravitational attraction could capture more matter directly. Jupiter's large mass could perturb the orbits of planetesimals in the asteroid belt sufficiently to prevent their cohering into a single planet.

**BOX 18.2**
*Gravitational Energy*

When the concept of energy was introduced (in chap. 8), it was as a property of moving objects. This form of energy is **kinetic energy**. If an object is raised above the floor and released, it will fall. By the time that it strikes the floor, it will have gained speed and have kinetic energy. This kinetic energy is a result of the gravitational attraction between the object and the Earth. On striking the floor, the falling object loses its kinetic energy, which is mostly converted to heat energy.

In a similar way, heat energy is produced when gravitational attraction causes a cloud of gas and dust to condense to form a star. As material is attracted toward the center of the cloud, it accelerates. During the acceleration, **gravitational energy** is being converted into kinetic energy. Collisions between the falling material and the growing core convert the kinetic energy to heat energy, some of which will be radiated away.

It is thought that before the Sun started to consume its nuclear fuel, it went through a brief phase in which gravitational energy was rapidly released as radiation, effectively sweeping the developing solar system clear of most of the uncondensed matter. After this, the Sun became a stable star and the planets were left to cool slowly. Their atmospheres evolved as a result of surface and internal geologic changes. Finally, residual rocks too large to be swept out by solar radiation were gravitationally attracted towards the Sun and produced the heavy cratering seen on planets and some of their satellites with which they collided. As material condensed to form the planets and their moons, chemical reactions took place. Some of the reaction products were solid, others were gaseous and escaped.

Many details in this type of model remain unsettled. For example, it is not yet possible to decide which of two processes is the faster: the gravitational attraction (accretion) or the chemical reactions. Both possibilities have been explored through calculations, but the observational data do not yet allow a clear choice to be made.

Angular momentum remains a problem. The rotation of the condensing solar system should speed up, and in any event it is surprising that so little of the final momentum should reside in the Sun. In the form outlined, the evolutionary model does not provide an answer. Solutions have been sought that use the Sun's magnetic field and rotation to transfer angular momentum to objects in outlying regions of the solar system, some of which have since escaped. There is disagreement among theorists about the size of the initial cloud—some models take an initial mass much larger than one solar mass, but then they have to cope with the problem of getting rid of much extra mass. The time scale for the various stages is not well established, except that it is known that there could be only a brief interval of no more than 100 million years for the meteorites to condense.

Although it is easy to imagine that condensation depends on distance from the Sun, a simple picture would suggest condensation into a circular ring, perhaps a ring of planetesimals. Computer simulations of hypothetical solar systems show that a collection of many planetesimals will gradually coalesce into a smaller number of larger planets. One particular model (Fig. 18.7) shows that after 79 million years, there will be eleven surviving planets out of an initial 100, and after 151 million years there will be only six remaining. This process depends on the relative speeds of smaller bodies when

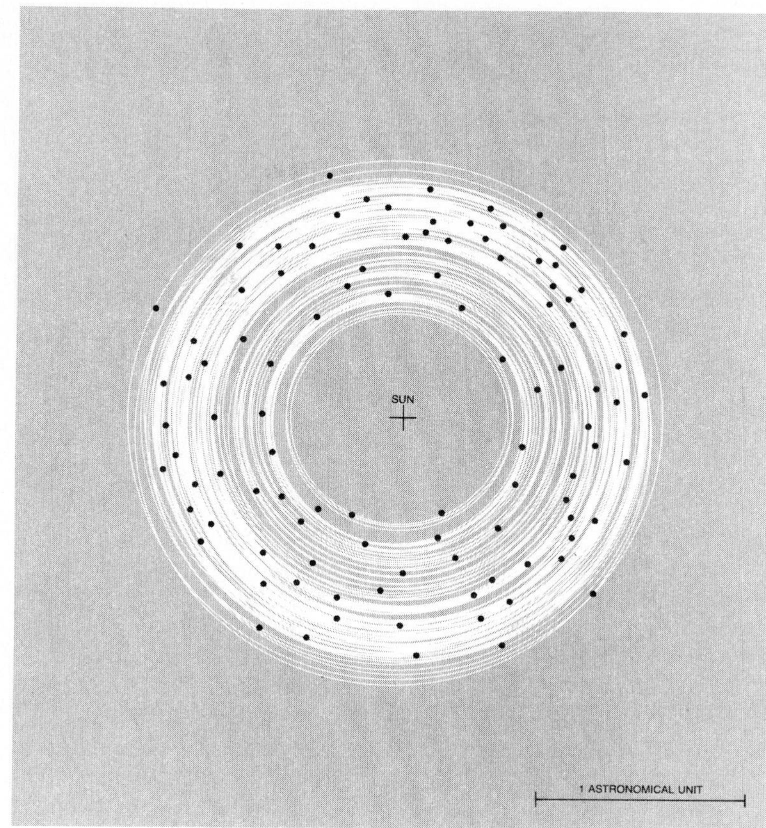


(a)

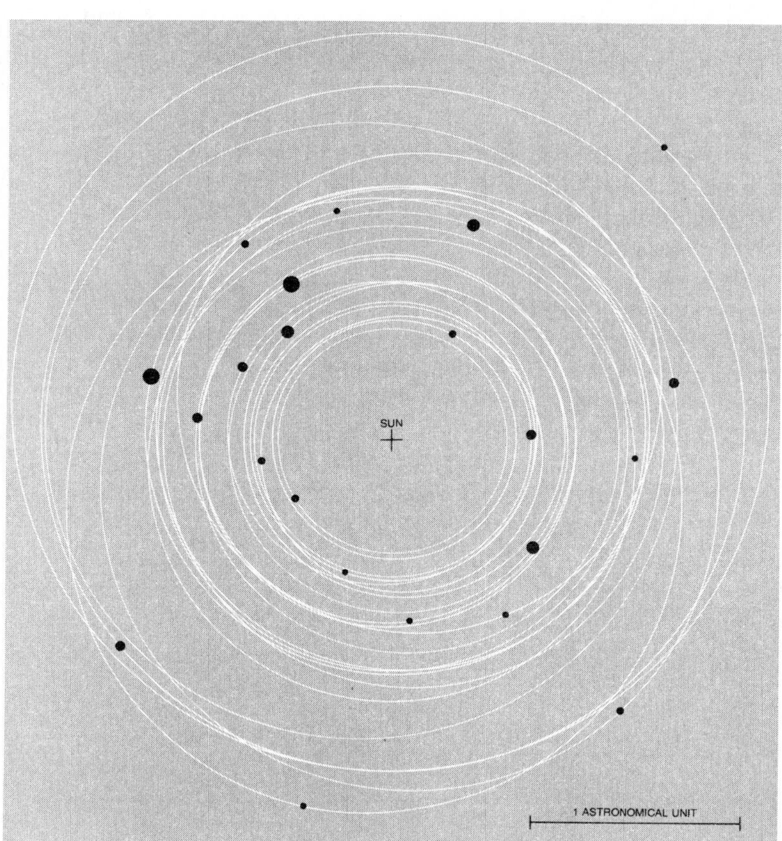


(b)

**Figure 18.7** Diagrams drawn from a computer simulation of the formation of the inner solar system, carried out by George W. Wetherill of the Carnegie Institution in Washington. (a) The simulation started with 100 planetesimals having a total mass equal to the present terrestrial planets and their moons. (b) After 30.2 million years, accretion through collision has reduced the number of orbiting bodies to twenty-two. Orbits are now noticeably elliptical (George W. Wetherill "The Formation of the Earth from Planetesimals," June 1981 *Scientific American*).

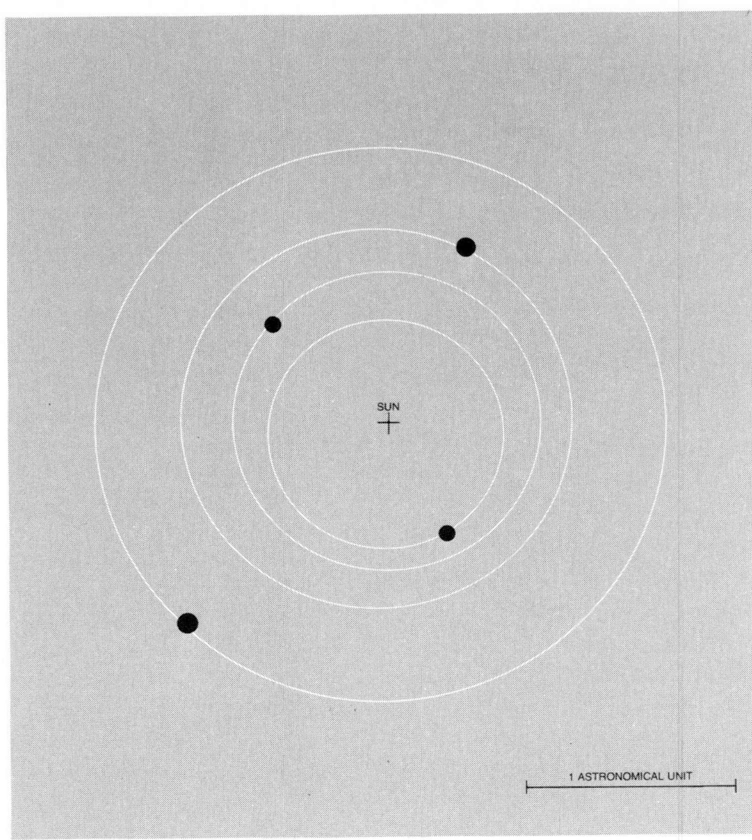


(c)

**Figure 18.7 (Cont.)** (c) After 441 million years four planets remain in almost circular orbits (George W. Wetherill "The Formation of the Earth from Planetesimals," June 1981 *Scientific American*).

they collide. If the collision speed is less than about twice the mutual escape speed, the bodies can stick together, but more violent collisions will probably result in fragmentation.

The formation of the smaller bodies of the solar system can be accommodated in the evolutionary model. The minor planets might represent pieces from a larger planet that was perhaps fragmented before reaching full planetary size, or perhaps they never grew any larger than they now are. More is known about the meteorites: Their structure shows that most probably once were parts of larger planetisimals. On the other hand, the great diversity in properties of planets and their satellites complicates the picture. How could the Earth and Moon have condensed close together in the evolving nebula yet have their observed differences? How could the Galilean moons differ so from one another and from Jupiter?

The evolutionary model incorporates reasonable explanations for dynamical properties 1 through 5 (Table 18.1), the chemical properties 1, 3, and 4, and the third of the miscellaneous items. This is sufficiently encouraging to be worth pursuing. The use of computers allows the consequences of different assumptions to be followed in great numerical detail, but progress may depend more on the availability of data, and that in turn will be remedied only slowly. Still, we can feel very pleased with what has already been accomplished—simply compare the current models with those of the sixteenth or seventeenth centuries.

## CHAPTER REVIEW

Properties of the solar system that a theory of origin must deal with:

**Dynamical properties:** orbits of planets are nearly circular and coplanar with the Sun's equatorial plane—

directions of motion in orbit are the same for the planets.

- angular momentum: mostly carried by the planets, very little by the Sun.

**Physical and chemical properties:**

- chemical composition of planets, atmospheres, moons.
- age of the solar system: 4.6 billion years.
- temperatures, pressures, and densities at solidification, as shown by structure of planets and meteorites.
- isotopic abundances.

**Other features:**

- systems of satellites and rings.
- minor bodies: comets, meteorites, and minor planets.
- bombardment phase in early solar system.

Models for origin and evolution of the solar system fall into three classes:

**Catastrophe model:** material for the planets came from the Sun.

**Capture models:** Sun and planets formed in different places; the planets were later captured by the Sun.

**Evolutionary models:** the bodies of the solar system all formed from the same original cloud of gas and dust.

Only evolutionary models are seriously considered today, and an evolutionary sequence can explain many of the observed features. The central idea in this model is the contraction and cooling of a large nebula. Materials with different melting points will solidify at different distances from the Sun. Problems include angular momentum and detailed compositional differences between various bodies.

## NEW TERMS

| | | | |
|---|---|---|---|
| **angular momentum** | **cosmogony** | **gravitational energy** | **refractory** |
| **capture model** | **cosmology** | **kinetic energy** | **volatile** |
| **catastrophe models** | **evolutionary models** | | |

## PROBLEMS

**1.** In the seventeenth and eighteenth centuries, scientists tried to construct theories for the origin of the solar system. What sorts of scientific information do we now have that they did not then know about and so did not include in their theories?

**2.** What are the weaknesses of theories of the origin of the solar system that assume that the planets formed from matter drawn out from the Sun?

**3.** What are the major features and properties of the solar system that must be included in a satisfactory theory of its origin?

**4.** Outline the present theory for the formation of the planets in the solar system.

**5.** The terrestrial and Jovian planets differ in many respects. How does the nebular theory explain their major differences in average density and composition?

**6.** What explanation does the nebular theory offer for the difference in masses between the terrestrial and the Jovian planets?

**7.** A large cloud of gas and dust rotates once in a month. It starts to contract. How will its rotation period have changed by the time the cloud has contracted to half its original size if the contraction is uniform?

**8.** Compare the angular momentum of two systems with the same mass:

a. a sphere of radius $R = 10^{11}$ cm that rotates once in a month and

b. an equal mass in a circular orbit with radius $10^{11}$ cm, taking 1 month to make a complete revolution.

**9.** In what ways do the terrestrial planets differ from the Jovian planets?

**10.** You are highjacked on a space flight and land on a cratered surface. Your luggage, however, has gone on to another destination, so you have no instruments with you. How could you tell whether you have landed on the Moon or on Mercury?

**11.** Which are the four largest satellites of the planets? Make a table of their diameters and masses, and compute their average densities.

**12.** Which planets have rings, and which have none?

**13.** Which planets have satellites, and which have none?

**14.** What are some of the differences between craters seen on Mercury, the Moon, and Mars?

CHAPTER

19

# The Stars: Distances, Motion, and Brightness

In choosing to study the solar system first, we have followed the historical sequence. It was the conspicuous mobility of the planets against the unchanging background patterns of the stars that attracted the attention that developed into scientific astronomy. Stellar astronomy languished. The stars did not seem to move, and they twinkled gently but inscrutably.

In a world where myths played a greater role than they do today and where the night skies were not brightened by city lights, the ancients saw patterns among the stars and gave them names that are still used as convenient designations for the regions of the sky. The original constellations still in use are those that could be seen from about 35° north latitude (Fig. 19.1). New constellations were added when European adventurers pushed south. In 1928 the International Astronomical Union tidied up the boundaries of the eighty-eight constellations. These names provide one of several ways in which stars' names are assigned. Thus the brightest star in the constellation Orion is designated α Orionis (the first of Orion). The second brightest is labeled β Orionis and so on, but there are too many stars for this method to be used for more than a tiny fraction.

Catalogs have been compiled for stars having various properties, and some stars are known by their catalog numbers. Some stars were individually named—α Ori is more widely known as Betelgeuse—and many of these names are still used, so that some stars have a name as well as several catalog numbers.

Some of the stars' names (such as Sirius) come from the Greeks. Others such as Betelgeuse, Aldebaran, and Fomalhaut survive in sometimes distorted forms from the Arabic. Some have been named for astronomers who made particular studies of them: Barnard's star and Ross 614 for example. Others are known by their places in catalogs of stars.* But the names of the stars tell us nothing of their nature. How can we study them, now that we have an arsenal of modern scientific weapons at our disposal? And, when we examine them, what do we find? These are the topics we shall cover in the next six chapters.

As the skies were mapped at radio, visible, IR, and X-ray wavelengths, one task has been the identification of stars from one catalog with their counterparts in other catalogs. There is a catalog of catalogs that helps with this. Identifying objects between catalogs is not always a simple task. Sometimes the position of an X-ray object, for example, might not be well defined, and within the uncertainties in its position there might be several stars. Further measurements would then be needed before a positive identification could be made.

Contrary to our first visual impressions, the night sky does *not* have millions of stars. The catalog that Ptolemy inherited from Hipparchus and then extended, listed only slightly more than 1000 stars, and there are only about 6000 stars bright enough for the eye to perceive. Now, by using large telescopes, we estimate that there are around 100 billion stars in the assembly we call our galaxy.

To make progress beyond simply noting the position and brightness of a star, we need to know the star's distance and we must examine its spectrum so that we can try to identify the physical processes that are taking place. We would like to know the star's mass and size, its surface temperature, and the composition of its outer layer whose radiation we can detect. Studying a large number of stars and combining our observations with theoretical calculations, we can try to understand the structure and evolution of stars of different types. We can study their groupings, in close pairs or in threes or fours, in clusters of hundreds or thousands, and in billions to make up a galaxy. Everywhere we shall be looking for systematic trends—in sizes, temperatures, composition—from which we can develop models.

* You may sometimes see an advertisement that carries the offer to name a star for you and provide a certificate—for a fee, of course. Do *not* be taken in. Star names cannot be bought.

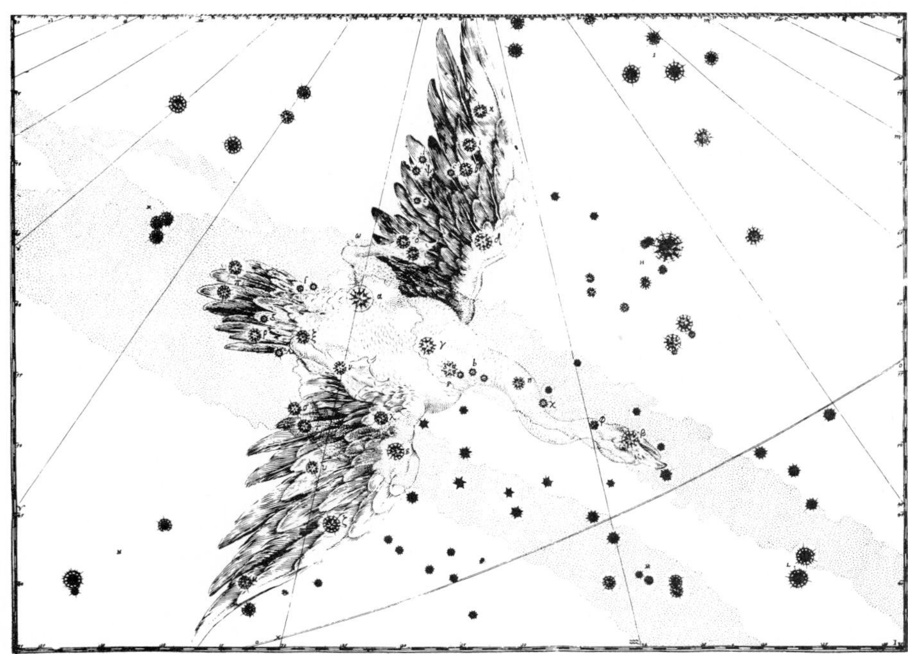

**Figure 19.1** Constellation Cygnus (the Swan), as depicted in Bayer's "Uranometria," 1655 (Yerkes Observatory).

Stellar astrophysics has been enormously productive, and the emergence of new techniques including space astronomy has revolutionized our understanding. We shall start our survey with a review of the techniques and then move to the pictures that have emerged.

## 19.1 STELLAR DISTANCES: MEASUREMENT OF PARALLAX

The simplest means of measuring the distance between two points is to use a ruler or tape measure, but this approach rapidly becomes impractical and then impossible when the distances are large and the places inaccessible. Surveyors, therefore, use a technique based on a few measurements of distance, together with the measurement of angles and the application of geometry and trigonometry. When the baseline *AB* and the angles at *A* and *B* (Fig. 19.2) have been accurately measured, the remaining dimensions of the triangle can be easily computed. The point *C* need not be reached in order to find its distance from *A* and *B*. This is basically the method that we use all the time with our eyes, although we rarely stop to think of it. Our stereoscopic vision is remarkable. When we look at an object, our eyes rotate so that the two lines of sight converge on the desired point, and automatically the image is kept in focus. By experience, we learn to judge distances even though we have probably never measured the baseline for this surveying (the distance between our eyes). The same system is used in cameras with rangefinders, with the lens having an engraved scale so that the distance can be read directly.

Clearly, the accuracy of this method depends entirely on the accurate measurement of the baseline and of the angles at *A* and *B*. In astronomy, the distance to any star (*C*) is much larger than any baseline that can be found on Earth. The angles at *A* and *B* are, therefore, each very close to 90° and hard to measure with accuracy. A different approach must be used based on the phenomenon of **parallax**.

Hold a pencil in your outstretched hand, and view it from each eye in turn against a more distant background. The pencil will appear to move to the left when you use your right eye, and to the right when you use your left eye. We can achieve the same effect astronomically if we photograph a star against a background of more distant stars, and do this from two widely spaced positions. The size of the Earth itself is too small to provide an adequate baseline; the best we can do is to take our photographs 6 months apart when the Earth passes through positions on opposite sides of its orbit around the Sun (Fig. 19.3). Viewed in this way, the closest stars will appear to move from side to side with an annual occupation of each extreme position. This method requires the very precise measurement of stellar images on a photographic plate so that the minute changes in position can be reliably detected.

The maximum length of the baseline for the celestial triangle is 2 AU, the diameter of the Earth's orbit. A long and very thin triangle is then defined (Fig. 19.4), and the angle at *C* is measured. By convention, *one half* of this angle is defined as the parallax of the star. It is usually measured in arc seconds and denoted by *p*. This definition uses the average Sun-Earth distance of 1 AU for the baseline.

For a triangle with 1 AU baseline (length *BD*) and an apex angle of 1 arc sec (angle *DCB*), the distance (*CD*) of the apex from the base is very

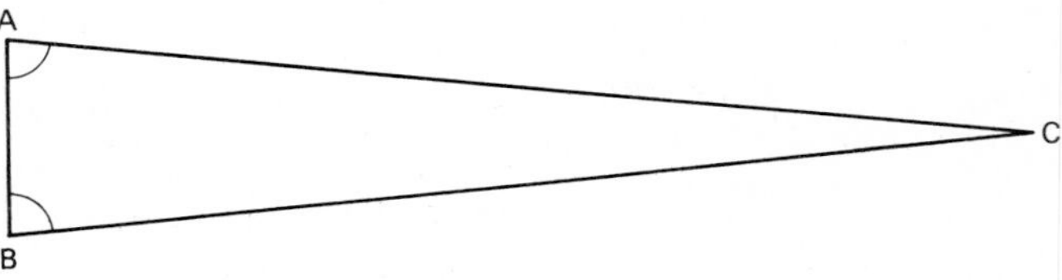


**Figure 19.2** Measurement of distance to a remote point (*C*) by triangulation. Distance can be determined from measurement of two angles and the length of the baseline.

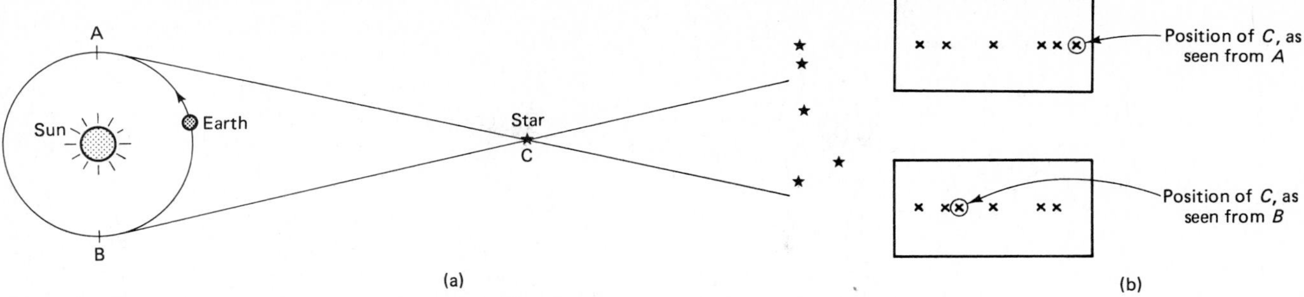


**Figure 19.3** Measurement of distance to a star by parallax (a) As the Earth moves in its orbit around the Sun, a closer star will appear (b) to move relative to more distant stars.

closely the same as the length of the long sides, and this length is defined as one **parsec**. For an apex angle of 1″, it can be shown that the length of those long sides is 206,265 times the length of the base. Thus

$$\text{distance } DC = 206265 \times \text{distance } DB$$

$$1 \text{ parsec} = 206265 \times 1 \text{ AU}$$

$$= 206265 \times 1.5 \times 10^{13} \text{ cm}$$

$$= 3.09 \times 10^{18} \text{ cm}$$

The parsec is a convenient unit of length for the distances between stars. Larger units are often used for greater distances. Thus 1000 pc = 1 kiloparsec = 1 kpc, and 1 million pc = 1 megaparsec = 1 Mpc.

There is also a popular unit, the light year, which is the distance traveled by light in 1 year, given by

$$1 \text{ LY} = (\text{speed of light}) \times (\text{number of seconds in 1 year})$$

$$= (3 \times 10^{10} \text{ cm/sec}) \times (3.16 \times 10^{7} \text{ sec})$$

$$= 9.48 \times 10^{17} \text{ cm}$$

Thus 1 pc = 3.26 LY.

In defining the parsec, an angle of 1 arc sec was used. Suppose we now consider a star that is farther away so that (for the same 1 AU baseline) the parallax (angle) is 0.5 sec. Clearly, the star is 2 pc distant. We can generalize this:

$$\text{distance in pc} = 1/(\text{parallax, in sec})$$

Thus a star that is 8 pc distance will have a parallax of $\frac{1}{8} = 0.125$ sec.

Measurement of stellar parallax is an exacting art that completely defeated the classical astronomers. Their inability to detect any parallax provided a major argument against the idea that the Earth was moving around the Sun. It was only in 1838 that the first parallaxes were measured inde-

**Figure. 19.4** The astronomical triangle. In practice, triangulation to stars involves very long and thin triangles, where the base is 2 AU in length (the diameter of the Earth's orbit) and the two long sides can be taken as equal. One half of the angle *ACB* is defined as the parallax. When angle *DCB* is 1 arc sec then, by definition, the distance *DC* is one parsec.

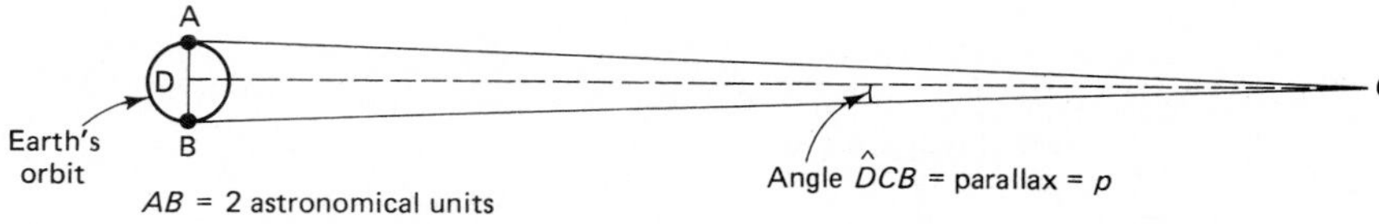

pendently by F. W. Bessel (61 Cygni; 0.35″) working in Konigsburg in Prussia, by Wilhelm Struve (Vega; 0.13″) in Russia, and by Thomas Henderson ($\alpha$ Cent; 0.74″) at the Cape of Good Hope.

**BOX 19.1**
*Measurement of Parallax*

In chapter 9, the scale of images produced by a telescope of focal length $F$ was given as

$$s = 1.75 \times 10^{-2} F \text{ cm/deg}$$

where $s$ is the plate scale, in centimeters per degree on the sky and $F$ the focal length in centimeters. For a parallax of one arc sec, $\frac{1}{3600}$ degree, the scale is

$$\frac{1.75 \times 10^{-2}}{3600} F \text{ cm/arc sec} = 4.8 \times 10^{-6} F \text{ cm/arc sec}$$

**Astrometric** measurement thus requires a long focal length ($F$), for example, around 15 $m$ = 1500 cm, so that each arc second corresponds to about 73 microns ($7.3 \times 10^{-3}$ cm) in the photographic image. With this scale, parallaxes as small as about 0.02″ can be measured.

Relatively few stars are close enough for reliable parallax measurements to be made. Of the 100 billion stars in our galaxy, less than 10,000 have measured parallaxes. For only about 30 is the parallax greater than 0.2″ (distances less than 5 pc), and for about 1000 it is greater than 0.05″. Direct measurement of parallax less than about 0.02″ is not presently reliable, and other methods must be used to deduce distances of the more distant stars.

A final comment is in order. We can use parallax *only* because some stars are much closer than others. If all stars were very far away (further than about 10,000 pc) or if all stars were at the same distance (no matter how far or close, such as on the classic sphere of the heavens), then there would be no stellar parallax to be observed. We would have great difficulty in measuring distances and would probably have made little progress towards understanding the structure and behavior of stars and more distant matter.

## 19.2 OTHER METHODS OF MEASURING DISTANCES

Several different methods have been developed for measurement of distances to stars that are too far to display a measurable parallax. Starting with the stars whose distances are known from their parallaxes, we can develop other methods and then extend them to greater distances, but without the parallax data for a base, we would have only a poor calibration for these other methods.

Some methods can be used only with certain classes of stars and we can review them further on. One type of measurement is so important, though, that we shall describe its main features here.

This method uses a standard light, that is, an object whose intrinsic brightness or luminosity is known. Suppose we take a standard light that radiates a known amount of energy, $L$ erg/sec. Placed at a distance $D$ cm from us, the measurable flow of energy will be $L/(4\pi D^2)$ erg/cm$^2$ sec (Fig. 19.5). What we actually measure is that composite quantity; if we already knew $L$, the only remaining unknown would be $D$, which could then be easily calculated.

The method hinges upon knowing the nature of the standard light and thus its luminosity, $L$. Astronomical standard lights come in several different types. They can be identified by spectral analysis or by the recognition of

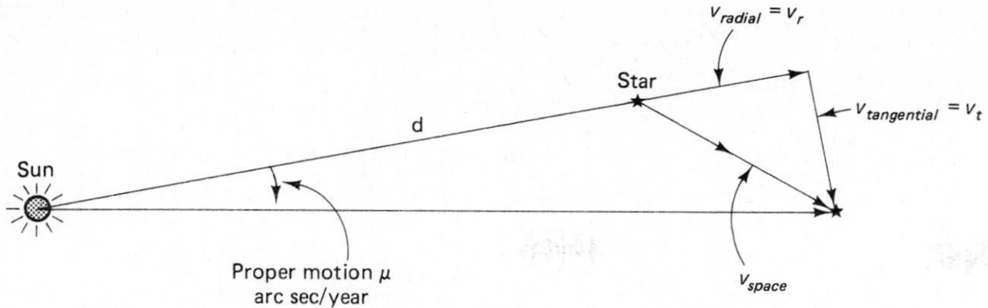


**Figure 19.5** Stellar motion. A star that moves with a true (space) velocity $v$ will appear to have two components of its movement. One component is along the line joining us to the star. This is the radial component, $v_r$. It can by measured by the Doppler shift in spectral lines (if the speed is large enough). The other component is $v_t$, the transverse component at right angles to the line of sight. This directly, cannot be measured but a related quality, can sometimes be measured, the proper motion, which is the rate at which the direction to the star seems to move. The proper motion is measured in arc sec/year and can be related to $v_t$ if the distance to the star is known.

the regular patterns of variations of luminosity that some stars display. Once the standard has been identified, the limit to the method is set by the faintness of the objects we can detect. Modern cosmology would not exist as a scientific discipline without this method.

## 19.3 THE MOTIONS OF A STAR

On many occasions, we have referred to the fixed stars and used their apparent immobility to gauge the conspicuous movements of the planets and the parallax of a relatively small number of stars. In general, however, even the *fixed* stars are moving. Because of their great distances, their changes in position are so slow as to be barely detectable. Over thousands of years, their movements become more noticeable and the appearances of constellations and groups slowly change. Far in the future, the Big Dipper will seem different.

Observations of stellar movements are made from an Earth that is always in motion around the Sun. When discussing the movements of stars, it is a simple task to remove this annually varying contribution to the apparent stellar motion, since the Earth's speed is well known. Accordingly, when we refer to stellar motions we shall be referring them to a stationary Sun. (The Sun itself is in orbit around the center of our galaxy. For some purposes, we will therefore need to make an additional correction to the speeds, but we can leave that until later.)

The movement of a star can be in any direction in space. From Earth the movement will appear to be made up of two components: the **radial** component along the line of sight towards or away from us, and the **transverse** or **tangential** component at right angles to the line of sight (Fig. 19.5).

Radial motion will not change the star's apparent position in the sky. If the radial velocity is large enough, a measurable Doppler shift will be imposed on the wavelengths found for spectral lines, so that we can tell both the speed and whether the star is approaching or receding from us (Chap. 8). We do not need to know the star's distance in order to measure its radial velocity, but that is not the case for transverse velocity.

Transverse movement appears to us as a change in a star's position. Whereas a star's position may change on an annual and cyclic basis because of parallax, a transverse velocity relative to the Sun produces a persistent drift in one direction. Experimentally, then, the transverse and parallax movements need to be disentangled. The transverse movement is termed the **proper motion** ($\mu$), and it is an angular speed measured in arc seconds

**Figure 19.6** Proper motion of Barnard's star is shown by photographs made on July 31, 1938 and June 24, 1939. The two plates were superimposed, then shifted horizontally to make this print. Other stars, which do not show any proper motion, show up as pairs all at the same separation. Barnard's star's movement is easily seen and amounts to 10.3 arc sec. At 1.81 pc, this is one of the closest stars. (P. van de Kamp and Sarah Lippincott, Sproul Observatory, Swarthmore College)

per year. Barnard's star has the largest known proper motion with a value of 10″.25/yr (Fig. 19.6). To translate this to a transverse velocity, we need to know the star's distance. If the parallax is p arc seconds and the proper motion is μ sec/year, then the transverse velocity is

$$v_t = [4.74\ \mu/\mathrm{p}]\mathrm{km/sec}$$

where the constant 4.74 comes from the units chosen.

The true (or space) velocity of the star (in km/sec or cm/sec) can be found by combining the radial and transverse components (Fig. 19.7). In

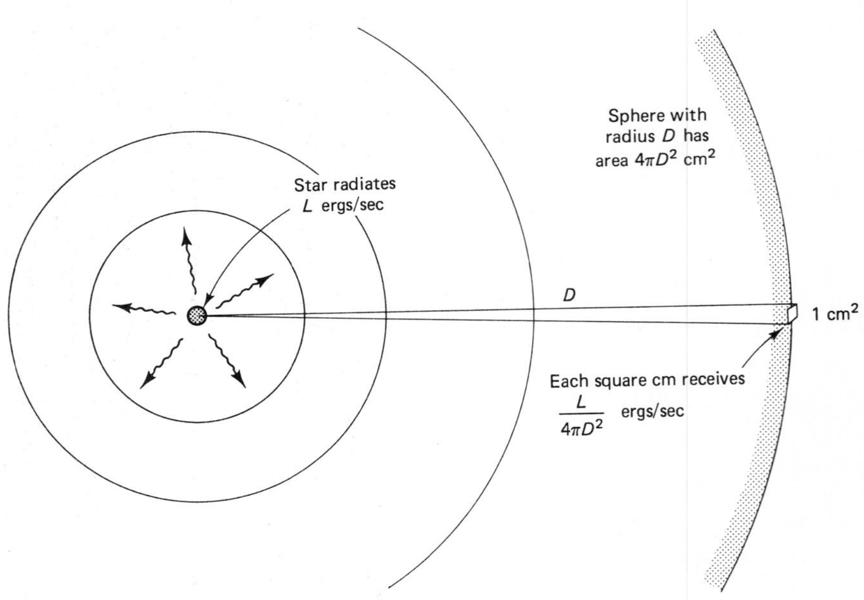


**Figure 19.7** Luminosity and radiation flux. If a star radiates $L$ erg/sec, then $L$ is termed the luminosity. This energy spread out as it is radiated. After traveling a distance $D$ cm, the radiated energy covers a spherical surface of $4\pi D^2$ cm$^2$, and the flux of the radiation at this distance $D$ from the source is thus $(L/4\pi D^2)$ erg/cm$^2$ sec. This is the amount of energy that arrives in each second on an area of 1 cm$^2$.

many cases, we may be able to measure only one component, $v_r$, so that we may not be able to calculate the space velocity.

The measurement of stellar distances and motions is affected by the Sun's own motion. Analysis of the redshifts and blueshifts observed in stars in different directions shows that the Sun is moving approximately in the direction of Vega (brightest star in the constellation Lyra). The speed of the Sun's movement is 20 km/sec or about 4 AU/year.

A result of this solar movement is that over several years parallax can be seen among stars that are otherwise too distant for direct parallax measurement. Those that are slightly closer appear to move steadily against those that are much farther away. Analysis of this effect for related groups of stars also involves measurement of their radial velocities and leads to an estimate of their **statistical parallax**, an average for the group. In this way, much greater distances can be measured than is possible with individual parallaxes.

## 19.4 STELLAR MAGNITUDES

Since the time of Hipparchus in the second century B. C., the brightest stars have been called first magnitude, and the fainter stars have been designated by larger numbers (third magnitude, sixth magnitude, and so on). This scale seems inverted, and if we were inventing the scale today, we would surely use a larger number for a brighter star. The use of the historical scale is so well established, however, that we do need to review it.

Hipparchus listed stars in six groups of brightness: there were fifteen stars of first magnitude; forty-five stars somewhat fainter at second magnitude; and larger numbers that were fainter down to sixth magnitude, the faintest he could see. With the use of telescopes much fainter stars can be seen. The early measurements, whether telescopic or by eye, were based on subjective comparisons until the introduction of the first photometers. By around the middle of the nineteenth century, techniques for measuring light intensity had improved, and it was found that the first magnitude stars seemed to be close to 100 times brighter than those of sixth magnitude (Fig. 19.8). It was then agreed to define the magnitude scale so that a difference

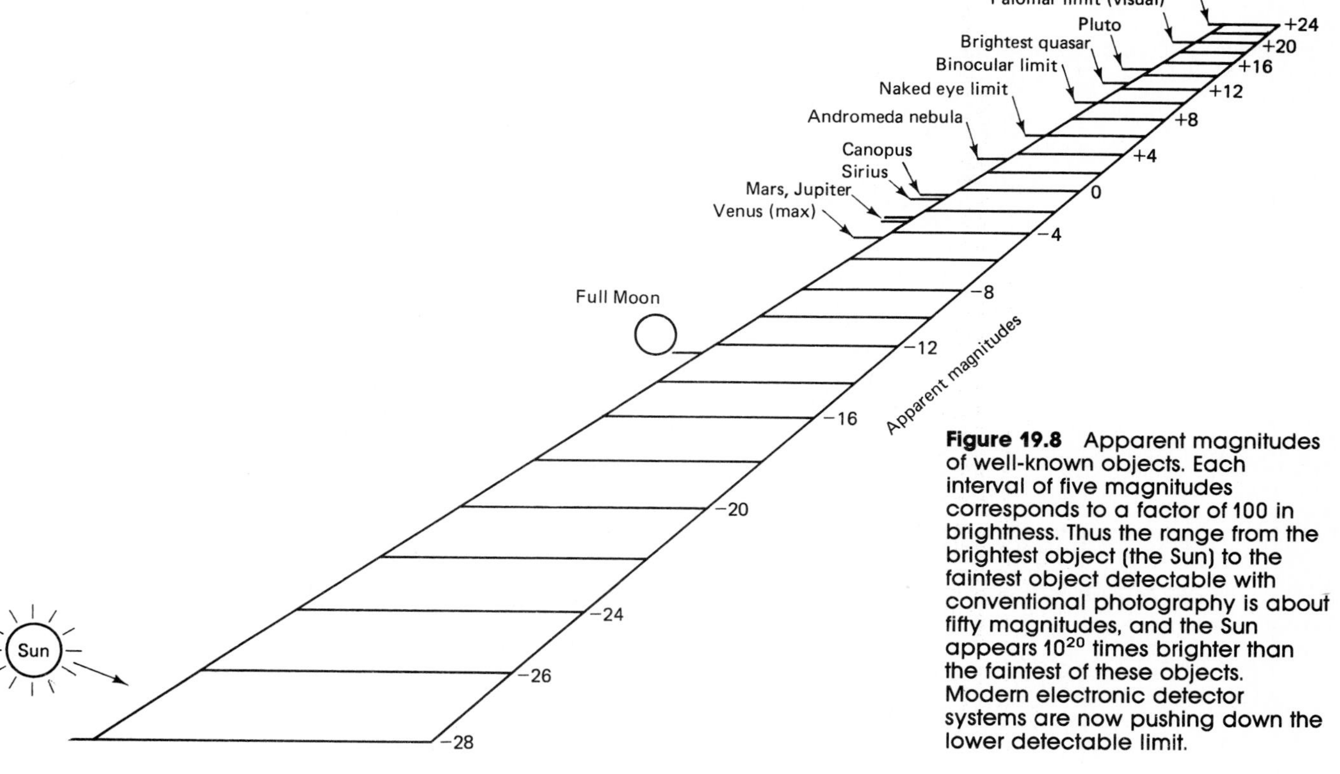


**Figure 19.8** Apparent magnitudes of well-known objects. Each interval of five magnitudes corresponds to a factor of 100 in brightness. Thus the range from the brightest object (the Sun) to the faintest object detectable with conventional photography is about fifty magnitudes, and the Sun appears $10^{20}$ times brighter than the faintest of these objects. Modern electronic detector systems are now pushing down the lower detectable limit.

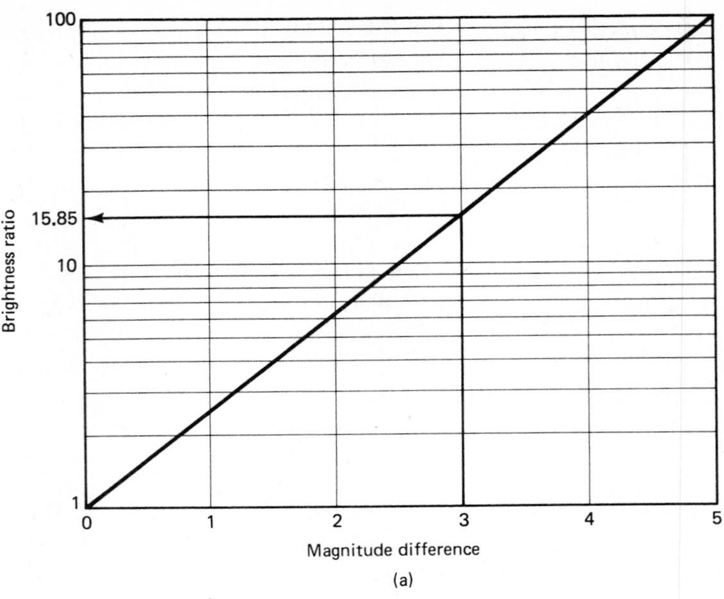


$10^{10}$
$10^9$
$10^8$
$10^7$
$10^6$
$10^5$
$10^4$
$10^3$
$10^2$
10
1
Brightness ratio
0 2 4 6 8 10 12 14 16 18 20 22 24 26
Magnitude difference
(b)

**Figure 19.9** Relation between brightness ratio and difference in magnitudes. (a) The magnitude scale is defined so that a difference of five magnitudes corresponds to a brightness ratio of 100:1. This is known as a logarithmic relation and is shown by the straight line on the graph. A difference of three magnitudes will, with this relation, correspond to a brightness ratio of 15.58:1. (b) For large brightness ratios and greater magnitude differences the line can be extended as shown.

of five magnitudes (first to sixth or eighth to thirteenth, for example) corresponded *exactly* to the ratio of 100 in brightness. It was also agreed to define the scale so that equal steps in magnitude always corresponded to equal ratios of brightness.

The conversion between the magnitude scale and brightness ratios can be accomplished by using the underlying formula (Box 19.2), by using the graph in Fig. 19.9, or by using the values in Table 19.1.

**BOX 19.2**
*The Magnitude Scale*

If we have two stars whose magnitudes have been measured to be $m_1$ and $m_2$, and their measured brightness are $l_1$ and $l_2$ (ergs/cm$^2$ sec), then the relation between the magnitude and brightness is

$$m_1 - m_2 = 2.5 \log(l_2/l_1)$$

This is shown as Fig. 19.9 and some representative values are also listed in Table 19.1.

Fractions of a magnitude or larger magnitude differences can either be calculated from the formula or (with less accuracy) read from the graph.

For example, if we find two stars whose magnitudes are $m_1 = 3$ and $m_2 = 7$, then their magnitude difference of 4 tells us that star 1 is forty times brighter than star 2.

Alternatively, if we know that one star is fifty times brighter than another, then we can use the graph to tell us that the corresponding magnitude difference will be 4.2.

When more accurate measurements were made, it was found that not all the first magnitude stars had precisely the same brightness, and the scale had to be extended into the negative numbers. Thus Sirius is the brightest star, with a magnitude of $-1.4$, and the southern star Canopus is next brightest at $-0.7$. Betelgeuse has a magnitude of $+0.41$ (usually the + is omitted but understood). Sirius is, therefore, $0.41 - (-1.4) = 1.81$ magnitudes brighter than Betelgeuse, and this corresponds to a brightness ratio of 5.3 : 1. The magnitudes of the stars are all positive with the only exceptions being Sirius, Canopus, α Centauri, and Arcturus. The planets can have magnitudes as negative (bright) as $-4$. The Moon ($-12.6$) and the Sun ($-26.7$) are obviously in a class by themselves.

*Table 19.1* MAGNITUDE DIFFERENCES AND BRIGHTNESS RATIOS

| *Magnitude Difference* | *Brightness Ratio* |
|---|---|
| 0.0 | 1 |
| 0.25 | 1.26 |
| 0.5 | 1.58 |
| 0.75 | 2.00 |
| 1.0 | 2.5 |
| 2.0 | 6.3 |
| 2.5 | 10 |
| 3.0 | 16 |
| 4.0 | 40 |
| 5.0 | 100 |
| 6.0 | 251 |
| 10.0 | 10,000 |
| 15.0 | 1,000,000 |
| 20.0 | 100,000,000 |
| 25.0 | 10,000,000,000 |

On first acquaintance, the magnitude scale seems both upside-down and unnecessarily complicated. It is, however, based on human perception, and it seems as though we are often more sensitive to ratios than to a scale of equally spaced steps. We can easily see that one object is twice as bright as another or hear that one sound is twice as loud as another. These differences are large enough for us to be able to remember, but much smaller differences cannot usually be recognized without a side-by-side comparison. So, when the magnitude scale came to be calibrated, it was found that the first magnitude stars were about twice as bright as those of second magnitude. Those of second magnitude turned out to be about twice as bright as those of third magnitude, and so on. The use of magnitudes is so deeply entrenched in astronomy that it is quite unlikely to be abandoned in favor of a more rational scale.

The magnitudes that we have been dealing with are more correctly termed **apparent magnitudes**, for they describe only the appearance of the objects and thus present a combination of intrinsic brightness and the dimming due to distance. For a catalog of stellar appearances that identifies the brighter and fainter objects, this is adequate. However, when comparing those objects' intrinsic properties in order to understand them, allowances must be made for the effects of their different distances. This can be done in two ways. The first introduces the idea of **absolute magnitude**.

Absolute magnitude is defined as the apparent magnitude that an object would display if it were located at a standard distance of 10 pc. Placing, in concept, all objects at the same standardized distance would permit a direct comparison (Fig. 19.10). An object that is actually closer than 10 pc would appear fainter if pushed back to 10 pc, and it would be assigned a larger

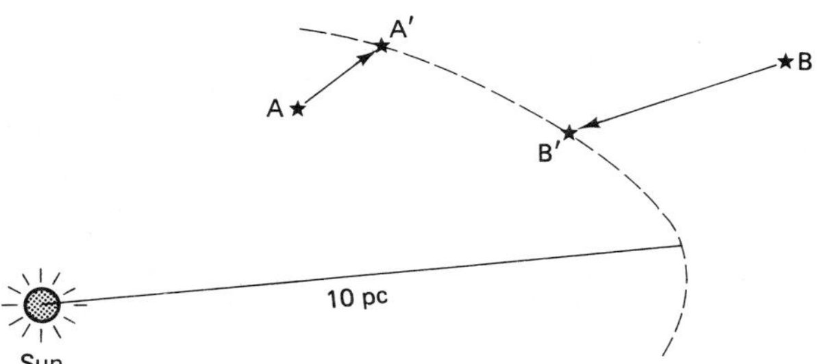


**Figure 19.10** Absolute magnitude represents the magnitude that an object would have if it were moved to be at a standardized distance of 10 pc. Thus an object at *A*, closer than 10 pc, would have to be moved back to be at 100 pc. It will thus appear fainter, and the numerical value of its absolute magnitude will be greater than its apparent magnitude. Another object *B* will have to be moved in making it appear brighter, and its absolute magnitude will thus be numerically less than its apparent magnitude.

absolute magnitude. Conversely, an object that is really farther than 10 pc will appear brighter (smaller absolute magnitude) if brought in to 10 pc.

The combination of inverse square law of brightness and the definition of the magnitude scale can be combined to compute the absolute magnitude, and the result is shown in Fig. 19.11. In this figure, distance (in pc) is plotted along the vertical axis, and along the horizontal axis we can read off the correction that must be applied to convert an apparent magnitude to its corresponding absolute magnitude. Thus Betelgeuse has an apparent magnitude of 0.41 and is at a distance of 150 pc. For 150 pc, we read off a correction of −5.88. The absolute magnitude is thus

$$M = 0.41 + (-5.88)$$
$$= -5.47$$

We recall that a negative number on the magnitude scale indicated a very bright object. Betelgeuse would appear to be as bright as −5.47 magnitudes if it could be placed at 10 pc from Earth.

We can make a similar calculation for the Sun, but it is so close (1 AU = $4.8 \times 10^{-6}$ pc) that it would need a separate conversion graph. If we make the necessary calculation shown in Box 19.4, we would arrive at the value of $M = +4.87$, far fainter than Betelgeuse if both were placed at the same 10 pc distance.

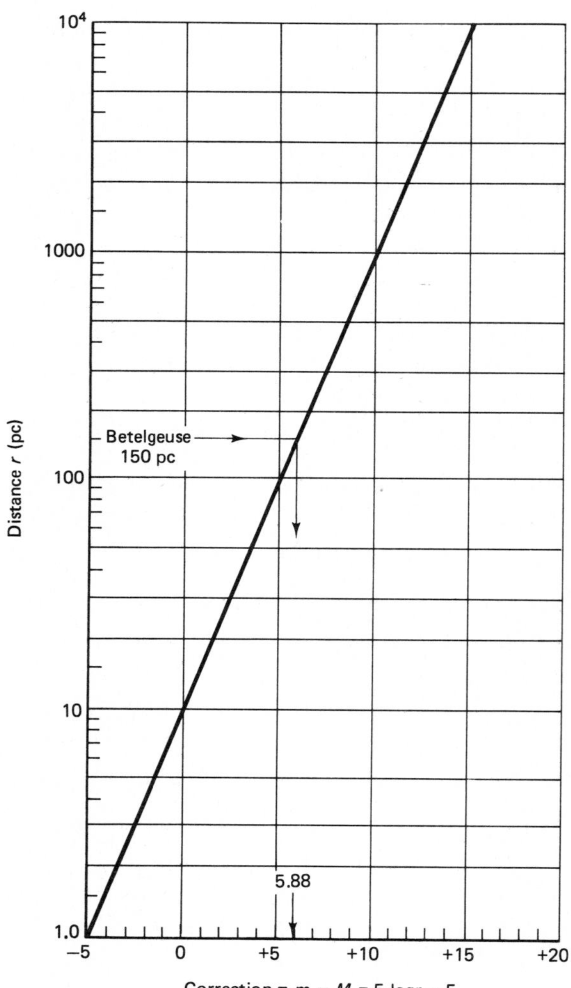


**Figure 19.11** The correction that must be applied to convert apparent magnitude to absolute magnitude requires knowledge of the star's distance. The straight line in this graph represents the distance modulus, the difference between apparent magnitude ($m$) and absolute magnitude($M$), in terms of distance $r$ (in pc). For example, Betelgeuse is 150 pc distant, and the distance modulus is thus 5.88 magnitudes.

**BOX 19.3**
*Absolute Magnitude*

The combination of the inverse-square law of brightness, with the definition of the magnitude scale leads to the relation

$$m - M = 5 \log r - 5$$

or

$$M = m + 5 - 5 \log r$$

where $m$ is the apparent magnitude of an object actually at $r$ pc distance, and $M$ is its absolute magnitude, that is, how bright it would appear if it were located 10 pc away. For example, Betelgeuse is at a distance of 150 pc and has an apparent magnitude of 0.41. Its absolute magnitude is thus

$$\begin{aligned} M &= (0.41) - 5 \log(150) + 5 \\ &= (0.41) - (5 \times 2.176) + 5 \\ &= -5.47 \end{aligned}$$

Similarly, the Sun's apparent magnitude is $-26.7$, but its distance of 1 AU is only (1/206525) pc $= 4.84 \times 10^{-6}$ pc. Its absolute magnitude then turns out to be

$$\begin{aligned} M &= (-26.7) - 5 \log(4.84 \times 10^{-6}) + 5 \\ &= 4.87 \end{aligned}$$

At their actual distances, the Sun appears much brighter than Betelgeuse. If both were moved to 10 pc, then Betelgeuse would be $5.47 + 4.87 = 10.34$ magnitudes brighter than the Sun.

**BOX 19.4**
*Conversion: Magnitude Difference to Brightness Ratio*

Comparing absolute magnitudes, Betelgeuse has been found to be 10.34 magnitudes brighter than the Sun. The ratio of their intensities can be computed as follows:

$$10.34 = 5 + 5 + 0.34$$

Each 5 magnitudes corresponds to a factor of 100 times, whereas the 0.34 magnitude corresponds to 1.37 (see Fig. 19.11 or Table 19.1). The overall ratio is thus $100 \times 100 \times 1.37 = 1.37 \times 10^4$.

The second way of comparing stars and allowing for their different distances is to use scientific units (ergs/sec) and to calculate the **luminosity**, which is defined as energy emitted per second. The magnitude scale is used very widely, but it does not tell us directly how much energy is detected. It is confined to comparisons, and even the absolute magnitudes do not introduce scientific units of energy.

If we measure the flow of energy from a star as E ergs/cm$^2$ sec at the Earth (Fig. 19.5), and if we know that the star's distance is $D$ cm, then the **luminosity** of the star can be calculated:

$$L = (4\pi D^2)\mathrm{E} \text{ ergs/sec}$$

We can use the Sun for an example. At the top of the Earth's atmosphere, at 1 AU from the Sun, the total flow of solar energy is $1.37 \times 10^6$ ergs/cm$^2$ sec. The luminosity of the Sun is thus

$$\begin{aligned} L_\odot &= 4\pi(1.5 \times 10^{13})^2(1.37 \times 10^6) \\ &= 3.8 \times 10^{33} \text{ ergs/sec} \end{aligned}$$

When observations at visual wavelengths are used for comparisons between stars or for the classification of stars, it is customary to use the magnitude scale. However, for other computations and especially when data from a much wider range of wavelengths is being used, it is generally more convenient to use luminosities.

**BOX 19.5**
*Distance Determination using a Standard Light*

Suppose that we measure the intensity of radiation from a distant star and obtain the value of $2 \times 10^{-3}$ ergs/cm$^2$ sec. Suppose, in addition, that the spectrum of the star allows us to identify it as one of the types which is known to have a luminosity of $3 \times 10^{39}$ ergs/sec, nearly a million times as luminous as the Sun. We can then calculate the distance to this star as follows:

$$\frac{L}{4\pi D^2} = 2 \times 10^{-3} \text{ ergs/cm}^2 \text{ sec}$$

so that

$$D = \frac{3 \times 10^{39}}{4\pi(2 \times 10^{-3})} = 3.45 \times 10^{20} \text{ cm}$$

$$= 112 \text{ pc}$$

## CHAPTER REVIEW

**Distances:** determined from measurement of parallax:

- distance (in pc) = 1/(parallax in arc sec)
- 1 pc = $3.09 \times 10^{18}$ cm
- other methods: use of known luminosity and measured brightness.

**Motions of stars:** radial motion—measured from Doppler shift in spectrum; transverse motion measured from changes in stellar position.

**Statistical parallax:** use of a group of stars and the Sun's own movement to measure distance to the group.

**Magnitudes:** 5 magnitudes difference corresponds to 100 times ratio in brightness.

**Apparent magnitudes** are those actually observed.

**Absolute magnitudes** are the magnitudes that would be seen if stars were at a standard distance of 10 pc.

**Luminosity** (ergs/sec) is the total radiant power of an object.

## NEW TERMS

**absolute magnitude**
**apparent magnitude**
**light year**
**luminosity**
**magnitude**
**parallax**
**parsec**
**proper motion**
**statistical parallax**

## PROBLEMS

**1.** What is the parallax of the following stars:

a. at a distance of 20 parsec?
b. at a distance of $3 \times 10^{19}$ cm?

**2.** What is the distance in centimeters to a star that has a parallax of 0.25 arc sec?

**3.** Suppose that no star were near enough to have a measurable parallax. How might we make a start in measuring distances to the stars?

**4.** Astronomical measurements of distance use the 2 AU diameter of the Earth's orbit as a base, and parallax angles as small as about 0.10 arc sec are measured. Suppose we could use the same method to measure distances on Earth, but with 3 in. (7.5 cm) distance between our eyes as the base. If we could measure a parallax as small as 1 arc sec, how great a distance could we measure?

5. Parallax is to be measured on photographs taken with a telescope that has a lens with a focal length of 10 m. How much will a stellar image move during the course of 6 months if the star has a parallax of 0.5 sec?

6. One parsec = 3.26 light years. The diameter of the galaxy is 30,000 parsec. How long does it take light to cross the galaxy?

7. Suppose a star at a distance of 2 pc from the Sun has a planet in an orbit with a 50 AU radius. What parallax would the Sun appear to have to an observer on that planet?

8. Suppose that stellar parallax were measured from Jupiter instead of from Earth. Barnard's Star is 1.83 pc from the solar system. What parallax would it appear to have?

9. What are apparent and absolute magnitudes? Why do we use both terms?

10. What are the absolute magnitudes of the following stars:

a. $m = 20$, $r = 100$ pc?
b. $m = 3.5$, $r = 5$ pc?

11. What are the distances for the following stars:

a. $m = 13.5$, $M = 15$?
b. $m = 5$, $M = 10$?
c. $m = -26.5$, $M = 5$?

12. At its brightest, Venus has an apparent magnitude of −4. Using Appendix I and list three stars that are brighter than $\frac{1}{100}$ of Venus' maximum brightness.

13. The faintest star visible to the naked eye has $6.5^m$. What is the apparent magnitude of a star that is 10,000 times fainter?

14. How far could the Sun be from the Earth and still be visible to the naked eye?

15. How much brighter is a $3^m$ star than one of $10.5^m$?

16. The Sun has an apparent magnitude of −26.5 and the full Moon −14. What is the ratio of their brightnesses?

17. Two telescopes have primary mirrors 1 m and 2 m in diameter. What is the difference in magnitude between the faintest objects that can be seen with each telescope during a given observing time?

18. What measurements are needed in order to separate proper motion from parallax?

19. Can a star have a measurable proper motion but zero radial velocity? Can it have a measurable radial velocity but zero proper motion?

20. A star has a transverse velocity of 200 km/sec and is at a distance of 25 pc. What is its proper motion?

21. Polaris moves by 0.9 arc sec in 20 years. Calculate its proper motion. At its distance of 300 pc, to what transverse velocity does this proper motion correspond?

22. Look at Fig. 19.6. If Barnard's Star did not have a measurable proper motion, how would its appearance be different on that photograph?

CHAPTER

# 20

# The Stars: Diameters and Masses

In the preceding chapter, we saw how the traditional observations of stellar positions and brightness have been refined. They now provide us with measurements of the distances to many stars and thus allow us to calculate the intrinsic brightness or luminosity. In the present chapter, we shall see how we can determine diameters and masses. Then in the next chapter we shall show how these various measurements can be combined, sometimes with the results of other observations, to lead to a systematic classification of the stars, an essential stage in the development of our understanding of these remote objects.

## 20.1 STELLAR DIAMETERS

We were able to determine the actual diameters of the Moon and planets by measuring their distances and their angular diameters. Then (Box 6.1), we have

$$\text{diameter (in cm)} = \frac{\text{(distance in cm)} \times \text{(angular diameter in arc sec)}}{206265}$$

This method cannot generally be applied to stars, however, for in all but a very few cases their angular diameters are too small to measure. Most stars appear to us as points of light, and determination of their diameters is based on the use of radiation laws we discussed in Chapter 9. Each square centimeter of a hot surface radiates $\sigma T^4$ ergs/sec; $\sigma$ is Stefan's constant, $5.67 \times 10^{-5}/\text{cm}^2 \text{ deg}^4 \text{ sec}$, and $T$ is the surface temperature. A star of radius $R$ will have a surface area of $4\pi R^2$ cm$^2$ and will therefore radiate

$$L_* = (4\pi R^2)(\sigma T^4)$$

$$\frac{\text{ergs}}{\text{sec}} = (\text{cm}^2)\,\frac{\text{ergs}}{\text{cm}^2\,\text{sec}}$$

where $L_*$ denotes the luminosity of the star (Fig. 20.1). We have already seen how luminosity can be calculated if a star's distance is known. This leaves two quantities to be determined, $R$ and $T$. There are several possible ways of determining the surface temperature $T$. Which method is used may depend on particular circumstances for each star.

If the star's continuous spectrum can be traced over a wide enough range of wavelengths (see Chap. 8), then the wavelength of maximum emission can be determined and Wien's Law can be used to calculate an effective temperature

$$\lambda_{\text{max}}\, T = 0.3 \text{ cm deg}$$

More generally, measurement of the apparent magnitude at several wavelengths can lead to a calculation of $T$ even if the peak in the spectrum cannot be observed (see Chap. 21). Other methods use more detailed spectral analysis.

The result is that $T$ can often be determined, and this leaves $R$, the star's radius, as the only unspecified quantity. It can then be calculated as

$$R = \sqrt{\frac{L_*}{4\pi} \cdot \frac{1}{\sigma T^4}} \text{ cm}$$

**BOX 20.1**

*Using the Radiation Law to Calculate the Radius of a Star*

Sirius, with an apparent magnitude of $-1.47$, is the brightest star. With a relatively large parallax of 0.377 sec, it is also one of the nearest stars. Its luminosity can be calculated to be about twenty-three times greater than the Sun or about $9 \times 10^{34}$ ergs/sec. From its spectrum, an effective temperature of 11,400 K can be derived. Its size can then be calculated

$$L = 4\pi R^2 \sigma T^4$$

so that

$$R = \sqrt{\frac{9 \times 10^{34}}{4\pi\sigma(11{,}400)^4}} \text{ cm}$$

$$= 8.6 \times 10^{10} \text{ cm}$$

Other methods of measuring stellar sizes make use of the wave nature of light. A characteristic feature of waves is their ability to **interfere** with one another. If the crests of two waves with the same wavelength arrive at

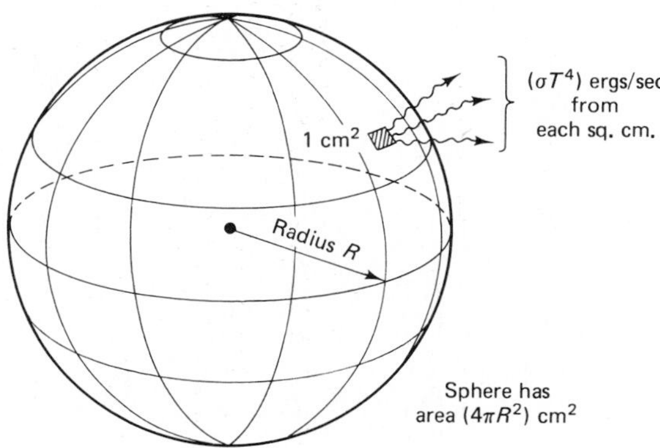


**Figure 20.1** Thermal radiation from a surface. Each square centimeter of a surface at a temperature $T$ radiates $\sigma T^4$ erg/sec. A spherical surface such as that of a star will, therefore, radiate $4\pi R^2 \sigma T^4$ erg/sec.

the same place at the same time, they can add to produce a wave of larger amplitude. On the other hand, if one wave's crest arrives at the same time as the other wave's trough, they can cancel totally or partially. Combinations of light waves can, therefore, produce complicated **interference patterns** of light and dark where different waves add or cancel.

Albert Michelson was the first to measure a stellar angular diameter using an interference method. Michelson had already had a distinguished career in optics, first at Annapolis, then in Germany, and later at the Case School of Applied Science (now part of Case-Western Reserve University). His measurements of the speed of light had set new standards of precision. Starting in the 1880s his search for small changes in the speed of light depending on the direction of motion of the Earth in its orbit around the Sun had failed to reveal even a minute change, and this null result was one of the major milestones along the path to Einstein's theory of relativity.

In the 1920s, Michelson and F. G. Pease devised a system to observe interference patterns produced by light waves coming from different regions across a stellar disc. Analysis of these patterns allows the determination of the angular diameter of a star, but the method is feasible in practice only for a few stars that are not very far away from Earth. Michelson and Pease were able to measure angular diameters for eight stars and found values in the range 0.02 to 0.056 arc sec.

A different type of interferometer is that designed by Hanbury-Brown and Twiss in Australia in the 1950s. Instead of using a single telescope they used two mirrors each 6.5 m in diameter moving on a circular track so that the separation between them could be adjusted up to a maximum of 188 m (Fig. 20.2). Each mirror focused the starlight onto a phototube, and the two sets of signals were combined electronically. Analysis of the resulting pattern of intensity variations allowed Hanbury-Brown and Twiss to measure stellar angular diameters as small as 0.001 arc sec.

Another interferometric method of measuring stellar sizes is based on occultations of a star by the Moon or a planet. As one of these relatively nearby bodies occults the star, the star does not disappear instantaneously. Instead, the light is reduced over several seconds, and during this brief interval an interference pattern of rapidly changing light and dark is observed (Fig. 20.3). The pattern depends on the angular size of a star.

**Figure 20.2** The stellar intensity interferometer at Narrabri Observatory, Australia. Each reflector consists of over 250 hexagonal mirrors nested to an overall diameter of 6.5 m. The reflectors can be moved around a circular track 188 m in diameter. Correlating the signals received at the two detectors permits stellar diameters to be measured (R. Hanbury Brown, University of Sydney).

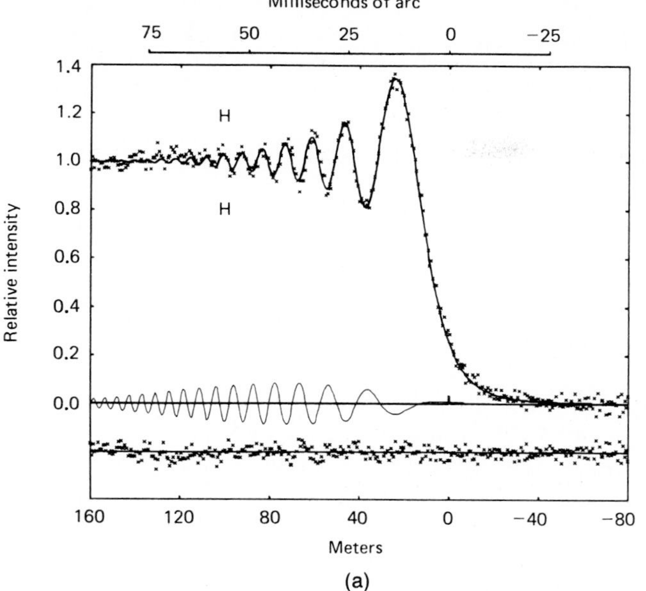


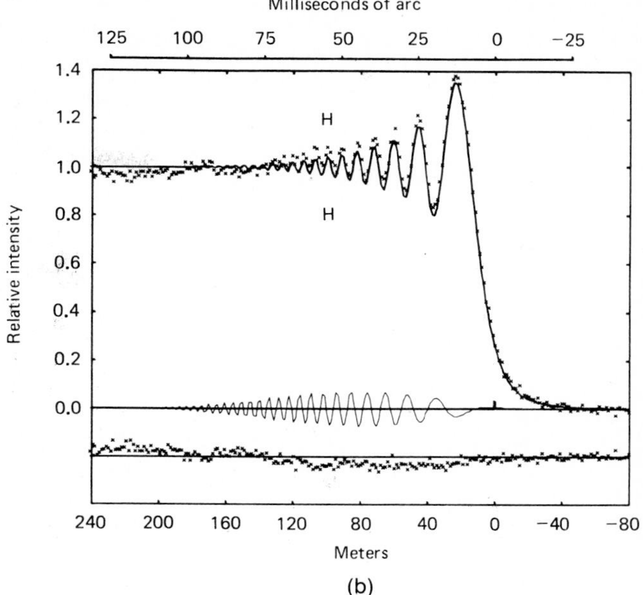


**Figure 20.3** Diffraction patterns produced by the lunar occulation of the stars SZ Sgr and $\gamma$ Lib, Analysis of these patterns indicates stellar diameters of 3.2 and 2.2 milli arc sec respectively (S. T. Ridgway, Kitt Peak National Observatory).

Speckle interferometry has already been described in connection with the measurement of the angular diameter of Pluto (Chap. 17). This method has also been applied to stellar measurements, and Betelgeuse's diameter has been found to be 0.08 arc sec.

Each technique may be applied to only certain stars or types of stars—for example, only a few stars lie close enough to the ecliptic to be occulted by planets or minor planets. On the other hand, these observations are sufficiently valuable that forthcoming occultations are now listed in astronomical tables so that with some advance planning particularly interesting objects can be scrutinized at many different wavelengths. As is so often the case in astronomy, we need to assemble the parts of our picture from a wide variety of techniques, and try to fit them into a coherent whole.

## 20.2 BINARY STARS

Contrary to appearances, most stars do not exist as isolated objects. They tend to be found in pairs or even in systems of three or four stars. Sometimes we might see two stars close together, but without further measurements we cannot tell how close together they really are. Their closeness might be only apparent, with a truly great separation in distance being masked by the closeness of their directions. Stars that only appear to be close are termed optical doubles and hold little interest for us. It is the genuine **binary stars** (i.e., gravitationally bound) that can provide us with a means of measuring stellar masses.

You will recall that we made repeated use of Kepler's third law, as expanded by Newton's Law of Gravity, in our study of the solar system. Noting the orbital sizes and periodic times of planets allowed us to determine the mass of the Sun; noting the orbital times of satellites, we derived values for the masses of their planets. All that was needed was the identification of one object bound to another by gravity and the measurement of the orbital size and period. We can apply this technique to the stars. We start by cataloging several types of binaries by the characteristics they display. Indeed, some binaries can meet more than one of these criteria.

**Visual binaries** are those that can be seen by telescope to be separate and are thought to be associated. With many the size of their mutual orbit is such that their apparent separation can be seen to change in a regular pattern. Sometimes the orbital motion is superimposed on a steady proper motion of their center of mass so that the two stars seem to be moving along irregular paths. Several years of observing can allow the two motions to be disentangled (Fig. 20.4). In **astrometric binaries**, one star of the pair is too faint to be seen, but its presence is signaled by the apparently wandering path of its visible companion. In this way in 1844 Sirius was found by F. W. Bessel to be an astrometric binary, but it was only in 1862 that its very faint partner, Sirius B, was accidentally discovered by Alvan Clark who was testing his new 18 in. refractor.

A **spectroscopic binary** is identified through the periodic variations in the spectrum of the system (Figs. 20.5 and 20.6). Depending on their relative masses, one or both of the two stars in the binary will be sometimes moving rapidly toward Earth or sometimes away. Lines in the binary's spectrum will be seen to be alternately blue- or redshifted and the shifts will repeat regularly. Finally, **eclipsing binaries** do just that: Each star in turn eclipses its partner so that the observed brightness is modulated in a regular fashion provided the line of sight lies close to the orbital plane of the pair (Fig. 20.7). The changes in brightness during eclipses depend on the relative sizes and surface brightness of the two stars. In some cases one of the two stars will be faint enough that its eclipse can scarcely be noted. The duration of the eclipse provides a measure of the sizes of the stars.

Just how a binary star system will look is the result of chance. Only those that are sufficiently close to the Earth and/or have large enough separations will be detectable as visual binaries. The number of identified visual binaries now approaches 100,000.

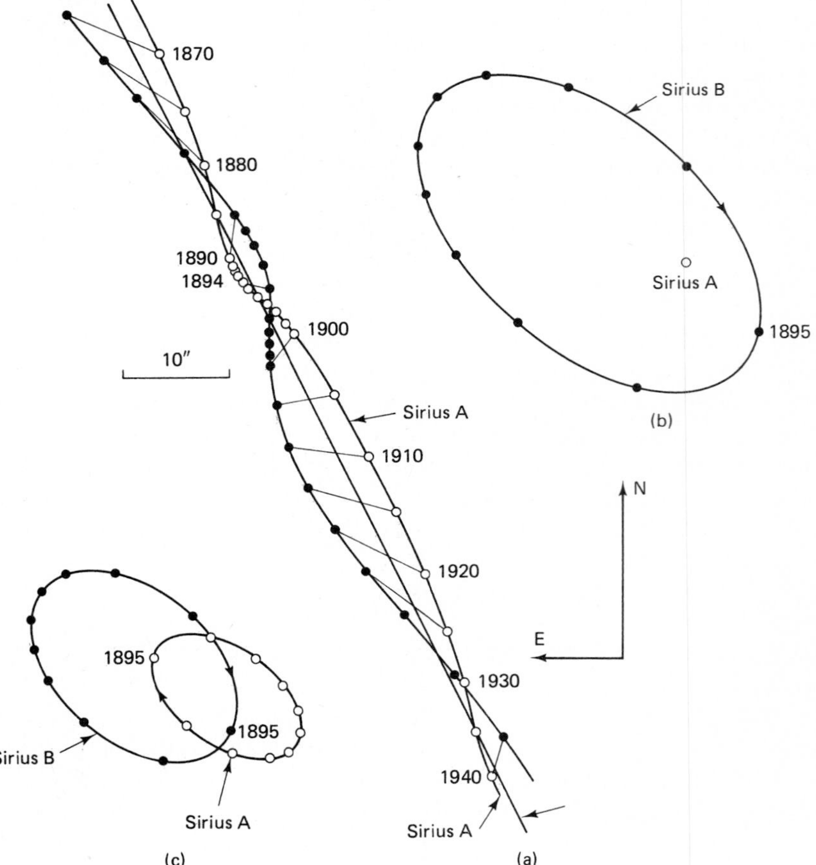


**Figure 20.4** Motions of Sirius A and its companion white dwarf, Sirius B. (a) As seen from Earth, the motion of Sirius A appears to wobble slightly about the general drift (down, in this diagram). The unseen path of Sirius B is indicated by the somewhat more curving line. The straight line represents the path of the center of mass of the system. (b) Computed orbit of Sirius B about Sirius A. (c) Computed orbits of the two stars about the common center of mass (P. van de Kamp).

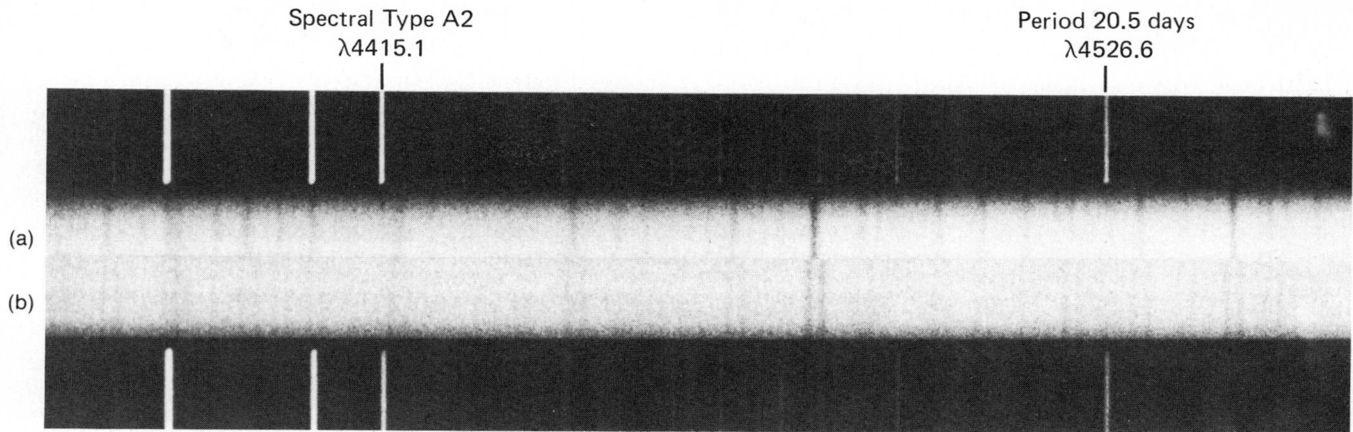


**Figure 20.5** Spectrum of a spectroscopic binary star, Mizar (Zeta UMa). When the two stars are moving at right angles to the line of sight (a), June 11, 1927, their spectral lines coincide. In spectrum (b), June 13, 1927, when the one star is moving towards us and the other away, there is a velocity difference of 140 km/sec between the two, and their spectral lines can be seen separately, one being blue-shifted and the other red-shifted (Mt. Wilson and Las Campanas Observatories).

**Figure 20.6** Spectrum of the spectroscopic binary star, α Gem. The Doppler shift in the spectral lines can be clearly seen, but with this star the spectrum of only one of the two components can be seen, the other being too faint to yield a separate spectrum (Lick Observatory).

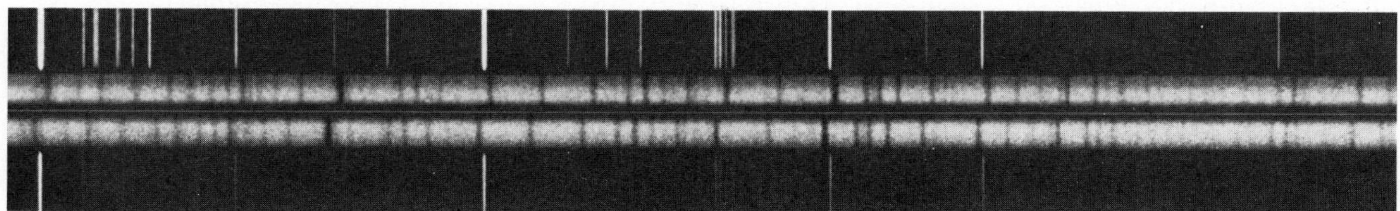

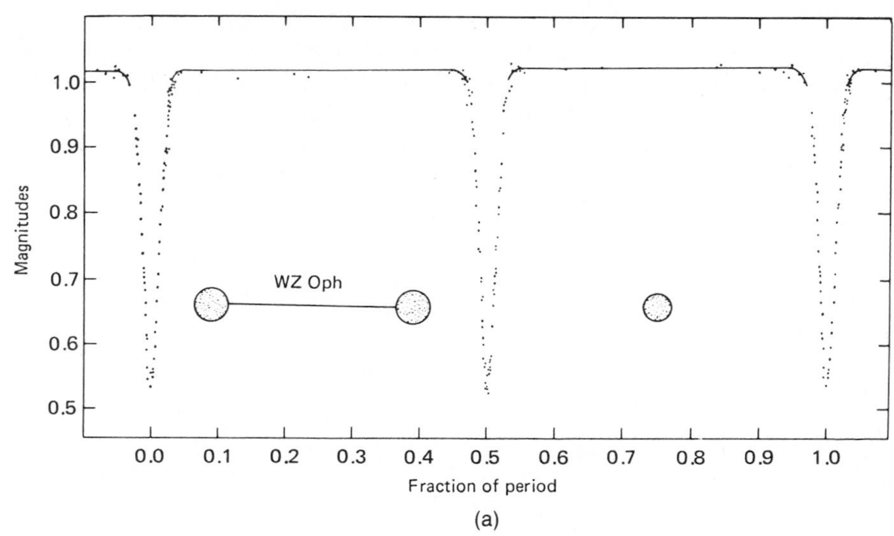


**Figure 20.7** Light curves for three eclipsing binaries: (a) WZ Ophiuchi, two widely separated stars with a period of 4.18 days. Equal depths of the successive eclipses show that the eclipses are central and that the two stars have equal surface brightness.

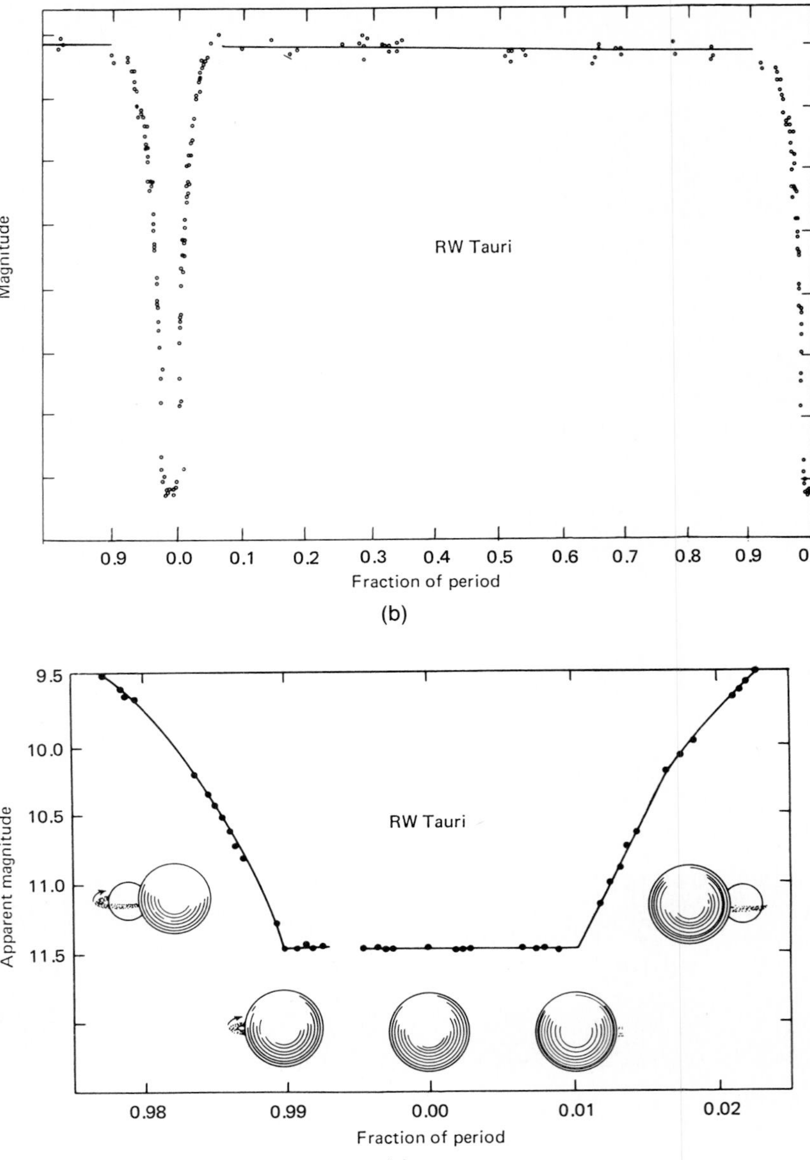


**Figure 20.7 (Cont.)**
(b) RW Tauri with a period of 2.77 days. The two stars are very different, one having a mass of 2.55 $M_\odot$ and the other 0.55 $M_\odot$. (c) Expanded picture of the eclipse of RW Tauri with sketches to show the geometry of the eclipse (Cecilia Payne-Gaposchkin and Katherine Haramundanis, *Introduction to Astronomy*, 2nd ed., 1970, reprinted by permission of Prentice-Hall, Inc.).

With the identification of a binary, what can we measure? The period is perhaps the easiest. For very slow visual binaries, we may have to deduce the period from only a part of a complete orbit, but for the other types we can observe many of the regular repetitions. From the changing positions of the two stars as they orbit one another in a visual binary and from their known distance from the Earth, the semimajor axis of the binary can be

*Table 20.1* SOME BINARY STAR DATA

| *Class of binary* | *Star* | *Period (P)* | *Semimajor axis a (AU)* | *Masses ($M/M_\odot$)* A | B | *Total* |
|---|---|---|---|---|---|---|
| Visual | Sirius | 50.1 years | 19.7 | 2.2 | 0.9 | 3.1 |
| | Ross 614 | 16.5 years | 3.90 | 0.14 | 0.08 | 0.22 |
| | δ Equ | 5.7 years | 5.02 | 2.0 | 1.9 | 3.9 |
| Eclipsing | Algol (β Per) | 2.87 days | 0.065 | 3.7 | 0.8 | 4.5 |
| | α Cru B | 17.36 days | 0.197 | 2.5 | 0.9 | 3.4 |
| | AH Vir | 9.8 hr | 0.0133 | 1.3 | 0.6 | 1.9 |

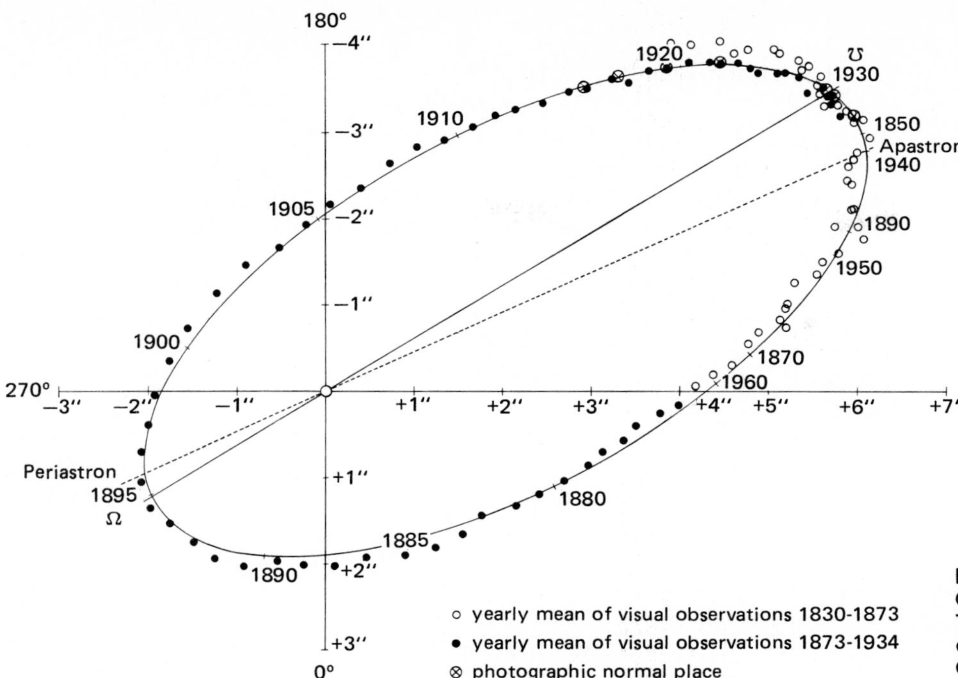


**Figure 20.8** Apparent orbit of 70 Ophiuchi, with an 88 year period. The two stars have masses of 0.9 and 0.65 solar masses. (Yerkes Observatory).

calculated (Fig. 20.8). We can then make use of Kepler's third law to determine the sum of the masses.

For an eclipsing system, the appearance at Earth depends sensitively on the inclination of the plane of the orbit to our line of sight. We have no initial way of knowing precisely what that inclination is, but we know it must be close to 0° in order for eclipses to be seen. We can determine the inclination from the combined use of velocity variations (deduced from observed Doppler shifts), light variations (produced by the eclipses), and estimates of the relative sizes of the stars (based on spectral analysis).

A survey of stars in the solar neighborhood has shown that only about 30 percent are singles, the remainder occurring in multiple systems with about 15 percent triples, fours or even larger groups, leaving about 55 percent in binaries. The periods of the binaries cover quite a range. As the separations of visual binaries are the largest, so too are their periods, typically tens of years. Most spectroscopic binaries have periods in the range 4 to 50 days, although some have periods as long as a few years. Eclipsing binaries are the closest, with typical periods between half a day and a few days. Separations between partners generally lie in the range of 0.01 to 20 AU. Some representative values are listed in Table 20.1.

## 20.3 STELLAR MASSES

If we have two masses $M_1$ and $M_2$ bound together by gravitational attraction and in orbit around their common center of mass (0), we can depict the motion as though one mass ($M_1$) were not moving while the other moves in an orbit with semimajor axis $a$ and period $P$. Then the appropriate form of the Kepler-Newton Law is

$$P^2 = \frac{4\pi^2 a^3}{G(M_1 + M_2)}$$

The period $P$ is in seconds, when the masses $M_1$ and $M_2$ are in grams and $a$ is in centimeters. Alternately, we can use $P$ in years, the masses in units of the solar mass ($M_\odot$) and $a$ in AU so that

$$P^2 = \frac{a^3}{(M_1 + M_2)}$$

The period $P$ and the angular size of the orbit can be measured. If the distance to the binary is known, then the size of the semimajor axis can be determined. The Kepler-Newton formula can then be used to deduce the combined mass of the two stars. If the two masses are not too different, we may also be able to determine the distance of each component from the center of mass and we can then calculate each mass separately. (Refer to the discussion of center of mass in Chap. 7). In many cases, though, we can find only the sum of the masses.

This application of Newton's version of Kepler's third law has been of enormous importance in astrophysics, for it is the only method of determining accurate stellar masses, but for isolated stars we have to resort to less direct methods.

The masses that we find range from about 10 percent of the solar mass up to several tens of solar masses. Modern theories of stellar structure (Chap. 23) now provide an understanding of why we rarely find masses far outside this range.

**BOX 20.2**
*The Mass of Sirius*

Sirius, the brightest star in the sky, is a visual binary. The parallax of the pair is 0.377 sec so that their distance from Earth is 1/0.377 = 2.65 pc. Sirius has a measurable proper motion, but careful measurements over many years have allowed the movements of the two components to be separated. In this way the angular size of the semimajor axis of the system has been found to be 7″.62. Since the distance from Earth is 2.65 pc, the actual size of the semimajor axis is 7.62 × 2.65 = 20.19 AU (see Box 19.2).

The measured period is 49.98 yr. Thus

$$M_1 + M_2 = 20.19^3/49.98^2 = 3.29M_\odot$$

Careful observations of the two components about their center of mass have allowed us to calculate the individual masses: for Sirius, 2.29 $M_\odot$ and for the companion, 0.98 $M_\odot$.

## 20.4 X-RAY BINARIES

The systems just discussed, display their binary natures through regular variations in their visible radiation. During the 1970s, a new class of binary star was discovered through satellite-based observations of X-ray emitters (Figs. 20.9 and 20.10). Some were found to have optical companion stars.

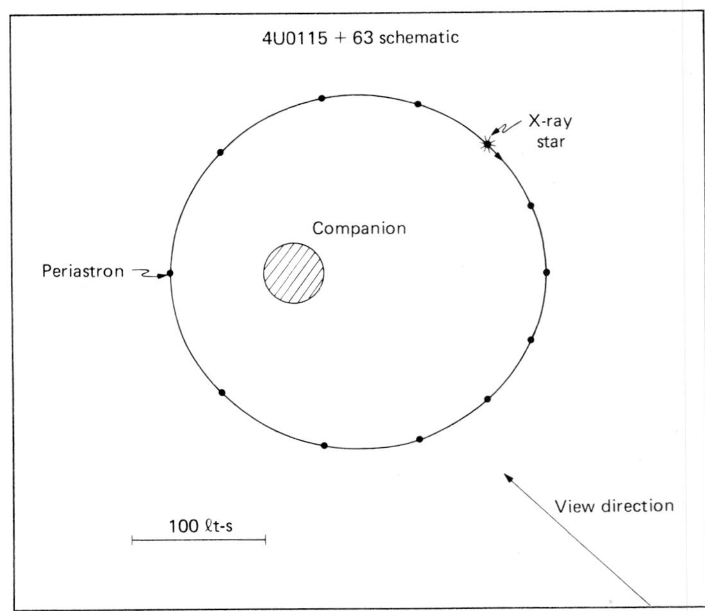


**Figure 20.9** Orbit of the X-ray binary star, 4U 0115 + 63. The orbital period is 24.31 days, and the positions of the X-ray star are shown by dots at intervals of 2 days. The eccentricity of the orbit is calculated to be 0.34 (Saul A. Rappaport, M. I. T.).

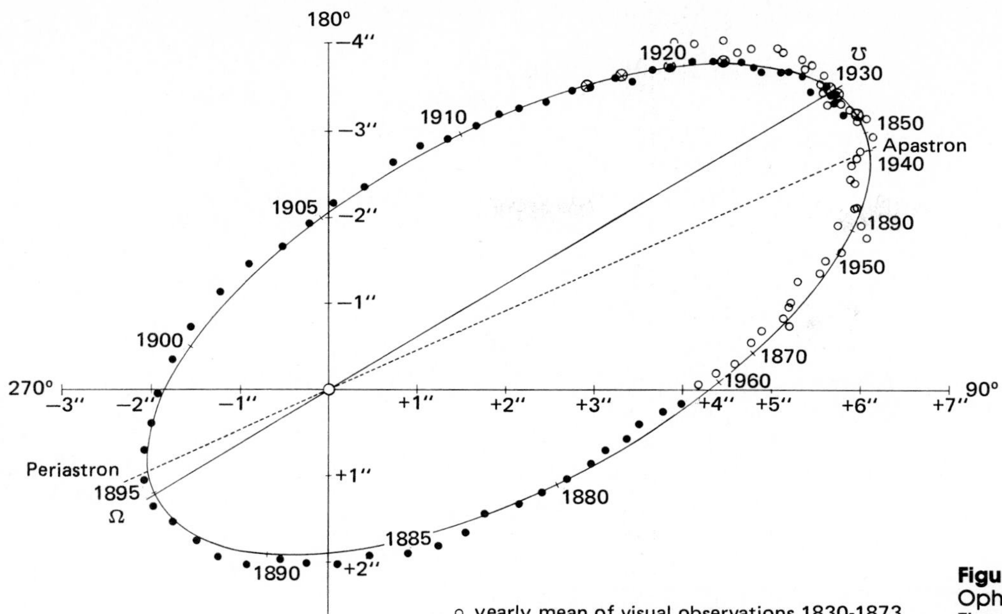

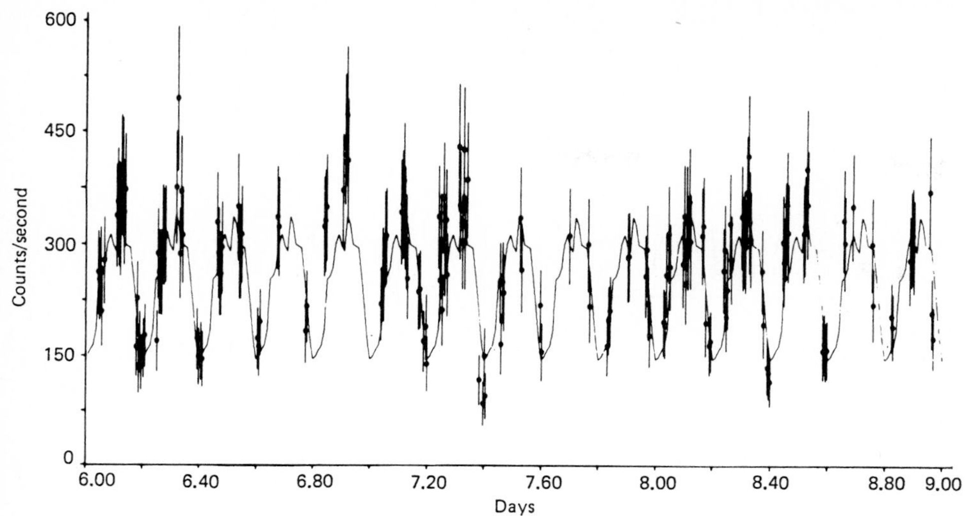


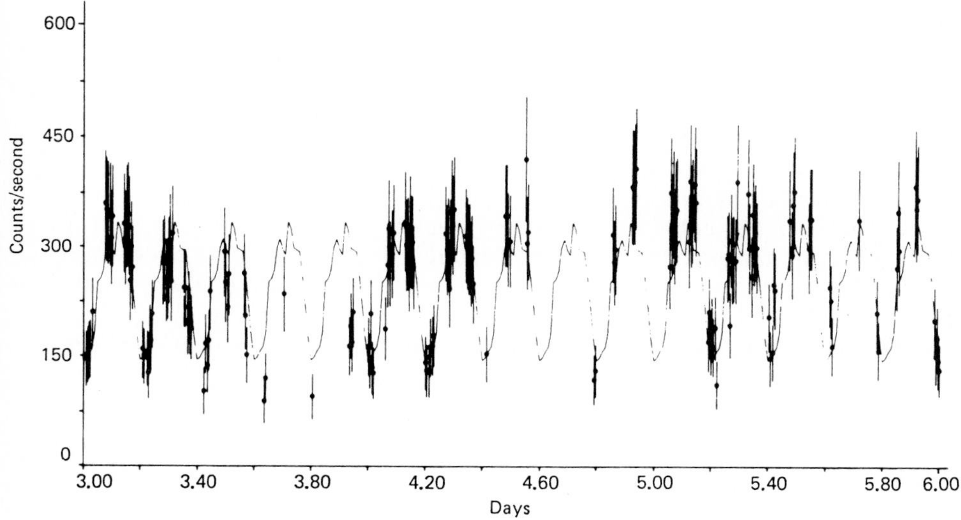


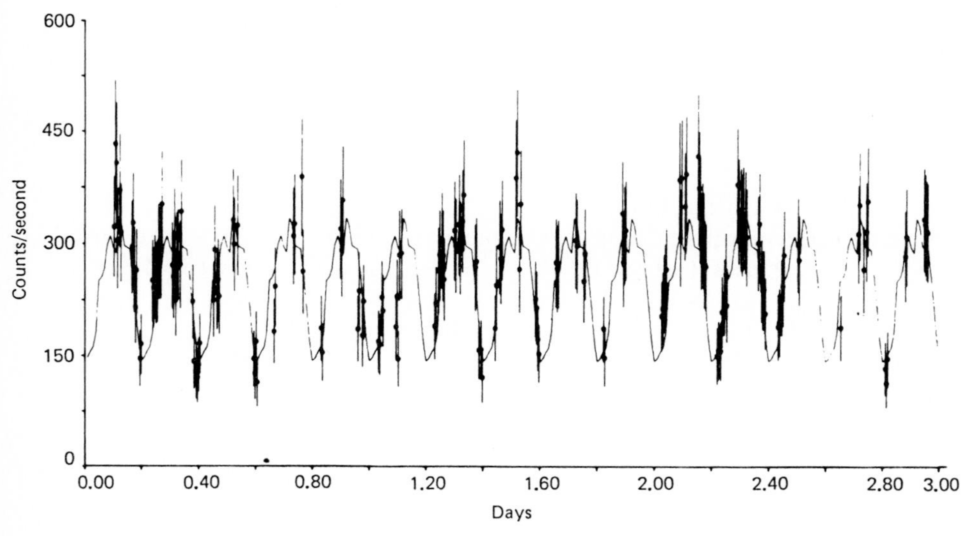


**Figure 20.10** Observations of the X-ray star Cygnus X-3. The intensities are indicated in counts/sec, the number of X-ray photons being detected each second. The period of this binary is 4.80 hr (H. Tananbaum, Harvard/Smithsonian Center for Astrophysics).

*Table 20.2* MASSES OF X-RAY BINARIES

| System | *Binary Period (days)* | *X-ray Emitter ($M/M_\odot$)* | *Optical Star ($M/M_\odot$)* |
|---|---|---|---|
| 3U 0900-40 | 8.97 | 1.0–3.4 | 19–32 |
| SMC X-1 | 3.89 | 0.5–1.8 | 13–22 |
| Cen X-3 | 2.09 | 0.7–4.4 | 17 ± 3 |
| Her X-1 | 1.70 | 0.4–2.2 | 1.4–2.8 |
| 3U 1700-37 | 3.4 | ≥ 0.6 | >10 |
| Cyg X-1 | 5.6 | 9–15 | > 8.5 |
| 4 U 0115+63 | 24.3 | ~1 | > 5 |

Very rapid and regular variations in the intensities of some X-ray sources suggested that they were components of eclipsing binaries similar to the familiar optical binaries.

We can apply the same type of analysis as we did for the conventional binaries, but the distances to these new X-ray binaries are not well determined and neither are the inclinations of their orbits. As a result, the observations do not yet permit precise mass values to be derived; only the ranges can be specified, as shown in Table 20.2.

It is notable that most of these X-ray binaries consist of a massive optical star with a much smaller X-ray emitter. Only Her X-1 has both components of moderate size. We shall return to this matter of stellar sizes in chapter 25 where we will see that the masses of the X-ray emitters provide an important test of a theory.

## 20.5 PLANETARY SYSTEMS

The detection of planetary systems around other stars would be a major discovery, and the techniques of binary star observing are being applied to this end. It is not expected that other planets will be luminous enough (in reflected light) to be easily detected, but the presence of a massive planet might be detectable through its influence on the movement of its parent star. In such a case, we would find an astrometric binary in which the unseen companion was far less massive then the wobbling visible star.

Just how difficult this observation will be can be seen if we consider the Sun-Jupiter system. Jupiter's mass is about $\frac{1}{1000}$ that of the Sun. The center of mass of the Sun-Jupiter system is very close to the center of the Sun (see Chap. 7). The Sun will, therefore, appear to an external observer to be moving in an orbit with a radius of about $\frac{1}{1000}$ of the Sun-Jupiter distance of 5.2 AU (i.e., $5.2 \times 10^{-3}$ AU or $2.5 \times 10^{-8}$ pc). An observer at a distance of 5 pc, for example, would thus see the Sun moving in an orbit of angular size $\frac{1}{1000}$ arc sec (Fig. 20.11). Such a small angle would be barely detectable, if we used an interferometer with widely spaced mirrors, and the Sun would be seen to wobble slightly with a period of 11.9 years.

Had the masses of the Sun and Jupiter been more nearly equal, the movement of the Sun would be larger and more easily noticed. A systematic search has been conducted for many years by Peter Van de Kamp at Swarthmore College's Sproul Observatory using a long-focus refractor so that 1 mm on his photographic plates corresponds to 18.9 arc sec. The stellar movements are measured to about $\frac{1}{100}$ mm so that an angular movement of almost

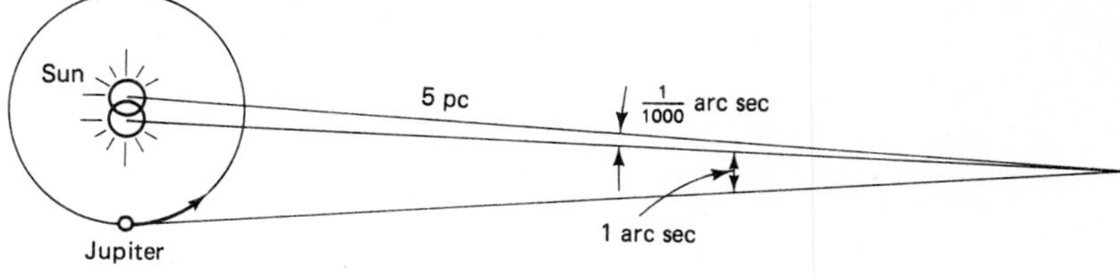


**Figure 20.11** The Sun-Jupiter system considered as a binary. If it were viewed from a distance of 5 pc, the Sun would be seen to move in an orbit with a radius of 1/1000 arc sec. Jupiter would move in a much larger orbit, but it would be difficult to observe against the great brightness of the Sun.

0.2 arc sec might be seen on a single photograph. Smaller movements can be detected by using many photographs. At one time, Barnard's star was thought to have a "planetary" companion. This star is the closest to Earth in the northern celestial hemisphere and thus most favorably placed for observing. Measurements of several thousand plates, mostly since 1938 but with a few as far back as 1916, seemed to indicate a wobble and suggested to Van de Kamp that Barnard's star has an unseen companion of about Jupiter's mass. Recalculations in 1977 have led to this claim being challenged and point again to the extreme difficulty of these observations.

## CHAPTER REVIEW

**Stellar diameters can be determined for some stars:**

— application of radiation laws if stellar temperature and distance are known.

— interference methods, in which light from different regions across a star produces an interference pattern. Angular diameters as small as about 0.001 arc sec can be measured in this way.

— occultation of a star by the Moon or a planet. Resulting interference pattern depends on stellar diameter.

**Binary stars:** optical doubles only seem to be close but are actually far apart.

— visual binaries: can be seen to be separate, in orbit around one another.

— astrometric binaries: one star too faint to be seen, but its presence is indicated by nonstraight-line motion of its bright partner.

— spectroscopic binary: stars not seen separately, but regular Doppler shift of spectral lines indicates orbital motion.

— eclipsing binaries: each star regularly eclipses the other, so that observed intensity varies in a regular way.

**Stellar masses:** application of Kepler's third law allows the sum of the masses of the two stars to be determined. If star masses are not very different, then individual masses can sometimes also be determined.

**X-ray binaries:** new class of binaries discovered in 1970—one star is an X-ray emitter, the other is an optical star, usually much more massive than the Sun.

**Planetary systems:** detection of planets around a star might be based on use of methods already employed for binary star observation. A number of stars has been examined carefully, but so far no clear evidence for the presence of planets.

## NEW TERMS

**astrometric binary**
**binary star**
**eclipsing binary**
**interference**
**optical double**
**spectroscopic binary**
**visual binary**

## PROBLEMS

**1.** What would be the angular diameter of the Sun if it were at a distance of 2 pc?

**2.** What kind of observations can indicate that a star is really a binary system even though we cannot see the two stars separately?

**3.** Describe a method for determining the diameter of a star.

**4.** A binary star system contains two stars of the same type and size. Suppose this system is viewed from a direction in which the plane of the orbit is seen edge on. Sketch how the total intensity of the observed light will vary through a complete cycle.

**5.** Most spectroscopic binaries have relatively short periods, and most visual binaries have longer periods. Why?

**6.** Figure 20.4 shows the motion of Sirius and its dwarf companions. Copy that diagram, and mark in the dates for the next cycle.

**7.** Figure 20.8 shows the elliptical orbit of the companion of 70 Oph about the larger star. When will the smaller star next be at periastron, that is, closest to the larger star?

**8.** If the surface temperature of the Sun doubled, how small would the Sun have to be in order for the luminosity to remain unchanged?

**9.** A star has a temperature three times that of the Sun, and a luminosity 0.01 times that of the Sun. Calculate the radius of the star (in solar radii).

**10.** What are the radii of the following stars:

a. $M = -4.1$, $T = 28{,}000$ K?
b. $M = +9.0$, $T = 3500$ K?
c. $M = +5.9$, $T = 4900$ K?
d. $M = -0.4$, $T = 3200$ K?
e. $M = -5.0$, $T = 4000$ K?

**11.** Verify by calculation that Kepler's third law holds for Ross 614. The two components of this binary have masses 0.14 and 0.08 solar masses. The semimajor axis of their orbit is 3.90 AU, and the period of the motion is 16.5 years.

**12.** A binary star system contains one star with 1 $M_{\odot}$ and one star with 3 $M_{\odot}$. What will be the period of revolution if the semimajor axis of the relative orbits is 10 AU?

**13.** A binary star is observed to have a parallax of 0.25 arc sec, and the orbital period is 40 years. The semimajor axis of the orbit has an angular size of 4 arc. sec. Calculate the combined mass of the two stars.

**14.** The X-ray binary 4U 0115 + 63 has an orbital period of 24.3 days, and the semimajor axis appears to be 0.28 AU. Calculate the sum of the two stellar masses.

**15.** Procyon is a double star in which the orbital period is 40.6 years and the separation is 15.8 AU. Calculate the combined mass of the two stars. One of the stars is a white dwarf. What is the minimum mass of the other star?

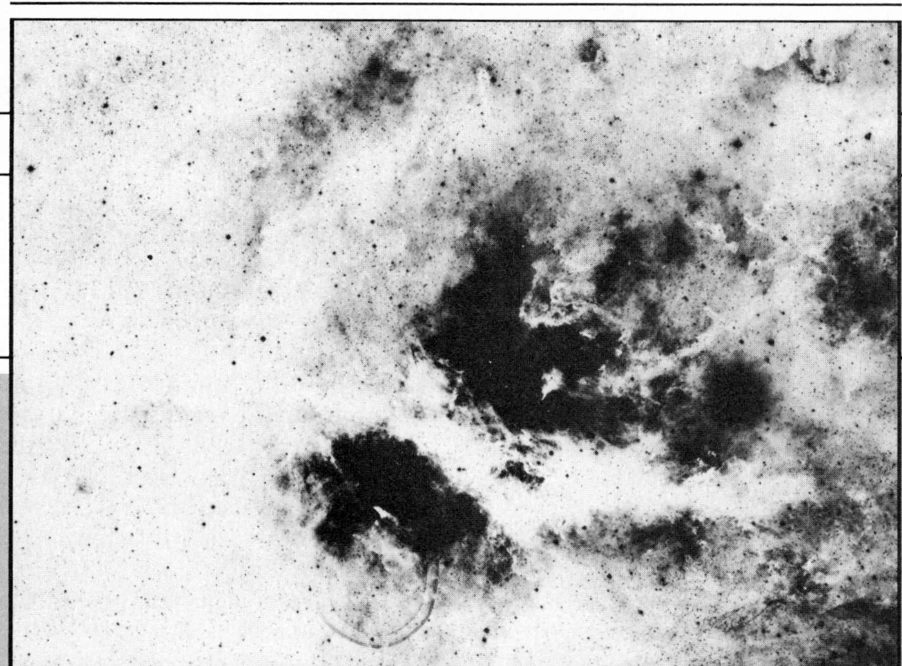

CHAPTER 21

# The Stars: Classification

Can we discover any order in the myriad of stars that our telescopes now reveal? Can we detect relationships or trends in their properties and thus be led to a general theory of stars—their compositions and behavior, their formation, structure, and evolution? We can examine the brightest stars or the nearest, but must be alert to the possibility that this sort of selection might be so unrepresentative that we could be quite misled in the theories we devise.

The ancient grouping of stars into constellations could lead to no insights. In contrast, the modern classification of the stars has been essential to the development of our understanding. With a large collection of data, we can try to arrange the stars in sequence according to some property such as temperature or magnitude and hope that careful classification will be as fruitful as it has been in biology, for example, where Linnaeus devised a system of classifying plants and animals in the eighteenth century.

## 21.1 SPECTRAL CLASSIFICATION

Our visual impression of the magnitude of a star involves the combined effect of all the wavelengths that our eyes can detect. We can gain deeper insights when we examine the spectra of stars.

Dark (absorption) lines were discovered in the solar spectrum by William Wollaston in 1802. They were cataloged in 1814 to 1815 by Joseph Fraunhofer, who later discovered similar lines in stellar spectra. It was not until 1860, however, that the identification of some of the solar lines with lines in laboratory spectra was made by Gustav Kirchhoff, and a few years later Fraunhofer made similar identifications in stellar spectra. This identification showed that the Sun and stars contained the same types of chemical elements as were known on Earth.

As more stellar spectra were obtained, classification schemes were developed. The first, devised by Father Angelo Secchi in 1863 had four classes: I: white or bluish; II: yellow stars like the Sun with many narrow lines; III: red stars like Betelgeuse and Antares with dark bands; and IV: faint and fiery-red stars with different dark bands. Refinements were gradually introduced, until we have the classification scheme in use today, in which the main classes are *O*, *B*, *A*, *F*, *G*, *K*, *M* and their subdivisions (see Fig. 21.1).

This strangely labeled sequence was devised by Annie J. Cannon at the Harvard College Observatory in the late nineteenth century before the radiation laws were fully known, and started as a straight alphabetical sequence *A* through *Q*. This sequence was thought to be followed as the stars age so that *A* and *B* types were termed "early" and those at the *K*, *M* end of the sequence were termed "late." Further work failed to confirm the existence of some of the spectral features, and a few letters/groups were eliminated. Then came the recognition that the color index (CI) (see Sec. 21.2) was related to temperature. Finally, the *O* and *B* classes were relocated

**Figure 21.1** Principal types of stellar spectra. Each spectrogram runs from short wavelengths (blue) to long wavelengths (red), left to right. Note how the blue intensity decreases as we move through the classes from hot (*O* and *B*) to cool (*M*). In addition, different absorption lines are emphasized at different temperatures (Palomar Observatory photograph).

*Table 21.1* SPECTRAL CLASSIFICATION FOR MAIN SEQUENCE STARS

| *Spectral Class* | *Color* | *Color Index (CI)* | *Temperature Range (K)* | *Main Spectral Features* | *Examples* |
|---|---|---|---|---|---|
| O | blue | <−.31 | >25,000 | Hottest stars; strong UV continuum; strong lines of ionized atoms HeII, OII, SiIV; hydrogen lines weak. | ζ Pup<br>10 Lac |
| B | blue | −.31 to 0.00 | 11,000–25,000 | Neutral helium (HeI) dominant; lines of OII, MgII; HI weaker than in O stars | Rigel<br>Spica<br>Achernar |
| A | blue | 0.00 to +0.27 | 7,500–11,000 | Hydrogen lines getting weaker; singly ionized metals Mg, Si, Ti, Ca, Fe[a] | Sirius<br>Vega<br>Castor |
| F | blue to white | +0.27 to 0.58 | 6000–7500 | Metallic lines stronger—CaII, FeII, CrII, FeI, CrI; HI lines weaker than A stars | Canopus<br>Procyon<br>Polaris |
| G | white to yellow | 0.58 to 0.89 | 5000–6000 | Solar type spectra; CaII dominant; ionized & neutral metals; H lines weaker; CH bands strong | Sun<br>Capella |
| K | orange to red | 0.89 to 1.45 | 3500–5000 | Neutral metal lines dominant; CH still present | Aldebaran<br>Arcturus<br>Pollux |
| M | red | >1.45 | <3500 | TiO molecular bands; neutral metals | Betelgeuse<br>Antares |

[a] *Metals* is a term often used to include all atoms heavier than H and He.

leaving the jumbled-letter sequence still in use. The defining properties of the spectral classes are listed in Table 21.1.

It was recognized that the classification also corresponded to a steady trend in temperature, from the hot stars (*O* and *B*) to the coolest (*M*) stars. Finally, each class is now divided into ten parts, with the sequence 0 to 9 running from higher to lower temperature. Thus *G5* is midway between *GO* and *KO*, and *AO* corresponds to a temperature of about 10,000 K.

## 21.2 COLORS OF STARS

The stellar classification into *O*, *B*, *A*, etc., is based on the lines in the spectra, which indicate the presence of the different chemical elements. Another but related method of classification is based on measurement of the thermal radiation from a star using several standardized wavelength bands. This can be done by using a **filter** that allows only some wavelengths to pass and absorbs or reflects all others. A filter may consist of two glass sheets with a colored dye or special coatings on one of the internal surfaces. Colored plastic, such as is used for wrapping candies acts like a filter, but its properties are not sufficiently standardized for scientific use. Filters can be constructed to pass wide or narrow bands of wavelengths.

For astronomical purposes, an internationally standardized series of filters has been defined. Most widely used is the *U*, *B*, *V* system with the *U* (ultraviolet) passing wavelengths around 3650 Å, the *B* (blue) centered on 4400 Å, and the *V* (visual) at 5500 Å, approximating the wavelength of maximum sensitivity of the human eye (Fig. 21.2). Other filters extend the range into the IR and are listed in Table 21.2.

The apparent magnitude of a star in one of these wavelength bands is denoted by *B* or *V* and so on for the other filters. The corresponding absolute magnitude will be $M_B$ or $M_V$.

The usefulness of this system becomes apparent when we examine the visual radiation that comes from the hot outer surface of a star. To measure the temperature, we could use Wien's Law ($\lambda_{max} T = 0.3$ cm deg), but if the temperature is too high or too low, $\lambda_{max}$ will not lie within the visual

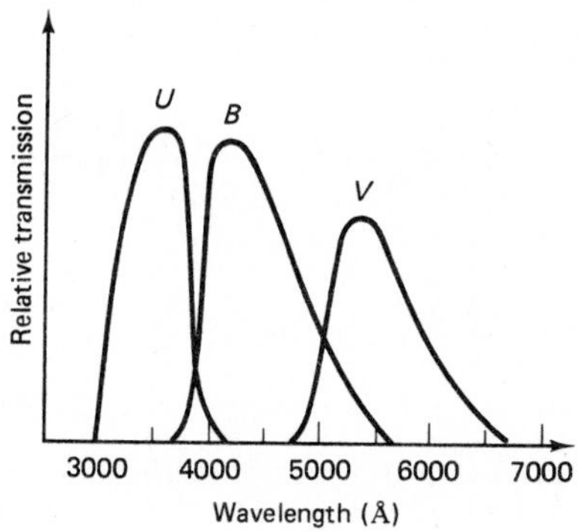


**Figure 21.2** Transmission of light through the *U*, *B*, and *V* filters. Each filter allows a small range of wavelengths to pass. The effective average wavelengths are listed in Table 21.2.

*Table 21.2* STANDARD FILTERS[a]

| Ultraviolet to near-IR | | | | | |
|---|---|---|---|---|---|
| Filter | U ultraviolet | B blue | V visual | R red | I infrared |
| Central wavelength (Å) | 3650 | 4400 | 5500 | 7000 | 9000 |
| **Middle-IR** | | | | | |
| Filter | J | K | L | M | N | Q |
| Central wavelength (microns) | 1.25 | 2.2 | 3.4 | 5.0 | 10.2 | 22.0 |

[a] Ref: C. W. Allen, *Astrophysical Quantities*, 3rd Ed. The Athlone Press, University of London, 1973.

part of the spectrum. We can still determine the temperature if we measure the magnitude in the $U$, $B$, $V$ regions. The **difference** $B - V$ is termed the **color index** (CI), and its diagnostic use can be seen by referring to the curves in Fig. 21.3. The difference $B - V$ is equal to the difference between the corresponding absolute magnitudes, $M_B - M_V$.

The curves in Fig. 21.3 represent the radiation from stars of the same size but with different temperatures, with $T_3$ the highest and $T_1$ the lowest. Star 3 will emit relatively more energy in the $B$ range than in the $V$; accordingly, its $B$ magnitude will be numerically *less* than its $V$ magnitude, and the color index $B - V$ will be negative. Visually, it will appear bluish. On the other hand, star 1 has a lower temperature, and it emits relatively less energy in the $B$ than in the $V$. It will appear reddish and yield a positive $CI = B - V$. Use of the CI provides a simple means of classifying stars by temperature.

The $B - V$ index ranges from about $-0.6$ for the hottest blue stars (with surface temperatures of 25,000 K) to $+2.5$ for the coolest red stars (with surface temperatures below 3000 K).

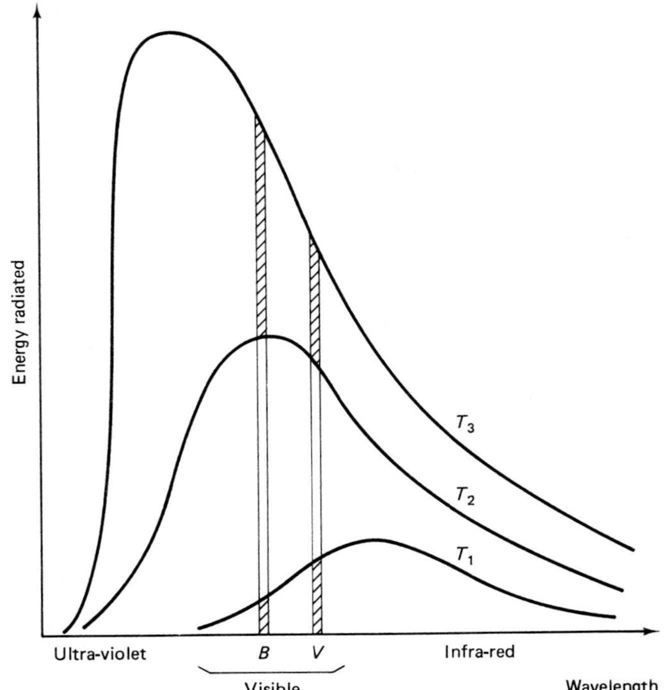


**Figure 21.3** Color Index. Schematic radiant energy curves are shown for three stars, having temperatures $T_1$, $T_2$ and $T_3$, with $T_3$ being the hottest. Star 3 radiates more energy in the $B$ band than in the $V$. Its $B$ magnitude will, therefore, be numerically less than its $V$ magnitude, and its CI, $B - V$, will be negative. Star 1 being much cooler radiates less in $B$ than in $V$, its $B$ magnitude will be numerically larger than its $V$ magnitude, and its CI will be positive. Star 2 is intermediate, and as shown will have a slightly negative CI.

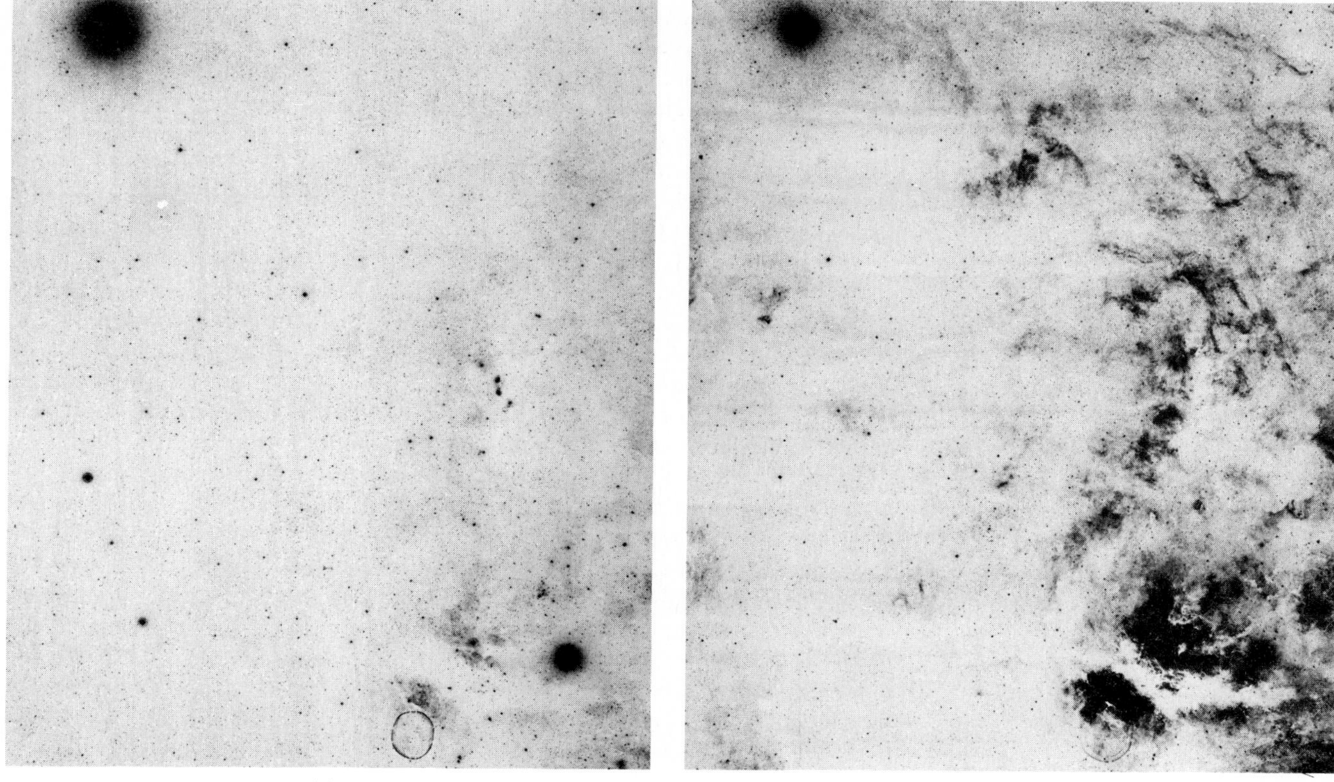


**Figure 21.4** Views of the same part of the sky, through (a) blue and (b) red filters. Note the great difference between the appearance of stars and clouds in these two views. The red filter transmits the wavelength of hydrogen Hα line, which is the strongest feature in the emission spectrum of the nebulosity. The very bright star at the top of the photo is Deneb, the star at the lower right is Gamma Cygni. (The oval-shaped ring at the lower right is a photo defect.) (© 1960 National Geographic Society—Palomar Sky Survey. Reproduced by permission of the California Institute of Technology).

The sky looks dramatically different when viewed or photographed through different spectral windows. Figure 21.4 shows two views of the same part of the sky, one photographed with a blue filter, the other with red.

The $U$, $B$, $V$ magnitudes and the $O$, $B$, $A$, etc., classes are related through temperature, and $AO$ defines the color index at $B - V = O$. At different temperatures, atoms will have different numbers of their electrons in different energy levels, and different spectral lines can thus appear. The full theory was worked out by the Indian physicist, M. N. Saha in 1922.

The color magnitudes defined by the $U$, $B$, and other filters refer to specified regions of the spectrum. Another magnitude system can be defined that would include the total energy, radiated at all wavelengths, and this is termed the **bolometric magnitude**, $M_{bol}$ or $m_{bol}$. Conversion between the $V$ and bolometric magnitudes involves a bolometric correction, which effectively takes into account the fraction of stellar energy not detected through the $V$ filter.

## 21.3 THE HERTZSPRUNG-RUSSELL DIAGRAM

In 1911 Ejnar Hertzsprung, a Danish astronomer, pointed out that stars with the same spectral classes could differ significantly in their luminosities (or absolute magnitudes). In 1913, quite independently, H. N. Russell at Princeton also examined the relation between luminosity and spectral class. The relation between these two quantities becomes apparent at once when we look at Fig. 21.5, in which absolute magnitude $M$ is plotted against spectral class for approximately 6700 stars. This form of display is now known as the Hertzsprung-Russell (H-R) diagram.

Absolute magnitude is usually plotted, but luminosity could just as well be used, and both scales are shown in the figure. The first diagrams used spectral classes in alphabetical sequence, and this convention has been retained even though it has the disadvantage that the temperature scale then increases from right to left.

Most stars fall somewhere along a band that runs diagonally from upper left to lower right. This **main sequence** (MS) shows the smooth progression of the characteristics from very hot stars (with high luminosity) to cool and far less luminous stars. The Sun, type G2, is on the main sequence. There are some stars that lie well above the MS, their luminosities being larger than an MS star of the same temperature. These stars are referred to as **giants**; **supergiants** are the few extremely luminous stars even further from the MS. In the opposite direction, there are stars whose temperatures are high, but whose luminosities are very small, placing them well below the MS. These are the **white dwarfs**. Figure 21.6 shows a schematic form of the H-R diagram with the locations of the various classes labeled.

There are additional subcategories among the giants and supergiants, as indicated by the almost horizontal lines labeled Ia, Ib, II, III, and IV in the H-R diagram. Category V refers to the main sequence. Thus the full designation for the Sun is G2 V, for Betelgeuse, it is M2 I, and for Aldebaran it is K5 III.

We can understand the basic idea behind the very descriptive *giant* and *dwarf* labels. The luminosity of a star depends on the fourth power of

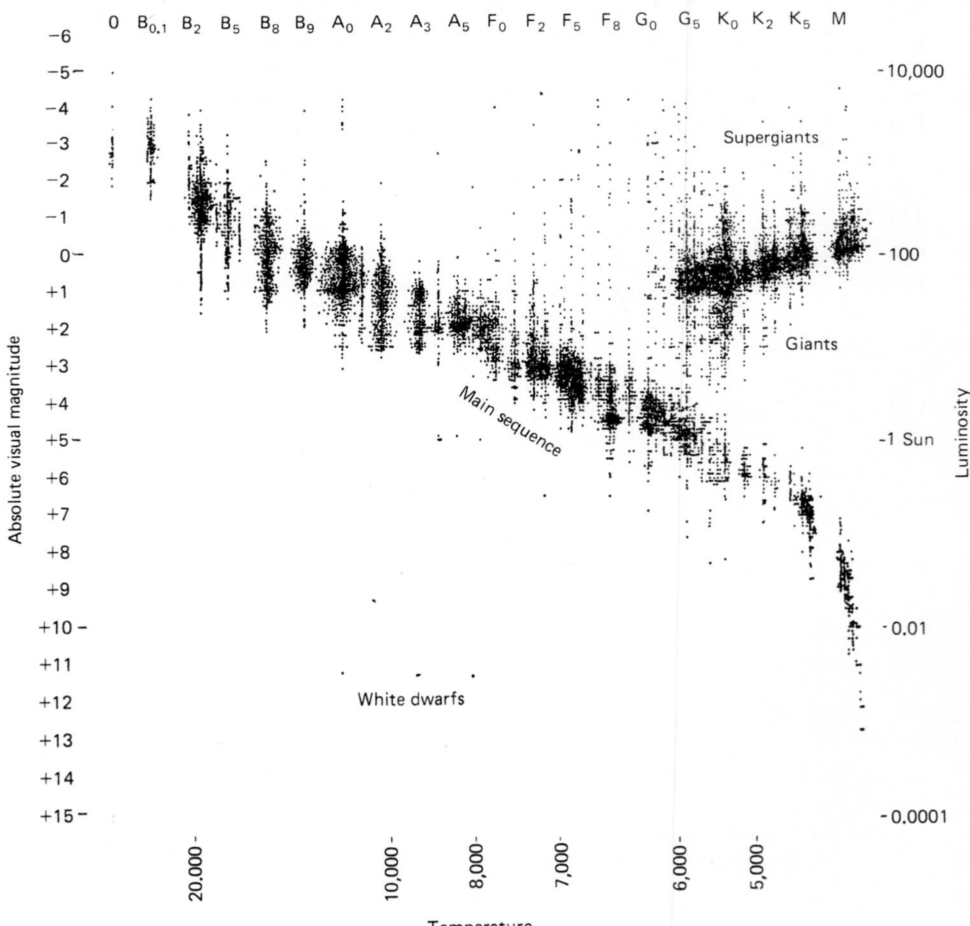


**Figure 21.5** H-R diagram for 6700 stars. The main sequence is strongly marked, as is the giants region (upper right). The apparent clumping of stars is an artifact of the measuring technique in this 1936 diagram. Modern surveys produce a smoother diagram (Yerkes Observatory).

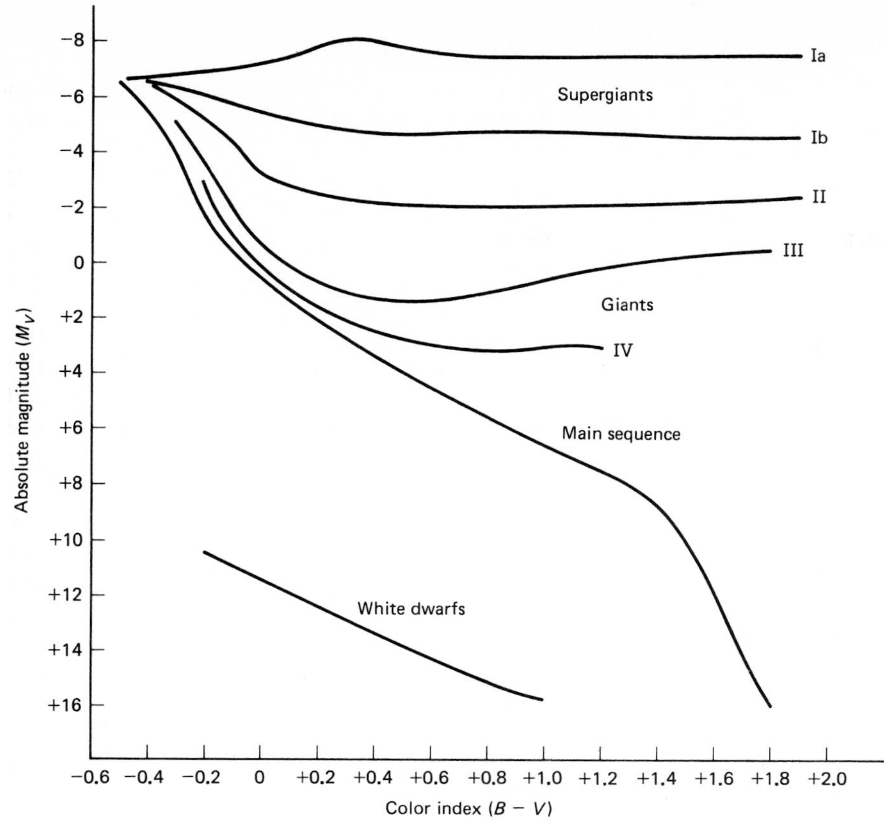


**Figure 21.6** Schematic H-R diagram showing the regions occupied by the different types of stars.

the temperature ($T^4$) and also on the surface area which in turn depends on the radius as $R^2$. Thus, among stars with the same temperature, those with larger radius have higher luminosities and lie above the MS; these are the giants. Those below the MS must be smaller or dwarf.

---

BOX 21.1
*Giants and Dwarfs*

Recall that the luminosity of a star is given by

$$L = 4\pi R^2 \sigma T^4 \text{ ergs/sec}$$

The Sun can be used as a standard denoting its luminosity by $L_\odot$, its radius by $R_\odot$ and its temperature by $T_\odot$. Then a comparison with another star will yield the ratios

$$\frac{L_*}{L_\odot} = \left(\frac{R_*}{R_\odot}\right)^2 \left(\frac{T_*}{T_\odot}\right)^4$$

For example, a star that has a surface temperature of 3000 K and a luminosity of $10^4$ times that of the Sun will have a radius given by

$$10^4 = \left(\frac{R_*}{R_\odot}\right)^2 \left(\frac{3000}{6000}\right)^4 \quad \text{(using 6000 K for } T_\odot)$$

leading to $R_* = 400\ R_\odot$; it will be classed as a supergiant.

A very different example is provided by a star that is white in color, has a temperature of 12,000 K (twice as hot as the Sun), but a luminosity only $\frac{1}{100}$ as much. Then

$$\frac{1}{100} = \left(\frac{R_*}{R_\odot}\right)^2 \left(\frac{12{,}000}{6000}\right)^4$$

from which

$$\left(\frac{R_*}{R_\odot}\right)^2 = \frac{1}{2^4}\frac{1}{100}$$

and

$$R_* = \frac{1}{40} R_\odot$$

Such a star is termed a white dwarf.

For these calculations, 6000 K was used for the Sun's temperature rather than the more precise 5800 K. For the broad identification of stars as giants or dwarfs, this approximation is clearly adequate.

The numbers of points representing the stars in different parts of the H-R diagram vary strongly with the method of selection. If we plot points for as many stars as we can, we find that all regions are populated. If, however, we plot the points only for those 100 stars that are close to the Sun, within about 6 pc, we find something different (see Fig. 21.7.). There

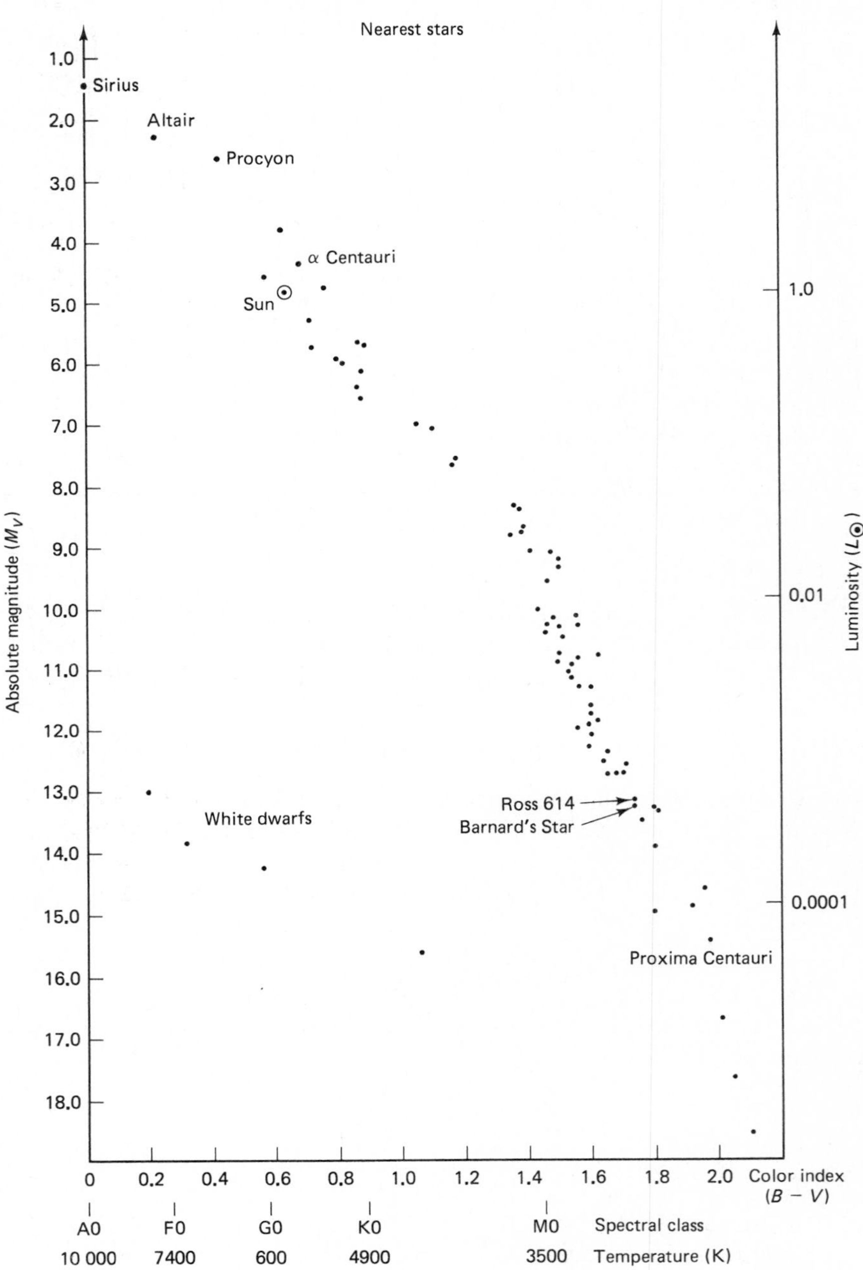


**Figure 21.7** Color-magnitude (H-R) diagram for the 100 stars nearest to the Sun. The main sequence is marked by the diagonal band. There are few high-luminosity stars and a few stars, the white dwarfs, that lie below and to the left of the main sequence.

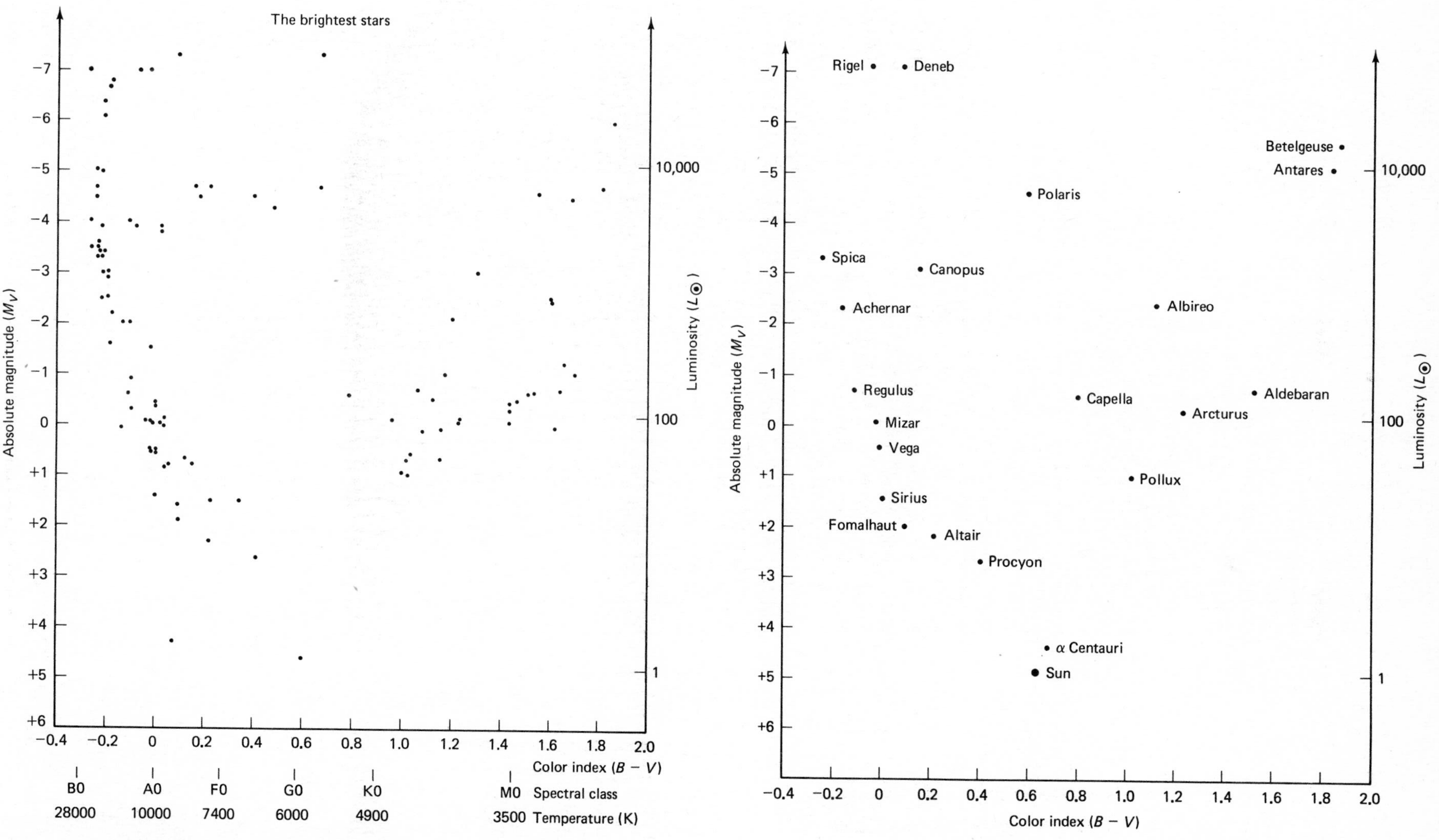


**Figure 21.8** H-R diagram for the 100 brightest stars. Stars that seem bright are in fact very luminous and not simply bright because of being nearby.

**Figure 21.9** H-R diagram for well-known stars.

are few white (hot) stars, no giants, and many stars at the lower (cooler) end of the MS. We find a different distribution if we select and plot data for the 100 brightest stars (Fig. 21.8): There are far more giants and supergiants. These are very luminous stars, which can be seen even at great distances. If we select stars that appear to be close together in clusters, we find distributions of points on the H-R diagram that can be quite different from those so far examined. Some have no giants, some have no stars along part of the MS.

As we shall see (chap. 24), the H-R diagram is a powerful and versatile tool in helping us to understand the evolution of stars. Before undertaking that analysis, it is necessary to find a way to include the data on stellar masses, for the H-R diagram makes use only of magnitudes and spectra.

## 21.4 THE MASS-LUMINOSITY RELATION

From binaries we find that in general the most luminous stars are also the most massive. The masses and luminosities can be displayed in the M-L diagram as shown in Fig. 21.10. The scale of masses turns out to be far more compressed than the scale of luminosities, which extends from about $\frac{1}{100}$ of the Sun's luminosity ($L_\odot$) to about $10^4\,L_\odot$, a range of $10^6:1$. In sharp contrast, the masses lie between slightly less than 0.1 $M_\odot$ to around 50 to 100 $M_\odot$, a range of only about 1000 : 1. This is an important observation that our theories of stellar structure have to explain. Those theories also have to explain why we see neither very small nor ultramassive stars.

In the H-R diagram, we found a well-defined band which contained most of the data points, and we find a similar grouping now with the M-L diagram. The main sequence stars lie in a band across the figure, and a few dwarfs lie clearly below the main group. The **mass-luminosity relation** can

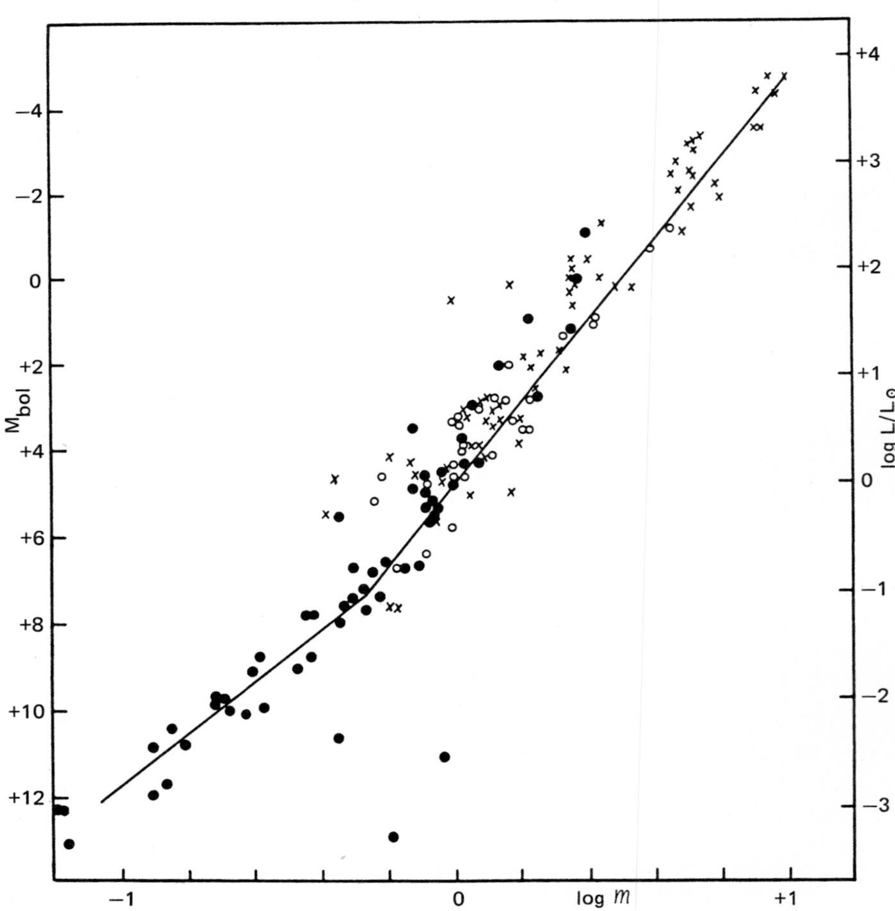


**Figure 21.10** Mass-luminosity relation. Data come from visual binaries (dots and circles) and eclipsing binaries (crosses). For the more luminous stars, the line shows the luminosity increasing as $M^{3.8}$, for the less luminous stars $L$ increases as $M^{2.4}$ (W. D. Heintz, Swarthmore College and B. Reidel Publishing Co.).

be expressed as

$$L/L_{\odot} = (M/M_{\odot})^{\alpha}$$

with the exponent $\alpha$ having an average value of 3.5 but varying somewhat with stellar mass. A better fit to the data uses two lines, as shown in this figure. Giants and white dwarfs do not follow this relation.

Now that we know this relation between mass and luminosity for main sequence stars, we can refer back to the H-R diagram. The most luminous stars are at the left-hand end of the main sequence, in the *O* and *B* categories, and these we can now recognize as the most massive. The least luminous stars are at the right-hand end of the sequence, the *M* stars, and these are the least massive. The main sequence can thus also be seen to correspond to a mass sequence.

## 21.5 SUMMARY

In this chapter and the two that preceded it, we have seen how different types of observations can be used to deduce many of the properties of stars. Not all techniques can be used for all stars, however, so that we are sometimes left with only fragmentary information. For some stars, we can measure their distances from the Earth and we can then convert apparent magnitudes to absolute magnitudes and luminosities. Measurement of stellar radiation in broad wavelength intervals allows us to calculate the temperature of the stellar surface, and the lines in stellar spectra allow us to identify particular elements and their states of ionization. The CI (or temperature) and the presence or absence of spectral lines then form the basis of the classification of the stars into the *O*, *B*, *A*, etc., groups. The H-R diagram uses the CI and the absolute magnitude (or luminosity) to produce an unsuspected order among the masses of stellar data. Finally, using binary star observations to calculate stellar masses, we find a relation between luminosity and mass for main sequence stars. This, in turn, shows that the stars along the main sequence are ordered not only according to CI but also by mass. The stars with the largest masses are at the left-hand end of the main sequence, in the *O* and *B* classes, whereas the lowest mass stars are to be found in the *M* class.

The classification that we have obtained, impressive as it is, does not yet provide answers to several major questions: we detect the radiation from the stellar surfaces, but what processes are taking place deeper down? What is the source of the energy that we detect as radiation? If the stars are radiating away their energy, how do they evolve? How do stars form, and what will they become? How does stellar behavior and evolution depend on stellar mass? What is the time scale for stellar evolution—hundreds of thousands, millions, or billions of years?

Answers have been found to most of these questions and we will review them in the next few chapters. We will first examine the Sun, the nearest star, and, as such, available for examination to a degree not possible for any other star.

## CHAPTER REVIEW

**Spectral classification:** Classes *O, B, A, F, G, K, M* are characterized by the appearance of different lines in stellar spectra.

**Colors of stars:** Spectral filters (defining wavelength bands) are used to measure magnitudes. Standardized bands are *U, B, V*, etc. These magnitudes are measurements of thermal radiation from a star, rather than indicative chemical elements (as is the case for lines used for spectral classification).

— color index CI $= B - V$ is a measure of stellar surface temperature: negative CI indicates a hot star, positive CI a cooler star.

— CI ranges between about −0.6 and +2.5.

**Hertzsprung-Russell (H-R) diagram:** data for stars displayed on graph of absolute magnitude versus spectral class or Color Index.

— most stars lie along the main sequence, a relatively narrow band.

— giants lie above the main sequence, white dwarfs below.

**Mass-luminosity diagram:** plot of luminosity versus mass shows a clustering of main-sequence stars around a line, expressed as the mass-luminosity relation. Giants and white dwarfs do not follow this line.

## NEW TERMS

| | | | |
|---|---|---|---|
| **bolometric magnitude** | **Hertzsprung-Russell (H-R) diagram** | **mass-luminosity relation** | **supergiants** |
| **color index** | **main sequence** | **O, B, A, F, G, K, M classification** | **U, B, V magnitudes** |
| **giants** | | | **white dwarfs** |

## PROBLEMS

**1.** For the color index CI = $B - V$, what are the central wavelengths of the two colors used?

**2.** Three stars have the following values for CI = $B - V$:

a. +0.5
b. 0.0
c. −0.5.

Which star is the coldest? Roughly what temperature does its color index indicate?

**3.** A star appears five times brighter in visual ($V$) light than in blue ($B$): Calculate the star's color index, $B - V$.

**4.** Three stars have the following absolute $B$ and $V$ magnitudes:

| star | $B$ | $V$ |
|---|---|---|
| A | −2.5 | −2.3 |
| B | −0.6 | −0.5 |
| C | +1.9 | +1.6 |

a. What is the color index for each star?
b. Which star is bluest? reddest?
c. Approximately what temperature would you assign for each star (use an H-R diagram)?
d. Which spectral classes would you place them in?

**5.** Three stars have the following apparent $B$ and $V$ magnitudes:

| star | $B$ | $V$ |
|---|---|---|
| A | 10.5 | 10.6 |
| B | 11.7 | 11.6 |
| C | 9.2 | 8.6 |

a. Which star is coolest?
b. If these are main sequence stars, what are their absolute visual magnitudes?

**6.** Sketch the H-R diagram indicating the location of the main sequence stars and the giants and white dwarfs. Indicate approximately the magnitude and CI or temperature values along the axes to your sketch.

**7.** Three main sequence stars have the same apparent magnitude of 9.7. Their temperatures are

a. 12,000 K
b. 5,800 K
c. 2,800 K.

Use a figure of the H-R diagram in the chapter to determine

a. What are the absolute magnitudes of the stars?
b. What are their distances (you will need to calculate these)?
c. Which is the most luminous star?
d. Which star is the reddest?

**8.** For stars represented on the main sequence of the H-R diagram

a. Which spectral classes are less massive than the Sun?
b. Which spectral classes contain stars that are older than the Sun? (First read Chap. 24).

**9.** Refer to Fig. 21.9, showing well-known stars on the H-R diagram.

a. Which is the third most luminous star shown?
b. Which star is the hottest? coolest?
c. Which star has the largest diameter?

**10.** In which spectral classes ($O$, $B$, etc.) are stars with the following spectral features?

a. ionized helium, but only weak hydrogen lines
b. strong lines of hydrogen
c. molecular bands of titanium oxide are strong
d. some hydrogen lines, but many lines of singly ionized calcium and iron.

**11.** A main sequence star has a surface temperature of 10,000 K.

a. What is its absolute magnitude?
b. What is its mass?
c. At what wavelength does it emit most radiation?

**12.** A star is 20 pc away and has a luminosity of $10^{35}$ ergs/sec.

a. What is the apparent magnitude?
b. If the star is on the main sequence, what must its surface temperature be?

**13.** Consider the following stars:

| star | apparent magnitude | spectral type |
|---|---|---|
| α Centauri | −0.72 | F0 |
| Arcturus | −0.06 | K2 |
| Vega | +0.5 | A0 |
| Spica | +0.9 | B1 |

a. Which star is the hottest?
b. Which star appears brightest?

**14.** A main sequence star of spectral class F5 is observed in a distant galaxy with an apparent magnitude of +9. What is its distance?

**15.** Betelgeuse has an absolute magnitude of −5.6, and its photospheric temperature is 2900 K.

a. What is its luminosity (in units of the solar luminosity)?
b. How large is it (in solar radii)?

**16.** Two stars that seem to be close together in the sky are Castor and Pollux. Their absolute magnitudes and parallaxes are:

Castor $M = +1.6$ $p = 0.072$ arc sec
Pollux $M = +1.0$ $p = 0.093$ arc sec

a. Which star is closest to us?
b. Which star appears to be the brightest?
c. What is the ratio of their luminosities?

**17.** Calculate the radius of a star that has an absolute magnitude $M = -5$ and a photospheric temperature of 4000 K.

**18.** A massive main sequence star has a luminosity forty times that of the Sun.

a. What is the star's mass in solar masses?
b. What is the luminosity in ergs/sec?
c. What will be its apparent magnitude if its distance is 100 pc?

**19.** Antares, one of the brightest stars, has an absolute magnitude $M = -4.5$, and its spectrum classifies it as an *M* star. Calculate the radius of Antares, relative to the Sun. If Antares were placed in the center of the solar system, how large would its angular diameter appear to an observer on Jupiter?

**20.** Use the mass-luminosity relation to find the luminosity of a star that has a mass of 10 $M_\odot$.

**21.** Describe the major differences between the H-R diagram and the mass-luminosity diagram and what types of observations are needed for them.

**22.** Use the mass-luminosity relation to find the masses of main sequence stars that are fifty times and one fiftieth as luminous as the Sun.

**23.** A main sequence star of 2 solar masses should have a CI $= B - V$ of how much?

**24.** The sequence of stellar classes is *O*, *B*, *A*, *F*, *G*, *K*, *M*. A famous mnemonic for this is Oh Be A Fine Girl, Kiss Me. Devise a better one.

# CHAPTER 22

# The Sun

Studying the Sun has immediate and practical results. Long before any written records were kept, the Sun's movements were followed to regulate the planting of crops, to signal the times of the year and the impending approach of winter. Their correlation with the seasons is still the basis of our calendar. In a more scientific way today, the study of the Sun will, we hope, lead to a better understanding of the weather patterns on Earth, for the Sun provides the driving energy, and we still do not understand many aspects of the solar-terrestrial relationship. The savings account of accumulated solar energy from past ages is now being drawn upon as we use fossil fuels, and we hope to be able to put the present flow of solar energy to equally good use as a clean, less expensive, and essentially inexhaustible supply.

The Sun also provides us with a **plasma** laboratory. (A plasma is a hot gas of ions.) Deep in its interior in a dense plasma, nuclear reactions are continually releasing the energy that is radiated away, and in the thin outer atmosphere there are other physical processes whose radiative output we detect. We can use our observations of the Sun to extend our knowledge of the behavior of plasmas and of fusion nuclear reactions.

There is, therefore, a very practical incentive for the study of the Sun. There is also a purely scientific basis for study, as the Sun is the star which is closest and thus easiest to examine in detail. The Sun is classified as a type G2 star lying in the main sequence about in the middle of the H-R diagram, an astronomical equivalent of the man-in-the-street. We might therefore hope that studying the Sun could give us guidance in understanding other much more remote stars. In this chapter, we shall describe many observable features of the Sun, and we shall reserve until Chapter 23 the discussion of models that attempt to describe the internal workings and the evolution of the Sun and other stars.

## 22.1 GENERAL FEATURES

Viewed from the Earth, the disc of the Sun has an average angular diameter of 31′59″, corresponding to a linear diameter of $1.39 \times 10^{11}$ cm. One arc sec resolution thus reveals detail on a scale of about 700 km, something to be remembered when looking at photographs of sunspots or eruptions silhouetted at the solar limb. Using orbital radii and periods for the planets, the Kepler-Newton Law yields a value of solar mass $M_\odot = 1.99 \times 10^{33}$ g. The average density of the Sun is thus 1.41 g/cm$^3$, about the same as Jupiter.

The visible parts of the Sun comprise its atmosphere, with the **photosphere** being the layer that we see as the bright disc. From this region, around 500 km thick, there emerges most of the Sun's radiation, and this is the layer that also contains the sunspots. The rotation of the Sun is clearly shown by the steady movements of the sunspots across the disc. (see Fig. 22.1). We find, however, that the Sun does not rotate at a rigid body. At its

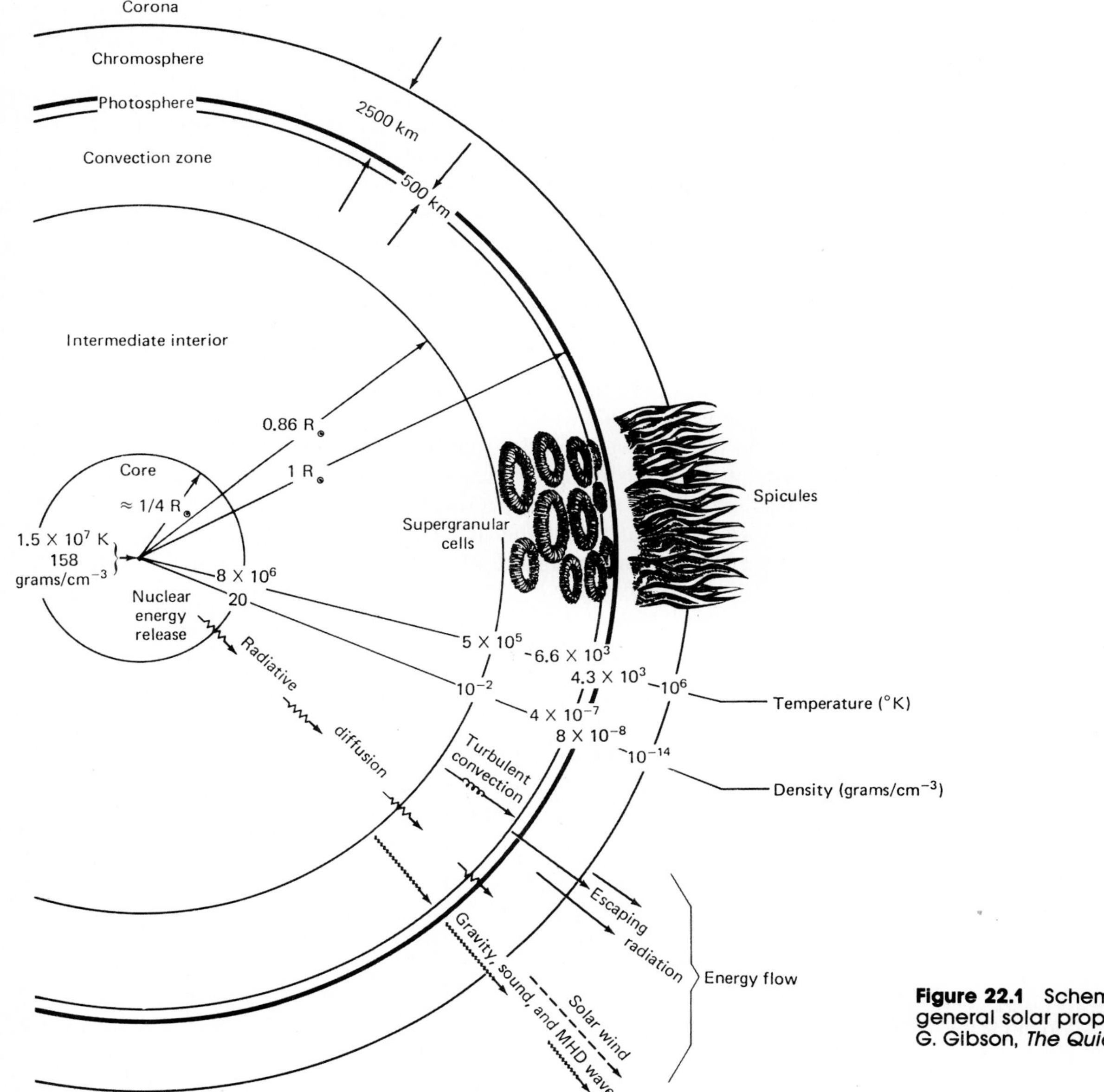


**Figure 22.1** Schematic diagram of general solar properties (Edward G. Gibson, *The Quiet Sun*, NASA).

equator, each rotation takes 24.8 days; at a solar latitude of 20° each rotation takes 25.4 days and at 40° latitude the rotation takes 26.7 days; these are sidereal periods. As seen from the moving Earth, the average rotation period is just over 27 days. The solar equator is tipped at 7°15′ to the plane of the ecliptic.

Above the photosphere are the **chromosphere** and **corona**, both emitting far less light than the photosphere. For a long time, both of these regions were visible only during eclipses, but they are now open to inspection at any time by using a specially adapted telescope, the **coronagraph**, which blocks out the light from the central disc. The chromosphere can also be studied by using special very narrow band filters that transmit light only at the wavelengths known to be emitted strongly in the outer layers. Even without occultation, the chromosphere can also be observed regularly at nonvisible wavelengths. The corona extends out for tens of millions of kilometers. The outer corona is gradually expanding through the solar system, where it is termed the solar wind.

The structure below the photosphere must be inferred from calculations and will be discussed in chapter 23 along with other stellar models. Here it will suffice to note that the source of the Sun's energy is in its central core where nuclear reactions are continually taking place. In these reactions, energy is released, carried by high-energy photons. This energy diffuses slowly up towards the photosphere from where it is finally radiated. The rate at which energy is released in nuclear reactions is equal to the rate at which the Sun's atmosphere radiates that energy. This luminosity ($L_\odot$) is $3.83 \times 10^{33}$ erg/sec and includes all radiation from radio through to X-rays, although the nonvisible wavelengths contribute very little to the total.

The temperature is not constant all over the solar disc or throughout the photosphere, but we can characterize the Sun by an effective temperature, $T_{eff}$. This is the uniform temperature that it should have in order to radiate as a black body the measured amount of energy from its known area

$$L_\odot = 4\pi R_\odot^2 \sigma T_{eff}^4 \text{ ergs/sec}$$

$R_\odot$ is the radius of the solar disc, $6.96 \times 10^{10}$ cm, and $T_{eff} = 5771$ K.

We should note that the radiation we receive from the Sun is a mixture of radiation that comes from different depths in the solar atmosphere (Fig. 22.2), where the temperatures differ significantly from the $T_{eff}$ we have just calculated. When we examine the solar spectrum from the near UV to the near IR, we find that the **shape** of the spectrum is best described by the radiation curve for a temperature of 5900 K (Fig. 22.3).

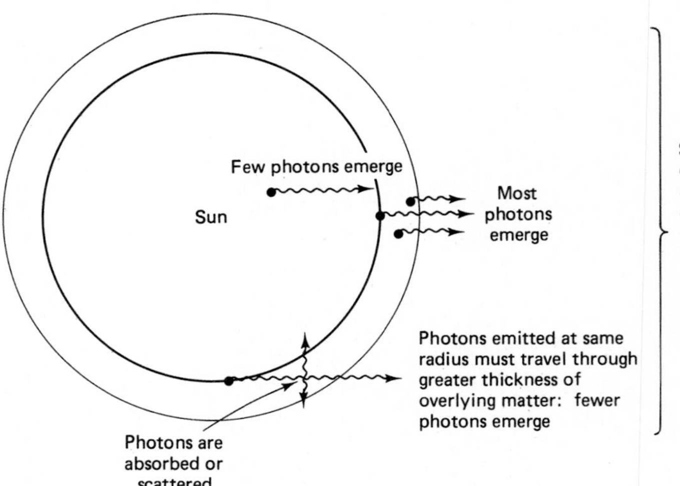


**Figure 22.2** Escape of photons from different depths in the solar atmosphere.

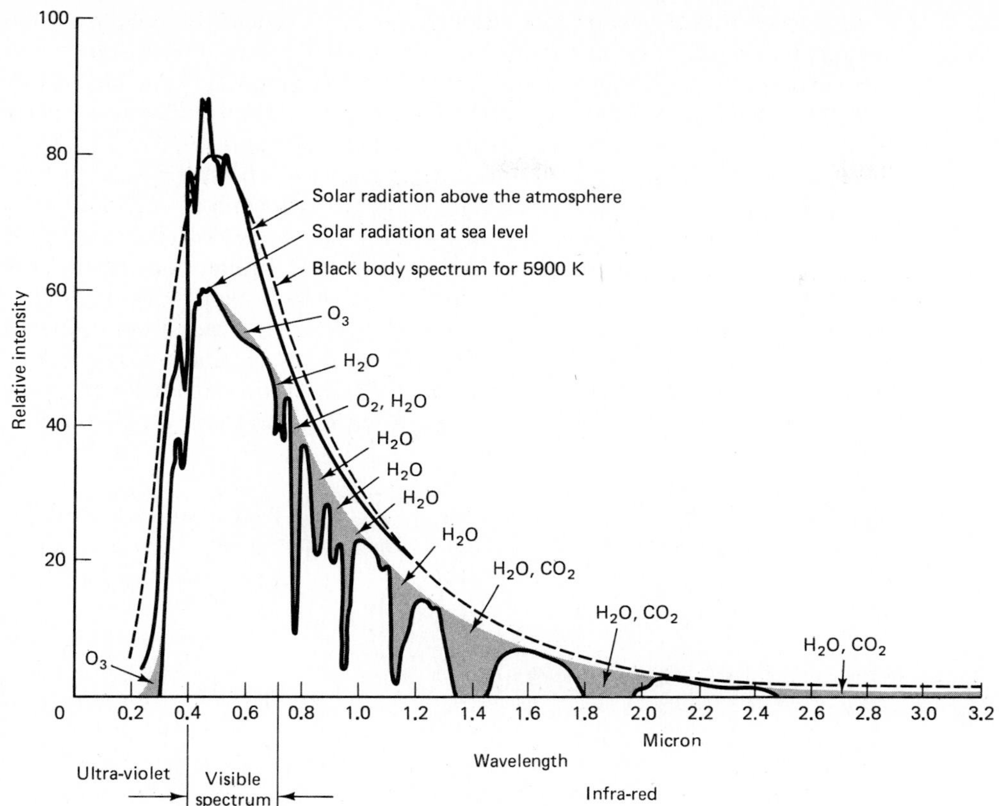


**Figure 22.3** The solar spectrum, from the near UV through the visible to the near-IR. Above the atmosphere, the solar spectrum is well represented by a black body spectrum for 5900 K. Absorption by the atmosphere especially by water vapor and carbon dioxide in the IR) reduces the total intensity and blocks some wavelengths completely (NASA).

Because of its extreme brightness, we do not usually think of the Sun in terms of magnitudes. Its apparent magnitude ($m = -26.7$) sets it apart from all other objects. This corresponds to an absolute magnitude $M = +4.8$ so that it is modestly embedded in the H-R diagram among the many G-type stars.

## 22.2 THE PHOTOSPHERE

Few photons can escape from great depths within the Sun. The thickness of overlying atmosphere absorbs or scatters most photons back into the interior. If we follow a path outward from the center of the Sun, we will find that the density, temperature, and pressure, decrease steadily. A stage will be reached where the density is sufficiently small so that an emitted photon has only a small chance of being reabsorbed or scattered and is instead much more likely to escape. This region is the photosphere, a relatively thin layer of about 500 km whose temperature drops from about 6600 K at its lower boundary to about 4300 K at the visible edge of the Sun.

The emission and absorption of visible photons within the photosphere is controlled in a complex way by the density of $H^-$ ions, hydrogen atoms that have each captured an extra electron. These ions are too fragile to exist at the higher temperatures below the photosphere, and above the photosphere there are too few electrons and hydrogen atoms to produce any appreciable number of ions.

The solar spectrum we observe is dominated by the light from the photosphere. Viewed from the Earth, the spectrum (Fig. 22.4) shows many absorption features that we can identify as arising from constituents in the

Earth's atmosphere—water and carbon dioxide especially. When allowance has been made for this atmospheric absorption, we find that not only the luminosity but also the spectrum shape is quite close to that for a black body with a temperature of about 5900 K. The spectrum shows a peak intensity around $5 \times 10^{-5}$ cm wavelength, and there are also thousands of dark absorption lines that must be attributed to absorption within the solar atmosphere. These lines were first cataloged in 1817 by Joseph Fraunhofer, who had earlier founded an Optics Institute in Munich, where he was responsible for major improvements in lenses and telescopes. Fraunhofer labeled the more prominent absorption lines $A$, $B$, $C$, etc., without being able to identify their origin. Some of these designations are still used. Fraunhofer's $D$ lines are now known to come from sodium and the $H$ and $K$ lines from calcium. Most of the lines come from iron and other metals. Hydrogen lines are strong: Fraunhofer's $C$ line (in the red) is now termed $H\alpha$ and the green $F$ line $H\beta$.

**Figure 22.4** The solar spectrum between 3900 and 6900 Å. Some of the absorption lines are produced by the Earth's atmosphere, but most are solar lines as identified (Mt. Wilson and Las Campanas Observatories).

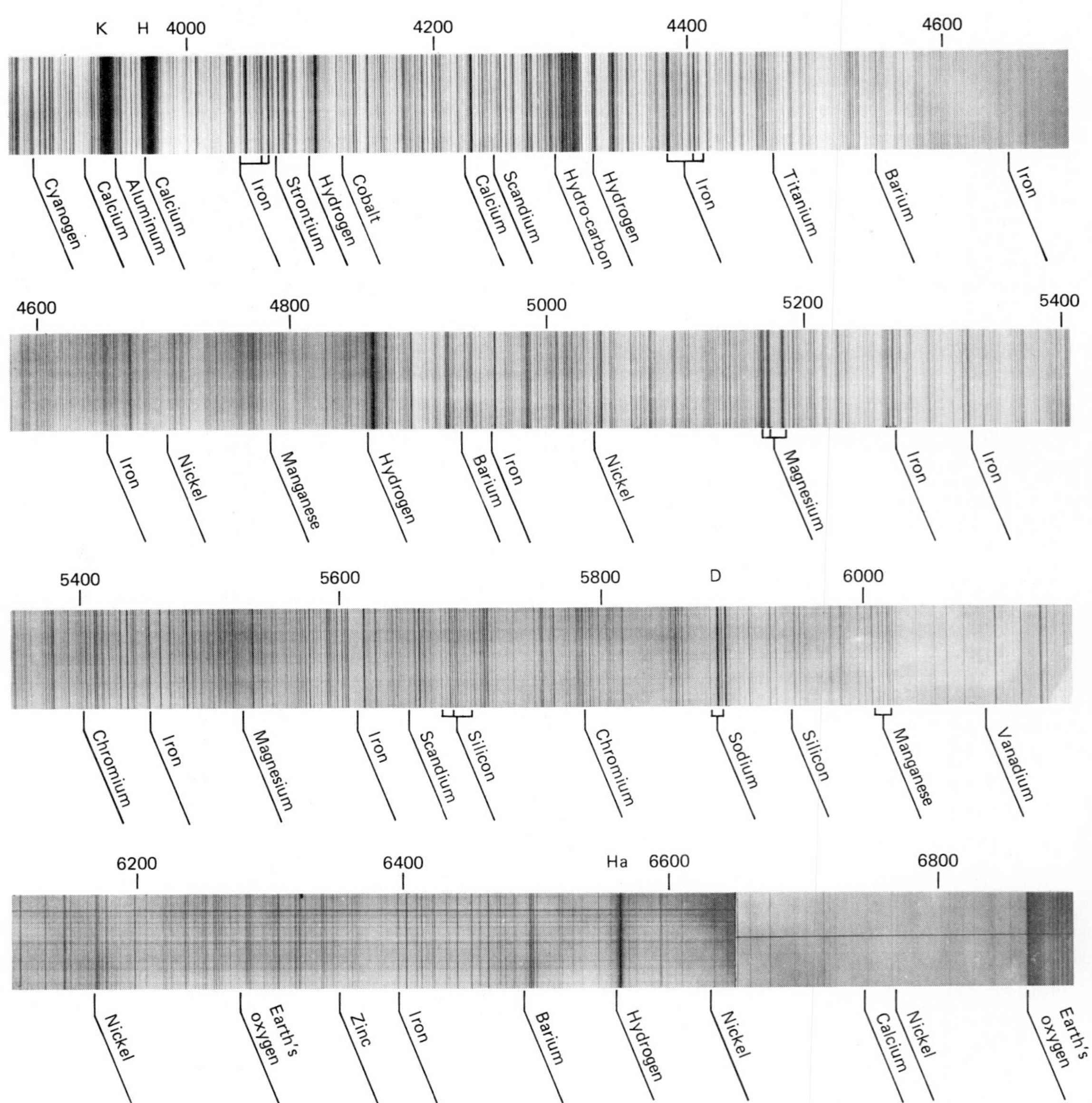

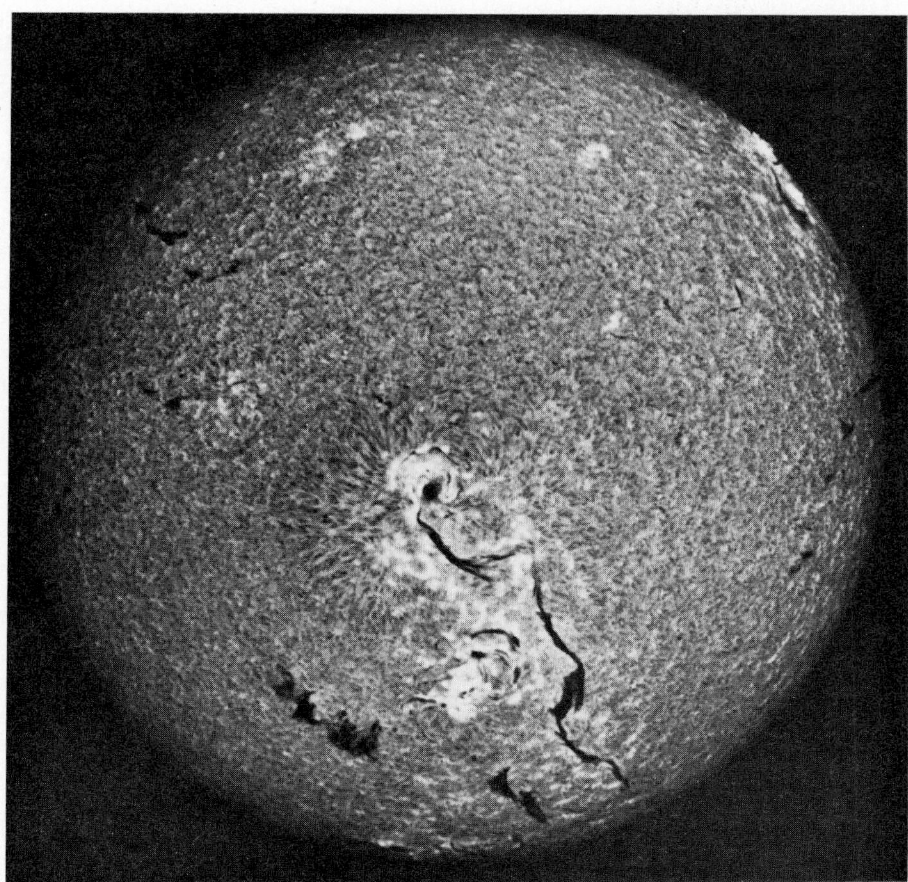

**Figure 22.5** Solar disc photographed in the hydrogen red Balmer H$\alpha$ lines (Sacramento Peak Observatory, A.U.R.A.).

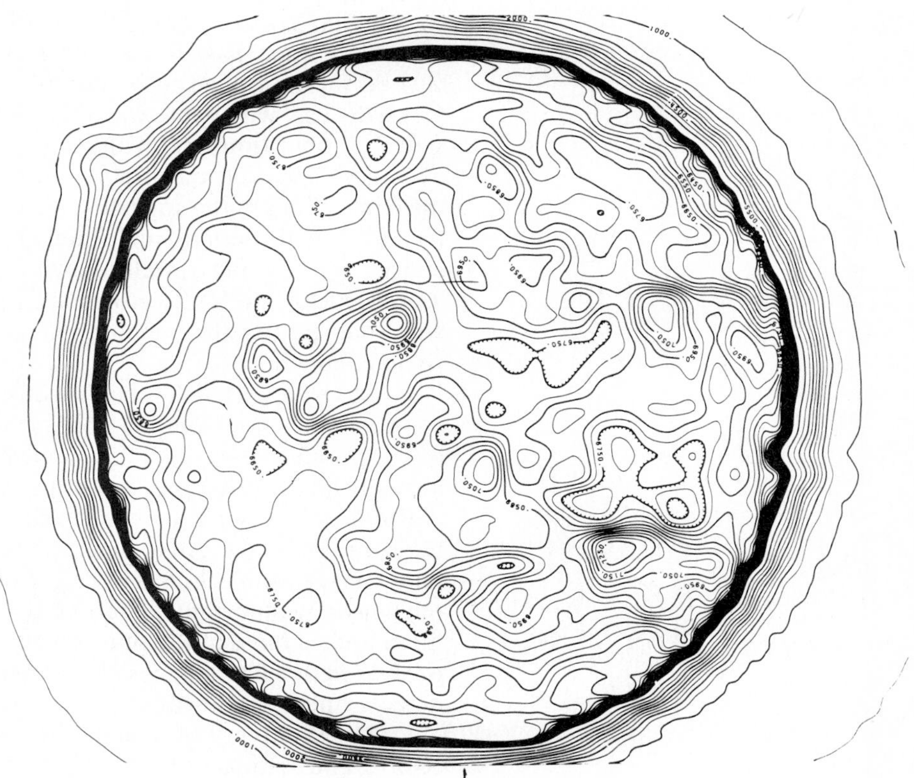

**Figure 22.6** The solar radio disc observed at a wavelength of 3.5 mm. Structure can be seen across the disc, and the radio brightness extends beyond the optical disc out into the corona (Mukul R. Kundu, University of Maryland).

***Table 22.1*** ABUNDANCES OF ELEMENTS IN THE SOLAR ATMOSPHERE[a,b]

| *Element* | *Atomic Number* | *Relative Number of Atoms* |
|---|---|---|
| Hydrogen | 1 | 1,000,000 |
| Helium | 2 | 67,700 |
| Carbon | 6 | 417 |
| Nitrogen | 7 | 87 |
| Oxygen | 8 | 692 |
| Neon | 10 | 98 |
| Sodium | 11 | 2.3 |
| Magnesium | 12 | 40 |
| Aluminum | 13 | 3.2 |
| Silicon | 14 | 38 |
| Phosphorus | 15 | 0.24 |
| Sulfur | 16 | 19 |
| Chlorine | 17 | 0.18 |
| Argon | 18 | 4.0 |
| Potassium | 19 | 0.13 |
| Calcium | 20 | 2.4 |
| Iron | 26 | 34 |

[a] A. G. W. Cameron, "Elemental Abundances in the Solar System," in *Essays in Nuclear Astrophysics*, (1982) Cambridge, England, Cambridge University Press; ed. by C. A. Barnes, D. D. Clayton and D. N. Schramm.

[b] Lithium, Beryllium, and Boron (atomic Numbers 3, 4, and 5, respectively) are present in minute quantities only, as are those in the range 21 to 25. Beyond Iron (26), the abundances drop off steeply.

The abundances of many elements can be deduced from their absorption lines, but the correspondence between these abundances and the spectral appearance is not simple. Each hydrogen atom has only one electron and the hydrogen spectrum is relatively simple and well known, but iron (Fe) has twenty-six electrons and can exist in many stages of ionization: FeII, which has lost one electron, FeIII, which has lost two electrons, and so on. Singly ionized calcium, CaII, produces the H and K lines. Each ionization stage has its own spectrum. Fortunately, no ionization stages beyond the first need to be considered in the relatively cool photosphere. Nevertheless, with approximately 100 chemical elements with many possible states of ionization, and within each many possible levels of excitation, the overall solar spectrum of absorption lines is extremely complex. Many of the fainter lines in the solar spectrum have not yet been identified, but based on Saha's calculations and later elaborations, the major elemental abundances are now fairly well known. They are listed in Table 22.1. Close to 93 percent of the Sun's atoms are hydrogen, nearly 7 percent are helium, and all others together amount to less than 0.1 percent. The helium deserves some comment.

Observations of the solar spectrum during the 1868 eclipse showed many emission lines not from the photosphere but close to it in bright extensions called **prominences** (Figs. 22.7 and 22.8). Some of these emission lines had never been seen before on Earth: A new element, Helium, was thus discovered by Norman Lockyer and named for Helios, the Sun. Even the 5900 K temperature of the photosphere is too low to excite helium atoms to emit their lines, and so helium's presence could only be seen in higher temperature regions above the photosphere. The helium abundance is still not as reliably measured as most of the others listed, and the present value is based in part on the relative abundances of hydrogen and helium among solar cosmic rays detected near the Earth.

The discovery of helium has an interesting sequel. The discoveries of X-rays and radioactivity by Henri Becquerel and the Curies in the 1890s provided the start of modern research into the structure of atoms. In 1902

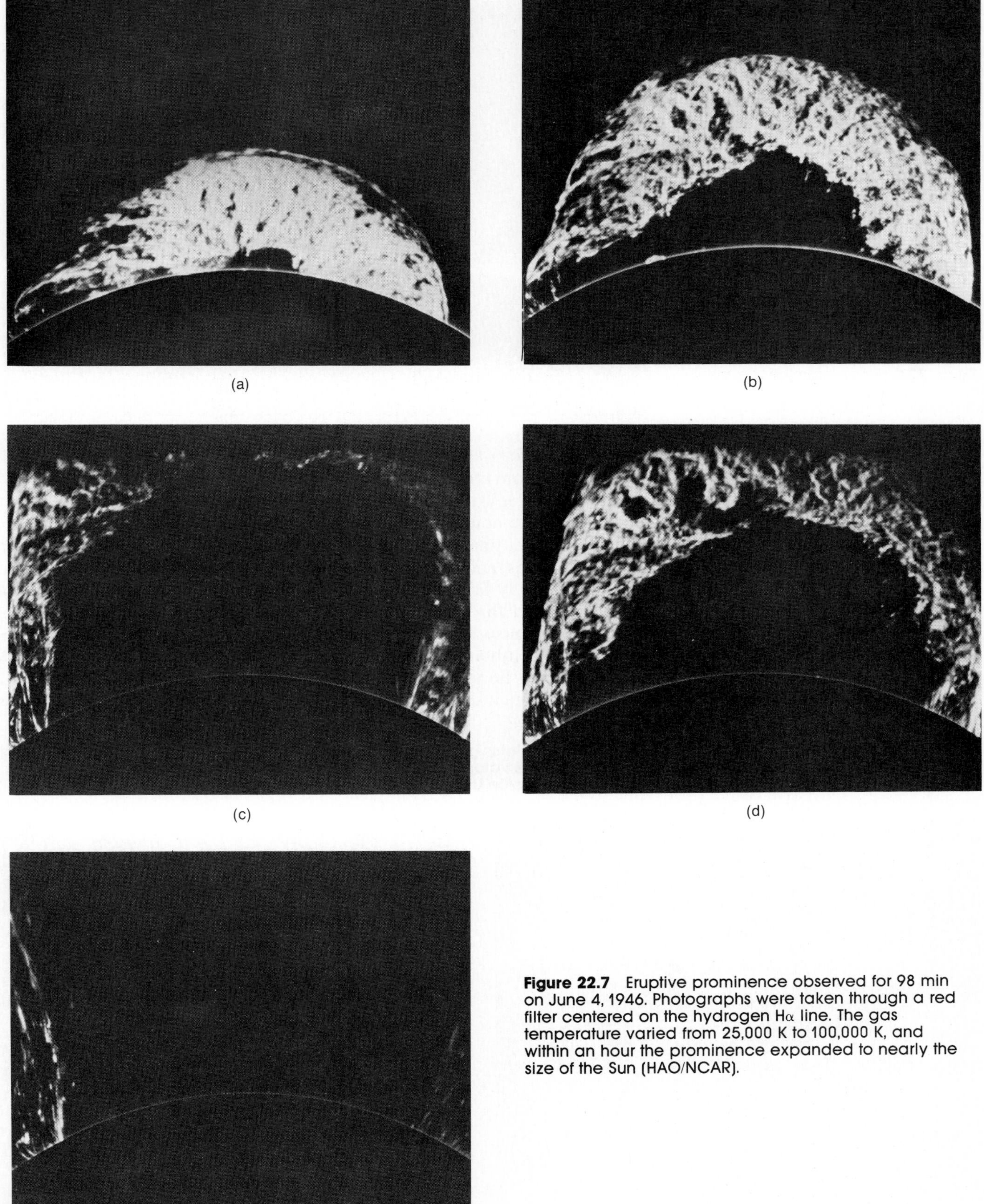

(a) (b) (c) (d) (e)

**Figure 22.7** Eruptive prominence observed for 98 min on June 4, 1946. Photographs were taken through a red filter centered on the hydrogen H$\alpha$ line. The gas temperature varied from 25,000 K to 100,000 K, and within an hour the prominence expanded to nearly the size of the Sun (HAO/NCAR).

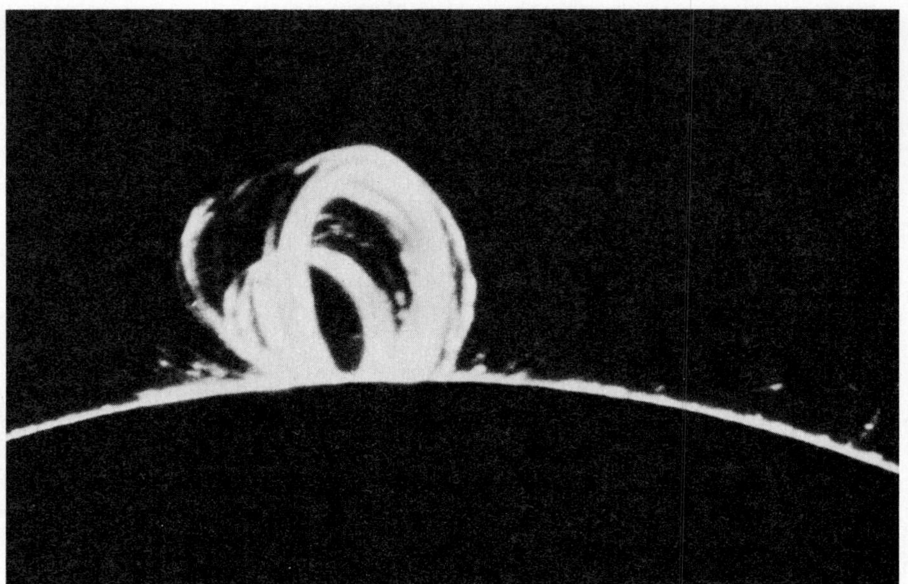

**Figure 22.8** Loop prominence, the result of a strong magnetic field (Sacramento Peak Observatory).

Ernest Rutherford and Frederick Soddy identified helium as the gas that accumulated from radioactive radium and thorium, and soon thereafter they found that the mysterious alpha particles given off by many radioactive substances were none other than nuclei of helium. Helium had escaped detection until that time for it does not normally combine with other atoms to form compounds, and it is only a trace constituent of the Earth's atmosphere.

The visually bright but bland solar disc shows considerable structure when examined through a telescope, especially when the Earth's atmospheric disturbances can be eliminated as when viewed from Princeton University's balloon-borne Project Stratoscope. We then see a surface that looks like a terrazzo floor, made up of many small and irregular bright areas, bordered by much darker lanes (Fig. 22.9a). This pattern of **granulation** is

**Figure 22.9(a)** Solar granulations shown at high magnification (Mt. Wilson and Las Campanas Observatories). (**b**) Solar granulations observed by balloon-borne telescope (Princeton University, Project Stratoscope).

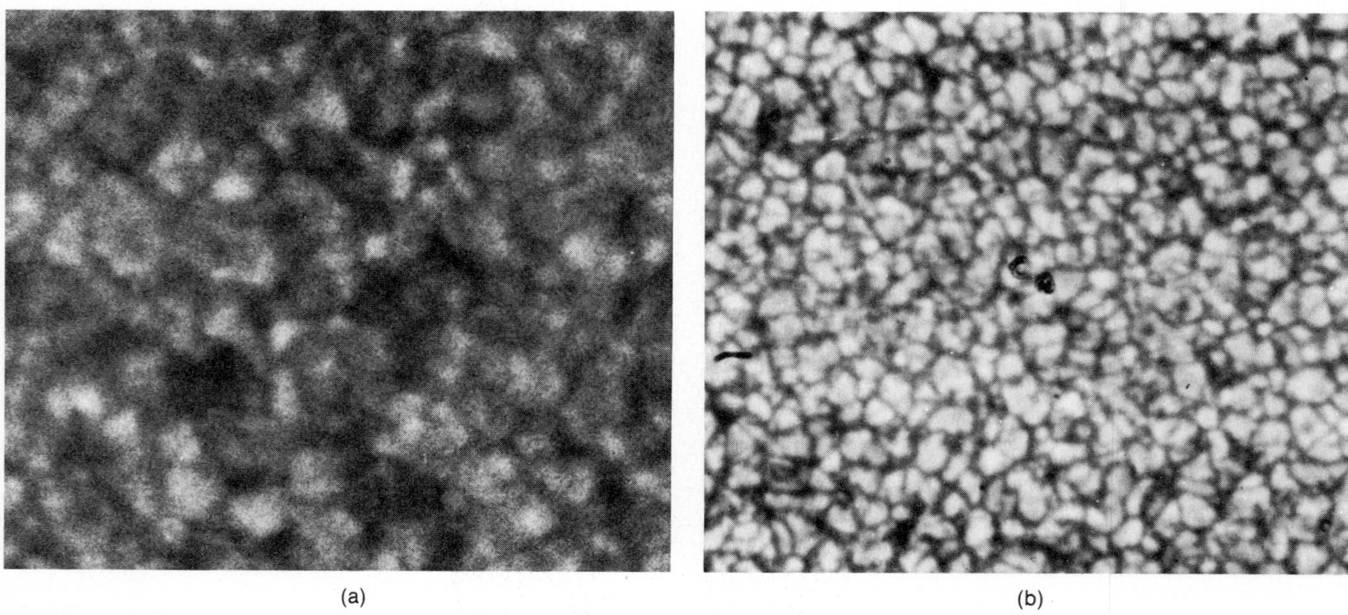

(a) (b)

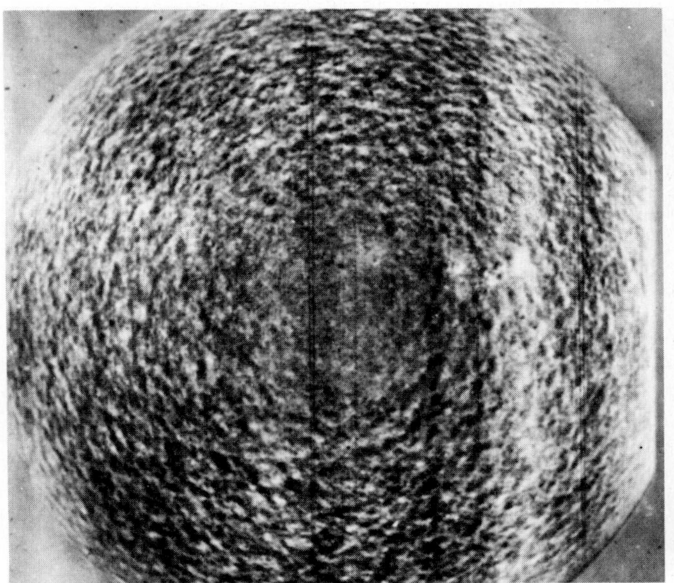


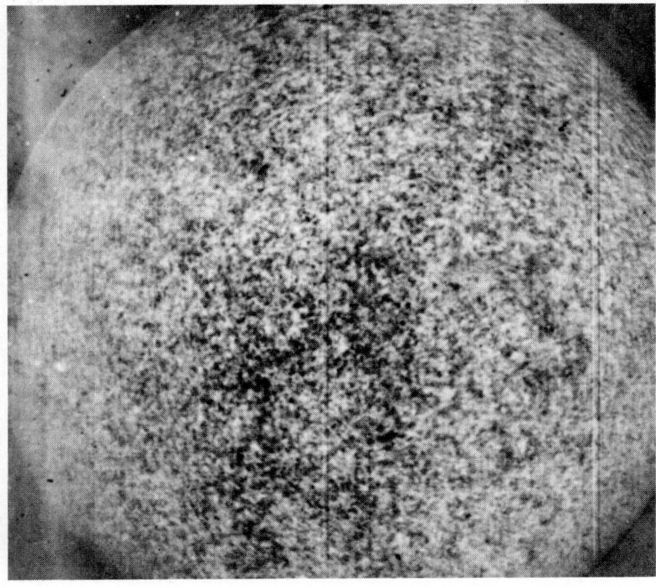


**Figure 22.10** Supergranulation observed at the wavelength of the calcium 6102.7 Å line. Photographs were taken 2.5 min apart, then photographically added (left image), and subtracted (right image). This analysis shows a slowly varying component (left) with a mean speed of 0.24 km/sec and an oscillating component (with a 5 min period and a velocity amplitude of 0.42 km/sec) (David K. Lynch, courtesy of San Fernando Observatory).

not steady. On a time scale of 5 to 10 min, individual cells come, go, and alter their boundaries. A typical granule is about 1000 km across, but granules as small as 300 km have been seen. The dark lanes are approximately 350 km wide.

The granular appearance is the result of considerable movement of gases in the **convection zone** beneath the photosphere. Hot gases are **convected** (carried) up and appear bright; some of their thermal energy is communicated to the photosphere, and the cooled gases then sink down in the lanes (Fig. 22.9b). Measurements show the grain centers to be about 50 to 100 K hotter than the darker lanes. Observations of Doppler shifts of the spectral lines show the upward and downward motions clearly. The vertical velocities are some hundreds of meters per second.

A larger-scale pattern of vertical movement has been seen by high-resolution spectroscopy where Doppler wavelength shifts can be measured to reveal a pattern of **super-granulation** (Fig. 22.10). The cells of the pattern have typical diameters of 30,000 km, and there are many smaller granules within. This is quite apart from the granulation that is at lower levels and lower temperatures.

## 22.3 THE CHROMOSPHERE

When the enormous brightness of the photosphere is blocked out, either by the Moon during an eclipse or with coronagraph, the spectrum of the **chromosphere** can be seen. Because of the very sudden appearance at the start of totality in an eclipse, its spectrum is known as the flash spectrum (Fig. 22.11). It shows a faint continuum (when compared to the photosphere) and bright emission lines. To the naked eye, the chromosphere has a pink-red color, mostly due to the hydrogen $H\alpha$ line at 6562 Å wavelength. In a spectrometer, numerous other and fainter lines can be seen.

The chromosphere is about 2500 km thick, with its temperature rising steadily from around 4300 K at its lower boundary to 100,000 K in its upper

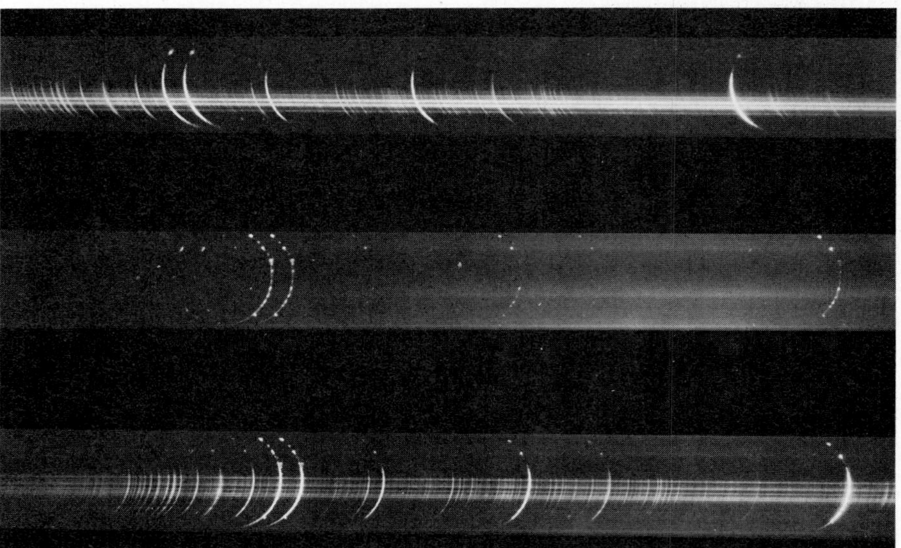

**Figure 22.11** Flash spectrum of the chromosphere photographed at the January 1925 total eclipse (Mt. Wilson and Las Campanas Observatories).

levels where it blends to a thin **transition region** joining the **corona**. Absorption lines will be seen in a spectrum if a hotter region is viewed through a cooler absorbing region. If, as we have here, a hot region is viewed against a dark background, then an emission line spectrum will be seen. This is the situation with the chromosphere. The temperatures in the chromosphere are high enough so that short-wavelength UV and X-ray photons will be produced.

The chromosphere is not a simple and uniform layer, but as can be seen along the solar limb (in silhouette) it consists of many narrow **spicules**, jets of glowing gas, typically 1000 km across and extending upwards for as much as 10,000 km (Fig. 22.12). These spikes last for about 15 min and form a constantly moving picture, with vertical speeds of around 25 km/sec.

**Figure 22.12** Solar spicules near the solar limb photographed in red ($H\alpha$) light (Big Bear Solar Observatory, California Institute of Technology).

## 22.4 THE CORONA

The brightness of the sky is due to sunlight that has been scattered in the Earth's atmosphere (Fig. 9.4). During an eclipse, not only is the direct sunlight from the photosphere blocked out, but so also is the light that can be scattered. As a result, we can see relatively faint glows, such as the spectacular streamers of the corona that extend outward for several solar radii. At one time, it was thought that the corona belonged to the Moon, but careful observation during the eclipses of 1851 and 1860 showed that this was not the case. The photosphere appears about a million times brighter than the corona, so that if even one millionth of the photosphere's light is scattered in the Earth's atmosphere, it will swamp the coronal light. Most of the coronal light actually originates in the photosphere, and comes to Earth after having been scattered by electrons in the inner corona and by dust particles in the outer corona. The coronal density of electrons decreases from $10^8/\text{cm}^3$ near the chromosphere to $10^5/\text{cm}^3$ at 4 to 5 solar radii, and $10/\text{cm}^3$ around 100 solar radii (Fig. 22.13).

The corona is heated to around $2 \times 10^6$ K by energy conveyed by waves through the lower levels but the details of the energy transport mechanism are not fully understood. Highly ionized atoms in the corona imprint their characteristic emission lines. As shown in Fig. 22.14, the stage of ion-

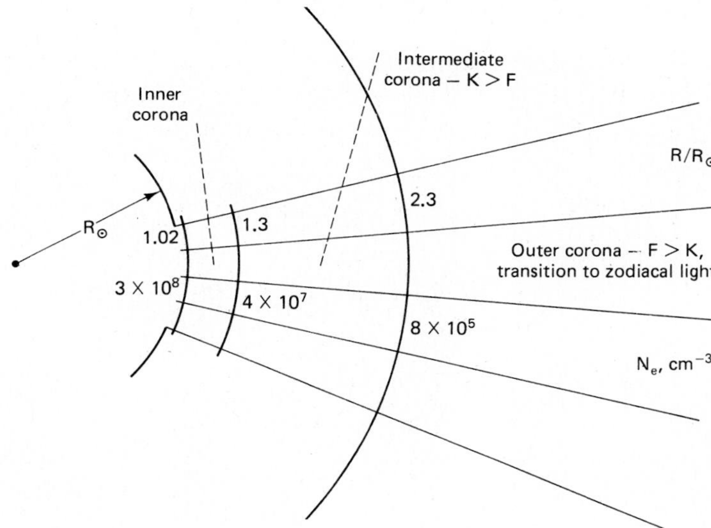


**Figure 22.13** Coronal regions, showing electron densities decreasing from inner to outer corona (NASA).

**Figure 22.14** Temperature profile across the boundary between the chromosphere and corona, showing the different stages of ionization at various temperatures. (From "Ultraviolet Astronomy" by Leo Goldberg. Copyright 1969 by Scientific American, Inc. All rights reserved.)

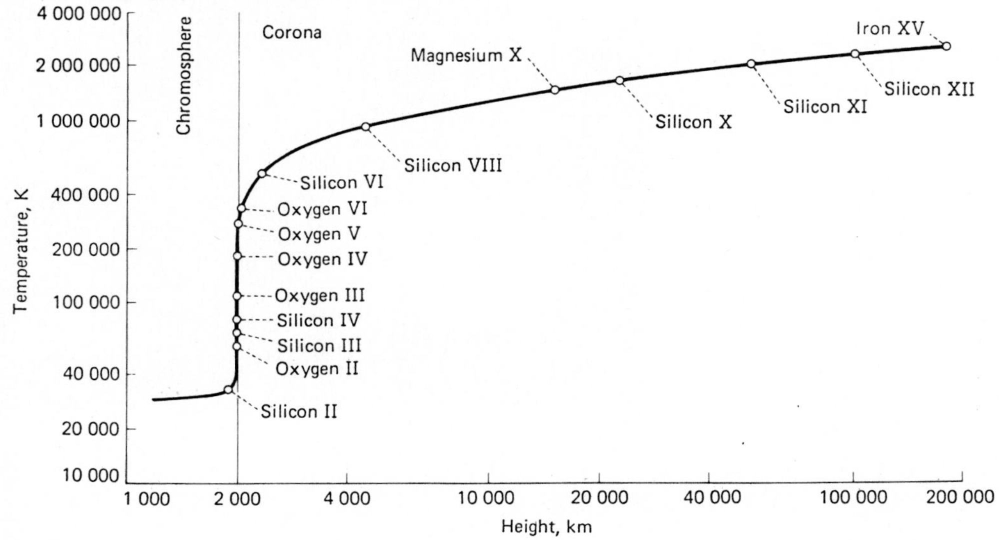

ization is very sensitively dependent on coronal temperature, and the identification of particular lines provides us with information on the temperature and density at different levels.

We can use Wien's Law to calculate that the corona should be radiating most strongly around wavelengths of $0.3/T = 0.3/(2 \times 10^6) = 1.5 \times 10^{-7}$ cm, in the X-ray region. This has been observed from rockets and satellites. The first X-ray observations of the Sun were carried out in 1948 by scientists of the U.S. Naval Research Laboratory, using V-2 rockets captured from Germany in WWII.

Early experiments detected the X-radiation and started to identify the ions responsible. Full X-ray images of the Sun were later obtained by a group at American Science and Engineering, Inc. Using a glancing angle telescope (chap. 10), they were able to show that the X-ray emitting regions are irregularly distributed across the disc. The X-ray emission is seen to be very strongly controlled by magnetic fields that extend up into the corona from active regions in sunspots.

The corona also radiates in the UV, undetectable from the Earth's surface but now well mapped from satellites and especially from Skylab (Figs. 22.15, 22.16 and 22.17). Some emission lines have been recognized as coming from very highly ionized states of multielectron atoms such as

(a)

(b)

(c)

**Figure 22.15a** Solar corona during the eclipse of November 12, 1966. The photograph was made through a filter that decreased in density from center to edges so that the brighter inner parts of the image would not be overexposed, yet the fainter outer parts could be seen. Three streamers can be seen coming from prominences. Venus can be seen at the left (HAO/NCAR). **(b)** Solar corona during the eclipse of March 7, 1970. This eclipse, near the time of maximum activity on the solar 11-year cycle, shows streamers nearer the poles than are seen in the 1966 eclipse (HAO/NCAR). **(c)** Solar corona, photographed from the lunar surface 1 min before sunrise July 31, 1971 during the Apollo 15 mission (NASA).

**Figure 22.16** Solar corona photographed with the Apollo Telescope Mount Coronagraph from Skylab December 27, 1973. The occulting disc of the coronagraph covers the solar disc and corona out to a distance of 1.6 $R_{\odot}$, permitting the corona to be observed out to about 6.5 $R_{\odot}$. Comet Kohoutek is visible shortly before its closest approach to the Sun (HAO/NCAR, NASA).

Ne VII 465

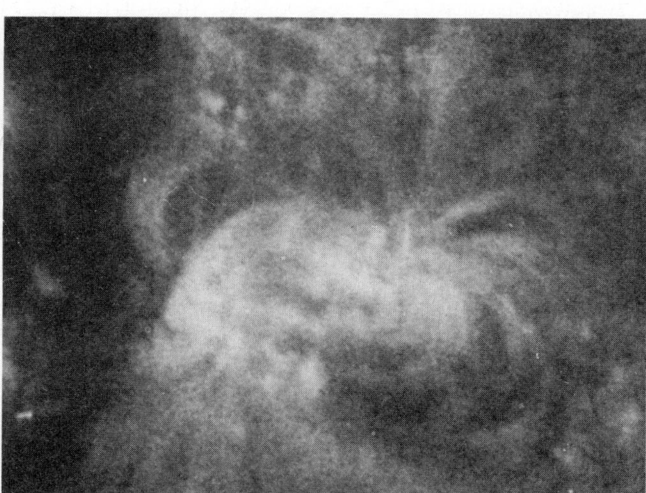

Mg IX 368

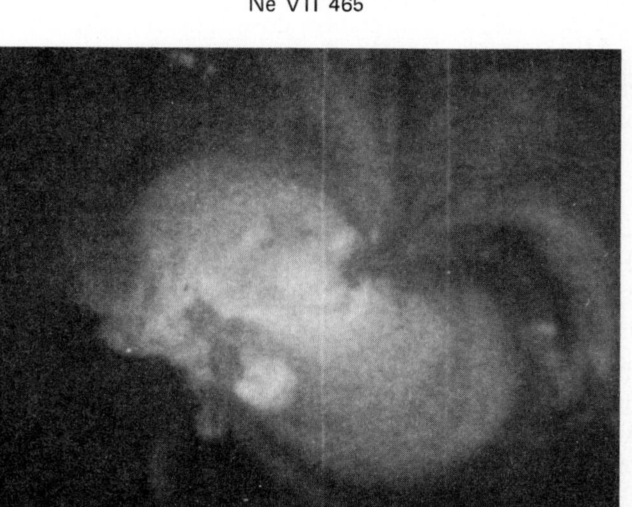

Fe XV 284

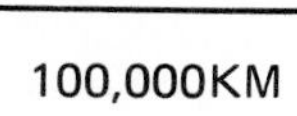

100,000KM

**Figure 22.17** Loops observed from Skylab at UV wavelengths. Temperatures between 600,000 K and 1,200,000 K are needed to produce highly ionized atoms such as NeVII, MgIX, and FeXV. These lines cannot be produced in the cooler photosphere. The loops extend from sunspots up into hotter regions. (Neil R. Sheeley, Naval Research Laboratory).

iron, but many spectral lines remain unidentified, for we cannot yet fully simulate solar conditions in the laboratory.

Some insights have been obtained unexpectedly in a fusion research program designed to investigate nuclear reactions in a plasma at very high temperatures. Some of the hot plasma briefly touched the stainless steel container walls, and produced spectral lines that had also been seen in the solar spectrum but not yet identified. These spectral lines were not observed in similar experiments where ceramic containers were used instead of steel. It is, therefore, inferred that some of the coronal spectral lines most probably came from very high-temperature iron (Fe) ions.

The corona and chromosphere are also sources of radio waves, and there has been extensive radio mapping of the Sun at wavelengths such as 3 nm, as shown in Fig. 22.6. The transparency of a hot plasma to radiation depends on the number of electrons per cubic centimeter (the electron density) in the plasma and the wavelength of the radiation. Because of the variation of electron density with depth in the corona, the transparency will vary with depth. Accordingly, observations at different wavelengths allow us to probe the electron densities and temperatures at different depths in the corona.

A major discovery from Skylab was the presence of **coronal holes**—dark regions in an otherwise bright corona. These holes, regions of lower density and temperature, are thought to be controlled by strong magnetic fields.

The high temperature of the corona causes it to expand, and a thin **solar wind**, mostly of high-speed protons and electrons, spreads out through the solar system at all times. The existence of the solar wind had been suggested by Ludwig Biermann in 1950 to explain some of the movements observed in comet tails. Direct observations came in 1959, first by the Russian space probe Lunik III and then by the U.S. Mariner II. In the vicinity of the Earth, the solar wind contains on the average about 80 protons and electrons per cubic centimeter, with fluctuations up and down by a factor of ten. Some heavier nuclei (for example, He, C, and O) have also been detected.

This solar wind sweeps by the Earth at a speed of about 500 km/sec. Because it consists of electrically charged particles, the solar wind acts as a partly transparent barrier to the cosmic rays (also electrically charged) coming from outside the solar system. The numbers of cosmic rays detected on Earth vary with the 11-year solar cycle under the influence of the solar wind. The solar wind also leaves its imprint on the lunar surface where the lunar materials steadily accumulate a radiation dose that diminishes with depth below the lunar surface. The surface layers are slowly churned up (''gardened'') by meteorite impacts, and measurements of the variation of accumulated radiation dose with depth have been used to estimate the rate of turnover.

## 22.5 THE ACTIVE SUN

The **active Sun** is the term used when very rapid and temporary, or major but slower variations take place. A number of these phenomena have been cataloged, and we start by considering the sunspots.

**Sunspots** Galileo not only used his new telescope to observe the sunspots, but also saw them move across the solar disc. He correctly deduced that this placed the spots on the Sun and not in the Earth's atmosphere (where they had earlier been thought to occur, to preserve the unblemished appearance of the Sun). Further, he was able to use their movement to measure the Sun's speed of rotation.

The spots appear dark against the brighter photosphere (Figs. 22.18 and 22.19). They have very dark centers (umbra) with gray edges (penumbra) and are often irregular in shape with a variety of sizes. With a temperature of about 4500 K, the sunspots are cool relative to the photosphere. The sunspots have very strong magnetic fields. On the Earth there are only two magnetic poles, but the Sun displays many. In a pair of sunspots, one will behave like a magnetic north pole and the other like a magnetic south pole. Through a solar cycle, the preceding spots in each pair will all be north, and in the next solar cycle they will all be south. The strength of their magnetic fields can be measured by the **Zeeman effect**, named for its Dutch discoverer. In a strong magnetic field, the energy levels in atoms are changed, altering the wavelengths of the photons that the atoms can absorb or emit. Precise wavelength measurements performed on the sunspot spectra for the first time in 1908 by G. E. Hale demonstrated the presence of the magnetic field.

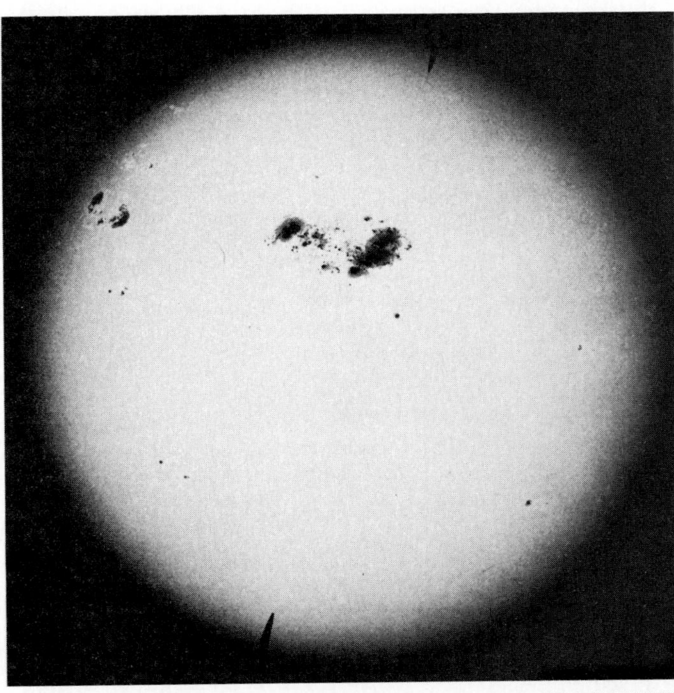

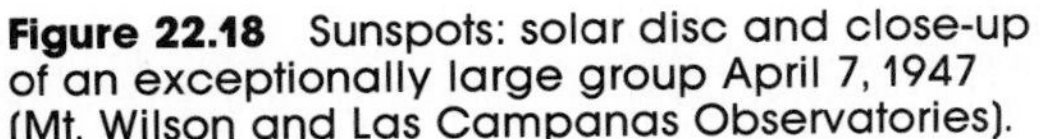

**Figure 22.18** Sunspots: solar disc and close-up of an exceptionally large group April 7, 1947 (Mt. Wilson and Las Campanas Observatories).

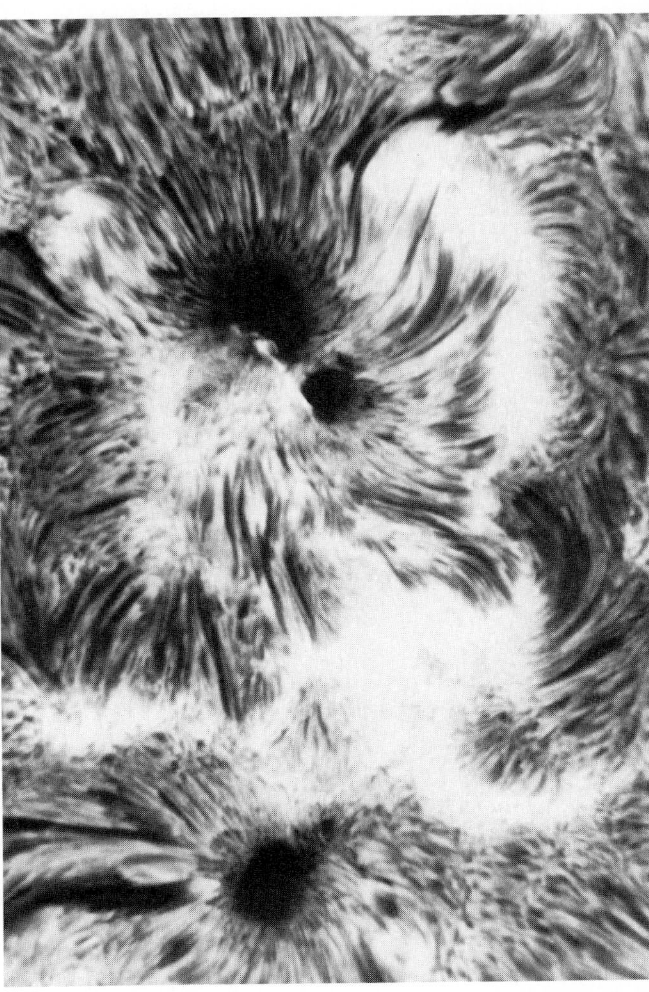

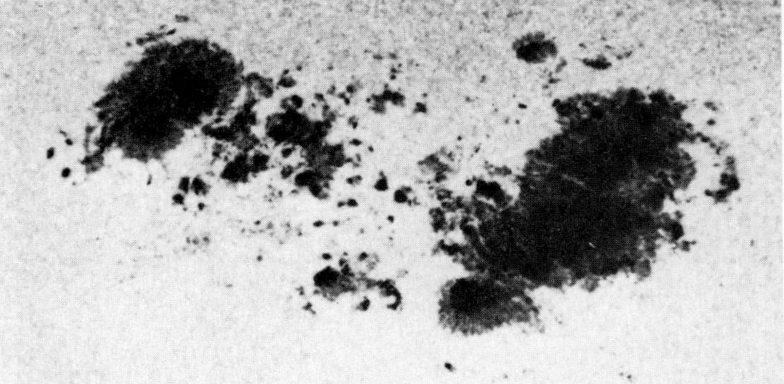

**Figure 22.19** Sunspot group photographed in red (H$\alpha$) light (Big Bear Solar Observatory, California Institute of Technology).

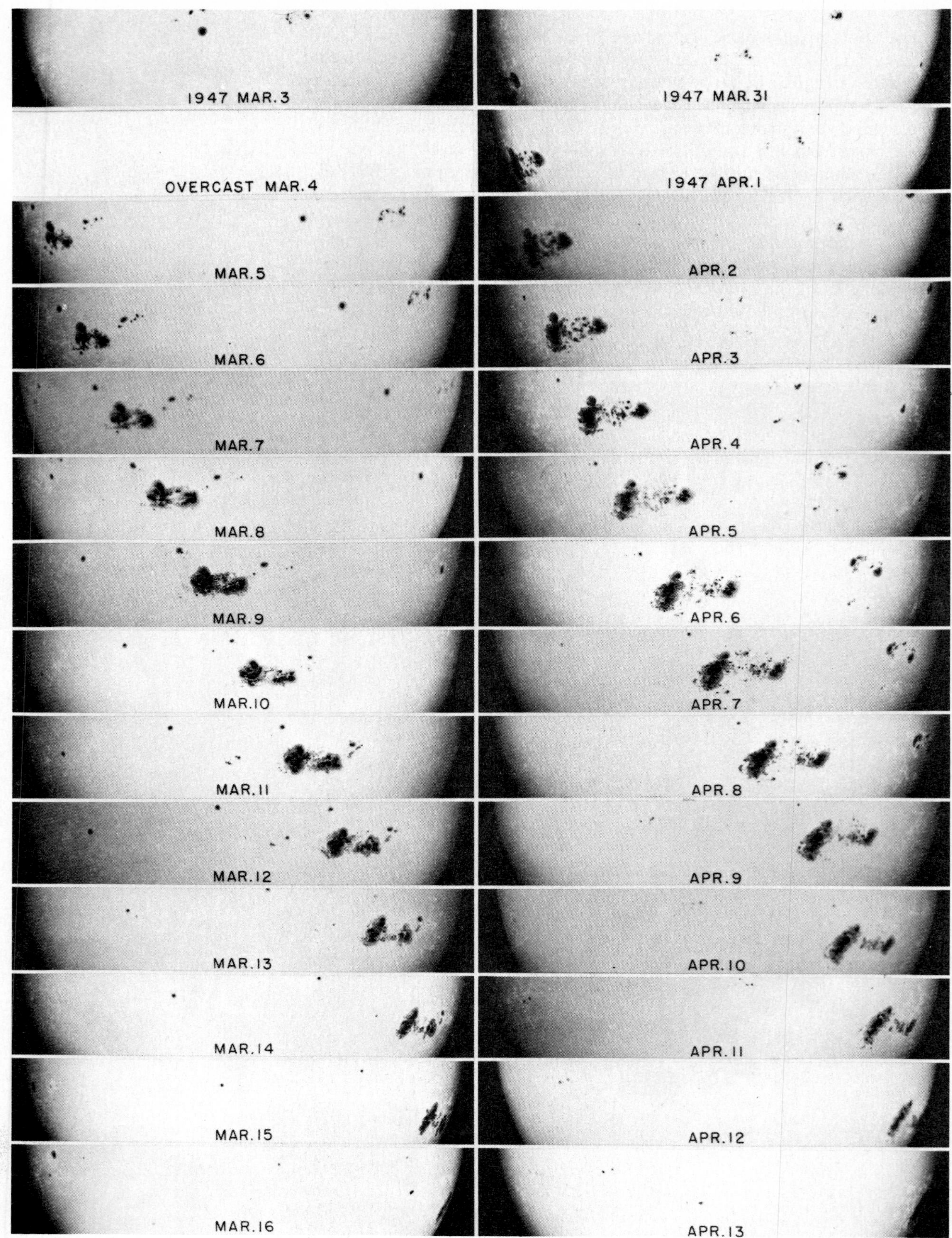


**Figure 22.20** Daily record of sunspot groups showing rotation of the solar disc. The group disappears around the limb on March 16 and reappears on March 31 (Mt. Wilson and Las Campanas Observatories).

After Galileo's discovery, records were kept of the number of sunspots that were visible. Some spots appeared alone, others in groups; some last long enough to be followed through several solar rotations (Fig. 22.20).

Around 1850 a recurrent pattern was noticed in the accumulated record: The number of sunspots was not steady but varied considerably. Records from about 1750 were fairly complete; they showed a very marked cycle with about 11 years between successive peaks of activity. Sunspots are now monitored regularly, and the sunspot number is a useful index of the solar activity. For example, a more variable and extended corona is seen when sunspot activity is at a maximum.

The cycles are not regular enough to be predicted with accuracy, but they are reliable enough to provide a schedule for coordinated observing programs, geophysical as well as astronomical, to try to extend our understanding of the ways in which the solar activity affects the Earth. The International Geophysical Year (IGY) was planned for 1957 to 1958 to cover the period of solar maximum, and the first artificial satellites, such as Sputnik, were launched as part of the IGY program. In 1964 to 1965 the International Year of the Quiet Sun was the umbrella program. More recently, the Solar Maximum Mission satellite, launched in February 1980, carried seven different instruments to cover the solar spectrum from the IR to gamma rays. In the scientific planning for these programs, scientists from many countries participate. Although only a few have the facilities to undertake satellite construction, launch, tracking and data telemetry, magnetic tapes of the data can be shared later so that participation may be extended to "Guest Observers." During these cooperative ventures, ground-based observations are correlated with those on spacecraft, and special periods are designated for simultaneous coverage over a wide range of wavelengths.

**Solar flares** Violent eruptions occur in the solar atmosphere. **Flares** (Fig. 22.21) are the most violent, starting without warning and lasting from minutes to a few hours. During this time, an enormous amount of energy can be released in the form of high-speed nuclei and electrons (solar cosmic rays) and radiation covering the spectrum from the radio to gamma rays (Fig.

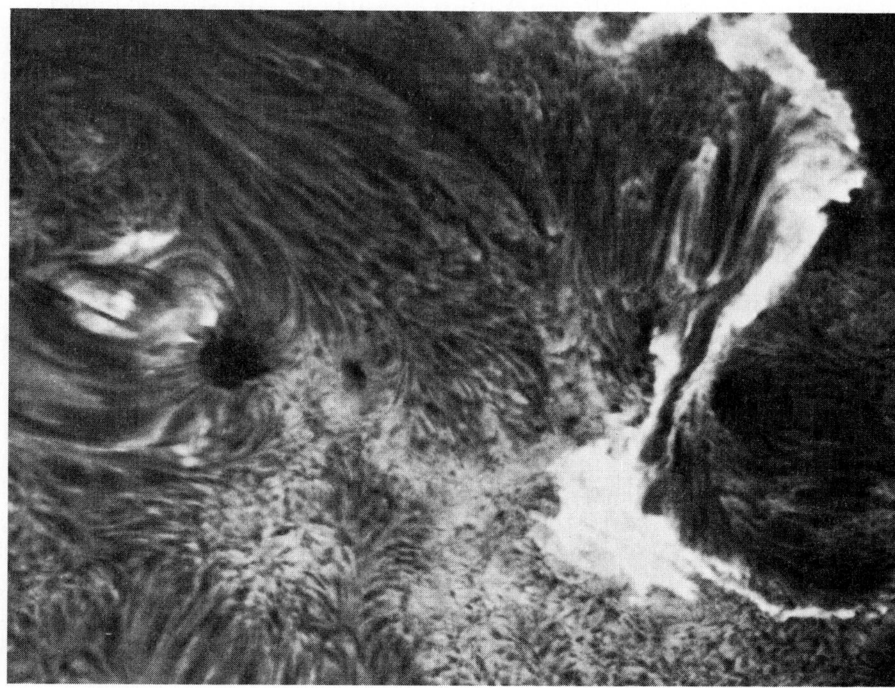

**Figure 22.21** Solar flare (right) observed near some small sunspots, as seen in red (Hα) light (Sacramento Peak Observatory).

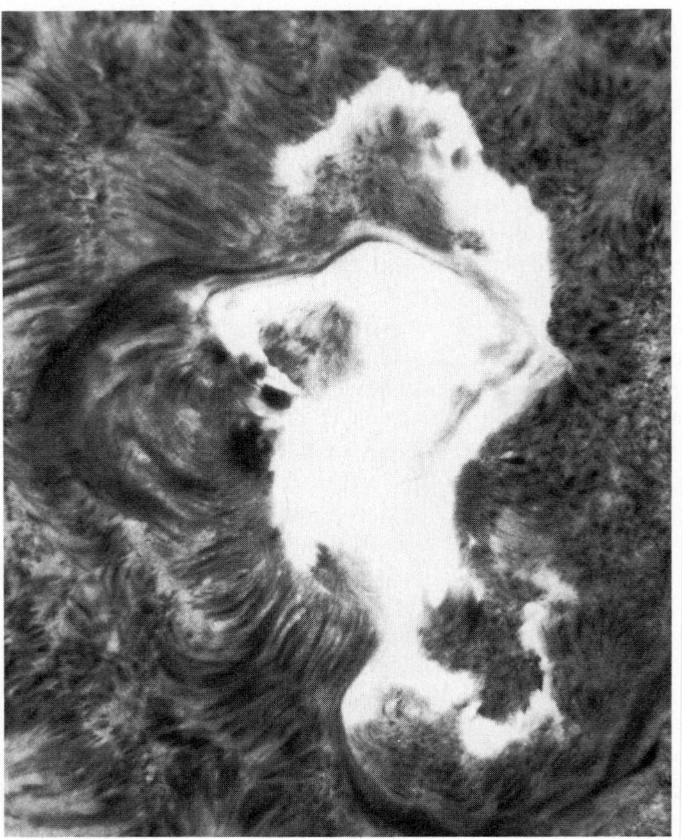

**Figure 22.22** Solar flare associated with the release of high-speed solar protons, photographed in red (Hα) light August 7, 1972 (Big Bear Solar Observatory, California Institute of Technology).

22.22). Within a flare, the temperature can exceed $5 \times 10^6$ K. The suddenness and the intensity of the solar flare particle radiation even when it reaches the Earth poses a radiation hazard to astronauts and to occupants in supersonic planes at very high altitudes. Supersonic airplanes need to be warned of solar flares, to be able to dive to low altitudes for protection. The arrival of flare particles triggers the auroral displays at polar latitudes and interrupts radio communications.

There are several other phenomena of solar activity. **Filaments** appear dark against the photosphere near sunspots, but seen at the solar limb they show up as bright turbulent flamelike streams of gas and are called **prominences**. They were shown in Fig. 22.8. Bright areas around the sunspot regions are called **plages**. Solar radio emission is variable, as are the coronal shape and the solar wind.

## 22.6 MAGNETIC FIELDS

The solar magnetic field is still a puzzle. We have a general and plausible model that provides a framework for further study, but many important details are far from being understood.

Strong magnetic fields have been detected in sunspots, through the Zeeman effect. Magnetic fields are also in evidence in the behavior of the glowing gases in flares, prominences, and other phenomena of the active Sun. We know that electrically charged particles (ions and electrons) will follow curved paths in a magnetic field. The complex shapes of prominences, like smoke being blown by a turbulent wind, provide a clear indication of the presence of a magnetic field.

Some ejected particles have sufficient speeds to get away from the Sun, and the paths they follow are defined by the solar magnetic field even as they spread over great distances through the solar system. Spacecraft have detected this extended magnetic field and the particles in it and monitored the ways in which this distorts the Earth's own magnetic field and even the magnetic field around the distant Jupiter.

The main question is still: What is the origin of the solar magnetic field? We think that the Earth's magnetic field comes from an internal electric current, a dynamo in effect, and that the Sun must also have an internal dynamo. Unlike the Earth, the solar rotational speed varies with latitude; we deduce that the rotation speed also varies with depth below the photosphere. As a result, an initially well-ordered field will after a few solar rotations be highly twisted. Some irregularities may be washed out but others may grow through the surface to form sunspots. These irregularities may extend beyond the surface if too much energy is suddenly concentrated in one region so that explosive release occurs. A plausible scenario for behavior of the solar magnetic field can be constructed, but it must be emphasized that much detail remains to be confirmed or modified.

## 22.7 LONG-TERM CHANGES

Variability in the Sun's behavior has shown up in several ways. Examination of the photosphere revealed the granulation and its rapid changes; monitoring of the sunspots drew attention to the existence of an 11-year cycle. Flares, prominences, and other outbursts display the Sun's activity on a relatively short time scale. It is, therefore, quite reasonable to ask whether there is any evidence for solar variability over much longer times. Once again, the sunspot record has given us the first clue.

The 11-year cycle was first noted in the sunspot record from around 1750. Around 1890 E. W. Maunder and G. Sporer independently drew attention to the historical record that *no* sunspots at all were seen between 1645 and 1715. No further notice seems to have been taken of this remarkably

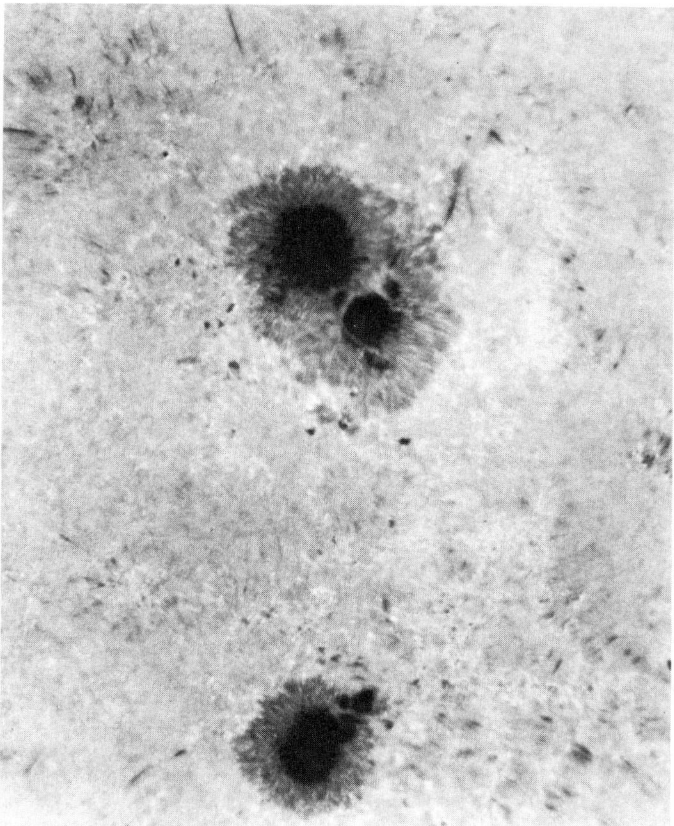

**Figure 22.23** Sunspot group photographed in white light May 21, 1972 (Big Bear Solar Observatory, California Institute of Technology).

quiet period until 1976 when John Eddy of the High Altitude Observatory in Boulder reviewed the older records and, going well beyond the sunspot analyses of Sporer and Maunder, showed that the 1645 to 1715 period was unusual in several ways. The period of reduced sunspot activity is now called the Maunder minimum, and it shows very clearly in Eddy's tabulation of the data (Fig. 22.24).

Suspecting that the absence of sunspots might be one indication among many of changed solar activity, Eddy examined the geophysical record for signs of possible effects on the Earth. These soon became obvious. Auroral displays in the night skies at northern latitudes are now known to be as-

**Figure 22.24** Annual mean sunspot numbers 1610 to 1979. Note the 11-year periodicity since about 1717, and the very small number of sunspots between about 1645 and 1715. This period is now termed the Maunder minimum (John A. Eddy, High Altitude Observatory).

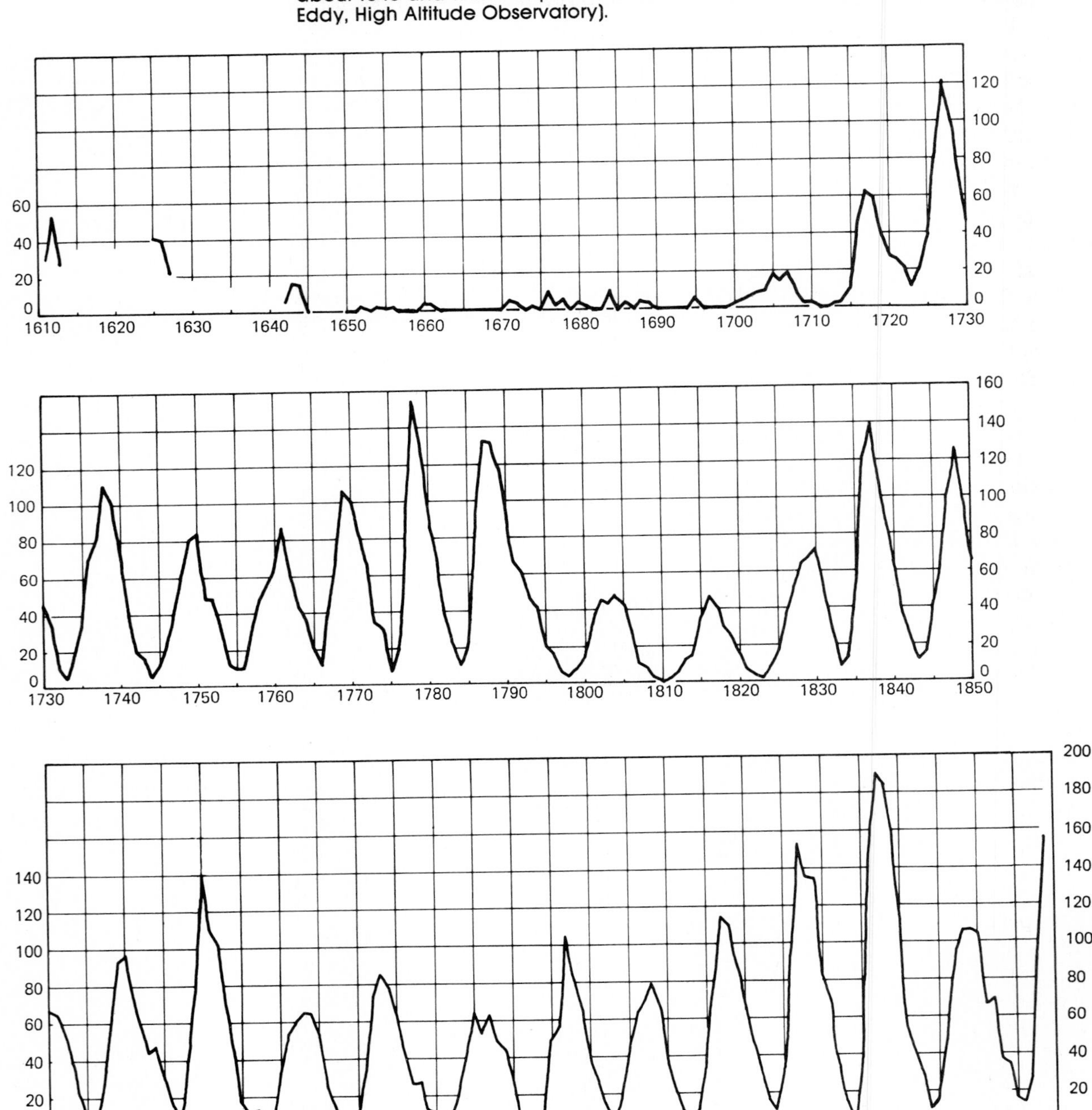

sociated with solar activity. Eddy found that there were very few reports of aurorae between 1650 and 1700 with a significant increase thereafter.

In years of high solar activity, the Earth is screened against some of the cosmic ray particles coming from outside the solar system so that the production of the radioactive isotope $C^{14}$ in the stratosphere is reduced. The $C^{14}$ mixes throughout the atmosphere and can be detected in the atmospheric carbon dioxide. In this form it is absorbed by trees and can be detected in wood. The analysis of $C^{14}$ in the growth pattern shown in datable tree rings, shows increased $C^{14}$ during the Maunder minimum. Contemporary accounts of the solar corona during eclipses visible from Europe between 1652 and 1715 provide additional confirming evidence—no report tells of coronal streamers; the reports mention only very small coronas, extending no more than 3 arc min from the Sun ($\frac{1}{10}$ of its diameter). It is very hard to escape the conclusion that solar activity was truly minimal for around 65 years.

These studies prompted a further examination of the sunspot records. Richard Herr of the University of Delaware has looked into the matter of the solar radiation, as deduced from old drawings of groups of sunspots. He has shown that the mean sidereal rotational period of the Sun was 27.1 days in 1612 and that there was a decided speed-up, to a rotation period of 24.2 days in 1643 as the Maunder minimum approached.

With this very persuasive evidence for solar variability, is there any evidence for change in the size of the Sun? Here the evidence does not go back so far. Eddy has examined the records of the Royal Greenwich Observatory dating from 1820. Each clear day, the time of passage of each edge of the Sun's image across a fixed north-south line was recorded. The angular diameter of the Sun can thus be calculated, allowance being made for the variable rate of rotation of the Earth and for the motion of the Earth around the Sun. Eddy found a clear trend, with the Sun shrinking by an average of 2.25 arc sec/century.

An independent approach for the measurement of the solar diameter was then taken by Irwin Shapiro at M. I. T. He reexamined the data on transits of the planet Mercury across the solar disc, which occur on the average about every 7 or 8 years. Mercury's diameter is 10 arc sec, and it takes about 5 hr to cross the disc. Shapiro's data show essentially no variation from 1740 until today. The effect detected by Eddy may be the result of steady changes in the clarity of the atmosphere above Greenwich, an industrial suburb of London. However, the required atmospheric optical change has to be very large to explain Eddy's result.

Several other sets of old records have been reexamined, to try to provide independent checks. Another set of transit observations with a meridian circle has come from the Naval Observatory in Washington, D.C.; another set of observations of transits of Mercury has been analyzed; and, finally, the duration of solar eclipses also gives a measure of the solar diameter and those data were examined. Ronald Gilliland of the High Altitude Observatory has compared all of these sets of measurements, and his summary of the data is shown in Fig. 22.25.

From his analysis, Gilliland has suggested that there might be a 76-year periodicity in the solar radius, with about 0″.4 between extreme sizes, and also a possible decrease of about 0″.2 over the past 265 years. As the figure shows, the data are quite scattered so that these trends should not be taken as firmly established.

Further observations with modern techniques and clearer skies than over London are in progress to try to resolve this question. The High Altitude Observatory has undertaken a project to make accurate measurements of transit times. Over a few years changes of less than 0.01 percent of the solar radius should be detectable.

**Figure 22.25** Observations of the solar diameter using three different methods. The meridian circle data were obtained by timing the passage of the solar disc across a vertical cross-wire in a telescope eyepiece. Timing the transits of Mercury across the solar disc provides some of the oldest data, but eclipse timing has yielded only a few data points. Each method shows a considerable scatter. Persistent trends are small, but there is a suggestion of a cyclic variation of about ± 0.02 percent (Ronald L. Gilliland, High Altitude Observatory, and *Astrophysical Journal*).

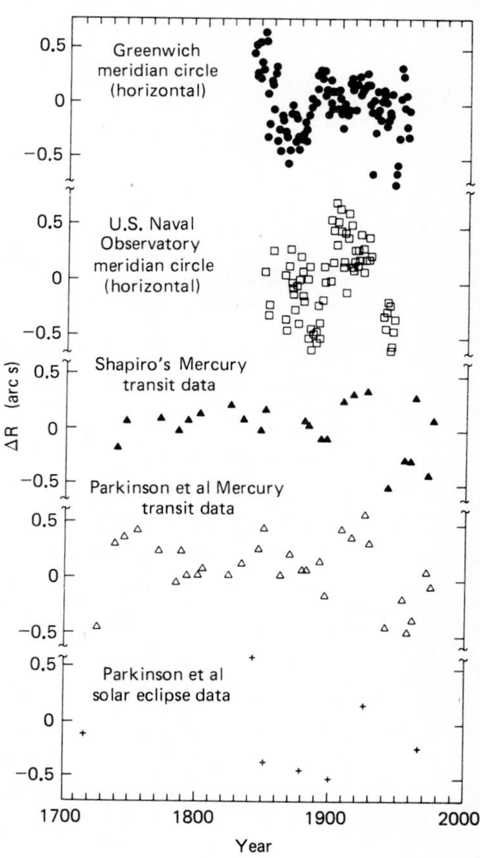

Finally, we need to ask whether total solar output, its luminosity, has changed over the years and if so by how much. The solar luminosity clearly controls the climate on Earth, but there are so many competing meteorological and geophysical effects in addition to man-made insults such as the combustion of wood and fossil fuels that there is still no agreed quantitative relation between climatic indicators and solar luminosity.

Direct and precise measurement of the solar luminosity is not easy; any small changes in the luminosity have to be extracted from long-term drifts in the sensitivity and calibration of the instruments. Until recently the accuracy was only about 1 percent. Variable atmospheric absorption reduces the reliability of ground-based measurements; it is necessary to get above the atmosphere.

The Solar Maximum Mission carried a precision photometer to monitor the solar luminosity. During the first 23 months of its operation, the record showed many fluctuations in the solar output generally within about 0.05 percent of the average but there were several periods of about a week where the output dropped by close to 0.1 percent and one drop as large as 0.23 percent was noted in July 1981. The largest drops seemed to correlate with the passage of large sunspot groups across the solar disc. In addition to the brief changes, there also seemed to be a systematic decrease in luminosity so that the average was down about 0.1 percent over the first 18 months. Thereafter, the luminosity started to increase again.

The significance of these changes is not yet known, and longer-term monitoring is needed before we will be able to see which, if any, of these phenomena correlate with other astronomical observations or with weather and climate on Earth.

## 22.8 SUMMARY

The Sun is still an object well worth further study. It has provided us with a new element (helium); it may still provide surprises when we understand more about the nuclear reactions that keep it glowing (Chap. 23). When we know more about it we may gain insights into the behavior of other stars. Further study of solar-terrestrial relations can be of great practical use for weather forecasting. We have learned much about the Sun, but we need to continue with our research.

## CHAPTER REVIEW

**General features:** angular diameter about half a degree

- linear diameter $1.39 \times 10^{11}$ cm.
- mass $1.99 \times 10^{33}$ g, average density 1.41 g/cm$^3$.
- luminosity $3.83 \times 10^{33}$ ergs/sec, leading to $T_{eff}$ = 5771 K.
- shape of spectrum very similar to black body of 5900 K (slightly different from $T_{eff}$, which describes luminosity).

**Photosphere:** (bright disc that is seen)

- temperatures range from 6600 K at lower boundary to 4300 K at top.
- spectrum shows many dark lines, which are used to identify atoms.
- prominences: bright extensions of photosphere.
- granulation and super-granulation: produced by upward/downward movement of hot gases.

**Chromosphere:** far less bright than photosphere, first seen during eclipses, now can be seen using coronagraph.

- temperatures range from 4300 K at lower surface to 100,000 K where it joins the corona.
- consists of many spicules, jets of glowing gas.

**Corona:** hot (million K) extended outer atmosphere of Sun.

- strong source of X-rays, also radio emission.
- solar wind—extension of corona, sweeping through the solar system.

**Active Sun:** sunspots—11-year cycle, but with significant variations such as in seventeenth century Maunder minimum.

— solar flares: increased emission of solar cosmic rays in addition to visible flare.

**Long-term changes:** solar rotation changed as Maunder minimum approached.

— solar diameter: some evidence for small change, perhaps cyclic.

— measurements not yet accurate enough to establish any long-term trend in luminosity, but satellite (SMM) observations show short-term decreases up to 0.23 percent.

## NEW TERMS

**active Sun**
**chromosphere**
**convection**
**corona**
**coronagraph**
**coronal holes**
**filaments**
**flares**
**Fraunhofer lines**
**granulation**
**hydrogen ions ($H^-$)**
**Maunder minimum**
**photosphere**
**plasma**
**prominences**
**solar wind**
**spicules**
**supergranulation**
**transition region**
**Zeeman effect**

## PROBLEMS

**1.** We are able to make many measurements of the Sun because it is so close. Suppose it were several parsec away. Describe some data that would no longer be obtainable.

**2.** What are the photosphere, the chromosphere, and the corona?

**3.** The intensity of the solar radiation at the top of the Earth's atmosphere is $1.37 \times 10^7$ ergs/cm$^2$sec. Suppose that 25 percent of this energy can reach ground level where it can be converted to electric power with 5 percent efficiency. What area of solar collector is needed in order to produce 1000 MW (typical of a large electric generator)? ($10^7$ erg/sec = 1 watt, 1 MW = $10^6$ watt.)

**4.** Why does the photosphere produce absorption lines but we see emission lines from the corona and the chromosphere?

**5.** What are Fraunhofer lines and how are they produced?

**6.** Why does the solar spectrum consist of a continuum with dark absorption lines? What conditions must exist for a star to show a spectrum of emission lines?

**7.** Why does the photospheric spectrum show many absorption lines, but the chromospheric spectrum shows emission lines?

**8.** What is the evidence for the corona being very hot?

**9.** Why are strong molecular lines or lines of ionized helium not seen in the spectrum of the solar photosphere?

**10.** What is the evidence for the solar corona being at a much higher temperature than the photosphere?

**11.** Why is it possible to use observations at different wavelengths to investigate different depths of the Sun?

**12.** The photosphere shows a pattern of granulation, with bright regions separated by dark lines. Why are those regions bright and the lines dark?

**13.** What is the *solar wind*?

**14.** The solar diameter has been measured by timing transits. How long would it take Mercury to pass across the solar disc if it crossed along a diameter? How long does it take the Sun's image to cross a vertical line in a telescope eyepiece?

**15.** How and why was helium discovered in the Sun before it had been found on Earth?

CHAPTER

23

# Stellar Structure

The accumulation of observational data has moved us from the earlier view that the stars were unchanging to the present picture of persistent change, usually slow but occasionally very rapid. The evidence for stellar evolution and its pace is compelling and comes in several ways.

There is direct observation of rapid changes as in the outbursts of supernovae. When we see the brightness of a star changing by a factor of more than a million, we know we must be witnessing major changes in its structure. Second, and on a longer time scale, we note that all stars persistently radiate energy that can only be drawn from some internal supply whose depletion must involve internal changes either in the structure or composition or both. Third, we note that the H-R diagram shows great patchiness in the density of stars. Diagrams for some clusters of stars have most stars on the main sequence, other diagrams (such as for some of the other star clusters) show many stars in the giant region. The best interpretation is that as stars age, they change their temperatures and luminosities, and thus they move on the H-R diagram.

Finally, we know from studies of radioactivity of the Earth, Moon and meteorites that the solar system and the Sun are about 4.6 billion years old. This sets a time scale for stellar evolution. Without this calibration, we could not tell whether the general scale of stellar change was in millions or billions of years. With the time scale established, we must seek energy supplies and stellar models that will last for such long times. Older theories based on short time scales and smaller reservoirs of energy are now known to be incorrect.

Now that we have large quantities of data and an understanding of the underlying physical processes (such as nuclear reactions and the behavior of radiation), we can construct plausible models that clarify our ideas and also guide us to new observations by which we can try to distinguish between competing models. We shall, therefore, in this chapter, describe the structure of a typical star, and in the following two chapters show how we can understand the evolution of stars, from their formation to their arrival (in some cases) at exotic stages such as neutron stars and black holes. We shall see that although there are places where our understanding is vague and our theories shaky, the general framework still seems to be well enough established, that we feel confident that we are moving in the right direction.

## 23.1 GENERAL SURVEY

To radiate appreciable amounts of energy at visible wavelengths, a star must have a surface temperature well above 1000 K. Actually, most of the known coolest stars have surface temperatures above 3000 K. The stellar interiors must be at even higher temperatures, and the stars must therefore be large balls of very hot gas, mostly hydrogen. Not only can solids not exist at these temperatures, but many of the atoms will be ionized. We must deal with a plasma of nuclei and electrons, and only in the outermost parts of stars (their atmospheres) can we expect to find appreciable numbers of electrons adhering to their parent nuclei.

The most important features of stars can be reviewed with a relatively simple treatment. A full theory must include the effects of rotation and magnetic fields within the stars, but this leads to such an increase in the complexity of the theory that progress has been slow. Fortunately, we can omit these complications in our survey.

Within most stars, competing forces act to keep an equilbrium allowing little collapse or expansion. A disturbance of this equilibrium can lead to very rapid and violent changes in a star's behavior. In most stars, the competing forces come from gravitation (inward) and gas pressure (outward). In some very luminous stars, radiation pressure (outward) also plays an important role.

The gravitational attraction on each atom, nucleus, and electron due to all the others produces a net inward force aimed toward the center of the star. At the center, the resulting gas pressure is enormous. The central temperature is therefore also very high, so high that the nuclei in the gas collide violently with each other, overcoming the repulsion due to their electric charges. Most nuclei rebound from their collisions, but an appreciable number can fuse to produce heavier nuclei and release energy. This energy moves away from the center and toward the stellar surface. The upward flow of radiant energy produces an upward force against the overlying matter, an effect that becomes important in very luminous stars. Energy that reaches the surface is radiated away, providing us with the observable quantities against which we can check our calculations of the internal behavior of the stars.

## 23.2 HYDROSTATIC EQUILIBRIUM

The still-accepted approach to the calculation of properties of stellar interiors was set out for the first time by A. S. Eddington in his 1926 book, *The Internal Constitution of the Stars*. By that date, knowledge of the properties of atoms and the developing quantum theory were sufficient to provide Eddington with the necessary theoretical tools.

We can follow the calculations in general outline by considering the star to be made up of spherical layers (Fig. 23.1). Each layer is held in equilibrium between the downward and upward forces. The downward force on the layer comes from the gravitational attraction of all the layers closer

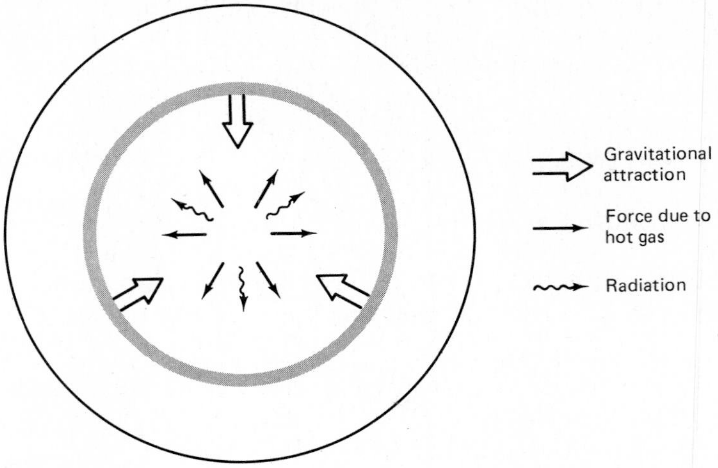

Gravitational attraction

Force due to hot gas

Radiation

**Figure 23.1** Hydrostatic equilibrium inside a star. Gravitational attraction produces an inward force, which is balanced by the outward force due to the very hot gas supplemented in some large stars by radiation pressure.

to the center. The upward force comes from the difference in the pressures at the upper and lower surface of the layer. The weight of each layer contributes to the pressure that increases steadily from the surface into the center, as more and more layers' contributions are added. Finally, at the center for a star like the Sun the pressure is about two billion times as high as the atmospheric pressure on the Earth's surface.

With the balance between the inward and outward forces, our star is said to be in **hydrostatic equilibrium**. It is in equilibrium because the forces are balanced, and it is called hydrostatic because the situation is (in some ways) similar to that in a standing liquid. With equilibrium, the pressure, density, and temperature remain steady for long enough times so that we can consider them as constant—a good starting point for our calculations. Catastrophic departures from equilibrium can be left until later.

## BOX 23.1
*Pressure at the Center of a Star*

Pressure can be defined as force per square centimeter. We are familiar with atmospheric pressure, and we talk of the pressure being 30 in. of mercury. More precisely, the atmospheric pressure is equal to the downward force exerted per square centimeter by a column of mercury 30 in. (76 cm) high. The units of pressure are dyne/cm$^2$, and one atmosphere corresponds to a pressure of about one million dyne/cm$^2$, equivalent to about 14 lb/in.$^2$

The pressure at the center of a star produced by the gravitational attraction that holds the star together can be calculated to be given approximately by

$$P_c = \left(\frac{GM\rho}{R}\right) \text{ dyne/cm}^2$$

where $G$ is the universal gravitational constant; $M$ is the total mass (in g) of the star of radius $R$ cm and average density $\rho$ g/cm$^3$. For the Sun, $M_\odot = 1.99 \times 10^{33}$ g, $R_\odot = 6.96 \times 10^{10}$ cm, $\rho = 1.41$ g/cm$^3$. $G$ is $6.67 \times 10^{-8}$ dyne cm$^2$/g$^2$ (Fig. 23.2). We then find $P_c = 2.7 \times 10^{15}$ dyne/cm$^2$ or two billion times the atmospheric pressure on the Earth's surface.

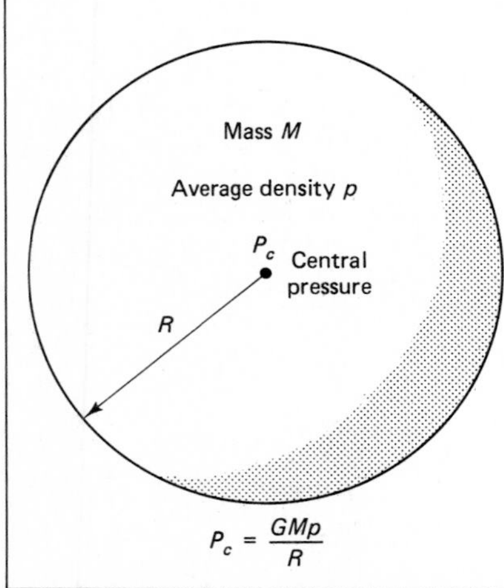


**Figure 23.2** For a star in hydrostatic equilibrium, the pressure at its center is simply related to the star's mass (*M*), radius (*R*), and average density ($\rho$). (*G* is the gravitational constant.)

**Plate 32** Montage of Saturn and six of its moons. Mimas partly overlaps Saturn and shows its giant crater; Dione is at lower left (NASA).

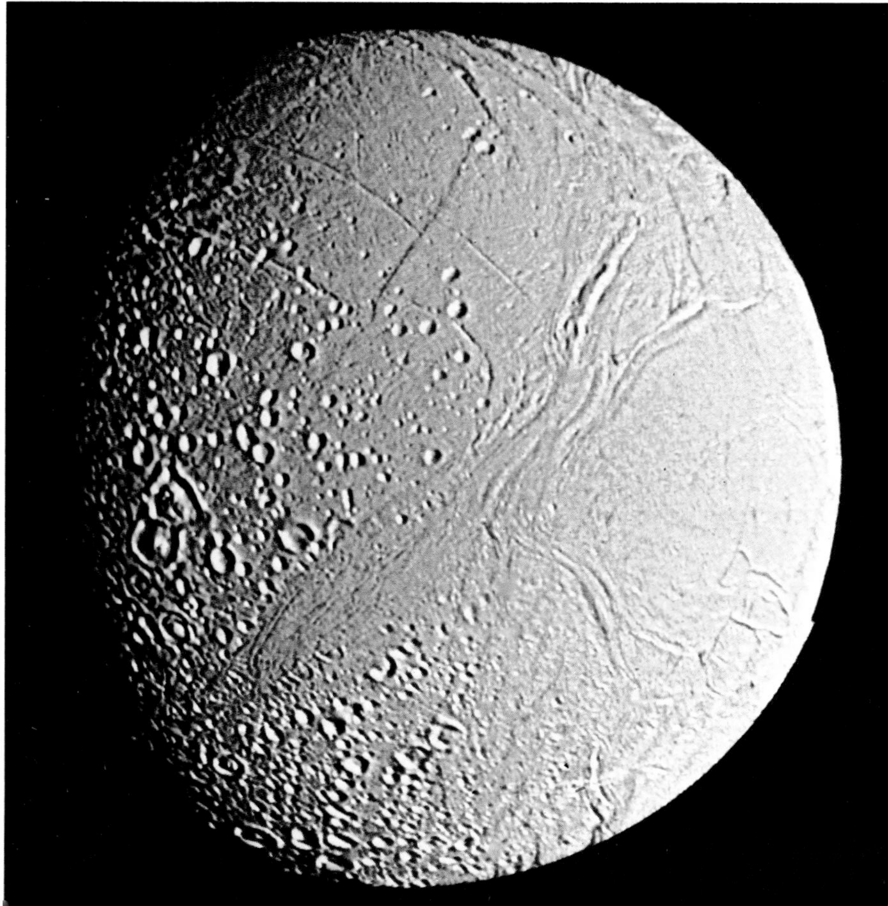

**Plate 33** Enceladus, one of Saturn's satellites. This view from Voyager 2 shows detail down to about 2 km (NASA).

**Plate 34** Iapetus, one of Saturn's moons, seen at closest approach of Voyager 2. Detail down to 21 km can be seen. Dark material (lower right) covers part of the cratered surface (NASA).

**Plate 35** Allan Hills Meteorite discovered in the Antarctic. Composition of this meteorite indicates a lunar origin (NASA/JSC).

**Plate 36** Line spectra for the elements indicated and spectrum of the sun, showing absorption lines. (Courtesy of Bausch and Lomb Incorporated)

**Plate 37** Solar spectrum obtained using a small aperture and a prism.

**Plate 38** X-ray photograph of the Sun, taken from Skylab in 1973. Active regions (bright), loops and coronal holes (dark regions) can be seen (American Science and Engineering, Inc., and NASA).

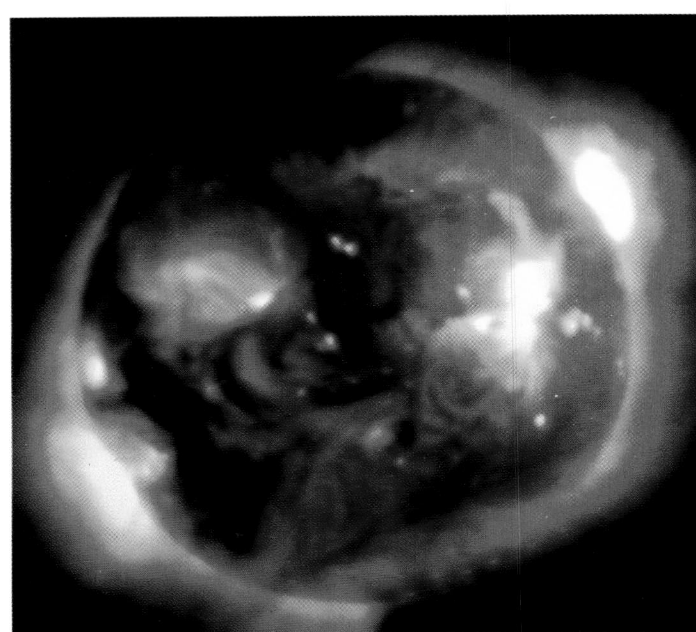

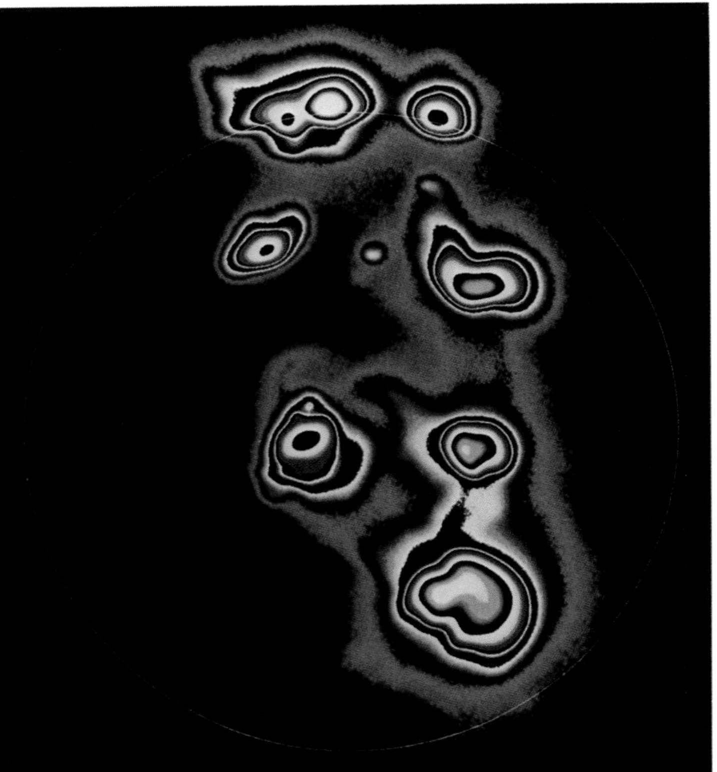

**Plate 39** The active Sun, showing regions of emission of high-energy X-rays. The highest temperatures (shown in white) are around 5 million K, 1000 times hotter than the underlying photospheric temperatures (NASA/Marshall).

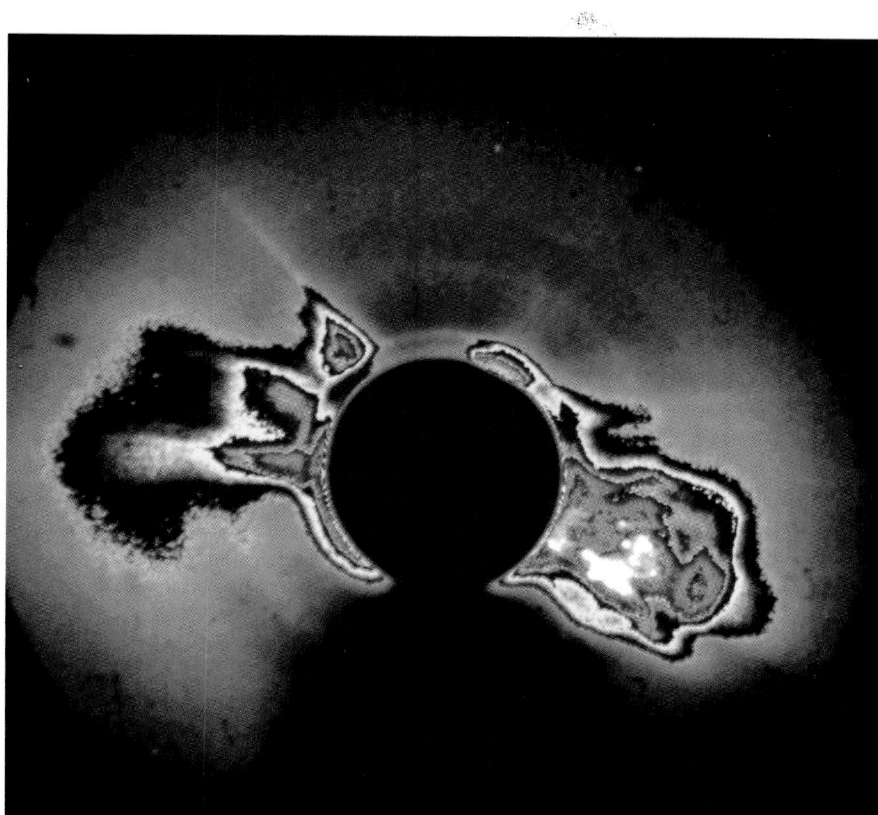

**Plate 40** Giant prominence, photographed from Skylab in the UV (Naval Research Laboratory).

**Plate 41** Coronal transients: part of the outer corona, expanding at speeds of 100 to 1000 km/sec, connected with solar disturbances that can be detected as far away as the Earth's orbit. Photographed from Skylab (HAO/NCAR and NASA/ Marshall).

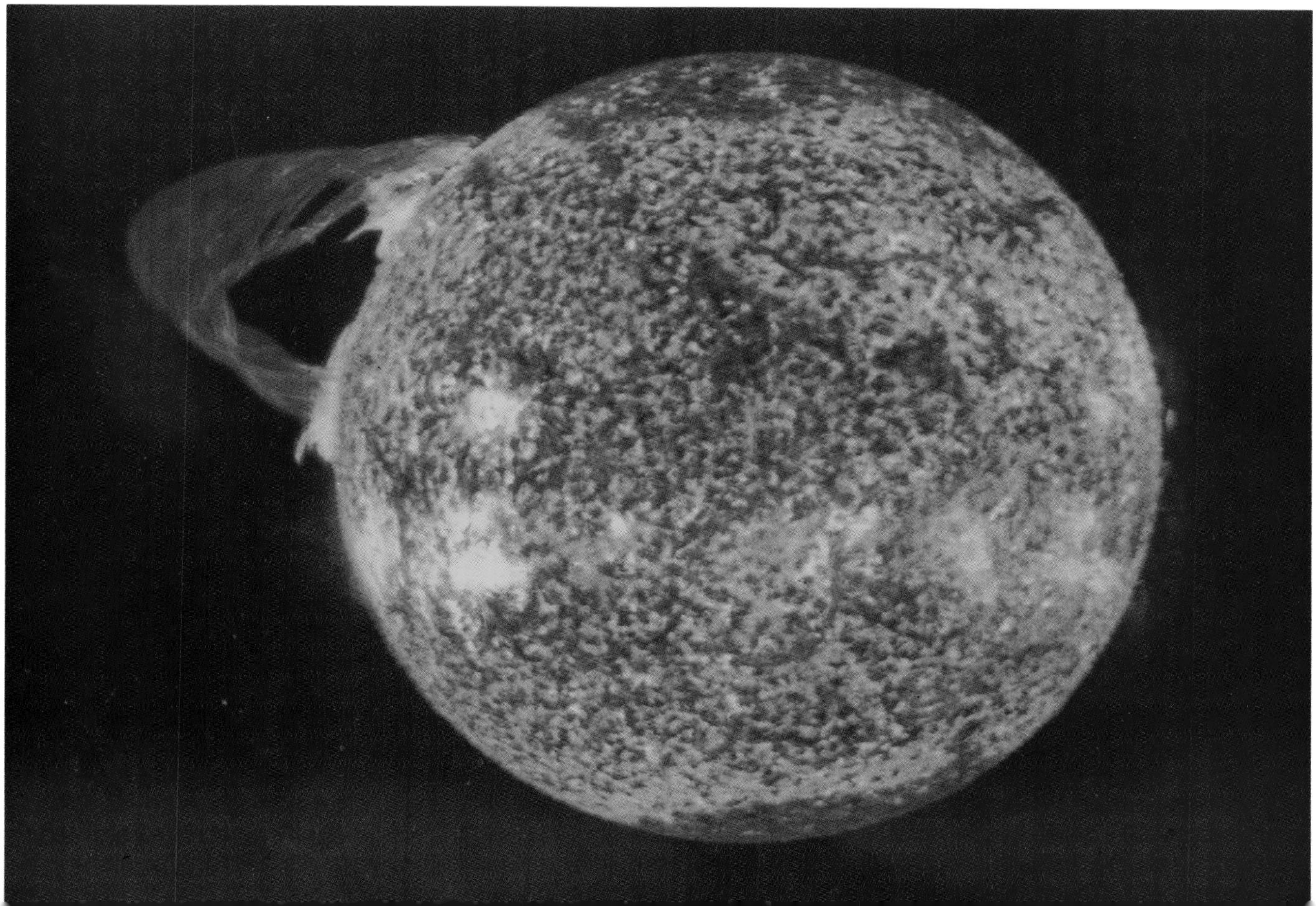

**Plate 42** X-ray and UV emission from an active region on the solar limb [HAO/NCAR].

**Plates 42 through 45** Coronograph observations of the Sun, made from the Solar Maximum Mission (SMM) spacecraft (Gordon Newkirk, Jr., HAO/NCAR). In all of these, a blank disc blocks direct light out to 1.5 solar radii to allow the fainter outer atmosphere to be observed.

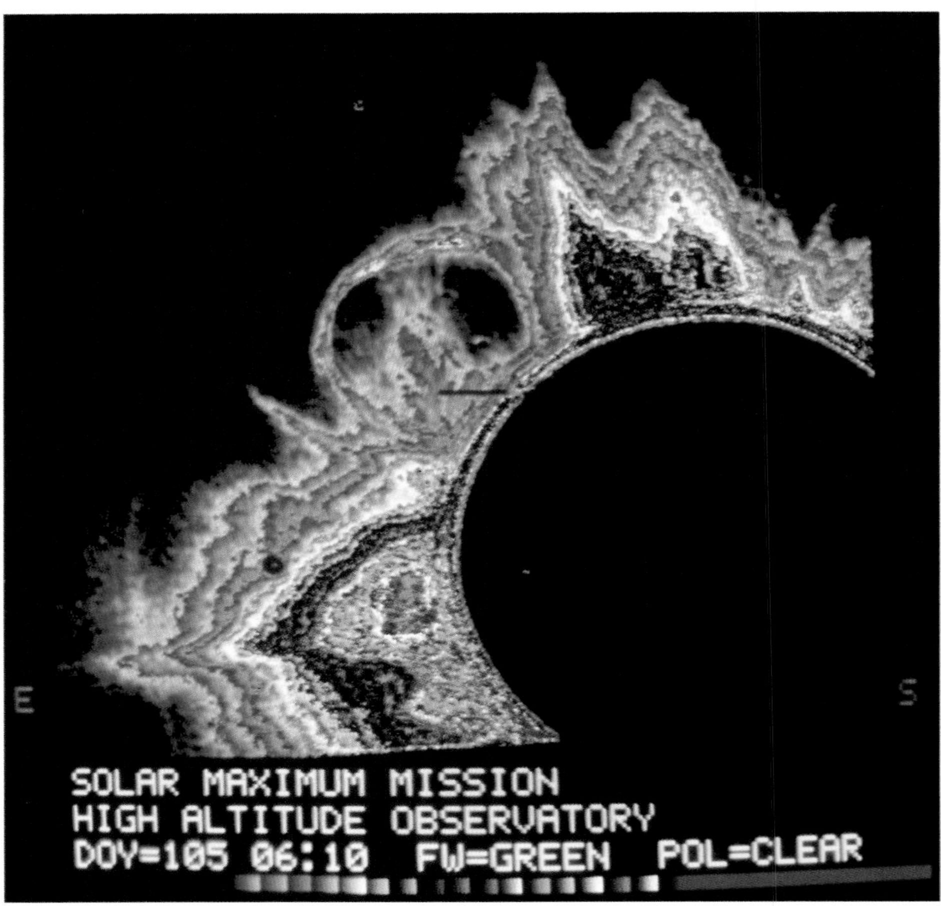


**Plate 43** Coronal streamers with color scale chosen to emphasize their structure [HAO/NCAR].

**Plate 44** Eruptive prominence that expanded out to three solar radii [HAO/NCAR].

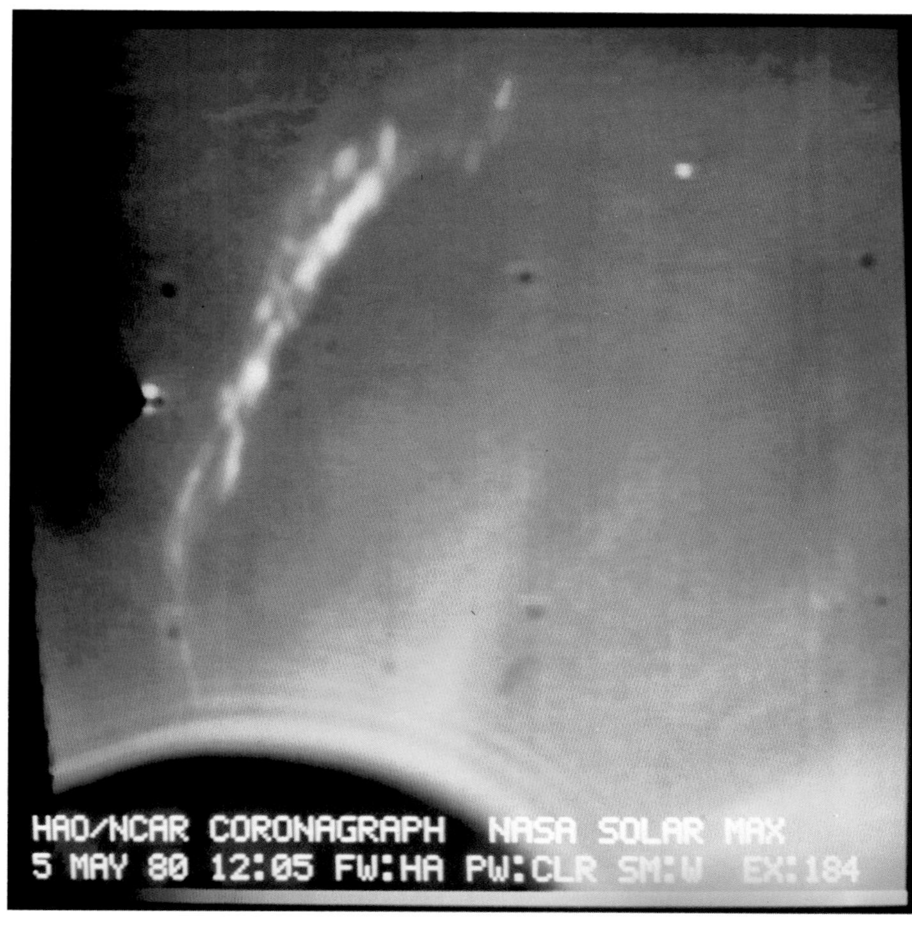


**Plate 45** Photograph taken to emphasize the hydrogen Balmer ($H\alpha$) line, shown here as bright yellow. Hydrogen atoms are ionized at a few thousand degrees, and it is not yet understood how such relatively cool gas can exist in the million degree corona [HAO/NCAR].

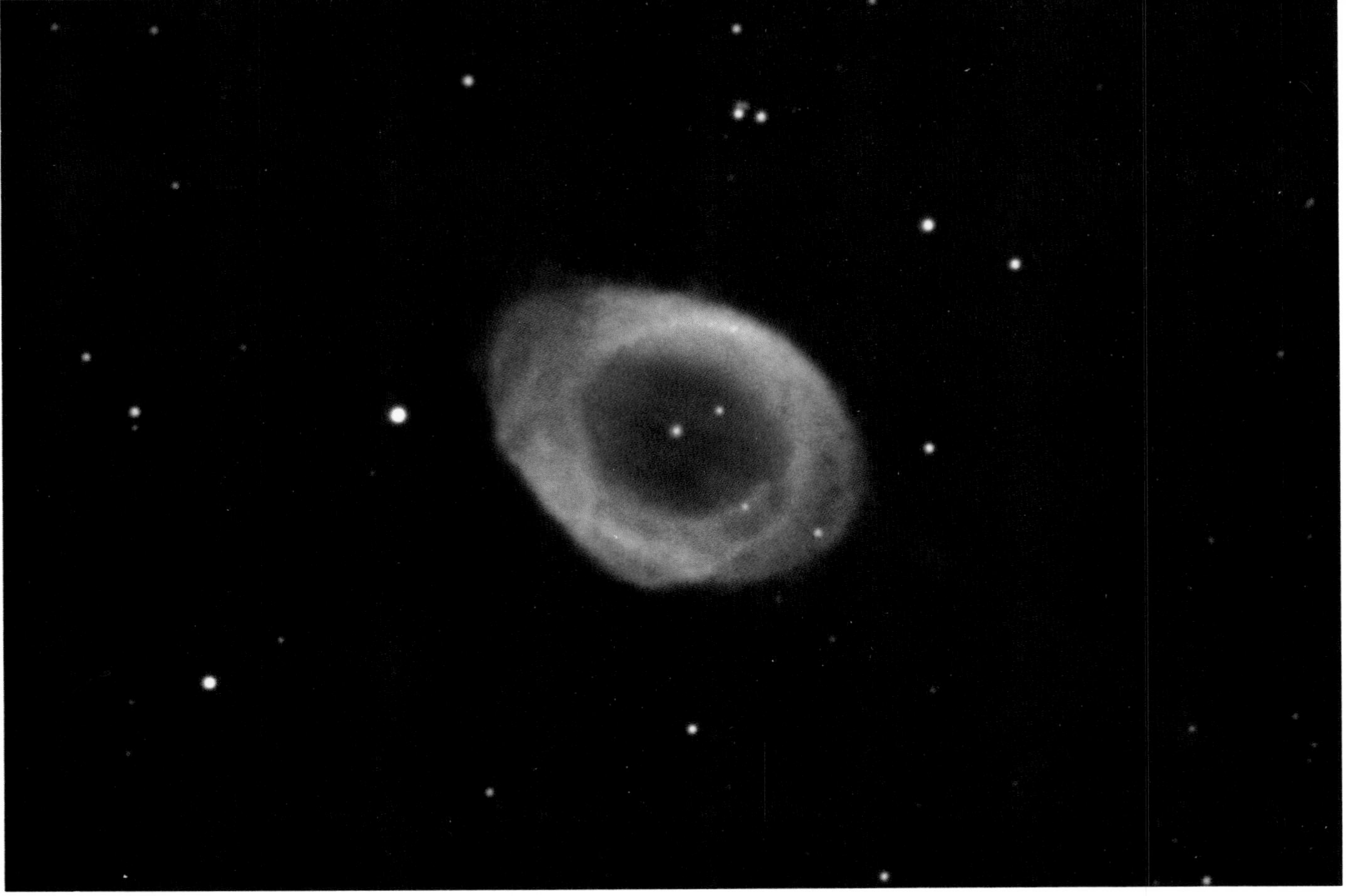

**Plate 46** Ring Nebula in Lyra (M57, NGC 7720). The red outer edge indicates the presence of hydrogen, and the green inner region indicates ionized oxygen. The central star is about fourteenth magnitude, and its spectrum indicates a 100,000 K surface. The diameter of the ring is about 3 arc min (KPNO).

**Plate 47** The Great Nebula in Orion (M42, NGC 1976). With binoculars, this nebula appears as a green glow in Orion's sword, but photographs with large telescopes show the greater extent of the nebula. The concentration of gas and dust and the number of young hot stars strongly suggests that this is a region of star formation (KPNO).

**Plate 48** Map of the Orion Nebula, based on radio observations with the VLA at 6 cm wavelength (M. Simon, S.U.N.Y., Stony Brook).

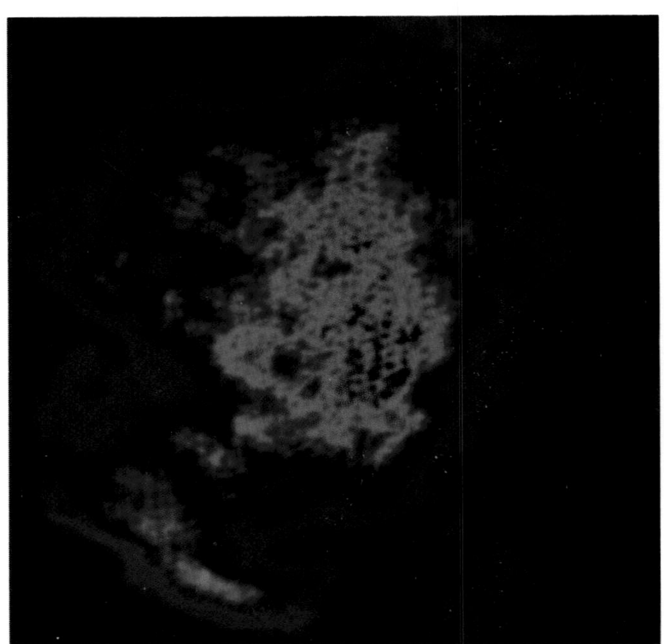

At such high pressures, the gas of nuclei and electrons will be at very high temperatures. The relation between temperature, density, and pressure is called the **equation of state** (Fig. 23.3). For a gas at ordinary temperatures and pressures, Boyle's Law relates the pressure ($P$) and temperature ($T$) of a gas at density $\rho$:

$$P = \text{constant} \times \rho T$$

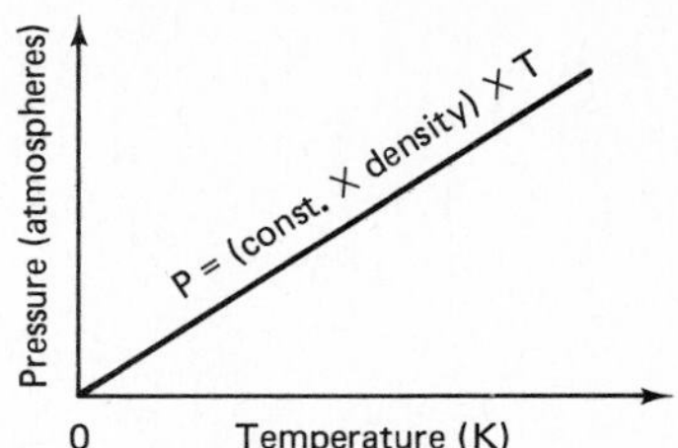


**Figure 23.3** Relation between pressure and temperature for a gas. If the gas density is kept constant, then pressure and temperature rise together (temperatures measured on the Kelvin scale).

More complex equations of state are needed for stellar interiors where we are dealing with electrically charged particles at such high densities that nuclear effects may have to be taken into account. In the case of the Sun, we are able to calculate the temperature at the center and we obtain $T_c \sim$ 15 million K.

From the nuclear reactions occurring in a stellar core at this high temperature, the released energy makes its way to the surface of the star. The way out is not direct; photons will be absorbed or scattered by the overlying layers. Absorbed photons can be reemitted or new photons can be emitted; scattered photons can be absorbed in subsequent collisions with atoms, and new photons will then be emitted. The result of all of these processes is that the energy moves steadily outward, but we find that it is being carried by an increasing number of lower-energy photons. The resistance offered by the star to the flow of radiation is termed **opacity**. The value of the opacity depends on the composition (the proportions of different nuclei and electrons), density and temperature of the gas. Many different physical processes contribute to the opacity. Its calculation is a complex but important factor in the development of models of stellar interiors.

At each level within the star, there is thus a condition of equilibrium between the downward gravitational force and the force due to gas pressure. If, for any reason, the pressure at some level should change, then the gas must move up or down to find a new equilibrium level. The temperature and flow of energy may then change to compensate. If the star cannot make the necessary changes, it may collapse or explode. This is a topic of great current interest discussed in Chapters 24 and 25.

## 23.3 SOURCES OF ENERGY

Without a supply of energy that can be continually released at its center, a star would collapse because it would have no way of resisting the ever-present inwardly directed gravitational force. As we know, large amounts of energy are required. An average star such as our Sun radiates $\sim 3.83 \times 10^{33}$ ergs/sec so that in 4.6 billion years it has radiated about $\sim 5 \times 10^{50}$ ergs. Where does this energy come from?

One possible source of energy is gravity. As the star contracts, first from a large diffuse and cold cloud of gas and dust and then as a hot denser gas, gravitational energy is released. The same principle is at work in a hydroelectric power plant on the Earth's surface where the gravitational potential energy of the falling water is used to run the electric generators. The release of gravitational energy during stellar contraction was first explored by Helmholtz and Kelvin in the late nineteenth century. They showed that for a star of mass $M$ that has contracted until its radius is $R$, the gravitational energy available is about $GM^2/2R$ ergs. Again, taking the Sun for our example, the available gravitational energy is

$$\frac{(6.67 \times 10^{-8})(1.99 \times 10^{33})^2}{2 \times (6.96 \times 10^{10})} = 1.9 \times 10^{48} \text{ ergs}$$

With energy being radiated at a rate of $3.83 \times 10^{33}$ ergs/sec, this supply will last for only about $5 \times 10^{14}$ sec or about 15 million years, far shorter than the now known age of the solar system. Therein lies the dilemma that was posed as soon as the age of the Earth became known from the measurements of radioactivity. The solution was not found for many years: the only adequate supply of energy is the fusion of nuclei. The application of the theory of nuclear reactions to stellar energy production was worked out only as recently as 1938 by Hans Bethe of Cornell University.

As long as temperatures are low, atoms are electrically neutral, and the outcome of each low-speed collision is a rebound with only a tiny fraction of the kinetic energy going into the excitation of the atomic electrons. In contrast, deep within the stars the nuclei are without their protective electrons and the repulsive forces between the positively charged nuclei keep them apart unless the temperature is very high. Then the nuclei will have speeds so great that even this electrical repulsion can be overcome, and the nuclei can collide to fuse and produce heavier nuclei.

The most abundant nuclei are protons, the nuclei of hydrogen atoms. At temperatures over about $10^7$ K they can fuse to form nuclei of helium, the next heaviest chemical element

$$4\text{H} \rightarrow \text{He}$$

The mass of each H is 1.0078 amu on the **atomic mass unit** scale (1 amu $= 1.66 \times 10^{-24}$ g), so that four of them (4 H) total 4.0312 amu. This is slightly (but crucially) more than the 4.0026 amu of their product, the helium nucleus, He. The difference ($0.0286 = 4.0312 - 4.0026$ amu $= 4.74 \times 10^{-26}$ g) is called the **mass defect** (Fig. 23.4). In the fusion reaction this mass defect is lost, transformed from mass into energy that is then released. This sort of conversion from mass to energy was first suggested by Einstein in 1905, in the now well-known relation

$$E = mc^2$$

where the energy $E$ in ergs is obtained from the conversion of mass $m$ g and $c$ is the speed of light, $3 \times 10^{10}$ cm/sec.

For $m = 1$ g, the energy available is thus $E$ ergs $= (1 \text{ g}) \times (3 \times 10^{10})^2 = 9 \times 10^{20}$ ergs. In the case of 4 H $\rightarrow$ He conversion, each fusion releases 0.0286 amu $= 4.74 \times 10^{-26}$ g and thus $4.27 \times 10^{-5}$ ergs. Thus a fraction $0.0286/4.0312 = 0.0071 = 0.71$ percent of the hydrogen mass disappears to reappear as energy.

Although most of the Sun consists of hydrogen, only about the central 10 percent is hot enough to react to produce helium. The overlying layers are too cool for nuclear reactions to proceed. The consumable hydrogen fuel thus amounts to about $\frac{1}{10}(2 \times 10^{33})$ g $= 2 \times 10^{32}$ g; the mass that can be converted into energy is thus $0.0071 \times 2 \times 10^{32} = 1.42 \times 10^{30}$ g. At the conversion rate of $9 \times 10^{20}$ ergs/g, this will produce $1.28 \times 10^{51}$ ergs.

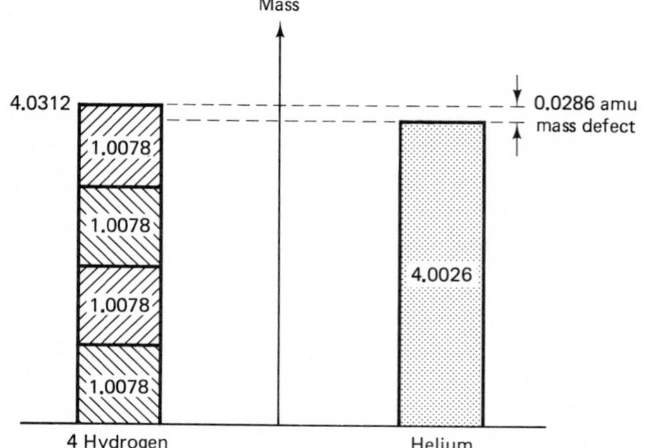


**Figure 23.4** Mass defect. When four hydrogen nuclei (protons) fuse to form a nucleus of helium, there is an apparent mass loss of 0.0286 amu. This "mass defect" is not truly lost but appears as energy released in the fusion reactions. The conversion between mass and energy follows Einstein's formula, $E = mc^2$, where $c$ is the speed of light.

As long as the Sun radiates energy away at the present rate of $3.83 \times 10^{33}$ ergs/sec, the supply can last for

$$\frac{1.28 \times 10^{51} \text{ ergs}}{3.83 \times 10^{33} \text{ ergs/sec}} = 3.34 \times 10^{17} \text{ sec}$$

or about 10 billion years—rather more than twice the known age of the solar system.

Even though collisions between hydrogen nuclei (protons) are frequent, the chance of four nuclei colliding simultaneously is so low as to be negligible. Instead, the fusion of hydrogen nuclei into helium nuclei proceeds through several steps, in each of which only two particles collide to produce a heavier particle and release some energy. The net result of successive reactions is equivalent to four hydrogen nuclei fusing to form one helium nucleus.

There are two different sequences of fusion reactions that produce helium from hydrogen: the proton-proton (p-p) chain and the carbon (CNO) cycle. At temperatures below $2 \times 10^7$ K as in the Sun, the p-p chain is the more important although both chains occur. In the CNO chain, heavier nuclei with larger electric charges are involved; higher temperatures are, therefore, needed before this fusion chain becomes important. The CNO cycle is unimportant for stars of type G (like the Sun), K, and M with low temperatures, but for types O to F where the central temperatures are greater than $2 \times 10^7$ K, the CNO cycle dominates the energy production. Details of the two cycles are described in Box 23.2.

## BOX 23.2
*Nuclear Reactions*

We are accustomed to the convenience of using abbreviations to represent the different chemical elements and their compounds. Thus Na (from the German **natrium**) is used for sodium, Cl for chlorine, and NaCl for their combination. When it comes to representing nuclear reactions, we want our shorthand to carry more information and so we add superscripts. We shall use a system in which a superscript represents the mass of the nucleus on the amu scale. So, for example, hydrogen nuclei of mass one are represented by $H^1$, helium of mass four by $He^4$, neutrons of mass one by $n^1$, and so on. Because hydrogen nuclei are often referred to as protons, you may sometimes find the alternative 'p' being used to represent these particles in the nuclear reaction formulas.

Electrons have so much smaller mass that we will dispense with a mass-superscript, but we will use a subscript of + or −, to indicate whether the electron carries a positive or negative electric charge. Other particles that we shall encounter are neutrinos, $\nu$, and gamma rays, $\gamma$.

The p-p (proton-proton) chain comprises the successive reactions

**a.** $H^1 + H^1 \rightarrow H^2 + {}_{+}e + \nu$

**b.** $H^2 + H^1 \rightarrow He^3 + \gamma$

**c.** $He^3 + He^3 \rightarrow He^4 + H^1 + H^1$

In the first reaction two protons fuse to produce a deuteron ($D$ or $H^2$), the nucleus of heavy hydrogen or deuterium. At the same time, a positron (${}_{+}e$) and neutrino ($\nu$) are released and carry away some energy. (The positron has the same mass as an electron but carries a positive charge. Neutrinos have no mass and no charge.)

With deuterons being formed in reaction **a**, some will be hit by other protons (reaction **b**) to produce an isotope of helium $He^3$ and a gamma-ray photon ($\gamma$), which carries away some energy. Finally, with enough $He^3$ produced, some will collide among themselves (reaction **c**) to produce $He^4$.

If we double all the quantities in *a* and *b* to provide the two $He^3$ needed for reaction *c*, and then add everything together, we have, for the complete cycle:

$$4[H^1] \rightarrow [He^4] + 2[{}_{+}e] + 2[\nu] + 2[\gamma]$$

There is an alternative p-p chain, in which 6 percent of the $He^3$ collide with $He^4$ to produce $Be^7$, but the complete cycle has the same products as the chain with the $He^3$-$He^3$ collisions.

Some of the early reaction products are consumed in later stages; the overall result is the conversion of four protons into a single helium nucleus, and the release of positrons, neutrinos, and photons. In addition, there is a mass defect at each stage, which appears as energy in gamma rays and as kinetic energy of particles. This energy is seen later as the star's luminosity. [The neutrinos also carry away some of the energy; see sect. 23.6.]

If the temperature is high enough, the CNO cycle can occur when sufficient $C^{12}$ nuclei are already present to act as seed nuclei. The sequence of reactions is

$$C^{12} + H^1 \rightarrow N^{13} + \gamma$$
$$N^{13} \rightarrow C^{13} + {}_{+}e + \nu$$
$$C^{13} + H^1 \rightarrow N^{14} + \gamma$$
$$N^{14} + H^1 \rightarrow O^{15} + \gamma$$
$$O^{15} \rightarrow N^{15} + {}_{+}e + \nu$$
$$N^{15} + H^1 \rightarrow C^{12} + He^4$$

Here again, some reaction products are used in later stages, and the overall result, once more, is the conversion of $4H \rightarrow He$ with the release of energy.

When the temperature is even higher (above about $10^8$ K), reactions take place between heavier nuclei, such as

$$He^4 + He^4 \rightarrow Be^8 + \gamma$$

followed by

$$Be^8 + He^4 \rightarrow C^{12} + \gamma$$

In this "triple-$\alpha$" reaction chain, the intermediate Beryllium is rapidly consumed, and three $\alpha$-particles fuse to form a carbon nucleus.

At yet higher temperatures, above about $10^8$ K, $He^4$ nuclei can fuse to produce $C^{12}$ nuclei, and more complex reactions can occur. As these reactions do not start until a star has aged considerably or unless the star starts out with enough heavier nuclei, we shall defer their discussion until chapter 25, when we come to review the later stages of stellar evolution.

A characteristic feature of all of the cycles of nuclear reactions is the release of energy. How does this energy display itself? Some of the energy is the kinetic energy of the moving particles. Some of the energy is carried away by neutrinos. Most of the energy, though, is carried off by gamma-ray photons. The next section is devoted to a discussion of the processes by which the energy from the nuclear reactions slowly reaches the surface of the star.

## 23.4 TRANSPORTATION OF ENERGY FROM THE CENTER OF A STAR

The radius of the Sun is close to $7 \times 10^{10}$ cm. We might thus expect that a photon released in a nuclear reaction at the center should reach the surface after a time given by

$$\frac{\text{distance}}{\text{speed}} = \frac{7 \times 10^{10}\text{ cm}}{3 \times 10^{10}\text{ cm/sec}} = 2.3\text{ sec}$$

but this does not happen. Why?

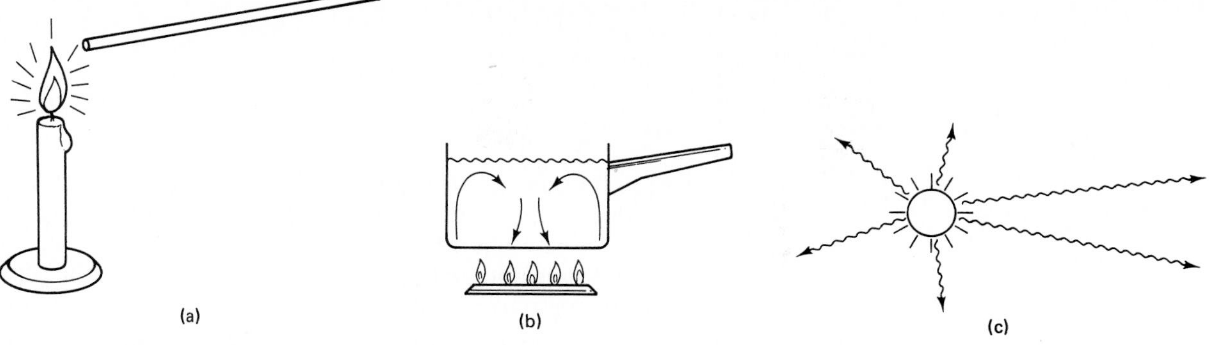


**Figure 23.5** Transport of energy. (a) Energy can be conducted along a metal rod through the internal vibrational collisions between atoms, but the rod itself does not move. (b) With convection, as in a pot of heated water, the hot water rises and carries the energy with it. Cool water descends and will rise after being warmed. Convection currents are set up. (c) Radiation carries energy without any matter being moved. We can feel the heat radiation from a fire and from the sun.

Energy can be transported by three means (Fig. 23.5). **Conduction** is familiar to us. If you hold a metal rod with the other end near a flame, you will soon notice an increase in temperature at the end you are holding. Energy is being conducted along the bar by internal collisions between the atoms, but no bulk motion occurs. **Convection** is also familiar. In a pot of water on the stove, heat is conducted through the metal base to the water at the bottom. This hot water rises and is replaced by colder water from higher up, setting up a convection current. The heat is carried around by the moving water itself. **Radiation** is the third means of transporting energy. Again, this is familiar: we can feel the heat of the direct rays of the Sun and the IR warmth of a fire. We discussed radiation in some detail in chapter 8. Electromagnetic radiation or waves carry energy with the speed of light, and we can also picture the energy as being carried by the photons emitted by atoms.

In most stars, conduction plays almost no role, and convection is limited to the outer regions. For example, in the Sun convection becomes important in the outer one-tenth of the radius. By far the most important process is radiation or **radiative transport** as it is often termed.

### BOX 23.3
*Speed of Energy Transfer*

The three ways of transferring energy depend on quite different physical processes. A consequence is that the speed with which energy is transferred will vary from one mode to the next and will depend on a number of properties such as density of the region, temperature, and composition.

There is a familiar example of different speeds. In a shower, the two faucets that control the hot and cold water will usually be several feet from the outlet. When you adjust the faucets, you will feel a very prompt change in the water flow: the change in pressure at the faucets is transmitted through the water with the speed of sound (about 500 m/sec). The change in water temperature, however, will not be noticeable until several seconds later, as this depends on the water flow (convection). Conduction is so much slower that it is unimportant under these conditions.

Instead of escaping directly from the center of a star, the flow of energy is strongly affected by **scattering** and **absorption** of photons. In a scattering process, a photon collides with an electron and is deflected into a new direction to continue until its next collision. Absorption may occur when a

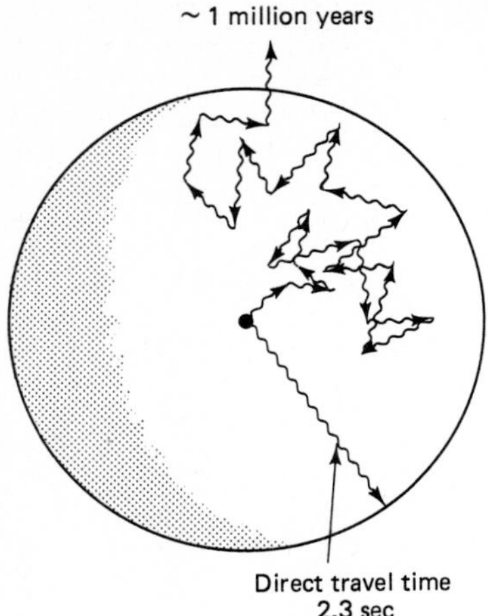


**Figure 23.6** Transport of energy from the solar interior to the surface. Energy is scattered, absorbed, and reemitted, and the path followed from the center of the Sun is, therefore, far from direct. The energy seen radiated today was produced in nuclear reactions in the core more than a million years ago.

photon collides with an ion and all of the photon's energy is transferred to one of the ion's electrons. Very quickly the electron returns to its original energy (orbit) and emits a new photon or several lower-energy photons.

After both the scattering and absorption processes, the emerging photons will usually be traveling in directions quite different from the incoming photon, and some of the originally upward-moving photons may even be completely turned around. After many collisions a general outward flow dominates, but the rate at which the energy moves out has been considerably slowed. Consequently, instead of the energy taking 2.3 sec to reach the surface, it takes more than a million years (Fig. 23.6). Thus the radiant energy that we see today coming from the Sun's photosphere was released in nuclear reactions at the center more than a million years ago.

## 23.5 STELLAR MODELS

After all of the discussion, theory, and computing, what does a stellar model look like? The Sun provides an excellent case study. What goes into such a calculation? The most accurately known ingredients are the radius, mass, luminosity, and surface temperature. The chemical composition is also needed for the calculation of the behavior of plasma at the very high temperatures and pressures in the core and for the calculation of the opacity of the overlying layers. It is usual to designate the proportions of hydrogen, helium, and the total of everything else (often misleadingly labeled as "metals") by the fractions X, Y, and Z, respectively. Thus a typical model for the Sun illustrated here is based on $X = 0.71$, $Y = 0.27$, and $Z = 0.02$; model calculations have been run for numerous different sets of values.

The major results for the present properties of the Sun are listed in Table 22.1 and displayed in Figs. 23.7 and 23.8. To show how these results can be understood, (for example) consider conditions at two distances on the way out from the center. Within the first 20 percent of the radius, 35 percent of the Sun's mass is contained; 94 percent of the luminosity originates within that region. Beyond 0.3 $R_{\odot}$, there is virtually no fusion taking

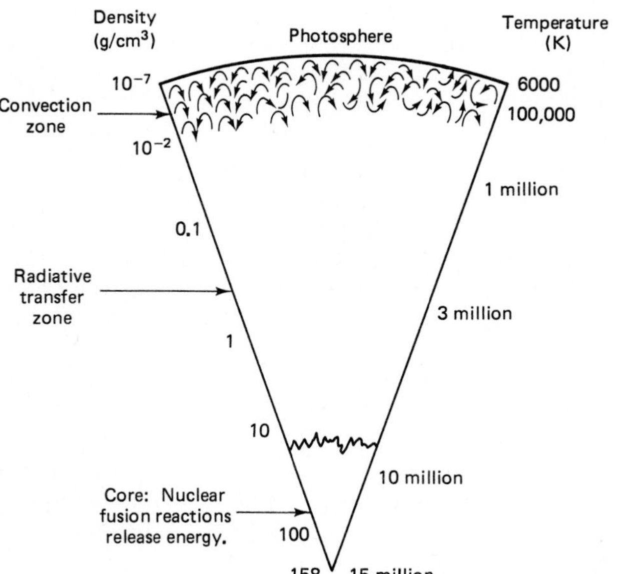


**Figure 23.7** Schematic model of the Sun. Density and temperature increase rapidly towards the center where nuclear reactions are taking place.

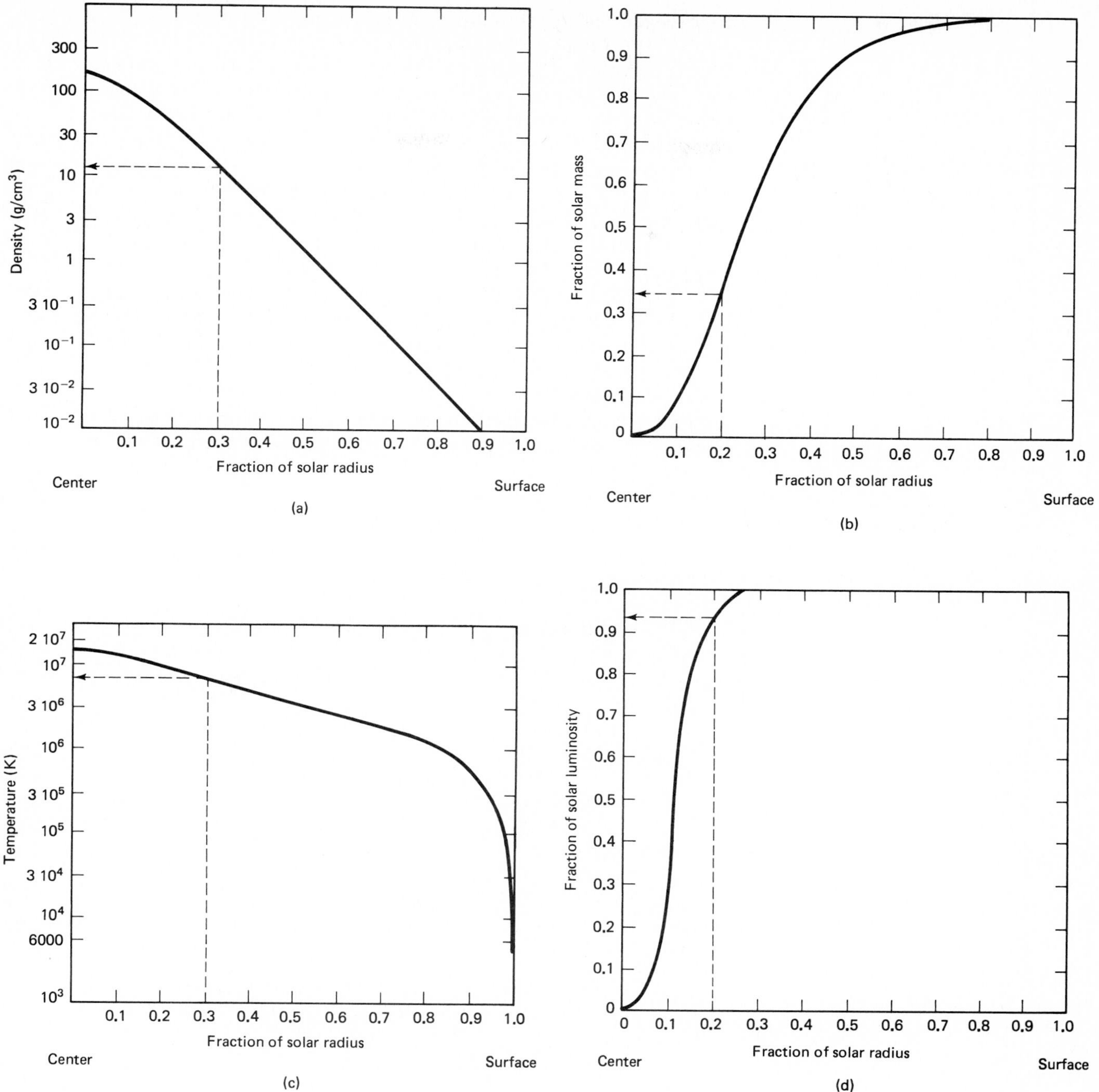


**Figure 23.8** Results of detailed model calculations for the Sun. (a) Density versus depth. The density decreases from a central value of 158 g/cm$^3$ to 0.01 g/cm$^3$ at nine tenths of the radius, then drops very rapidly. (b) Fraction of the solar mass versus distance from center. Because the density is so high in the inner regions and decreases rapidly with increasing distance, most of the solar mass is very concentrated. Thus, 35 percent of the mass is contained within the inner 0.2 of the radius. (c) Temperature versus distance from the center. The temperature drops steadily for the first 80 percent of the radius, and then decreases very rapidly towards the photosphere which is seen. Only in the central region is the temperature high enough for fusion reactions to take place. (d) Fraction of solar luminosity produced within different regions. Fusion reactions take place only in the very central region where the temperature is high enough. So, for example, 93 percent of the solar luminosity comes from reactions in the central 20 percent of the radius, and virtually nothing from beyond about 0.3 $R_\odot$ (based on diagrams in *The Quiet Sun*, NASA).

place, since the temperature and density are too low. At 0.3 $R_\odot$ the temperature is 6.7 million K (compared to the central temperature of 15.4 million K). At this radius, hydrogen comprises 71 percent of the mass (compared to 73 percent at the surface and 38 percent at the center where it fuses to form helium). The average density of the Sun is 1.41 g/cm$^3$, but local values run from 158 g/cm$^3$ near the center to 13.3 g/cm$^3$ at 0.3 $R_\odot$ to 0.74 g/cm$^3$ at 0.95 $R_\odot$.

In contrast (Fig. 23.9), a main sequence star of 10 $M_\odot$ has a radius four times larger than the Sun and a luminosity 6000 $L_\odot$. Its temperature drops from 28 million K near the center to 28,000 K at its surface. The density is much lower than that of the Sun—only 0.4 g/cm$^3$ on average—and it has a surprisingly low 7.8 g/cm$^3$ near the center.

The calculation of stellar models is a significant research area in modern astrophysics. Models for stars of different mass and composition have provided considerable insight into the probable structure and the way in which externally measurable quantities depend on the hidden properties, but a word of caution is necessary. While we think that the main features of the models are probably correct, even for the Sun there is a very serious confrontation between theory and experiment to be described in the next section.

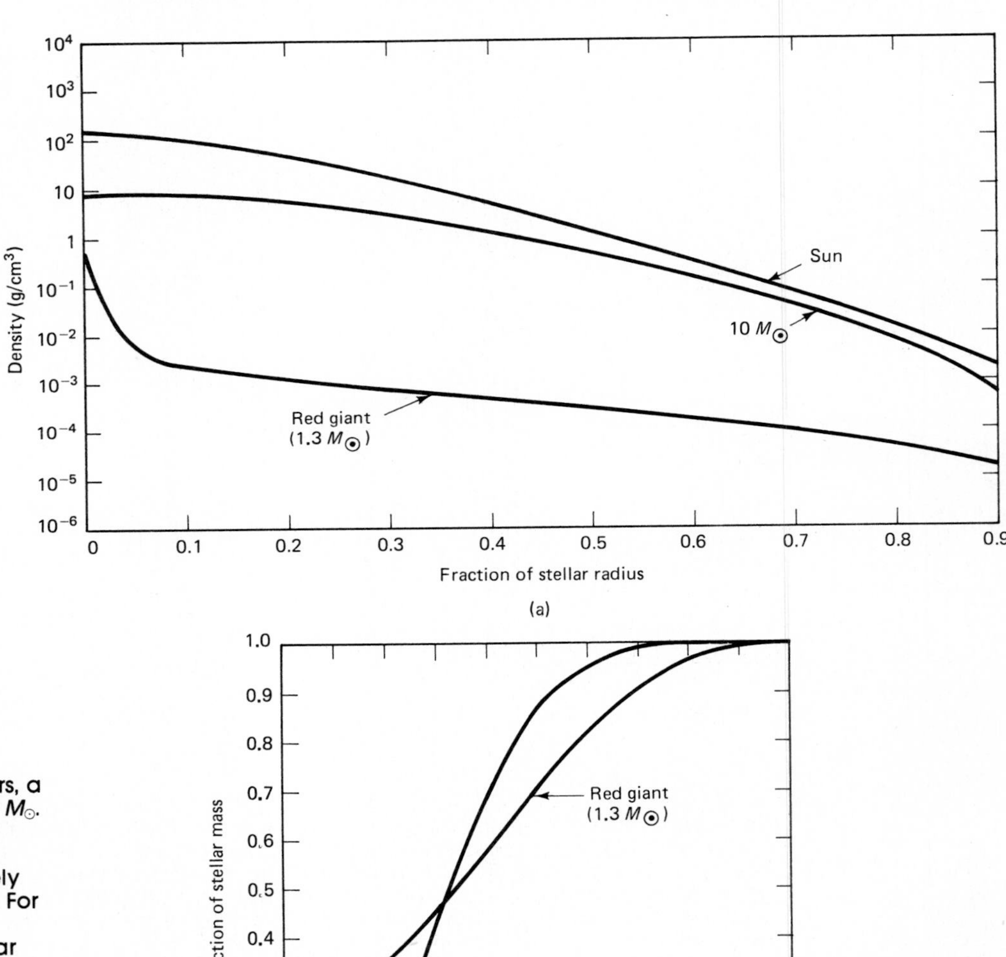


**Figure 23.9** Models for two stars, a red giant and a star of mass 10 $M_\odot$. (a) Density. The red giant has a small compact core, and the density then decreases relatively slowly for most of the star's size. For the Sun and a 10 $M_\odot$ main sequence star, there is no similar drop in density near the core. (b) Fraction of stellar mass. The 10 $M_\odot$ star, like the Sun, shows a great concentration of mass towards the center with about 50 percent of the mass within the central 30 percent of the radius. The red giant has 27 percent of its mass within the first 1 percent of radius.

## 23.6 THE SOLAR NEUTRINO PROBLEM

Nuclear reactions at the center of the Sun provide the energy that we ultimately detect in the radio, visible, and other regions of the electromagnetic spectrum, but our measurements tell us directly only about the outer regions of the Sun. We have only one direct probe, the neutrinos, for examining the deep interior in a way that is not influenced by the outer layers.

Some of the nuclear reactions release neutrinos. These particles have bizarre properties. In some ways, they are like photons: They have no mass, they carry no electric charge, but they do have momentum and energy. But whereas scattering and absorption prevent the photons from making a straight line exit from the center, the neutrinos can and do go cleanly through most atoms they encounter. Most neutrinos escape promptly and traveling at the speed of light reach the surface of the Sun after 2.3 sec and arrive at the Earth about 8 min later.

If the neutrinos can escape from the Sun without being absorbed or scattered, how can we hope to detect them on Earth? The answer is that we *can* detect them but very inefficiently by providing a special target for them. Some neutrinos that collide with atoms of an isotope of chlorine, $Cl^{37}$, can react to product an isotope of argon, $Ar^{37}$

$$\nu + Cl^{37} \rightarrow Ar^{37} + {}_{-}e$$

The argon atoms are radioactive with a half-life of 35 days and can be detected by their radioactivity: $Ar^{37} \rightarrow Cl^{37} + {}_{+}e$.

The experiment, therefore, consists of a very large tank containing chlorine atoms in the form of a compound $C_2Cl_4$, perchlorethylene, a commercial cleaning fluid that is obtainable in very pure form in large (and relatively cheap) quantities. The tank, holding 400,000 liters (about 100,000 gallons) is located 1.5 km underground in the Homestake Gold Mine in Lead, South Dakota, and is connected to a special plumbing system (Figs. 23.10

**Figure 23.10** General layout of the Brookhaven Solar Neutrino Experiment. Radioactive argon atoms ($Ar^{37}$) are produced by solar neutrino reactions in the fluid in the large tank of perchlorethylene. These atoms are removed by bubbling helium gas through the system, and the number of radioactive $Ar^{37}$ atoms can then be counted (Raymond Davis, Jr., Brookhaven National Laboratory).

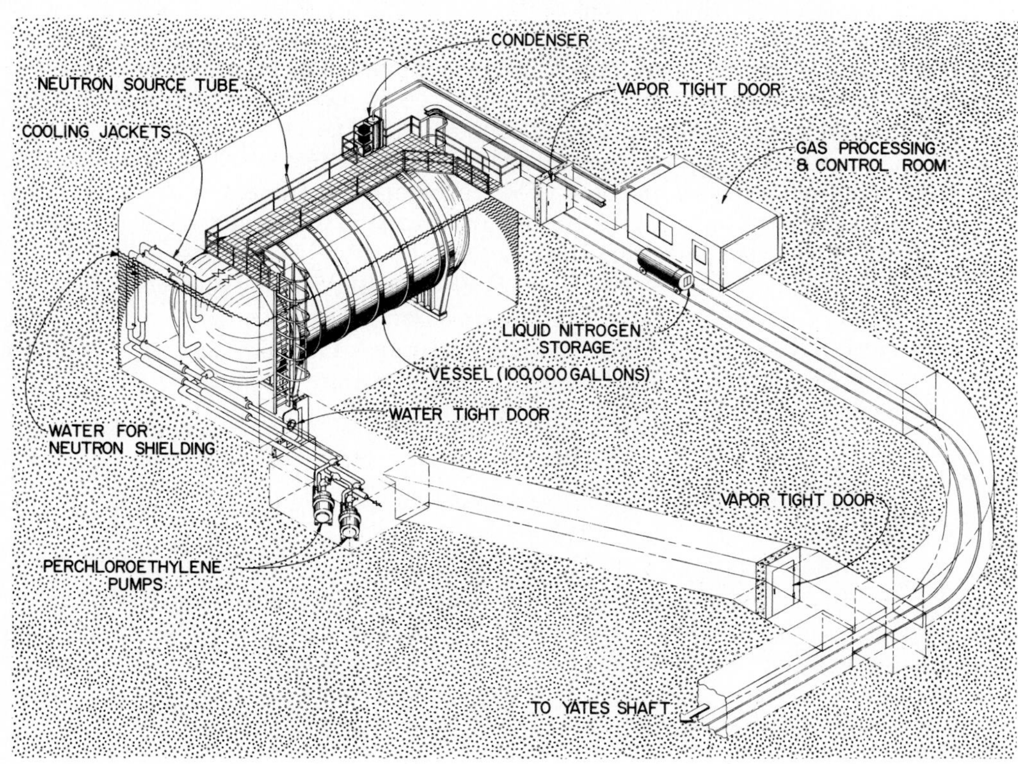

**Figure 23.11** Tank containing the neutrino detector, perchlorethylene, in the Homestake Gold Mine, South Dakota (Raymond Davis, Jr., Brookhaven National Laboratory).

and 23.11). The choice of the underground site and the additional shielding provided by covering the tank with water are designed to absorb most of the cosmic-ray particles that would penetrate even that far underground. Cosmic-ray particles would react in the cleaning fluid and swamp the solar signal if there was not enough shielding. Solar neutrinos interact in the $C_2Cl_4$, and so a small amount of $Ar^{37}$ gradually builds up. Every 100 days helium gas is bubbled through the system. This removes the argon atoms that are carried along as a trace gas with the helium. The radioactivity of the argon is then measured providing an index of the number of solar neutrinos that have interacted in the tank. This ingenious experiment was designed by Raymond Davis, Jr. of the Brookhaven National Laboratory in Long Island, and it has been operating since the late 1960s.

At the Earth, the flow of neutrinos from the Sun is expected to be about $10^{10}$/cm$^2$ sec. Most come from the reaction $H^1 + H^1 \rightarrow H^2 + {}_{+}e + \nu$, but these have too low a kinetic energy to trigger the chlorine reaction, which is much more sensitive to higher-energy neutrinos from the other reaction chain in which there is the reaction

$$Be^7 + {}_{-}e \rightarrow Li^7 + \nu$$

There is only about one of these higher-energy neutrinos for every $10^4$ of low energy. For every $10^{21}$ neutrinos that penetrate Davis' tank of perchlorethylene, only *one* will react to produce an $Ar^{37}$ atom so that Davis will be looking for only about three $Ar^{37}$ atoms produced per week.

Calculations of the expected number of neutrinos depends on several assumptions regarding the solar interior and the behavior of neutrinos. The predictions are expressed in **solar neutrino units** or SNUs: the number of captured neutrinos for each $10^{36}$ target atoms. Over the years the predicted number has changed, as improved calculations have been made. The present prediction is 7.6 SNUs with an uncertainty of 1.1. The observed number is $1.8 \pm 0.3$ SNUs (Fig. 23.12). Our only direct probe of the solar interior thus shows an embarrassing discrepancy from theory by a factor of about four.

Can the experimental result be in error? It is easy to get too large a result if one has not been careful to exclude "background" signals, but too low a result is harder to explain. Careful checks have failed to show any systematic experimental effect that would result in too low an answer. How about the theory? Theoreticians have tried, but they have not been able either to reduce their predicted value sufficiently or to pinpoint the source of disagreement.

All sorts of possibilities have been considered. Since the neutrinos tell what is happening now at the center of the Sun, and the luminosity tells what was happening a million years ago, perhaps the neutrinos are simply giving advance warning of a coming reduction in the luminosity with the possibility of another ice age on Earth. Perhaps the neutrinos' properties differ from those assumed. Some of the neutrino properties assumed in the calculations have not been tested directly; rather they have been inferred from other experiments with the aid of theory. It has been assumed that the neutrinos have zero mass, but if instead they have any mass, even a very little, the consequences would be far-reaching. Experiments have shown that the neutrino mass must be less than 1/10,000 of an electron mass, and further experimenting is in progress. Perhaps calculations of the solar opacity

**Figure 23.12** Yearly averages of the $Ar^{37}$ production rate in the solar neutrino detector. Although there are considerable variations, the combined average rate is about one quarter of the predicted value (Raymond Davis, Jr., Brookhaven National Laboratory).

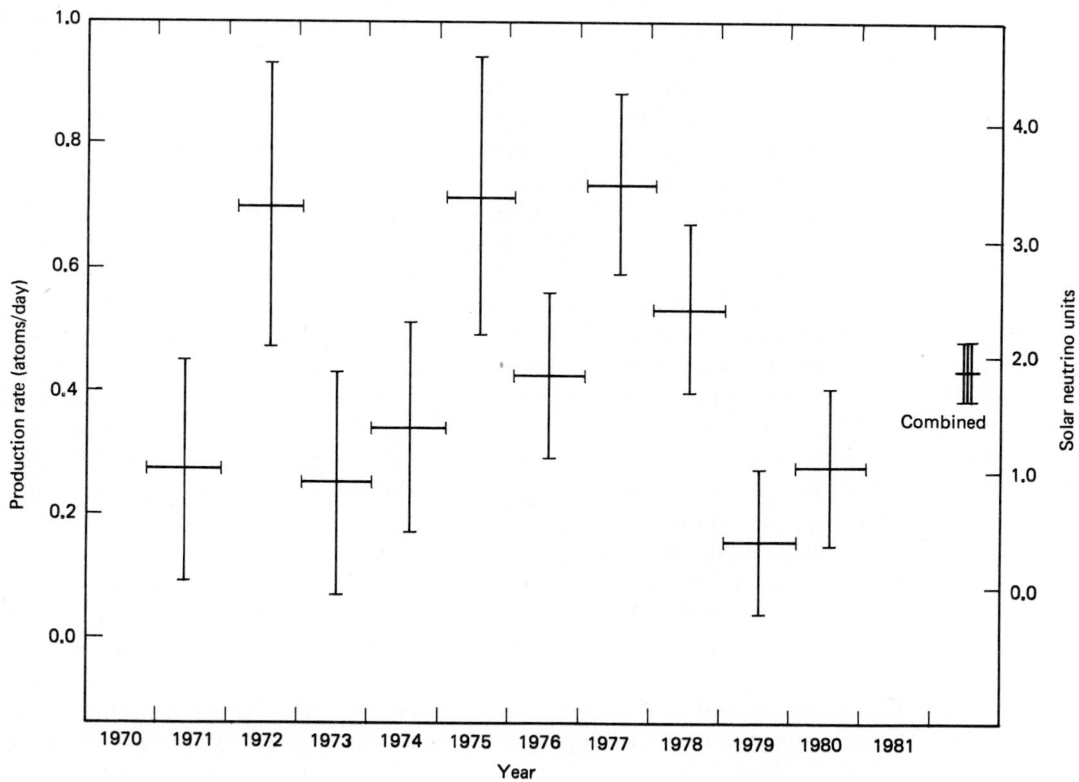

need revision. For each of these and many other possibilities, there seems to be a plausible refutation—but the end result is still a major dilemma.

Other experiments are being designed to detect the more plentiful low-energy neutrinos from the p-p chain, but it will be several years before any of these experiments can start. If we cannot get better agreement between theory and observation for the Sun, is it possible that we could be seriously wrong for other less accessible stars? We think the theory is basically correct, but . . . .

## 23.7 COMMENTARY

After the detailed discussions in the preceeding pages, it may be useful to stand back and take a broad view of the role of stellar models in astrophysics. Our objective is the understanding of the universe: its present construction, its origin, and probable further development. We now know that the time scale for stellar evolution is measured in billions of years, but our observing base is a matter of only about 100 years. The inaccessibility of the stars imposes another major difficulty. The magnitude of the task might be gauged from a comparison.

Suppose we wanted to study an animal that lived for only a few years, but our observing lasted at most a few seconds. During those seconds, we could observe a number of those animals and guess that some seemed to be old, others young, but we could not yet follow them for long enough times to see any appreciable changes. Further, we could not lay hands on any of them for dissection, so that our ideas about their internal structure and functioning—skeletons, digestive systems, etc.—could come only from their external appearance.

We might seek other animals that aged much faster and study them in our laboratories, hoping that our findings could be generalized. We would almost certainly have to make some simplifying assumptions, and introduce complicating factors as new observations forced them on us.

This is close to the way we proceed with stellar models. We know the external appearances: size, surface temperature, composition, (from spectral analysis), and luminosity. We know a lot about the basic physics of nuclei and radiation, and so we construct models whose internal structure leads to the observable quantities. We find that there is some leeway: Perhaps several different models can provide equally good agreement with the observations. However, the models lead to different predictions so that we might have a means of selecting some and rejecting others. This can guide our observing programs.

The inherent complexity of model calculations forces us to make some simplifying assumptions at the start. For instance, we usually assume that our model stars are perfectly spherical with quantities such as density, temperature, and opacity depending only on distance from the center but not on direction (stellar latitude or longitude). Rotation complicates the picture in many ways. Stellar model calculations have been revolutionized by the improvements in computers, which allow theorists to work numerically with equations that could not previously be solved exactly. The high speed with which computers carry out their computations has opened the way to examining models far more complex than could have been considered even as recently as the 1960s.

Newer models, for example, for extremely massive or dense stars require the inclusion of speculative processes in nuclear physics that have not yet been verified under laboratory conditions. To the extent that these models provide and survive observable tests, we might be able to add to our knowledge of the properties of nuclei and their even smaller constituent particles. The scientific difficulties are great, but the prospects are exciting.

## CHAPTER REVIEW

**Evidence for stellar change:** observations of supernova outbursts; stars radiate energy—depletion of energy reservoir requires changes; H-R diagrams are different for various groups of stars. As stars age, their locations on the H-R diagram will change. Time scale for stellar evolution is billions of years, as indicated by age of Earth and solar system.

**Hydrostatic equilibrium:** steady state in which a star radiates energy released in its interior, while there is a balance between forces that would either make it collapse or expand. Inward force comes from gravitational attraction between constituent atoms of the star. Outward forces come from hot, high-temperature gas trying to expand and radiation that is traveling from the center out. (Radiation exerts an appreciable force only in large stars.) At the center of a star, pressure and temperature are very high. Nuclear reactions take place releasing energy.

**Nuclear reactions:** p-p chain of reactions in which four hydrogen nuclei fuse to form helium with release of energy. Alternative is CNO chain, which requires higher temperatures. In the Sun the p-p chain is responsible for most of the energy, but some comes from the CNO chain, which dominates in massive stars.

**Transportation of energy:** in stars most of the energy is carried by radiation. Under some conditions convection carries energy, and conduction is generally negligible. Energy radiated today from surface of the Sun was released in nuclear reactions at the center about a million years ago.

**Stellar models:** theoretical schemes that try to match the observed mass, radius, luminosity, and composition by suitable internal structure.

**Solar neutrinos:** released in nuclear reactions in the Sun. Numbers detected at Earth are significantly less than expected.

## NEW TERMS

**absorption**
**Boyle's law**
**conduction**
**deuterium**
**equation of state**
**fusion**
**gravitational energy**
**hydrostatic equilibrium**
**mass defect**
**neutrino**
**nuclear reactions**
**opacity**
**positron**
**radiation pressure**
**radiative transfer**
**scattering**
**solar neutrino unit (SNU)**

## PROBLEMS

**1.** Why are higher temperatures required for the CNO cycle than for the p-p cycle, although both result in the fusion of hydrogen into helium?

**2.** What sources of energy are available for stars, and what is the evidence that persuades us that nuclear reactions must be the source of the Sun's luminosity?

**3.** How did the discovery of radioactivity lead to the realization that the energy radiated by the Sun could not be coming from gravitational contraction?

**4.** Using the formula given in this chapter calculate the pressure in the center of the Earth.

**5.** What are the different mechanisms by which energy can be transported?

**6.** Assume that the Sun's luminosity remains steady at its present value. How long will the Sun remain on the main sequence, that is, until it has converted one tenth of its total mass of hydrogen into helium? (Assume that initially 72 percent of the Sun consists of hydrogen.)

**7.** The solar luminosity is $3.84 \times 10^{33}$ ergs/sec, derived from the fusion of hydrogen to helium. What mass of hydrogen is consumed each second to produce this energy, and how many hydrogen nuclei are reacting each second?

**8.** A star in the Eta Carina nebula is thought to have a mass of 100 times the solar mass, and it has a luminosity of about $10^{40}$ ergs/sec. How long will it take at the present luminosity before it has consumed 10 percent of its hydrogen and converted it into helium? (Assume that the star is initially 100 percent hydrogen.)

**9.** Why do fusion reactions take place in the center of the Sun but not on the surface?

**10.** After you strike a match, the hot gases in the flame soon disperse. The Sun also consists of hot gases. Why do they not disperse? The Sun also does not collapse. Why not?

**11.** The fusion reaction that powers the Sun involves the joining of four hydrogen nuclei to form helium. However this takes several steps, and other nuclei are formed and then consumed in a chain of reactions.

Identify these intermediate nuclei in the p-p reaction chain.

**12.** Stars can live for millions or billions of years, far longer than the span of any of our observations. What evidence do we have that stars do not remain constant but evolve?

**13.** The triple alpha reaction involves the fusion of three helium nuclei into a single carbon nucleus. The mass of a helium nucleus is 4.0026 amu, and the mass of a carbon nucleus is 12.000 amu. Calculate the amount of energy released in this process. Compare this to the energy released in the p-p reaction that fuses hydrogen to form helium.

**14.** In the solar neutrino experiment $Ar^{37}$ atoms are produced. These are radioactive with a half-life of 35 days. These atoms are collected every 100 days. If all of the $Ar^{37}$ atoms were produced 100 days before collection, how many would not yet have decayed by the time of collection?

**15.** Why is the solar neutrino experiment important?

CHAPTER 24

# Star Formation and Evolution

Now that we have a generally satisfactory understanding of the structure of stars, we can move on to explore the successive stages through which a star will pass as it ages. As we shall see, both the rate of aging and the end result depend on the mass of the star. The formation of a star starts in a region of gas and dust. Stars with masses between about a tenth of a solar mass and some tens of solar masses condense and evolve onto the main sequence where they spend most of their lives. Evolution off the main sequence is strongly influenced by mass. For example, a star like the Sun is expected to expand to become a red giant, lose much of its outer envelope, and finally settle down as a white dwarf. A much heavier star may end as a neutron star or black hole.

We think that we know enough of the basic physics to construct reasonable models and then follow their progress with computers. But what seems a reasonable model may turn out to be so complicated that some simplifying assumptions must be made to keep the calculations to a manageable size. As a result, we should expect the main features of the calculations to be correct, but the fine details may be less reliable. Many different calculations have been made, starting with alternative assumptions about the proportions of hydrogen, helium, and the rest (the $X$, $Y$, and $Z$ mentioned in the last chapter). In this chapter we shall review the early stages of stellar evolution, and leave for the next chapter the description of the exotic final stages such as neutron stars and black holes.

## 24.1 STAR FORMATION

In this section we will review, in general terms, the evolution of a star onto the main sequence, and in the next section we will show how these changes can be represented on the H-R diagram.

In many regions around and between the stars, we can see extensive clouds of gas and dust, which can supply the ingredients for new stars (Figs. 24.1 and 24.2). At the typical temperatures of pre-stellar clouds (around 10 K), most of the gas is molecular hydrogen ($H_2$) with traces of heavier com-

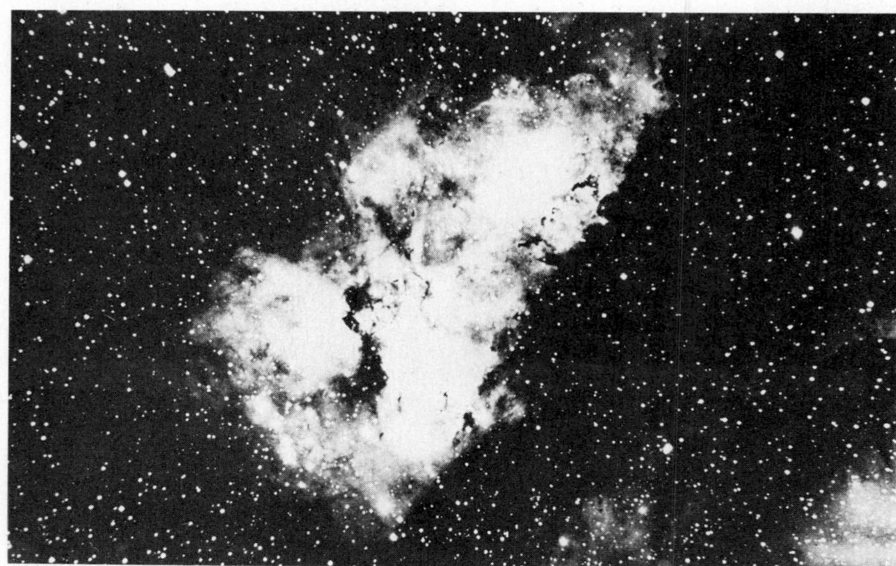

**Figure 24.1** The Eta Carinae nebula. This region contains a large mass of gas and dust with many very bright and hot young *O* type stars, suggestive of a region of recent and perhaps current star formation (Yerkes Observatory).

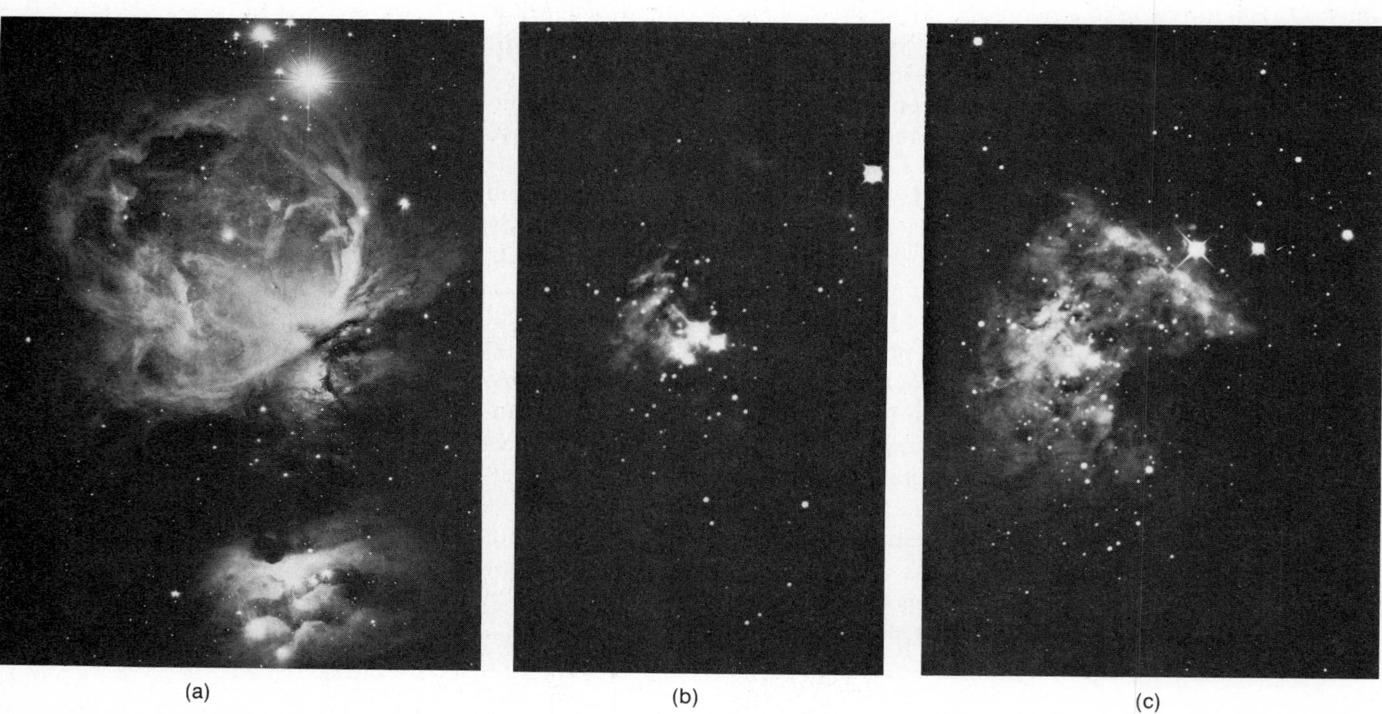

(a) (b) (c)

**Figure 24.2** (a) The Orion nebula photographed in deep blue light. This region also contains much gas and dust and many young stars (Lick Observatory). (b) The Orion nebula, photographed in IR light, showing the Trapezium, the group of four stars near the center (Lick Observatory). (c) The Orion nebula, shown in a longer IR exposure. Fainter stars and more of the nebulosity can now be seen (Lick Observatory).

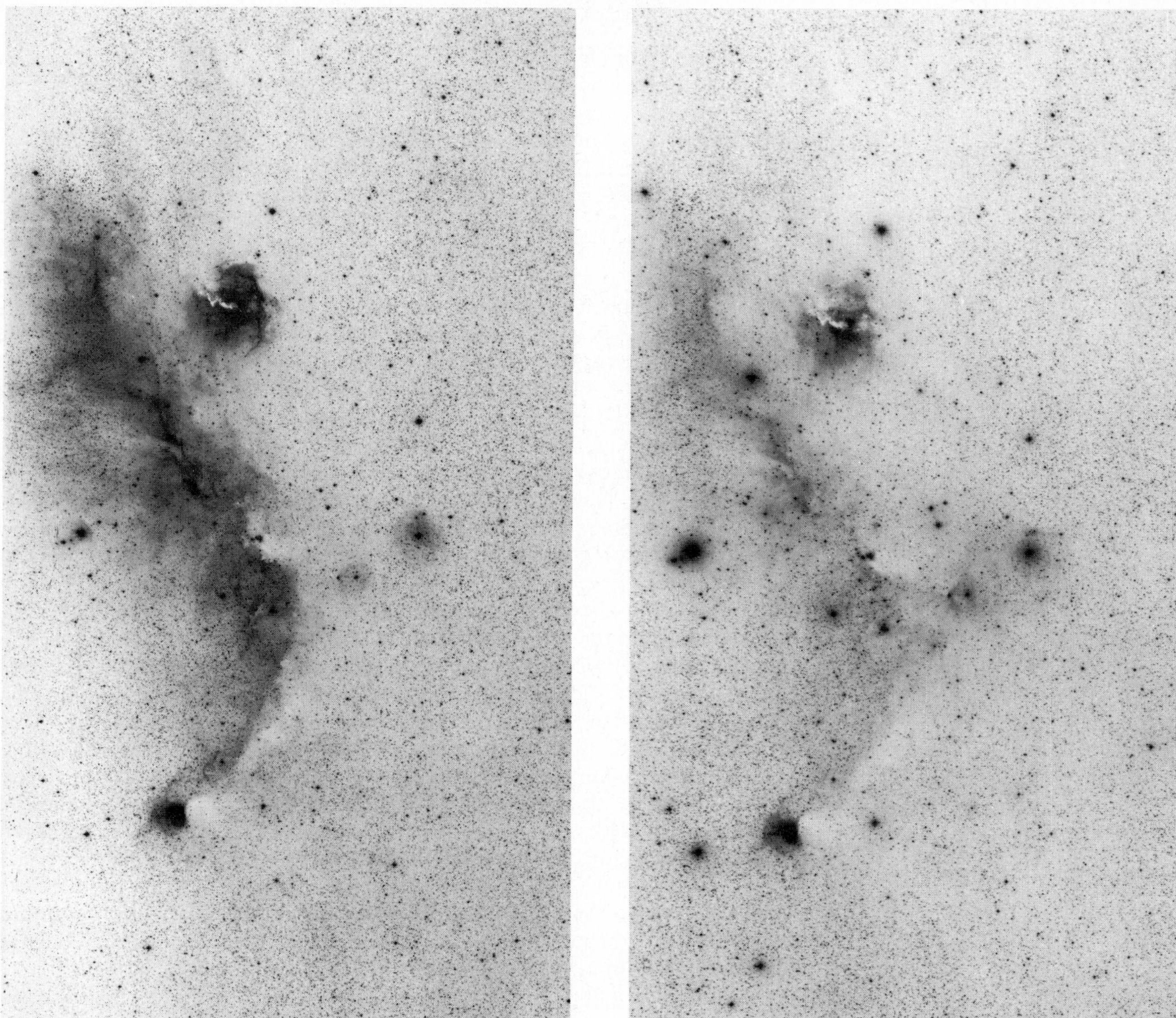

**Figure 24.3** The formation of stars might be triggered when the local density of the interstellar medium is significantly increased. The dark region in the photograph on the left (taken through a red filter) marks the boundary of the expanding gas and dust shell, probably from an old supernova. Along the boundary, there are many young stars seen more clearly in the photograph on the right taken through a blue filter. Light from these young stars is reflected by the dust and gas to produce the nebula, which is about 30 pc in length (W. Herbst and G. E. Assousa; © 1960 National Geographic Society—Palomar Sky Survey. Reproduced by permission of the California Institute of Technology).

ponents. The very first steps of star formation are not yet well understood. We think that a star starts to form when a local density within the gas/dust cloud becomes appreciably larger than that of the surrounding region, but the minimum required density is not known.

Present theoretical speculation tends to focus on two possibilities for the mechanism that will produce the higher densities and thus initiate star formation. The **density wave** theory assumes that the critical densities can be produced in the gas and dust by a wave that travels around the galaxy, much as a sound wave compresses air without moving it in bulk. The other theory starts with a supernova explosion and follows the passage of the shock wave that spreads out compressing the material it encounters (Fig. 24.3).

Stages in Stellar Evolution

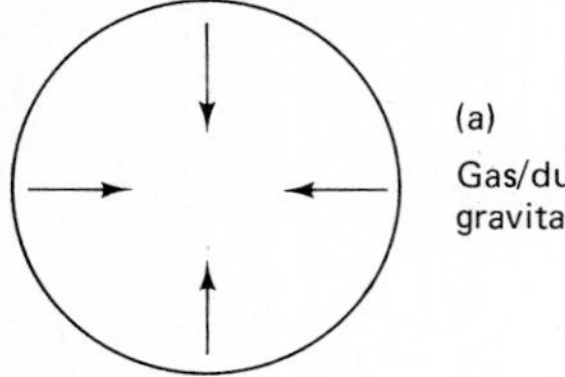

(a)

Gas/dust cloud contracts, gravitational energy released.

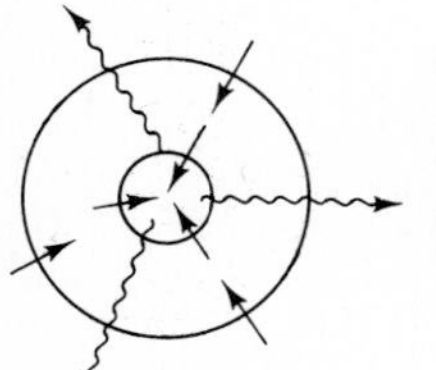

(b)

Material closer to center falls faster, heats up, radiates. Radiation can still escape as collapsing outer cloud is not dense.

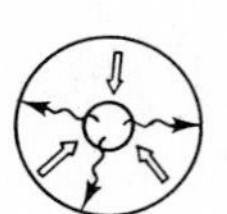

(c)

Inner region heats up as energy can no longer easily escape through dense outer region; collapse slows.

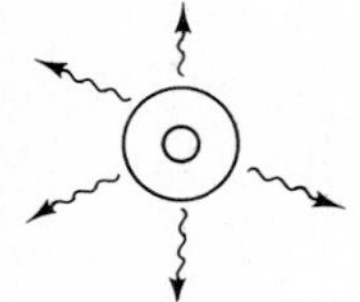

(d)

Star in hydrostatic equilibrium: central temperature high enough to sustain nuclear reactions and central pressure prevents further collapse.

**Figure 24.4** Stages in stellar evolution shown schematically.

Each theory provides an explanation for some observations, but neither is yet firmly established. Whatever the mechanism, once the density is high enough and the condensation starts, the behavior of the cloud can be followed in theory ("modeled") with computers.

In many parts of the galaxy there is (on the average) only one hydrogen atom or molecule in each cubic centimeter. A prestellar cloud might have a density of 100/cm$^3$ and a diameter of a few light years ($10^{18}$–$10^{19}$ cm). The gas and dust in a cloud are pulled inwards towards the center by gravity in the same way that matter in a condensed star is pulled to the center. In its later stages, a star will be supported against collapse by the pressure of its hot gas core, but in a cold cloud there is no central release of energy at the start, and so the collapse is initially unopposed. Because of the very large cloud dimensions, the time of collapse is typically several hundred thousand years to produce a star of one solar mass. Later stages of evolution last for much longer.

As the collapse proceeds, material close to the center falls faster than that further out. The density in the central region increases and so does the frequency of collisions between molecules and dust grains. Meanwhile the outer region is still collapsing but not as rapidly so that its density increases slowly (Fig. 24.4).

Gravitational energy released in the contraction appears as increased kinetic energy of the molecules. During the early stages, the interior warms up. Its thermal radiation is mostly at IR wavelengths and escapes easily through the outer region, which is still very diffuse.(Compare this to the later hydrostatic equilibrium in stars where the opacity is large enough to play an important role by slowing the escape of the radiant energy.)

Further collapse leads to greater trapping of the radiation through two effects. First, the collapse causes the inner core to heat up more. The higher temperature has its peak radiation at shorter wavelengths. Opacity depends on the wavelength of the radiation, and in general, opacity is larger for shorter wavelengths. Second, the collapse causes the density to increase, and opacity increases with increasing density. With these two reinforcing effects, the core therefore heats up to several hundred degrees, and the increased radiation pressure acts as a brake slowing further contraction.

The outer material continues to fall in towards the core, adding to its mass and heating it further. By the time that the temperature reaches 2000 K, collisions between hydrogen molecules are violent enough to break them into their constituent atoms; at around 5000 K the electrons are stripped off the atoms. Because each of these changes requires energy, the core pressure temporarily drops, but the resulting drop in the upward radiation force is promptly followed by further gravitationally driven contraction leading to yet higher pressure and temperature. The cycle continues until the central temperature is high enough to ignite the nuclear fusion reactions. Until the nuclear reactions start, the only source of thermal energy is the gravitational contraction. After the fusion reactions have started in the core, the contraction slows and there is an interval in which the luminosity does not change much. During this interval, the nuclear reactions steadily increase in rate until they are providing about 98 percent of the energy. The star then takes its place on the main sequence where it is termed a zero age main sequence (ZAMS) star. At this stage, the ZAMS star is in hydrostatic equilibrium.

Not all of the cloud will have contracted onto the star by the time it switches on its nuclear reactions and reaches the main sequence. Radiation and shock waves from the new star can then heat residual dust which emits in the IR. Such objects have indeed been seen. Some termed Herbig-Haro objects (Fig. 24.5) also emit appreciably at visible wavelengths. Others have

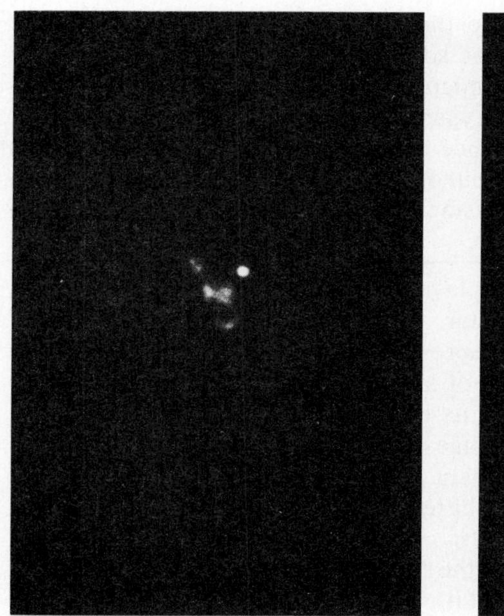

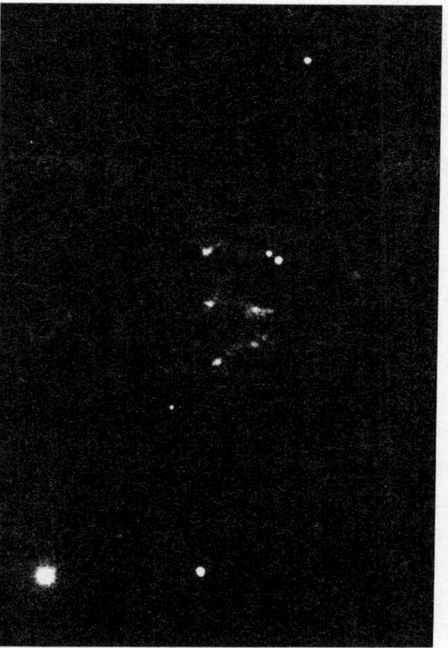

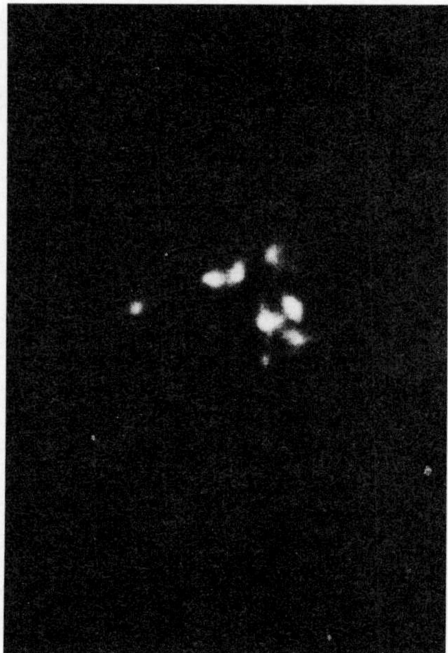

**Figure 24.5** Herbig-Haro objects (Lick Observatory).

been detected at IR wavelengths in regions such as the Orion nebula, which are known to be rich in gas and dust and, therefore, quite likely sites for star formation.

In general the original cloud will be much more massive than the resulting star, and most of the cloud must be blown off at some stage although it is not clear just how or when. It is likely that the star goes through unstable periods in which its luminosity varies irregularly before settling to the main sequence value. For example, the visible luminosity of T Tauri stars (Figs. 24.6 and 24.7) can vary by several magnitudes, perhaps to be expected for

**Figure 24.6** T Tauri with the wispy nebula NGC 1555 just to the right (G. Herbig, Lick Observatory).

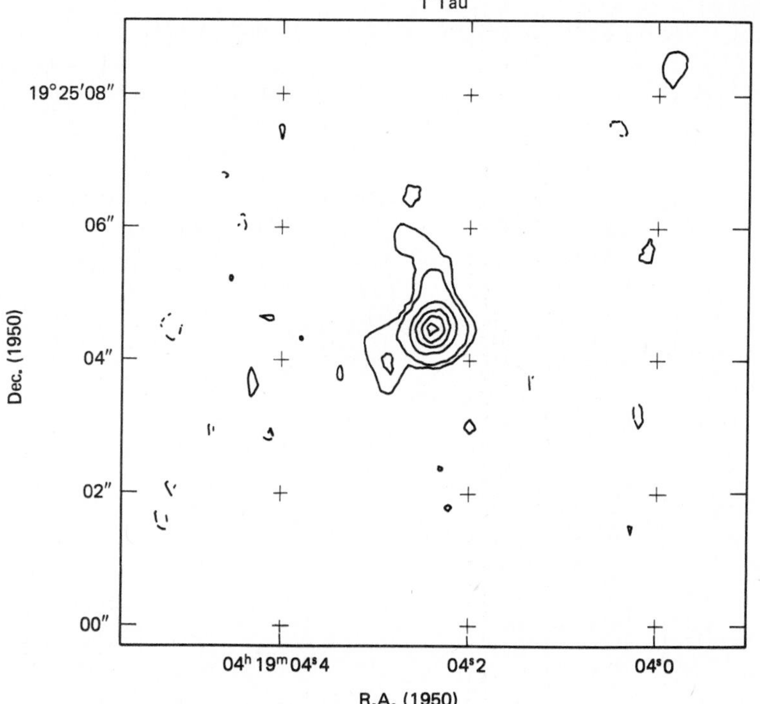


**Figure 24.7** T Tauri observed at 6 cm wavelength using the VLA. The central core is unresolved with a diameter of less than 1 arc sec. The outer contours show a considerably extended region of ionized gas with the irregular shape probably due to a stellar wind (Martin Cohen, NASA Ames Research Center).

a star that has not yet settled onto the main sequence. In addition, the T Tauri radiate strongly in the IR, and some of them show a strong emission feature at the IR wavelength of 10 microns ($10^{-3}$ cm). This feature is generally associated with the silicon-oxygen (SiO) radical and is seen in many dusty regions. The overall appearance of the T Tauri, therefore, suggests radiation from a dust coccoon surrounding a variable luminosity star, possible indicators of an early stage of star formation.

## 24.2 USE OF THE H-R DIAGRAM

We can use the H-R diagram to display the changes that take place during the formation and evolution of a star. The H-R diagram we used earlier showed a number of points, each representing a star as we now see it. At each stage of its life the luminosity of a star depends on the physical processes releasing energy and thus on its effective (or apparent) surface temperature and total surface area. Changes in luminosity, temperature, and/or radius will thus show up as movement on the H-R diagram, and the path traced out by a developing star as these quantities evolve is termed an **evolutionary track**.

For examples, we will follow the tracks of a few stars with different masses, and see how their representative points move on the H-R diagram as these stars evolve. We start with a star of one stellar mass.

During the earliest phase ($A$ to $B$, Fig. 24.8), the cold gas and dust are contracting. The gravitational energy that is released warms the cloud, and the luminosity increases. A maximum luminosity is reached (at $B$); after this, more of the energy from contraction goes into internal heating, while the opacity increases and the luminosity drops. During the ensuing gravitational contraction (along $C$ to $D$, Fig. 24.8), there is little change in the effective surface temperature, but the luminosity continues to drop. Recalling that $L$ is proportional to $R^2T^4$, we see that if $T$ remains fairly steady, then the stellar radius $R$ must decrease by a factor of about thirty while the luminosity drops by a factor of 1000. From $D$ to $E$ the luminosity hardly changes but the temperature increases by about a factor of 1.5. Using the same formula, we see that $R$ must decrease by a factor of about two as the star progresses from $D$ to $E$.

During the gravitational contraction ($C$ to $D$), our evolving star is actually more luminous than it will be later even when the nuclear reactions have started. The gravitational energy released in the contraction is able to

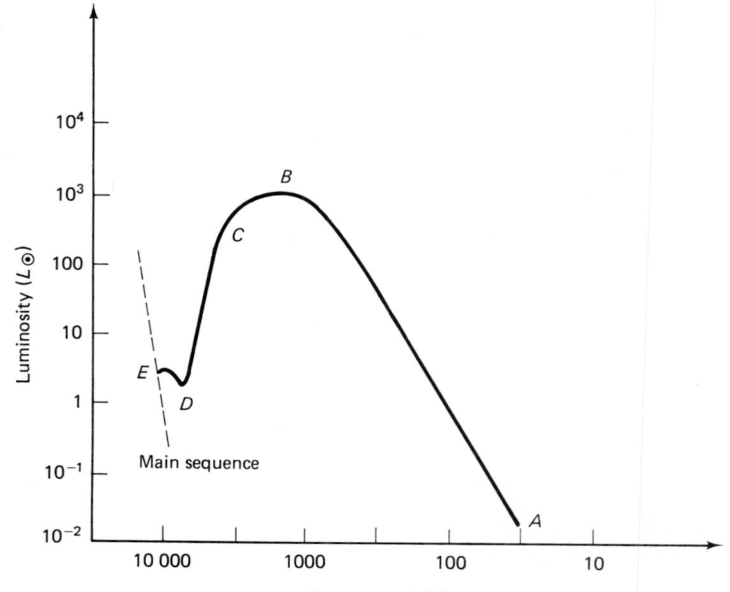


**Figure 24.8** Schematic evolutionary track of a solar mass star on the H-R diagram. Initially ($A$ to $B$) there is a contraction, warming, and luminosity increase. This is followed by a phase ($B$, $C$ to $D$) while the gravitational contraction continues, but the increasing density of the nebula prevents the radiation from escaping: the luminosity drops. The temperature of the central region is now increasing and nuclear reactions start, and the star is now ($E$) on the main sequence. The surface temperature is far lower than that in the core.

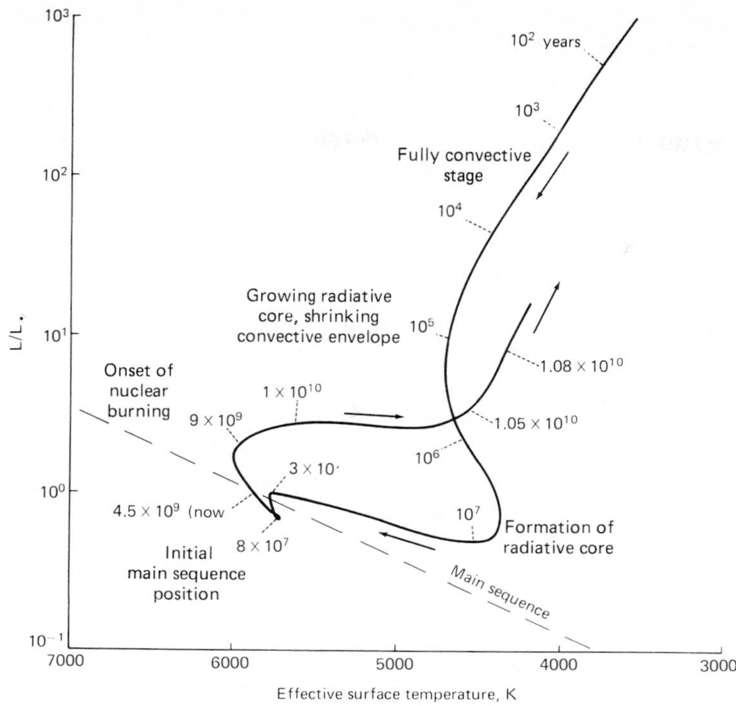


**Figure 24.9** Evolutionary track on the H-R diagram for the Sun. Times marked indicate times from start of condensation. Evolutionary stages before this diagram are shown in Fig. 24.8.

escape through the thin collapsing cloud, but (as can be noted from the time markers) it is very bright for only a relatively short time, and most of the radiated energy will be in the IR. The rate of contraction slows down as the star approaches point *D*. By point *D* the opacity has increased to such an extent that it is much harder for the energy to escape, and (as shown by the *DE* track) the star must, therefore, heat up.

The release of energy in nuclear reactions becomes increasingly important along the *DE* part of the track. By *E*, almost all of the energy comes

**Figure 24.10** Evolution of the temperature, luminosity, and size of the Sun. The various quantities are shown as ratios to the present values (NASA).

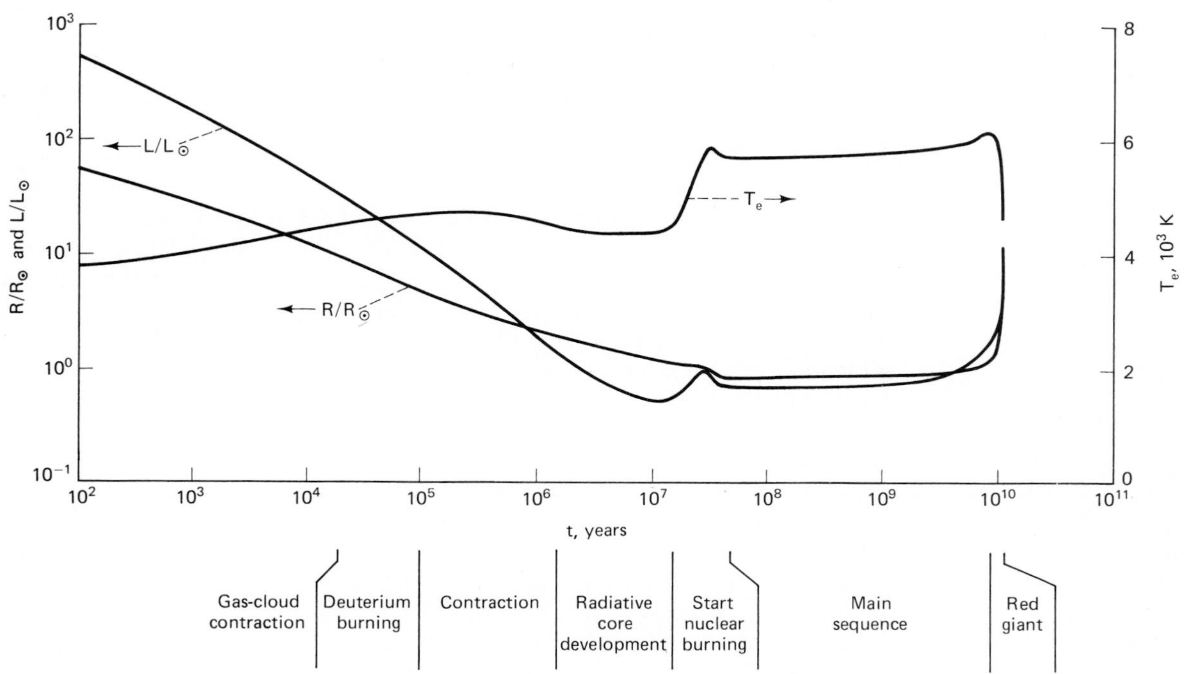

from nuclear reactions, and the contraction has almost ceased. The main sequence is reached after $5 \times 10^7$ years.

For the more massive stars the evolutionary tracks bear a general resemblance to the track for 1 $M_\odot$, but with increasing stellar mass the temperatures change by larger factors and the duration of the successive phases becomes much shorter. Where the gravitational contraction (section *CD*) takes about 10 million years for a star of 1 $M_\odot$, the corresponding times are just over $10^5$ years for 3 $M_\odot$, and just over 1000 years for the 10 $M_\odot$ star. The tracks give a misleading impression for the *D* to *E* segments: In the diagram these lines seem to increase with increasing mass, but the actual duration (in years) is decreasing. More massive stars arrive on the main sequence much more rapidly.

Along the main sequence the position of a star is determined by its mass. At the top left-hand end of the sequence we find the most massive stars with the highest luminosities, highest temperatures, and (as we will see) the shortest times on the main sequence. As we move down the main sequence, we are moving through stars of smaller mass with lower luminosities and temperatures. These stars are consuming their nuclear reserves so slowly that they evolve over very long times.

## 24.3 STAR MASSES

The scenario set out in the preceding sections does not seem to place any restrictions on the sizes of the stars that can be formed. Given a cloud sufficiently large or small, can we produce very large or very small stars? Present calculations indicate lower and upper limits of about 0.08 $M_\odot$ and 150 to 200 $M_\odot$ respectively for stable main sequence stars formed in this way. The theoretical limits are not firm.

Below 0.08 $M_\odot$ somewhat smaller stars can still form, but they will not be on the main sequence. They will be stars of very low luminosity, radiating away only the energy released by gravitational contraction since their central temperatures will not be sufficiently high for nuclear reactions to take place. Our planet Jupiter, which radiates twice as much energy as it receives from the Sun, might be considered to be in this category. These small stars will presumably be very hard to detect unless they happen to be close to the solar system. No main sequence star of mass definitely lower than 0.08 $M_\odot$ has been found.

At the top end of the mass scale, the limit may be uncertain by as much as a factor of two. Some calculations have indicated that when the central condensation gets as large as about 100 $M_\odot$, its luminosity will be so great that the resulting radiation pressure will prevent infall of further matter, and the mass should not increase. If, however, a star of greater mass should somehow form, then its life on the main sequence would be quite brief. Infrared and visual surveys in recent years have given evidence for the existence of a number of stars with masses up to around 200 $M_\odot$. Stellar model calculations suggest that such massive stars will not age through the red giant phase but instead may heat up and move to the left on the H-R diagram while losing mass steadily by their stellar winds. There is no clear evidence for stars with masses above 200 $M_\odot$ in our galaxy, but recent observations of an extremely luminous object (R136A) in the Large Magellanic Cloud suggest that its mass is about 2000 $M_\odot$.

## 24.4 EVOLUTION ON THE MAIN SEQUENCE

Once established on the main sequence, a star will be in hydrostatic equilibrium, more or less gently consuming its nuclear fuel in the reactions in its core. How long the star remains on the main sequence depends very sensitively on its mass.

It might be thought that massive stars would live longer than small stars because they have more fuel available. A calculation shows that this is not the case, since the large stars consume their fuel so much more rapidly. Thus the very luminous *O* and *B* stars will age rapidly and move off the main sequence more quickly than the cooler and less luminous *K* and *M* main sequence stars.

**BOX 24.1**
*Lifetime and Stellar Masses*

We can use a simple calculation to get approximate values for relative times that stars of different masses will remain on the main sequence:

$$T = \text{time on main sequence}$$
$$= \frac{\text{total energy available}}{\text{rate of consuming (radiating) energy} = L}$$

The total energy available is proportional to the mass $M$, and from the mass-luminosity relation we know that $L$ is proportional to $M^{3.5}$. Thus

$$T \sim M/L \sim M/M^{3.5} \sim M^{-2.5}$$

This tells us that a star of mass 10 $M_\odot$ will be on the main sequence for about $10^{-2.5} = \frac{1}{300}$ of the time that our Sun will be in its place.

A complicating factor enters in the case of massive stars. At the center of such a star, the temperature will be high enough for the CNO cycle to operate—but only if there are sufficient carbon nuclei to start and sustain the reactions. Thus the evolutionary rate can depend on the composition of the star through both the opacity and the core composition.

As the hydrogen in the core is consumed, the core contracts slowly and thus produces a higher temperature. The luminosity increases slightly, and the star's position on the H-R diagram changes slowly. As long as hydrogen is available for conversion in the core, the star can stay on the main sequence moving only slowly. We might expect that this would continue until all of the hydrogen is used up, but instead we find more catastrophic changes taking place. Some will be described in the next section, others in the next chapter.

## 24.5 EVOLUTION OFF THE MAIN SEQUENCE

The consumption of the core hydrogen finally leads to an all-helium core (Figs. 24.11 and 24.12). For a 1 $M_\odot$ star, this will take close to 10 billion years, for a 10 $M_\odot$ star only about $10^7$ years. Hydrogen will still fuse in a surrounding shell, thus adding to the helium in the core. Initially, the helium core is dense enough to resist further compression, but as no fusion of the helium is taking place, the helium core cools slowly. In addition, the core is gaining mass as helium is added by continuing fusion of hydrogen in the immediately surrounding layer. A stage will, therefore, come when the core contracts suddenly and heats up again. This contraction in turn causes a collapse of the surrounding hydrogen, which also gets hotter, and the rate of hydrogen fusion thus increases. The increased energy output forces the outer regions of the star to expand; with this expansion comes a cooling and the end of the time on the main sequence. The star's position on the H-R diagram moves rapidly to the right away from the main sequence.

Another important change takes place during the expansion. Convection becomes more important for carrying heat to the surface. As a result the luminosity increases; the star moves up on the H-R diagram to the red giant stage. Meanwhile, although the surface has expanded, the core temperature is rising. When the core reaches $10^8$ K, the helium ignites in the

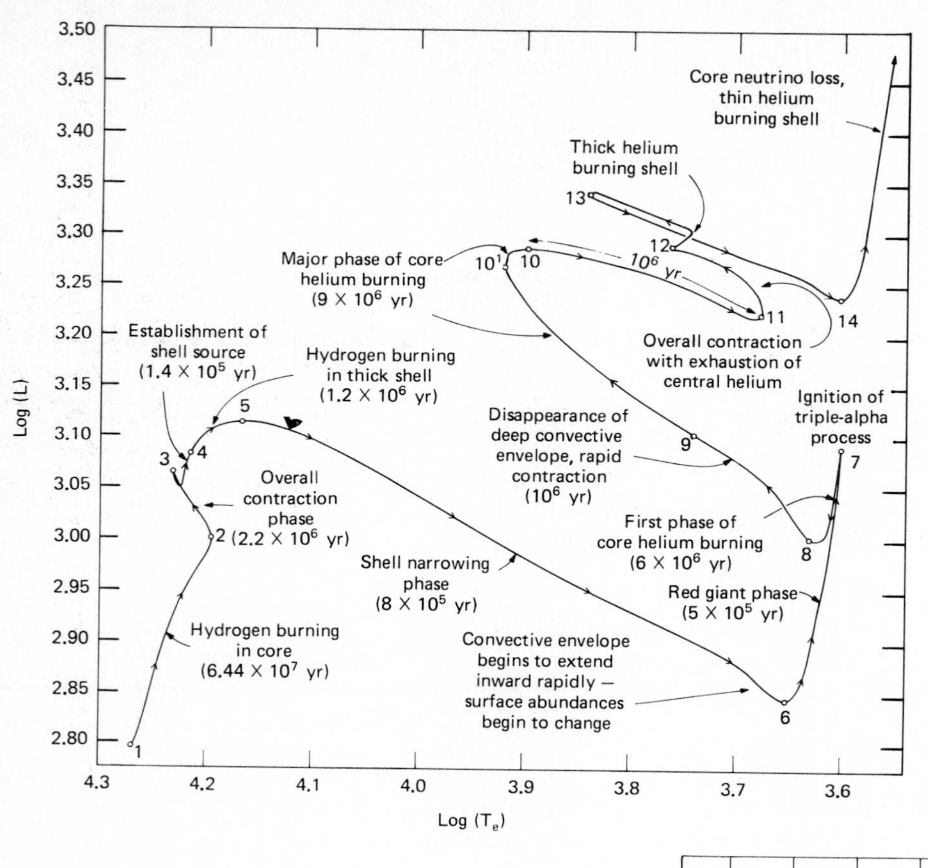


**Figure 24.11** Evolutionary track on the H-R diagram for a 5 $M_{\odot}$ star after it leaves the main sequence. Some phases are rapidly passed through, but the star lingers in others (Icko J. Iben, Jr. and *Annual Reviews of Astronomy and Astrophysics*).

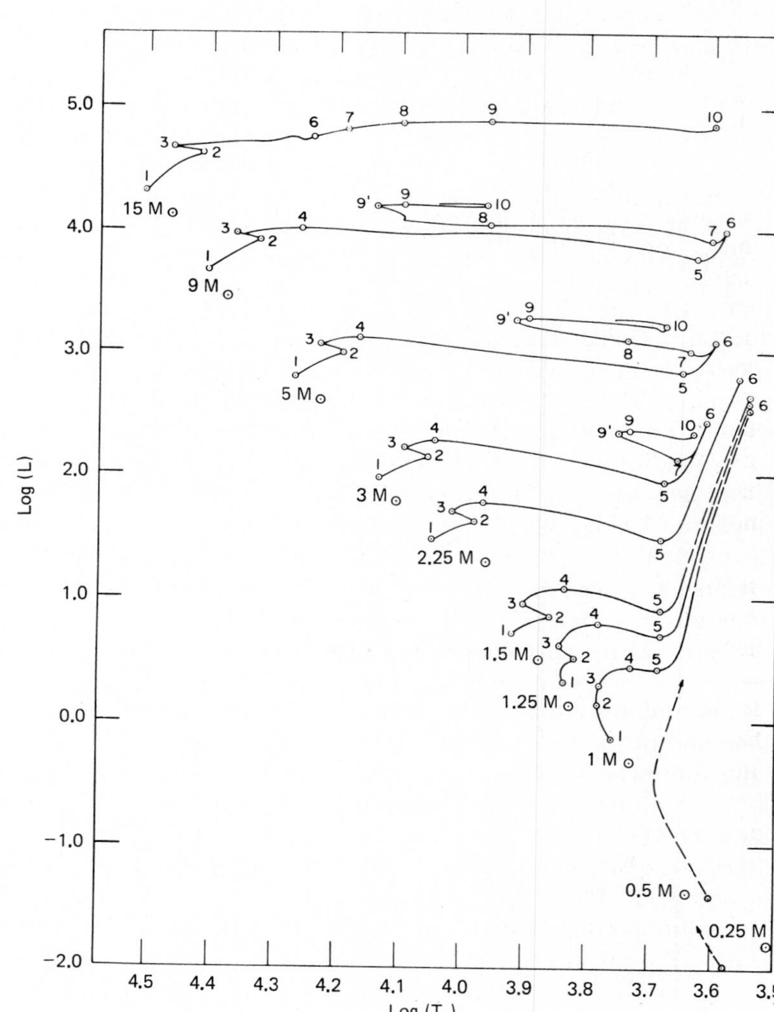


**Figure 24.12** Evolutionary tracks for stars of different masses leaving the main sequence. Numbered points along each track correspond to the same stages as shown in Fig. 24.11 (Icko J. Iben, Jr. and *Annual Reviews of Astronomy and Astrophysics*).

triple alpha process (Chap. 23) and starts to react to form carbon. There may be an initial contraction and drop in luminosity, followed by a steady rise as more of the helium reacts. Hydrogen fusion then begins to play a less important role in the total luminosity. (Fusion reactions are sometimes termed "burning": thus hydrogen burns to form helium and at sufficiently high temperatures, the helium burns to produce carbon.)

By this stage the evolution is increasingly sensitive to the mass and to differences in stellar composition. The relative proportions of hydrogen, helium, and metals determine just which reactions dominate at any temperature and also control the opacity which in turn controls the energy flow. Detailed predictions vary from model to model, but there is general agreement that a sequence of steps will follow in which contraction is followed by increased heating in the core or the region immediately around it leading to more (or, later, different) nuclear reactions until a highly unstable condition is ended either by a readjustment to a new burning phase or by a rapid implosion and then a violent explosion. When this occurs, some (perhaps much) of the outer layer of the star may be blown off to form a shell that shows up (when cooled) as a **planetary nebula**. (This name is an historic relic—no planets are involved, but the appearance is certainly nebulous. See Sect. 24.7.) In the last violent stages of massive stars (those with masses between 4–8 $M_\odot$) there is probably a very rapid series of nuclear reactions in which the fusion of nuclei proceeds far beyond the helium and carbon we have considered so far. This explosive stage is thought to occur in supernovae, and we will return to review some of its interesting details in the following chapter.

Once again, we can follow the progress of the evolutionary changes by using H-R diagrams. The successive stages are shown in great detail in Fig. 24.11 for a star of 5 $M_\odot$, and in lesser detail for several different mass stars in Fig. 24.12.

For the 5 $M_\odot$ star, we can trace its aging from the main sequence with its hydrogen-burning core, through to hydrogen burning in a thin shell. A minimum luminosity is reached, followed by a rapid expansion in the red giant phase, which leads to the start of the triple alpha reaction. The helium burning is accompanied by an increase in both surface temperature and luminosity. As the core helium becomes exhausted, the star goes through another cooling-contraction cycle before experiencing another rapid increase in luminosity.

The tracks for stars with other masses once again show family resemblances. It should again be noted, though, that the H-R diagram can be misleading: the length of any line is not proportional to the time taken for that stage, and the time scales differ significantly from one stellar mass to another.

Of all the stages we have considered by far the longest is that spent on the main sequence, and this is true for all masses. This explains why we generally see more stars on the main sequence than elsewhere in the H-R diagram.

## 24.6 RED GIANTS

Betelgeuse and Antares are among the twenty brightest stars in the sky, and their reddish colors clearly distinguish them from Sirius or Vega, for example. Once their distances were known, spectral measurements and the application of the radiation laws showed both Antares and Betelgeuse to be giants. Now, after our review of stellar models, we can understand much more about their structure. We can also see them as representatives of a late stage in stellar evolution.

With surface temperatures around 3000 K and luminosities around $10^4$ $L_\odot$, they must have radii around 400 $R_\odot$ (see Box 22.2). Since $R_\odot \sim 7 \times$

$10^{10}$ cm, these giants have radii of about $2.8 \times 10^{13}$ cm or about 2 AU. When the Sun expands in its red giant phase, it will push out past the orbits of the Earth and Mars.

The structure of a red giant is quite different from solar-mass main sequence stars, which typically have densities that run from over 100 g/cm$^3$ in the core down to 1 g/cm$^3$ by about half way out. Model calculations for a red giant show a high central density ($10^5$ g/cm$^3$ or about 1 ton/in.$^3$) surrounded by a very rapidly decreasing density that drops below 1 g/cm$^3$ by one tenth of the radius from the center. The major part of the star is a gas that is less dense than Earth's atmosphere, although it will be above 30,000 K. When it evolves to the red giant stage, the Sun's expansion past the Earth will, therefore, appear like a very hot wind.

Stellar models indicate that the red giant phase may last for millions of years rather than billions, for a star of around one solar mass; then, quite rapidly, it will pass through the planetary nebula stage on its way to becoming a white dwarf.

## 24.7 PLANETARY NEBULAE

The growing instability in a red giant will lead to the blowing off of its outer envelope leaving behind a central core. The exploded shell may contain 10 to 20 percent of a solar mass of material, expanding at a velocity of 20 to 30 km/sec. The density of atoms in the shell is low, perhaps 100/cm$^3$, so that our view of the remnant star is clear, and the shell takes on the appearance of a ring where we view it through its greatest thickness (Fig. 24.13). A good example of a planetary nebula is the Ring nebula in Lyra, the fifty-seventh object in Messier's catalog (Fig. 24.14). At ninth magnitude, it is bright enough for its features to be seen in a 6 in. telescope.

For a typical diameter of about 1 light year, the age of a planetary nebula (since its explosion) can be calculated assuming that the expansion speed has been constant

$$\text{Age} = \frac{\text{present radius}}{\text{expansion speed}} = \frac{0.5 \text{ LY or } 5 \times 10^{17} \text{ cm}}{20 \text{ km/sec or } 2 \times 10^{6} \text{ cm/sec}}$$

$$\sim 2.5 \times 10^{11} \text{ sec} \sim 8000 \text{ years}$$

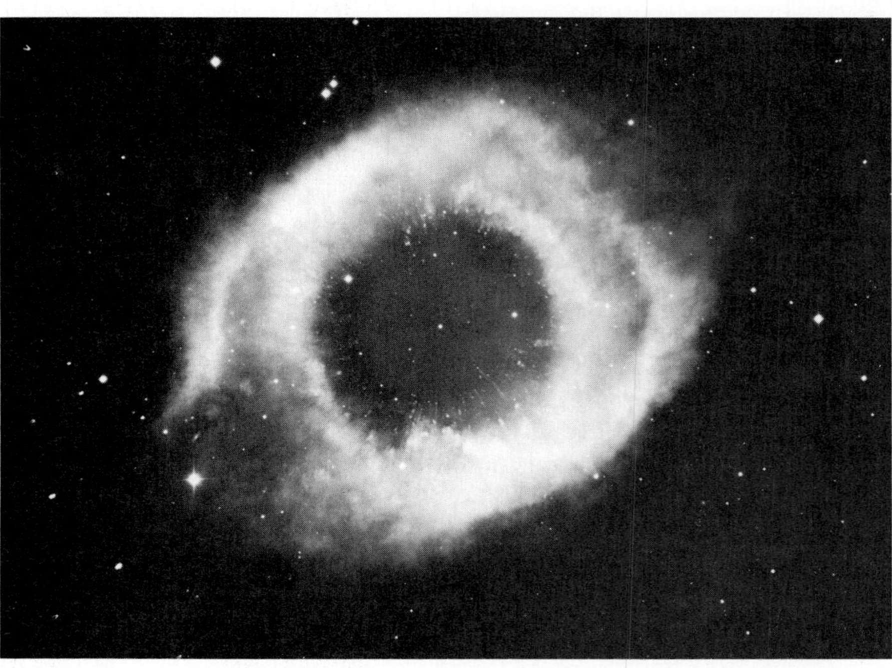

**Figure 24.13** Planetary nebula (NGC 7293) photographed through a red filter to select the hydrogen Balmer H$\alpha$ and doubly ionized nitrogen lines near 6500 Å. Located at a distance of 140 pc and having a diameter of 0.5 pc, this nebula appears to be nearly 1/4° in diameter (Palomar Observatory Photograph).

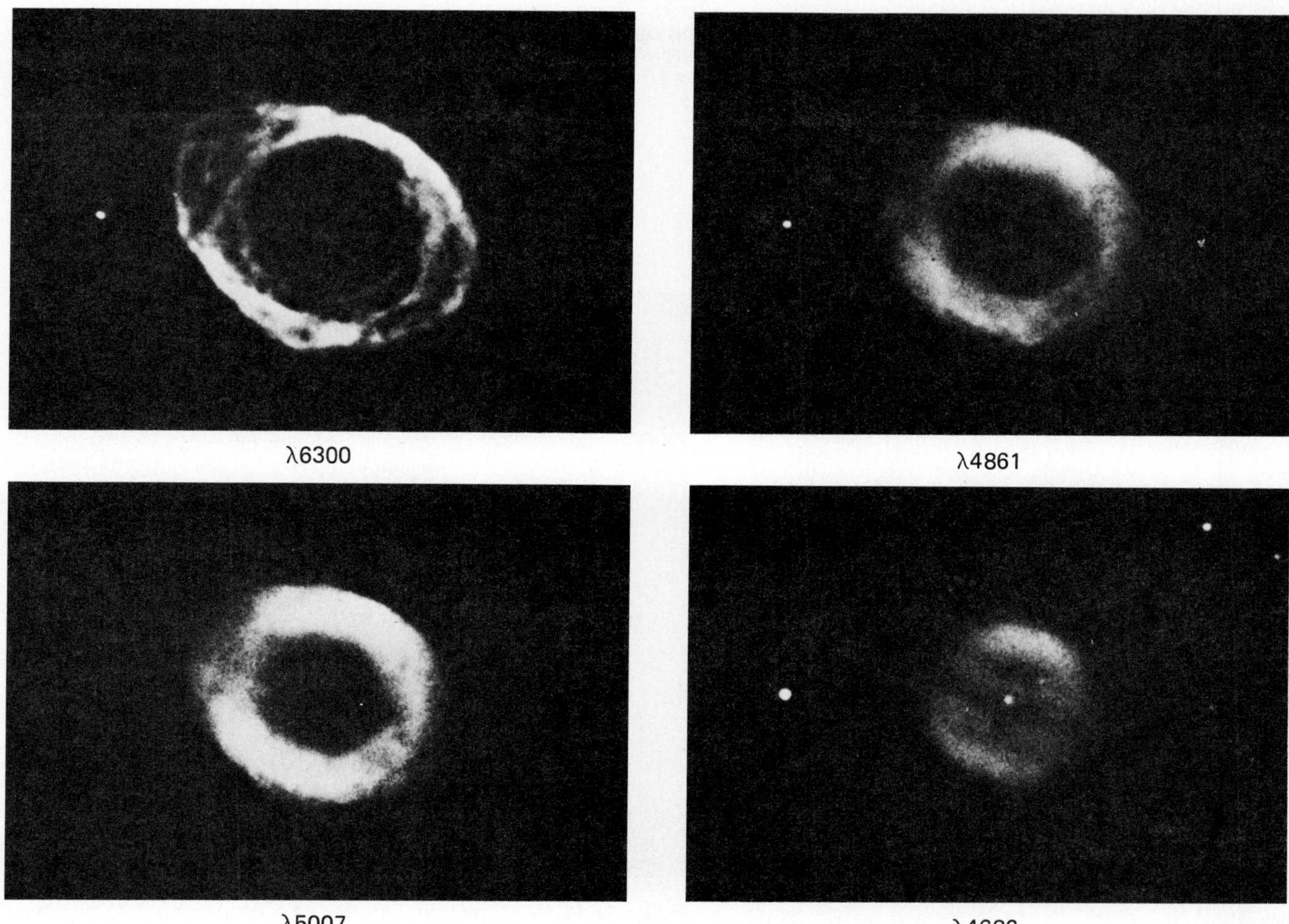


**Figure 24.14** Ring nebula shown at different wavelengths. The filters center on spectral lines of neutral oxygen (6300 Å), doubly ionized oxygen (5007 Å), and neutral hydrogen (4861 Å, the Balmer Hβ line). The hot central star is emphasized in the blue filter (4686 Å) centered on singly ionized helium. This nebula is 700 pc from Earth. Its diameter is 0.26 pc, and its angular diameter is 1.25 arc min (KPNO).

A few as old as approximately 50,000 years have been found. These very low ages tie in with our ideas on the evolution of planetary nebula. At the present expansion velocity, the shell will soon disperse (like a smoke ring) beyond the point where we can see it, and it owes its light to energy absorbed from the central star.

Central stars in planetary nebula have extremely hot surfaces—20,000 to 100,000 K—among the hottest known. These remnants consist mostly of carbon and oxygen in their cores and are not massive enough to sustain temperatures of $6 \times 10^8$ K to burn those nuclei. It is thought that these stars become white dwarfs, gradually cooling and becoming less luminous as they radiate away their energy without being able to replenish it from nuclear reserves. The properties of the resulting white dwarfs are so different from regular stars and giants that we shall discuss them separately in Chapter 25.

## 24.8 STAR CLUSTERS

In Chapter 20 it was noted that most stars occur in binary or multiple systems. Closer examination shows that sometimes one of a pair is itself binary, and there are even systems with five or six stars moving in close and complex

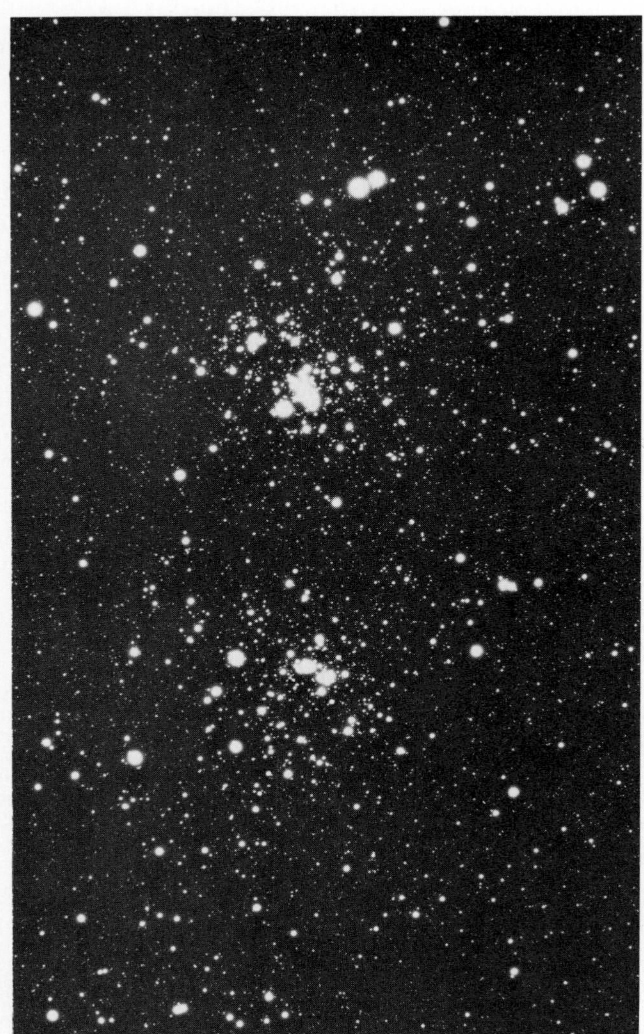

**Figure 24.15** The open double cluster, h and x Persei. These clusters are about 2400 pc distant with angular diameters about 22 arc min (Lick Observatory).

**Figure 24.16** M67, an open cluster, at a distance of 830 pc and with an angular diameter of 18 arc min (Palomar Observatory Photograph).

orbits with the dimensions of the system still no more than around 20 AU. Groups of stars have also been found to occur on a much larger scale where the separations are in the parsec rather than AU range and the number of associated stars run from tens to hundreds of thousands. Messier's catalog is rich in these **clusters** (Fig. 24.15).

**Galactic** or **open clusters** contain tens to some hundreds of stars (Figs. 24.16 and 24.17). Some clusters such as the Hyades (the closest cluster) and the Pleiades are sufficiently bright that they attract attention even without the telescope. Most clusters, though, cannot be seen unaided. With a cluster diameter of 5 to 10 pc, the average separation between stars is about 1 parsec or about the distance of the closest star from the Sun. The number of known open clusters is more than 1000.

**Globular** clusters have many more stars—tens to hundreds of thousands mostly concentrated into a central region of 10 to 20 pc diameter but with outlying members as far as several times this distance (Fig. 24.18). Only through a large telescope do individual stars become resolved.

When we introduced the H-R diagram, we noted that the number of stars represented in different parts of the diagram could depend on the way in which they were selected. This becomes clear when we display the H-R diagrams for some clusters, for example, the open Pleiades, Hyades, and M67, and the globular cluster, M3 (Fig. 24.19).

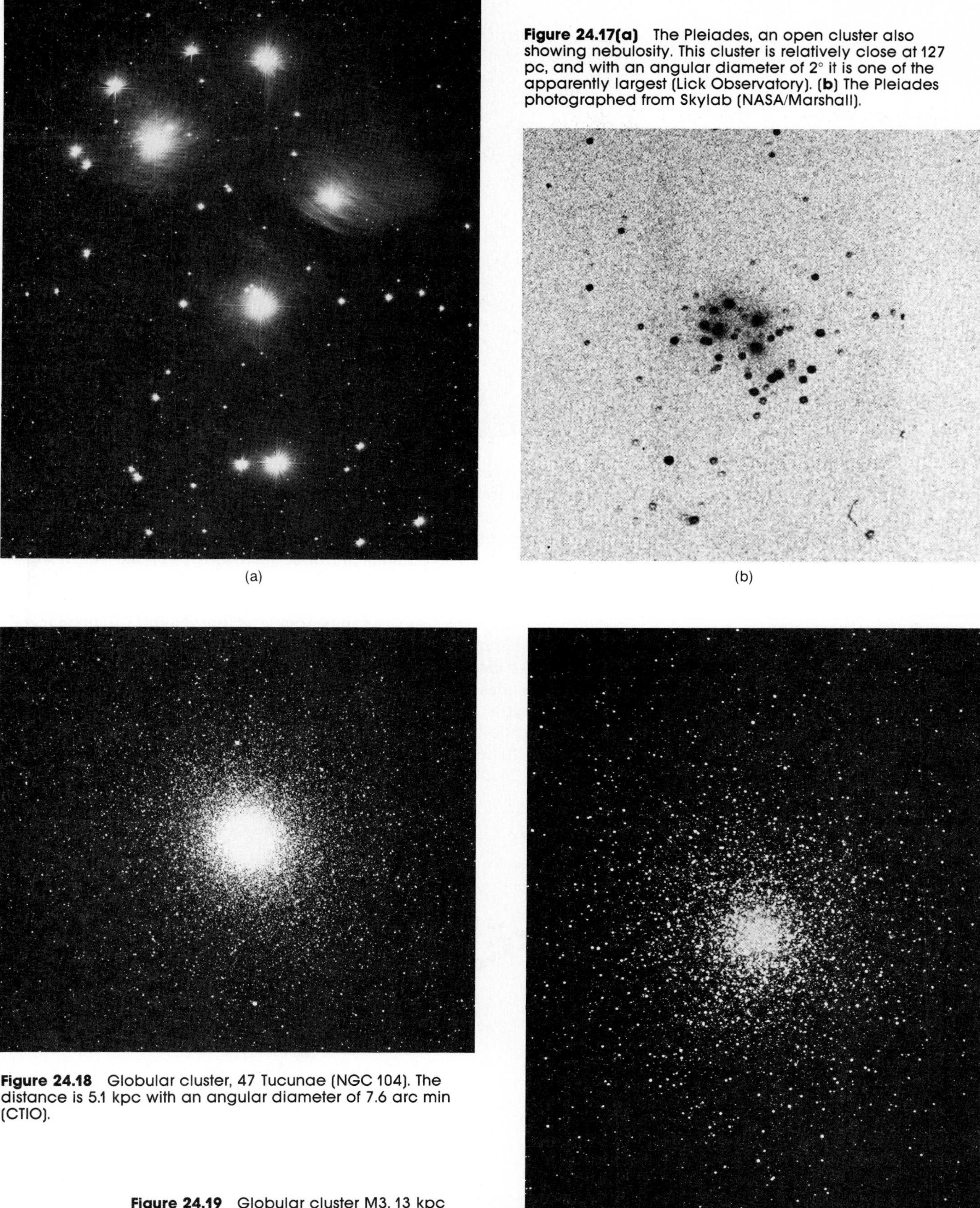

(a)

(b)

**Figure 24.17(a)** The Pleiades, an open cluster also showing nebulosity. This cluster is relatively close at 127 pc, and with an angular diameter of 2° it is one of the apparently largest (Lick Observatory). (**b**) The Pleiades photographed from Skylab (NASA/Marshall).

**Figure 24.18** Globular cluster, 47 Tucunae (NGC 104). The distance is 5.1 kpc with an angular diameter of 7.6 arc min (CTIO).

**Figure 24.19** Globular cluster M3, 13 kpc distant and with angular diameter 3.4 arc min (Lick Observatory).

**Figure 24.20** Two globular clusters (NGC 6522 and NGC 6528) near the center of our galaxy (KPNO).

In the H-R diagram, we plot absolute magnitude (or luminosity) against CI or temperature. To derive absolute magnitudes we must know the distances. In the case of star clusters, the stars are much closer to one another than they are to us. The conversions from apparent to absolute magnitudes would, therefore, involve about the same factor for each star. In this case we can just as well plot apparent magnitudes instead of absolute magnitudes. The omission of the apparent/absolute conversion amounts to a simple shift of the magnitude scale, but the appearance of the diagram is unchanged. This use of apparent magnitudes has the great advantage that we can examine clusters whose distances are not known. The results are striking.

The Pleiades (Fig. 24.21) show no giants, and only a few stars, the subdwarfs below the main sequence. The Hyades (Fig. 24.22) have no subdwarfs but do have some giants (top right area of the H-R diagram). M3, the globular cluster, however, has very few bright stars along the main sequence (Fig. 24.23). Instead, it has a horizontal giant branch that meets a

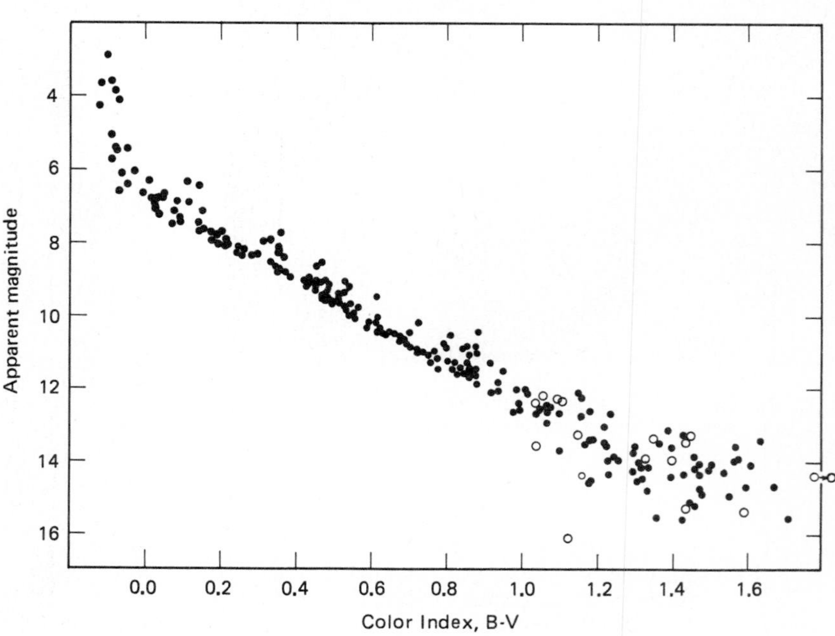


**Figure 24.21** H-R diagram for the Pleiades. (Cecilia Payne-Gaposchkin and Katherine Haramundanis, *Introduction to Astronomy*, 2nd ed., 1970 reprinted by permission of Prentice-Hall, Inc.)

heavily populated band that turns up from the main sequence at about $B - V = 0.4$. Other clusters also display populations that are very different from the first H-R diagrams we saw (Fig. 24.24).

We can now use our knowledge of stellar structure and evolution to understand the differences between the H-R diagrams for different clusters. We can make the reasonable assumption that the stars in a cluster probably formed at about the same time, but as the various stars in a cluster will probably have different masses, they will evolve at different rates.

As we have seen, the more massive stars arrive soonest on the main sequence, consume their hydrogen cores most rapidly, and thus are the first to start to move off the main sequence. Less massive stars take longer to go through each successive stage. As a result, the massive stars will be leaving the main sequence while their lighter contemporaries are still burning hydrogen. If we were able to make measurements on the stars every 100,000 years (for example) and plot the data on a succession of H-R diagrams, we would find the population shifting. Initially, all stars would be collapsing, and their points would be above and to the right of the main sequence. Somewhat later, we would find many stars strung out along the main sequence with a few low-mass stragglers not yet quite on the main sequence. Still later, we would find stars like the Sun moving off toward the

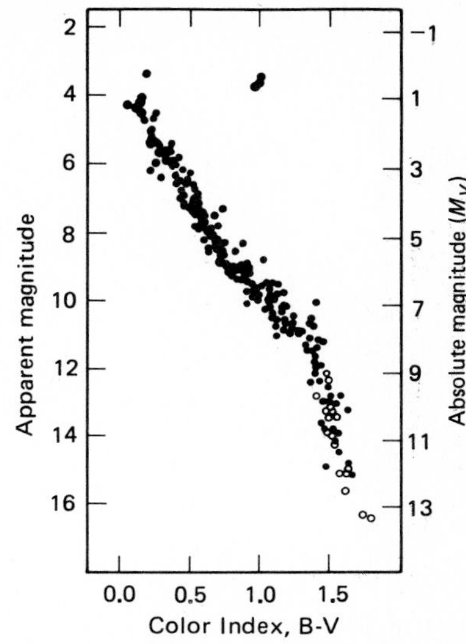


**Figure 24.22** H-R diagram for the Hyades cluster. (Cecilia Payne-Gaposchkin and Katherine Haramundanis, *Introduction to Astronomy*, 2nd ed., 1970 reprinted by permission of Prentice-Hall, Inc.)

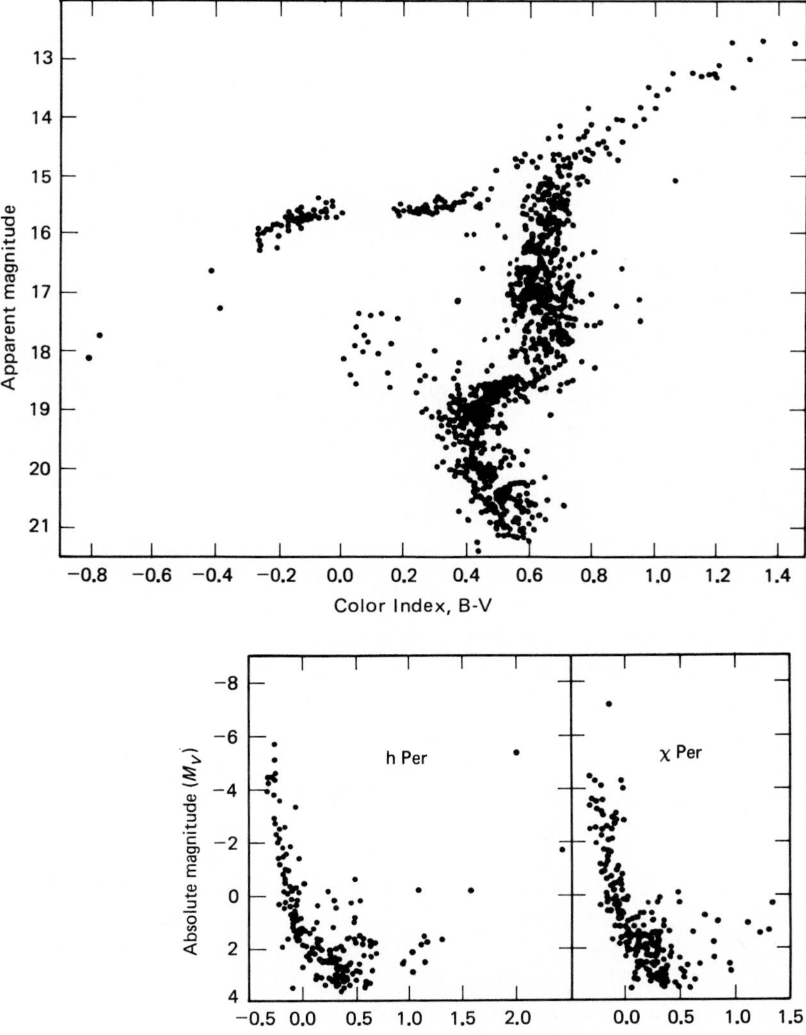


**Figure 24.23** H-R diagram for M3. (Cecilia Payne-Gaposchkin and Katherine Haramundanis, *Introduction to Astronomy*, 2nd ed., 1970 reprinted by permission of Prentice-Hall, Inc.)

**Figure 24.24** H-R diagram for the double cluster h, χ Persei. (Cecilia Payne-Gaposchkin and Katherine Haramundanis, *Introduction to Astronomy*, 2nd ed., 1970 reprinted by permission of Prentice-Hall, Inc.)

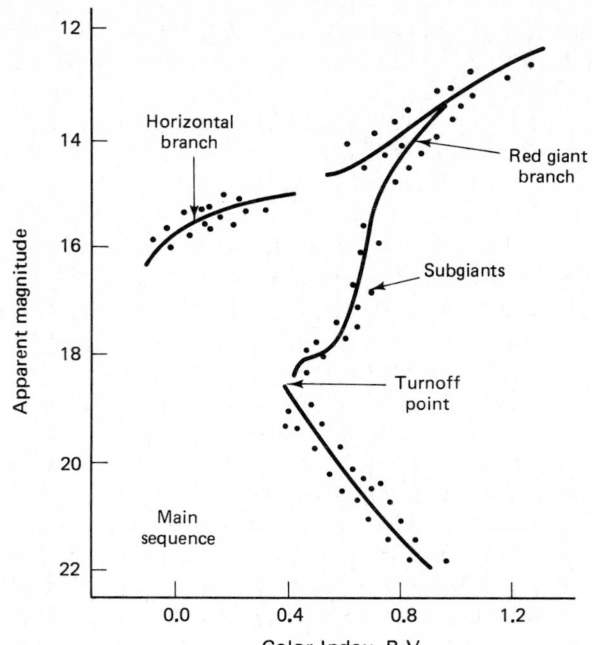


**Figure 24.25** Schematic H-R diagram for an old globular cluster. The heavier stars will be the first to leave the main sequence so that the location of the turnoff point is an indicator of the cluster's age.

red giant region. Much later, more stars would have moved from the main sequence (Fig. 24.25); many would now be spread through the giant region and some might even have reached the planetary nebula and white dwarf stages. The stars of very low mass will, however, stay on the main sequence for a very long time.

Because stellar theory seems generally well established, we know the length of time needed to reach these different stages. Accordingly, we can use the appearance of an H-R diagram to help us gauge the age of a cluster.

The open cluster M67 shows numerous giants and a large number of stars that have moved off the main sequence. There are very few stars with CI less than about 0.4 on the main sequence. From the theory, it is estimated that the age of this cluster must be about 10 billion years in order for its stars to have evolved so far. The globular cluster M3 also shows many stars that have evolved from the main sequence. The turn-off from the main sequence is slightly below $B - V = 0.4$, and there are still main sequence stars down to $B - V \sim 0$. The age of this cluster is estimated to be slightly less than that of M67, perhaps 8 to 9 billion years.

The Pleiades show no giants, and the Hyades have only four. These open clusters must be very young, for they show virtually only the main sequence. The age of the Pleiades is estimated to be only about $10^8$ years, and this is supported by the appearance there of dust regions as expected where star formation is recent. Two distant star clusters in the constellation Perseus also show most of their stars on the main sequence, and their ages are estimated to be only about 10 million years.

As might have been suspected, the analysis of the ages of clusters is not quite as simple as has been described. One problem is the selection of stars as cluster members. Where a cluster is seen with background or foreground stars ("field stars"), how can cluster membership be decided? One method is to measure the proper motions of the stars. Genuine cluster members will have similar proper motions, and precision measurements can allow unrelated stars to be identified. As an example, Fig. 24.27 shows the cluster NGC 2506, and Fig. 24.28 shows the H-R diagrams (a) for all measured stars and (b) for those whose membership of the cluster has been determined with at least 90 percent probability. The effect of careful selection is dramatic.

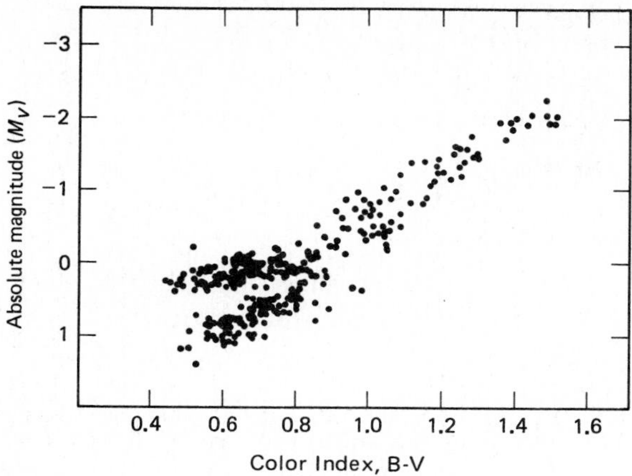


**Figure 24.26** H-R diagram for 47 Tuc. The age of this cluster is estimated to be 15 to 18 billion years. (Cecilia Payne-Gaposchkin and Katherine Haramundanis, *Introduction to Astronomy*, 2nd ed., 1970 reprinted by permission of Prentice-Hall, Inc.)

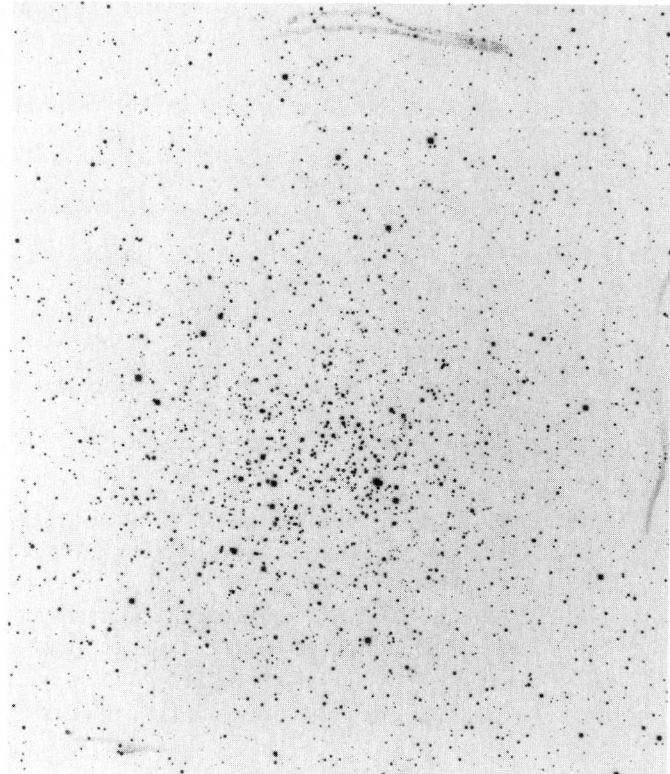

**Figure 24.27** Open cluster NGC 2506 photographed with the Kitt Peak 4 m telescope (William F. van Altena, Yale University Observatory).

**Figure 24.28** H-R diagrams for NGC 2506 (left) all stars in the field, and (right) only stars whose proper motions indicate cluster membership with high probability (stars with proper motion membership probabilities ≥ 90%) (William F. van Altena, Yale University Observatory).

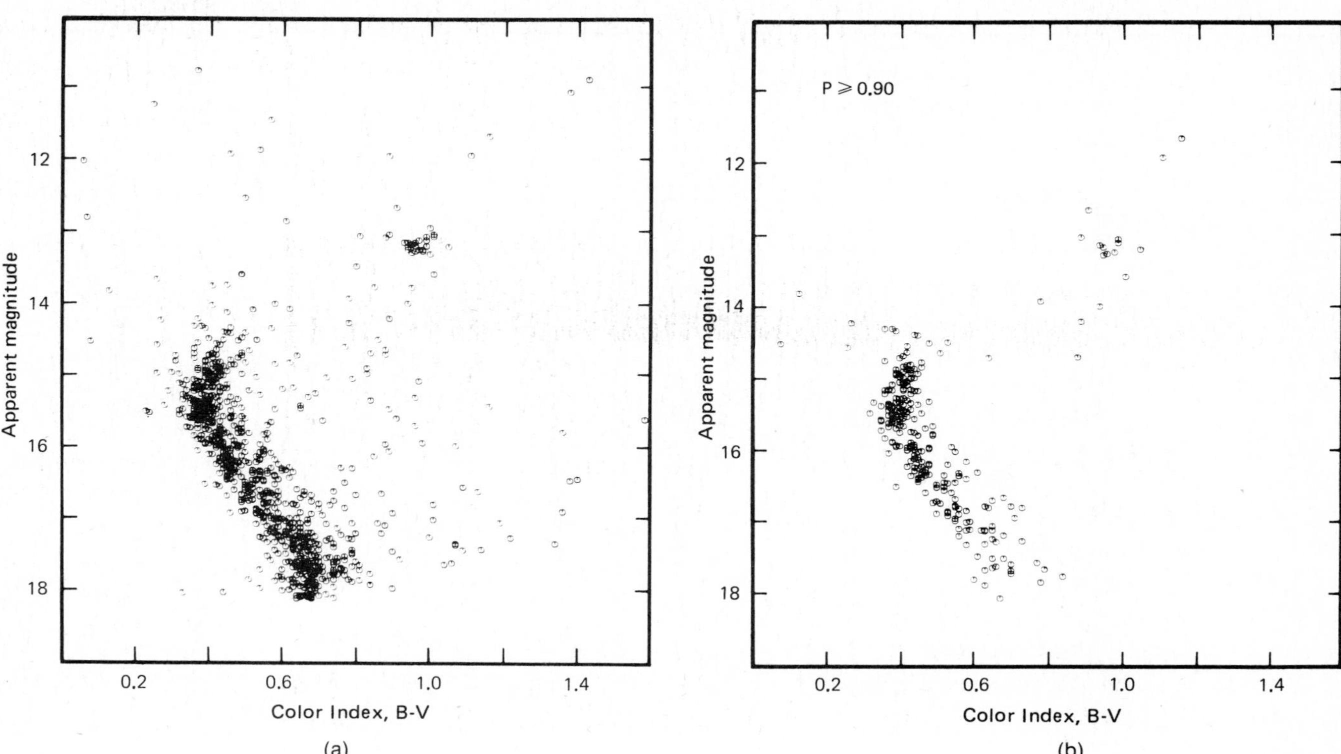

We have found a considerable spread in the ages of clusters, as deduced from their H-R diagrams. Although we can tell a single star's age only within wide limits, we can use the clusters for a population age estimate and this is important when we come to examine the structure of our galaxy and, later, when estimating the age of the universe.

## CHAPTER REVIEW

**Star formation:** extensive cloud of gas and dust at low temperature and much more massive than final star. Star formation triggered by density wave or shock (perhaps from supernova explosion).

- — contraction of cloud, with warming from released gravitational energy.
- — radiation can initially escape, because cloud is not dense.
- — as temperature rises and cloud density increases, opacity increases leading to radiation being trapped and further temperature rise.
- — contraction slows, core temperature increases until nuclear reactions start, star is on main sequence.
- — excess gas/dust blown off, but when/how not certain.

**Main sequence:** star is powered by fusion of hydrogen to helium in core.

- — main sequence stage is longest in evolution of star.
- — more massive stars burn hydrogen fastest, evolve soonest from main sequence.

**Further evolution:** depends strongly on stellar mass:

- — solar mass star will go through stages of core contraction, increasing temperature, ignition of additional nuclear reactions, then expansion through red giant stage, probably followed by planetary nebula stage in which some outer material is blown off and remaining star is initially very hot, then cools to become white dwarf.
- — more massive stars have different end points—Chapter 25.

**Star clusters:** groups of stars that probably formed about the same time.

- — within a cluster, stars of different masses evolve at different rates so that H-R diagram for a cluster can be used to determine cluster age.

**Clusters:** galactic or open clusters—loose structure; globular clusters—very tight concentration with tens or hundreds of thousands of stars.

## NEW TERMS

| | | | |
|---|---|---|---|
| **burning** | **evolutionary track** | **open cluster** | **zero age main sequence (ZAMS)** |
| **clusters** | **galactic cluster** | **planetary nebula** | |
| **density wave** | **globular cluster** | | |

## PROBLEMS

**1.** What mechanisms are thought to provide the conditions for the start of star formation?

**2.** It is thought that stars condense from clouds that initially have about 100 hydrogen atoms in each cubic centimeter. If such a cloud has a radius of $10^{19}$ cm, what would be the mass of that cloud, and how does this compare to the Sun's mass?

**3.** Why are early stages of star formation not observable at visible wavelengths?

**4.** Nuclear reactions provide the energy for stars after they have reached the main sequence. What is the source of energy before that time?

**5.** Why do more massive stars evolve much more rapidly than low-mass stars?

**6.** A star on the main sequence has a mass of $2 \times 10^{35}$ g and a luminosity of $4 \times 10^{39}$ ergs/sec. Suppose it starts as pure hydrogen. How long will it take shining at its present luminosity before 10 percent of the hydrogen has been converted to helium?

**7.** Hydrogen can be consumed to form helium in two main processes: directly via the p-p reaction chain, or via the CNO cycle.

a. Which process is most important in the Sun?
b. Which process could not have occurred in the first stars formed in the galaxy?

**8.** During which part (or parts) of a star's evolution is each of the following energy sources important?

a. nuclear fusion of hydrogen to form helium.
b. gravitational potential energy released by falling matter.
c. nuclear fusion of helium to form carbon.

**9.** As a star forms, it goes through several stages before arriving on the main sequence. Refer to Fig. 24.8.

a. During this phase *C* to *D* the temperature of the outside is not changing much. What is happening?
b. During the next phase *D* to *E* the temperature changes but the luminosity does not change by much. What is happening?
c. Sketch an H-R diagram to show roughly where you would expect to find the evolutionary track of a heavier star than shown in Fig. 24.8.

**10.** Describe in general terms what changes take place as a star evolves, and sketch an H-R diagram to show how this evolution can be represented.

**11.** Why and when does a main sequence star of 1 $M_\odot$ start to move off the main sequence?

**12.** As a star evolves away from the main sequence, there will be a stage when the star is less luminous even though the temperature in the core is higher than when it was on the main sequence. Why is this so?

**13.** Which stage in a star's evolution lasts the longest?

**14.** Three main sequence stars have masses of 20, 1, and 0.5 solar masses. What differences will there be in their evolution, and what is the reason?

**15.** Why do main sequence stars lie in a band on the H-R diagram rather than along a narrow line?

**16.** The Ring nebula (Fig. 24.14) is at a distance of 700 pc and has a diameter of 0.26 parsec and an angular diameter of 75 arc sec. The expansion appears to be about 0.005 arc sec/year. What is the expansion speed (in km/sec), and how old is the nebula?

**17.** The central star of the Ring nebula has a temperature of 90,000 K. It is at a distance of 700 pc, and it has an apparent magnitude of 14.6. Compute the radius of this star.

**18.** Why does the Ring nebula have a reddish outer boundary and greenish inner region?

**19.** The planetary nebula NGC 7293 (Fig. 24.13) is relatively close at 140 pc and has an angular diameter of 800 arc sec. What is its diameter in kilometers?

**20.** What is an important difference between globular and open clusters (apart from their appearance)?

**21.** How does the H-R diagram for a cluster provide us with information on the cluster's age?

**22.** Compare the H-R diagrams for the clusters M3, 47 Tuc, and the Pleiades. Note the differences in appearance. Which of these clusters is the youngest, and which the oldest?

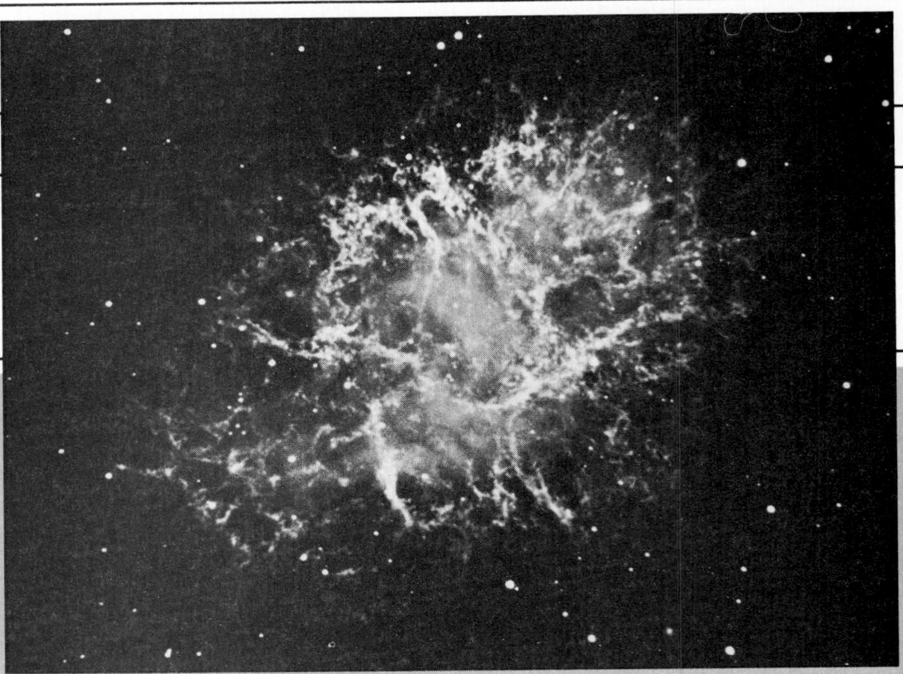

# CHAPTER 25

# White Dwarfs, Neutron Stars, Pulsars, and Black Holes

The final stages of stellar evolution can produce objects whose bizarre properties we have great difficulty in imagining, so far are they from anything we can duplicate in our laboratories. Why do we think that such objects exist? How are they produced and how can we recognize them? In some cases, they turned up unexpectedly when observational data were being interpreted. In others, we have been led to propose their existence by following the evolutionary tracks of stars, applying the laws of physics that we have found on Earth and assuming that these laws can be applied everywhere. When radically new theories are produced to try to explain some strange effect, the test of the correctness of the theory is not its agreement with common sense—the only test is the confrontation with further observation. If the picture that we then assemble seems to be consistent, we may fee encouraged, but we may still find it difficult to understand or visualize these new objects. This is especially the case with the objects we are about to describe.

## 25.1 DENSE MATTER

We know that some matter in stars is at very high densities, far higher than anything we can personally encounter. Everyday materials (solids and liquids) have a very small range of densities: The density of water is 1 g/cm$^3$ (about 62 lb./cu. ft), iron 7.8 g/cm$^3$, gold the heaviest at 19.3 g/cm$^3$, styrofoam perhaps the lightest at 0.05 g/cm$^3$ (Fig. 25.1). Yet our stellar models (chaps. 23 and 24) have led us to the conclusion that far greater densities occur in the interiors of the stars. For example, the density at the center of the Sun is over 150 g/cm³. White dwarfs are even denser. From the known luminosities and temperatures of a few white dwarfs, we calculate central densities of more than $10^6$ g/cm³ (more than 10 tons/cu. in.).

These densities result from the weight of overlying material, but what changes must take place in the matter that is being compressed? As you may have found out, styrofoam compresses easily. By pushing or using a hammer, you can force the air out of the spaces and finally produce a density of around 1 g/cm³. But once the air is out, there is little more than you can do. Wood can also be compressed somewhat (with a vise), but metals seem totally incompressible. Under the most extreme laboratory conditions, the maximum pressure attainable is slightly over a million atmospheres, and even then a tiny, solid sample can be compressed to little more than twice its original density. Clearly, something more is needed to reach densities in the hundreds or millions of grams per cubic centimeter.

At normal temperatures, each atomic nucleus is surrounded by its electrons, but most of the volume is empty. The nucleus has a diameter of around $10^{-12}$ to $10^{-13}$ cm, while the electrons occupy orbits at about $10^{-8}$ cm. To push the atoms closer together requires forcing the electrons into smaller orbits where they have more energy than usual, and this extra energy must be supplied from the outside. At the center of a star, the pressure is large enough to do this. In the Sun, for example, the central pressure is more than one billion atmospheres. More importantly, though, the stellar interiors are hot enough to ionize virtually all the atoms. The electrons and nuclei in the resulting gas can then be forced very much closer together.

Why does the gas of electrons and nuclei not simply collapse further? The answer was not known until the theory of quantum mechanics had been developed in the 1920s. The simple gas laws of everyday experience do not hold for electrons in the stellar interiors; modifications must be made in those laws to include quantum theory. The result, unexpectedly, is a severe restriction on the behavior of the electrons and the degree to which they can be compressed. An electron gas under these extreme conditions is then termed **degenerate**, and the theory indicates that degenerate electron gas exists in the white dwarfs and in the cores of some giants.

The application of quantum mechanics to stellar interiors was first explored in detail by A. S. Eddington, the English mathematical physicist, in 1926, and the critical concept of electron degeneracy was introduced in that same year by another English theorist, R. H. Fowler. Some electrons in the stellar cores have very high speeds so that additional corrections are required

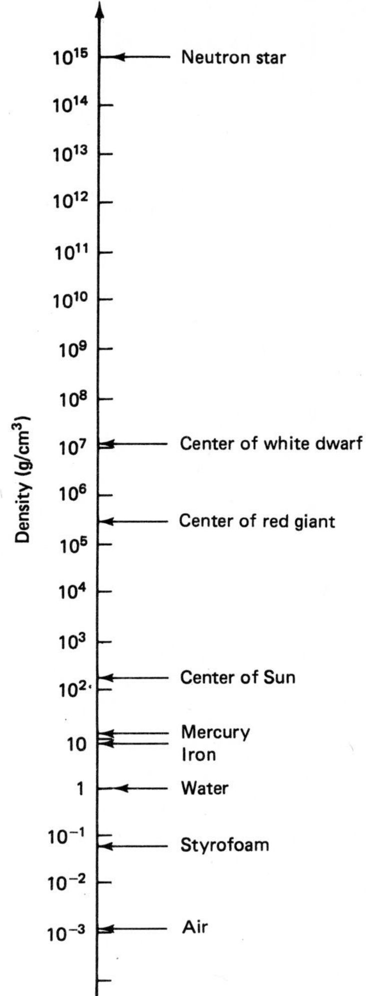


**Figure 25.1** Densities of common materials and the interiors of some stars.

*Table 25.1* CENTRAL TEMPERATURES AND DENSITIES IN STARS

| *Star* | *Mass (Solar Masses) $M/M_\odot$* | *Spectral Type* | *Central Temperature $T_c$ ($\times 10^6$ K)* | *Central Density (g/cm³)* |
|---|---|---|---|---|
| Main sequence | 20 | B0 | 34 | 5.0 |
| | 10 | B3 | 31 | 8.9 |
| | 5 | B6 | 27 | 20 |
| | 2 | A6 | 20 | 68 |
| | 1 | G2 | 15 | 100 |
| | 0.5 | M0 | 8 | 63 |
| Red Giant | 1.3 | | 40 | $3.2 \times 10^2$ |
| White Dwarf | 0.9 | | 8 | $1.6 \times 10^7$ |

in the theory involving relativity. This necessary elaboration came in 1932 from S. Chandrasekhar, at that time at Cambridge University but thereafter for many years at the University of Chicago. In Russia Lev Landau developed the same ideas independently. These important theoretical advances marked a complete departure from the older astrophysics which was in essence based solely on Newton's Laws and the nineteenth century theory of electromagnetism.

Since the 1930s and especially since the introduction of computers, the theoretical investigation of dense stars has continued with great vigor. How well do the calculations agree with the observations?

## 25.2 WHITE DWARFS

The strange nature of white dwarfs could not become apparent until their masses and dimensions were known. The best known white dwarf is the faint companion to Sirius discovered 1862, but it was not until 1915 that Walter S. Adams, using the then new Mt. Wilson 2.5 m (100 in.) reflector, obtained the first spectrum from which the surface temperature could be determined. The stellar radius could then be calculated, using the relation $L = 4\pi R^2\sigma T^4$ and the known small luminosity (Box 21.2). A typical radius is about one fortieth of the solar radius or about three times the radius of the Earth (Fig. 25.2).

With the mass of Sirius B already known from the observations of its movements around its binary partner, its density could be calculated. In fact, this was the second white dwarf to be identified. The first (but not as well known) had been 40 Eridani B. When the white dwarfs' density was calculated, the high value was the cause of concern, for in the context of the physics of 1915 there was no known way of producing densities of 1 million g/cm$^3$.

One early suggestion was that these stars were not dwarfs but rather normal-sized and that they were luminous over only a small part of their surfaces. (The small luminosity and high surface temperature together imply that the emitting surface is small. This can be due either to the star being small or to the star being normal size but having only a part of its surface actually emitting radiation.) This explanation did not gain much support, perhaps because a hot-spot should rotate with the star, and the brightness should change every rotation, and this had not been seen.

The alternative idea of very compact objects with sizes close to that of the Earth found rather skeptical acceptance until Eddington's calculations

**Figure 25.2** Scale and properties of white dwarfs and neutron stars compared to the Sun.

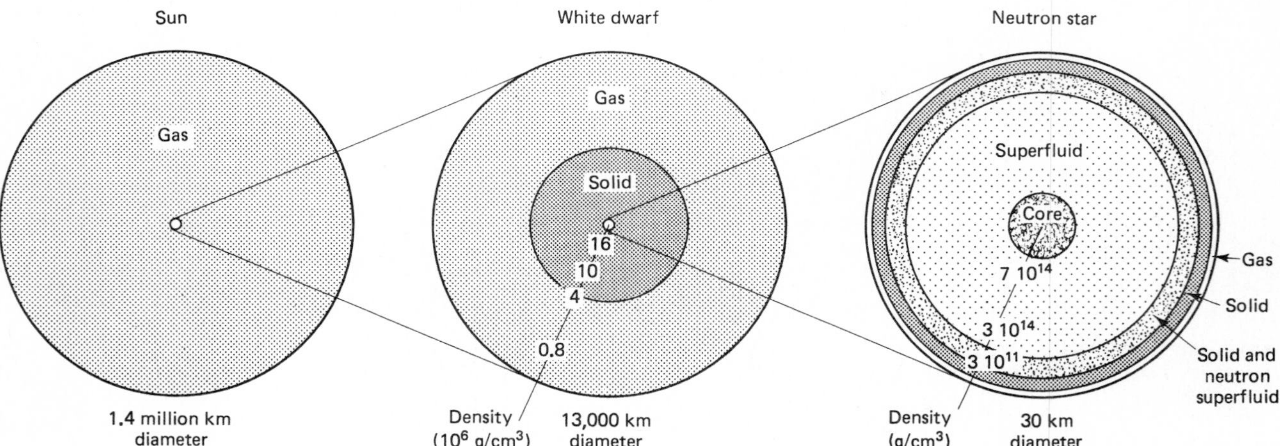

(and the rapid development of the new quantum theory) provided a physical explanation. Chandrasekhar's theory went much further. He was able to show that the radius of the dwarf should be smaller for more massive dwarfs and that there should be a maximum mass for a star to be able to exist as a white dwarf. This **Chandrasekhar limit** is 1.44 $M_\odot$. More modern calculations show that a definite upper limit is hard to determine with accuracy and it may be somewhat closer to 1.40 $M_\odot$.

Even today of the several hundred white dwarfs known only a few have masses that have been accurately measured. The well-determined masses all fall below Chandrasekhar's limit. For example, Sirius B has a mass of 1.05 ± 0.03 $M_\odot$, Procyon B has a mass of 0.63 $M_\odot$, and 40 Eri B has a mass of 0.43 ± 0.02 $M_\odot$.

Although most of the mass of a white dwarf will be at or near the calculated high densities, a thin surface layer must have much lower density. It is the light from this photosphere of hot but nondegenerate matter that is actually seen.

**Box 25.1**
*Gravitational Redshift*

From Einstein's general theory of relativity, it can be calculated that light leaving a massive object has its wavelength shifted to longer values, for the gravitational pull on the photons causes them to lose energy. The change in wavelength is given (to a close approximation) by

$$\frac{\Delta\lambda}{\lambda_0} = \frac{GM}{Rc^2}$$

where $\Delta\lambda$ is the change in wavelength, $\lambda_0$ is the original wavelength, $R$ is the radius of the star, $M$ its mass, $c$ the velocity of light, and $G$ the universal gravitational constant. For example, for the case of light leaving the Sun, we have $R_\odot \sim 7 \times 10^{10}$ cm, and $M_\odot \sim 2 \times 10^{33}$ g. Thus,

$$\frac{\Delta\lambda}{\lambda_0} = \frac{(6.7 \times 10^{-8})(2 \times 10^{33})}{(7 \times 10^{10})(3 \times 10^{10})^2} \sim 3 \times 10^{-6}$$

For a white dwarf, the mass will be about the same as for the Sun, but with $R \sim 10^9$ cm, the effect is much larger,

$$\frac{\Delta\lambda}{\lambda_0} \sim 2 \times 10^{-4}.$$

Surface temperatures of white dwarfs range from about 50,000 K down to around 5500 K, comparable to the Sun. Examination of the absorption lines in their spectra shows that the white dwarf stars' atmospheres are overwhelmingly either hydrogen or helium and almost never a mixture.

Redshifts have been measured in prominent lines and have been attributed to gravitational effects after corrections have been made for Doppler shifts (see Box 25.1). Gravitational redshifts predicted by Einstein have been measured on Earth in very accurate experiments. For a compact object such as a white dwarf, the redshift is a few parts in 10,000, well within present observational capabilities, and the agreement with predictions from relativity theory is good.

It might be thought that we could use a measured redshift to calculate the radius of a white dwarf if the mass is independently known. However, as we know from our studies of the Sun, a stellar photosphere can have vertical convection at quite high speeds. The observed redshift may, therefore, be the result of a complicated mixture of effects due to the gravitational redshift and the Doppler effect due to convection. We cannot, therefore, reliably determine the dwarfs' radii in this way. In addition, the star may be moving away from the Earth, and the resulting Doppler redshift can be larger than that due to gravity.

Because of their low luminosities, white dwarfs are hard to find. There is a bias towards finding those that are larger and hotter or are in binaries with prominent companions. Estimates have been made of the total number of white dwarfs allowing for these effects, and rough values of around 5 percent of all stars seem reasonable.

Some theorists have suggested that white dwarfs represent a final stage in the evolution of stars originally less massive than about 4 $M_{\odot}$. With degenerate electrons resisting further compression and energy loss, any further cooling has to be at the expense of the nuclei amongst the electrons. The nuclei do not have the same restrictions on their behavior as the electrons. As the nuclei give up their energy, the star will slowly cool and move down and to the right on the H-R diagram ending up as a black dwarf. Here the observational bias intrudes again—black dwarfs will be extraordinarily hard to detect, and none have yet been identified.

Theory and observation for white dwarfs seem to be in general agreement. This does not mean that they should no longer be studied. More refined measurements can give us further insights into their dense structure, composition, and role in stellar evolution. For example, we still do not know just how they evolve from planetary nebulae, but they are well understood compared to the next class of objects we shall discuss.

## 25.3 NEUTRON STARS

Chandrasekhar's calculations for stellar structures showed that there was a maximum mass that a star could have and still exist as a white dwarf supported by its degenerate electrons. What happens to a star whose mass exceeds the Chandrasekhar limit, that has consumed its hydrogen and helium, and collapsed to a compact remnant? The core continues to collapse to still higher densities. This was the subject of calculations by J. Robert Oppenheimer who predicted in 1938 the existence of even denser objects, neutron stars. (Neutrons had been discovered only in 1932 after Chandrasekhar's calculations on white dwarfs.)

When the central pressure is more than the electrons can support, they are forced so close to the protons that they combine to produce neutrons and neutrinos

$$_{-}e + p \rightarrow n + \nu$$

This reaction absorbs energy, which the neutrinos remove when they escape. The neutrons remain, constituting a degenerate gas, similar in many ways to the electron gas of an earlier stage. We now have a neutron star, far more compact than the white dwarfs with the inward gravitational force being held in balance by the outward force from the dengenerate neutron gas. The diameter of the neutron star will be about 20 km. In its central core, the density will exceed $10^{15}$ g/cm$^3$. Even the average density (more than $10^{14}$ g/cm$^3$) far exceeds that of the white dwarfs (Fig. 25.2).

The surface is expected to be extremely hot, initially around 10 million K and cooling rapidly. Its emitted radiation will be mostly in the X-ray region. The interior is expected to consist of several regions, some even behaving like solids despite their high temperatures. Because we have had no experience with matter under these conditions, it is by no means clear just what physical effects need to be included in the calculations, and the predictions accordingly differ in response to assumptions made by different theorists. It may be several years before a consensus develops.

There are two effects of which we are far more certain. Most of a star's angular momentum is probably retained as it collapses, so that its rotation will speed up considerably. Thus a star like the Sun, which rotates once in about a month, would take less than $\frac{1}{100}$ sec for a rotation if it were to contract to neutron star dimensions. The magnetic fields on the surface of some stars are far stronger than the Earth's magnetic field. Contraction to a radius of 10 km will result in very intense fields; this together with a high rate of rotation are just the conditions that we need for the acceleration of electrons to almost the speed of light. We might expect to detect the resulting synchrotron radiation from the high-speed electrons.

In contrast to white dwarfs where the theory gave an explanation for existing but puzzling observations, in 1938 there were no immediate candidates for the theoretically predicted neutron stars, and the prediction lay unconfirmed until the discovery of pulsars nearly 30 years later. But before we review those newer objects, we need to ask where the neutron stars come from and how they fit into the general evolutionary sequence.

## 25.4 THE NEW STARS: NOVAE AND SUPERNOVAE

Before the invention of the telescope, the sudden appearance of a very bright star not previously been known was recorded by the Chinese and other Eastern astronomers, along with their other "guest stars" (Chap. 4). In western Europe, the novae of 1572 and 1604 were seen by many and were discussed by some of the leading scientists of those days—Tycho Brahe, Kepler, and Galileo. We now use the term supernova to describe the very few novae with extreme brightening, and we know that they represent a separate class of phenomena from the more plentiful novae. Both novae and supernovae represent late stages in stellar evolution. Because far more is known about the supernovae and also because they are more important in several ways, we will devote more space to them. First, though, we will take a brief look at the novae.

The signature of a nova is a brightening by more than 9 magnitudes in less than a few days (Figs. 25.3 and 25.4), together (now) with spectra that

**Figure 25.3** Nova Herculis, before and after its 1934 outburst. At its peak, this nova had an apparent magnitude of $1.^m4$ (Yerkes Observatory).

**Figure 25.4** Nova Herculis in 1951, photographed through UV, green, and red filters showing how the ejected material has spread out in 17 years. The distance to the nova is about 300 pc (Palomar Observatory Photograph).

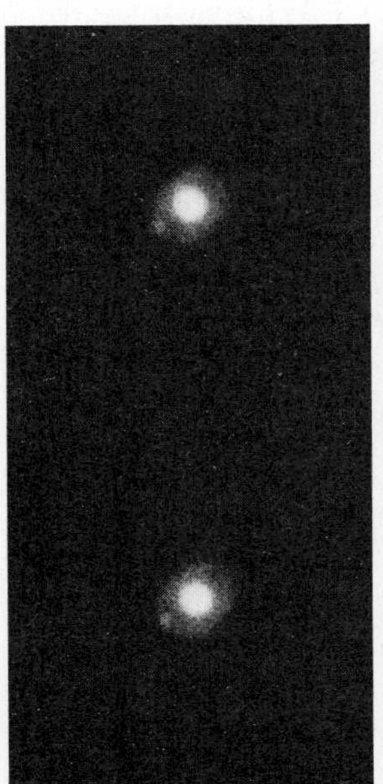

λ3727 (OII)
103a-O Emulsion
plus UG-2 filter

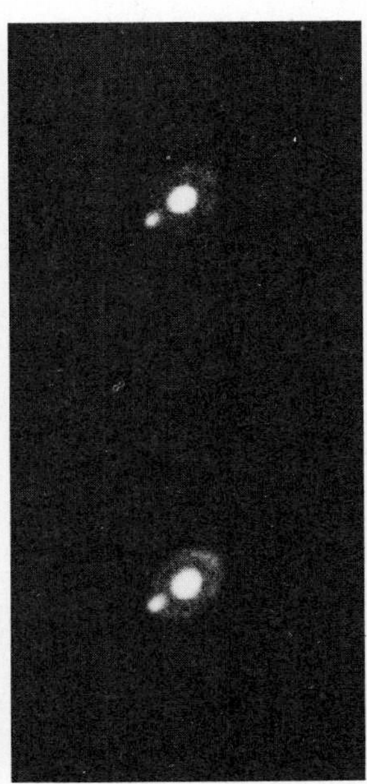

λ4986 + λ5007 (OIII)
103a-J Emulsion
plus GG-11 filter

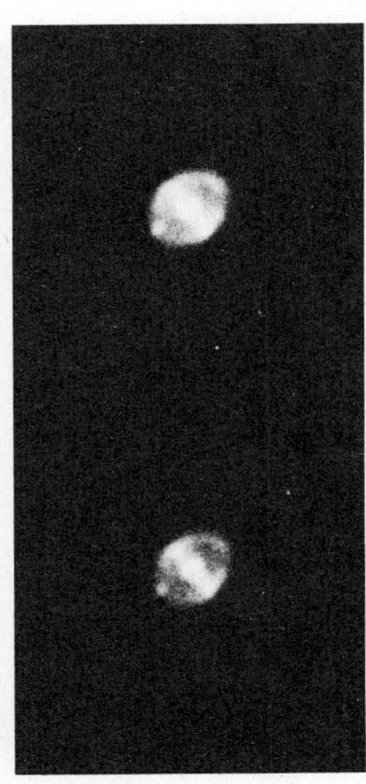

6563 Hα and
λ6548 + λ6584 (NII)
103a-E Emulsion
plus RG-2 filter

show Doppler shifts corresponding to speeds of 100 km/sec to a few thousand km/sec. There are no reliable estimates for the masses of these stars; it is thought that novae occur in close binaries where the two stars have masses of about $\frac{1}{2}$ and 1 $M_{\odot}$. The more massive of the pair may be a white dwarf that is attracting matter from the very extended atmosphere of its less massive companion. Two mechanisms have been proposed for the source of the explosive energy that we see. One theory considers the transferred mass to build up on the surface of the white dwarf until the temperature at that boundary is so high that thermonuclear reactions suddenly start. The other theory postulates a sudden increase in the mass transfer leading to the explosion. In either case, material is ejected at high speed. The outburst is relatively short-lived, but the peak luminosity can reach $10^5$ $L_{\odot}$ and as much as $10^{45}$ ergs can be radiated in a few months (Figs. 25.5 and 25.6). Some estimates have been made of the rate of mass transfer needed to trigger this event, and values of $10^{26}$ g/year seem possible. The frequency of novae is hard to estimate reliably, and a value of 40/year has been suggested for our galaxy, only a few being visible to us.

The more spectacular outbursts of supernovae were recorded relatively few times before the invention of the telescope. Since 1604 the only supernovae seen have been in distant galaxies, so that we have not been lucky enough to get a close-up view. Our modern data, therefore, come from observations of supernovae at great distances (Fig. 25.7), and from the remnants of supernovae in our own galaxy. We can now try to see how these

**Figure 25.5** Nova Cygni 1975, showing decline from magnitude 2 to magnitude 15 (Lick Observatory).

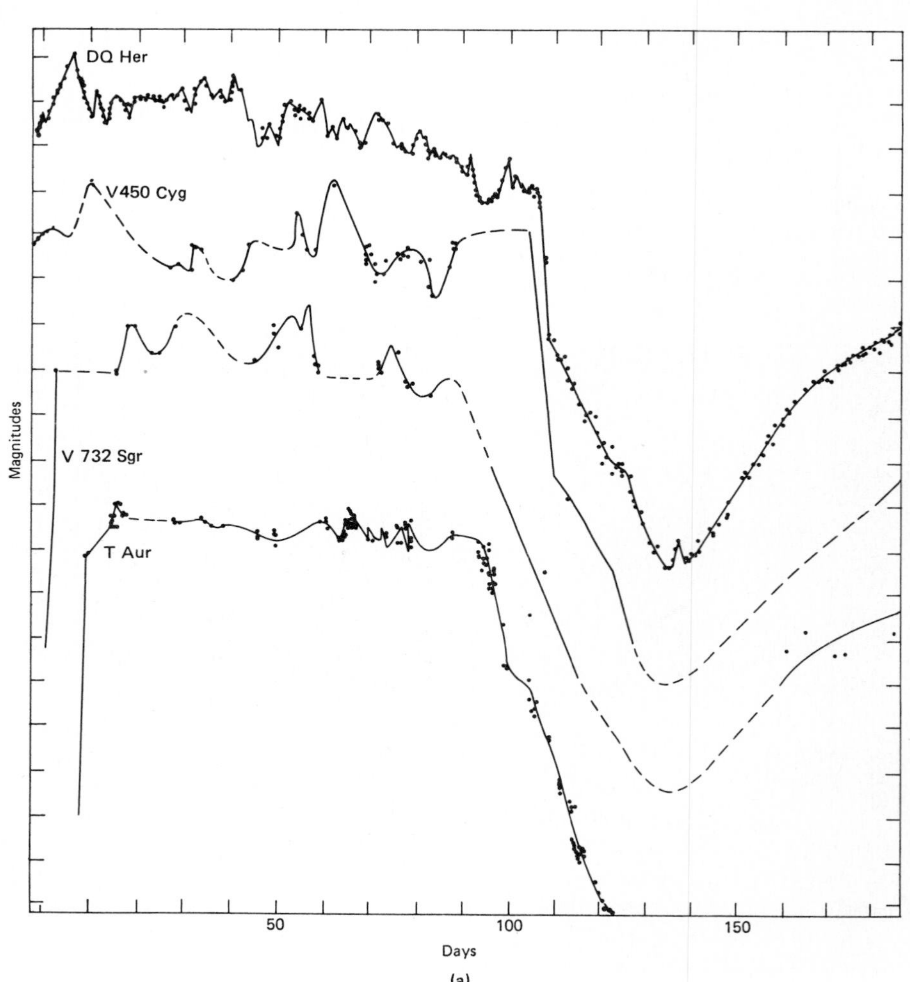


**Figure 25.6** Light curves of novae showing variations within two broad groups. (Cecilia Payne-Gaposchkin and Katherine Haramundanis, *Introduction to Astronomy*, 2nd ed., 1970 reprinted by permission of Prentice-Hall, Inc.)

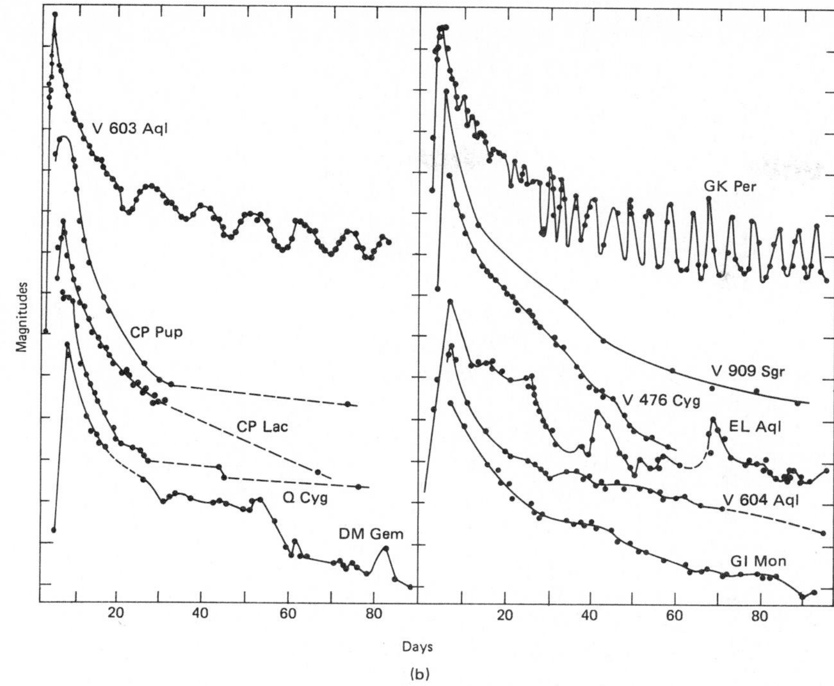


**Figure 25.6** continued.

**Figure 25.7** Supernova in the spiral galaxy NGC 4303 observed in 1961 (Lick Observatory).

observations can be understood in terms of our ideas of stellar structure and evolution.

The observational signature of a supernova is an increase in luminosity by as much as a factor of $10^8$ in a very few days followed by a relatively slow decline in brightness. Fritz Zwicky devised a classification scheme for supernovae based on their light curves. Most supernovae can be classified as Type I or II, (Fig. 25.8) but we do not know whether other suggested types are really distinct classes or simply part of a broad range of variability.

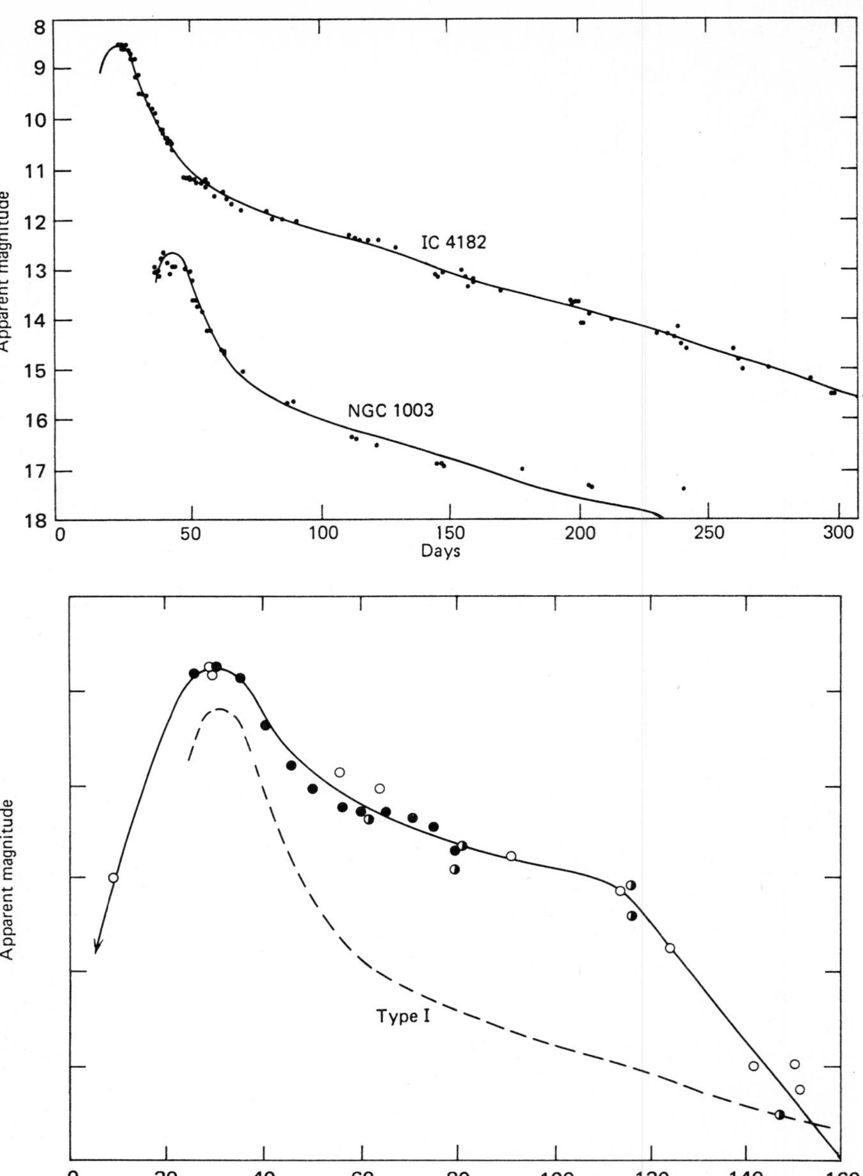


**Figure 25.8** Light curves for supernovae: (top) for Type I, (below) for Type II. In the Type I events there is a rapid initial decrease in brightness and then a very slow and steady decline. In Type II events the initial drop is not great, but there is a rapid decline after a shorter slow decrease. These curves are composite curves from several supernovae, scaled to show similarities in shape within each type. (Compare to the light curve for Tycho's supernova, Fig. 4.6)

Peak luminosities may exceed $10^{43}$ ergs/sec, and the total energy radiated is calculated to be as much as $10^{52}$ ergs. Estimates of the frequency of supernovae are insecure for several reasons. There is a very small data base: No supernova explosion has been seen in our galaxy since 1604, and we do not know how complete the older pretelescopic records really are. (Remember that there is no Western record of the Crab nebula's outburst.) Among the Eastern records of "guest stars," we have been able to identify many supernovae, but we remain uncertain about possible observational or recording biases. In addition, our view to much of the galaxy is obstructed by interstellar dust so that supernovae cannot be seen in many regions. Attempts have been made to correct for these factors, and an average rate in the range of one supernova every 30 to 100 years in our galaxy is generally accepted.

The Crab nebula, first object in Messier's catalog and the remnant of the 1054 supernova, has been studied more intensively than almost any other object outside the solar system. Although it provides an excellent specimen for us to examine, we should be careful about generalizing from a single object. The nebula today is decidedly nonstellar and quite irregularly shaped

**Figure 25.9** The Crab nebula (M1) photographed in yellow light. The Crab is about 2 kpc distant and by now has an angular diameter of about 5 arc min (Lick Observatory).

(Fig. 25.9) with its longest dimensions now about 6 arc min. At its distance of 2 kpc, this translates into a diameter of about $10^{19}$ cm or 10 LY. Photographs taken at intervals of tens of years show that the nebula is still expanding at a speed of about 1400 km/sec. At this speed, it need only have started in A. D. 1140 in order to reach its present size. This suggests that there is something in the nebula that produces an acceleration of the expanding material.

The visual apparent magnitude is 8.6, but the nebula looks very different at different wavelengths. Through a red filter, which centers on the hydrogen H $\alpha$ line (Fig. 25.10), it takes on the appearance of a glowing ball of tangled steel wool. At wavelengths that avoid the major spectral lines, the Crab shows a general glow but without the filamentary features. In the early 1950s as radio astronomy developed, the Crab was found to be one of the strongest radio sources. Iosef Shklovsky suggested that this radiation could be coming from very-high-speed electrons, spiraling in a magnetic field. (See chap. 8 for a discussion of this type of radiation.) A characteristic feature of synchrotron radiation is its polarization; Shklovsky went on to

**25.10** The Crab Nebula, in red (H$\alpha$) light (Lick Observatory).

suggest that some of that the visible light might be synchrotron radiation and thus be polarized. Walter Baade, using the Mt. Palomar telescope, discovered the visible polarization in 1956. Figure 25.11 shows quite clearly how the Crab's appearance varies as the polarizing filter is rotated.

Shklovsky used the synchrotron radiation theory to calculate the rate at which electrons were radiating away their energy, and it turned out that a typical electron should be able to radiate for only about 10 years. Compared to the known age of the Crab (about 900 years) there could be only one conclusion—that within the Crab nebula there had to be some mechanism that was continually replenishing the energy emitted by the electrons. This was a serious problem: The total luminosity of the Crab is about $2 \times 10^{38}$ erg/sec; if most of this comes from electrons and the electrons are losing their energy on a time scale of about 10 years, what is keeping the electrons going? No reservoir of energy for the electrons was known until the discovery of pulsars in 1967.

Walter Baade and Rudolph Minkowski in 1942 had suggested that the Crab was a remnant of the 1054 event recorded by the Chinese. Shklovsky was able to persuade the Chinese to examine their records for other supernova remnants. Several lists of guest stars have now been ransacked and radio searches have been conducted for remnants. The result is a number

**Figure 25.11** The Crab Nebula photographed through a polarizing filter set at different angles. The major changes observed demonstrated that much of the light is synchrotron radiation, produced by high-speed electrons in a magnetic field (Palomar Observatory Photograph).

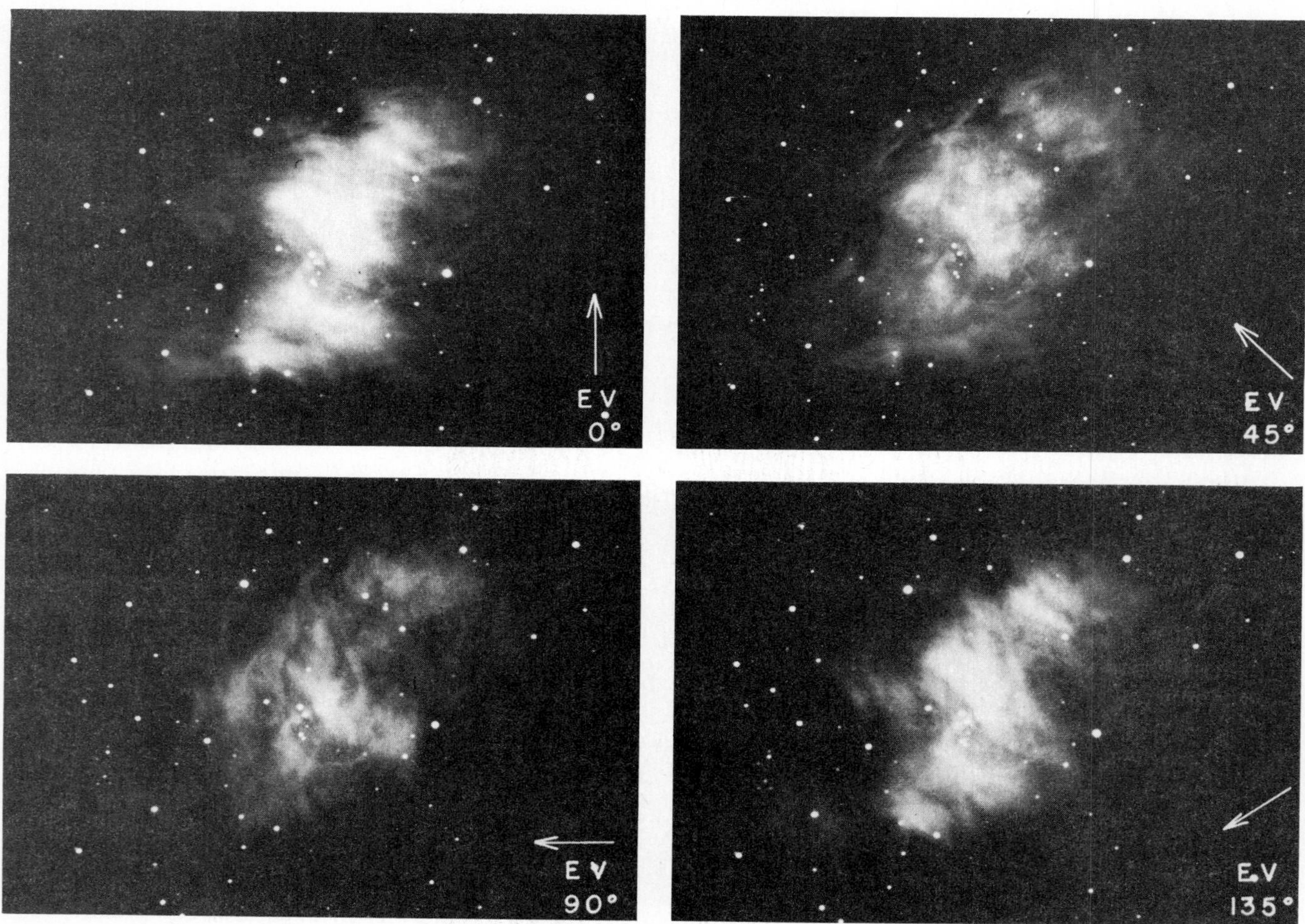

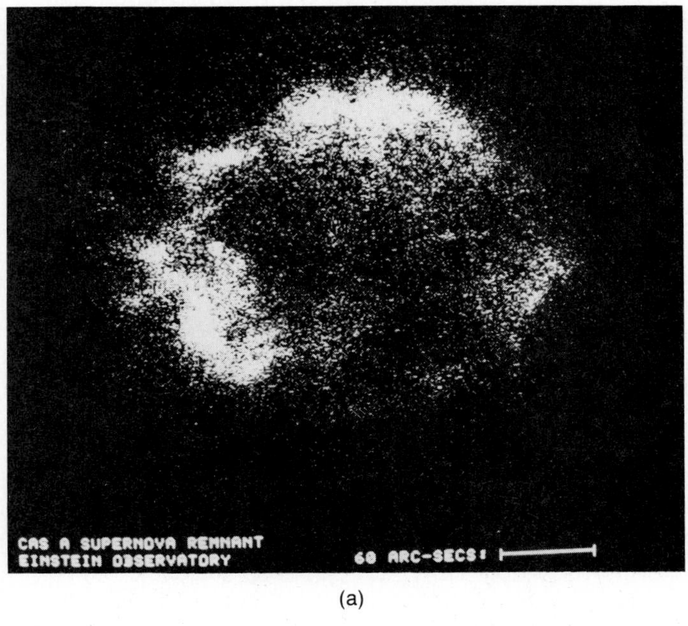


(a)

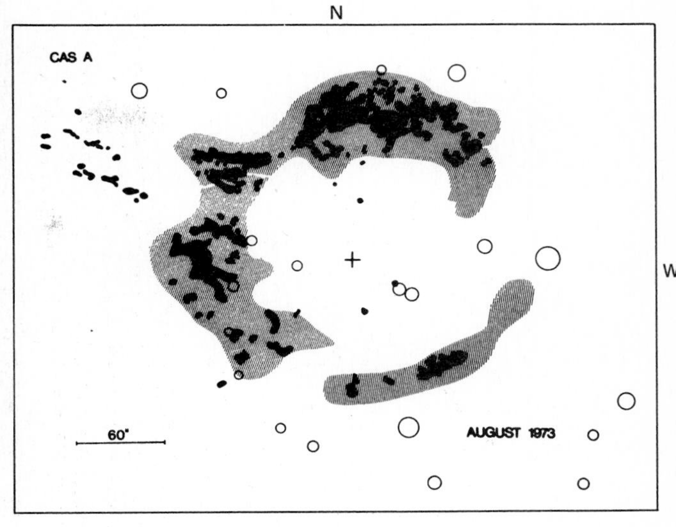


(b)

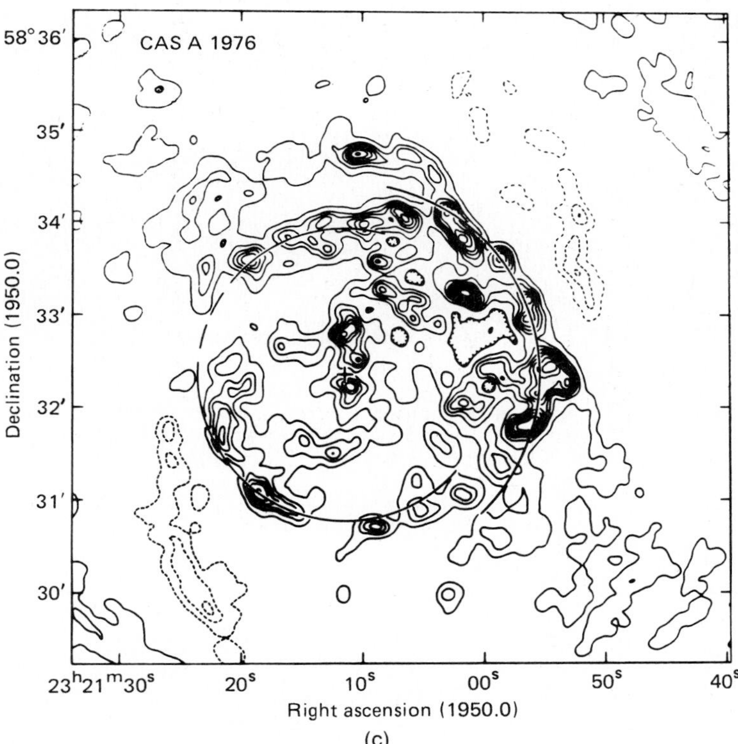


(c)

**Figure 25.12** Supernova remnant, Cas A, observed (a) by the Einstein X-ray observatory (HEAO-2 satellite), (b) optically, and (c) at radio wavelengths (S. S. Murray, Harvard/Smithsonian Center for Astrophysics).

of supernova remnants (usually abbreviated to SNR) (Fig. 25.12), where the structures have the appearance of an expanding shell but where any stellar remnant is too faint to be seen on photographs (Fig. 25.13). This is the case of the Brahe (1572) (Fig. 25.14) and the Kepler (1604) supernovae. The SNR in Vela has been discovered by radio astronomers; its present size and expansion speed suggest an age of 11,000 to 20,000 years. How does all of this fit with the theories of stellar evolution?

We saw that a star of about 1 $M_\odot$ probably evolves through the red giant and planetary nebula stages. Those are clearly quite unlike the supernovae for they show no strong radio or synchrotron radiation. Some cal-

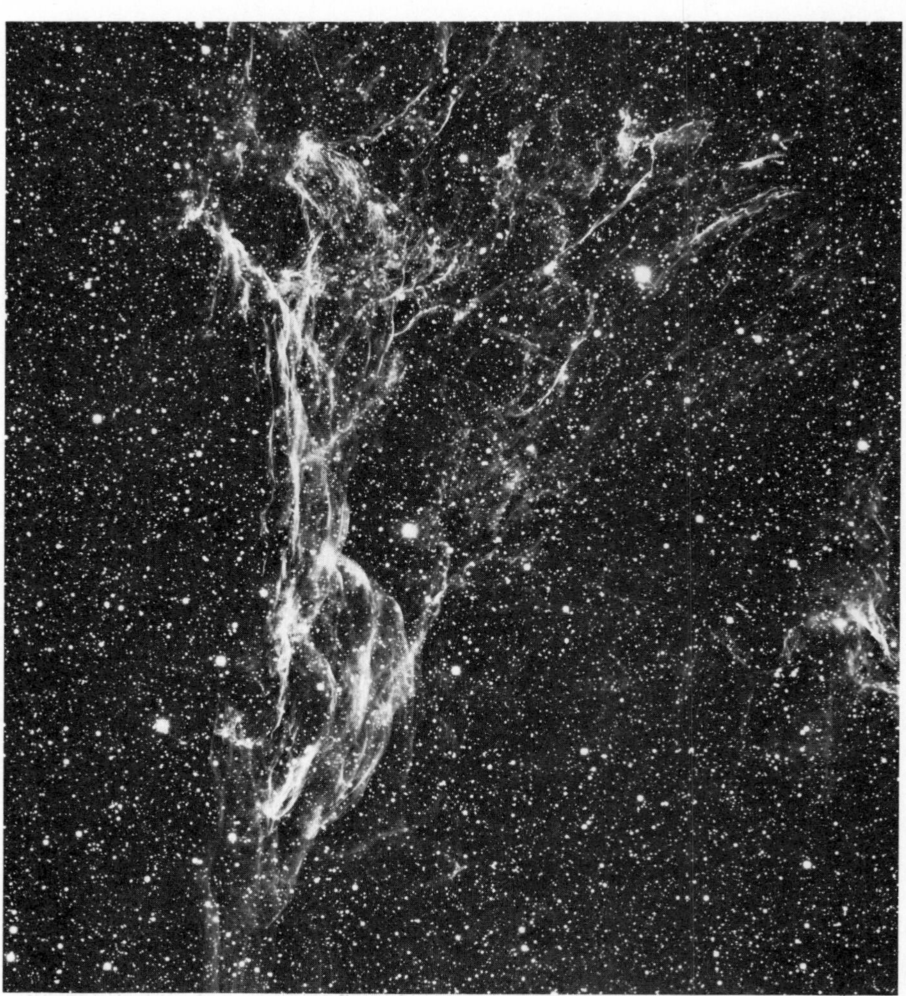

**Figure 25.13** Central section of the Veil Nebula, a supernova remnant (CTIO).

**Figure 25.14** Maps of Tycho Brahe's 1572 supernova (**a**) at 50 cm wavelength and (**b**) observed with X-rays of energy 0.15 to 4.5 keV. This supernova remnant is thought to be 3 kpc distant (P. B. Reid, and *Astrophysical Journal*).

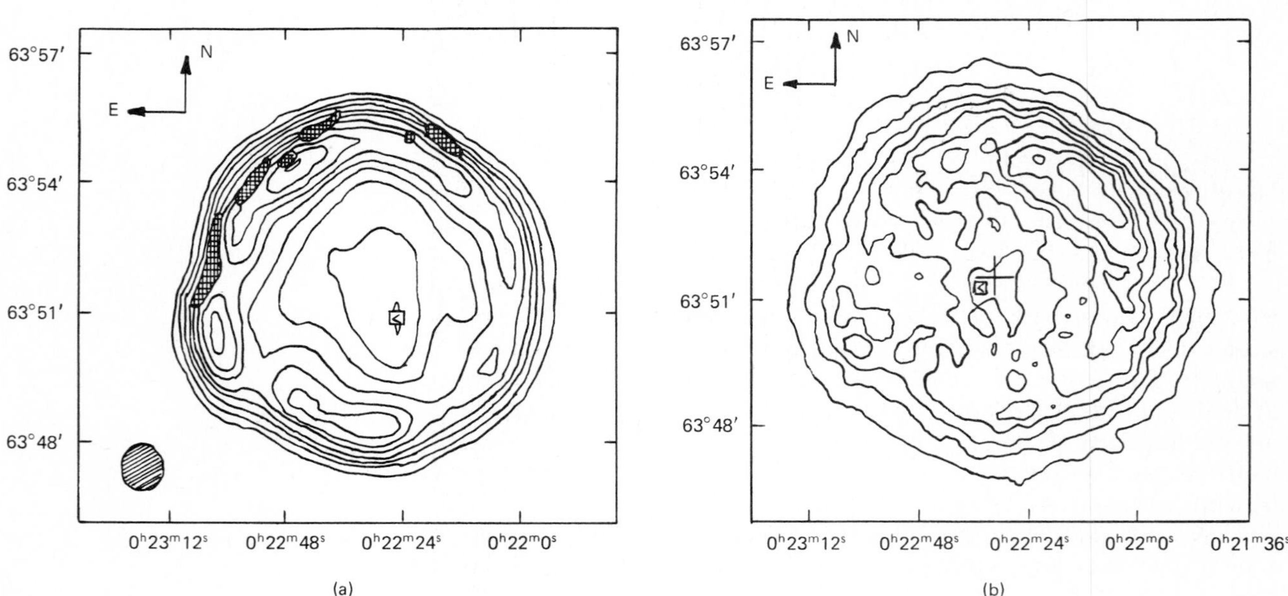

culations seen to indicate that a supernova results from the evolution of a star originally in the mass range 4 to 8 $M_\odot$, but this result is not yet well established. At the center of such a massive star, burning of hydrogen and helium is probably followed by carbon burning so that heavier nuclei are rapidly formed. Collapse and subsequent heating lead to nuclear reactions in which many neutrons are produced. These neutrons, in turn, may be captured by nuclei to produce yet heavier nuclei.

In our discussion of stellar energy sources (Chap. 23), we saw how the fusion of hydrogen nuclei to produce helium was accompanied by the release of energy. Energy is also released as other light nuclei fuse to make heavier nuclei until iron (Fe) is produced. Thereafter, in order to produce yet heavier nuclei, energy must be added. In the development of a supernova, therefore, there comes a stage where heavier nuclei are being produced in the core, but this consumes some of the core energy and the core collapses. As usual, a collapse releases gravitational energy and the core heats up again. In this ultimately explosive process, there is a further release of energy as lighter nuclei fuse in the outer layers.

In the few seconds needed for all of this to occur, heavier nuclei are being produced initially through the combination of lighter nuclei but later by the addition of neutrons to existing nuclei. Neutron addition produces heavier isotopes, and can proceed along two main routes termed the **slow** or **s-process** and the **rapid** or **r-process**. (Other processes are generally less important.)

If the neutrons are arriving in a slow trickle—intervals of months or even years between neutron arrivals—then a newly-formed nucleus will have time to decay if it is radioactive and has a short half-life. If, on the other hand, there is a torrent of neutrons—perhaps a million colliding with each nucleus every second—then the newly formed nucleus will be able to add several more neutrons before it undergoes its radioactive decay.

Neutron capture does not stop with Fe or Co. Successively heavier nuclei are produced as more neutrons are captured, but just which particular nuclei are produced depends on the competition between neutron capture and radioactive decay. We can understand this by noting that the lifetimes of some radioactive nuclei are long (years or days), others very short ($Fe^{62}$ has a half-life of 0.19 sec.) We find that the r- and s-processes produce different nuclei. During fusion in a main sequence star, the production of heavier nuclei is governed by the slow (s-) process. In a supernova, the collapse is part of a runaway in which the number of neutrons increases very rapidly and the r-process takes place probably in a few seconds.

We see the brilliant flash of a supernova while the nuclei are being rapidly transformed. The debris of a supernova will thus contain many heavier nuclei that are thrown out into the interstellar medium where they provide material out of which the next generation of stars will form. A star like the Sun, with its attendant planets, was formed with atoms that had been processed in stars and then explosively distributed.

The theory of **nucleosynthesis** was first worked out in its present form by Margaret and Geoffrey Burbidge, William Fowler, and Fred Hoyle in 1957 (in an article usually referred to as $B^2FH$) (Figs. 25.15 and 25.16). This theory provides a quantitative understanding for the relative abundances of most of the different chemical elements found in nature.

The theory is based on extensive data from nuclear physics and is now widely accepted as a generally reliable explanation for the evolution from a universe that was initially mostly hydrogen with some helium (from the early stages of the Big Bang—see Chap. 31) to the present universe with 100-plus elements. Our solar system and even we, ourselves, consist of atoms that have been recycled from earlier generations of stars.

**Figure 25.15** B²FH: The authors of the historic paper on nucleosynthesis gather 24 years later to celebrate F's seventieth birthday. E. Margaret Burbidge, Willie Fowler, Fred Hoyle, and Geoffrey Burbidge.

**Figure 25.16** B²FH, as it appeared in the *Reviews of Modern Physics* (American Physical Society).

REVIEWS OF

MODERN PHYSICS

VOLUME 29, NUMBER 4 OCTOBER, 1957

**Synthesis of the Elements in Stars***

E. MARGARET BURBIDGE, G. R. BURBIDGE, WILLIAM A. FOWLER, AND F. HOYLE

*Kellogg Radiation Laboratory, California Institute of Technology, and Mount Wilson and Palomar Observatories, Carnegie Institution of Washington, California Institute of Technology, Pasadena, California*

"It is the stars, The stars above us, govern our conditions";
(*King Lear*, Act IV, Scene 3)

but perhaps

"The fault, dear Brutus, is not in our stars, But in ourselves,"
(*Julius Caesar*, Act I, Scene 2)

**BOX 25.2**
*Production of the Heaviest Nuclei*

Among the important and observable differences between the nuclei produced by the r- and s-processes, there is one which we should discuss, for we have made use of it before.

With the s-process it is possible to produce nuclei as heavy as lead and bismuth. Nuclei that are slightly heavier have very short half-lives. Any that are produced in the s-process will decay quickly. With the r-process, however, the very rapid addition of neutrons carries the production of nuclei far beyond lead before they have a chance to decay. If sufficient neutrons can be added to produce uranium or thorium, we suddenly find a very different situation. Uranium and thorium have isotopes with very long half-lives, so long in fact that we have used them to date the age of the solar system. The very long-lived isotopes that were used for dating the Earth, Moon, and meteorites, were all produced before the formation of the solar system, presumably in supernova explosions. They are our most striking evidence for the operation of the r-process, since the s-process and other fusion reactions that occur in normal stars are quite unable to produce uranium and thorium.

There is additional evidence that nucleosynthesis is now taking place. The spectra of some stars show lines corresponding to the element technetium. This is an unusual element, for it is not found on Earth. Technetium has no stable isotopes—all are radioactive, and the longest-lived has a half-life of less than a million years. Technetium has been produced on Earth in nuclear physics experiments. With its half-life far shorter than the ages of the stars in which it has been found, the only conclusion we can draw is that this element is still being produced in those stars probably through the s-process.

Another product of the supernovae is the release of energy that shows up in magnetic fields and high-speed electrons as observed for example in the Crab nebula. In addition, the supernovae probably produce the high-speed nuclei that we detect as cosmic rays (Chap. 27). The origin of the elements that we can see today can thus be understood; we still need to inquire into the fate of the stellar core, presumably left behind after the explosion.

## 25.5 PULSARS

If research led to only another decimal place or to slight extensions of our current knowledge, astronomy would be a very dull subject indeed. One of the fascinations of astronomy is the total unexpectedness of many of its discoveries, and a good example of this is the discovery of pulsars in 1967. This came in the course of a radio astronomy survey being carried out at Cambridge University in England in one of the pioneering radio observatories. The group under Anthony Hewish was searching for very rapid variations in radio signals and had devised a special radio telescope. If you tune your radio away from the strong local stations, you will probably hear the hiss or crackle of atmospheric and man-made "noise." To overcome this undesirable background, radio astronomers normally "smooth" their signals by rejecting rapid variations and recording only the averages over adjustable intervals of time. (In the same way, a photographic time exposure smooths over any variations in the light intensity.) This technique of course eliminates the possibility of detecting very rapid real fluctuations in the signal from a distant object.

Hewish set out to look for rapid variations that occur when cosmic radio waves pass through the electrically charged solar wind. His equipment by design was uniquely able to detect very rapid variations in signal strength. One of his research students, Jocelyn Bell, noted a strange signal that was seen repeatedly each night, not at the same local time (which would indicate man-made interference) but at the same sidereal time. This showed that the source of the signals moved with the stars and had to be outside the solar system. Another puzzling feature of the signals was their intermittent quality—a series of very evenly spaced radio pulses. So regular was the spacing that one needed ultraprecise clocks to measure them. The first of

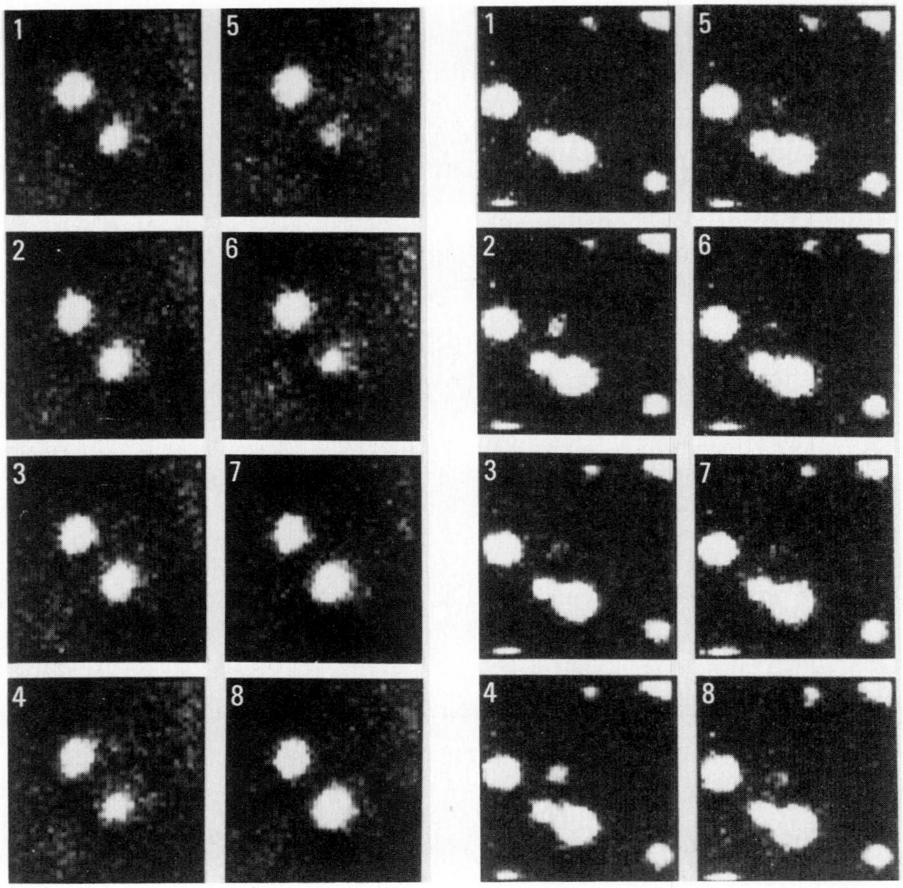

**Figure 25.17** The Crab and Vela optical pulsars photographed in blue light. Each field of view covers only 12 arc sec × 12 arc sec, and the photographs were spaced evenly at intervals of one eighth of each pulsar's period. For the Crab, the greatest brightness occurs in phases 7 and 8, and the intermediate brightening (the *interpulse*) is seen in phases 2 and 3. For Vela, the main double pulse is seen in phases 2 and 4 (Anglo-Australian Telescope Board, and *Nature*).

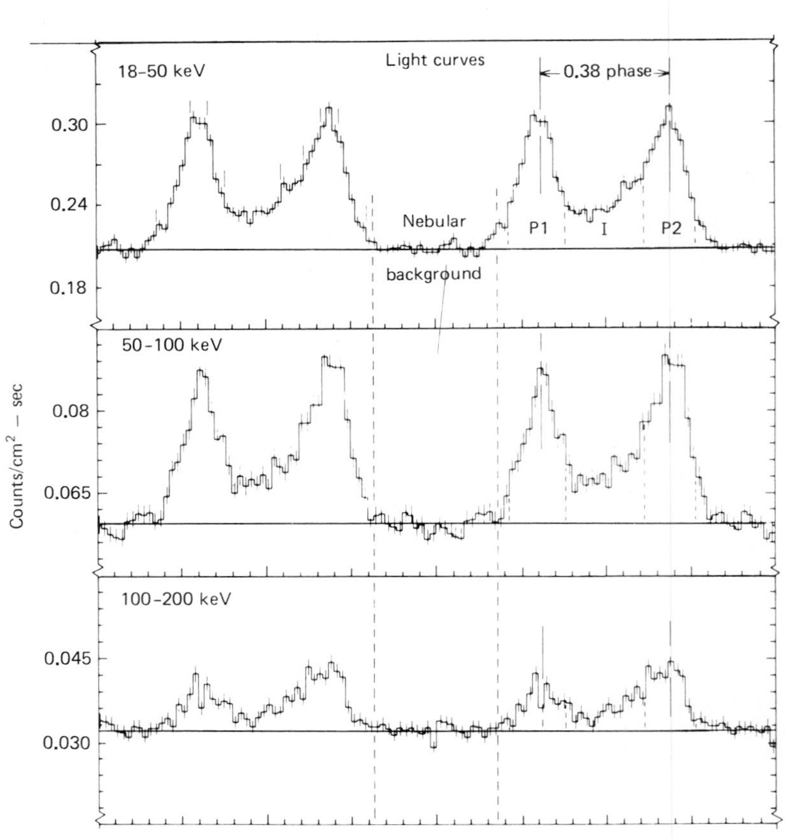


**Figure 25.18** Light curves for the Crab pulsar for X-rays of various energies (F. K. Knight, Naval Research Laboratory).

these **pulsars** gave one pulse every 1.33733 sec, and the next three discovered had repetition periods of 0.253065 sec, 1.187911 sec, and 1.2737635 sec. No celestial object was known to pulse either as rapidly or as regularly, and local sources were ruled out by the nightly return of the signals each at its correct sidereal time. So regular was the frequency of the pulsing that the Doppler shift due to the Earth's orbital motion could be detected. Correcting for this, the true period of the first pulsar was found to be

$$1.3372795 \pm 0.0000020 \text{ sec}$$

After briefly entertaining the idea that the celestial transmitters were operated by Little Green Men, the Cambridge group suggested that the underlying source was the rapid pulsation of a neutron star. Earlier, there had been theoretical calculations that showed that white dwarfs could certainly not pulse as rapidly as once per second. Even neutron stars were marginal, but this mechanism seemed the best option at the time.

The neutron star hypothesis was taken a step further by Thomas Gold of Cornell University, who first suggested (in 1968) that the extreme regularity of the pulsars came from rotation rather than from pulsation. The rotating neutron star model for pulsars is now accepted and has been elaborated in great detail in many theoretical papers that have since been published. Details of the model are still controversial and uncertain.

Searches have led to the discovery of over 300 pulsars. The pulse periods range from the shortest at 0.0015 sec up to the present maximum of almost 4 sec.

Once again the Crab nebula has played an important role in extending our understanding of stellar evolution. Pulsed radio signals were found to come from the Crab, and a search was quickly carried out for pulsed optical signals. Normal photography could not be used, since only the total number of photons is recorded, not their actual arrival times. With an electronic phototube, the arrival of each photon can be timed. When this was done, it was found that there was an optical pulsar in the Crab (Fig. 25.17). A star that had long been seen embedded in the nebula was now seen to switch on and off every 0.033 sec exactly as did the radio signals. Further experiments have shown the Crab to be pulsing in X- and gamma rays as well (Fig. 25.18). All of this pulsed radiation comes from the central star, quite separate from the long-known surrounding nebula whose radiation is not pulsing.

Pulsar patterns are complex. Each shows a regular repeated pulse, and in some this is followed by a smaller "interpulse." The shape of each pulse—showing the speed with which the signal increases and decreases—varies from pulsar to pulsar and occasionally changes within the signals from a single pulsar.

Our present model to explain all of this (and much more) is based on Gold's proposal of a rotating neutron star with a strong magnetic field that is (like the Earth's field) not aligned with its rotation axis. As the neutron star rotates, the magnetic field sweeps around accelerating electrons whose radiation we see every sweep—something like the beaming from a lighthouse (Fig. 25.19). Whether we see any pulse or none from a particular object will depend on the accident of the direction of the beam in space. (The same geometrical accident determines whether a binary system is or is not seen to eclipse.)

The details of the magnetic field and particle acceleration are still being argued over. Ultimately, the energy of the system depends on the rotational energy of the neutron star. As the star slows down, kinetic energy is fed into the surrounding region and converted to radiation. Timing of the Crab over several years has shown a steady slowing down in its rotation rate

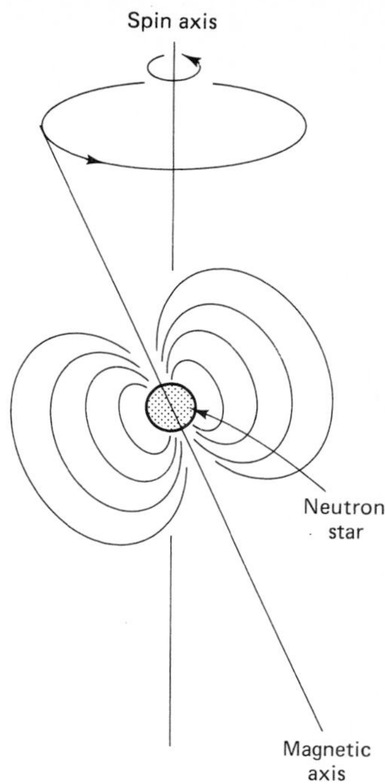


**Figure 25.19** Rotating neutron star model: with the magnetic field not aligned along the rotation axis, the magnetic poles (and the regions where the radiation is being produced) rotate, creating a lighthouse effect. A distant observer will receive only pulses of radiation, the precise time pattern depending on factors such as the orientation of the system and the angle between the spin and rotation axes.

punctuated by some abrupt changes ("glitches"). It has been suggested that the glitches come from "star-quakes"—sudden changes in the structure of the neutron star that affect the distribution of its mass and thence its rotational speed.

The steady slowing down of the Crab pulsar corresponds to a rate of loss of rotational kinetic energy of about $2 \times 10^{38}$ ergs/sec. The known luminosity of the Crab is also about $2 \times 10^{38}$ ergs/sec. This seems to confirm the idea that the source of the radiated energy is actually the rotational kinetic energy of the very compact central star, but it has not yet been possible to model in detail the physical processes involved in the transfer of the energy.

The pulsars we have been discussing were discovered through surveys at radio wavelengths. Visible pulsing as in the Crab and Vela SNRs could be examined once an optical source was identified for the radio pulsar, but most pulsars still do not have visible counterparts.

## 25.6 X-RAY BINARIES

A distinct class of nonradio variable sources has been discovered among the X-ray stars that have been located during surveys from the satellites Uhuru (1970) and Einstein (1978). Optical searches then showed that some of the X-ray sources were located in close binaries where the optical and X-ray intensity variations provide the usual information on the masses. In addition, the X-ray emission is pulsed with the observed periods of the different X-ray sources covering the range 0.7 sec to over 835 sec. Present evidence suggests that all of these X-ray sources are members of binaries. Constructing models for these sources has been a vigorous branch of astrophysical theory; the models differ in detail but agree in the main properties.

In the binary, the optically visible star is generally normal, on the main sequence, but (as deduced from the binary features) often much more massive than the Sun (Fig. 25.20). The other star, which is emitting the X-ray pulses, is not optically detectable. In normal binaries the companion stars circle around each other with the only variation in their observable radiation coming from their eclipses. In X-ray binaries, matter is continually drawn off from the main sequence star and falls to its companion neutron star. Upon arriving at the neutron star, this matter (now traveling at high speed) slams into the neutron star's surface producing the X-rays that we detect. This process of **accretion** is thought to draw the matter either from an atmosphere of a main sequence star or from the stellar wind (if the star is very massive) (Fig. 25.21). The overall picture, then, is of a severely distorted large star, having its mass steadily transferred to its compact partner. The X-ray pulsation probably results from the orbital motion in the binary, in which the area heated by the impact of the mass transfer is sometimes facing toward Earth, sometimes away from Earth.

In two special cases (Cyg X-1 and LMC X-3) the collapsed star is thought to be something more exotic than a neutron star—a black hole, to which we now turn our attention.

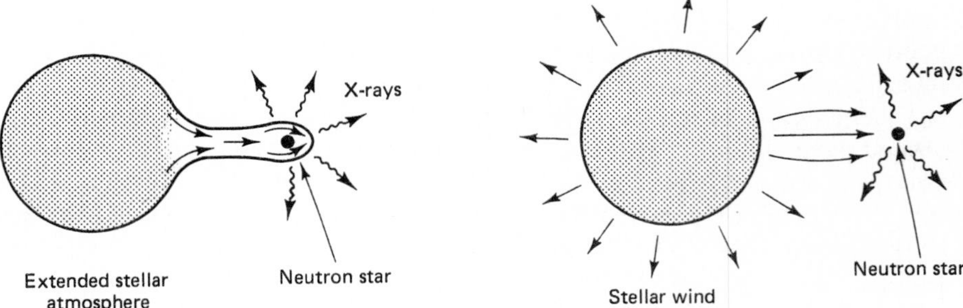


**Figure 25.20** Models for the production of X-rays in a binary system. In one model (**a**), matter is drawn from the extended atmosphere of a large main sequence star, and the X-rays are produced when this matter strikes the surface of a very compact companion, probably a neutron star. In the other model (**b**), the attracted matter comes from a star that is not so extended but which has a strong stellar wind.

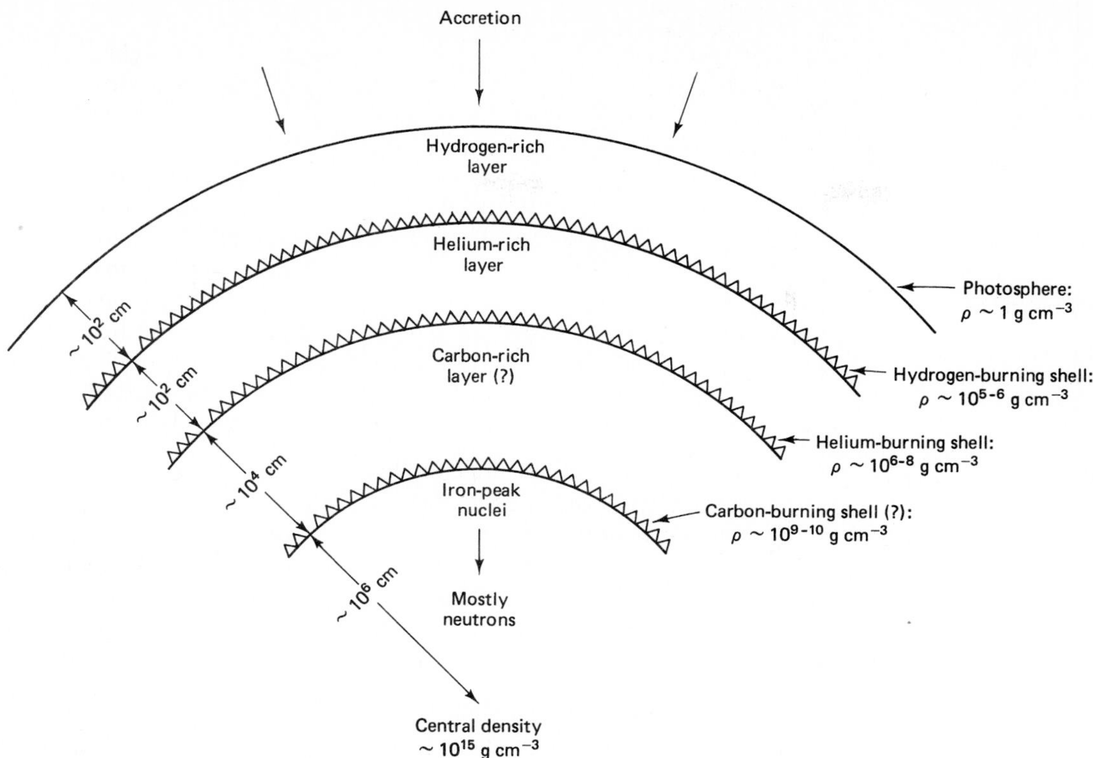


**Figure 25.21** Schematic sketch of the successive layers of a neutron star that is accreting matter from its companion (Paul Joss, M. I. T., and Gordon and Breach Publishers).

## 25.7 BLACK HOLES

Even further from our everyday experience than the white dwarfs and neutron stars are the black holes, objects of fascination for the scientifically interested public as well as for the professional community. What are these objects, how do they form and evolve, and what are their observable properties?

We have seen that white dwarfs are collapsed remnants of stars originally less massive than about 4 $M_\odot$; neutron stars seem to be the corresponding end stages for stars originally with masses about 4 to 8 $M_\odot$. The white dwarfs with $M < 1.4\,M_\odot$ are supported by degenerate electrons against further collapse; in the neutron stars, the cores of mass $< 2\,M_\odot$ have degenerate neutrons that prevent collapse. Stars that start with mass greater than about 8 $M_\odot$ may evolve through an explosive stage to a remnant core, depleted of hydrogen and helium and more massive than about 2 to 3 $M_\odot$. Even the degenerate neutrons cannot stop the collapse which then continues to the black hole stage.

The most extreme property of the black hole is that which gives its name. The black hole is so much more compact than a dwarf that any photon leaving its surface loses *all* of its energy: No radiation escapes. This effect was first calculated in 1916 by Karl Schwarzschild, using Einstein's new equations of general relativity. The critical dimension is now known as the Schwarzschild radius. A star of this size (or smaller) will appear *black*—that is, no light can leave its surface. The actual Schwarzschild radius depends on the mass of the star. For a star of 3 $M_\odot$, it is about 9 km, and for 5 $M_\odot$ it is 15 km.

### BOX 25.2
*The Schwarzschild Radius*

In Box 25.1, we saw that the gravitational redshift for light coming from the surface of a star was given by $\Delta\lambda/\lambda = GM/Rc^2$. This expression holds for normal stars and even for such compact stars as white dwarfs, but it is not

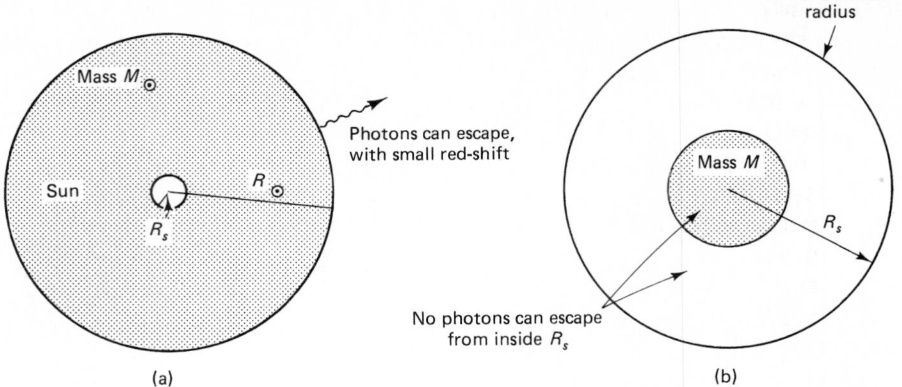


**Figure 25.22** Escape of photons from massive objects. **(a)** Photons can escape from an object such as the Sun, whose Schwarzschild radius is far below the photosphere. Photons will be slightly redshifted. **(b)** For a compact and massive object, the Schwarzschild radius will be at the surface of the star or farther out. Photons released from the surface are inside the Schwarzschild radius and cannot, therefore, escape.

valid for neutron stars or black holes. Those objects are so compact and their gravitational fields so intense that general relativity theory is then needed in order to compute how light will behave near them.

This calculation defines the Schwarzschild radius: It is the radius of a spherical surface surrounding an object of mass $M$, and it is given by

$$R_s \equiv \frac{2GM}{c^2}$$

Light emitted from that surface cannot escape—it will return to that surface (Fig. 25.22). Since no light can escape, none will be seen, and the object and its surface will be black.

For a star with a mass of 1 $M_\odot$, we find $R_s$ has the value of $3 \times 10^5$ cm or 3 km, roughly 2 mi.

The spherical surface at the Schwarzschild radius is called the **event horizon**. Within this surface, the collapsed object can condense even further, but we would not know that this had occurred. The minimum density of the unseen remnant is even higher than in a neutron star. For example, a 3 $M_\odot$ black hole will have a density of about $10^{16}$ g/cm$^3$ (about 10 billion tons in a cube a quarter of an inch on the side). The collapsed remnant can be smaller than its Schwarzschild radius, which is determined by mass alone.

The properties of black holes are by no means settled. Some calculations suggest that black holes could lose energy (evaporate) by creating particles (converting mass to energy), and another theory has raised the possibility that black holes might evaporate so quickly as scarcely to exist at all; this suggestion aroused considerable interest but later work seems to have satisfactorily refuted it. How can one theory clearly refute another? Apart from calculational errors, how can a mathematical theory be wrong? All theories must be based on some assumptions: For example, which physical processes are important and must be included, and which can be ignored? Different theorists can thus disagree on particular assumptions and on the approximations that might be used for the numerical calculations. The complexity of the theories can thus result in differing predictions.

Although it cannot radiate in the usual way, the black hole's mass can make its presence felt on the neighboring matter. For example, a binary system consisting of a black hole and a normal star would show an unexpectedly variable proper motion as Sirius did before we found its small companion. From careful measurements of the proper motion, the orbit of the binary could be deduced, and the sum of the masses of the two stars then estimated.

With this theoretical background, what is the observational evidence? Regretfully, it must be admitted that none of these exotic objects has yet been clearly identified. Since the theory places limits on the possible black hole mass, and since our best way of estimating stellar masses is through observations of binaries, we would expect our strongest evidence for the existence of a black hole to come from a binary star. This may be the case for our best candidates so far.

A ninth magnitude star, HDE 226868 has been found at the location of the X-ray source Cyg X-1 (Fig. 25.23). This star's spectrum is that of a blue supergiant (09 type) and (if it is on the main sequence), it has a mass of about 5 $M_{\odot}$. Spectral shifts recur with a period of 5.6d, indicative of a binary system. The unseen partner responsible for the X-rays must have a mass appreciably about 3 $M_{\odot}$, the limit for a neutron star, and the best present value is 8 $M_{\odot}$. Observation of X-rays requires the source to be a compact object. The only candidate thus seems to be a black hole. The X-rays then probably come from the matter that is being drawn to the black hole from its supergiant partner, and the matter is accelerated as it approaches the black hole. Some of this matter might go into orbit around the black hole outside the Schwarzschild radius, and its X-rays can thus escape. The details of such a binary system are largely speculative; the X-ray emission might be pulsed, but this has not yet been seen and more sensitive instruments are needed.

In 1983 it was discovered that the X-ray emission of Cyg X-1 had a 294-day variation in addition to the well-established 5.6-day period. Further examination has shown that the 5.6-day cycle changes shape during the longer cycle. Based on these and optical observations, it has been suggested that Cyg X-1 contains a third star, either a small *F* or *G* type in orbit around the other two. The mass of this third star would have to be no more than 2.6 solar masses, and it is suggested that accretion from the supergiant in the binary to the black hole is sometimes hampered by the third star to produce the observed changes in the 5.6-day emission.

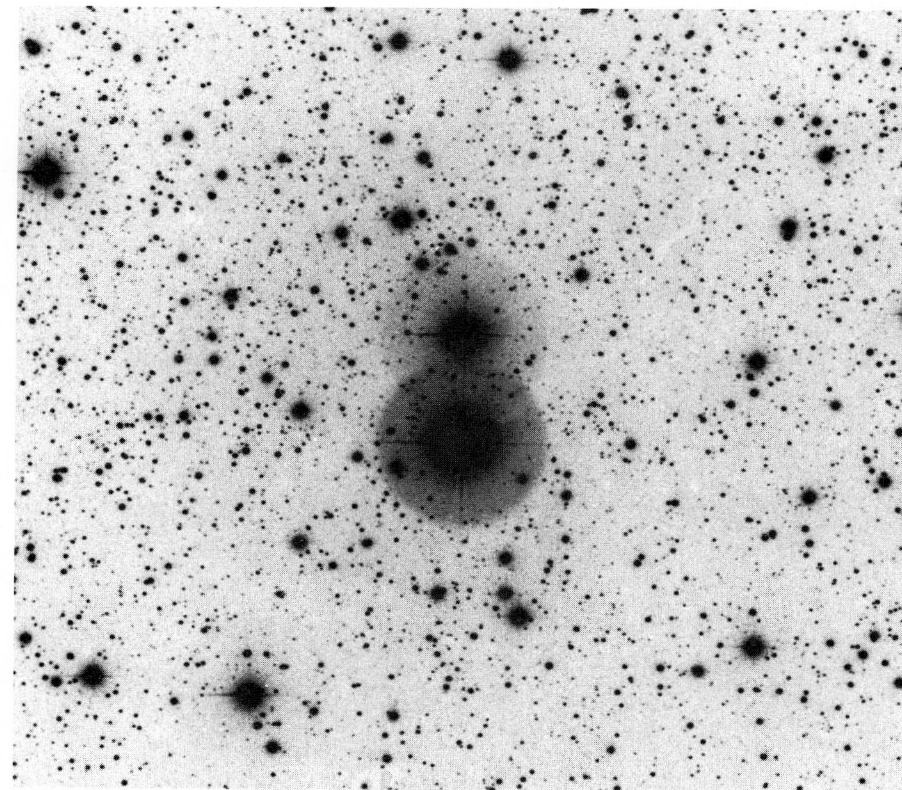

**Figure 25.23** Cygnus X-1. One of the strongest X-ray sources, possibly containing a black hole. At the position of the X-ray source there is a ninth magnitude star, HDE 22686, appearing in this photograph as the larger of the two stars in the center. The other star, not as bright and just above Cyg X-1, is unrelated and it is *not* the partner of Cyg X-1 in the binary system. Note that the large sizes of some star images are the results of the time-exposure and are not indications of stellar size (Jerome Kristian, Palomar Observatory).

A second candidate black hole has been located in the X-ray source LMC X-3. In this case, the structure appears to contain a black hole of about 10 solar masses in orbit with an ordinary star. The orbital period is 41 hr, and the separation of the two stars is about 11 million km, less than 0.1 AU. The spectrum of the visible star indicates a main sequence B3, with a mass of about 6 solar masses.

Theory also indicates that, given appropriate conditions, very much smaller and very much larger black holes might form. Large black holes might occur in the centers of galaxies, and small black holes might be very old remnants left over from an early stage in the evolution of our universe. As you have probably realized by now black hole theory is a good example of Mark Twain's comment

> There is something fascinating about Science. One gets such wholesale returns of conjecture out of such trifling investment of fact (*Life on the Mississippi*).

## 25.8 GRAVITATIONAL RADIATION

The catastrophic collapse of a supernova to a black hole releases a very large amount of energy. Theory suggests that this should be accompanied by a burst of radiation in gravitational waves. Attempts have been made to detect these, so far without clear success.

How can gravitational waves be detected? In the 1960s Joseph Weber at the University of Maryland devised and built the first gravitational wave *telescope*, now on permanent exhibit at the National Museum of American History in Washington, D.C. (Fig. 25.24). Weber used an aluminum cylinder, 150 cm long, 61 cm in diameter, and weighing about $1\frac{1}{2}$ tons. This was delicately suspended and had extremely sensitive sensors attached at the midpoint of the cylinder's length. The arrival of a gravitational wave would be signaled by a submicroscopic vibration of the cylinder. A displacement of $2 \times 10^{-14}$ cm, less than the diameter of an atomic nucleus, could be detected.

Since all sorts of man-made and other natural causes such as thunderstorms might well give spurious vibrations, Weber operated two detectors, one at the Argonne National Laboratory near Chicago and the other

**Figure 25.24** First gravitational wave detector, constructed by Joseph Weber and now on permanent display at the Museum of American History in Washington (Joseph Weber and the Smithsonian Institute).

at the University of Maryland. It is highly improbable that spurious signals would be recorded simultaneously at two detectors so far apart; most coincidences between signals at the two detectors should be genuine.

Weber detected many apparent coincidences. What was particularly notable was the increased coincidence rate every 12 hr in sidereal time, when the two detectors were pointing toward or away from the galactic center. (If the sources of the vibrations were on Earth, one might expect to find repeated coincidences in solar rather than sidereal time.) Weber extended his observations in collaboration with another group at the University of Rome, and they have published reports of coincidences between detectors in Maryland and in Rome.

Stimulated by Weber's results, other groups have also undertaken these demanding and delicate observations. In some individual detectors events have been recorded but have been attributed to natural causes (on Earth), and so far no other group has reported signals that they think are due to gravitational waves.

Despite the lack of confirmation for his results, Weber's pioneering experiment had the very positive result of stimulating the intensive study of gravitational wave detector design. More sensitive detectors are now being used and the search continues.

There has, however, been indirect confirmation for the radiation of gravitational energy. A pulsar, PSR 1913 + 16, has been found to be a component of a binary system. According to gravitational theory, the orbital period of the components in such a system should change slowly as energy is radiated away via gravitational waves. The predicted effect is about two parts per billion per year—far too small to be seen in conventional binaries. With a pulsar, things are different because the pulses come with such regularity. PSR 1913 + 16 has an orbital period of 7.75 hrs and a pulse period of 0.059 sec, and red- and blueshifts can be detected in the frequency of its signals due to its orbital motion in the binary. From the very slow change in these frequency shifts, the rate of change of the binary orbit can be calculated, and it is found to be in excellent agreement with the theory.

## CHAPTER REVIEW

During stellar evolution on the main sequence, nuclear reactions in the core release enough energy to allow the star to resist gravitational contraction. At late stages of stellar evolution, the rate of energy production decreases, and gravitational contraction dominates until a new equilibrium is reached. Properties of this stage depend on mass of star.

**White dwarf:** probably end stage of star of original mass less than 4 $M_\odot$ after passing through planetary nebula stage.

- white dwarf has average density around $10^6$ g/cm$^3$, achieved when atoms are ionized and the resulting gas of nuclei and electrons is compressed. Electrons are degenerate. Maximum mass of white dwarf about 1.4 $M_\odot$ as calculated from theory of Chandrasekhar.
- size of white dwarf comparable to the Earth.

**Neutron star:** star originally probably 4 to 8 $M_\odot$ ends as neutron star—compression preceeds beyond density of white dwarf until electrons are so densely packed with protons that they combine to form neutrons.

- density in neutron star core may exceed $10^{15}$ g/cm$^3$.
- surface will be hot, initially around $10^7$ K, so X-rays are emitted.
- radius of neutron star about 10 km.
- because angular momentum is retained, neutron star will rotate very much more rapidly than original star.
- magnetic field of original star is also compressed when neutron star forms, and magnetic field at surface of neutron star will be very high.

**Novae:** great brightening of a star, perhaps nine magnitudes in a few days.

- probably due to transfer of mass from one star

to another in a binary with build up of mass on one until temperature rises sufficiently to ignite nuclear reactions on the surface of the star to produce observed outburst.

**Supernovae:** now known to be different from novae, not simply larger version.

- peak supernova luminosity around $10^{10}$ times solar luminosity.
- mechnism of outburst: runaway fusion reactions when a massive star has consumed most of core hydrogen, then most of core helium, and has collapse that ignites further reactions.
- star then explodes, blows off a large amount of matter, and leaves a neutron star remnant.
- during supernova runaway stage, heavy atomic nuclei are produced rapidly in nuclear reactions.

**Pulsars:** neutron stars emit radio pulses that recur at very regular and short intervals (typically less than a second).

- extreme regularity of pulse emission can be explained only by rotating neutron star, probably left over from supernova.

**X-ray binaries:** transfer of mass between stars in a binary in which one star is massive, the other not optically detectable, probably neutron star.

- mass accretes onto neutron star with emission of X-rays.
- stellar masses can be determined in some cases (using Kepler's Law), and binary usually found to contain one very massive star.

**Black holes:** stages of contraction beyond neutron star—more massive stellar remnant continues to collapse to become black hole.

- black hole is so compact that gravitational redshift prevents emitted photons from escaping.
- two candidates for black holes have been identified, but evidence is not conclusive.

**Gravitational radiation:** expected to be emitted in supernova collapse.

- detectors first designed and constructed by Weber, later by many others, but no clear uncontested evidence yet for detection of gravitational waves.
- evidence for radiation of gravitational energy comes from the binary pulsar PSR 1919 + 16. Orbit of binary changes exactly as predicted by gravitational theory.

## NEW TERMS

**accretion**
**black hole**
**Chandrasekhar limit**
**degenerate**
**event horizon**
**gravitational radiation**
**neutron star**
**nova**
**nucleosynthesis**
**pulsar**
**r-process**
**s-process**
**supernova**
**Schwarzschild radius**

## PROBLEMS

**1.** The density of lead is 11.5 g/cm$^3$, that of air 0.001 g/cm$^3$, and the density at the center of the Sun is almost 160 g/cm$^3$. Calculate the volume of your bedroom and then the mass of the air in it. Now calculate the volume of lead and stellar core matter that would have the same mass as the air in your room.

**2.** What is the maximum mass of a white dwarf star? What happens if a collapsing star has more mass than this special value?

**3.** For a white dwarf, the gravitational redshift amounts to a change of wavelength of about 2 parts in 10,000. What speed must a white dwarf have approaching us to produce a blueshift that exactly cancels its gravitational redshift?

**4.** What will be the gravitational redshift of light emitted from the surface of a neutron star? If light is emitted at a wavelength of 5000 Å, at what wavelength will it be detected at the Earth? (Assume 2 $M_\odot$ for the star's mass.)

**5.** The Sun takes about a month to rotate once on its axis. If the Sun were to collapse first to the size of a white dwarf and then to a neutron star's size, what would its rotation period be assuming that it keeps its angular momentum?

**6.** A supernova radiates about $10^{52}$ ergs of energy. If all of this came from the fusion of hydrogen into helium, how much hydrogen would be needed?

**7.** A supernova might radiate $10^{52}$ ergs during its outburst. How long does it take the Sun to radiate this amount of energy?

**8.** Why do we think that supernova explosions are essential for life on Earth?

**9.** The Crab nebula is about 2 kpc distant and is expanding at about 1000 km/sec. What will its angular diameter be after another 1000 years?

**10.** What observational evidence leads us to believe that there are many very-high-speed electrons in the Crab nebula?

**11.** What are the major differences between the r- and s-processes of nucleosynthesis and their products?

**12.** What evidence is there that atomic nuclei are being formed today in nuclear reactions? (An alternative is that the nuclei were all formed by some initial time and none since.)

**13.** A large star might lose mass via its stellar wind (like the solar wind) at a rate of $10^{-5}$ solar masses per year. Suppose 10 percent of this accretes onto a neutron star companion and generates X-rays. If this matter forms a layer on the surface of the neutron star at the same average density of that star, how thick a layer is built up in a year? (Assume 10 km radius for the neutron star.)

**14.** Which has the higher surface temperature, a white dwarf or a neutron star?

**15.** What is thought to be the mechanism by which pulsars emit radiation at such regular intervals?

**16.** How do pulsars differ from Cepheids and other variable stars? (See Sec. 26.3)

**17.** What is thought to be the mechanism that produces X-rays in X-ray binary systems?

**18.** Suppose that the Sun collapsed suddenly to become a black hole. How would the orbital period of the Earth be affected?

**19.** Calculate the Schwarzschild radius for an object having the mass of the Sun. How does this compare with the solar radius?

**20.** Suppose the Sun collapsed to become a black hole. Without the input of solar radiant energy life would not last long. In the short remaining time, what astronomical differences might you see?

**21.** A black body can be too hot to touch, yet invisible to our eyes. Explain this, and contrast it to the behavior of a black hole.

**22.** To date, the best candidates for identification as black holes are members of binary star systems. Why is this?

# CHAPTER 26

# The Milky Way Galaxy

Viewing the night sky from within a city is usually not very rewarding. Apart from the moon, a few of the planets, and a relatively small number of bright stars, we are likely to see little more than a hazy glow. Light from street lamps and advertising signs is reflected back to us by the ever present auto and industrial smog, which we can see more clearly from an airplane as we come in to land. The view is very different if we can get away into the countryside. Thousands of stars are then visible, even without binoculars, and a dim but glowing band can be seen to stretch across the sky. This Milky Way is made of stars, gas, and dust—the equatorial region of the **galaxy** in which the solar system is located.

It was one of the major discoveries of this century to find that other galaxies exist, separated from our own Milky Way galaxy by great distances. As we examine those galaxies, we can sometimes recognize features that we find in our own galaxy, but sometimes we also find almost totally different systems. We have been able to extend our knowledge of the Milky Way galaxy by these studies, and we have also been able to assemble the results of these galactic studies into a sweeping model of the large-scale structure of the universe. These are the topics we shall cover in this and the next few chapters.

It is very difficult for us to measure, let alone get a feeling for, distances that are vastly greater than those we encounter in our eveyday activities. What we find in the development of astronomical knowledge is a succession of spurts, following each innovation in distance measurement and each opening up of a new wavelength window. Classical astronomy had no means of measuring distances so that the scale of the starry heavens was for long quite unknown. Later, with improvements in the measurement of small angles, the exploration of the solar system could proceed without involving the obviously more remote stars, which could still be relegated to a distant, crystal sphere. Even after the first observations of stellar parallax in 1838, the full dimensions of our galactic system remained unsettled until the 1920s.

Early astronomers, recognizing their inability to separate stars by distance, had been content to display their starry maps as strictly two-dimensional; modern star charts still have this form and are of great use. The first serious attempt at measuring the three-dimensional extent of the distant heavens was due to Sir William Herschel, discoverer of Uranus. His starting idea was that if two apparently close stars were actually at very different distances, they might appear to separate with time because of parallax and motion by the Earth or by either star. In searching for close stars it was his lucky observation that revealed Uranus.

Failing to detect any stellar parallax, Herschel proceeded with a survey based on an ingenious idea. If we can see as far as the ends of the galaxy and if the stars are distributed roughly evenly in all parts, then (as Herschel reasoned) the number of stars counted in a given direction must be proportional to the extent of the galaxy in that direction (Fig. 26.1). In 1785 Herschel published the results of his counts of stars in 683 regions of the heavens. From these *star-gauges* he constructed the first observational map of the galaxy that clearly showed a nonspherical shape with a decidedly irregular boundary. Later he added another 400 gauges. This was Herschel's universe; the recognition of external galaxies did not come for another 150 years. Herschel's method is still used for mapping the extent of the radio halo that surrounds our galaxy: The dimensions of the halo are deduced from the strengths of the radio signals in different directions.

In this chapter, we will start with a broad review of the present picture and then describe in more detail the methods of discovery and the component parts of the galaxy.

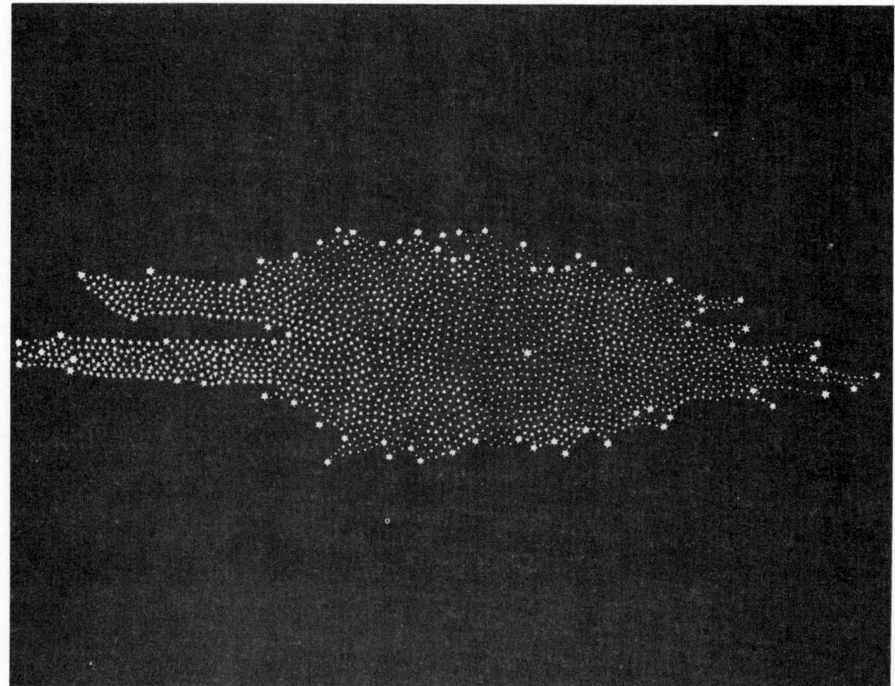

**Figure 26.1** The Milky Way galaxy, as modeled by Herschel using his star gauges (Yerkes Observatory).

## 26.1 GENERAL FEATURES OF THE GALAXY

The stars are separated from their closest neighbors by distances on the scale of parsecs (or light years), and all are confined within the galaxy whose overall diameter is about 30,000 pc ($10^{23}$ cm). The distribution of stars is far from uniform. Most stars lie within a broad equatorial **disk** surrounded by the galactic **corona** or **halo** whose full extent is still not settled. The solar system is almost on the galactic equator and 8.7 kpc from the center of the galaxy (Figs. 26.2 and 26.3).

The galactic disk has a thickness of about 1 kpc ($3 \times 10^{21}$ cm) tapering down toward the outer edges from a broad central bulge about 2 kpc in radius that surrounds the galactic nucleus, which is about 5 pc across. Within the disk, the stars are concentrated along **spiral arms**, which are wrapped around the galaxy from near the nucleus out past the solar system.

The halo consists of a very thin gas, mostly hydrogen, so that the combination of the halo and disk comprise a roughly spherical system of diameter $10^{23}$ cm, somewhat flattened at the galactic poles. Dotted fairly uniformly in the disk and halo are groups of stars, the globular clusters.

We cannot see all of these features in our galaxy, but we have been able to construct this picture from our fragmentary direct evidence and from observations of other galaxies such as Andromeda (M31). The stars are the most immediately observed constituents; in the previous chapters we reviewed present theories of their formation, structure, and evolution. The distribution of different types of stars in different parts of the galaxy is something that we will note in Sect. 26.2. We shall also need to discuss variable stars (Sect. 26.3), which provide us with important means of measuring distances.

Interstellar space was once thought to be largely empty and of little interest. Recent observations at nonoptical wavelengths have shown that this is far from the truth: The **interstellar medium** (ISM) contains a rich variety of atoms and molecules whose study has developed into a vigorous branch of modern astrophysics. The ISM is now known to play an important role in star formation, and our present knowledge of the structure of the galaxy is heavily dependent on observations of the ISM. We will review the ISM briefly in this chapter and at greater length in the next chapter because of its importance.

The galaxy is not static. Its constituents are rotating about the center in a complex pattern. At our distance from the center, the solar system takes about 210 million years to complete one revolution. During the solar system's

**Figure 26.2** A section of the Milky Way from Sagittarius to Cassiopeia (Palomar Observatory photograph).

(a)

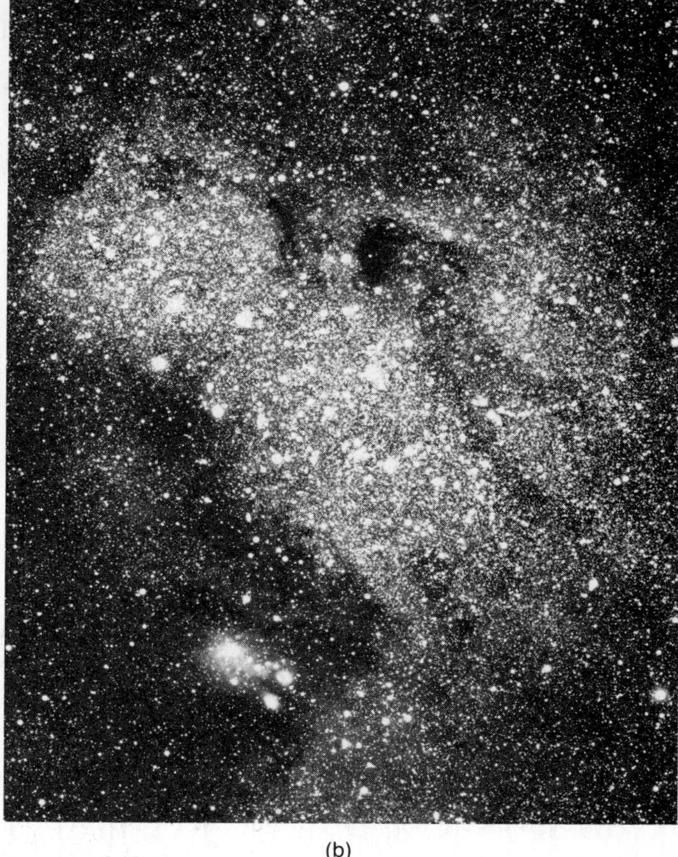

(b)

**Figure 26.3** Enlarged views of parts of the Milky Way: (a) in Sagittarius, south of the galactic center and (b) Small Star Cloud in Sagittarius (Lick Observatory).

4.6 billion years, the Sun and the planets have thus completed over twenty galactic revolutions. It is hard to escape the conclusion that the evolution of the solar system must have been influenced by this galactic rotation, which slowly alters the environment of the solar neighborhood.

This picture of our galaxy is by no means obvious from telescopic observations. Around and between the stars there are enormous quantities of dust that seriously limit the range of our vision in the disc. Only with the introduction of radio, IR, and X-ray astronomy have we been able to detect the radiation from more distant parts of the galactic disc including the galactic center. Looking away from the plane of the galactic disc, we can observe at visible wavelengths to far greater distances (Figs. 26.4 and 26.5). Indeed, we can then observe other systems of stars (galaxies), and the picture we construct for our own galaxy has been strongly influenced by the appearance of those distant systems.

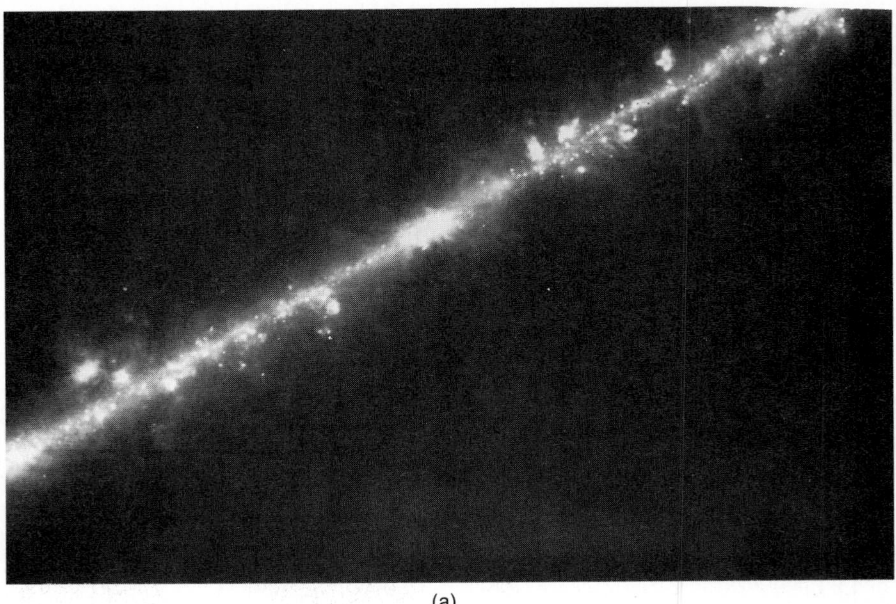
(a)

**Figure 26.4** a) Infrared view of the central region of the Milky Way galaxy, as observed from the IRAS satellite. The Milky Way stretches across this field of view which is about 48 × 33 degrees. Regions of more intense infrared emission from interstellar gas and dust (warmed by nearby hot stars) appear as knots and blobs. The position of the galactic center is shown by the bright feature in the middle (NASA/JPL).
b) The same region of the sky as shown in (a), seen here in optical wavelengths. In this view, the dust along the galactic plane obscures the more distant stars and therefore appears as an irregular dark band. The bright regions at greater distances from the galactic equatorial plane show that the stars are not as concentrated to the plane as are the dust clouds (NASA/JPL and Palomar Observatory photograph).

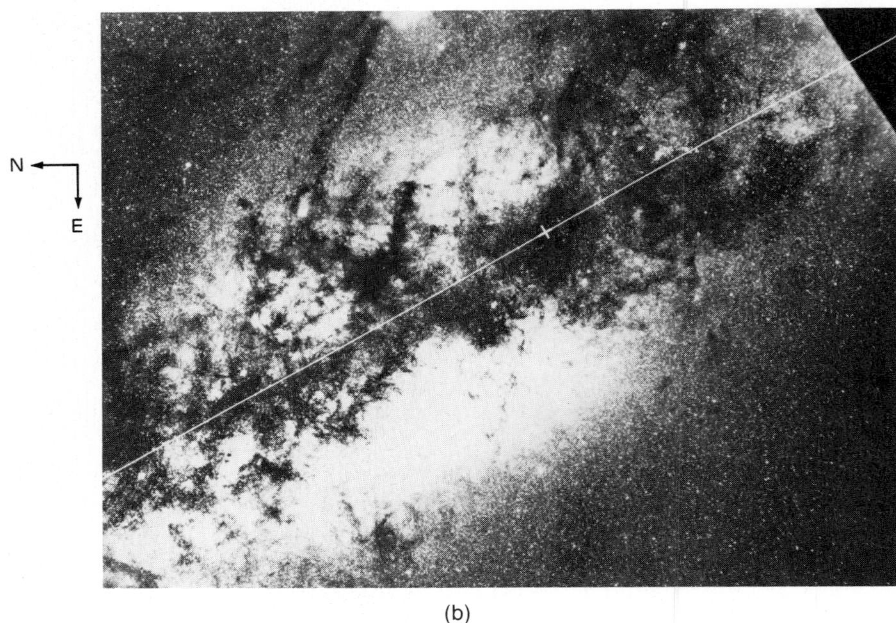

(b)

**Figure 26.5** The Milky Way as seen in gamma rays from the satellite COS-B (Mayer-Hasselwander et al., Max-Planck Institut für Astrofysik, and *Astronomy and Astrophysics*).

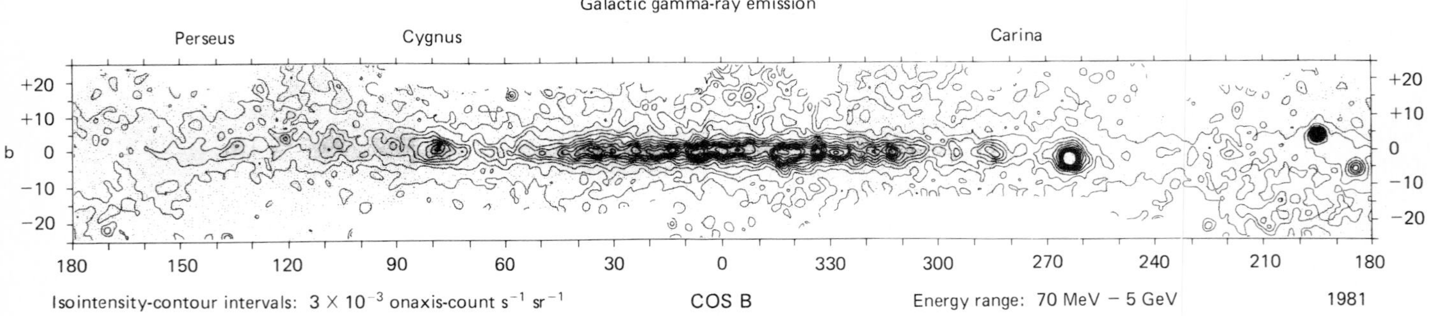

## 26.2 THE STELLAR POPULATION

When we first examined the stars, we concentrated on their individual properties—their structure, formation, and evolution. These are the trees of the celestial forest (our galaxy), and we now need to survey the forest itself. Are the same kinds of trees to be found in all parts of the forest, or are there regional differences? Are there clearings in the forest, with overgrown tangles elsewhere? And what about the celestial undergrowth, the interstellar medium?

We can make a start by looking at two well-defined groups of about 100 stars, those that seem brightest to us and those that are nearest. Earlier (Chap. 20), we used these groups to demonstrate the use of the H-R diagram in classifying the stars. We can now tabulate some of the data for these same stars in a different way, as in Fig. 26.6. The absolute magnitudes of these stars are displayed in histograms. We see, at once, that this straightforward means of selecting stars produces very different groups.

The nearest stars are generally not very bright—our part of the forest is not well illuminated. Our Sun is one of the intrinsically brightest stars in this neighborhood, although it has a very modest luminosity when compared to those in the second group. Those stars that seem bright really are very luminous. (Only Sirius and Procyon are in both groups, near and bright.) The nearest stars include a number of white dwarfs and many *late type* (*K* and *M*) stars of moderate size and low surface temperature. In contrast, although the bright stars include some *K* and *M* giants and supergiants, most are hot *B* or *A* types.

When we extend our survey to include fainter and more distant stars, we find we can divide the stellar population into two broad groups, which are known as Population I and Population II. This classification came from Walter Baade in an article published in 1944. He noted the great differences between the H-R diagrams of the open and globular clusters and pointed to other major differences between the properties of the two types of clusters.

As shown by their spectra stars in the open clusters such as the Pleiades (Fig. 24.17) and Hyades are metal-rich. On the other hand, the spectra of the globular clusters such as M3 in the constellation Canes Venatici (Fig. 24.19) are found to be metal-poor. We have already noted that the Pleiades and Hyades stars occupy locations in the H-R diagram typical of young groups of stars (Figs. 24.21 and 24.22). Most stars in the globular cluster M3 are off the main sequence (Fig. 24.23) following a branch line through

**Figure 26.6** Absolute magnitudes for (a) the 100 brightest stars, and (b) the 100 nearest stars.

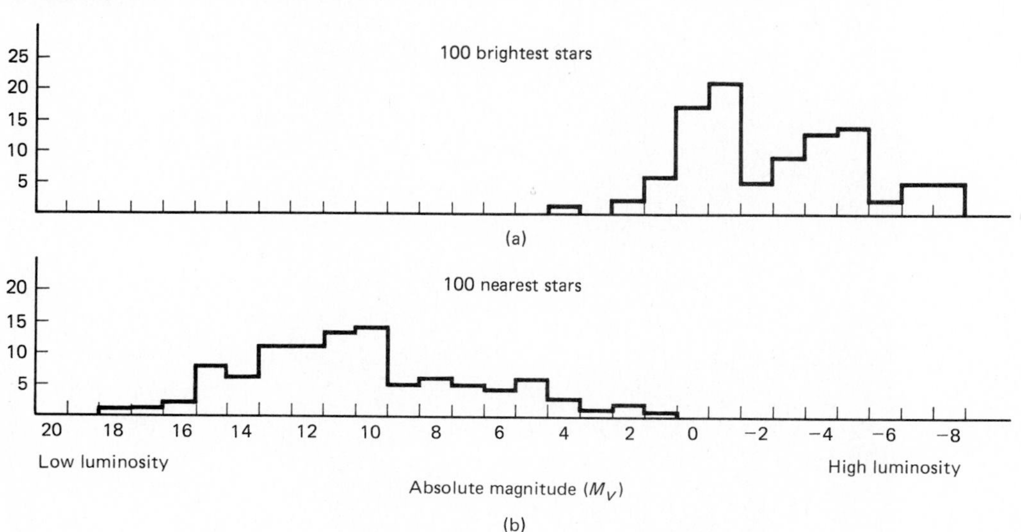

**Figure 26.7** Globular cluster M2 (Yerkes Observatory).

the giants. Those M3 stars still on the main sequence have low luminosities and can thus burn for a long time. We deduced that the globular clusters (Fig. 26.7) were much older than the open clusters (Fig. 26.8) (although there are a few exceptions).

The open and globular clusters differ in another important way: The open clusters are strongly concentrated in the equatorial region of the galaxy, whereas the globular clusters are distributed fairly uniformly in, above, and below the galactic disc (Fig. 26.9). Baade noticed that there also appeared to be a division between types of stars in the Andromeda nebula M31. Stars in Andromeda's spiral arms seemed similar to stars in open clusters, and stars in Andromeda that lay between its spiral arms or close to the galactic nucleus bore a closer resemblance to stars in the globular clusters. Based on these comparisons, Baade proposed the classification that is still in use with the addition of some subgroups.

**Population I** stars are those in the spiral arms. These are young stars, and very luminous. They consist of hydrogen (more than 90 percent), with some helium (close to 9 percent), and a little (barely 1 percent) of the heavier elements, the metals.

**Population II** stars comprise the rest: those that lie between the spiral arms, or out of the plane of the galaxy, or in the galactic nucleus. They are older than the Population I stars, and they are metal-poor, with typically 10 to 100 times less of the heavier elements than those in Population I.

At first sight this might seem to be inverted: Surely the old stars should have had more time to produce heavy elements through nucleosynthesis? The explanation is that the old stars formed long ago from an interstellar medium that was still almost all hydrogen, and any heavy elements that these old stars are producing are contained within them in their cores. The young stars, on the other hand, formed more recently by which time the ISM had been enriched with heavier elements from the explosions of older fully evolved stars. Their metal-enrichment can be seen in the spectra of their photospheres. The fact that the young Population I stars are in the spiral arms indicates that these are the regions where stars are probably still forming: The density of gas and dust is sufficient.

**Figure 26.8** Loose globular cluster Abell 2 (Lick Observatory).

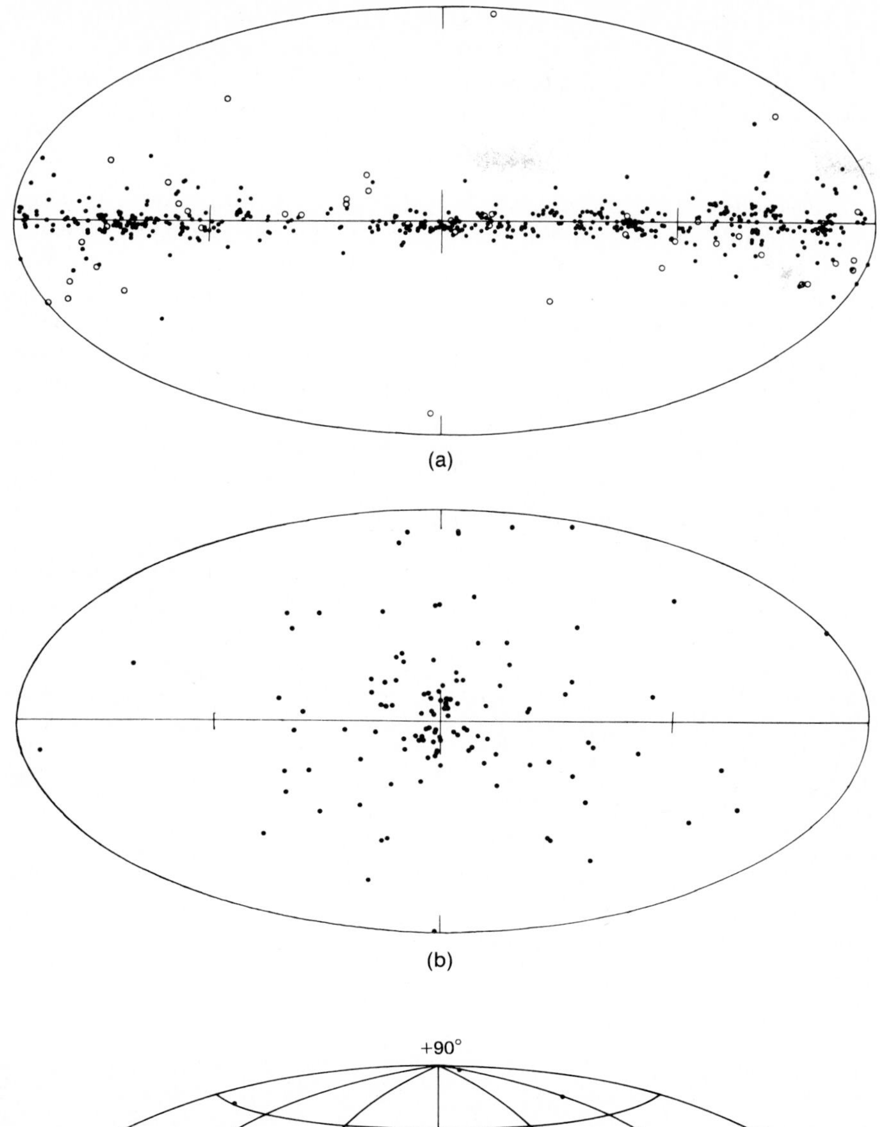


**Figure 26.9** Spatial distribution of star clusters: (a) open clusters and (b) globular clusters. The strong concentration of the open clusters towards the galactic equatorial plane is quite different from the widely dispersed distribution of the globular clusters (Cecilia Payne-Gaposchkin and Katherine Haramundanis, *Introduction to Astronomy*, 2nd ed., 1970, reprinted by permission of Prentice-Hall, Inc.).

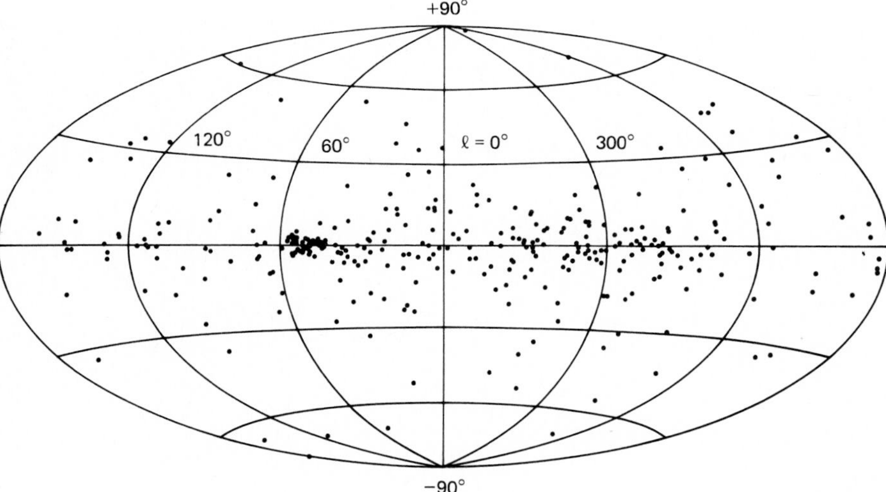


**Figure 26.10** Distribution of pulsars in the galaxy showing the concentration around the equatorial plane (Joseph H. Taylor, Princeton University).

As we saw in Chapter 24, ages can be estimated for clusters from stellar theory and the distribution of stars on the H-R diagram. The very short main sequence of globular clusters indicates extreme ages with individual estimates ranging from 8 to 13 billion years. This sets a lower limit to the age of galaxy at about two to three times the age of the solar system.

## 26.3 VARIABLE STARS

The dismantling of the ancient view of the constant heavens continues. The fixed stars are no longer fixed—they move even if slowly. In addition, they do not always shine with the same brightness. We have already come across the novae and supernovae, but even on a less spectacular scale we find variations in the intensity of the radiation from stars. Some variations are irregular, others are cyclic and have turned out to be of great but unexpected use. Among the regularly varying stars, the binaries have already been discussed. From them, we have been able to deduce the combined masses of their components. Those stars that vary regularly but are not binaries have played an equally important but different role. The cause of their brightness variations is a real physical change taking place in the star.

Some of the regular variables change their brightness noticeably in hours, others over many days. The variations over successive cycles trace out the light curve. Although there may be minor differences between successive cycles for a given star, the main shape of the curve persists. The combination of light curve, period (time to repeat), and spectrum permit classification of the variables into major groups (Fig. 26.11).

The fourth brightest star of Cepheus, δ Cephei, varies in brightness between $3^m.6$ and $4^m.3$ every 5.4 days. During this cycle its surface expands and contracts with a speed that reaches a maximum of 10 km/sec so that its

**Figure 26.11** Light curves of representative Cepheids. These have all been standardized to show their variations through each cycle and so emphasize the differently shaped light curves. Some (top left) display a gentle cycling in brightness while others show sharp increases and slower declines. Actual periods in days are marked (Adapted from a diagram by Cecilia Payne-Gaposchkin and Katherine Haramundanis, *Introduction to Astronomy*, 2nd ed., 1970, reprinted by permission of Prentice-Hall, Inc.).

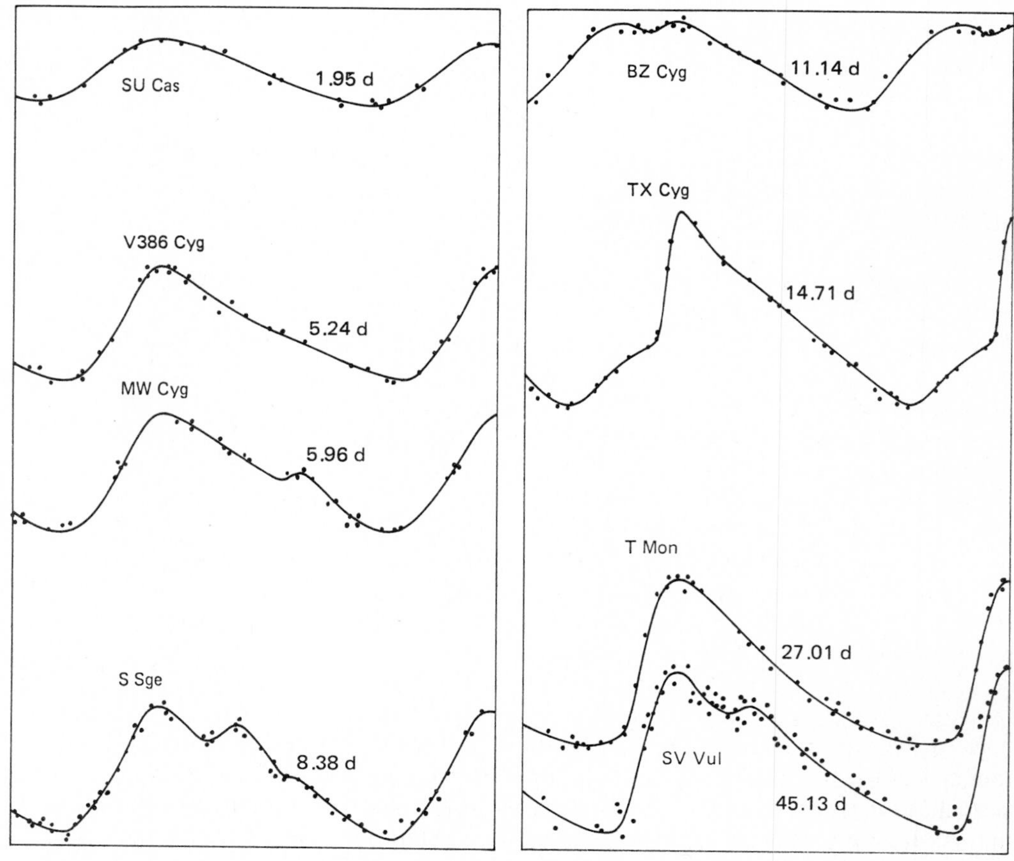

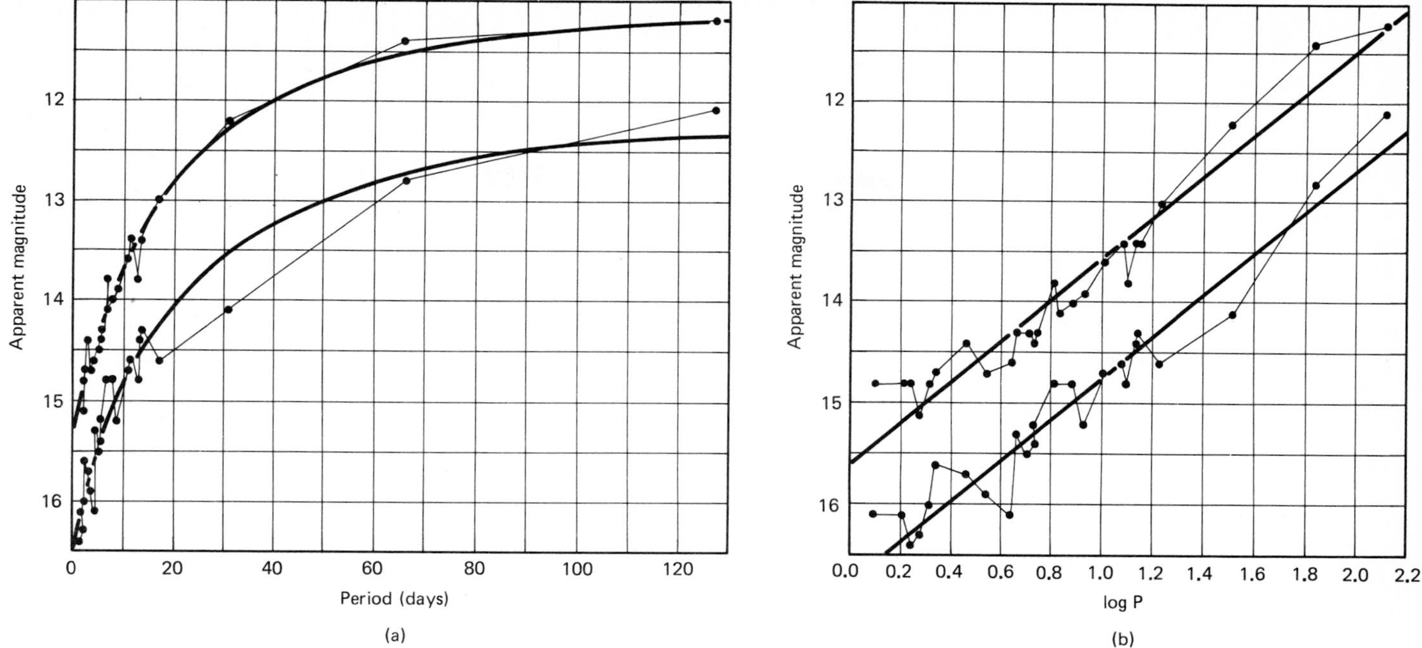


**Figure 26.12** Original diagrams of Henrietta Levitt, showing relation between apparent magnitudes and periods for Cepheid variables in the Magellanic clouds. In each diagram the maximum and minimum apparent magnitude is plotted against the period for that star. The diagram on the left uses a linear scale for the periods (in days), and the diagram on the right has the logarithms of the periods plotted. This is the form now generally used (Harvard College Observatory Circular 173, March 3, 1912).

radius changes by more than 1 million km (about 7 percent of its average radius). At the same time, the surface temperature swings up and down between about 5500 and 6500 K. The class of similar variables is referred to as Cepheids; other members have periods as short as a day or as long as 100 days, and all show a characteristically regular pulsing. Only about 700 Cepheids have been found in our galaxy, but because of their great intrinsic brightness they can also be seen in distant galaxies. In fact, this is the way in which they first attracted persistent attention.

Henrietta Leavitt at the Harvard College Observatory examined the light variations of some stars in the Large and Small Magellanic Clouds. These clouds, found in parts of the southern sky that cannot be seen from northern observatories, were discovered by Magellan during his voyage of exploration in 1520. Miss Leavitt found an unexpected regularity in the light variations for the Cepheids in the Small Magellanic Cloud (SMC): The bright stars had systematically longer periods for their variations. Figure 26.12 shows Miss Leavitt's data for the first twenty-five Cepheids published in 1912. Since the distance to the SMC was not known at that time, only the apparent magnitudes could be measured.

Further analysis required knowledge of absolute magnitudes and, therefore, knowledge of distance, but no Cepheid is close enough to provide a measurable parallax. Statistical methods more complicated and less direct were developed, and by 1917 Harlow Shapley (then at the Mt. Wilson Observatory and later at the Harvard College Observatory) had calculated the distances to the Magellanic Clouds and to some globular clusters that contained Cepheids. Once the distances to some Cepheids were known, the period-luminosity relation could be recast in terms of absolute magnitude or luminosity rather than in terms of apparent magnitude (Fig. 26.13). The

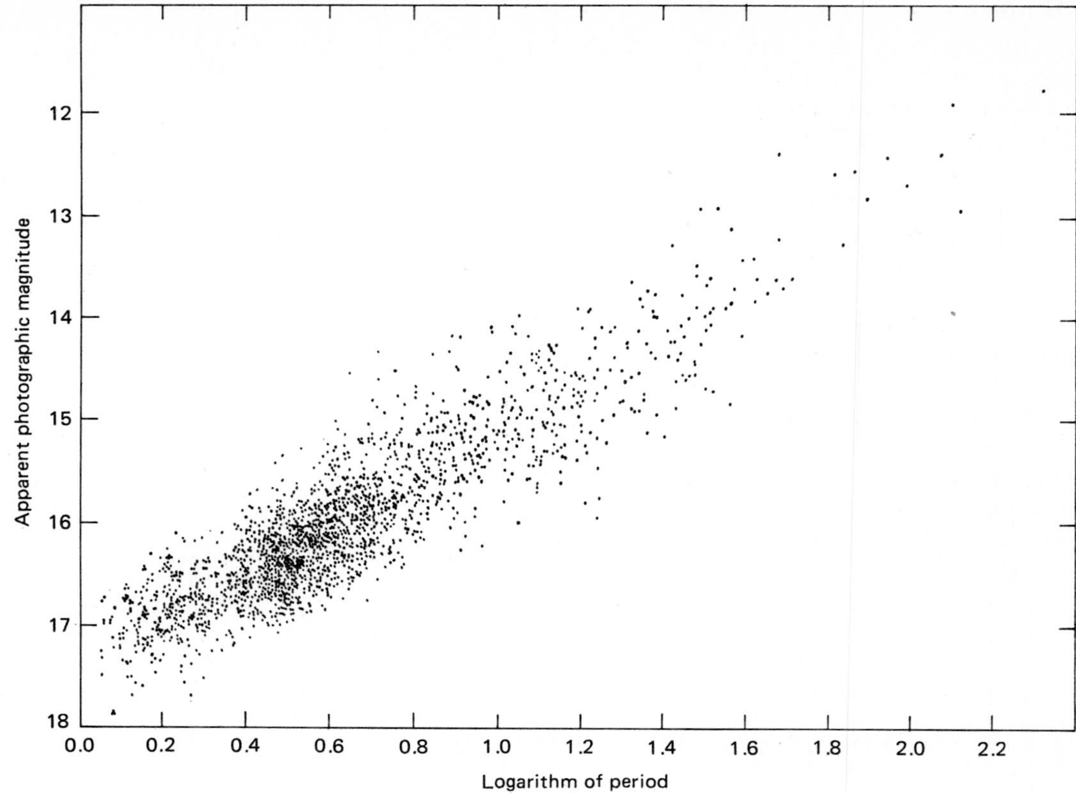


**Figure 26.13** Modern data showing the period-luminosity relation for Cepheids in the Magellanic Clouds. Data have been slightly corrected for the small difference in distances to the two clouds (Data obtained by Sergei Gaposchkin and Cecilia Payne-Gaposchkin; reprinted from Cecilia Payne-Gaposchkin and Katherine Haramundanis, *Introduction to Astronomy*, 2nd ed., 1970, reprinted by permission of Prentice-Hall, Inc.).

relation could then be applied to Cepheids anywhere (see Box 26.1) to determine distances. Shapley was able to show that the globular clusters lay within our galaxy whose overall size was far larger than had previously been suspected. Further, the Magellanic Clouds were shown to be galaxies in their own right, smaller than our own galaxy and well-separated from it.

## BOX 26.1
### *Calibrating the Cepheids*

Knowing the distance to some Cepheids, their absolute magnitudes can be calculated and the period-luminosity diagram can be redrawn using absolute magnitudes (or luminosities) rather than apparent magnitudes (Fig. 26.14).

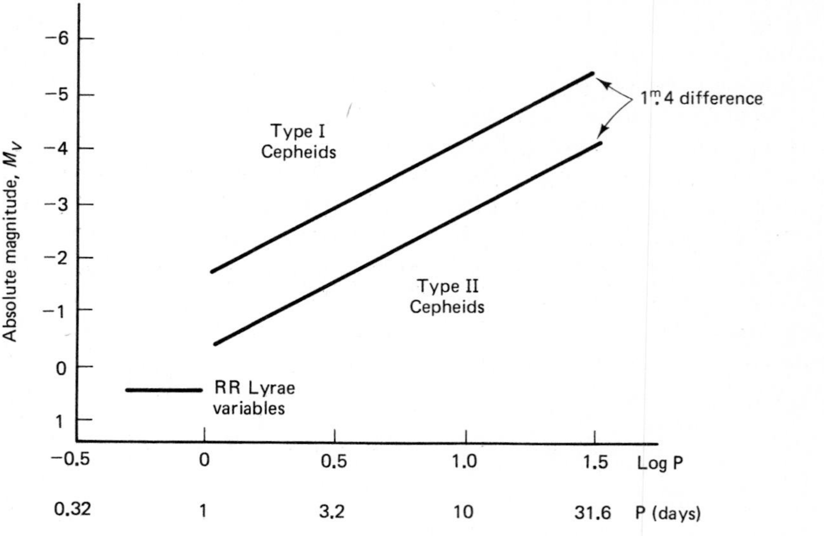


**Figure 26.14** Schematic diagram, showing period-luminosity relation for Cepheids (Types I and II) and RR Lyrae variables. Data points are actually spread in narrow bands about half a magnitude wide on each side of the average lines shown.

If a new Cepheid is now discovered, its period can be easily measured; reference to the period-luminosity relation tells us the absolute magnitude. Knowing the absolute magnitude and the apparent magnitude we can calculate the distance. For example, suppose a Cepheid is detected in a globular cluster and is found to have an apparent magnitude of 13 and a period of 3 days. From the period-luminosity diagram, we can then see that its absolute magnitude must be $M = -2.2$. (Fig. 26.14). We can then use the relation $m - M = 5\log r - 5$ to obtain $r = 1.1 \times 10^4$ pc as the distance to that cluster.

Until 1950 all Cepheids were grouped together, and a single period-luminosity relation was used for the distance calculations. It was Walter Baade's discovery that there are two types of Cepheids. For the same period the classical or Type I Cepheids are about 1.4 magnitudes brighter than those of Type II, which are found in some globular clusters and also in the galactic halo. More careful tracing of the light curves through their cyclic variations shows other differences between the two types. Type I are sometimes termed classical Cepheids, and Type II are termed W Virginis variables. For distance measurement it is, therefore, important to identify the type of variable. Cepheids play a critical role in the measurement of distances to other galaxies (Chaps. 28 and 31), and the scale of distances required a major revision following Baade's discovery.

In principle other types of stars can also be used as standards for distance measurement provided that they can be unambiguously identified and are bright enough to be seen even at great distances. The short period RR Lyrae variables have been used in this way because they have easily observed periods between an hour and a day. Their absolute magnitudes (0.5 to 1.0) are so much less than the Cepheids, however, that their use has been confined mainly to relatively nearby stars, such as those connected with globular clusters in our galaxy.

The main behavior pattern and the physical mechanism of these and other types of variable stars are well understood. Their cyclic brightness and spectral changes have been well documented. When a new type of variable star, the pulsar, was discovered, there was no way in which the extreme regularity and very high frequency of repetition could be explained in terms of conventional variable star behavior. Only a very tiny and very rapidly rotating object could produce the radiation patterns observed.

### BOX 26.2
*Naming the Variable Stars*

As might be expected, the names assigned to variable stars are a mixture of the historic and the systematic. Some (such as Algol) are bright enough that individual names had long been attached. Others such as δ Cephei had been cataloged among the brightest in a constellation. With the increase in photometric accuracy, which showed that stellar variability was widespread, a more sytematic approach was adopted. Each variable star carries the name of its constellation and a capital letter to indicate the order of its discovery. The first variable found in a constellation is denoted by *R*, thus *R* Monocerotis was the first found in Monoceros. Subsequent variables are designated by *S* through *Z*, then pairs of letters *RR*, *RS*, *RT*, . . . , *RZ* and *SS*, *ST*, . . . , *ZZ*. After exhausting these combinations, the sequence continues with *AA*, *AB*, . . . , *QZ*, but omitting *J* to avoid possible confusion with *I*. This use of single and double capital letters accounts for the first 334 variables in a constellation. Thereafter, the lists continue with *V335*, *V336* (*V* for variable). . . . Some variables may carry both a new listing as well as an older name. For example, Nova Aquilae 1918 was the 1918 nova in Aquila and is now listed as *V603* Aquilae.

The continued monitoring of the light variations of these stars is a major project for many amateur astronomers. Data are collected, compiled, and made available through the American Association of Variable Star Observers (AAVSO).

Far less regular variations in brightness also occur. We have previously discussed the violent eruptions of the novae and supernovae. Less spectacular are flare stars. We see flares in the Sun because we are so close, but the flare energy is so much less than the regular radiation that solar-type flares cannot be seen in most other stars. For fainter stars such as *M* type dwarfs, however, a large flare can double the total luminosity. Some flare stars have been seen to brighten by several magnitudes in a few seconds before fading slowly. The underlying causes are complex but generally understood.

A completely new type of variable was discovered in the early 1970s, when bursts of X-rays and gamma rays were detected by instruments carried by satellites. The nature and even the location of most of these objects remain a puzzle, and they will be discussed in Chapter 29 along with other perplexing objects. Perhaps, when they are better understood, they might be used as standards for distance measurements, as we use the well-known RR Lyrae and Cepheid variables.

## 26.4 NEBULAE AND THE INTERSTELLAR MEDIUM

As telescopes improved, it was found that some fuzzy objects remained fuzzy even when seen with improved resolution. Their shapes were distinctly nonstellar, and often they extended over considerable distances. These objects were termed nebulae; we now know they are clouds of gas and dust. Some (the **reflection nebulae**) (Fig. 26.15) glow because they scatter light,

**Figure 26.15** Reflection nebula surrounding the star Merope, in the Pleiades (KPNO).

**Figure 26.16** Emission nebula: the Orion Nebula (Lick Observatory).

**Figure 26.17** Diffuse nebula in Cygnus (IC 1318) (Lick Observatory).

usually from nearby hot *O* and *B* stars and sometimes from very luminous *M* stars. Others (**emission nebulae**) (Fig. 26.16) glow because they are heated by hot stars within. It might be thought that this distinction is trivial if both draw their energy from the same kind of source (hot stars), but closer inspection shows important differences.

Hot stars radiate at all wavelengths and especially strongly in the UV. Short-wavelength radiation is scattered more strongly than long wavelengths. Reflection nebulae, therefore, look bluish for the same reason that the sky looks blue. Emission nebulae, on the other hand, surround their hot stars and are dense enough to absorb much of the short-wavelength radiation. The more energetic photons can eject electrons from hydrogen atoms to produce an ionized (HII) region around the star. This region, termed a **Strömgren sphere**, can extend for 10 pc or more around a star that has a surface temperature of 40,000 K (Fig. 26.18). Some of these nebulae will glow in the IR because of radiation from the dust they contain.

Because hydrogen is the major component of most interstellar gas, hydrogen lines and especially the red H$\alpha$ are conspicuous indicators in the visible spectrum. Photographs of the sky taken with red and blue filters will often show marked differences, which permit the hydrogen clouds to be traced easily.

Reflection and emission nebulae are conspicuous because of their proximity to hot stars and also because of their fairly high densities, typically $10^3$ to $10^4$ atom/cm$^3$. At greater distances from the stars, the gas density tails off so that the average density between stars is down around the level of 1 atom/cm$^3$. For a long time the vast volume between the stars attracted

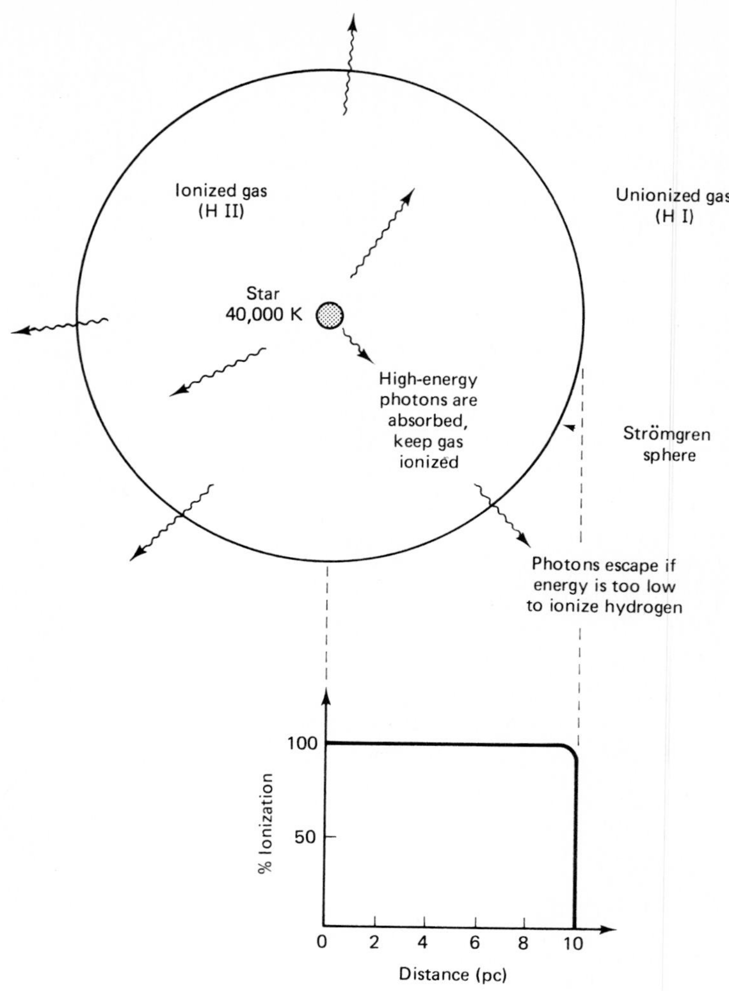


**Figure 26.18** Structure of the HII region surrounding a very hot star. Out to 10 pc, the hydrogen is ionized through absorption of the high-energy UV photons. Lower-energy photons escape. Beyond the Strömgren sphere, the gas remains neutral.

little attention, but the importance of the ISM has become recognized. Because we shall discuss it in detail in the next chapter, we will give here only a few of its properties.

The presence of matter in the interstellar regions was noted as stellar spectra were examined in increasing detail. The continuous spectra of stars tended to depart from what was expected from theory; some of the light was apparently being lost along the way. This **extinction** depended on the wavelength and could best be interpreted in terms of absorption and scattering by dust grains, tiny particles with diameters of about $10^{-5}$ cm or less. Absorption lines were found that were identified as being produced by gaseous atoms and molecules such as CH and CN between ourselves and the stars. With the increasing sensitivity of radio receivers and the opening up of the millimeter-wavelength region to radio astronomy, many molecules have now been identified. Giant molecular clouds have been found. Clouds of hydrogen and carbon monoxide have been found and have provided important clues to the distribution and movement of matter in the galactic disc. Most of the hydrogen is in the atomic form (HI), some (especially near hot stars) is ionized (HII), and some if cold enough is molecular ($H_2$).

## 26.5 SPIRAL ARMS

Even though we think that there may be planetary systems around other stars, we know of only our own for certain. Although we are located well inside it, we know a great deal about its structure partly because there is very little dust to obstruct our view. The situation is very different when we turn our attention to the galaxy, which is so dusty that we cannot see to its center. In contrast to the case of the solar system, we know that there are other galaxies, and we have to draw on their appearance to guide us in the construction of a model of our galaxy. We can see that many galaxies have some sort of spiral structure with arms coiling outwards from a central core, much like a pinwheel. With this clue, we can see evidence for a similar pattern in our galaxy when we plot the results of surveys at visible and radio wavelengths. But without the clues provided by external galaxies we would have only very hazy ideas about the structure of our galaxy.

Hot *O* and *B* stars of great luminosity can be seen at larger distances than other stars, but the general obscuration still confines our visual observations to the nearby parts of the galaxy. These hot stars are consuming their fuel supplies so rapidly that their lifetimes cannot be long; accordingly, they must have condensed relatively recently from gas and dust. We find them concentrated along the spiral arms. Their surrounding HII regions show the same concentration along spiral arms in other galaxies and in our own. Another indicator of the spiral structure is provided by neutral hydrogen (HI) regions whose presence the radio astronomers can trace using the characteristic hydrogen line at 21 cm wavelength (frequency 1420 MHz). These cold clouds of hydrogen are in motion so that the observed wavelengths are slightly shifted from 21 cm to longer and shorter wavelengths. These Doppler shifts tell us the radial speeds of the HI clouds. The temperature of the gas within a cloud determines the average speeds of the hydrogen atoms and thus, in turn, the spread of wavelengths on either side of the average. Observation of the line shape thus gives information on the gas temperature.

## BOX 26.3

### *Mapping Galactic Hydrogen by Its 21 cm Radiation*

Because of the way in which the Sun and the spiral arms rotate about the center of the galaxy, there will usually be a difference in speed between the Sun and any part of a spiral arm that we wish to observe (Fig. 26.19). As a result, when 21 cm radiation from hydrogen in a spiral arm is detected, a Doppler red- or blueshift may be seen depending on the direction of relative motion. Receivers attached to radio telescopes can be swept in frequency

**Figure 26.19** Schematic model of a spiral galaxy. Radio observations made from some location *O* will received signals from hydrogen clouds in spiral arms located at different distances. Because the rotation speed varies with distance from the galactic center, the different regions will be moving with different radial speeds relative to the observer. Some will be approaching and others receding. As a result, the nominal 21 cm hydrogen lines will be red- or blue-shifted. The intensities will depend on the amount of hydrogen as well as on the distance. The observed spectrum (shown in (*b*)) will have a series of peaks, one for each spiral arm with its hydrogen. Different spectra (not shown) would be found when observing in other directions, and sometimes the peaks will not be clearly separated. From observations like these, a picture of the galaxy can be constructed.

O A B C D E F

(a)

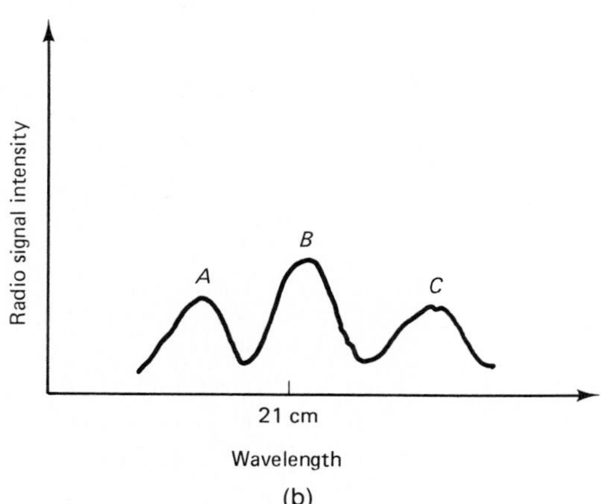


(b)

(just as we do when tuning a radio), and the intensity can be measured at different frequencies so that the Doppler-shifted signal can be traced out.

From our position in the galaxy, many directions of observation will cut across one or more spiral arms (Fig. 26.19a). The signals we receive will, therefore, be the combinations of signals from hydrogen in the different spiral arms that the line of sight intercepts, and each component will probably have a different Doppler shift. There is an added factor that complicates the shape of the signal. Within each hydrogen cloud, the atoms will be in motion in random directions. At the instants that those atoms emit the photons we later detect, some of the atoms will be approaching us, some moving away, and many will be moving in diagonal directions. Each spiral arm's hydrogen, therefore, emits not a sharp 21 cm line but a line that has been broadened by the motion of the hydrogen atoms within the cloud, and this broadening depends on the cloud temperature. The central frequency is shifted by the bulk motion of the cloud as it moves with the spiral arm.

The composite spectrum actually observed (Fig. 26.19b) must, therefore, be unscrambled to separate the contribution from each spiral arm. Already having some knowledge of the spiral structure from optical observations, we can use the radio observations to supplement and extend our knowledge of the spiral structure. Further, the width of each broadened line can be used to measure the temperature in the hydrogen.

Putting all of this together, drawing on observations of other galaxies, we end up with a picture of our own spiral galaxy (Fig. 26.20). The limited observations and the necessity to make assumptions lead to uncertainties in the picture. For example, it is not certain whether there are three or four spiral arms. Older observations identified three, but more recently it has appeared that might be a fourth. The Sun is located in the Orion arm. Further from the galactic center is the Perseus arm; closer in is the Saggitarius arm, which may be connected to the Carina arm (Fig. 26.21).

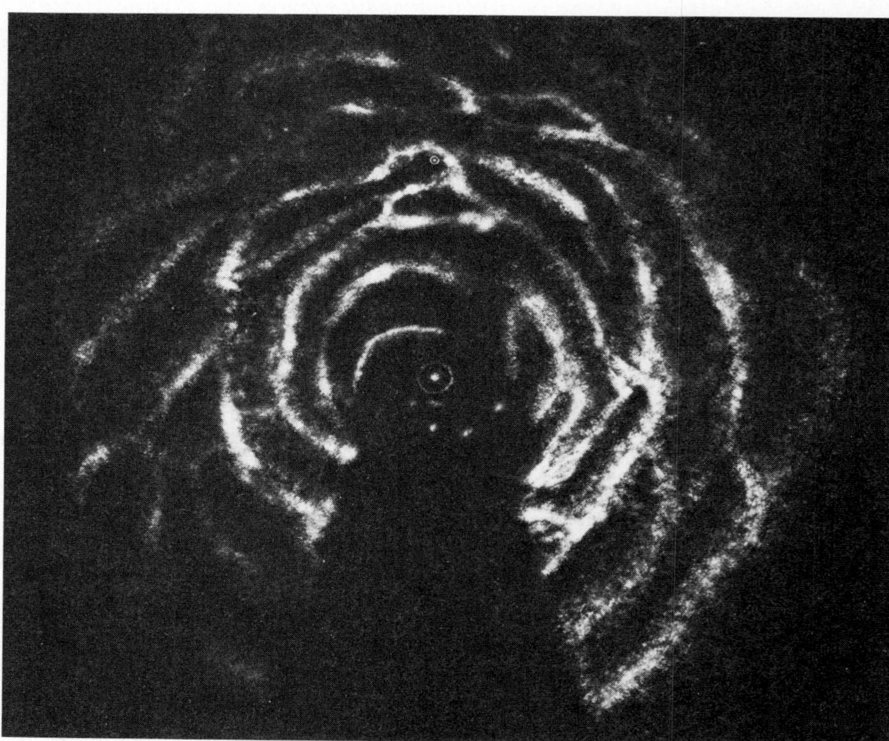

**Figure 26.20** Artist's impression of the structure of the Milky Way galaxy as it would be seen from above the galactic north pole. In this view the galaxy rotates in a clockwise direction. This diagram is based on observations of the 21 cm line emitted by neutral hydrogen, made with radio telescopes in Leiden, Netherlands and Sydney, Australia. This should be considered as a vivid impression rather than an accurate representation. Other and more recent observations lead to the diagram shown in Fig. 26.21 (Gart Westerhout, U.S. Naval Observatory).

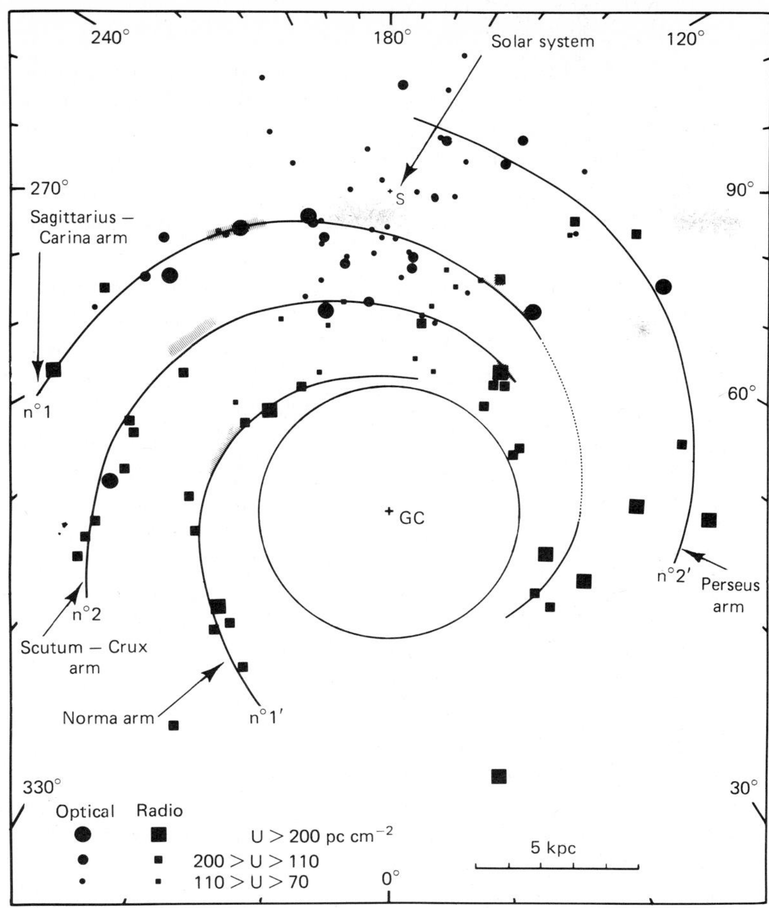


**Figure 26.21** Spiral structure of our galaxy based on optical and radio data. Circular data points are for optical HII regions, square points for radio HII regions. Regions of strongest radio emission are shaded. No data are shown for the innermost 4 kpc. The solar system is located between the Perseus and Sagittarius-Carina arms (Y. M. Georgelin and Y. P. Georgelin, Observatoire de Marseilles, and *Astronomy and Astrophysics*).

## 26.6 ROTATION OF THE GALAXY

The spiral structure is not static. Observations show that the arms are rotating about the galactic center and that the rotation speed depends on the radial distance from the center. The rotation (Fig. 26.22) is not well known close to the galactic center. If the galactic spiral structure were rigid and rotated like a phonograph record, the angular speed (degrees/million years) would be the same at all distances from the center. In that case, the linear speed (km/sec) would increase smoothly with distance. This behavior seems to exist only close to the center out to about 1 kpc. Further out, even at the greatest radial distances, the linear speed shows little spread around an average of 250 km/sec. This means that angular speeds are higher at smaller distances and that the spiral structure is coiling tighter assuming that the arms are trailing. At the Sun's position the linear speed is about 250 km/sec and the angular speed about 1.7°/million years. It takes about 210 million

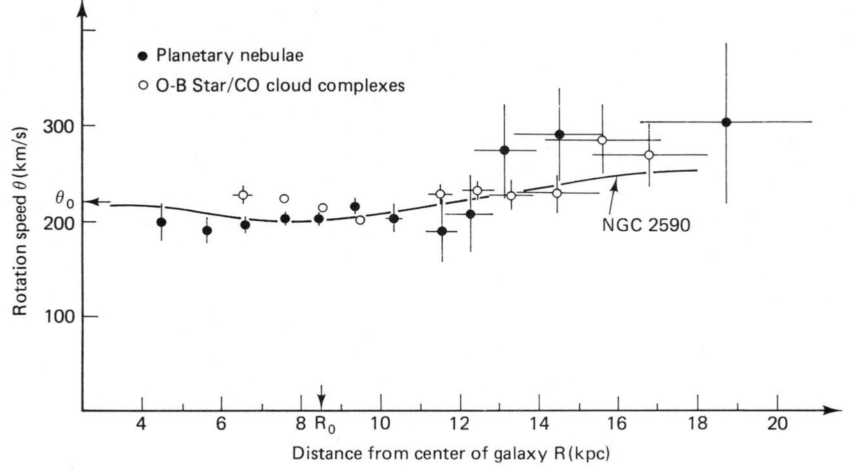


**Figure 26.22** Rotation curve for the galaxy. In the central region (not shown in this diagram) the rotation speed increases with distance from the center. At distances greater than about 4 kpc from the galactic center, there is little variation of the rotation speed, but there appears to be a slow increase towards a value of about 300 km/sec at the outermost parts. Data shown in this diagram come from measurements of young O-B star associations, CO cloud complexes and planetary nebulas. (Courtesy of Y. Terzian and the *Astrophysical Journal Letters*.)

years for the solar system to make an orbit around the galaxy. We can use this information to determine the mass of the galaxy by applying Kepler's third law on a large scale (Box 26.4).

This simple calculation can be somewhat improved by making allowances for the actual distribution of the matter in the galaxy, but the result is probably good to within a factor of two; the total mass of our galaxy is several hundred billion solar masses.

Evidence is accumulating that the extent of the galaxy and its total mass may be several times larger than the preceding discussion has indicated. If most of the galactic mass is contained within a radius of 12 to 15 kpc, then the rotation speed should be seen to decrease at greater distances. This, however, is not seen. Measurements on globular clusters suggests that the rotation curve might be flat even as far out as 50 kpc.

The mass of the luminous components of the galaxy is reasonably known. If the total mass is several times larger, where is it and what form does it take? A variety of observations seem to rule out regular stars, gas, dust, and molecules. It has been suggested that the missing mass fills the halo, making a slightly flattened-spherical system, and that small subluminous stars might be the unseen component. Further observations of rotation speeds are clearly needed.

**BOX 26.4**
*The Mass of the Galaxy*

The relation $P^2 = 4\pi^2R^3/GM$ is applied to the motion of the Sun in its orbit around the galactic center. The distance from the center $R$ is 8.7 kpc or $2.7 \times 10^{22}$ cm. The time for one complete orbit (deduced from the speed and radius) is 210 million years or $6.7 \times 10^{15}$ sec. We then have

$$\begin{aligned} M_{gal} &= \frac{4\pi^2R^3}{GP^2} \\ &= \frac{4\pi^2(2.7 \times 10^{22})^3}{(6.67 \times 10^{-8})(6.7 \times 10^{15})^2} \\ &= 2.6 \times 10^{44} \text{ g} \\ &\sim 10^{11} M_\odot \end{aligned}$$

This is the total mass of stars and ISM that lie between the Sun and the galactic center.

More elaborate calculations take into account the distribution of the mass in the galaxy. In addition, instead of basing the calculation on the motion of the Sun, similar calculations can be performed using the observed motion of matter much further from the galactic center, as shown in the rotation curve (Fig. 26.22).

For some time, it appeared as though the mass of the galaxy was about $4 \times 10^{11} M_\odot$. However, more recent measurements have extended the rotation curve beyond 15 kpc and have shown that the rotation speed is either constant or even rises slowly. This rotational behavior implies a galactic mass of around $10^{12} M_\odot$ and this poses a problem because the visible mass is far smaller. The source of this apparent discrepancy is not understood.

During the life of the galaxy, the spiral structure should have made many revolutions and should apparently be fully coiled up by now. Why do we still see structure? The present theory to explain this came from an idea of the Swedish astronomer, Bertil Lindblad, in the 1920s. The mathematical details were elaborated by the M. I. T. mathematicians C. C. Lin, Frank Shu, and their colleagues in the 1960s.

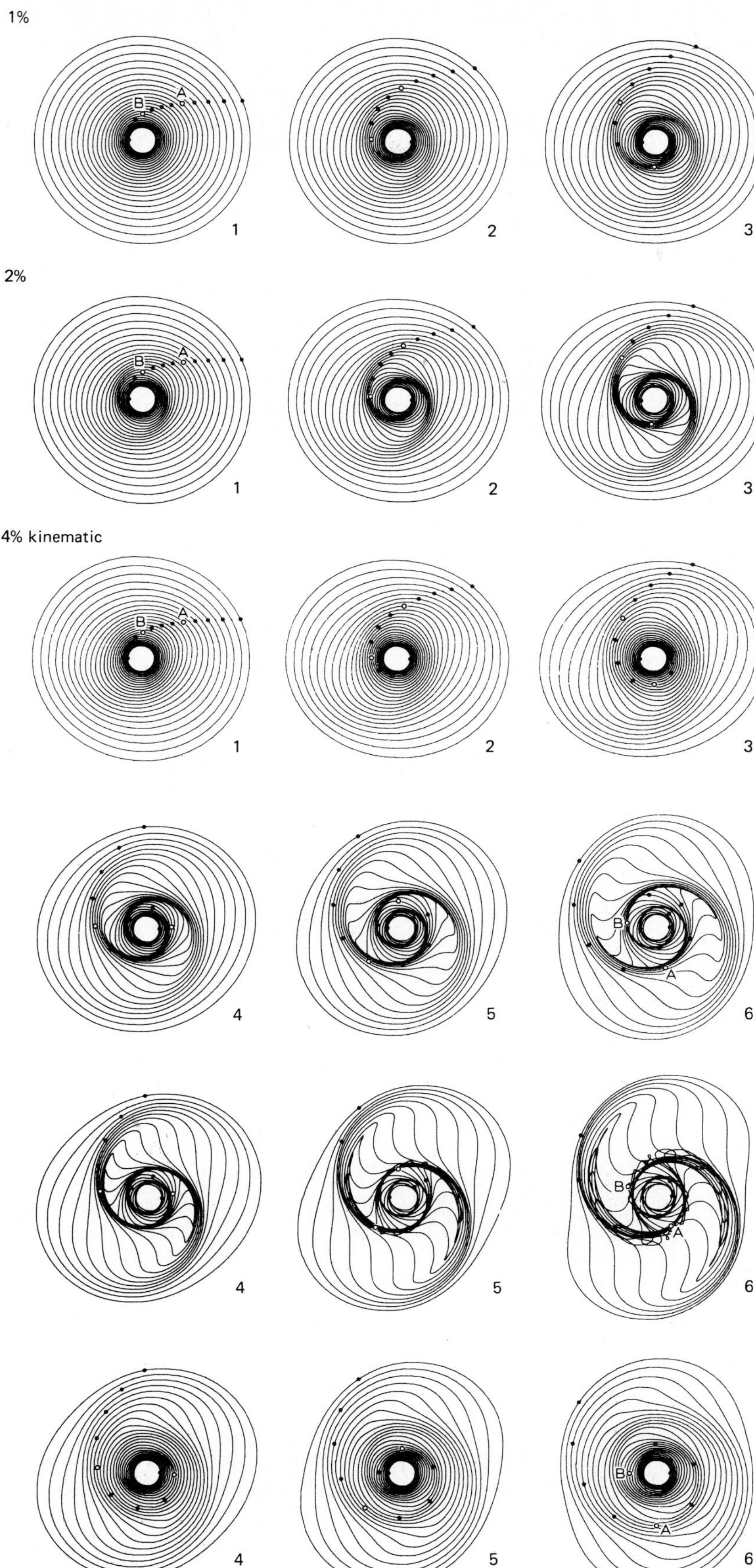


**Figure 26.23** Computer simulation of development of spiral structure in a galaxy. The starting point is a thin disc of stars with a spread of radial velocities in circular orbit around the center. A temporary tidal force is introduced in addition to the central force. The figure shows the results, for the cases where the tidal force is 1, 2, and 4 percent of the central force (Alar Toomre, M. I. T. and Cambridge University Press).

This model is based on spiral density waves. We can get an impressionistic understanding of the model by considering the matter in the galactic disc as though it were made up of concentric strips. Within each strip, the stars, gas, and dust follow a nearly elliptical orbit around the galactic center. The gravitational attraction of other matter, both closer and further from the center, will cause the major axis of this ellipse to precess slowly, much as Mercury's orbit precesses around the Sun. The ellipses of different strips will precess at different rates; the net effect will be a complicated rotational pattern in which the local density will be occasionally high in some places (the spiral arms) and low in others. As the density wave moves round, it behaves like a shock from an explosion, compressing and heating the gas so that conditions become ripe for star formation.

Galactic patterns with similar appearances have been seen by Alar Toomre and his colleagues at M. I. T., who have used computers to follow the consequences of a momentary force being applied from outside a galactic disc (Fig. 26.23). An external force amounting to no more than 1 or 2 percent of the forces already present within the galaxy can lead to the production of a spiral structure as the galaxy continues to rotate. It is possible that the additional force is supplied by the passage of another galaxy; computer modeling can provide insights into these large-scale phenomena, which are still not well understood (Fig. 26.24).

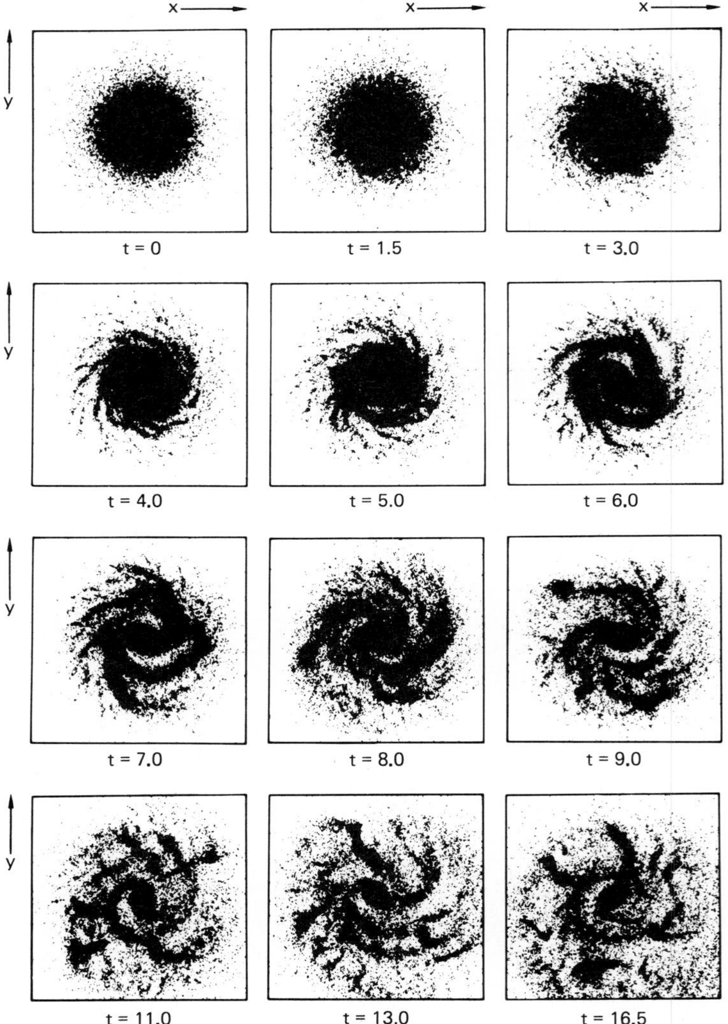


**Figure 26.24** Computer simulation of galactic evolution. At the start, the stars are distributed with small velocity spread in a galactic disc that is rotating uniformly. As the simulation proceeds, collisions between stars will change their distribution. Energy loss is represented by removing the random velocities of some stars each revolution. After four revolutions, the spiral structure is apparent, and very marked by six revolutions (Frank Hohl, NASA Langley Research Center).

## 26.7 THE GALACTIC NUCLEUS

Does our galaxy have a bright core, as we see in many other galaxies? The center of our galaxy (located in the direction of the constellation Sagittarius) has been mapped in great detail at IR and radio wavelengths (Figs. 26.25, 26.26 and 26.27). These wavelengths penetrate the dust in the disc, and the use of very long baseline radio interferometry (VLBI) gives angular detail down to about one thousandth of an arc sec, corresponding to distances of about 10 AU. Figures 26.25 to 26.27 show maps obtained in surveys at wavelengths 2.2 microns, 10 microns, 69 microns, and 3.75 cm. The galactic center is clearly a complex region.

Radio observations show that at the center there is a very compact source (Sgr A) with two components: an intense HII region (Sgr A West) at the very center, and a weaker component (Sgr A East) about 10 pc away. Another source, Sgr B, comprises several HII regions in a large molecular cloud. In the general vicinity of the center, there are many supergiant *M* stars (identified by IR radiation from their relatively cool atmospheres) and some hot *O* stars (which show up via their surrounding HII regions). Star density increases towards the center. On the average there is one star per 100 cubic pc in the whole galaxy, but close to the center the density is over a million times higher and the distance between stars is only light weeks.

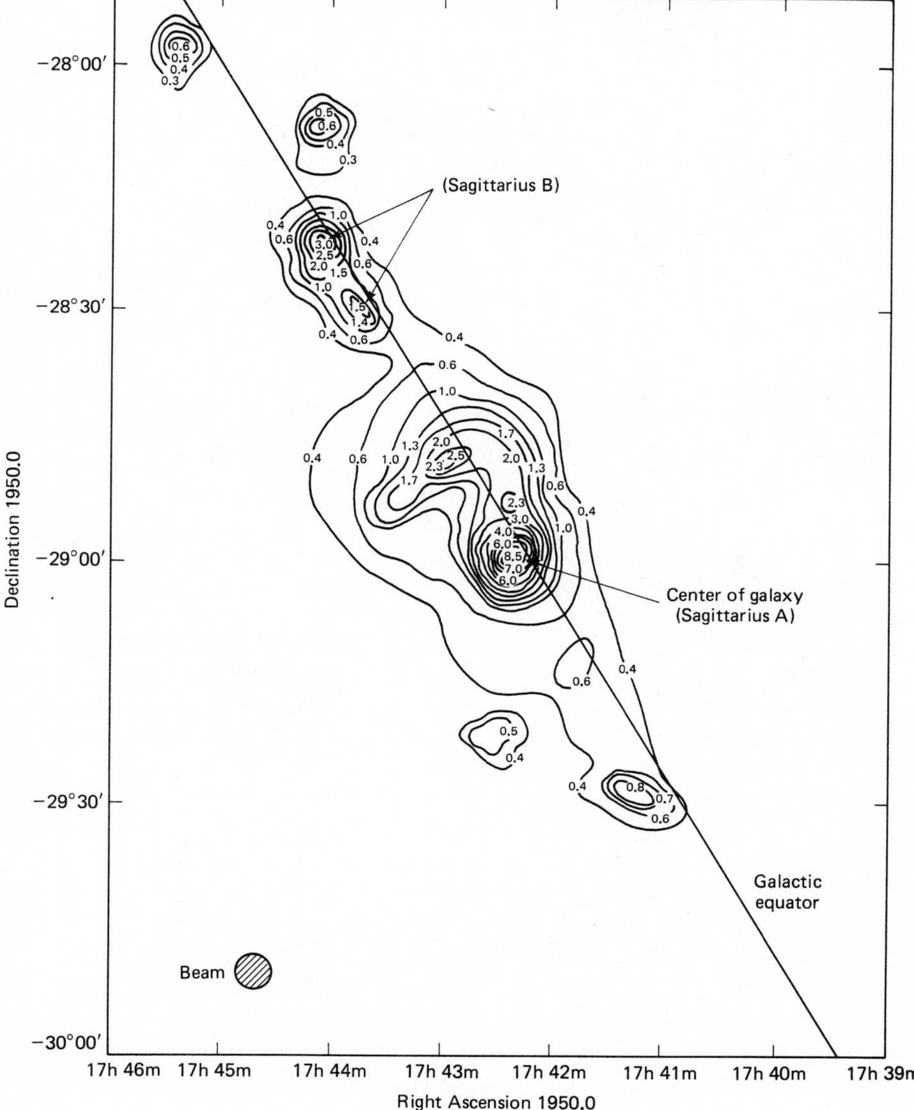


**Figure. 26.25** The center of the Galaxy: radio map from observations at 3.75 cm wavelength. The galactic equator appears as a diagonal line in this equatorial coordinate map. Closely spaced contours identify separate sources. This map is based on observations with a single radio antenna with a beam size as indicated by the circle at lower left. Higher resolution can be obtained, by using a radio interferometer or shorter wavelengths (A. Maxwell, Harvard Radio Astronomy Station).

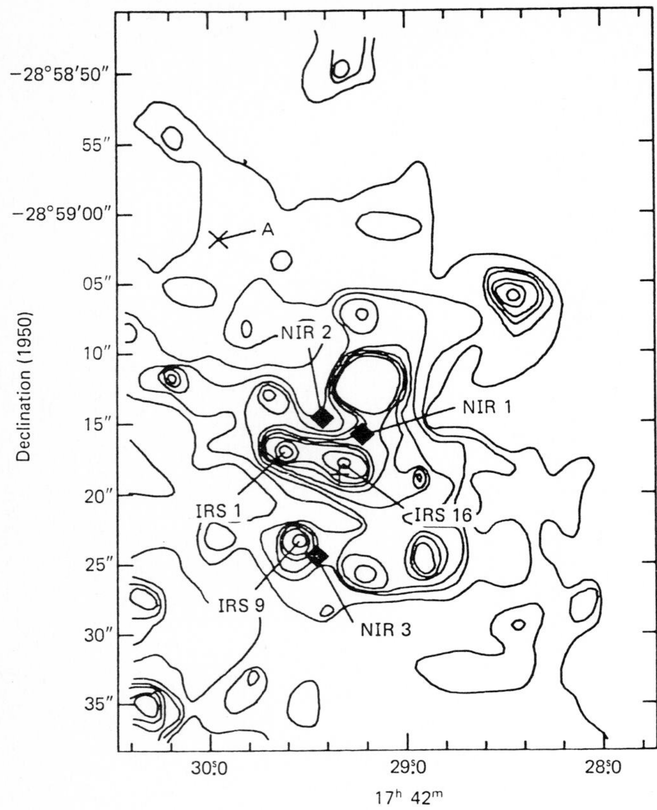


**Figure 26.26** Galactic center, based on IR observations at 0.9 and 2.2 microns. This map corresponds to only the central region shown in Fig. 26.25. The multiplicity of separate but close IR sources can now be seen. Most of these sources are probably stars or groups of stars (G. R. Ricker, M. I. T., and *Astrophysical Journal*).

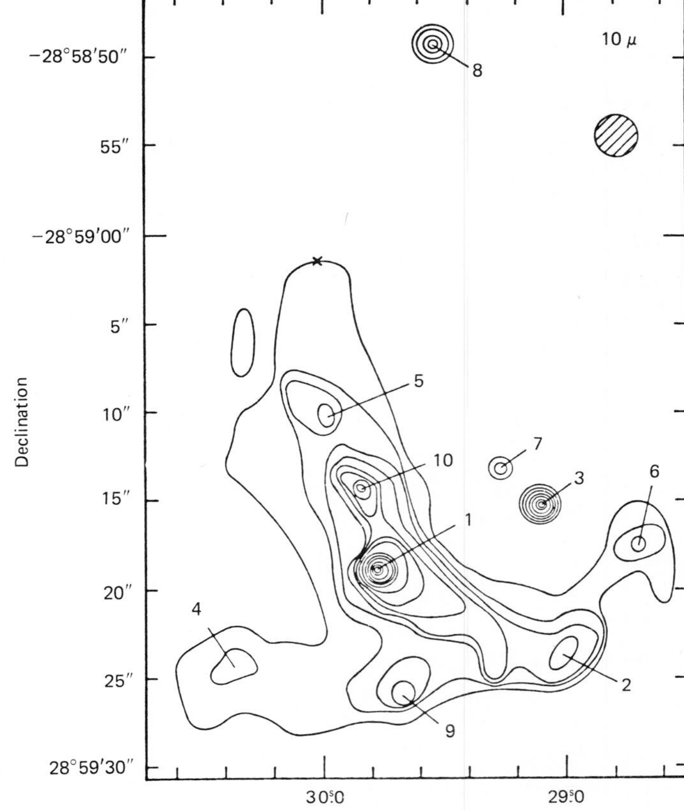


**Figure 26.27** Galactic center based on IR observations at 10 microns. At this wavelength, there should be little radiation from stars, and most of the radiation is thought to come from dust (G. Neugebauer, California Institute of Technology, and *Astrophysical Journal*).

Observing at IR wavelengths up to about 20 microns, we can see several separate and very luminous objects (each brighter the $10^6\ L_\odot$). At longer wavelengths (around 100 microns), the radiation seems to come from an extended region of gas and dust probably heated by stars within. At yet longer (radio) wavelengths, we find seaprate sources as well as broad emission regions. Both the polarization and the shape of the radio spectrum identify synchrotron radiation and thus the presence of very-high-speed electrons in a magnetic field of Sgr A East. Hydrogen (HI) has been surveyed by its 21 cm line. The distribution of the CO molecules has been mapped by their characteristic 2.6 mm line. This gas appears to be moving away from the center, almost as though from the site of an explosion.

When we add together the energy in all parts of the spectrum, we find that the central region (out to about 150 pc) is emitting more than $10^8\ L_\odot$, mostly at IR wavelengths. This is close to $\frac{1}{1000}$ of the radiation from the entire galaxy. We do not yet have any satisfactory explanation for the source of the large amount of energy emitted from such a small region. Models have been proposed that require massive black holes or the annihilation of matter and antimatter, but the question is far from being settled.

## 26.8 SUMMARY

Optical astronomy and the study of the Cepheid variables showed the size of the galaxy. There is a spiral structure whose full extent we do not yet know. Different stellar populations are found inside the spiral arms and outside. We have discovered extensive molecular clouds and a pervasive interstellar medium with gas and dust. The energy content in cosmic rays and magnetic fields is comparable to the energy in starlight. In the galactic nucleus there are numerous IR sources and an unexplained powerhouse at the center.

## CHAPTER REVIEW

First attempt to map the Milky Way Galaxy was that of William Herschel, who counted the numbers of stars in different directions. Separate existence of other galaxies was not recognized.

**General features:** most stars concentrated in a disc, about 30 kpc ($10^{23}$ cm) diameter.

- — typical thickness of disc 1 to 2 kpc.
- — surrounding halo with few stars and little gas.
- — stars in disc are mostly in spiral arms with whole structure rotating around galactic center with angular speed that varies with distance.
- — sun about 8.7 kpc from galactic center.

**Stellar population:** Population I stars—in spiral arms, young, very luminous, containing about 1 percent heavy elements.

- — population II stars: between spiral arms, out of plane of disc or in galactic nucleus—older stars, 100 times less heavy elements composition than Population I stars.

**Variable stars:** Cepheids and RR Lyrae—show regular variations in brightness.

- — light curves and periods differ for different types of stars, with periods between about a day and 100 days.
- — Cepheids and RR Lyrae variables are important for distance determination, since their absolute magnitudes are known.

**Nebulae:** emission and reflection nebulae, illuminated by nearby bright stars.

**Spiral Arms:** mapped from radial velocity measurements of stars, from positions and movements of hydrogen clouds mapped by 21 cm radio line, and by comparison with other spiral galaxies.

- — galaxy has three or four spiral arms (number not yet firmly established).

**Mass of the galaxy:** computed from measured rotation and application of Kepler's Law.

- — mass of galaxy is about $4 \times 10^{11}$ solar masses, but rotation curve now suggests much higher mass. Problem: What is the "missing mass"?

**Nucleus:** not visible at optical wavelengths, but surveyed at radio and IR wavelengths.

- — contains a large number of highly luminous objects; ultimate power source not identified.

## NEW TERMS

**Cepheid variable**
**disc**
**emission nebula**
**extinction**
**galaxy**
**halo**
**Interstellar medium (ISM)**
**nebula**
**Population I, II stars**
**reflection nebula**
**RR Lyrae variable**
**spiral arms**
**spiral density wave**
**star gauges**
**Strömgren sphere**
**Type I, II Cepheids**
**W Virginis variable**

## PROBLEMS

**1.** What are two mechanisms that have been suggested for the initiation of star formation?

**2.** What are the major differences between stars in Populations I and II?

**3.** Calculate the average density of matter in our galaxy using a simplified model. Take the mass as $4 \times 10^{11}\,M_{\odot}$. For the volume assume a cylinder of diameter $10^{23}$ cm, and height 2 kpc $= 6 \times 10^{21}$ cm.

**4.** The disc of the galaxy has a volume of about $10^{68}$ $cm^3$. If the mass of the galaxy is about $4 \times 10^{11}$ $M_\odot$ and the average star has about the same mass as the Sun, calculate the average distance between stars.

**5.** How do we know that the Sun is far from the center of our galaxy?

**6.** What is the observational evidence for the age of the galaxy?

**7.** How can you use the H-R diagram to determine the age of a star cluster?

**8.** What can you deduce about the relative ages of two clusters, if cluster A has no *O* stars on its main sequence but cluster B has types *O*, *B*, and *A*?

**9.** How did the discovery of Cepheids in the Magellanic Clouds help to determine their distances?

**10.** There are Cepheids variables in our own galaxy as well as in other galaxies. Why was the period-luminosity relation first recognized for Cepheids in the Magellanic Clouds?

**11.** What is the absolute magnitude of a Type I Cepheid with a period of 10 days? What is the period of a Type II Cepheid of the same absolute magnitude?

**12.** Why can Cepheids be used for distance measurements to much greater distances than for RR Lyrae stars? Roughly how much farther can the Cepheids be used (for the same limiting magnitude and ignoring any effects of interstellar absorption)?

**13.** How do these nebulae differ from one another?

a. emission nebulae
b. reflection nebulae
c. planetary nebulae

**14.** About $2 \times 10^{-11}$ erg of energy are needed to ionize a hydrogen atom, that is, to remove its electron. The HII region around a hot star contains ionized hydrogen. What wavelength photons are needed to ionize hydrogen? Use Wien's Law to calculate the temperature of the stellar surface that will radiate photons of this wavelength.

**15.** What is the evidence for our galaxy having a spiral structure?

**16.** How can the mass of the galaxy be estimated?

**17.** The Sun is 8.7 kpc from the center of the galaxy and appears to be moving in a circular orbit with a speed of 250 km/sec.

a. Calculate how long it will take to complete one orbit around the galaxy.
b. Using Kepler's Law calculate the mass of the galaxy within the circular orbit.

**18.** Suppose the Sun were located at the outer edge of the galaxy. How would the sky appear at night?

**19.** What wavelengths are used for studies of the galactic center?

**20.** What major discoveries have come from observations of the galactic center?

**21.** Stars are not as constant as used to be thought. Stars whose brightness varies include Cepheids, novae, supernovae, and binaries. Use the index to locate descriptions of these stars, and summarize the major differences. Which types can be used as distance indicators for clusters or galaxies?

CHAPTER 27

# The Interstellar Medium

The interstellar medium (ISM) might easily be thought of as only the empty space between stars. When investigating the ISM we encounter problems that are quite different from those confronting us when we examine the stars and planets. How can we measure the properties of regions which have fewer atoms in each cubic centimeter than the best vacuum we can produce in the laboratory? Look across the room in which you are sitting. You can see a window or door, perhaps a picture on the wall. How can you tell that there is any air? Perhaps you can feel a breeze, even though you cannot see the air moving. There are too few air molecules along your line of sight to produce an easily measured optical effect. How about any dust? Perhaps you can see dust specks dancing in a strong beam of sunlight. The dust particles are not luminous, but you can *see* them by the sunlight they scatter.

When we look through interstellar distances, there are enough gas atoms, molecules, and dust particles along our lines of sight to produce measurable effects on the spectra of the stars. In some cases, we can even detect radiation emitted from the gas and dust clouds themselves. Because the methods differ from those we have relied on so far, it will be useful to review them briefly before describing the results that have been obtained. As we shall see, the ISM is far more than just a good vacuum.

## 27.1 OBSERVATIONAL METHODS

Most of the gas and dust clouds are too cold to radiate strongly at visible wavelengths, although they can shine in reflected light, usually when close enough to hot and luminous *O* and *B* stars. Thermal emission from the clouds can be detected at radio, millimeter, and IR wavelengths. Some of the

**Figure 27.1** Horsehead Nebula in Orion photographed in red light. Located about half a degree south of Zeta Ori, the eastern-most star in the belt of Orion. This nebula is difficult to see by eye, even with a telescope, and long photographic time exposures are needed. This dense cloud stands out against a background largely illuminated by Zeta Ori, an *O* star of 1.79 magnitudes. Distance about 300 to 400 pc (Palomar Observatory photograph).

**Figure 27.2** Cone Nebula in Monoceros photographed in red light. The stars are part of the open cluster NGC 2264. The brightest star in this view, just above the top of the cone has magnitude 7.14, and the star to the right has magnitude 9.10. The cluster is about 740 pc distant from the Earth (Palomar Observatory photograph).

most spectacular astronomical photographs show giant clouds of interstellar gas and dust, sometimes bright in their reflections or emission, sometimes dark with their outlines silhouetted against a brighter background (Figs. 27.1 and 27.2).

Dark nebulae obscure distant stars whose light would otherwise come to us. Sometimes the obscuration is total; little or no light emerges from the cloud so that a part of the sky seems devoid of stars. Sometimes the obscuration is selective. Just as atoms emit at characteristic wavelengths, so they absorb at those same wavelengths. As a result, the continuous spectrum from a distant star will be punctuated by dark absorption lines. We need to distinguish between the absorption lines that come from interstellar material and those caused by the star's own atmosphere.

This effect was first noted in 1904 by J. F. Hartmann, who correctly attributed a dark absorption line in the spectrum of the star δ Orionis to the element calcium. Soon other lines from Na, Ti, K, and Fe were identified. None of these atoms is really plentiful, and it might seem strange that the very abundant hydrogen should not have been detected. But the interstellar gas is cold so that the atoms are almost all in their lowest energy levels. For cold hydrogen the chief absorption lines lie in the UV and could not be seen until observations were made above the Earth's atmosphere in the 1970s. In contrast, heavier atoms have many electrons; those electrons in outer orbits are easily excited. The resulting absorption lines are in the visible and can be seen even though the abundances of heavy atoms are much lower than that of hydrogen.

There are, of course, regions of the ISM whose emission can be detected; characteristic lines then permit identification of the radiating particles. This has been especially fruitful in identifying molecules through their millimeter wavelength emissions.

The ISM leaves its imprints in other ways as well. In a vacuum all wavelengths (radio, IR, visible, etc.) travel with the same speed, but in all transparent materials (whether glass, plastics, or interstellar gas) the speed depends slightly on the wavelength. This has an interesting result. If, for example, a flare occurs on a distant star, the signals at different wavelengths will reach us at slightly different times. The delay in arrival times can be measured, and we can use this **dispersion** to deduce the average density of electrons along the line of sight. Further, interstellar electrons together with any interstellar magnetic field can affect the polarization of the radiation. Ordinary light such as that from an incandescent lamp is unpolarized; if it is viewed through Polaroid glasses, there will be no change in brightness as the glasses are rotated. Some stars (pulsars) emit polarized radiation; the signal strength at Earth depends on the direction of the radio antenna or Polaroid filter. Interstellar electrons and magnetic field combine to rotate the plane of polarization with the degree of rotation also depending on the wavelength. Measurement of the direction of polarization at different wavelengths thus allows us to estimate the strength of the interstellar magnetic field. (Note: very-high-speed electrons will produce synchrotron radiation. In the interstellar space, there are many free electrons that came from ionized atoms and have very low speeds. These slow electrons do not produce synchrotron radiation, and it is these low-speed electrons and not their high-speed relatives that are responsible for the rotation of the polarization.)

Finally, we can probe the ISM through investigations of cosmic rays at the Earth. Cosmic rays are high-energy protons, electrons, and atomic nuclei that travel throughout the galaxy. We can identify them, measure their abundances and energies and note their arrival directions. Cosmic rays provide the only matter that we can detect whose origin lies outside the solar system.

## 27.2 DUST AND INTERSTELLAR REDDENING

Only a minute fraction of the radiation emitted by a star will initially be aimed in the right direction for us to receive it. Moreover, as the starlight traverses the ISM, some of the light will be scattered or absorbed by interstellar dust. The combination of scattering and absorption is **extinction**, a reduction in the amount of light that we finally detect. The degree of extinction depends on the wavelength of the light and on the size and nature of the dust particles. In general, short-wavelength light suffers more extinction than does long wavelength so that stars appear **reddened**. This will show up when we measure the $B$ and $V$ magnitudes to calculate the color index, $B - V$.

Suppose we were examining a star for which $B = 0.76$, $V = 0.35$, and there was no extinction. Then the color index would be $B - V = 0.76 - 0.35 = 0.41$. Now suppose we examine an identical star located in a dusty region. The interstellar extinction might reduce the $B$ magnitude by 0.3 (in this hypothetical example) to make $B' = 1.06$. (Remember that larger numbers denote fainter stars.) The light in the longer wavelength $V$ interval will not be dimmed as much, perhaps 0.2 magnitudes. (Fig. 27.3). The $V$ magnitude will then appear as $V' = 0.55$, and the new color index will be $B' -$

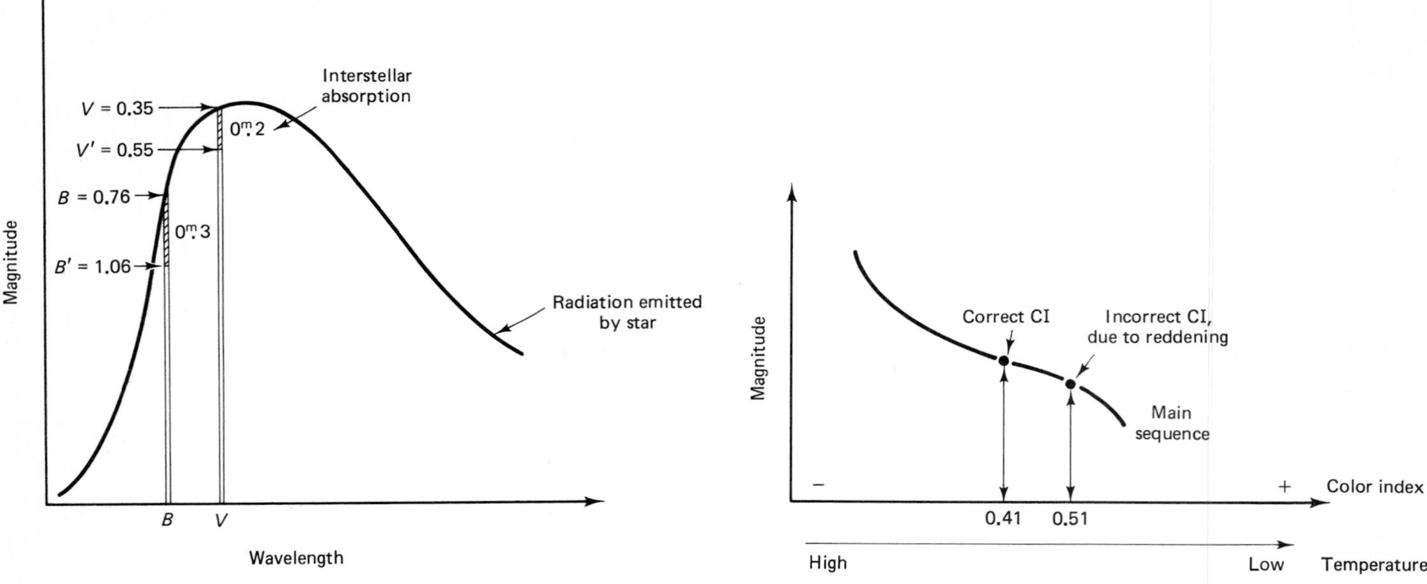


**Figure 27.3** Color excess. Short (blue) wavelengths are scattered more than long (red) wavelengths. As a result, when light from a star passes through interstellar dust, the blue magnitude ($B$) will be changed by a larger amount than the magnitude corresponding to a longer wavelength such as $V$. The color index, CI = $B - V$, will be increased and the star appears to be reddened. This can lead to errors in the calculation of the star's distance and its position on the H-R diagram.

$V' = 1.06 - 0.55 = 0.51$. A larger color index indicates a redder spectrum. Unless we knew that this star's light had been reddened and we corrected for it, we would use the value 0.51 to position the star incorrectly on the H-R diagram and we would also calculate too low a surface temperature and too great a distance (see Box 27.1).

One method of measuring interstellar extinction requires observation of a star of known spectral classification identified by its spectral lines. The star's intrinsic color index will thus be known. The measured color index will be different because of extinction; the difference between the measured and intrinsic values of the color index is the **color excess**, and this can be used to calculate a correction factor for the distance measurement.

How do we know that the extinction is produced by dust grains? The theory of scattering and absorption of light by small particles has been worked out in great detail. Atoms, molecules, and electrons also scatter and absorb light, but their effects differ markedly in wavelength dependence from those due to dust. By *dust*, we mean tiny particles, typically around $10^{-5}$ cm in diameter comparable in size to the wavelength of light. (Being so small, they will not affect the long-wavelength radio and IR, which is why we can see for greater distances with those long wavelengths.) What are the dust grains made of? We are not completely sure. Strong extinction in the UV, especially for wavelengths around 2200 Å, suggests graphite grains. Infrared evidence on circumstellar dust shows a distinctive absorption for wavelengths near $10^{-3}$ cm (10 microns). A very similar absorption feature has been seen in spectra from meteorite grains. This has been attributed to silicates, molecules which include the SiO group. Interstellar silicate grains probably also include Mg, Al, and Fe. Another IR absorption feature near 3 microns can probably be identified with ice. From these very different observations, a composite picture of the interstellar grains emerges: a core of silicates, perhaps with a coating of ice (Fig. 27.4). The formation of grains probably takes place in the outer and cooler atmospheres of stars that are losing mass where the densities are thought to be high enough for the silicates or other materials to condense.

**Figure 27.4** Schematic drawing of an interstellar dust grain. The presence of silicate material is inferred from the observation of an absorption feature near 10 microns in many spectra. This wavelength is characteristic of the chemical bond between silicon and oxygen. Another interstellar absorption feature often seen is at 3.1 microns characteristic of ice. An interstellar grain might thus consist of a silicate core surrounded by a thin layer of ice. The surface is probably very irregular, and the typical size range is between 0.1 and 1 micron ($10^{-5}$ to $10^{-4}$ cm).

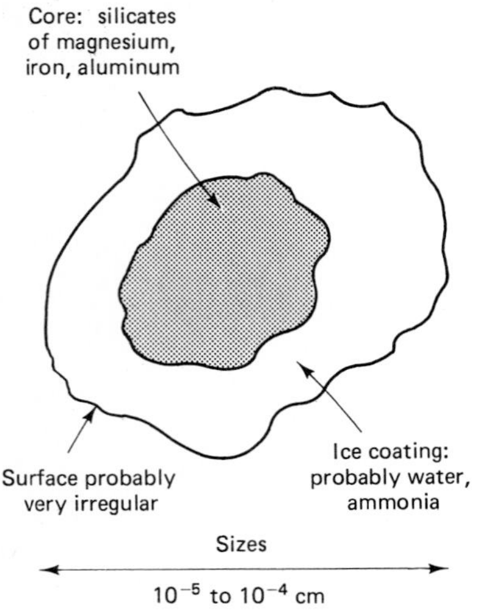

Interstellar grains are thought to play an important role in the formation of molecules. Generally, the density of the ISM is so low that simple collisions between atoms in space will be too infrequent for molecule production. Instead it is thought that some interstellar atoms will be absorbed when they collide with the grains. Once on a grain surface, atoms can combine to produce molecules some of which may remain on the grain, while others escape into the ISM where they are detected by their characteristic spectral lines.

We still know very little about the behavior of grains and their surfaces at very low temperatures and pressures. Laboratory studies can provide important clues. So, for example, J. Mayo Greenberg in Leiden has deposited thin films of $NH_3$ and CO on very cold surfaces and irradiated them with UV for several hours. Infrared spectra have shown that more complex molecules are formed: HCO, $H_2CO$, HNCO, and $HCONH_2$, for example.

The interstellar grains are almost certainly not spherical, as is often assumed to make the calculations easier. Complicated shapes and the presence of surface contaminants make it difficult to calculate just how a grain will behave when it absorbs starlight or atoms.

The extinction varies across the sky, but a good average value in the visible is one magnitude per kpc in the galactic plane. A star at 1 kpc will appear one magnitude fainter than it really should because of extinction caused by dust. Unless we correct for this, we will tend to overestimate distances, for we will think that the dimming is due to greater distance rather than being caused by dust (Box 27.1).

**BOX 27.1**
*Interstellar Extinction*

Suppose a star is found to have an apparent magnitude of 6 and is known to be a main sequence star with absolute magnitude $M = -5.5$. Its distance can then be calculated

$$m - M = 5 \log r - 5$$

$$\log r = \frac{m - M + 5}{5}$$

and we obtain $r = 1995$ pc.

If now extinction of $A$ magnitudes occurs, then a correction term must be included

$$m - M = 5 \log r - 5 + A$$

and to solve for $r$ use

$$\log r = \frac{m - M + 5 - A}{5}$$

With the same numbers as before, and taking $A = 2$ magnitudes as an example, we obtain $r = 794$ pc.

Thus, if we do not include the extinction correction, we will overestimate the true distance to the star. Another way of looking at it is that a star of given absolute magnitude will appear fainter because of the extinction; failure to recognize this causes us to think the faintness is due to greater distance, whereas it should really be attributed to extinction. The estimation of distances should, therefore, include correction for extinction.

Because the dust produces extinction that depends strongly on wavelength, there can be very noticeable changes in stellar spectra, but the actual density of dust is very low. Typically, there is 100 times as much mass in gas as in the dust. Dust grains are about 100 m apart, and we detect the dust only because of the very long sight lines.

## 27.3 GAS: HYDROGEN

Hydrogen is the most abundant chemical element, and it comprises over 90 percent of the interstellar gas. Observations in different wavelength windows are needed in order to locate the different forms in which the hydrogen exists: molecular ($H_2$), atomic (HI), and ionized (HII).

When we described the spectral lines that characterized each element, we were concerned only with the switching of electrons between orbits of different energy; this effect dominates the elemental line spectra. Out in the cold regions of interstellar space and away from hot stars, almost all hydrogen atoms will have their electrons in the lowest energy orbit allowed by the Bohr theory. The only emission feature will then be the 21 cm line in the radio region (chap. 8). That line comes from hydrogen atoms that switch between two close energy levels that involve the spins of the proton and electron in the atom.

The strength of the emitted 21 cm signal depends on the number of hydrogen atoms in the upper of these two closely spaced energy levels. What determines that number, that is, how many hydrogen atoms have extra energy that can be radiated at 21 cm? The excitation from lower to higher levels comes from collisions, but collisions between atoms in interstellar space are very infrequent so that few atoms will have their spins parallel. Worse still, once a hydrogen atom gains the extra energy and has the spins parallel, it can remain in that state for 11 million years on average before switching back to the lower level and emitting its photon. This time is so long that most of the excited atoms return to their lowest energy level through further collisions (with no radiation resulting). The overall likelihood of an atom first gaining the energy and then radiating before collision is thus very, very small. The only reason we can detect the 21 cm line is that our astronomical fields of view cover such large volumes that sufficient hydrogen atoms are included.

The idea of looking for the 21 cm line came from H. C. van de Hulst. During WWII there were few opportunities for astronomy in the Netherlands under German occupation, and van de Hulst's research was confined to theory. After the war, with the improvement in radio techniques, it still took until 1951 before E. M. Purcell and H. Ewen at Harvard detected the 21 cm line (Fig. 27.5). Since that time, it has been a basic tool of radio astronomy.

A hydrogen cloud at rest will emit the 21 cm line. If the cloud moves, the line will show a Doppler shift, proportional to the speed of the cloud. Movements of the atoms within the cloud depend on the cloud's temperature and show up as a broadening of the line. We can thus probe both the temperature within the cloud and the bulk speed. A typical cloud temperature is 100 K. Surveys of hydrogen in the galaxy have shown how the gas is distributed along the spiral arms (chap. 26). This information is complementary to that based on stellar observations and helps us to unravel the pattern of galactic rotation.

At the low temperatures of interstellar clouds, one would expect to find some hydrogen in molecular form. This proved difficult to observe. No spectral lines from $H_2$ fall in the visible range. The first observations came during a rocket flight in 1970 when George R. Carruthers of the Naval Research Laboratory found $H_2$ absorption lines in the UV spectrum of $\epsilon$ Persei. A typical rocket flight produces only about 5 min of useful data. Far more data were obtained 2 years later through an experiment on the Orbiting Astronomical Observatory (OAO), Copernicus (Fig. 27.6). Its reflecting telescope with a 90 cm mirror could be aimed by ground control at selected stars for far longer than in any rocket flight. In addition to detecting $H_2$, the Copernicus experiment also found traces of HD, the molecule formed by combining a hydrogen atom with one of its isotopes, deuterium. Deuterium is rare (about 1 D atom for every $10^4$ H atoms), and its abundance in the

**Figure 27.5** H. I. Ewen with the radio horn antenna with which he and E. M. Purcell made the discovery of the 21 cm radiation from the Milky Way (NRAO).

**Figure 27.6** Preparation of the Orbiting Solar Observatory (OSO)-3 which carried instruments with which George Carruthers made the observations of molecular hydrogen (NASA).

galaxy provides an important indicator for testing theories of galactic formation and evolution, as we shall see in Chapter 31.

It is hard to estimate the total mass of hydrogen in the ISM, but it now seems as though there may be comparable quantities in atomic and molecular forms, each amounting to about 1 percent of the mass of the galaxy. Densities of hydrogen clouds vary, the highest observed being no more the $10^6$ atom/cm$^3$, for example, in Orion.

Conspicuous regions of ionized hydrogen have been detected near hot *O* and *B* stars. The strong UV radiation from these hot stars ionizes the hydrogen out to about 30 LY from a star with a surface temperature of 40,000 K. The ionization is maintained by the absorption of the stellar UV photons that have wavelengths shorter than 912 Å. The resulting plasma is very hot, typically $10^4$ K. Some electrons and protons do recombine so that hydrogen atoms can again exist briefly until being torn apart by the arrival of the next UV photon. In the recombination process, an electron is captured into an orbit around a proton. The electron may then change orbits, skipping quickly to lower energy levels and emit a photon with each jump to a new orbit. The stellar radiation at wavelengths longer than 912 Å is not absorbed in the ionized hydrogen, and so this relatively long-wavelength radiation escapes together with the photons emitted in hydrogen recombination. The HII region is said to **fluoresce**, through the process of converting the short-wavelength UV radiation to longer wavelengths, and the region appears bright, especially in the red Balmer H$\alpha$ line.

## 27.4 GAS: OTHER ATOMS AND MOLECULES

A component of the interstellar gas at temperatures even higher than the circumstellar HII regions was discovered from the Copernicus satellite. Observations at UV wavelengths near $10^{-5}$ cm identified a pair of lines from OVI (Oxygen 6). For oxygen atoms to lose five of their eight electrons in collisions requires temperatures close to 1 million K so that the collisions between atoms will be violent enough. One explanation for such high temperatures proposes that the energy comes from supernova explosions, whose debris carries an expanding wave (a *shock front*) that rapidly compresses the interstellar gas as it passes through. After a long enough time and many supernovae, the ISM will have a tangle of interwoven shock fronts. We know the approximate rate of supernova explosions, and there is much evidence for shock fronts that travel at speeds above 20 km/sec so that the model is attractive but not yet proven.

Hydrogen molecules have already been mentioned, but long before their discovery other interstellar molecules had been found. Between 1937 and 1941, CN, CH, and $CH^+$ had been identified by optical astronomers using the Mt. Wilson 100 in. reflector (Fig. 27.7). The presence of these molecules in the ISM had shown up in absorption in some stellar spectra. Other absorption lines were also noted in the visible spectra, but many remain unidentified.

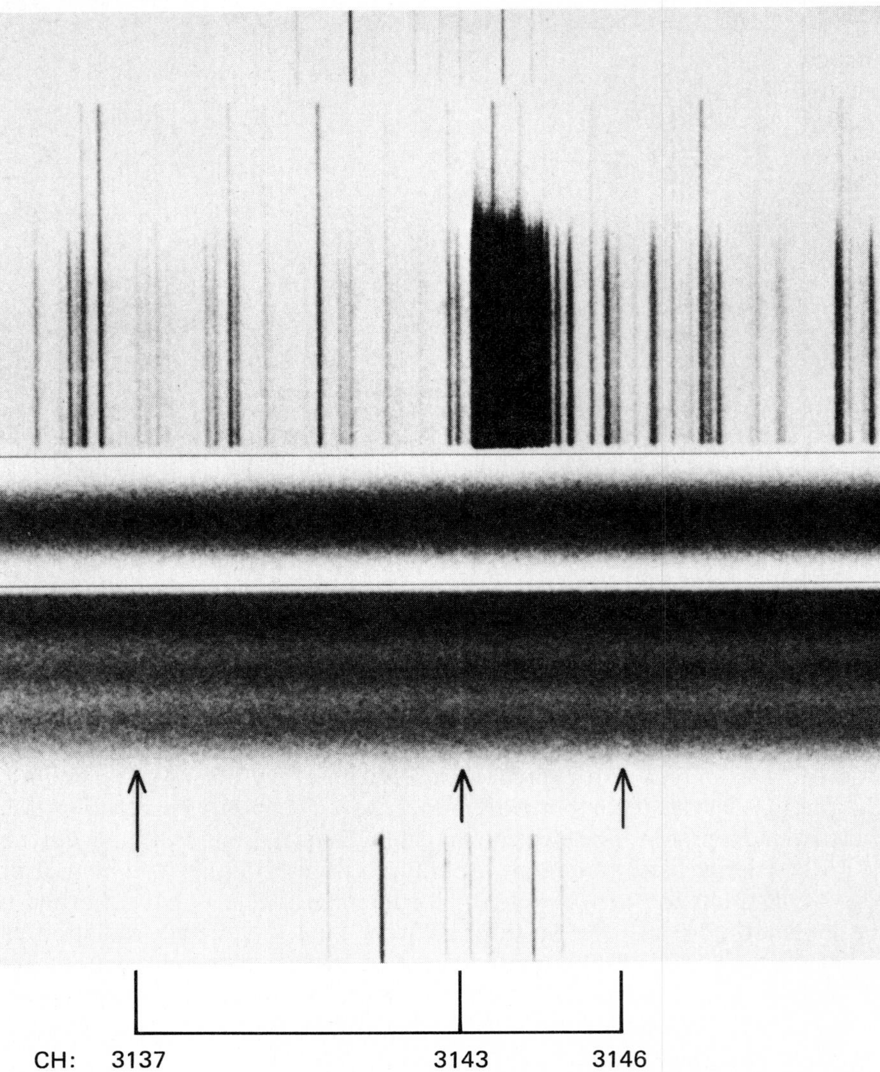


**Figure 27.7** Spectrum of ϵ Ophiuchi, displaying interstellar absorption lines of CH. Laboratory calibration spectra of CH and OH are shown at top. In this print, the CH absorption lines appear as very thin white lines in the continuous stellar spectrum (Lick Observatory).

**Figure 27.8** A 36 ft radio telescope used for observations at millimeter wavelengths primarily for the study of interstellar molecules (NRAO).

Next to hydrogen and helium, the most abundant elements are carbon, nitrogen, and oxygen so that combinations of these elements with hydrogen are to be expected. (Helium does not generally combine to form compounds.) In 1963 the OH (hydroxyl) radical was detected by radio astronomers at 18 cm wavelength. Another 5 years passed before the next discoveries were made: ammonia ($NH_3$) discovered at 1.3 cm, and water ($H_2O$) at 1.4 cm. Thereafter, with improved electronic techniques and the use of radio telescopes specifically for molecular searches (Fig. 27.8), a flood of discoveries followed. Some of the molecular lines are seen in absorption, some in emission. Table 27.1 lists a representative selection of the molecules discovered.

*Table 27.1* SELECTION OF INTERSTELLAR MOLECULES OBSERVED

| Number of Atoms in Molecule | Name | Chemical Formula |
|---|---|---|
| 2 | hydrogen | $H_2$ |
| | hydrogen deuteride | HD |
| | hydroxyl radical | OH |
| | methylidine radical | CH |
| | cyanogen | CN |
| | carbon monoxide | CO |
| 3 | water | $H_2O$ |
| | hydrogen sulfide | $H_2S$ |
| | sulfur dioxide | $SO_2$ |
| | hydrogen cyanide | HCN |
| 4 | ammonia | $NH_3$ |
| | formaldehyde | $H_2CO$ |
| 5 | formic acid | HCOOH |
| | cyanamide | $H_2NCN$ |
| 6 | methyl alcohol | $CH_3OH$ |
| | formamide | $HCONH_2$ |
| 7 | methylacetylene | $CH_3C_2H$ |
| | vinyl cyanide | $H_2CCHCN$ |
| 8 | methyl formate | $HCOOCH_3$ |
| | methyl cyanoacetylene | $CH_3C_3N$ |
| 9 | ethyl alcohol | $CH_3CH_2OH$ |
| | dimethyl ether | $(CH_3)_2O$ |

Most of the molecules appear to be concentrated in giant molecular clouds (GMC), where the density is typically $10^3/cm^3$ in the outer regions, increasing to probably $10^6/cm^3$ in the middle. These molecular clouds, mostly hydrogen, may be 10 LY across, each with a mass of molecular gas $10^5$ to $3 \times 10^6$ $M_{\odot}$. Most of the GMC are located in a ring of radius 4 to 8 kpc around the galactic center. One particularly notable cloud is the Orion molecular cloud near the Orion Nebula whose greenish glow can be seen by the naked eye on very clear winter nights. The density of dust is so high that molecules in the interior of that cloud are well shielded from the intense UV radiation from nearby hot stars. The Orion molecular cloud has proven a celestial gold mine for molecular studies.

The surprising aspect of this new field of molecular radio astronomy is not that molecules should have been found but rather the complexity of many of the molecules. One can imagine one carbon atom and one hydrogen atom combining, whether in space or on the surface of a dust grain, but the chances of successive attachments that will ultimately form molecules of eight or nine atoms such as methyl cyanoacetylene ($CH_3C_3N$) or ethyl cyanide ($CH_3CH_2CN$) seem vanishingly small.

## 27.5 COSMIC RAYS

Cosmic rays are not *rays* like light or any other electromagnetic radiation. They are very-high-speed particles, mostly protons, traveling at speeds close to that of light. (The name *cosmic rays* was given before their nature was established. It would have been more accurate to term them *cosmic particles*.) Although some cosmic rays come from the Sun, the majority have their origins outside the solar system but still within our galaxy and are usually termed galactic cosmic rays.

About 89 percent of all galactic cosmic rays are hydrogen nuclei (protons); 9 percent are helium nuclei (alpha particles); close to 1 percent are electrons, and the remaining 1 percent are heavier atomic nuclei completely stripped of all of their electrons. All cosmic rays carry electric charges. They cannot penetrate through the Earth's atmosphere; at sea level, we detect secondary products of their collisions in the atmosphere. To detect the primary cosmic rays themselves, we must use balloons or satellites.

In cosmic-ray research, we encounter some experimental problems that are quite unlike those we find in optical or radio astronomy. With enough interstellar gas or dust, there will be measurable absorption or emission signals, but we do not take note of each gas molecule, each dust particle or each photon. The protons and heavier cosmic-ray particles, however, can be detected only when they pass through special detectors, and we then literally count the cosmic rays one by one. Because all of the cosmic-ray particles are electrically charged, they produce electrical signals in the detectors. From our measurements, we can determine for each particle its electric charge and this serves to identify it: Alpha particles each carry two units of charge, carbon has six, iron has twenty-six and lead has eighty-two. In addition, we can measure the speed of each particle.

About 1000 cosmic-ray protons arrive on a square meter each second; the arrival rate of iron nuclei is only about $0.2/m^2$ sec, and the heavier particles are far less frequent. The best data on the heaviest particles have come from experiments on satellites where the largest experiments have used two detectors on HEAO-3. During that satellite's $2\frac{1}{2}$ year life, about 10 million cosmic rays were detected and identified.

One of the prime tasks of cosmic-ray research has been the comparison of the relative abundances of the different particles, listed in Table 27.2 with the general abundances in the solar system (as deduced from meteorites, the Earth, and Moon, and the Sun). Some striking similarities and differences

## *Table 27.2* COSMIC RAY ABUNDANCES[a]

| Elements | Atomic Numbers (Z) | Cosmic Ray Abundance | Solar System Abundance |
|---|---|---|---|
| hydrogen | 1 | $1 \times 10^{12}$ | $1 \times 10^{12}$ |
| helium | 2 | $1 \times 10^{11}$ | $6.7 \times 10^{10}$ |
| lithium, beryllium, boron | 3–5 | $1.4 \times 10^{9}$ | $2.6 \times 10^{3}$ |
| carbon, nitrogen, oxygen | 6–8 | $5.8 \times 10^{9}$ | $3.1 \times 10^{7}$ |
| florine to manganese | 9–25 | $1.8 \times 10^{9}$ | $4.0 \times 10^{7}$ |
| iron | 26 | $2.8 \times 10^{8}$ | $3.4 \times 10^{7}$ |
| beyond iron | >26 | $1.4 \times 10^{7}$ | $2.0 \times 10^{6}$ |

[a] Cosmic ray data represent averages, since precise numbers depend on particle energies.

can be seen. There are relatively more iron (Fe) nuclei in the cosmic rays than in the solar system. We think that many of the cosmic rays have come from highly evolved stars, perhaps injected into the ISM from supernova explosions. The lithium, beryllium, and boron are produced when the more plentiful carbon, nitrogen, and oxygen collide with interstellar atoms and are fragmented into lighter nuclei. Similar processes occur in the fragmentation of heavier particles (Fig. 27.9).

We can use these abundances to calculate how many collisions are needed to produce the observed numbers of Li, Be, and B; these three elements are rare in stars, since they are very rapidly consumed in the thermonuclear reactions in stellar cores. Present calculations show that the cosmic rays have been traveling on the average for several millions years in the galaxy. During this time, sufficient C, N, and O particles will collide with the interstellar hydrogen to produce the observed abundance of the Li, Be, and B nuclei.

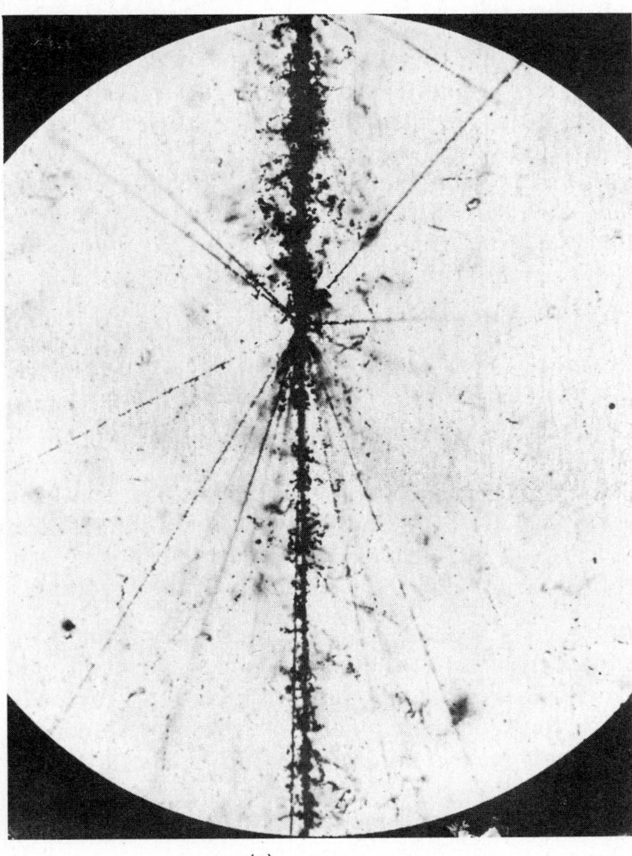

(a)

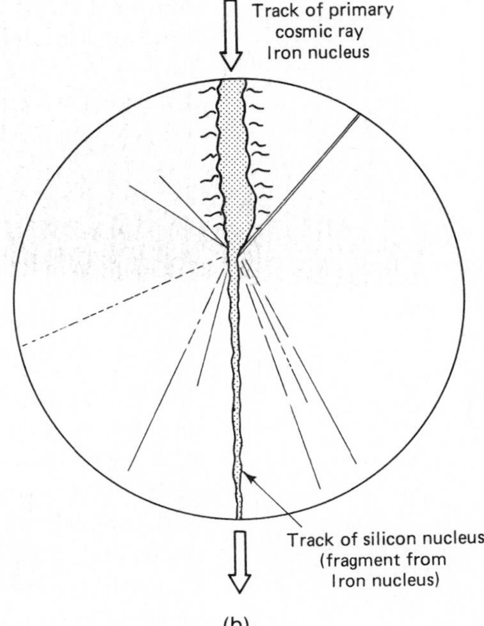


(b)

**Figure 27.9** Collision of cosmic ray iron (Fe) nucleus in a photographic emulsion. A silicon nucleus and numerous lighter particles emerge from this collision. It is this type of process that occurs as the cosmic rays travel through the interstellar regions and produce the otherwise rare nuclei observed.

Unlike electromagnetic radiation, which travels in straight lines across the galaxy, the cosmic rays are easily deflected by magnetic fields so that their arrival directions at Earth do not give us any clues to their points of origin. Indeed, when the arrival directions are noted and allowance made for the effects of the Earth's magnetic field, we find the arrival directions are almost isotropic, that is, they arrive evenly from all directions. There are several reasons for thinking that very few cosmic rays come from other galaxies; almost all must be produced in and confined to our own galaxy.

## 27.6 MAGNETIC FIELDS IN THE GALAXY

The magnetic field at the Earth's surface has a strength of about 1 gauss. This is sufficient to aim a compass needle; much stronger fields are needed to lift heavy objects. For example, electromagnets that produce fields of thousands of gauss are used to lift scrap iron. We now know that there are magnetic fields that extend over large distances in our galaxy. These fields are very weak, typically $10^{-5}$ gauss or a bit less, but they cover such great regions that their total stored energy is comparable to the energy of all starlight photons. There are four main lines of evidence for the interstellar magnetic field.

**Polarization of starlight** Starlight that is scattered by interstellar dust shows a polarization of a few percent (Fig. 27.10). A model proposed by Jesse Greenstein and Leverett Davis of Caltech in the 1950s attributes this effect to the alignment of the dust grains by an interstellar magnetic field. It has been suggested that the dust grains are somewhat elongated and tend to be aligned by a magnetic field, but the alignment is disrupted by impacts upon the dust grains by cosmic rays and gas molecules. Nevertheless, a general alignment will persist, and the light scattered from grains that are aligned will show a slight polarization. Qualitatively, the theory seems reasonable, but it still lacks quantitative confirmation partly because the properties of the dust grains are not well established.

**Synchrotron radiation** As we have mentioned earlier, electrons follow curved paths when traveling in a magnetic field, and they then radiate. The resulting synchrotron radiation can be identified by its characteristic spectrum and also by its polarization. Radio surveys of the galaxy have clearly shown the presence of this synchrotron radiation which is so different from the thermal radiation from stars and HII regions. We can calculate the magnetic field strength and the electron kinetic energies needed to produce the observed synchrotron radio signals. The field strength turns out to be $10^{-5}$ gauss or slightly less, close to that deduced from starlight polarization. The electron kinetic energies turn out to be in the hundreds of millions of electron volts, typical cosmic-ray energies. The synchrotron signals are not confined to the galactic disc: there is evidence for a halo that the electrons seem to fill. This large volume extends far from stars and the obvious sources of particles such as supernovae.

For comparison, we recall that synchrotron radiation has also been seen in supernovae remnants such as the Crab nebula, where the stronger magnetic field (about $10^{-3}$ gauss) and higher electron kinetic energies combine to produce synchrotron radiation in the visible part of the spectrum.

**Cosmic-ray confinement** The cosmic-ray electrons come from two sources. Some must be injected directly where the cosmic-ray protons and nuclei originate, and some are secondary products of the cosmic-ray collisions in the interstellar gas. Between them, enough electrons are produced so that synchrotron radiation is seen in all directions, including those well out of the plane of the galactic disc. It seems as though most cosmic rays are confined to the galactic disc by the magnetic fields, but some electrons do leak out to form a halo above and below the disc.

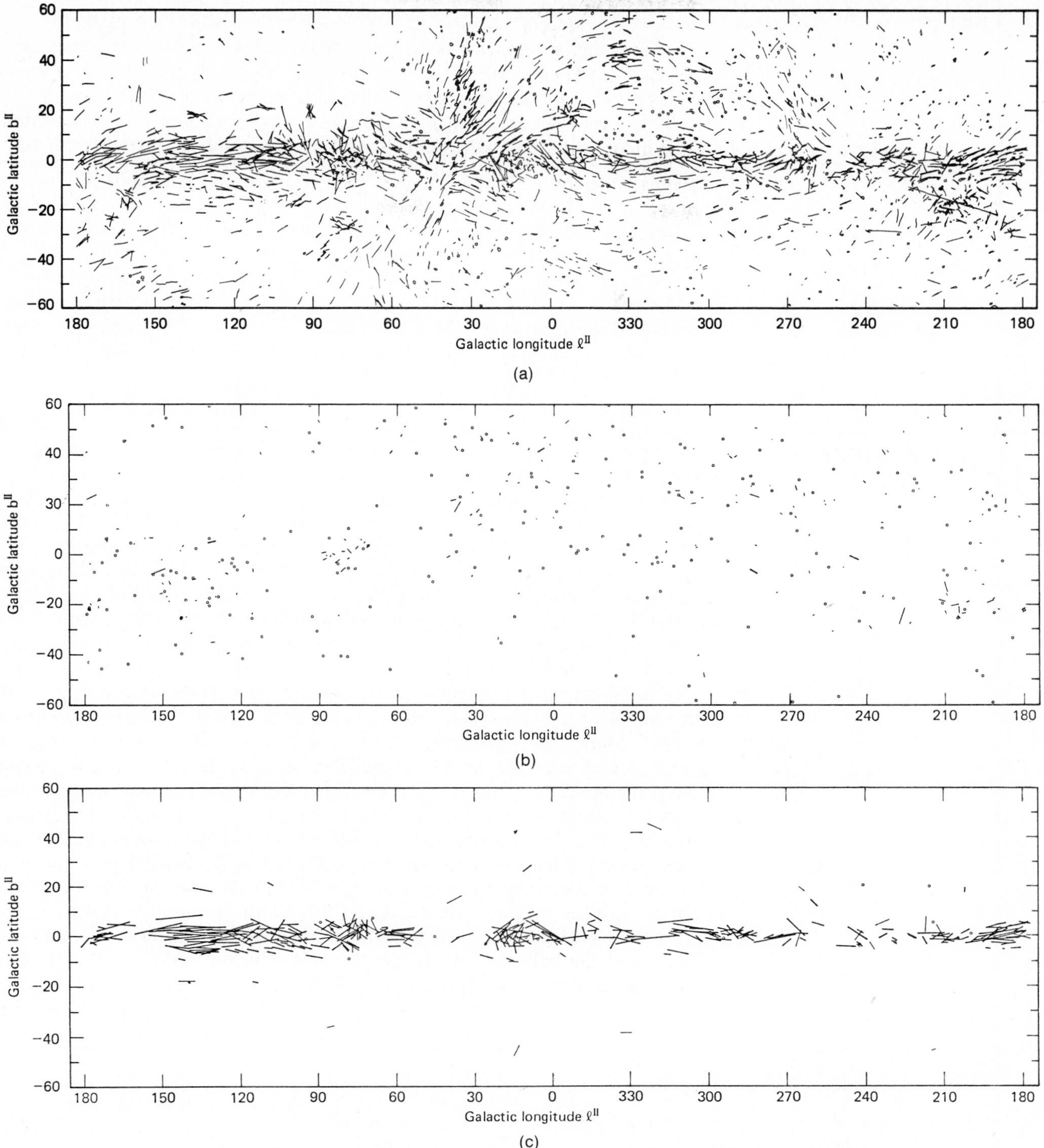


**Figure 27.10** Polarization of starlight, produced by scattering of light from interstellar dust grains. The measured polarization for each star is shown as a line. The length of the line corresponds to the degree of polarization, and the angle of the line indicates the orientation of the polarizing filter for maximum effect.

a) Data for 7000 stars, for all distances. The pattern of the galactic magnetic field can be seen.

b) Data for stars closer than 50 pc from the Sun.

c) Data for stars between 2000 and 4000 pc from the Sun.

The dependence of the scattering on the amount of dust and thus on distance shows very clearly when (b) and (c) are compared. (D. S. Mathewson and V. L. Ford, Monthly Notices of the Royal Astronomical Society, Vol. 74, (1970), pp. 139–182; reprinted with permission.)

Some features of the cosmic rays can be explained if there is a magnetic field that confines most of them to the disc; this does not give us a very accurate value for the strength of the field, but fields of around $3 \times 10^{-6}$ to $10^{-5}$ gauss are consistent with the cosmic-ray confinement. Once again, this is in agreement with the field strengths calculated from the other effects.

**Faraday rotation** The rotation of the polarization by a magnetic field is known as the Faraday effect. Observation of the direction of polarization at different wavelengths for pulsars again leads to values for the interstellar magnetic field of $10^{-5}$ to $10^{-6}$ gauss.

The picture of the galactic magnetic field as shown by different methods is very patchy, but its general features have emerged. Since the total energy content of the magnetic field is comparable to the energy residing in other forms—kinetic energy of gas clouds, starlight, cosmic rays—its influence must be included in any overall model of the behavior of the galaxy.

## 27.7 SUMMARY

The interstellar medium can no longer be considered as simply a region where there happen to be no stars. The first indication of the presence of interstellar matter came from Hartmann's observation of absorption lines in stellar spectra, but it has required the new observational techniques to show the rich variety of the ISM constituents. The composition of the interstellar gas and dust is an indicator of stellar evolution past and of the influence on stellar evolution to come. Evolved stars have enriched the ISM with heavy nuclei; in turn, those heavy nuclei will form part of the next generation of stars and play a major role in the way those stars evolve.

The complex molecules that are now being discovered by radio astronomers may provide clues to the origin of life. These fragile molecules are unable to survive the high temperatures of clouds that are condensing to form stars. The only molecules that will survive are those that are far from the new stars. From these molecules we might be able to find out more about the conditions that allow their formation and in this way we might be guided to those sites where the more complex biological molecules are assembled. The very-high-temperature component of the interstellar gas and the cosmic-ray particles are probably related to stellar evolution as they can be produced in supernova explosions and their expanding shock fronts. Together with the magnetic fields, they store a large amount of energy.

Even though the density of the ISM is very low, the total energy is large, and the behavior of the galaxy is greatly affected by the ISM. The importance of its study is firmly established.

## CHAPTER REVIEW

Absorption lines in stellar spectra gave first indication for presence of interstellar material.

- dispersion depends on wavelength, produced by interstellar electrons.
- rotation of plane of polarization produced by electrons and magnetic field.

**Reddening:** produced by dust, affects short wavelengths more than long.

- reddening changes the color index for a star, can lead to incorrect location on H-R diagram and also to error in distance computed.

**Dust grains:** some graphite, some silicates, perhaps with ice coating.

- interstellar magnetic field produces some alignment of grains, causing polarization of scattered starlight that depends on distance of star from Earth.

**Gas:** mostly hydrogen, with perhaps equal amounts of atomic and molecular forms.

- atomic: detected in 21 cm radio surveys;
- molecular: detected in UV.

**Molecules:** large number of complex molecules detected, using radio methods at millimeter wavelengths.

— formation of molecules thought to take place on surface of dust grains.

**Cosmic rays:** high speed nuclei and electrons, probably originating in supernovae explosions.

— relative abundances are used to determine history and origin of the radiation: cosmic rays travel on average several million years between acceleration and detection at Earth.

**Magnetic fields in the galaxy:** field strength about $10^{-5}$ gauss indicated by several different types of observations: polarization of starlight, synchrotron radiation, cosmic rays, rotation of plane of polarization of radiation from pulsars.

## NEW TERMS

**color excess**
**dispersion**
**Faraday effect**
**fluorescence**
**giant molecular cloud (GMC)**
**reddening**

## PROBLEMS

**1.** If the Sun has so much hydrogen, why was molecular hydrogen not discovered in the solar spectrum?

**2.** Why are calcium absorption lines seen in stellar spectra but not lines from the much more abundant hydrogen?

**3.** How can interstellar dust be detected? What evidence do we have for the composition of this dust?

**4.** What kinds of interstellar matter have been identified, and what is the evidence for each kind?

**5.** How does the presence of interstellar dust affect the spectrum we observe of a distant star? What erroneous conclusions might we draw if we did not correct for this effect?

**6.** A star is observed to have an apparent magnitude $m = 10$ and is known to be the type of star with $M = +5$. Suppose it is determined that there are two magnitudes of absorption due to dust. Calculate the distance to the star without taking the dust into account and then including the correction due to the dust.

**7.** If interstellar dust dims the light from a distant star by three magnitudes, what fraction of the light gets through the dust?

**8.** Why can we see for greater distances using radio and IR wavelengths than we can with visible light?

**9.** What is thought to be the composition of the interstellar dust, and what is the evidence for this?

**10.** Why is the 21 cm hydrogen line only observed from cold regions?

**11.** How is observation of the hydrogen 21 cm line used for helping to map out the structure of the galaxy?

**12.** A radio astronomer is surveying the neutral hydrogen in the galaxy observing in the region of 21 cm wavelength. Two hydrogen clouds are detected at wavelengths of 20.998 cm and 21.001 cm. How would these results be interpreted?

**13.** Sketch the appearance of the graph of rotation speed (km/sec) versus distance from the galactic center for the situation of the galaxy rotating as a rigid body. Compare this to the actual rotation curve. (Fig. 26.22)

**14.** Which region of the spectrum is used for studies of molecules in giant clouds?

**15.** What kinds of evidence do we have for the existence of a magnetic field in the galaxy?

**16.** Why would we expect to find molecular clouds near *O* and *B* stars?

**17.** What evidence do we have that there is matter between the stars?

**18.** Why do we think that the Orion nebula is a site of current and recent star formation?

**19.** What is a probable origin of the cosmic rays?

**20.** Cosmic rays that we detect at the Earth have traveled on the average for about 10 million years. If a typical cosmic-ray speed is about $\frac{2}{3}$ of the speed of light, what distance have these cosmic rays traveled? How does this compare to the diameter of the galaxy, and what conclusion can be drawn from these numbers?

CHAPTER

# 28

# Galaxies

These days, we take the existence of galaxies so much for granted and galactic research is such an important part of astronomy that it is easy to overlook the fact that the nature of galaxies was not clearly proven until as recently as 1925.

Simple positional observations of the planets, stars, and an occasional comet occupied the early astronomers until the invention of the telescope. Ptolemy's catalog did contain lists of nebulous or cloudy objects, but telescopic inspection showed them to be clusters of stars. Although the blurred shape of the Andromeda nebula is bright enough to be seen by the naked eye, its first recorded mention was only in 1612. Increasing use of better telescopes added rapidly to the list of nebulous objects whose nature continued to remain obscure for so long.

During the eighteenth century, several catalogs of nebulae were published. Most of the objects listed were actually star clusters, and the only galaxies known before 1764 were Andromeda (Fig. 28.1), its small elliptical companion, and a spiral galaxy in the constellation Hydra. The first catalog that contained a substantial number of galaxies was compiled by the French astronomer, Charles Messier, in 1781 (Fig. 28.2). Messier was mainly interested in comets. In those days, searching for comets involved looking for the hazy, non-stellar appearance of the comet's head, and Messier found that he was continually finding the same hazy but noncometary objects. He therefore published his list of 103 nebulae as an aid to other comet hunters, who would thereby be able to avoid wasting their time and so concentrate on better candidates. Messier's catalog includes mostly star clusters and

(a)

(b)

**Figure 28.1** (a) Andromeda Nebula, M31. At a distance of 680 pc, its angular diameter is 1.7°. Its spiral structure is taken as a model for our own Milky Way galaxy (Lick Observatory). (b) High-resolution image of the central part of the Andromeda nebula, obtained with the Einstein X-ray satellite (HEAO-2). The positions of the X-ray sources have been compared with those of optically known objects. One X-ray source coincides with a supernova remnant and several with globular clusters (L. Van Speybroeck, Harvard/Smithsonian Center for Astrophysics).

**Figure 28.2** Charles Messier comet-hunter, whose catalog of nebulous objects is still widely used (Yerkes Observatory).

**Figure 28.3** Whirlpool galaxy (M51) with a satellite galaxy NGC 5195. At a distance of 3.8 Mpc, this galaxy has an angular diameter of 9 arc min (Lick Observatory).

nebulae within our own galaxy, as well as thirty-five objects that are now known as separate galaxies. The catalog is widely used, and many amateurs make a point of systematically working their way through the complete list, which is given in Appendix K. (The faintest of the Messier objects has a visual magnitude of about 12, which can be seen through a 6 in. telescope.) Objects are often designated by their Messier catalog numbers: Thus, the Andromeda nebula is M31, and the spectacular spiral galaxy in Canes Venatici is M51 (Fig. 28.3 and 28.4).

Messier's catalog was quickly supplemented by William Herschel, whose first catalog of 1000 nebulae was given to the Royal Society of London in 1786. Herschel followed this with 1000 more nebulae in 1789 and another 500 in 1802. John Herschel's *General Catalogue of Nebulae* with 5079 entries appeared in 1864. These were superseded by J. L. E. Dreyer's *New General Catalog* (NGC) in 1888 with its 7840 objects, and his two *Index Catalog* (IC) supplements. Dreyer's catalogs are still in general use. Thus, Andromeda (M31) is also listed as NGC 224, and the Crab nebula (M1) is NGC 1952.

Most of the elder Herschel's work was done with his 19 in. telescope. Although giant telescopes such as Herschel's 58 in. and Lord Rosse's 72 in. were constructed, they produced few results of astronomical use (Figs. 28.5 and 28.6). Their metal mirrors tarnished easily, did not keep a sufficiently accurate shape, and suffered from condensation. As a result, galactic surveys reverted to refractors but did not truly take off until the arrival of the next generation of reflectors with large glass mirrors. The largest refractor, inaugurated at the Yerkes Observatory in 1897 and still in use today, has an objective of 40 in. diameter. At the Mt. Wilson Observatory, the 60 in. reflector started work in 1908, and the 100 in., for many years the world's largest, was put into use in 1918. To it we owe the recognition that many of the nebulae lie beyond our galaxy, that they are indeed the "island universes" so speculatively named by Immanuel Kant in 1755.

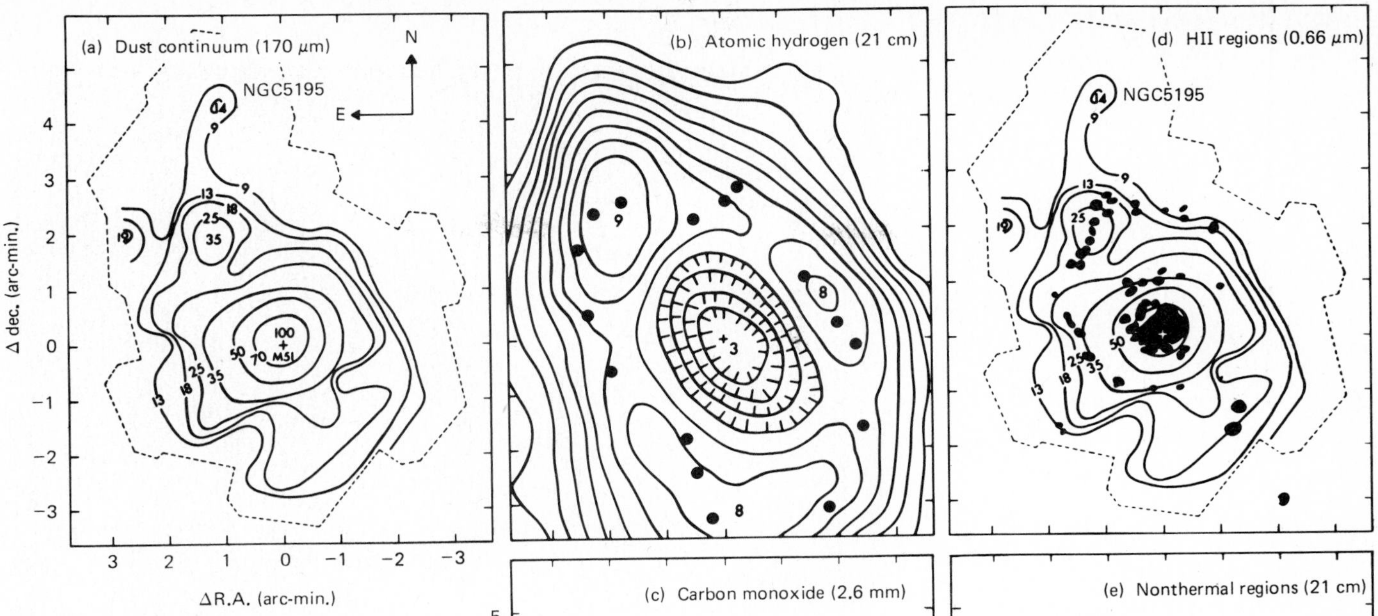

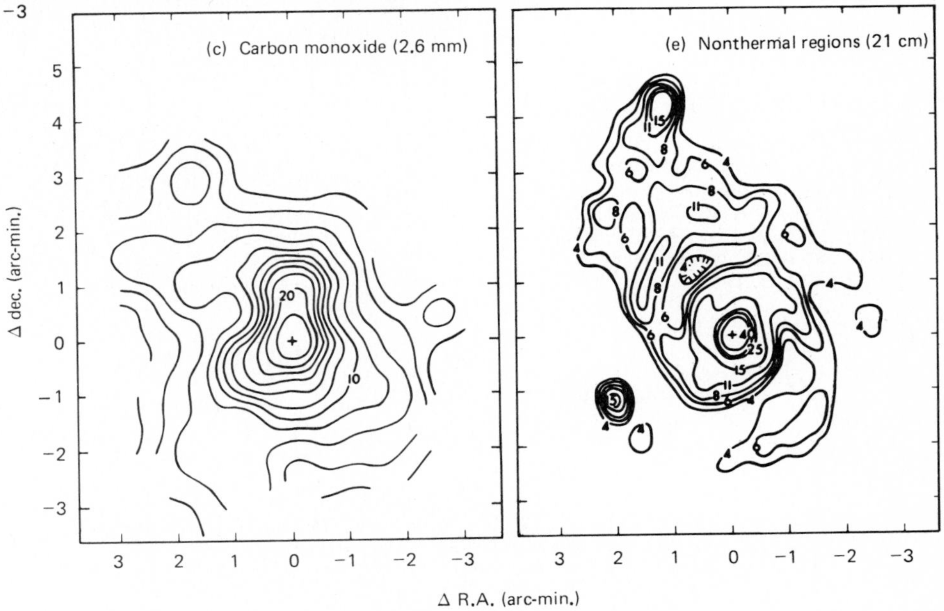

**Figure 28.4** Maps of the Whirlpool galaxy at different wavelengths. (a) at 170 microns in the far IR showing the distribution of heated dust, (b) at 21 cm, showing the distribution of atomic hydrogen, (c) at 2.6 mm, showing interstellar carbon monoxide (CO), (d) at 6600 Å (0.66 microns) showing the HII regions around very hot stars, and (e) at 21 cm showing the nonthermal (synchrotron) radiation (J. Smith, Yerkes Observatory).

**Figure 28.5** Sir William Herschel's 40 ft telescope. This dimension refers to the tube length. The primary mirror had a diameter of 1.2 m. Built between 1785 and 1789, the telescope was cumbersome to use and seems to have been used for little beyond some planetary observations (Yerkes Observatory).

**Figure 28.6** The Earl of Rosse's telescope, known as the Leviathan of Parsonstown. With this telescope built in the early 1840s, the spiral shape of the galaxy M51 (Figure 28.3) was seen for the first time (Yerkes Observatory).

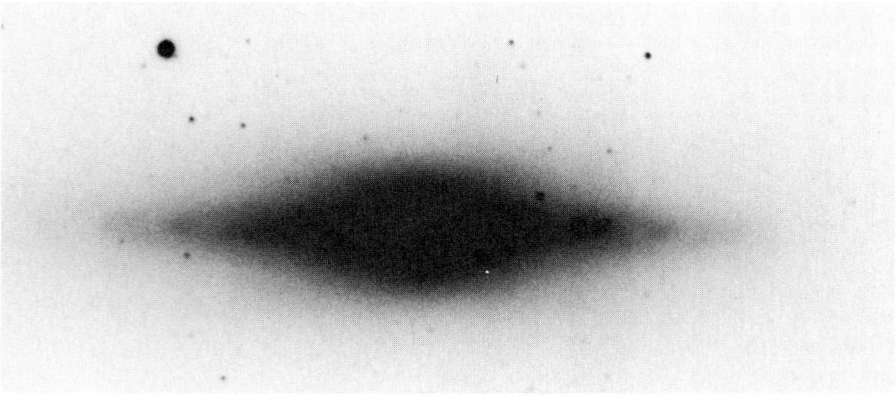

(a)

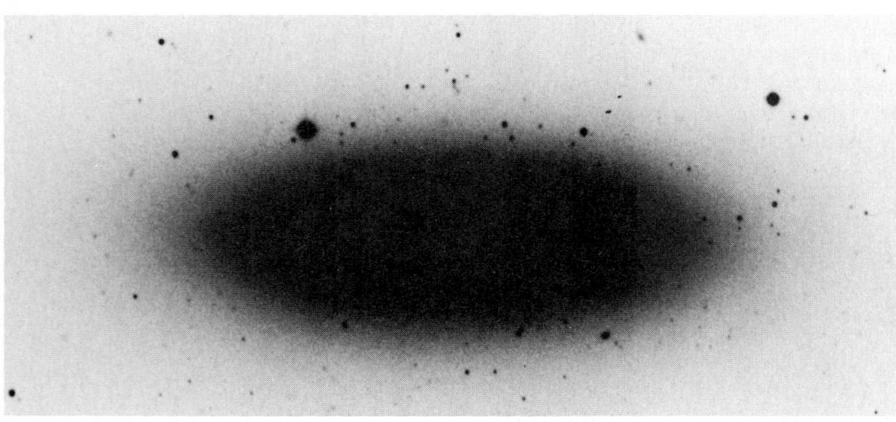

(b)

**Figure 28.7** Two photographs of the galaxy NGC 3115 showing how the appearance of a galaxy can depend on the method of photographing, (a) using a *U* filter and (b) using a *B* filter, which shows the extended halo. This galaxy has an angular diameter of about 1.5 arc min and is at a distance of 10 Mpc (S. E. Strom, Kitt Peak National Observatory).

## 28.1 DISTANCES

The nature of Messier's and Herschel's nebulae remained a puzzle as long as their distances could not be measured and as long as details of their structures could not be resolved. If they were parts of our galaxy, then their haziness could presumably be due to glowing clouds of gas, but if they were well outside our galaxy, then the fuzziness could be due to our inability to resolve their separate stars.

A reliable distance indicator did not emerge until the discovery of the Cepheid variables' period-luminosity relation (Chap. 26). Through their Cepheids, the globular clusters were shown to be part of our galactic system but the LMC was found to be much further away. The nature of the spiral galaxies remained a subject of dispute. It was perfectly understandable that very different views could be and were held by such major astronomical figures as Harlow Shapley at the Mt. Wilson Observatory and Heber D. Curtis at the University of California's Lick Observatory, who debated the subject before the National Academy of Sciences in 1920. Shapley held that the spiral galaxies were relatively near and were smaller than the Milky Way galaxy which he thought had a diameter of 100 kpc. Curtis' position was that our galaxy was only 10 kpc across and similar in form to the spirals that were separate and much farther away. With the unfailing clarity of hindsight, we can see how both Shapley and Curtis were partly correct and partly wrong, for both right and wrong reasons.

Their errors came from the lack of information which did not become available until several years later. Shapley erred in considering interstellar absorption to be negligible. Further, he based some of his conclusions on rotation measurements made by Van Maanen for some of the spirals; these were later shown to be in error. Curtis noted novalike explosions in spirals and used them as his standard of luminosity not knowing that the outbursts were really of supernovae and thus much brighter but farther away.

**Figure 28.8** Triangulum nebula, (M33), distance 720 kpc, and with an angular diameter of 35 arc min. Spiral arms can be seen, but they are not as pronounced in M81 (Figure 28.10) (Lick Observatory).

By 1923, Edwin Hubble was able to identify twelve variable stars in M31 (Andromeda) and twenty-two in M33 (another spiral galaxy) and show that they followed the Cepheid light curves. Using Shapley's period-luminosity relation, Hubble then calculated the distances to those spirals to be about 300 kpc, placing them clearly well outside the galaxy. The results were published in 1924. Although most astronomers accepted them fairly promptly, traces of the debate continued for another 10 years.

Identification of Cepheids remains our best and most important method of measuring galactic distances, but fewer than forty galaxies are close enough for even the brightest Cepheids to be recognizable. The distances to these galaxies range from around 50 kpc (for the LMC and SMC) to around 7 Mpc at most (Fig. 28.9).

For greater distances and therefore for most galaxies, it is necessary to find and calibrate other luminosity standards. Outbursts of novae are brighter than the Cepheids and can be seen to greater distances; their max-

**Figure 28.9** The extragalactic distance scale. Different phenomena are used for measuring different ranges of distances. At the greatest distances, there is still some disagreement between different determinations of the Hubble constant.

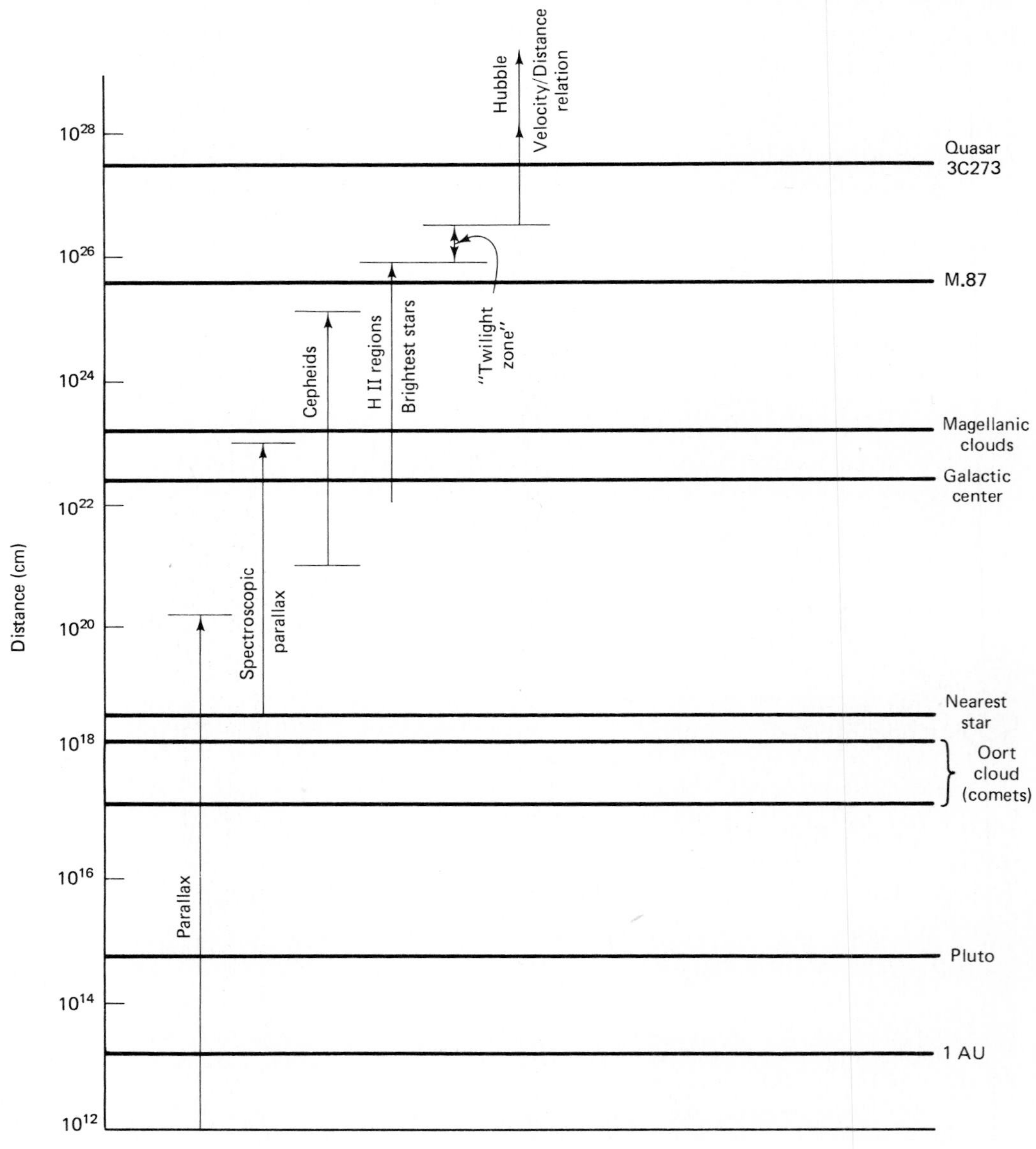

**Figure 28.10** Spiral nebula M81, distance 3.2 Mpc and angular diameter 20 arc min. This galaxy is one of the nearest beyond our Local Group. It is receding from us with a speed of 77 km/sec (Lick Observatory).

imum brightness is around $M = -9$. Bright globular clusters can also reach $M = -9$ and occasional supergiant stars can be as bright. These indicators can be used for galaxies out to about 25 Mpc. For yet larger distances, other indicators are needed. Supernovae are bright enough to be useful in principle, but the maximum brightness varies so much from one supernova to another that they are not reliable indicators. The HII regions around very hot *O* stars provide a different approach based on size rather than brightness. Surveys of the angular sizes of HII regions in galaxies whose distances we already know, show that the largest HII regions are about the same in galaxies of the same type. Measurement of the angular diameter of an HII region in a distant galaxy can thus allow its distance to be calculated (Box 28.1).

### BOX 28.1
*Use of HII Regions as Distance Indicators*

An *O* star with a surface temperature of 40,000 K can produce a surrounding HII region with 20 pc diameter (Fig. 28.11). If the angular diameter of this HII region appears to be one arc sec, the distance can be calculated

$$\frac{206265 \times 20 \text{ pc}}{1 \text{ arc sec}} = 4.1 \text{ Mpc}$$

With a long baseline radio interferometer, an HII region of 0.1 arc sec angular diameter can be identified so that distances of 40 Mpc can be measured in this way (see Sec. 19.1).

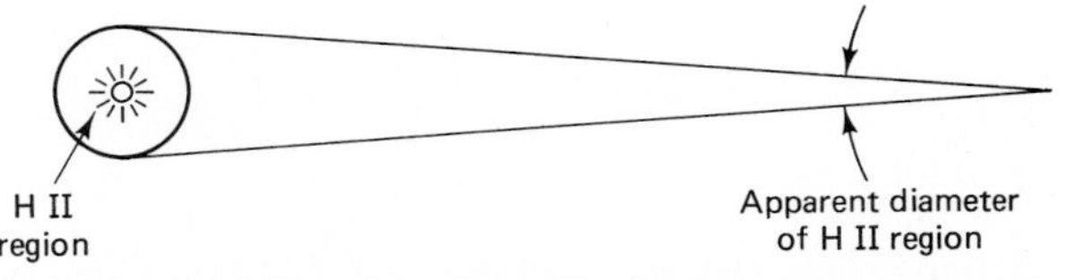


**Figure 28.11** Use of the angular size of an HII region for distance measurement. Since the physical size of an HII region is well known, measurement of its angular diameter can be used to determine the distance.

To go to still larger distances, we have to use less direct and less accurate methods. Although a random selection of galaxies will display a wide range in total brightness, some specific types cover only a small brightness range and we can look for them, either singly or among clusters. We can also note that the brightest galaxies in clusters tend to be similar, and we can use the brightest (or second or third brightest) as our galactic luminosity standard. In this approach, the same types of galaxies must be used. For example, supergiant spiral galaxies, which have an absolute magnitude of about $-21$, can be seen to about 400 Mpc.

Finally, there is an indirect method of assigning distances that can be used with caution. We shall describe it briefly here and will need to review it more fully in the next two chapters.

At the Lowell Observatory in Arizona, V. M. Slipher found from his spectroscopic measurements that most galaxies were moving away from us, as shown by their spectral lines being shifted to longer (redder) wavelengths. Very few galaxies showed a blueshift. When these measurements were made (1912 to 1925), the nature of the galaxies was not yet established, nor, of course, were their distances known. Once Hubble was able to measure some galactic distances, he went on to show that the speeds of those galaxies (as measured by Slipher) were proportional to their distances. This relation, based on the measurement of redshifts, is now known as the Hubble Law or the velocity-distance relation. It can be expressed as

$$v = HD$$

where $v$ is the speed of the galaxy in km/sec, $D$ is the distance in Mpc, and $H$ is the Hubble constant (in km/sec/Mpc). The precise value of the constant $H$ is not yet agreed. For the present we will use a value of 75 km/sec/Mpc, but we shall return to this topic in Chapter 31.

The Hubble Law can be used to determine distances. If a galaxy's redshift has been measured and its recession velocity thus calculated, the distance can be computed assuming that the galaxy in question does indeed follow the Hubble Law.

**Figure 28.12** M104, the Sombrero galaxy. The dark line of dust encircles the galaxy, which is 12 Mpc distant, and has an angular diameter of about 6 arc min. This galaxy was in the first group whose recession velocities were measured by V. M. Slipher in 1912 (CTIO).

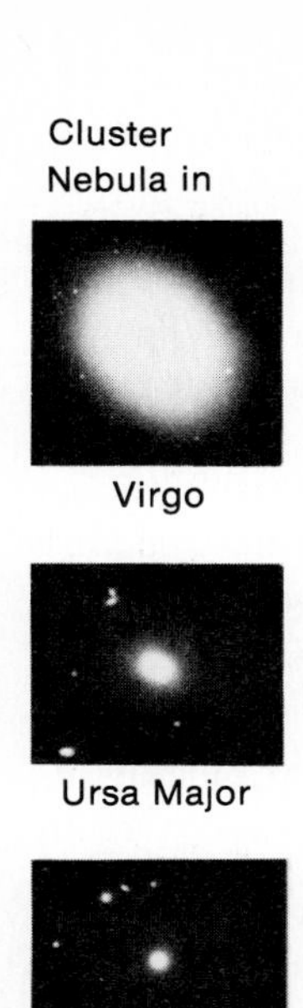

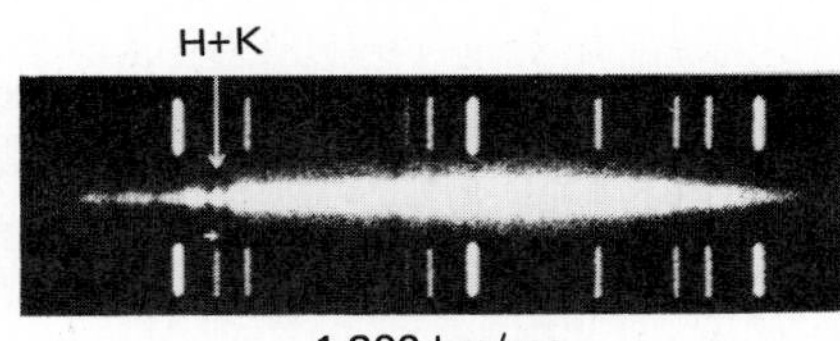
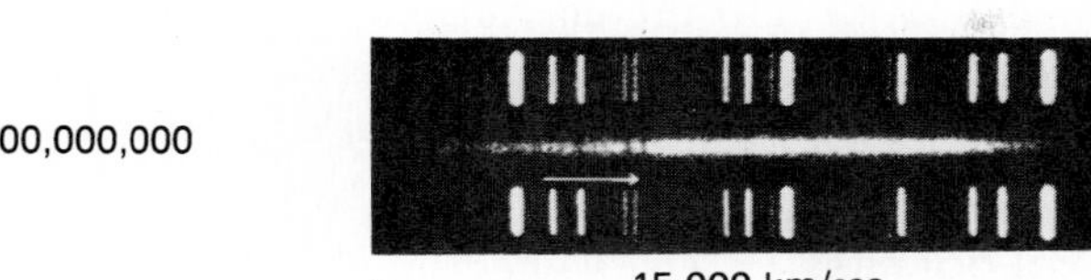

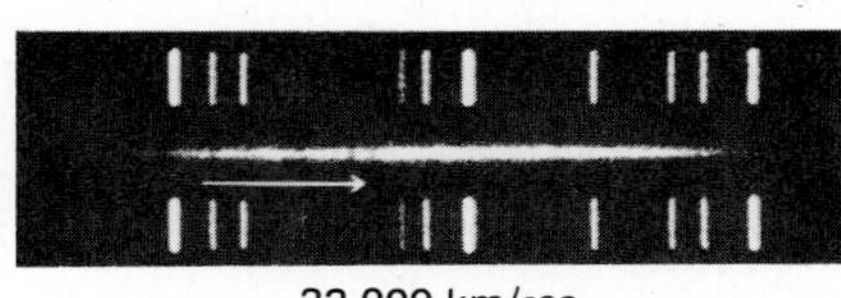
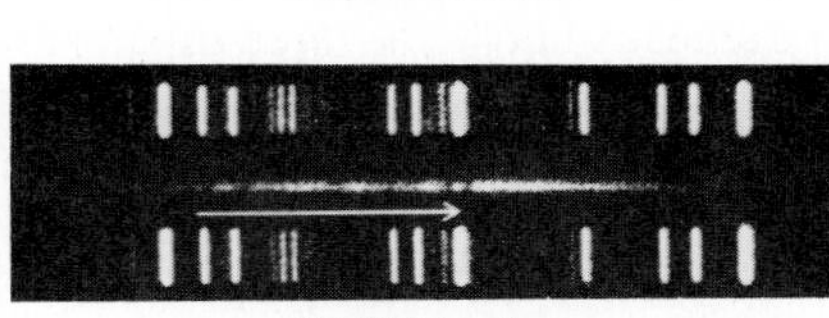
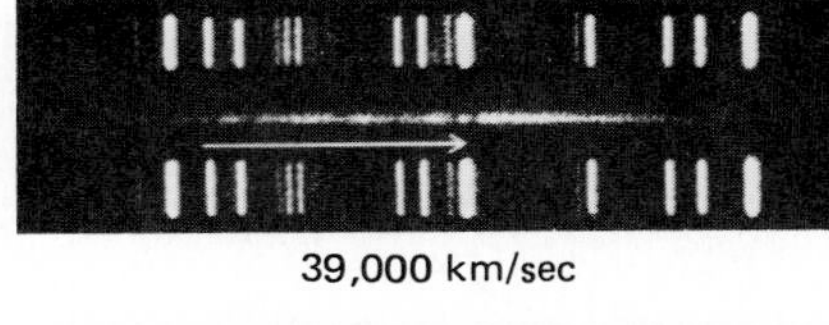


**Figure 28.13** Velocity-distance relation for external galaxies. The Hubble Law, experimentally discovered, showed that the velocity with which a distant galaxy recedes from Earth is proportional to its distance. These spectra show how the *H* and *K* lines of calcium are redshifted for each of the galaxies shown. The distances (in light years) have been calculated for a value of the Hubble constant, $H = 50$ km/sec Mpc. Redshifts are expressed as velocities, $cd\lambda/\lambda$. Arrows indicate shift for calcium lines *H* and *K*. One light-year equals about 9.5 trillion km, or $9.5 \times 10^{12}$ km (Palomar Observatory photograph).

**BOX 28.2**

*Calculation of Galactic Distance from its Redshift*

Suppose that we observe a redshift of $z = 1.003$ for a galaxy. This is related to the speed of recession by

$$(z - 1) = \frac{\Delta\lambda}{\lambda_0} = \frac{v}{c} \equiv 0.003$$

so that we find

$$v = 0.003\ c = 0.003 \times 3 \times 10^{10} = 9 \times 10^{7}\ \text{cm/sec}$$
$$= 900\ \text{km/sec}$$

Applying the Hubble Law, we then have

$$900 = HD$$

and, using $H = 75$ km/sec/Mpc, we find

$$D = \frac{900}{75} = 12\ \text{Mpc}$$

## 28.2 TYPES OF GALAXIES

Scientific progress has been made when objects were grouped together on the basis of some criterion and their other common properties were later recognized. Comparisons between groups might then point to evolutionary trends. Progress in astronomy was made in this way when the moving objects (planets) were separated from those that did not seem to move (the stars).

In studying the stars, the primitive groupings by magnitudes were supplemented by $B$ and $V$ measurements to lead to the H-R diagram. And so with galaxies, we need a classification scheme.

Galaxies can be classified on the basis of their appearance. Some have their stars tightly grouped towards the center with spiral arms extending out in a variety of patterns. Some galaxies show no spiral structure but rather an oval distribution of stars, sometimes punctuated by dark (presumably dust) lanes. There are some galaxies of quite irregular shape. Observational bias can enter into the classification: A flat spiral will show its structure when viewed from directions roughly perpendicular to the galaxy's plane, but if the galaxy is tilted at a large angle to the line of sight, only an elongated glowing strip will be seen.

**Spiral galaxies** The wonderful shapes of the spiral galaxies quickly attract attention (Figs. 28.14 and 28.15). Spirals comprise almost 70 percent of the brightest 1000 galaxies. Ellipticals, though actually more abundant among all galaxies, are smaller and fainter.

Typical spirals are our own (observed from within and not by any means fully mapped) and Andromeda, M31. In a spiral, we can identify a nucleus and surrounding disc with several arms in the characteristic pattern. The spiral arms show the sites of star formation where young and luminous stars can be seen. Star clusters and bright emission nebulae can also be seen. The concentration of matter in the arms is much more marked in some galaxies than in others. Hubble denoted the spirals by *S*; *Sa* have arms coiled close to the nucleus, *Sb* more open, and *Sc* barely coiled. While most spirals show the arms coiling from the nucleus, about one third show the arms bending from a straight bar of glowing stars, gas, and dust. These barred spirals were denoted by *SB*, with subgroups *SBa*, *SBb*, and *SBc*.

**Figure 28.14** Spiral galaxy NGC 4565 seen edge-on. The distance is not well determined but is probably about 8 Mpc. The angular diameter is 15 arc min, and the central region has a thickness of 1.4 arc min. The redshift of its spectrum indicates a radial velocity of about 1200 km/sec (Lick Observatory).

**Figure 28.15** Spiral galaxy NGC 4622, in Centaurus. Distance is 70 Mpc (CTIO).

Rotation can be estimated from Doppler shifts of spectral lines in light coming from opposite sides of a galaxy. Application of Kepler's Law (as with our galaxy) then leads to a measurement of the mass. A variation in this approach uses the hydrogen 21 cm line instead of visible lines. For spirals, the masses range from $10^9\ M_\odot$ to $4 \times 10^{11}\ M_\odot$.

**Elliptical galaxies** Elliptical galaxies show no signs of any spiral structure and are denoted by *EO*, *E3*, and so on, where the index is given by

$$n = 10\,\frac{(\text{largest diameter} - \text{smallest diameter})}{\text{largest diameter}}$$

Thus, *E0* is spherical and *E7* is very elongated. The brightest of ellipticals tapers off steadily from the center, often extending for large distances so that the dimensions are hard to define with precision. Most of the ellipticals are small, usually described as dwarf, with masses $10^6$ to $10^8\ M_\odot$. Rather fewer are giant ellipticals, such as M87, which have masses up to $10^{13}\ M_\odot$ or 100 times that of our galaxy.

**Irregular galaxies** As implied by their name, these galaxies show no regularity in their shapes, no appearance of any sort of symmetry. Irregular galaxies of Type I, such as the Large and Small Magellanic Clouds (LMC and SMC) are those in which *O* and *B* stars, clusters and emission nebulae can be identified. Irr II such as M82 show no such features: no stars, no clusters. Their spectra are similar to those of *A5* stars, which must be so numerous and so close together that we cannot see them individually. Unlike the Irr I galaxies, the Irr II show considerable evidence for thick, dusty regions that appear dark against a bright background.

**The Hubble classification** Hubble's classification system, introduced in 1925, is usually displayed in the form of a diagram (Fig. 28.16). More elaborate classification systems have been devised as more examples have been found which do not fit exactly into this scheme. There are no sharp divisions, and examples can be found representing all stages of appearance between adjacent types (Fig. 28.17).

We expect galaxies to evolve. Their constituent stars and interstellar material will change as they do in our own galaxy. Stars will form from the ISM, then age, and contribute to the ISM. Dynamical effects will alter the spatial distribution of the stars: Spiral arms may wind tighter and elliptical galaxies may flatten somewhat as they contract depending on their rotation. But we have no clear indication that the evolution of galaxies involves changes in galactic type. Spirals probably remain spirals, ellipticals probably remain ellipticals. In this sense, the classification scheme should not be compared to the H-R diagram where the evolution of a star can be followed along a track. Rather, the classification of galaxies should be seen as a useful

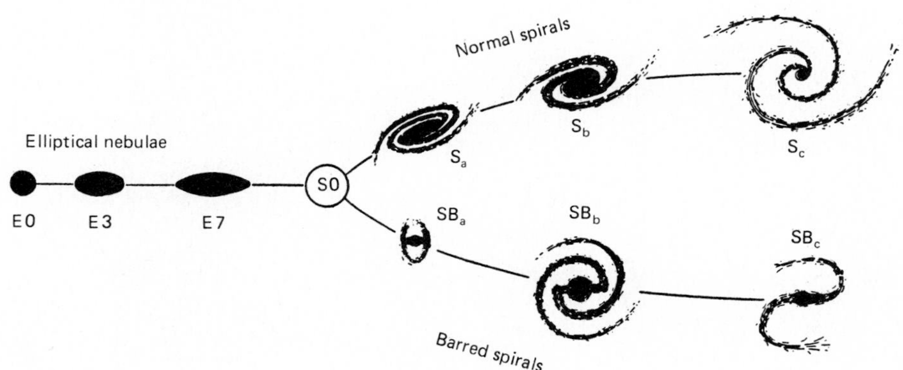


**Figure 28.16** The Hubble scheme for classifying galaxies according to their shapes. Elliptical galaxies are designated by a number between 0 and 7. Spirals fall into two broad groups, the normal spirals and the barred spirals (Yerkes Observatory).

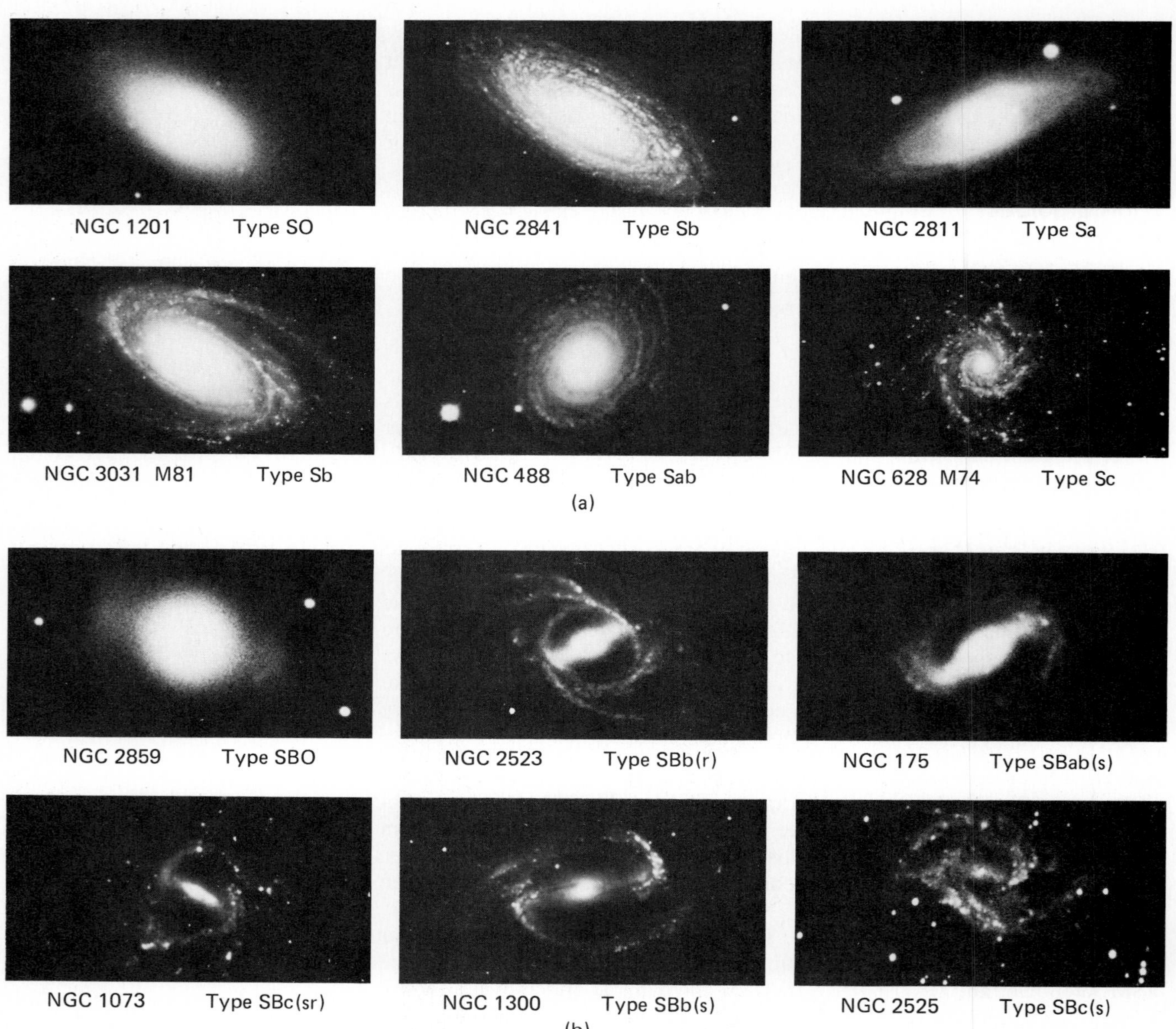


**Figure 28.17** Examples of the various types of galaxies (Palomar Observatory photograph).

*Table 28.1* PRINCIPAL CHARACTERISTICS OF CLASSES OF GALAXIES

| | *Spirals* | *Ellipticals* | *Irregulars* |
|---|---|---|---|
| Mass ($M_\odot$) | $10^9$–$10^{12}$ | $10^6$–$10^{13}$ | $10^8$–$10^{11}$ |
| Diameter (kpc) | 7–50 | 1–170 | 2–10 |
| Optical Luminosity ($L_\odot$) | $10^8$–$10^{10}$ | $10^6$–$10^{11}$ | $10^7$–$2 \times 10^9$ |
| Types of stars | Populations I, II (old and young) | Population I (old) | Populations I, II (old and young) |
| Mass/luminosity ratio | 2–20 | 100 | 1 |
| Dust | yes | very little or none | little or none in Irr. 1; much in Irr. II |
| Gas | yes | very little | much in Irr. II |
| Radio luminosity typical (erg/sec) | $10^{38}$–$10^{40}$ | up to $10^{45}$ | up to $10^{45}$ |
| Examples | M51, M101, Milky Way | Cen A, M106, M87 | M82, Cyg A, LMC, SMC |

*Table 28.2* COMPARISON OF OPTICAL AND RADIO LUMINOSITIES

| | *Optical Luminosity (erg/sec)* | *Radio Luminosity (erg/sec)* |
|---|---|---|
| Sun | $4 \times 10^{33}$ | $10^{12}$ |
| Flare star | $10^{32}$ | $10^{16}$ |
| Supernovae | $10^{36}$ | $10^{35}$ |
| Normal galaxy | $10^{44}$ | $10^{37}$–$10^{39}$ |
| Radio galaxy | $10^{44}$ | $10^{42}$–$10^{45}$ |

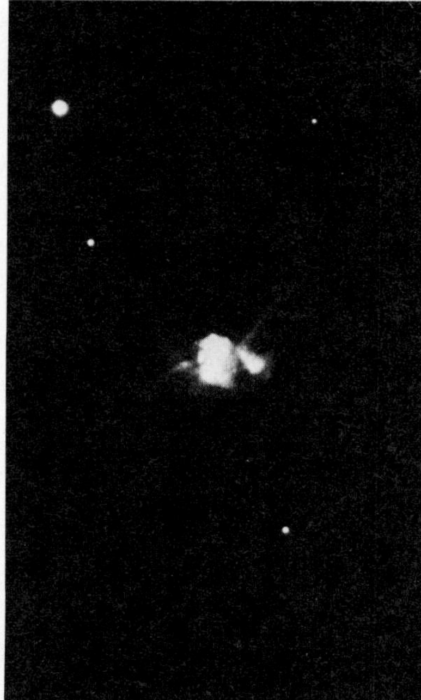

**Figure 28.18** Peculiar galaxy, NGC 2623, type Sc pec, a source of radio noise (Palomar Observatory photograph).

way of organizing the observational data. Properties of the main types of galaxies are listed in Tables 28.1 and 28.2.

Some galaxies fit quite well into the Hubble scheme but still have spectral features that seem to set them apart (Fig. 28.18). These galaxies are often grouped together and some are described in the following section.

**Seyfert, Radio, and X-ray galaxies** Seyfert galaxies, named after Carl Seyfert who first cataloged them in 1943, have very small, bright nuclei showing broad emission lines from hot gas. Many show strong IR emission from their nuclei, and one has been identified as an X-ray source. Perhaps 1 percent of all spirals are Seyferts (Figs. 28.19 and 28.20). The source of the intense radiation from a Seyfert core is not yet understood; observed luminosities of $10^{45}$ ergs/sec (ten times larger than that of our own galaxy) from a compact core requires a mechanism that we have not yet been able to model successfully.

**Figure 28.19** Seyfert galaxy NGC 1275, also known as the strong radio source Perseus A. Distance about 100 Mpc; this galaxy shows a Doppler shift of about 5000 km/sec (KPNO).

**Figure 28.20** Seyfert galaxy, NGC 4151, a strong source of IR radiation (Palomar Observatory photograph).

*Table 28.3* LUMINOSITIES OF SOME RADIO GALAXIES

| | *Type of Galaxy* | *Radio Luminosity (ergs/sec)* |
|---|---|---|
| Small Magellanic Cloud | Irr | $10^{36}$ |
| Large Magellanic Cloud | Irr | $10^{37}$ |
| M33 | Sc | $10^{38}$ |
| M31 (Andromeda) | Sb | $10^{39}$ |
| NGC 1275 (Perseus A) | Seyfert | $2 \times 10^{42}$ |
| M87 (Virgo A) | El | $10^{42}$ |
| NGC 5128 (Centaurus A) | Irr | $2 \times 10^{42}$ |
| Hercules A | Irr | $10^{44}$ |
| Cygnus A | Irr | $10^{45}$ |

Until this point, all of the classification of galaxies has been based on the visible part of the spectrum. Most spiral galaxies including our own, emit weakly at radio wavelengths: about $10^{38}$ to $10^{40}$ ergs/sec, compared to a typical $10^{44}$ ergs/sec in the visible. The development of radio astronomy showed that in some galaxies the reverse is the case—far more energy is radiated at radio wavelengths than in the visible. Some giant ellipticals have very strong radio emission: $10^{44}$ to $10^{45}$ ergs/sec.

A classification of radio galaxies can be made on the basis of their radio appearance. Some show a strongly concentrated core with or without a somewhat extended halo. The majority, however, show a *double-lobe* structure in which an extended region of radio emission is placed roughly evenly on each side of an optical galaxy. More recent observations with increased sensitivity show that in some cases there is a connecting bridge of weakly emitting material between the regions of strong emission (Table 28.3).

**Figure 28.21** Cygnus A, second strongest radio source in the sky. This object was discovered by radio telescopes before its optical counterpart had been identified. At one time it was thought to consist of two galaxies that were colliding, but for many reasons this model is not now accepted. The two visible blobs are separated by about 2 arc sec. Two regions of strong radio emission do not coincide with the visible object but are about 50 arc sec away on each side (Palomar Observatory photograph).

The extended sources display a range of shapes. Some have their radio-brightest parts at the far ends of a thin emitting region, but others have two bright regions with long tails extending further out. Several have emitting regions that bend behind a bright core looking something like a comet with its tail (Fig. 28.21). The sizes of the radio sources vary enormously. They range from being not much larger than their associated optical galaxies (with diameters of a few hundred parsecs) up to the giant radio galaxy 3C 236 whose length is 4 Mpc.

Among the earliest identified radio objects in the sky was Cygnus A (Fig. 28.22). With a radio power of about $10^{45}$ erg/sec, it is one of the strongest radio sources, and its radio contours show two bright regions equidistant from a visible galaxy of irregular shape. At one time, Cyg A was thought to consist of two colliding galaxies. This is no longer accepted, but neither has any other model proved satisfactory.

M87, a giant elliptical galaxy, has long been known to have an optically bright jet protruding from one side (Figs. 28.23 through 28.27). Radio observations have now shown that the jet is also a strong radio emitter. The visible light from the jet shows the polarization characteristic of synchrotron radiation, and the radio spectrum supports this picture. The jet of M87 must contain a large number of electrons and a strong magnetic field. There also seem to be extended but weaker radio regions roughly symmetric about M87. Recent photographs have been manipulated by computer to reveal more detail of the jet than show with normal photographic printing, and they display quite unexpected structure in the jet. It appears to be a row of connected condensations rather than a single smooth and bright jet. Observations over more than 20 years show small but definite changes. The core of M87 is less

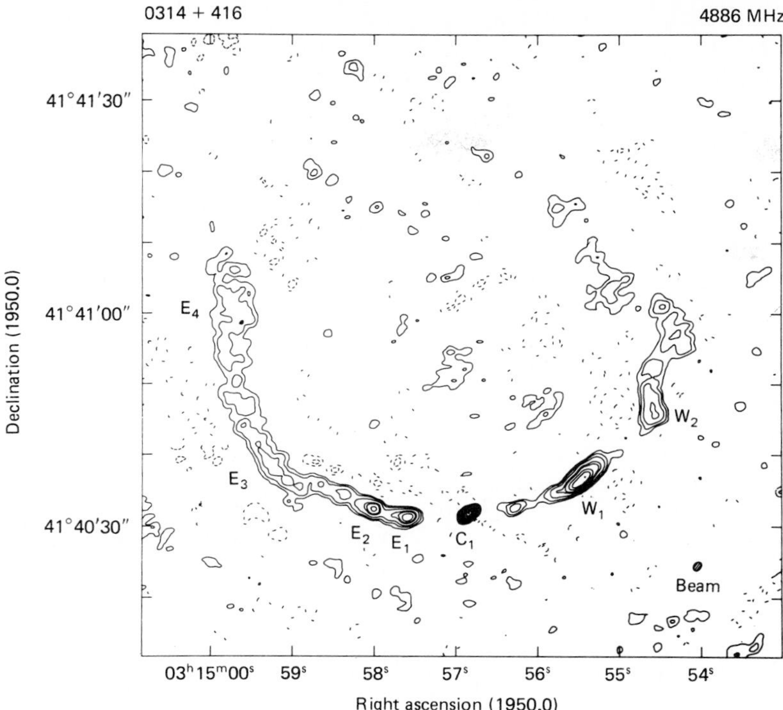


**Figure 28.22** Radio map of NGC 1265, from observations at 6.1 cm wavelength using the VLA. On each side of the galaxy, there is a narrow tail that extends for a considerable distance many times the galactic diameter. Each tail shows small bright regions. This sort of structure is thought to be the result of the galaxy moving through the hot intergalactic medium found in clusters of galaxies, but there is not yet a clear model for the physical processes involved (NRAO, and the *Astrophysical Journal*).

**Figure 28.23** Figures 28.23 through 28.27 show different views of M.87 (NGC 4486), also known as the radio source Virgo A. The spectrum of this giant elliptical galaxy shows a recession velocity of 1200 km/sec. A short exposure shows the bright core with a jet of condensation to one side (Lick Observatory).

**Figure 28.24** In a longer exposure, the large size becomes apparent, and globular clusters can be seen in the outer regions (Lick Observatory).

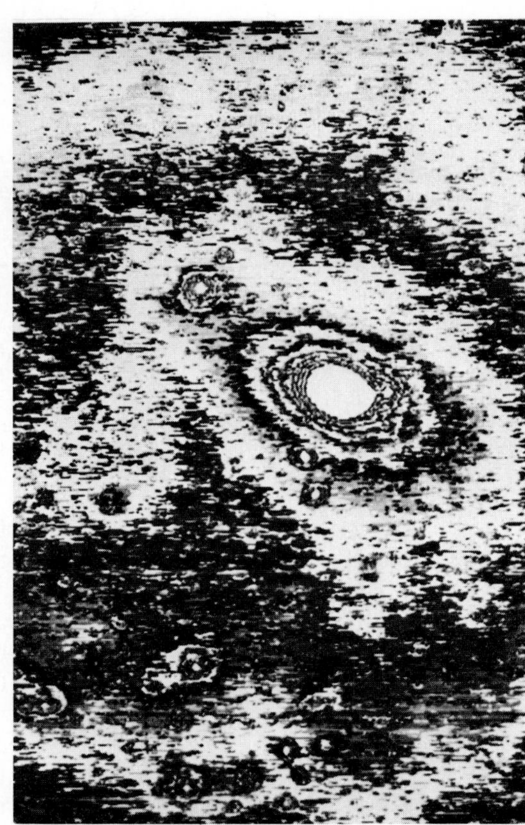

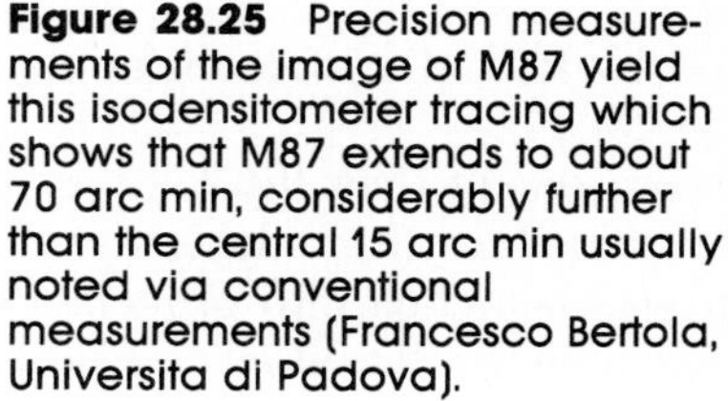

**Figure 28.25** Precision measurements of the image of M87 yield this isodensitometer tracing which shows that M87 extends to about 70 arc min, considerably further than the central 15 arc min usually noted via conventional measurements (Francesco Bertola, Universita di Padova).

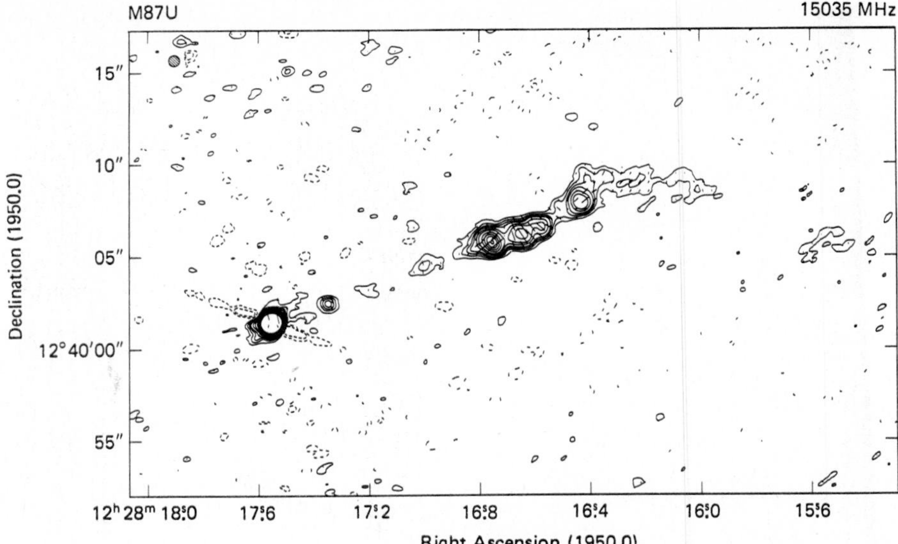


**Figure 28.26** Radio map made at a wavelength of 2 cm using the VLA. The core stands out strongly as does the jet with its condensations (NRAO).

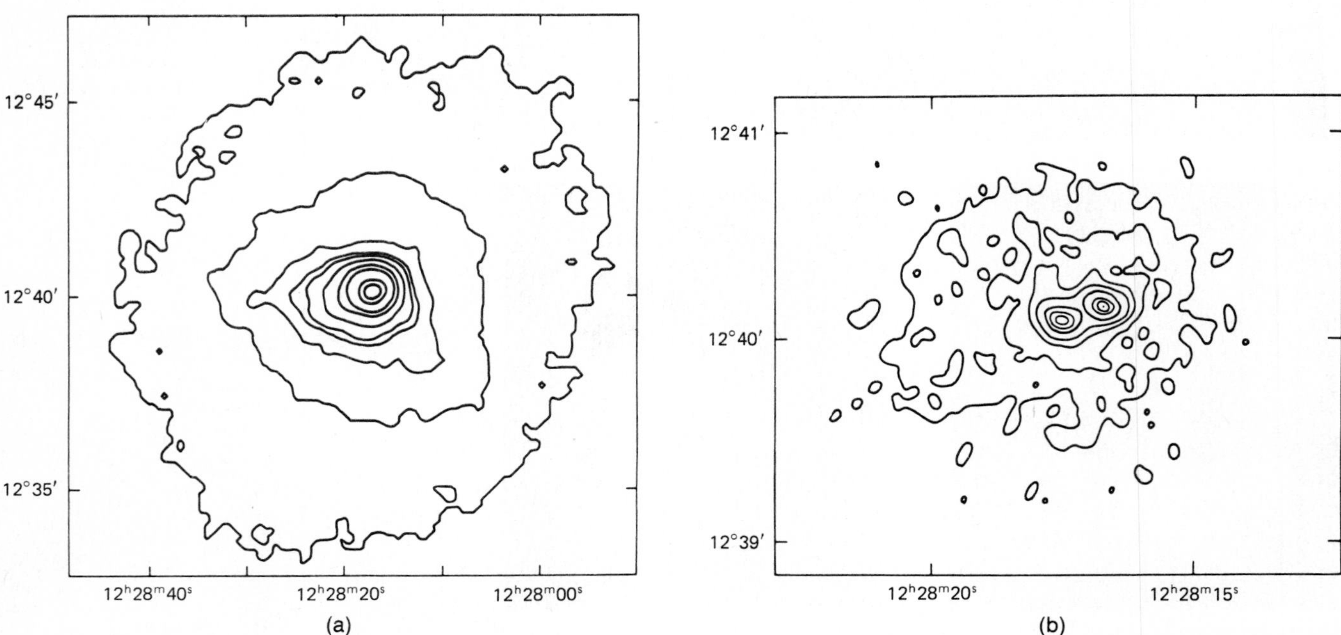


**Figure 28.27** X-ray contour maps of M87 based on observations made with the Einstein satellite. (a) Entire galaxy, showing strong central concentration but with extensions to the east and southwest. (b) Central regions of 2 arc min showing two peaks about 12 arc sec apart. The positions of these peaks agree well with the positions of the optical core and the brightest concentration in the jet (Paul Gorenstein, Harvard/Smithsonian Center for Astrophysics).

than a parsec across and, according to some models, must contain a compact mass of about $5 \times 10^9\ M_\odot$.

Centaurus A (also listed as a peculiar elliptical galaxy, NGC 5128) has a dense dust band across its bright glow. Its optical diameter of 17 kpc is dwarfed by its radio lobes, which extend for about 500 kpc on each side (Fig. 28.28).

More recent than the radio exploration of the galaxies have been the X-ray observations from the X-ray astronomy satellites Uhuru and Einstein. Uhuru was launched into orbit in 1970, from Kenya (hence the satellite's name, Swahili for "freedom"). Einstein, second in the series of High Energy

**Figure 28.28** Centaurus A, NGC 5128. A strong radio source extends about fifty times farther than the visible object, seen here with its thick dust band across the center. The optical galaxy has a diameter of about 15 kpc, and it is at a distance of 5 to 8 Mpc (CTIO).

Astronomical Observatories (HEAO), followed 8 years later. Together, these satellites have provided a remarkable survey of the X-ray sky both inside and outside our galaxy. Many galaxies were found to be strong X-ray emitters, and some of these are listed in Table 28.4. It should be noted that the listed X-ray luminosities are only lower limits, since the X-ray observations have been confined in the case of Einstein to wavelengths between $3 \times 10^{-8}$ and $5 \times 10^{-7}$ cm (3 to 50 Å).

*Table 28.4* X-RAY LUMINOSITIES OF SOME ACTIVE GALAXIES

| *Object* | *Type* | $L_x$ *(erg/sec)* |
|---|---|---|
| NGC 4151 | Seyfert | $4 \times 10^{42}$ |
| Cen A | Radio | $2 \times 10^{42}$ |
| M 87 | Elliptical | $5 \times 10^{42}$ |
| NGC 1275 | Seyfert | $2 \times 10^{44}$ |
| 3C 273 | Quasar | $7 \times 10^{45}$ |

Seyfert, radio, and X-ray galaxies together comprise a class that can be termed *active* galaxies. They exhibit similar features: intense radio and/or X-ray emission in addition to the usual optical radiation, and very compact components are often seen. Much of the radio emission and some of the optical emission show polarization and spectral shapes that can be identified as synchrotron radiation indicative of a region populated by many high-speed electrons. The X-ray emission requires high-speed particles or a very-high-temperature plasma. The overall picture they present is certainly one of great activity. Models that have been proposed include black holes that rotate and attract surrounding matter releasing gravitational energy. Such models are plausible, but the details are far from being fully worked out.

## 28.3 CLUSTERS OF GALAXIES

Just as we find clusters of stars, so we find galaxies congregating in clusters. However, whereas most stars are not members of clusters, it now seems as though the reverse is true for galaxies. Regular clusters consisting mainly of elliptical galaxies show a superficial similarity to globular clusters of stars—a generally spherical assemblage with a concentration to the center. With a population of 1000 galaxies or more, these clusters are often strong sources of X-rays.

Irregular or open clusters show neither symmetrical structure nor increased central density; some have local groupings randomly spaced. Some, such as the Virgo cluster, include large numbers of galaxies, whereas others may have only a few tens of members.

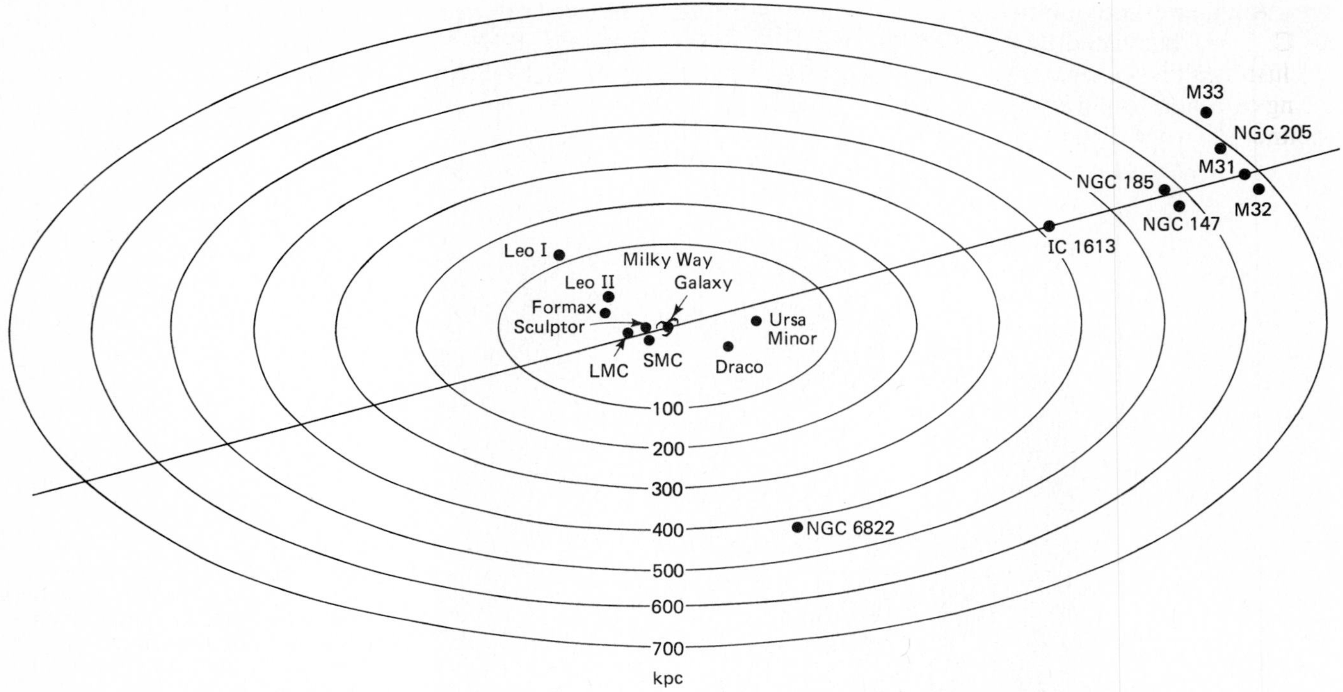


**Figure 28.29** The Local Group of galaxies, closest galactic neighbors of our Milky Way galaxy. In this diagram, the positions of the galaxies are shown projected onto the plane of our galaxy. There is one subgroup relatively close by, and another subgroup in the direction of Andromeda (M31).

**The Local Group** Our galaxy belongs to the Local Group which extends out for about 1 Mpc and has about two dozen identified constituents. Recognizing new members of the Local Group is not easy, because our own galaxy obscures some viewing directions. Members of the Local Group are listed in Appendix J and shown schematically in Fig. 28.29. More distant galaxies are all moving away from us, but within the Local Group there are several galaxies that are moving toward us.

Andromeda is the largest member of the Local Group, followed by the Milky Way and M33; others are considerably smaller. The Magellanic Clouds are our closest neighbors at just over 50 kpc. The group includes all types except barred spirals and giant ellipticals. The center of mass of the Local Group is roughly midway between the two largest members. We can use velocity measurements to determine the rotational speed and thence (with Kepler's Law) calculate the mass of the group. In this way, we obtain the value of about $5 \times 10^{12} M_{\odot}$.

Within the members of the Local Group, we have been able to identify many features we have found in the Milky Way: hundreds of Cepheids in the LMC, large amounts of hydrogen, clusters of hot *O* and *B* stars, and their emission nebulae.

**Other clusters of galaxies** Most galaxies seem to occur in clusters. In the nearest clusters, distances can be estimated from the brighter Cepheids; *O* stars and HII regions are the standards for further clusters such as the Virgo cluster which contains thousands of galaxies spread over 10° of the sky and is about 16 Mpc away (Fig. 28.30). Our investigation of the large-scale properties of the universe is heavily dependent on the measurement of the distances to the closest clusters, so that the Virgo, Coma, and Perseus clusters have received much observational attention.

A major survey of clusters of galaxies carried out by George Abell of U. C. L. A. has listed over 2500 rich clusters. Analysis of the distribution of clusters shows a clustering of the clusters; how far this hierarchy of clustering continues is uncertain, for the tabulations are limited by the increasing faintness of very distant galaxies.

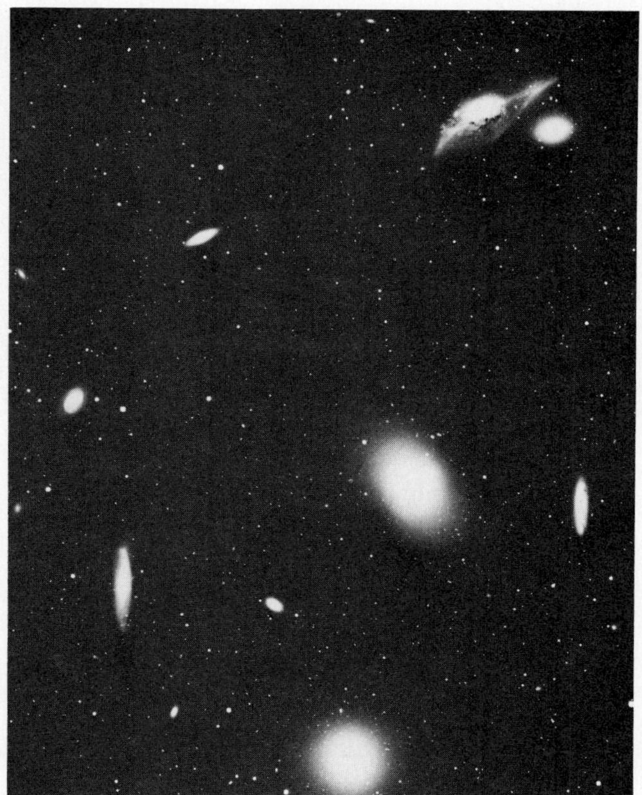

**Figure 28.30** Cluster of galaxies in Virgo, including M84 (NGC 4374) and M86 (NGC 4406). Determination of the distance to the Virgo cluster of about 2500 galaxies is an important step in the establishment of the extragalactic distance scale. The distance to the cluster is about 16 Mpc (KPNO).

**Figure 28.31** Cluster of galaxies in Hercules: At a distance of about 175 Mpc, this cluster comprises about 300 galaxies in a region with an angular diameter of about 6 arc min (Palomar Observatory photograph).

**Figure 28.32** Cluster of galaxies in Leo. There are about 300 members of this cluster, at a distance of 300 Mpc (Palomar Observatory photograph).

**The missing mass** Within a cluster the velocities of individual galaxies are randomly directed. Some might be aimed away from the center of the cluster and have speeds greater than the escape speed (see Chap. 11) so that one might expect a cluster to lose some members, much as a liquid evaporates its fastest molecules. If this loss continued for a long time, it would be surprising to find galaxies as efficiently grouped into clusters as is actually observed. It therefore seems reasonable to assume that clusters persist for a long time in which case something must prevent the 'evaporation' of the fastest galaxies. Presumably that something is the total cluster mass.

When calculations are carried through for the Virgo cluster, it turns out that the needed total mass is about fifty times the observed mass. We are encountering a similar problem with the mass of the Milky Way galaxy where the rotation curve indicates a much larger mass than is visible. This is a serious problem: How does one hide a mass many times the visible mass and prevent it from radiating any signal that would disclose its presence? Black holes could provide a convenient but unconfirmed solution; large amounts of gas (hot or cold) should have shown up. For the Hercules cluster, the visible mass is only about one fifth of what is needed. Recent X-ray observations indicate more hot gas inside the cluster than was previously recognized. Infrared surveys may locate cool stars or regions of cold gas, but that may simply raise new problems relating to the evolution of galaxies.

## 28.4 INTERGALACTIC MATTER

We have interstellar matter; do we have anything between the galaxies? So far, the evidence is negative—all that we have been able to do is set upper limits to the density of different possible constituents. Intergalactic dust would redden the light from distant galaxies; but this effect has not been seen. (The effects of dust in the Milky Way have to be considered and can be minimized by observing in directions well out of the plane of the galaxy along lines of sight that contain minimal galactic dust.) Hydrogen gas should be observable by its absorption of UV at 1216 Å. Failure to find this line leads to an upper limit in the density of intergalactic hydrogen of 1 atom in each 10 m cube. X-rays have now been detected (by satellite) from over forty clusters, but no X-rays have yet been detected from between clusters. Hot, ionized hydrogen, about 1 ion per 1000 $cm^3$, will suffice to produce these X-rays, but the mass is still far short of the missing mass needed to hold each cluster together.

## 28.5 FORMATION AND EVOLUTION OF GALAXIES?

From the wide range of observations, theorists have tried to construct models for the formation and evolution of galaxies. It has been easier to pose the questions than to find acceptable solutions. How do galaxies form and why are some galaxies spirals, others elliptical? How does a spiral evolve? What happens if galaxies collide? Do different types of galaxy represent stages of evolution?

Computer modeling allows us to examine some of these questions. In Chapter 26, we mentioned the results obtained by Toomre, which showed how a spiral structure might evolve from a gravitational force applied by a passing galaxy. Toomre has also modeled a galactic encounter between the spiral galaxy M51 (NGC 5194) and its irregular companion NGC 5195 (Fig. 28.33). Although these simulations can reproduce many of the observed

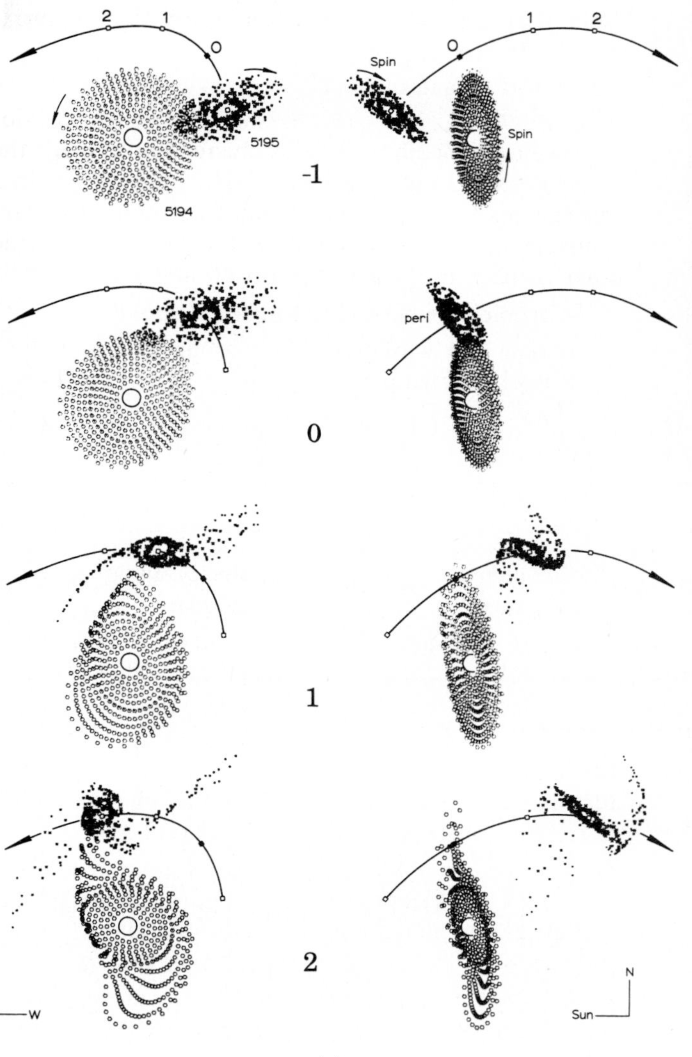


(a)

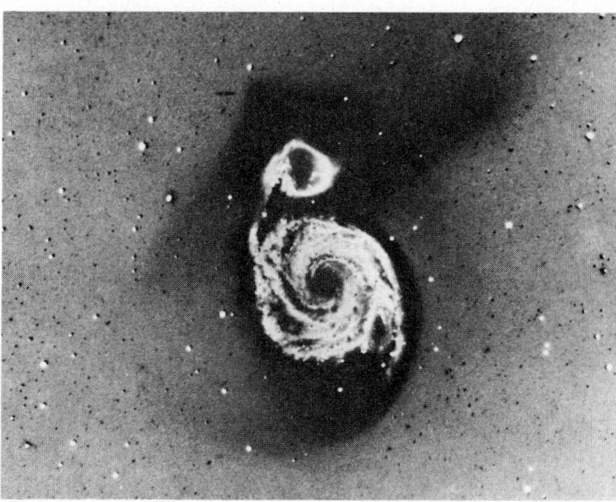

(b)

**Figure 28.33** (a) Computer simulation of the tidal encounter between M51 (NGC 5194) and its smaller companion NGC 5195. (Alar Toomre, M. I. T., © 1978 by the I.A.U.) (b) Specially processed photograph of M51, in which a standard image has been superposed on the sum of five other specially exposed photographs, all taken with the Palomar 5 m telescope. This composite photograph brings out the extended regions of the two galaxies and especially the long streamers from NGC 5195. Compare these features to the computer simulation and to the standard print (Fig. 28.3) in which these features cannot be seen (Martin Burkhead, Indiana University).

features, collisions between galaxies are expected to be so rare that we should not look to them to provide answers to all or even most problems of galactic structure.

There is a measure of agreement that galaxies probably formed when the universe was denser than it is now and that turbulence in the hot gas led to local fluctuations of density. Such fluctuations would attract more mass and while cooling condense into what would later be recognized as a galaxy. The theory of turbulence in hot, ionized gas is complicated, and though progress is being made, we really know little about the early stages of galaxy formation.

The scale of our observable universe is now so large and observational methods so newly expanded that it is not surprising that there are still so many unanswered questions. Even the present knowledge, however, allows us to narrow the scope of cosmological speculation as we shall see in Chapter 31.

## CHAPTER REVIEW

Messier's nebulae have turned out in many cases to be systems of stars that are separate from the Milky Way galaxy.

**Distances:** Cepheid variables gave first reliable distance measurements and are still valuable up to about 7 Mpc.

— at greater distances, other standard sources are used: globular clusters, HII regions, redshifts.

**Types of galaxies:** spirals and barred spirals.

— ellipticals.

— irregulars.

— special classes have their own designations, for instance, Seyfert galaxies with very bright core.

**Velocity-distance relation (Hubble law):** Distant galaxies are receding from us with speed given by $v = HD$, where $v$ = velocity of a galaxy at distance $D$ Mpc from the Sun, and $H$ = Hubble constant, approximately 75 km/sec/Mpc.

**Clusters of galaxies:** Local Group within about 1 Mpc of the Sun.

— galaxies seem to exist in clusters, with hierarchy of clusters of clusters.

**Missing mass:** Clustering of galaxies indicates gravitational force that keeps cluster together; required mass is often far larger than observable mass.

— problem: where is the missing mass?

**Formation and evolution:** not certain, but computer modeling is showing how galaxies can respond to specific forces.

## NEW TERMS

**elliptical galaxy**
**Hubble Law**
**irregular galaxy**
**Local Group**
**Seyfert galaxy**
**spiral galaxy**
**velocity-distance relation**

## PROBLEMS

**1.** Describe the method in which Cepheid variables are used to determine the distance to a galaxy.

**2.** What was the evidence that showed that the Andromeda nebula lay far outside our galaxy?

**3.** In a distant galaxy, a supernova is seen to explode and reach an apparent magnitude of +17. Assuming that the absolute magnitude of a supernova is −18, calculate the distance to that galaxy.

**4.** Suppose a Cepheid variable is found in a distant galaxy and is used to determine the distance. Later it is found that the Cepheid has been incorrectly identified and that its absolute magnitude is $2^m$ larger (fainter) than first thought. How does this change the distance that was calculated for the parent galaxy?

**5.** A Type I Cepheid has an absolute magnitude of $M = -2$. A Cepheid of this type is identified in a distant galaxy and found to have an apparent magnitude $m = +20$. How far away is that galaxy?

**6.** A Type II Cepheid is seen to have a period of 5 days and an apparent magnitude of 18.5. How far away is it?

**7.** Cepheid variables cannot be used for distance determinations beyond about 7 Mpc. What indicators are used for greater distances?

**8.** Apart from their shapes, what are some major differences between spiral and elliptical galaxies?

**9.** Use the illustrations to identify an elliptical galaxy, a spiral galaxy, and a barred spiral galaxy.

**10.** The Hubble relation involves the distances and speeds of galaxies. How are those quantities measured?

**11.** What is the distance to a galaxy that has a redshift of $\Delta\lambda/\lambda_0 = 0.25$?

**12.** What is the radial velocity of a galaxy that is 2.5 Mpc distant?

**13.** An HII region is identified in a distant galaxy. If that galaxy is known to be 10 Mpc distant, what will be the angular diameter of the HII region? (Assume that the HII region is 20 pc in diameter.)

**14.** In which type of galaxy can no individual stars be seen?

**15.** A distant galaxy is rotating and is viewed edge-on. The light we receive shows a Doppler shift. Light from the edge that is approaching us shows a redshift of 300 km/sec, and light from the other edge shows a redshift of 700 km/sec. Light from the center of the galaxy shows a redshift of 500 km/sec. The angular diameter of the galaxy is 400 arc sec.

a. How far is the galaxy (use $H = 75$ km/sec/Mpc)?
b. What is the diameter of the galaxy?
c. What is the mass of the galaxy (in units of solar masses)?

**16.** A line in the spectrum of hydrogen normally has a wavelength of 4861 Å. It is identified in the spectrum of a distant galaxy but at a wavelength of 5560 Å.

a. How fast is that galaxy receding from us?
b. If another line in the spectrum is observed at a wavelength of 5727 Å, what wavelength would be observed if that galaxy were not moving?

**17.** The irregular galaxy M82 has an apparent magnitude of 8.4 and is at a distance of 3 Mpc. It has a diameter of 7 kpc and a measured speed away from us of 400 km/sec.

a. Calculate the absolute magnitude.
b. By how much are its spectral lines redshifted?
c. If it contains a Type I Cepheid with an absolute magnitude of $-2$, would this be detectable on Earth?

**18.** What is the parallax of the Andromeda galaxy? Is this measurable?

**19.** How can the rotation rate of galaxies be measured?

**20.** What is the Local Group, and how large is it?

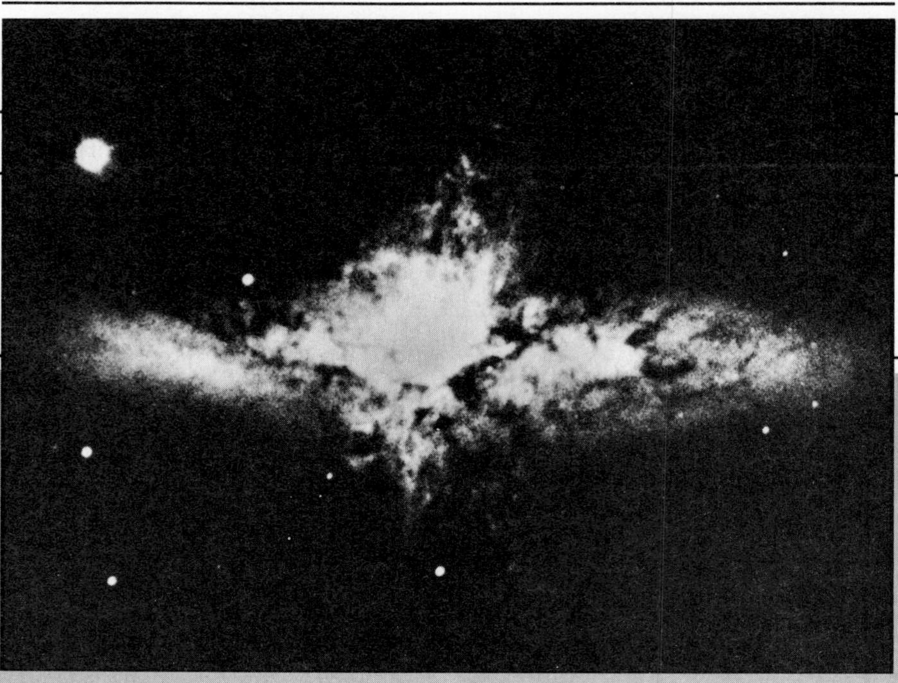

# CHAPTER 29

# Quasars and Other Puzzles

In the preceding chapters, we have described many astronomical objects and have been able to outline generally satisfactory theories. The main scheme of things seems to be reasonably well understood, even though nagging doubts may persist at a more detailed level. There have also been observations for which no satisfactory theory or model has yet been proposed. There are phenomena where the observations have resisted attempts at rationalization, despite major theoretical and observational efforts on the part of a large number of very able astronomers. Sometimes solutions may attract partisan support and the debates display the fervor of a good theological battle.

In this chapter, we shall review some major outstanding puzzles. Many of them require very technical discussion for different models to be compared in detail. We can, however, gain a general impression of the kinds of objects and problems that provide much of the fascination and scientific challenge at the very frontiers of astronomy. We shall start with the quasars discovered in 1963, with a caution that was well expressed by Geoffery and Margaret Burbidge. In their book, published only a few years after the discovery, they commented that 95 percent of the ideas for quasars must be wrong, but no one knows which 95 percent. The comment could probably just as well be applied to other objects we shall describe.

## 29.1 QUASARS

**Discovery** In optical astronomy, arc second positional accuracy has been available for many years, but the relatively large ($\lambda/d$) factor for even the largest single radio telescopes meant that the positions of radio sources were not nearly as well tied down.

**BOX 29.1**

***Angular Resolution of Radio Telescopes***

At the University of Manchester's Jodrell Bank radio telescope, the dish diameter is 250 ft or 7500 cm. Operating at a wavelength of 75 cm (400 MHz frequency), the angular resolution is:

$$\frac{206525 \times \text{wavelength}}{\text{diameter}} = \frac{2.06 \times 10^5 \times 75}{7500} \sim 2000 \text{ arc sec} \sim 0.6 \text{ deg.}$$

For a radio interferometer with a baseline of 100 mi ($1.5 \times 10^7$ cm), the resolution is improved by a factor ($1.5 \times 10^7/7500$) or 2000. The resolution attainable for other baselines can be computed by proportion (see Sec. 29.3). For the same baseline, resolution can be improved by using shorter wavelengths.

The use of radio interferometers, first by Ryle at Cambridge University in England and Mills at Sydney University in Australia, improved the situation and both of these radio observatories compiled extensive catalogs of radio sources. Successive Cambridge catalogs listed their objects as 1C, 2C, 3C, 4C. The very widely used 3C catalog was based on a survey at 178 MHz ($\lambda \sim 168$ cm). Many sources could be identified with known visible objects, some with supernova remnants, and some with galaxies.

In 1960 Matthews and Sandage (at the Mt. Wilson and Palomar Observatories) seemed to have identified the visible counterpart of the radio object 3C48. At the corresponding position in a photograph, there appeared to be a star, rather bluish and only $16^m$ (Fig. 29.1). In itself this identification

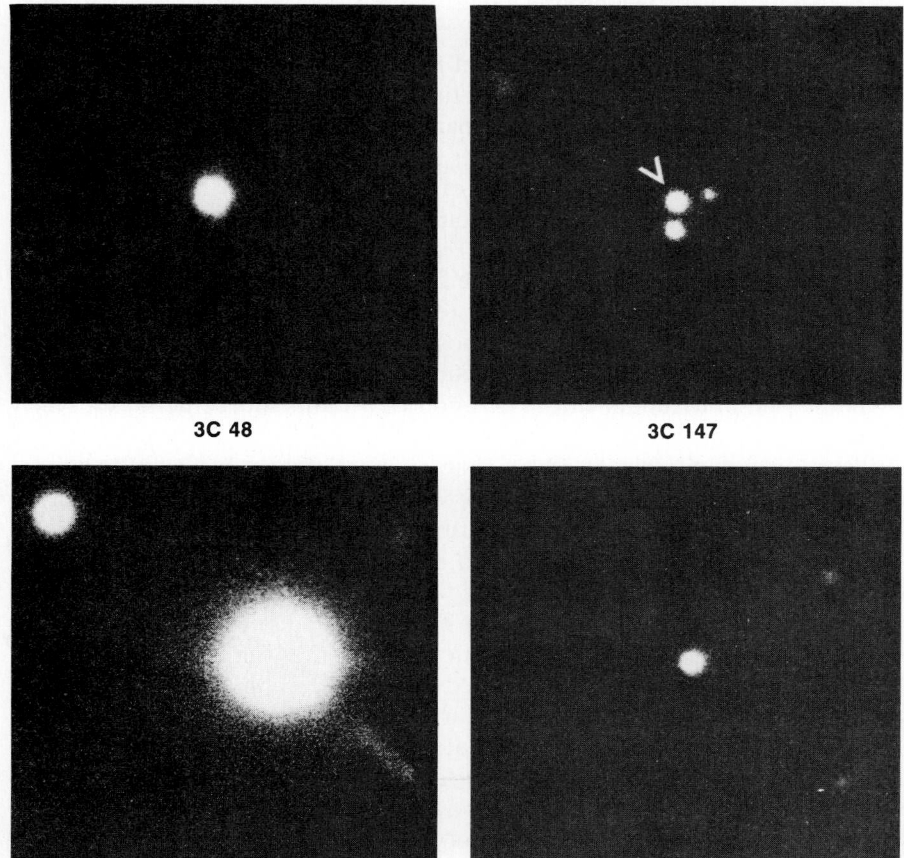


**Figure 29.1** Four quasars photographed with the 5 m telescope (Palomar Observatory photograph).

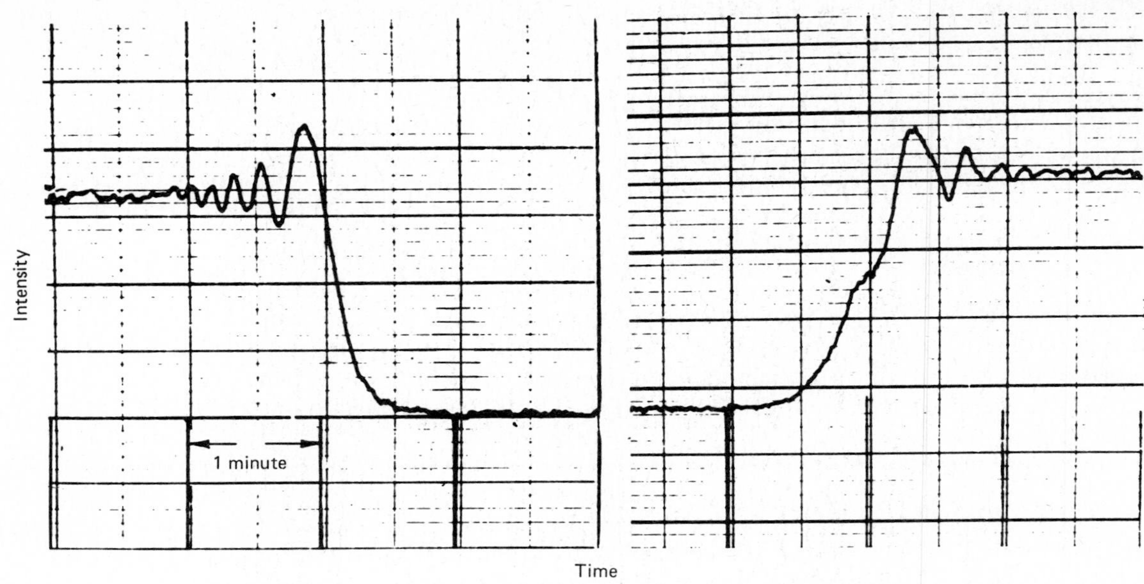


**Figure 29.2** Lunar occultations of 3C273 observed in 1962. Analyses of these diffraction patterns allowed Hazard and his colleagues to make precise determinations of the quasar's structure and coordinates (C. Hazard, University of Pittsburgh, and *Nature*).

was strange, for until that time no stars (other than the Sun) were known as radio sources. The spectrum of this star could not be understood, for it showed broad emission lines, none of which coincided with any well-known wavelengths. No further progress was made until 1963 when a new observational technique came into use.

Cyril Hazard at the University of Sydney used the Parkes 210 ft (64m) radio telescope to observe the occultation of radio sources by the Moon. At any given observatory only a small part of the sky can be covered by the Moon in its orbit, but for radio objects in that region more precise mapping becomes possible. The start and end of the occultation by the Moon's limb can be noted; since the Moon's position is very accurately known, the star's position can also be well determined (Fig. 29.2). In the earliest observations, the major source of uncertainty came from lack of knowledge of the exact profile of the Moon. Hazard used this technique to observe the radio source 3C273, and was able to pinpoint its position to within about 1 arc sec.

This great improvement in positional accuracy had important consequences. The uncertainty in the position of a radio source defines a small region on the sky that is sometimes termed an *error box*. If the error box is too large, it will contain so many visible objects that it becomes quite impractical to examine all of them to see which one might be the radio source. However, now that Hazard had been able to locate 3C273 to such accuracy, an optical search was conducted and showed a 12.8 magnitude star to be located at the radio position. The radio mapping also showed that 3C273 was not a simple radio source: It consisted of two emitting regions, 3C273 *A* and *B*, separated by 19.5″ and located astride the star.

Maarten Schmidt of Caltech used the 5 m telescope at Mt. Palomar to get a photographic spectrum of the star (Fig. 29.3). Like 3C48, its spectrum showed emission lines at no immediately recognizable wavelength. It was Schmidt's inspired idea that the lines actually included the well-known Balmer lines of hydrogen, but Doppler shifted so that all wavelengths were increased by exactly the same fraction. Table 29.1 lists the identified lines

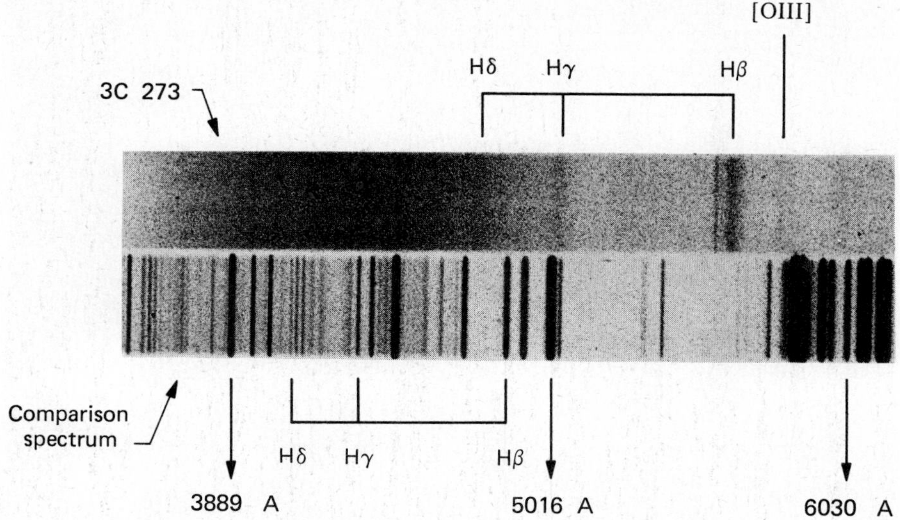


**Figure 29.3** Spectrum of quasar 3C273. Laboratory calibration spectrum of hydrogen and helium appears along the bottom row. The three Balmer lines (Hβ, Hγ, Hδ) in the hydrogen spectrum appear also in the quasar, but shifted by 16 percent (Maarten Schmidt, California Institute of Technology).

in the spectrum of 3C273, which has also more recently been observed at x-ray wavelengths (Fig. 29.4).

With measured wavelengths so different from the usual values and with so many chemical elements and lines to choose from, how can this sort of identification be secure? The answer is that spectra of galaxies tend to show certain lines with much greater regularity than other lines. Further, if the conditions of temperature, pressure, and density are right for the production of some lines, then (from theory) we know generally which other lines should also appear and, just as importantly, which other lines should *not* appear. The greater the number of lines there are observed in a spectrum, the easier it is to make an identification. Certainly, the identification of the lines in 3C273 show remarkable agreement in the wavelength shift. From the ratio of (measured wavelength)/(standard wavelength) = 1.158, the Doppler formula tells us that 3C273 must be moving away from us with a speed of 0.146 times the speed of light. Jesse Greenstein of Caltech then showed that the lines in the spectrum of 3C48 could also be reconciled with known lines, this time if a Doppler shift of 1.367 were used (Table 29.1), corresponding to a recession speed of 0.303 times the speed of light.

***Table 29.1*** EMISSION LINES IN QUASARS[a]

| | *3C273* | | *3C48* | |
|---|---|---|---|---|
| *Rest Wavelength* ($\times 10^{-8}$*cm*) | *Measured Wavelength* ($\times 10^{-8}$*cm*) | *Ratio* | *Measured Wavelength* ($\times 10^{-8}$) | *Ratio* |
| 2798 MgII | 3239 | 1.1576 | 3832 | 1.3697 |
| 2975 NeV | — | | 4066 | 1.3667 |
| 3346 NeV | — | | 4575 | 1.3673 |
| 3426 NeV | — | | 4685 | 1.3676 |
| 3727 OII | — | | 5097 | 1.3676 |
| 3869 NeII | — | | 5288 | 1.3668 |
| 3970 Hε | 4595 | 1.1574 | — | |
| 4101 Hδ | 4753 | 1.1590 | — | |
| 4340 Hγ | 5032 | 1.1594 | 5935 | 1.3675 |
| 4861 Hβ | 5632 | 1.1586 | 6646 | 1.3672 |
| 5007 OIII | 5792 | 1.1568 | — | |
| Mean red-shift | | 1.1581 | | 1.3675 |

[a] Adapted from a paper by Jesse L. Greenstein and Maarten Schmidt, *Astrophysical Journal*, 1964, Vol. 140, pp. 1–34.

**Figure 29.4** X-ray image of 3C273, obtained by the Einstein Observatory (HEAO-2). The bright spots dotted around are not associated with the quasar image but were produced by background photons. The sharp edge to the pattern shows the limit of the field of view (Harvey Tananbaum, Harvard/Smithsonian Center for Astrophysics).

**BOX 29.2**
*Doppler Shifts: Reminder*

When the relative speed ($v$) between a source of radiation and the detector is less than about one tenth of the speed of light, then the Doppler shift is given with sufficient accuracy by

$$\frac{\lambda - \lambda_0}{\lambda_0} = \frac{\Delta\lambda}{\lambda_0} = \frac{v}{c}$$

However, when the relative speed is greater than about 0.1 c, and especially when it exceeds 0.5 c, the full formula must be used

$$\text{redshift} = z = \frac{\lambda}{\lambda_0} = 1 + \frac{\Delta\lambda}{\lambda_0} = \sqrt{\frac{1 + v/c}{1 - v/c}}$$

In these formulas $\lambda_0$ is the wavelength measured at rest and $\lambda = \lambda_0 + \Delta\lambda$ is the wavelength measured when the source is moving at a speed $v$ away from the observer (see Box 8.2).

These Doppler shifts were large, larger in fact than for any other galaxy known at that time. What and where could these objects be? After a few more had been located, it was realized that only redshifts were being found. No blueshifts were detected. All of these objects had to be moving away from the Earth. Comparison of old sky photographs with new showed no signs of any proper motion, so the objects could not be nearby. The only known objects which showed Doppler shifts even approaching these high values were galaxies and that immediately raised a serious problem. If these new objects represented some new class of galaxy and if they too followed the Hubble velocity-distance relation (several "ifs"), then they would have to be very far away.

The calculation of luminosity from the measured intensity and the distance showed the luminosity to be enormous. Therein lay the problem, which has persisted. We shall go into this in more detail later; here it will suffice to say that the energy dilemma was quickly recognized and this, together with other strange properties, attracted considerable research activity. By now, some hundreds of quasars have been identified, and it is almost impossible to find an issue of the *Astrophysical Journal* without several articles on this topic. About eighty articles on quasars appear each year in just this one journal.

A word on the name. Because their photographic appearance was starlike yet they did not behave like stars, they were termed *quasi-stellar radio sources* or *QSS*; some objects were found which showed large redshifts but weak radio emission, hence *quasi-stellar objects* or *QSO*; **quasar** is now generally used to include all objects with very large redshift and with a compact starlike appearance.

**Properties** For many years, the following properties provided the criteria for labeling an object as a quasar:

1. starlike photographic appearance, often (but not necessarily) identified with a radio source.
2. variable light output. These variations are not as regular as for a pulsar but are quite uneven. 3C48 was found to vary by nearly $0\overset{m}{.}4$ (nearly 50 percent) over about 1 year. In 3C446 the change was a factor of two in only 1 day.
3. conspicuous UV brightness: *blue-stellar objects* was another early name. This feature provided one way of searching for quasars, photographing the sky through near UV and blue filters to locate objects with considerable ultraviolet excess.
4. broad emission lines. Later, many quasars were found with absorption lines that tend to be much narrower than the emission lines.
5. large redshifts. In general, the absorption redshifts are not the same as those of the emission lines. Numerous quasars show more than one redshift in their absorption lines.

As more quasars were discovered, exceptions to these requirements were found. For example, when the quasars were being found primarily because of their radio emission, their optical luminosities were found to vary. However, when quasars were selected through optical searches, much less variability was found. The strong UV excess (compared to main sequence stars) is not always seen and is no longer a requirement. No list of sharply defined criteria can now be applied to all objects, apart perhaps from the initially starlike optical appearance and the large redshift.

**Distances** Distances are central to the quasar puzzle. Most astronomers believe that quasars are at the distances indicated by their measured redshifts and the use of the Hubble relation. There are several lines of evidence. For example, some nearby quasars have been found close to regular galaxies that show the same redshift. What are the alternatives? Could the quasars be inside our galaxy? In that case, one might expect some to be approaching us and others receding, but *all* quasars have been found to recede. An early suggestion came from James Terrell of the Los Alamos Scientific Laboratory who suggested that quasars were high-speed objects ejected from our galaxy and that by now all were located outside the galaxy (Fig. 29.5). In this way, no quasar would be approaching us and only redshifted spectra would be seen. There was, however, a reason for rejecting it. With the high speeds indicated by their redshifts, the quasars are moving very rapidly away from us. Their speeds are many orders of magnitude greater than the speeds of any stars in the galaxy. The proper motion of nearby quasars should thus have been detectable, but none was found.

Could the redshifts be due to something other than the Doppler effect? Gravitational redshifts were considered very early and were rejected on the basis of detailed analysis of the spectra. Could the redshifts be Doppler, but the quasars not follow the Hubble relation? This has been forcefully argued by Halton Arp of the Mt. Wilson and Palomar Observatories who has studied

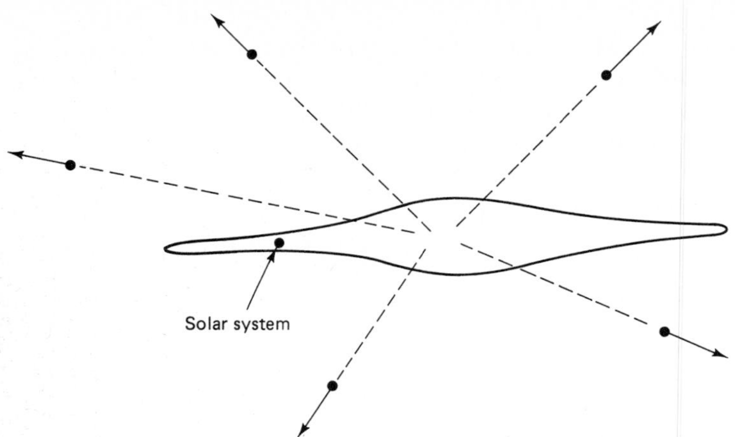


**Figure 29.5** Terrell's model. An early explanation for the observation that all quasars show redshifts in their spectra: Quasars were thrown out of our galaxy by a gigantic explosion, and by now all are receding from us. Increasing evidence, however, now rules out this model and suggests that most quasars are at the cosmological distances which their large redshifts would suggest.

galaxies that appear to be associated (Fig. 29.6). In many photographs, wispy connections may appear between some galaxies suggesting that they are linked. For example, the two galaxies NGC 4319 and Markarian 205 appear to be close together—only 42 arc sec separates them. They also seem to be connected by streamers of material, yet the spectrum of NGC 4319 indicates a recession velocity of 1800 km/sec, and the spectrum of the other galaxy shows a recession speed of 21,000 km/sec.

Arp has also found cases where two quasars are separated by a regular galaxy which has a different redshift. At the least, these results suggest that some quasars are not at distances given by the usual velocity-distance relation. Arp has not so far succeeded in converting most astronomers away from the view that most quasar redshifts are normal (that is, that they follow

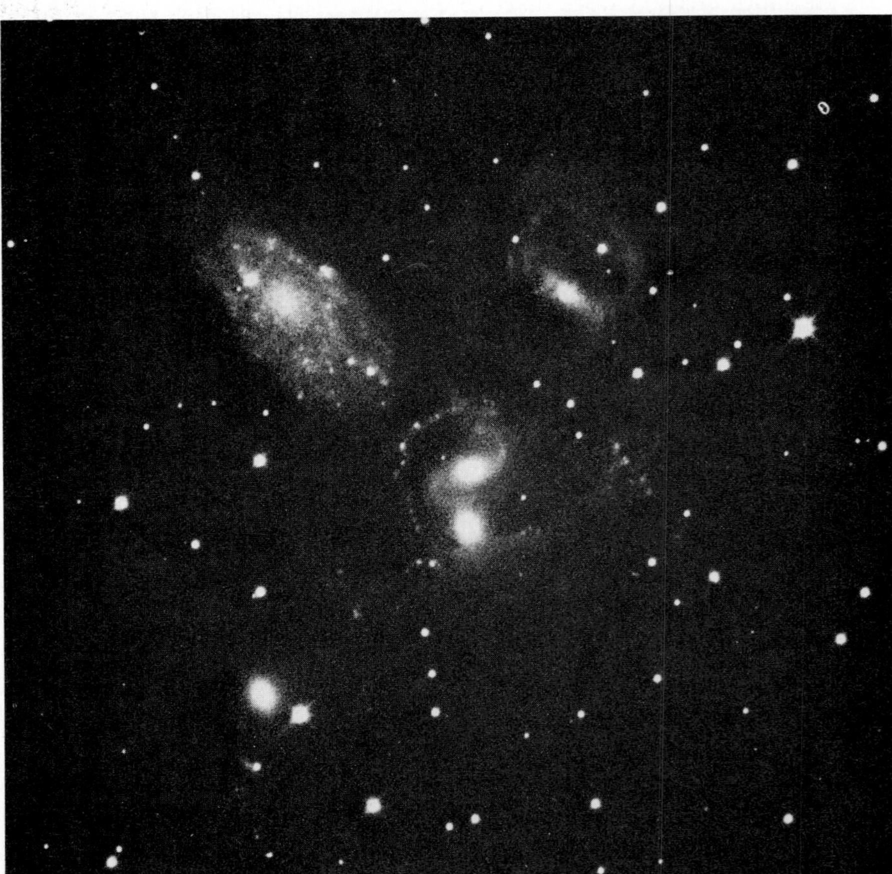

**Figure 29.6** Stephan's quintet: a cluster of five galaxies that seem to be connected by faint streamers. The redshifts shown by the different galaxies are not, however, the same. One galaxy has a spectrum indicating a recession speed of 800 km/sec, while the others have speeds of about 5700 km/sec. There are some indications that the quintet is also associated with another galaxy (NGC 7331) about half a degree away, which is at a distance of 15 Mpc and shows a red-shifted spectrum with a speed of 1050 km/sec (Lick Observatory).

the Hubble Law), nor has he provided an alternative distance indicator for the quasars. So, while the weight of professional opinion is against Arp, the question should not be considered as closed, for a number of major astronomers have oscillated in their views over the years. The accepted view certainly carries a high price in quasar energy, as we shall see in the next section. The first quasars had redshifts of 1.158 and 1.367. By now, redshifts as large as 3.5 have been found. The corresponding recession speeds must then be computed with a more complex formula than is adequate at low redshifts (Box 29.1) before the Hubble relation can be used to obtain the distance of the quasar.

**Energy** The energy problem was very quickly recognized. Data from the Greenstein and Schmidt 1964 paper are summarized in Table 29.2.

Here are two objects which in a single line of the spectrum (Hβ) are radiating more than a billion times as much energy as the Sun does at *all* wavelengths. The problem is compounded when another item is included, the variability. We can see how the time scale for light variations can be used to set limits to the size of the emitting region.

Consider, for example, an object 1 LY across (Fig. 29.7). If one side started to brighten, the "flame" could spread no faster than the speed of light and so reach the other side in no less than 1 year. Presumably, for the total output to increase noticeably, most of the object must brighten, and thus the time observed for a change indicates the maximum diameter.

*Table 29.2* QUASARS: ENERGY RADIATED

| | *3C48* | *3C273* |
|---|---|---|
| Redshift ($\lambda/\lambda_0$) | 1.367 | 1.158 |
| Distance (Mpc)[a] | 1100 | 474 |
| Luminosity: (erg/sec) | | |
| hydrogen Hβ | $6 \times 10^{42}$ | $9 \times 10^{43}$ |
| total visual | $10^{45}$ | $4 \times 10^{45}$ |

[a] These distances were based on the value of the Hubble Constant accepted at that time, $H$ = 100 km/sec/Mpc.

**Figure 29.7** Size of a distant object, inferred from intensity variations. If the one side of our galaxy (*A*) brightens first, its light will reach a distant observer before the light from the other side, *B*. The interval between the arrival of light from *A* and *B* is the time taken for a brightening to spread across the galaxy. With a diameter of 100,000 LY, the interval will be at least 100,000 years. For a small object, more rapid intensity variations are possible.

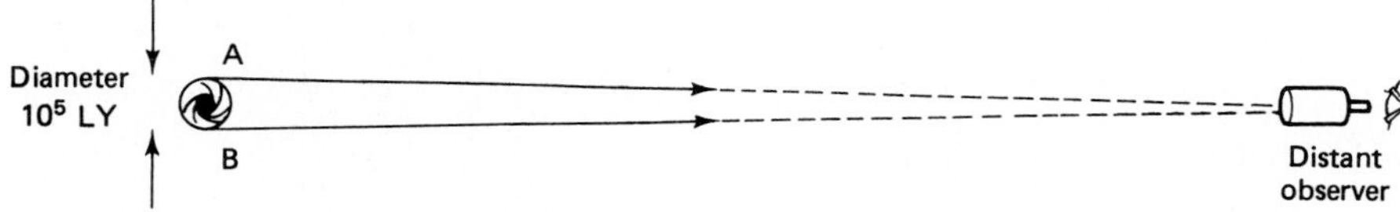


**BOX 29.3**
***Quasar Light Variations and Size***

Our galaxy is about $10^{23}$ cm or 100,000 LY in diameter. If the total emission from the galaxy were to brighten sufficiently to be seen by a distant observer, it is most likely that almost all of the galaxy will have to brighten—extreme brightening will most likely not be confined within a small region. If the brightening starts at one edge of the galaxy, the soonest it can reach the other edge is 100,000 years later.

Accordingly, when we observe quasar brightening in a week or a month, we can infer that the maximum size of the luminous regions must be 1 light week or 1 light month.

Significant variations of quasar brightness in much less than a year, therefore, imply that the size of the source region is much less than 1 LY across. This is a tiny fraction of the 100,000 LY diameter of our galaxy, yet from that small region there emerges more energy per second than from our entire galaxy. The compactness required for the energy supply is orders of magnitude different from anything astronomers had yet encountered.

Not only must intense energy sources be identified, but additional physical processes need to be considered. The density of photons in those source regions is so high that there will frequently be collisions between electrons and photons, changing the spectrum of the radiation that we will observe.

**Models** With so much data by now accumulated during more than 20 years, what sort of model can we construct for quasars? What we will describe represents a general consensus at the present time.

The redshifts are interpreted in terms of the Doppler effect, and the distances are then calculated from the Hubble relation. Luminosities of $10^{44}$ to $10^{47}$ ergs/sec have been found. Light variations tell us that source regions can be smaller than a parsec in diameter, and the emission line spectra then require about 100 $M_{\odot}$ of ionized gas within that volume at an average density of about $10^4$ protons/cm$^3$, but probably very patchy with some parts being as dense as $10^9$/cm$^3$. The absorption lines may be produced in a surrounding envelope of cooler gas or, as is now considered more likely, somewhere along the very long distance traveled by the light before it reaches us. Multiple absorption redshifts can be produced by the quasar light passing through the halos of distant galaxies whose own redshifts are thus imprinted on the spectra we detect.

What powers a quasar? There is some agreement that the most probable source is an extremely massive black hole, $10^8$ to $10^9$ $M_{\odot}$, into which surrounding matter (stars, gas, dust) is attracted, thus releasing gravitational energy. While it is relatively easy to conjure up a black hole as an energy source, it is far harder to specify the dimensions and details of the physical processes involved. We are, therefore, still far from being able to predict a spectrum which could be compared with observation. For example, many quasars have been found to be X-ray emitters. It is beginning to appear as if quasars fall into two categories: Type I with radio luminosities less than $10^{41}$ ergs/sec (about 90 percent of all quasars), and Type II with luminosities

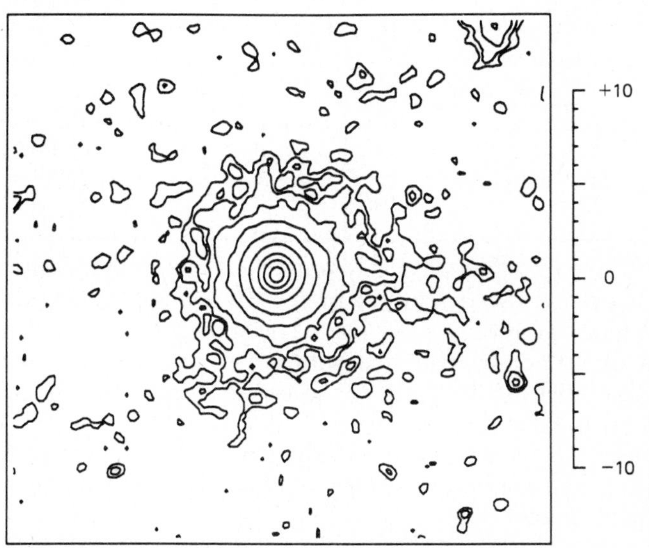


**Figure 29.8** Extended image of quasar PKS 0736 + 017. New techniques have been developed for the measurement and computer analysis of images on photographs. When applied to quasar images, it is found that the images are no longer starlike points, but that in many cases they show extended halos. The quasars thus appear to be the very bright cores of galaxies, where the surrounding halos could not be seen with conventional photographic techniques. (S. Wyckoff, P. A. Wehinger, T. Gehren, Arizona State University and Max-Planck Institut für Astronomie, Heidelberg; observations made with the European Southern Observatory 3.6 m telescope.)

greater than $10^{41}$ ergs/sec. Type I quasars appear less luminous in X-rays, and their radio spectra differ from those in Type II. There are also suggestions that Type I quasars are related to Seyfert and spiral galaxies, Type II to BL Lac objects (see Sect. 29.3) and elliptical galaxies. This classification may turn out to be related to an evolutionary sequence, but the evidence is still very sketchy.

In recent years, there have been considerable improvements in the efficiency of photon detection and in the computer enhancement of photographic images. Applied to quasars, these techniques have now shown that many quasars are not simply bright pointlike objects. They show angular dimensions that are consistent with galaxies at the great distances implied by their redshifts, and some show faint halos (Fig. 29.8). It is beginning to appear that quasars are very luminous and distant galaxies, admittedly with extremely energetic cores but still possibly part of a general sequence of galaxy types.

**Double quasars and gravitational focusing** There are so many stars in our galaxy that there will be many occasions when we find two that appear to be close together although they are actually unrelated and far apart. Quasars are much less abundant, and we would not expect to find close pairs. Nevertheless, the unexpected does sometimes happen (and provides the incentive in lotteries), and two quasars were indeed found with only 5.7 arc sec between them. The quasars (0957 + 561 A, B) were faint objects (seventeenth magnitude). Spectra were obtained using the 2.1 m telescope at Kitt Peak and the Multiple Mirror Telescope, both in Arizona. These spectra showed almost identical redshifts for the two objects, with $\Delta\lambda/\lambda_0 = 1.405$ for the emission lines.

It was proposed that these were not two identical quasars that happened to be close, but rather that what was observed were two images of a single quasar. Between the quasar and the Earth, it was suggested, there was a galaxy that was acting as a lens so that light rays passing close to the edges of the galaxy would be bent to produce the images that we observe (Fig. 29.9).

The gravitational bending of starlight predicted by Einstein in his general theory had been observed during the 1919 solar eclipse. What we are now seeing, on a much larger scale, is the same physical effect.

Confirmation came from photographs of the quasar pair, taken by Alan Stockton of the University of Hawaii, who was able to see a slight extension to the image of one of the pair (Fig. 29.10). Computer analysis of the photograph allowed the one quasar image to be subtracted to reveal an image of a galaxy. It had been noted that the absorption lines in the quasar spectra were also redshifted, but the $\Delta\lambda/\lambda_0 = 1.39$. This could now be understood: The absorption lines were produced when the quasar light passed through the thin outer region of the galaxy, and the 1.39 is really the redshift of the galaxy not the quasar. This also indicates that the galaxy is not as far as the

**Figure 29.9** Gravitational focusing of light by a galaxy to produce multiple images of a distant quasar. The spectra observed for the quasar images are almost identical and show almost the same redshift, highly improbable if the quasars were truly unrelated and just happened to be only a small distance apart.

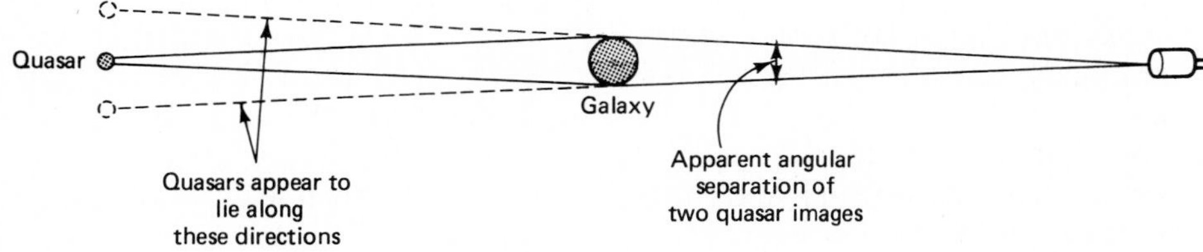

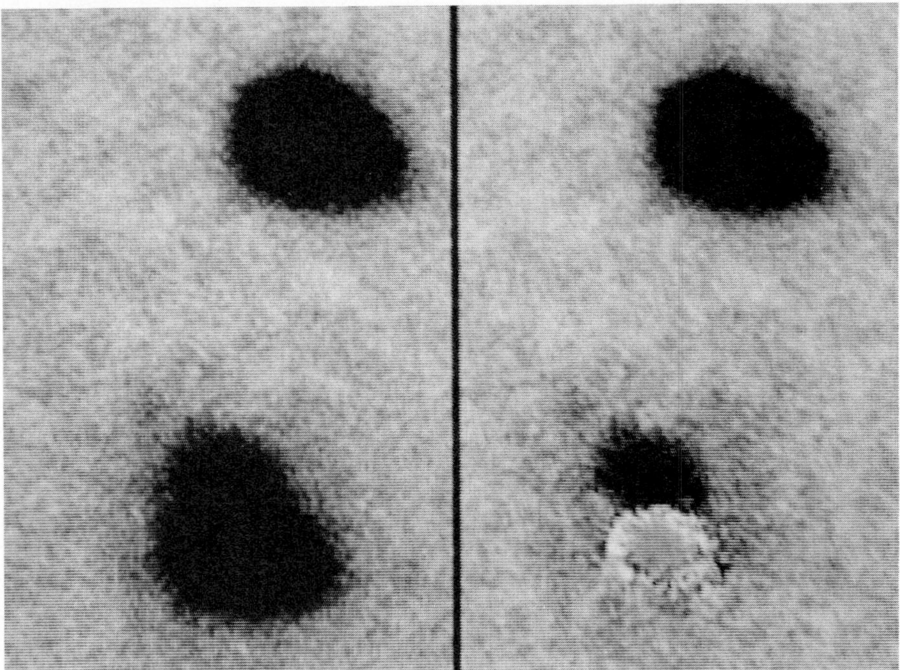

**Figure 29.10** Double quasar 0957 + 561. (a) Five separate exposures have been added to produce this image. (b) Subtraction of the image of the upper quasar from the lower yields the small remaining image, that of the galaxy whose gravitational focusing has produced this double image (Alan F. Stockton, Institute for Astronomy, University of Hawaii).

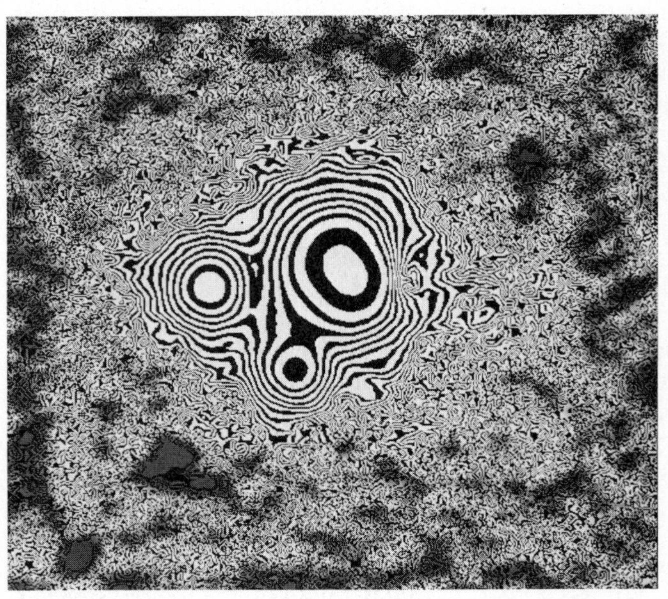

**Figure 29.11** Triple quasar, Q 1115 + 08: image obtained with one of the six telescopes of the Multiple Mirror Telescope, using a vidicon to obtain 1 sec exposures, which could be superimposed. The images of these quasars are approximately 2 arc sec apart and have been produced by the gravitational focusing of light from a single quasar (KPNO).

quasar. Since these observations, reported in 1979, additional examples of the gravitational lens effect have been found. In one case, three quasar images have been produced by the intervening galaxy (Fig. 29.11).

**Another quasar oddity** As if quasars had not already provided us with sufficient new observations, yet one more must be mentioned. Repeated observations have been made of the double radio quasar 3C273, and it has been found that the two bright radio cores are moving steadily apart. From measurements of the rate of angular separation, the apparent speed of separation (in km/sec) can be calculated, since the distance of 3C273 is already known (from its redshift). The surprising result is that the separation speed seems to be ten times the speed of light (Fig. 29.12). Speeds greater than $c$ are termed superluminal, and such effects have been detected in seven quasars since 1970. At first glance this phenomenon would seen to contradict a basic assumption on which Einstein built his theory of relativity that nothing could travel faster than the speed of light in a vacuum. A simple calculation (Box 29.4) shows that this apparently revolutionary discovery is the consequence of the accident of the direction of our line of sight relative to the direction of motion of the radio cores (Fig. 29.13). Other strange visual effects have long been known to result from very-high-speed travel, but this one had not been anticipated.

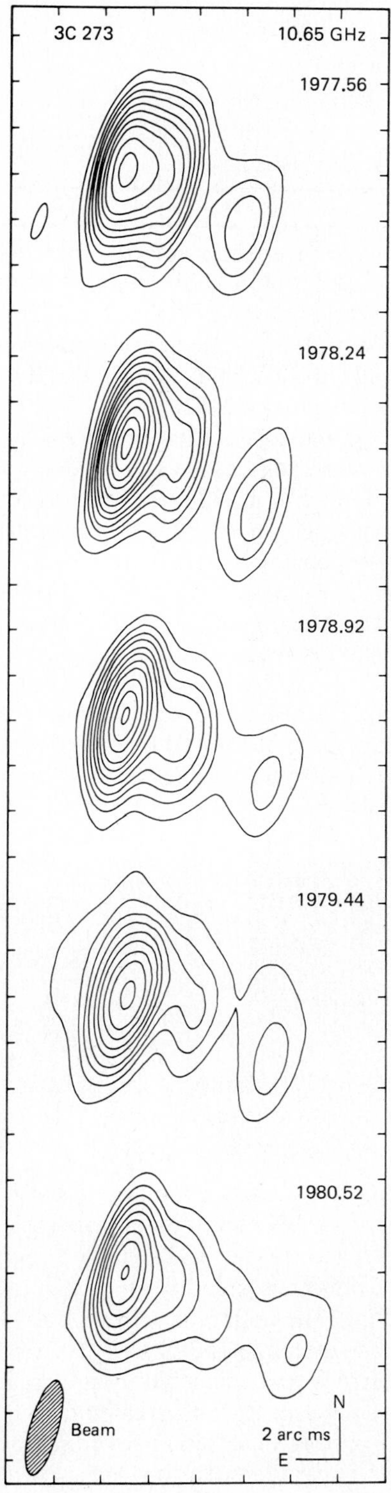


**Figure 29.12** Maps of the quasar 3C273B, constructed from radio observations at 2.8 cm wavelength, over a period of 3 years. Five radio telescopes in California, Texas, Massachusetts, West Virginia, and West Germany constituted the VLBI array. The steady expansion of the quasar, by just over .002 arc sec, can be seen. The system is expanding at a rate of 0.76 milli arc sec/year, corresponding to an apparent speed of 9.6 times the speed of light (T. J. Pearson, Owens Valley Radio Observatory, California Institute of Technology, and *Nature*).

**Figure 29.13** A possible model for the superluminal quasar. The quasar core ($A$) ejects a high-speed beam of electrons, which are seen at $B$ and $B'$ at later times. To a distant observer, the separation between $A$ and $B$ changes by a distance $d$. This appearance is deceptive because the time difference between $B$ and $B'$ depends on the angle $\theta$ as well as on the speed of $B$ and the speed of light. The *apparent* speed of $B$ can exceed $c$, the speed of light.

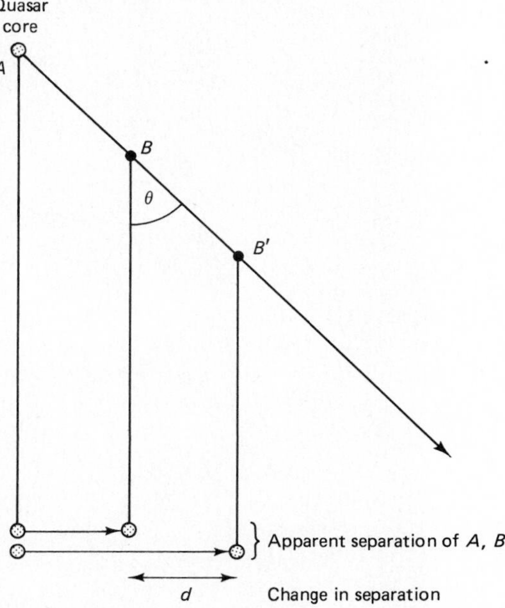

**BOX 29.4**
*Superluminal Velocities*

Observation of radio quasar shows two cores ($A$ and $B$, Fig. 29.13). Another observation at a later time, finds the cores at $A$ and $B'$ where the separation has increased by a distance $d$. ($A$ might be a massive core, with $B$ and $B'$ the positions of a beam of ejected electrons, moving with speed $v_o$.) The light signals from $B$ and $B'$ reach us with a time delay that depends on the angle between the direction of the electron beam and the line of sight to the Earth. If the separation distance between $B$ and $B'$ appears to be $d$, then it can be shown that the apparent speed of separation of $A$ and $B$ is given by

$$v_{\text{app}} = v_o \frac{\sin \theta}{1 - \dfrac{v_o}{c} \cos \theta}$$

and this can easily exceed $c$, even though $v_o$ is less than $c$.

## 29.2 ACTIVE GALAXIES

Quasars display the highest densities of energetic cosmic activity, but increasingly we are finding that many other objects also show evidence for great turmoil. We have moved very far from our earlier view of the heavens being populated by fixed stars twinkling sedately at us. Some of this evidence for active galaxies has already been reviewed, in Sec. 28.2. Another active galaxy from the Messier catalog is the irregular galaxy M82.

M82 provides an interesting example of the difficulties faced by astronomers as they piece together the information relating to such a distant object and try to construct a model for its behavior. M82 is classed as an Irr. II galaxy (Figs. 29.14 and 29.15). It shows an elongated blur of light with no individually resolved stars. The overall spectrum is similar to those of small spiral galaxies, but no spiral structure can be seen. Without any bright stars of its own to serve as luminosity standard, the distance to M82 is based on measurements of Cepheids in nearby galaxies with which it appears to form a group. The distance is 3 Mpc.

Glowing filaments extend out for about 3 arc min, corresponding to a length of 3 kpc. Bright parts of irregular shapes are separated by dark lanes probably due to large amounts of dust. The hydrogen H$\alpha$ line is so intense ($2 \times 10^{40}$ ergs/sec or $\sim 10^7\ L_\odot$) that it must originate in about $10^6\ M_\odot$ of

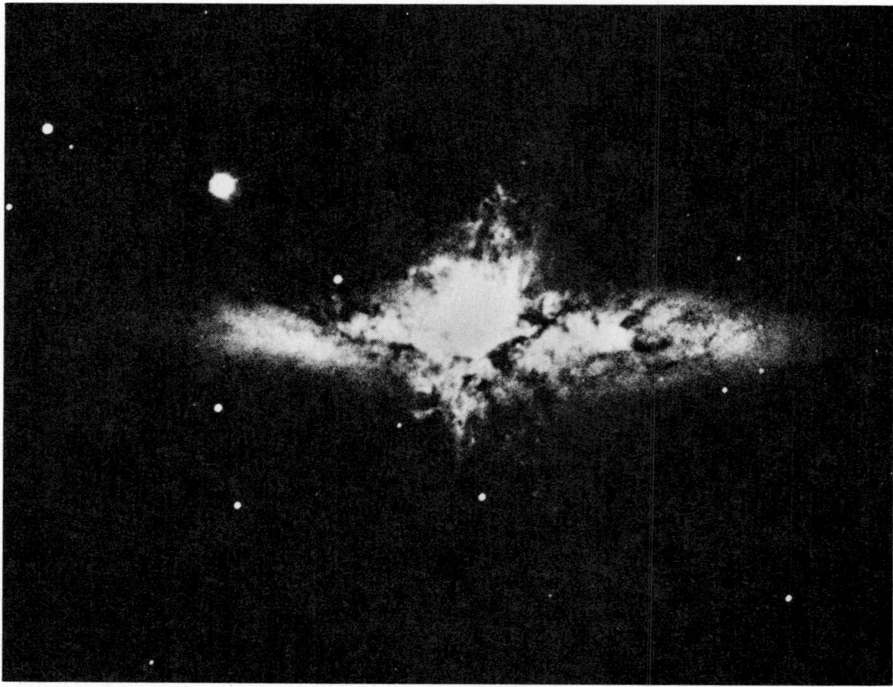

**Figure 29.14** Irregular galaxy M82, photographed in red (H$\alpha$) light. Strong polarization and radio emission have been observed, indicative of high-speed electrons in a magnetic field. Dark lanes are caused by absorption of light by dust. Glowing filaments expand to very large distances from the galaxy. No individual stars can be resolved. The distance is 3 Mpc, and the angular diameter is about 8 arc min (Palomar Observatory photograph).

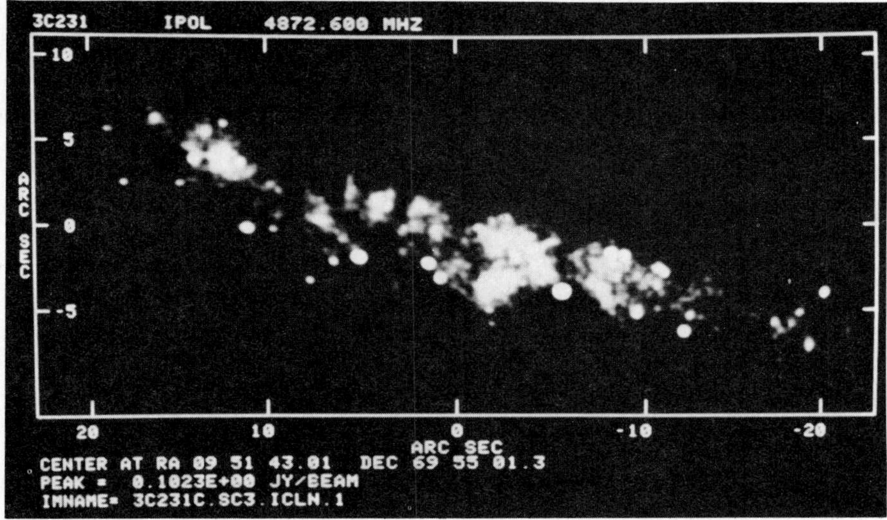


**Figure 29.15** VLA map of the central region of M82, at a wavelength of 6.2 cm. This map shows the central 700 pc, (one twelfth of the galaxy) with numerous hotspots (NRAO).

hydrogen. Spectral analysis shows a very complex shape for the Hα line indicating great turbulence.

Much of the light is polarized, indicative of synchrotron radiation. The redshift of the Hα line indicated that gas was streaming out of the galactic center at a very high speed. From some features, it appeared that we were observing the galaxy at an angle of about 10° to the plane of its disc, and that the speed of the gas streams was 1000 km/sec. This very high speed taken together with the observed polarization led many to believe that we were viewing a galactic explosion with many features like a much larger-scale version of the Crab nebula. More recent observations have led to a different model being considered.

Long-exposure photographs have shown much more detail. Infrared images show bright patches at longer wavelengths, and newer spectra show lines normally seen in very hot *O* and *B* stars. The polarization has been reexamined, and it now seems that it is not produced by synchrotron radiation but by the scattering of light from dust particles. The great quantity of dust is consistent with the observation of hot and very young stars, just as we find in regions of star formation in our own galaxy, such as Orion and the Carina nebula. Long-wavelength IR can penetrate even the thick dust, and measurements at wavelengths near 12 microns have been used to trace out the rotation curve of the galaxy.

The model that emerges from these newer observations is that of a spiral galaxy surrounded by much gas and dust and viewed almost edge-on. The dust scatters and polarizes the light that we see, and the movement of the surrounding gas is part of a much larger system that has been revealed by radio mapping, which shows M82 and its larger spiral neighbor M81 enveloped in an enormous hydrogen cloud.

The explosion model was first considered when quasars were new and active galaxies were very much the center of attention. It now seems as though M82 is a less-active but still interesting object, but further observations might well reveal new features that will require us to reconsider our model again.

## 29.3 BL LAC OBJECTS

The variable object BL in the constellation Lacerta (the Lizard) is obviously the scene of great turmoil, but we do not yet know for certain whether or not it belongs to our galaxy. Absence of proper motion sets a lower limit to its distance at only 200 pc, well within the galaxy, but other observations including some of the interstellar absorption and a disputed redshift of about $Z = 1.05$ place it well beyond at about 300 Mpc.

What is so peculiar about BL Lac (and another thirty similar objects) is the total absence of any emission lines from its bright central region, coupled with rapid variations in brightness seen in visible, radio, and IR. Most of its energy is radiated in the IR with a distinctly nonthermal spectrum. Polarization is strong, with rapid variations. These spectral signatures indicate a synchrotron source in which the direction of movement of the electrons and the controlling magnetic field are changing rapidly.

Some structure shows up in the VLBI radio observations at 8 GHz. There are two components about 1 milli arc sec apart, and a rate of increase in separation of about 1 milli arc sec/year has been seen. The brightness of the components shows such rapid variations that it is very difficult even to determine the shape of the radio spectrum. In visible light, BL Lac shows peaks of brightness around $13^m$ but is usually within the range $14^m$ to $16^m$. Night-to-night variations of $0.^m3$ have been seen. In contrast to many active objects, no X-rays have been detected thus far.

Coordinated observations have been made by observers in the U.S., Europe, and Israel. With such a wide spread in longitudes, BL Lac could be kept under observation for up to 12 hr. Polarization changes are large and rapid; up to 30° rotation in polarization was observed in one 6-hr period.

## 29.4 SS 433

In 1977 Bruce Stephenson and Nicholas Sanduleak (of Case Western Reserve University) published a list of stars showing strong hydrogen emission lines. Object SS433 had actually been discovered as far back as 1960, but its bizarre nature, even by standards that include such things as quasars and BL Lac, was not noted until 1978 when Bruce Margon (then at U. C. L. A.) first examined its spectrum in great detail. Since then an intensive observing program, carried out at many laboratories, has confirmed the strange nature of this object's radiation. The hydrogen spectrum was observed—it displays red and blue Doppler shifts simultaneously. The red- and blueshifts were seen to change in synchronism (Fig. 29.16), varying significantly from night to night, with repetition through a 164 day cycle.

**Figure 29.16** Red and blue Doppler shifts observed in SS 433, showing the regular 164 day periodicity (Bruce Margon, University of Washington).

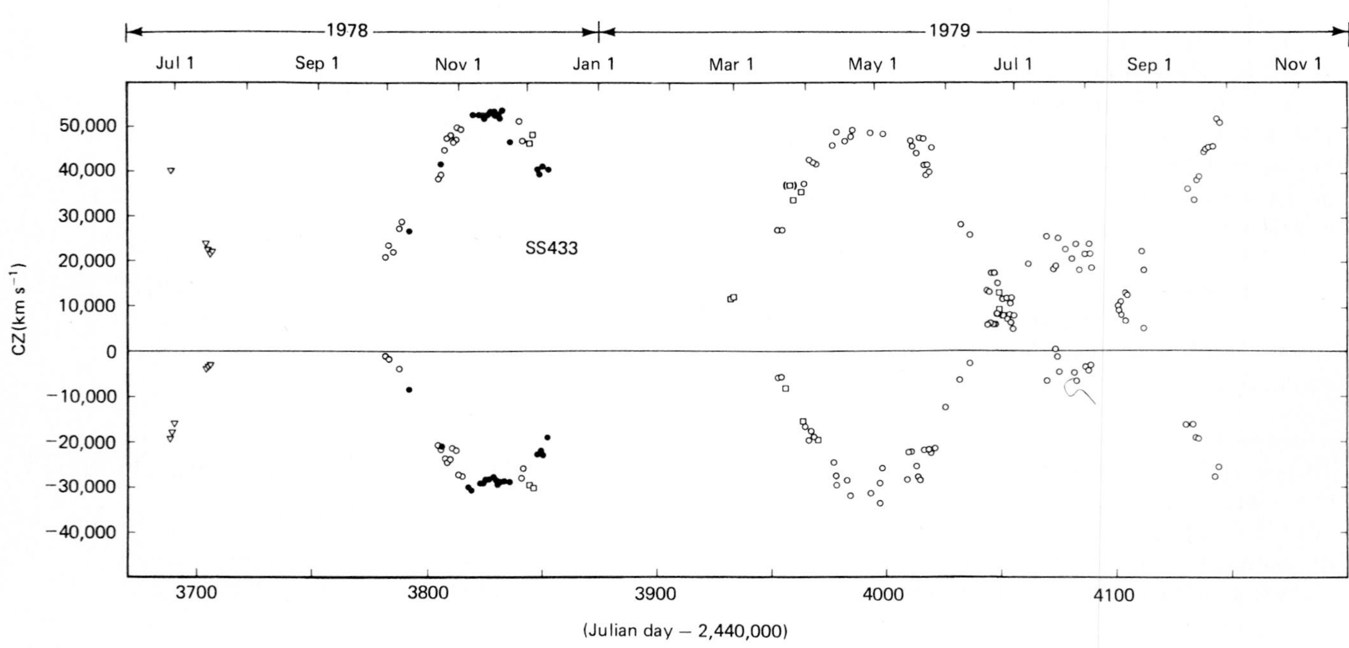

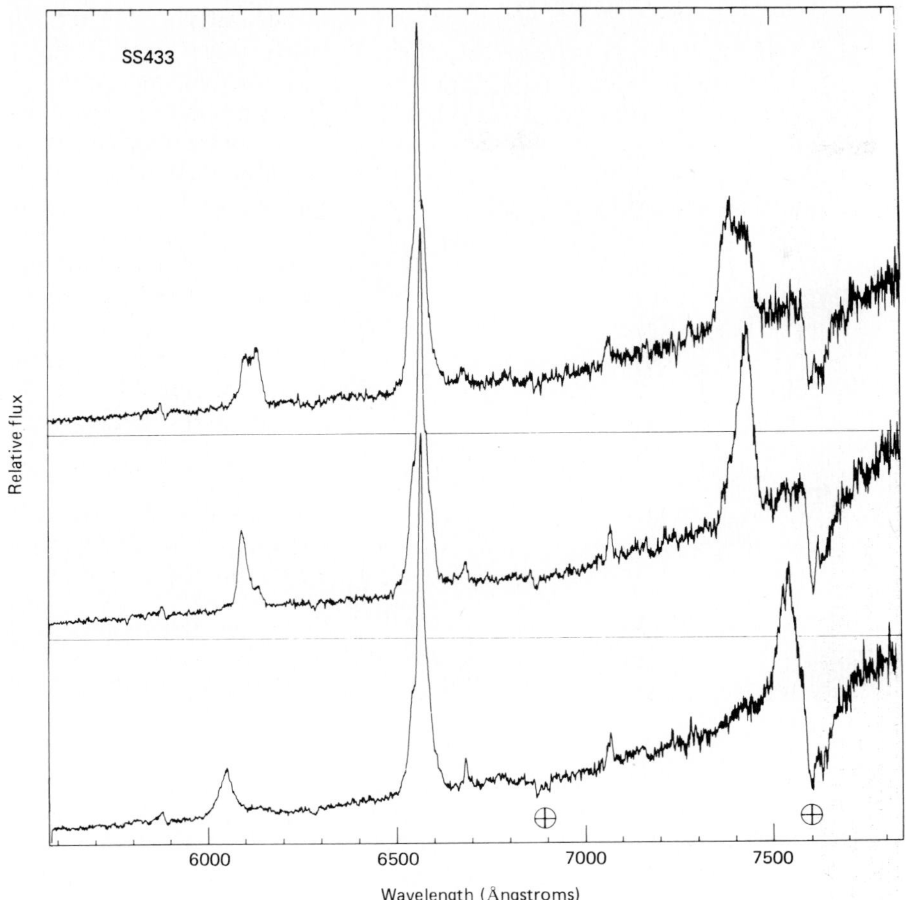


**Figure 29.17** Spectrum of SS 433 in the region of the red hydrogen Hα line (6562 Å). The spectra were taken over a period of 4 days. The strong peak at 6562 Å is unchanged, but the feature on the blue side moves by about 100 Å and changes shape. At the same time, the IR feature near 7500 Å moves out by about 150 Å. (Bruce Margon, University of Washington, and the *Astrophysical Journal*).

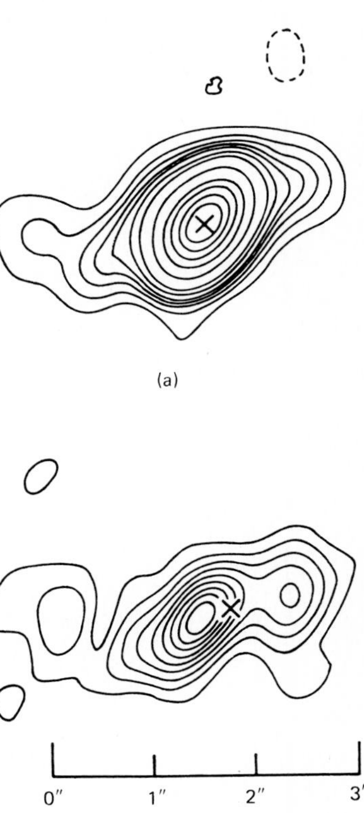


**Figure 29.18** VLA map of SS 433 at 6 cm wavelength. (a) total intensity, and (b) the linearly polarized intensity. The cross marks the optical position of SS 433 (NRAO).

One model involves a rotating pair of jets of gas, moving away from a central object at 80,000 km/sec (about the quarter of the speed of light) (Fig. 29.19). The line of sight from the Earth is at 78° to the rotation axis. Model details (dimensions, total energy) depend on distance, and this has not yet been determined. Interstellar absorption lines lead to a value of about 3.5 kpc, almost twice as far as the Crab. The Einstein X-ray satellite (HEAO 2) has detected X-rays coming from SS 433 but rather less intense than is usually seen in X-ray binaries. There seem to be connections between SS433 and W50, a nearby HII region that resembles a supernova remnant.

In addition to the 164 day cycle, there is some evidence for a 13 day periodicity in the average wavelength of the hydrogen lines, showing a maximum redshift velocity of 75 km/sec.

What sort of model can be constructed? Jonathan Katz, Margon's colleague at U. C. L. A., has suggested a binary system with transfer of matter from one star to a disc around the other. Perhaps so much matter is being transferred that the resulting X-rays blow some of the matter out into two jets whose radiation gives SS433 its unique pair of Doppler shifts. The size of the emitting region is estimated to be between 0.1 and 30 AU; about one millionth of a solar mass is being ejected per year and the energy radiated by the two beams is about $10^{40}$ ergs/sec. But much of this is conjecture, and more observations may well force us to change our model.

**Figure 29.19** Artist's drawing of a model of SS 433. Matter is drawn from the extended atmosphere of the large main sequence star into an accretion disc that surrounds a neutron star. The observed red- and blue-shifted features are emitted from high-speed jets (Bruce Margon, University of Washington).

## 29.5 X AND $\gamma$ BURSTS

Very short and intense bursts of X-rays and $\gamma$-rays have been detected by satellite-borne detectors (Figs. 29.20, 29.21 and 29.22). Faint astronomical objects need long time exposures and this effectively rules out the photographic detection of rapid changes. Electronic detectors can respond much more rapidly to changes in light intensity and are particularly well suited to X- and $\gamma$-rays where each photon carries enough energy to be detectable.

Between 1969 and 1972, the U.S. Vela satellites detected sixteen very short bursts of $\gamma$-rays, photons with energies of 0.2 to 1.5 MeV. The bursts lasted from less than 0.1 sec to about 30 sec. The Vela satellites (Fig. 29.23) were not primarily astronomical satellites but were designed to detect nuclear explosions and act as monitors for the Test-ban Treaty. Simultaneous observations by several Vela spacecraft could attest to the reality of the bursts but give only a very poor indication of the arrival direction of the pulse of photons.

$\gamma$-ray bursts occur at a rate of about ten per year; in a typical event, the flow of energy is $10^{-4}$ erg/cm$^2$ sec or twice as intense as from the brightest visible star. If the sources of these bursts were distributed evenly throughout the galaxy, some might be close and some far. A careful search using large-area detectors on high-altitude balloons found few small bursts. This result is taken to indicate that the sources are probably confined to our galaxy, since distant sources should produce weak bursts that were not detected.

After the initial report of the Vela observations by the group at Los Alamos Scientific Laboratory in 1973, further bursts were noted but no progress was made in identifying the sources. It was only after 1978 that specifically designed $\gamma$-ray burst detectors were available: the Soviet Union's Venus probes Venera 11 and 12 and the International Sun-Earth Explorer 3 (ISEE-3). Their observations have been coordinated with those from other scientific spacecraft: the solar orbiter Helios-2, the Pioneer-Venus Orbiter, Prognoz-7, HEAO-2 and the Solar Maximum Mission (SMM). The result is

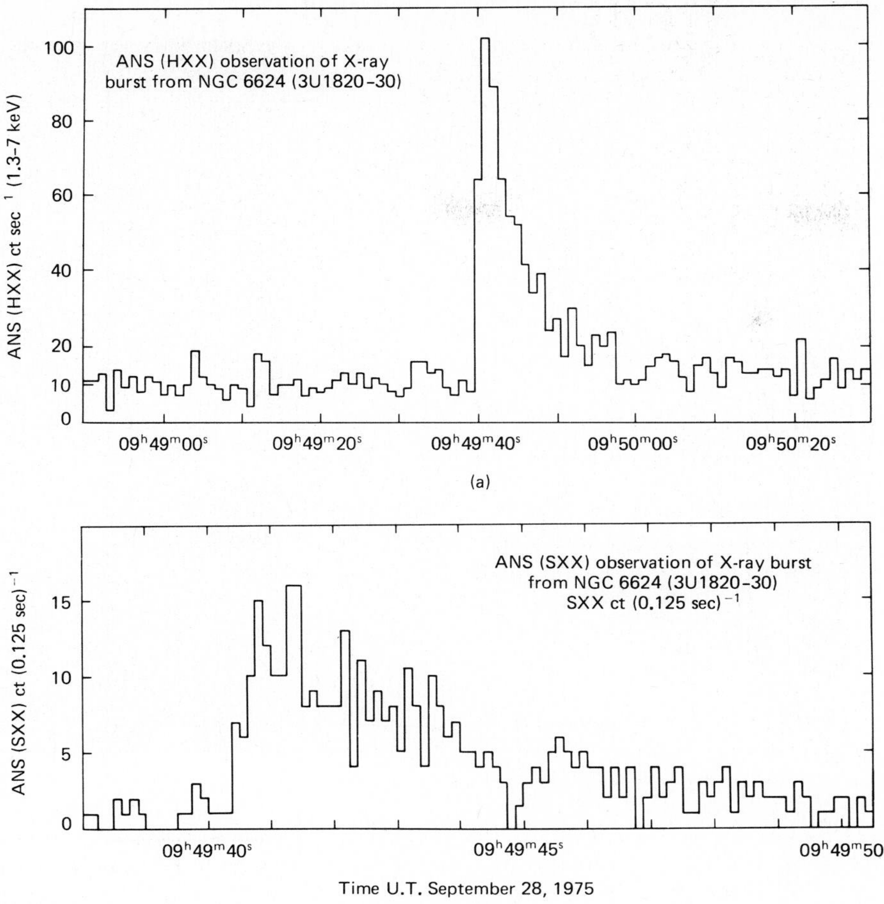


**Figure 29.20** X-ray burst from NGC 6624. Note the very sudden increase in intensity, and the rapid decay, with most of the activity occurring in only 10 sec. Observations were made with the Astronomical Netherlands Satellite (Jonathan E. Grindlay, Harvard College Observatory).

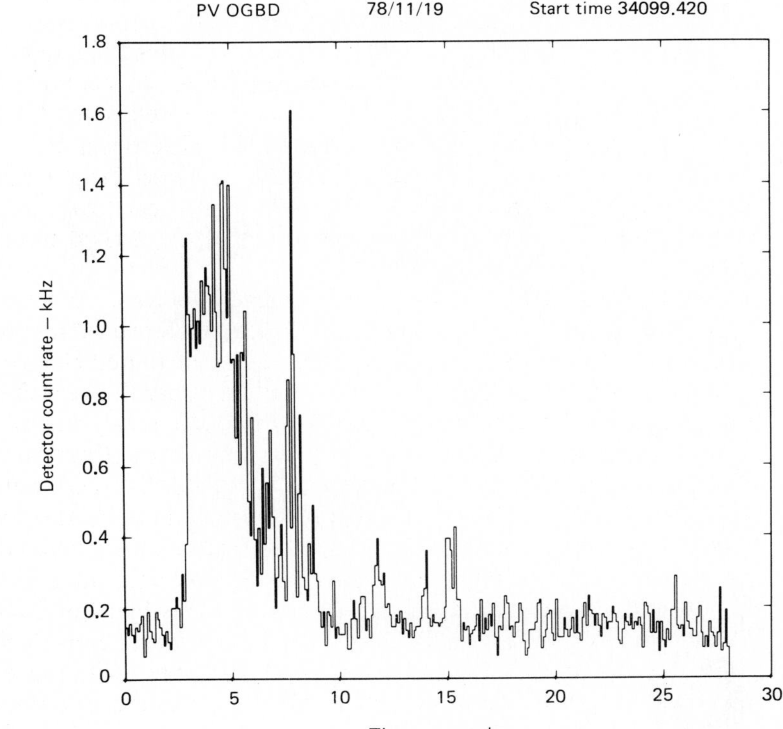


**Figure 29.21** Gamma-ray burst, observed in 1978 from Pioneer Venus spacecraft. This was one of the most energetic events recorded (Ray Wm. Klebesadel, Los Alamos National Laboratory).

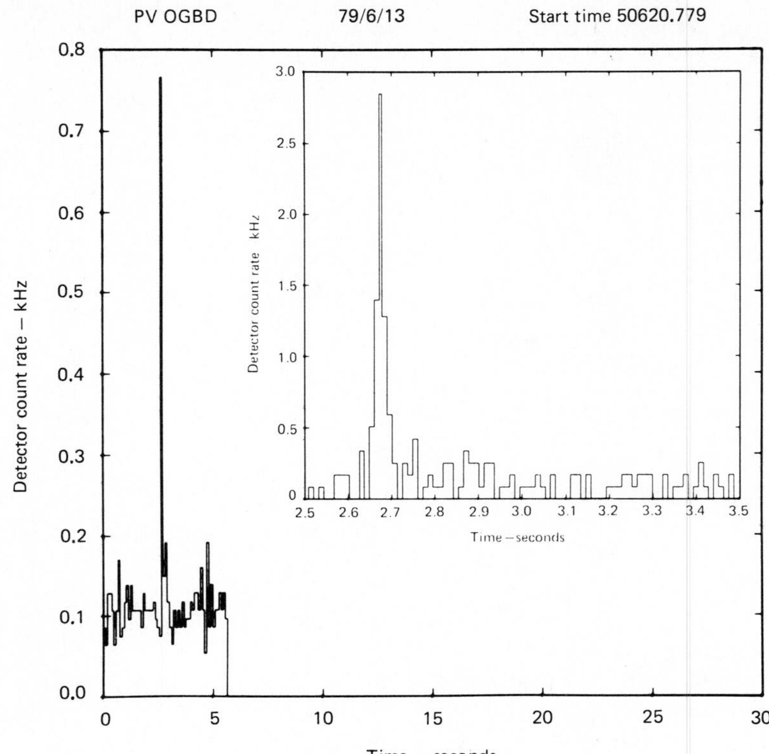


**Figure 29.22** Gamma-ray burst, observed in 1979 from Pioneer Venus spacecraft. This event is one of the shortest yet observed, lasting for only about one tenth of a second (Ray Wm. Klebesadel, Los Alamos National Laboratory).

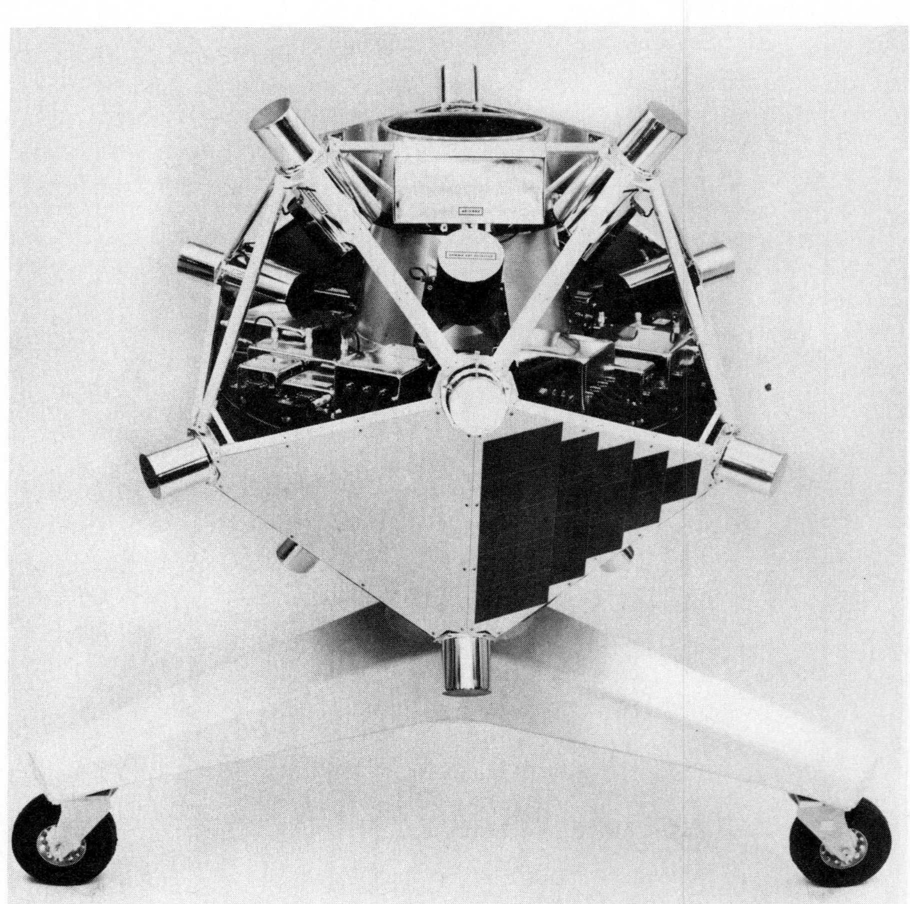

**Figure 29.23** Vela Satellite, with which the gamma-ray bursts were discovered (TRW Inc.).

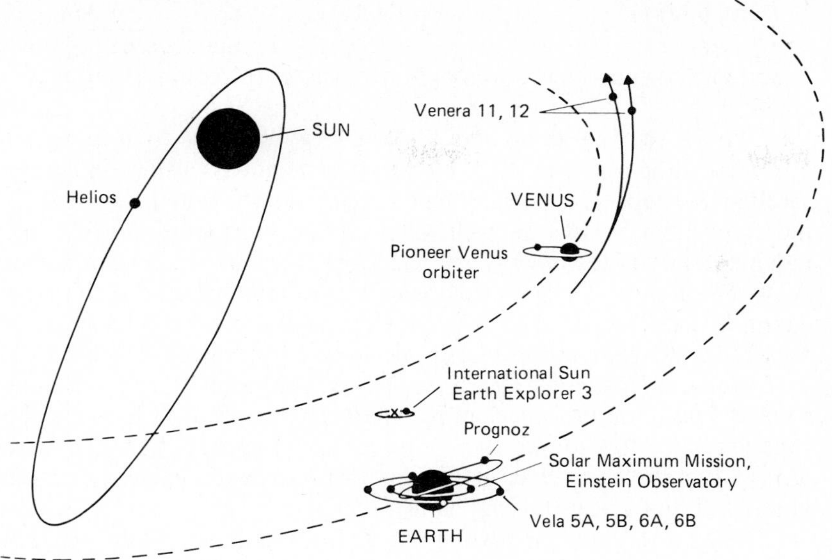


**Figure 29.24** Spacecraft network used for determining the arrival directions of gamma-ray bursts during 1978 to 1980. There was one group of vehicles in orbit around the Earth, three in orbit around the Sun, and one around Venus. By now many of these are no longer operating, but a similar network can be made up of other vehicles, and the system operates in some ways like a long-baseline radio interferometer on Earth (Ray Wm. Klebesadel, Los Alamos National Laboratory).

a network in space (Fig. 29.24), in which time delays between arrival of gamma rays at the different detectors lead to directional information for the sources of the bursts. This is somewhat like the widely separated radio receivers comprising the VLA and VLBI on Earth.

The first identification came in March 1979 at N49, a supernova remnant in the LMC (Fig. 29.25), but then came a number of events for which an error-box as small as 1 arc min$^2$ could be specified but without finding suitable candidates in any of them: no supernova remnants, no X-ray emitters, no quasars. Some source boxes were close to the galactic plane and contained many optical objects but none that seemed unusual; other source boxes contained no optical object at all.

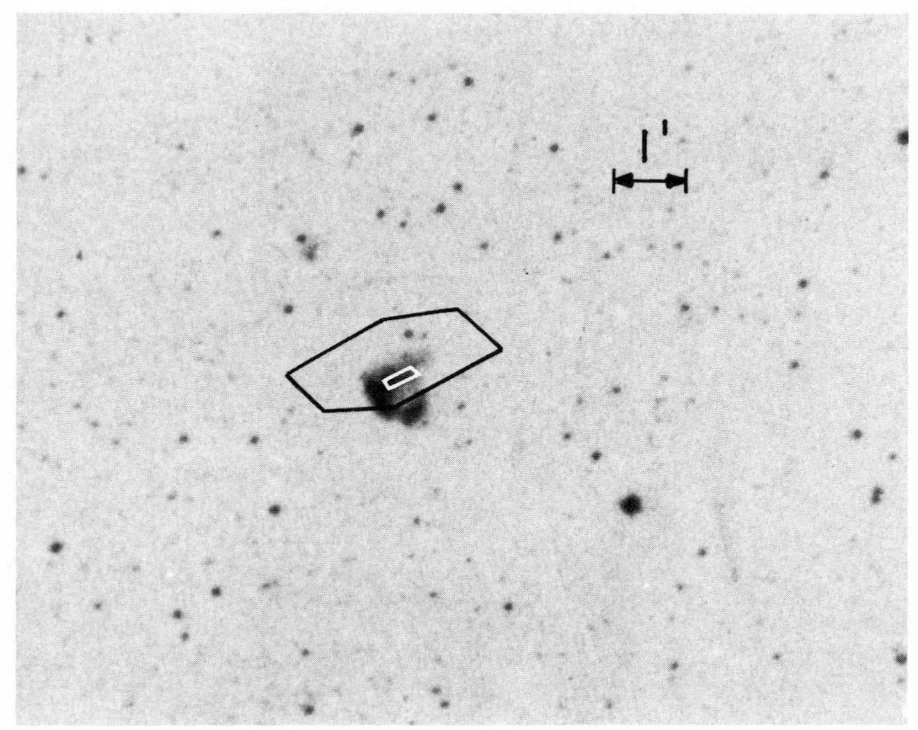


**Figure 29.25** Location of March 5, 1979 burst. The large box shows the source location as determined from the time-of-arrival of the data at three spacecraft. The error box contains the expanding cloud of gases of N49, a supernova remnant in the Large Magellanic Cloud. Using data from ten spacecraft, the error box was reduced to the small area marked within the supernova remnant (Ray Wm. Klebesadel, Los Alamos National Observatory).

As might be expected, theoretical speculation has been energetic, covering a range of physical processes but, in the absence of source identification and distance measurements, the situation remains confused and fascinating.

The X-ray bursts involve lower-energy photons than the gamma rays and were first noted in 1971 by Russian scientists using detectors on the satellite Kosmos 428. These results were not published until 1975 at about the same time as Jonathan Grindlay of the Harvard College Observatory independently noted a burst of 1 to 20 keV X-rays in data obtained by Dutch A. N. S. satellite. Many more observations have by now been obtained, for example with the M. I. T. X-ray experiment on the Small Astronomical Satellite 3 (SAS-3) and the Orbiting Solar Observatory-8 (OSO-8).

Some bursters last for 10 to 20 sec, and tend to recur at intervals of several hours, other repeat more rapidly. The arrival of higher-energy photons tends to stop sooner than those of lower energy. Several sources emit X-rays steadily as well as in bursts, with the steady emission amounting to about 100 times more energy than in bursts.

Unlike the gamma rays, good directional information was obtainable for the X-ray bursts. The globular cluster NGC 6624, already known as a regular X-ray emitter, was identified as a source of bursts. Interestingly, when the regular X-ray output of NGC 6624 is high, no bursts are seen, but when the regular output is low, bursts are observed. Intensive hunting has detected about thirty X-ray bursters, about a third of which are in globular clusters.

A *rapid burster* was discovered in 1976. It emitted several thousand bursts per day for 2 months, then stopped completely, resumed after a year and again stopped after a few weeks. This remarkable object seems to be located in a globular cluster that had not been previously known from optical astronomy (Fig. 29.25). Most bursters lie within about 30° of the galactic plane and if they are at typical distances of 10 kpc, then their luminosities must be in the range of $10^{38}$ to $10^{39}$ erg/sec.

The spectrum of X-rays within a burst seems to be thermal; for example, combining data from several bursts from two bursters, the M. I. T. group could deduce that the radiation came from ionized gas that heated to 30 million K in the first 7 sec, and then cooled steadily to 5 million K, after which it was too cool to be detected.

If these objects are at 10 kpc, then their sizes must be about 25 km, consistent with the size of neutron stars, but the limits of error are large, in distances and in the interpretation of the spectrum.

With the present small observational base, it is no surprise that a completely satisfactory model has not yet been proposed to account for the X-ray bursters. The best model (except for the rapid burster) involves nuclear burning on the surface of a neutron star. Another model involves a neutron star that attracts matter from its surroundings. This accelerated matter slams into the surface of the neutron star releasing energy as X-rays. Variations on this theme assume accretion from the companion star in a binary system or from a disc of matter that surrounds the neutron star and rotates rapidly with it. Irregularities in the accretion rate will show up as irregularities in the X-ray emission. No eclipse-type variations have been seen, and no model yet tells us why such a large fraction of the bursters are in globular clusters. An alternative model involves accretion onto a black hole instead of onto a neutron star. For each model, detailed calculations (with computers) can be traced out to predict further characteristic radiation signatures that can be searched for. In this way, we should be able to reject some models or refine others, but we have no final answers yet.

## CHAPTER REVIEW

**Quasars:** problem of identification of radio with optical objects

- use of lunar occultations and radio interferometers.
- spectra show large redshifts in emission lines, often also large but different redshifts in absorption lines.
- if quasars are at distances given by velocity-distance relation for galaxies, then power radiated is very large.
- subclass show connections with galaxies that have different redshifts.
- sizes of some quasars given by speed of brightness variations.
- gravitational focusing by galaxies to produce two or more images of a quasar.
- ejection of high-speed electrons and viewing direction can produce appearance of speed greater than the speed of light.

**Active galaxies:** very luminous, with synchrotron emission indicating presence of magnetic field and many high-speed electrons.

**BL Lac objects:** absence of emission lines, rapid variations of emission at all wavelengths.

**SS 433:** simultaneous and synchronized red- and blueshifts in spectrum, with 164 day period.

- probably two high-speed electron jets, from a rotating source.

**X- and gamma-ray bursts:** short duration bursts of high-energy photons.

- origin not well-established, mechanism probably transfer of mass onto a condensed object.

## NEW TERMS

| | | | |
|---|---|---|---|
| **active galaxy** | **superluminal velocity** | **gravitational focusing** | **quasar** |
| **error box** | **BL Lac object** | **bursts (X and gamma)** | |

## PROBLEMS

**1.** Why is it sometimes difficult to match an optical object to a radio or X-ray source?

**2.** How has the Moon been used to determine a more accurate position for the quasar 3C273?

**3.** What is meant by superluminal speeds? How are they explained?

**4.** The data listed in Table 29.2 were taken from the original paper by Greenstein and Schmidt. They used a value of $H$ = 100 km/sec/Mpc for the Hubble constant. Calculate the luminosities using the value of $H$ = 75 km/sec/Mpc.

**5.** Suppose a quasar is at a distance of 1,000 Mpc and has an angular diameter of 0.001 arc sec.

a. What is its linear diameter?
b. What is the shortest time in which one might expect to see it brighten significantly?
c. What redshift would be seen in its spectrum?

**6.** The quasar 3C48 has a spectral line at a wavelength of 5097 Å. From an analysis of other lines, we know that this line was emitted at a wavelength of 3727 Å.

a. At what wavelength would we expect to observe a line emitted at 4101 Å?
b. How fast is 3C48 moving away from us (with $H$ = 75 km/sec/Mpc)?
c. Assuming that the Hubble Law (with $H$ = 75 km/sec/Mpc) can be used for quasars, how far away is 3C48?
d. How long does it take light to reach us from this quasar?

**7.** A quasar is observed to have a redshift of $z = 1.15$. What is its recession velocity, and what is its distance (assuming the Hubble relation applies)?

**8.** A quasar is seen to brighten by 50 percent over the course of a day. How large is the quasar?

**9.** What is the evidence that supports the idea that quasars are at distances given by their redshifts and the Hubble relation? What evidence argues against this idea?

**10.** Telescopes on the Earth's surface cannot observe the Lyman series of lines in the hydrogen spectrum, even from the Sun, yet some of these lines can be seen in quasar spectra. Explain this.

**11.** The Balmer lines of hydrogen are some of the most prominent lines features of many astronomical objects. Why are these lines not observed in some quasars?

**12.** How large a redshift is needed in order for the Balmer H$\alpha$ line to be shifted out of the visible, to an IR wavelength of 10,000 Å, for a quasar?

**13.** Why were quasars given that name? What more recent observation suggests that they are the bright central cores of galaxies?

**14.** What is meant by gravitational focusing, and how was it discovered?

**15.** The Astronomical Society of the Pacific sells bumper stickers with these legends

a. Black Holes are Out of Sight.

b. Quasars are Far Out.

Explain those statements, and devise some other legends that are also astronomically correct.

**16.** What is so unusual about the spectrum of SS 433?

**17.** How are X-ray and gamma-ray bursts detected?

**18.** X-rays and gamma rays have very energetic photons. Their energies are usually measured in thousands and millions of electron volts. 1 electron volt $= 1.6 \times 10^{-12}$ ergs. Calculate the wavelength of a gamma ray that has 1 MeV energy (= 1 million electron volts).

CHAPTER

30

# Life in the Universe

Until this point, we have been describing the physical universe as we infer its properties from our astronomical observations, laboratory experimenting and calculations. In this chapter, we will turn our attention to the still unanswered question: Are we, on Earth, the only living creatures in this universe?

There has long been speculation about the possibility of life elsewhere in the universe or intelligent beings who could respond to our messages, perhaps even visit the Earth. With our present scientific knowledge, can we do more than speculate? We can describe the conditions in which recognizable life can exist. We can try to estimate the likelihood of those conditions existing elsewhere. More concretely, we can use these studies to guide our searches for that life. New areas of scientific activity have thus developed: **Exobiology** is the study of life away from the Earth; the search for extraterrestrial intelligence is designated **SETI**, as distinct from **CETI**, communicating with extraterrestrial intelligence.

The study of life beyond the Earth can very quickly pass from the consideration of other habitable planets and communications with other civilizations to the planning of space travel, to the establishment of space colonies and to the social structure of small communities that are isolated during many-generation journeys. In this chapter, we shall confine our attention to topics that make use of astronomical methods or information. Even though they have their own intrinsic interest, projects such as the establishment of a space colony or the exploitation of the mineral resources of the Moon very quickly raise the clearly nonastronomic matters of relief of population pressure on Earth, budgetary priorities, and social and political preferences. Endlessly debatable, those topics are probably better discussed outside an astronomy course.

We should note that despite some enthusiastic efforts, no clear signs of extraterrestrial intelligence or recognizable life have yet been found. A brief description of life on Earth is useful as a starting point.

## 30.1 LIFE ON EARTH

The life that we too often take for granted is very fragile. The environmental concerns of recent years have alerted some of us to many man-made biological insults, but there other factors that have just as important an influence on the existence of life on Earth.

Living organisms on Earth—plants and animals—are made up of molecules. While life continues and the plants and animals reproduce, the molecules are rearranged. At too low a temperature, chemical reactions will stop or proceed very slowly; at too high a temperature, the complex biological molecules will dissociate into simpler molecules or into atoms (Fig. 30.1). Life as we know it is, therefore, confined to temperatures roughly in the range 230 to 370 K (−40 to 210°F).

Essential molecular building blocks are the complex molecules DNA and RNA; we do not know whether alternative combinations of atoms could provide a basis for life. Inside biological molecules, most atoms are hydrogen (about 63 percent), about 24 percent is oxygen, 10 percent carbon, and about 1 percent nitrogen with traces of heavier atoms that are needed. In a sense, this composition is no surprise, for these are the most abundant atoms in the solar system (apart from the chemically inert helium and neon). We know that the solar system has these elemental abundances because the sun formed well after the birth of the galaxy. Very old stars have different elemental

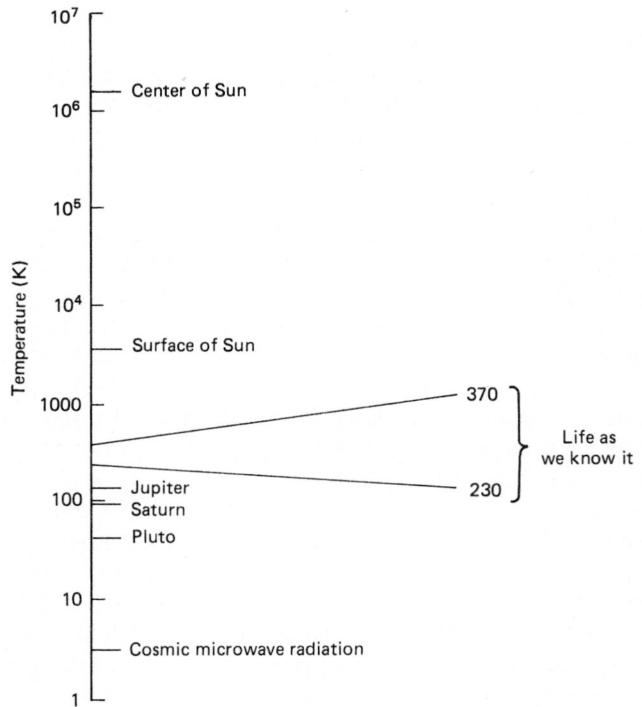


**Figure 30.1** Temperatures in the universe. Between the low temperature of the cosmic microwave radiation (3K) and the solar temperature, there is only a very narrow band in which life can exist as we know it. Above about 370 K, biological molecules will separate into their constituent atoms, below about 230 K chemical reactions needed for life will proceed very slowly or stop.

abundances with fewer heavier elements because they formed before the ISM had been enriched by heavier nuclei generated in rapidly evolving stars. Old stars would, therefore, not seem to be good candidates for searches for life.

Life on Earth needs a nourishing atmosphere. This can be liquid (rivers or oceans) or gaseous (the atmosphere). We ingest or inhale this atmosphere and extract vital nutrients. Photosynthesis in plants and solubility of chemicals in water are essential processes in the chemistry of life. We do not know whether intelligent life could develop or exist in an atmosphere other than air and use liquids other than water. Our atmosphere serves us in another important way: It shields us from the harmful effects of solar UV radiation and from cosmic rays. Again, we do not know how life would develop in a different radiation environment.

Whatever the nature of the surrounding atmosphere, animal survival requires the development of the means of sensing enemies and food. Thus our eyes are sensitive to radiation of certain optical wavelengths, and some snakes have IR sensors. Other atmospheres might be transparent at different wavelengths, and living creatures would adapt accordingly. This might affect their ability to receive our messages.

Life requires a steady supply of energy. In our case, the reservoir is the Sun. Conceivably, life could develop and survive with more or less energy than we receive from the Sun. Overall, then, we know the requirements for the support of life on Earth, but we cannot exclude the possibility that other forms of life may exist elsewhere under different physical conditions.

## 30.2 THE ORIGIN OF LIFE

Even with all of the ingredients that we think are needed, will life develop or is there an element of chance? What is needed to trigger the process of evolution? We do not know. The fossil record tells us that the earliest living organisms emerged about 1.5 billion years after the solidification of the solar system, and that more than another billion years elapsed before cells with nuclei existed. Recognizable hominids go back to only a few million years, and the interval during which man has been sufficiently advanced to send, receive, and identify intelligent signals is vanishingly small.

How did life start? During the 1930s, the English scientist J. B. S. Haldane and the Russian A. I. Oparin independently had the idea that the primitive atmosphere of the Earth would have had little oxygen, and that organic molecules might form in an atmosphere that consisted predominantly of ammonia ($NH_3$) and methane ($CH_4$). A crucial experiment was carried out by Harold Urey and Stanley Miller in the 1950s at the University of Chicago (Fig. 30.2). They fired an electric spark through a mixture of ammonia and methane and discovered that certain biological molecules called amino acids were produced. One can imagine lightning bolts in the Earth's primitive atmosphere producing such an effect. Miller and Urey's experimental results are suggestive but not conclusive. Recent geological evidence from very old Greenland rocks points to an atmosphere that had much more carbon dioxide than methane 3.8 billion years ago. In 1960 Philip Abelson at the Geophysics Laboratory of the Carnegie Institution of Washington produced amino acids from mixtures of nitrogen, carbon monoxide, and hydrogen.

What this evidence shows is that some biological molecules can be produced under conditions that might have existed long ago on the Earth. Equating those molecules with life requires far more evidence than we now have. Synthesizing processes are necessary but not sufficient.

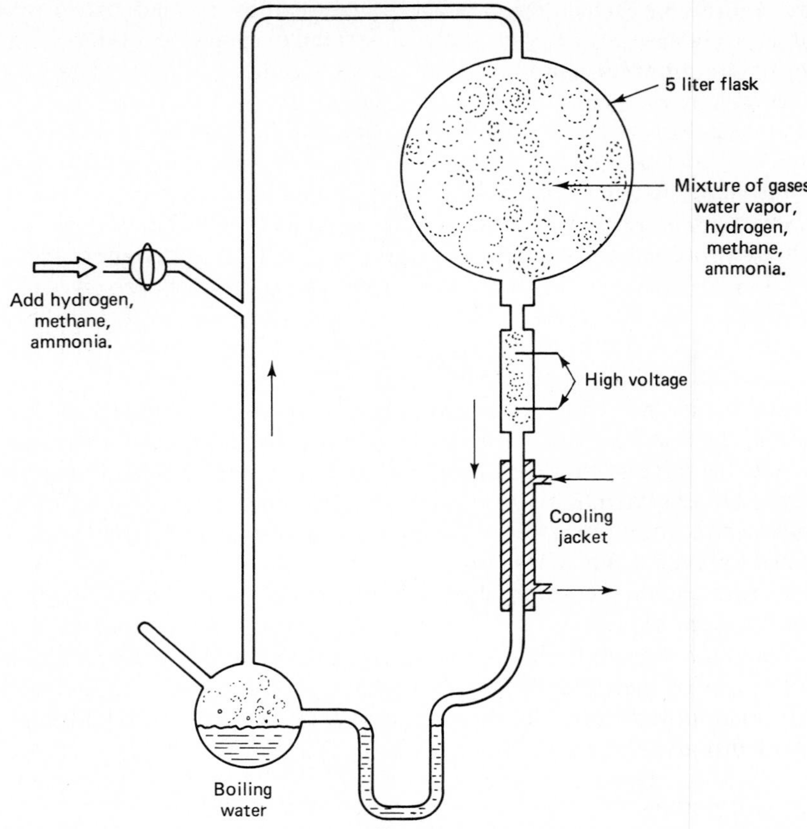


**Figure 30.2** The Miller-Urey experiment. The system was sealed after water had been boiled. The air was pumped out and metered quantities of hydrogen, methane, and ammonia added. The system was again sealed, and the water kept boiling for a week. A high voltage was applied across electrodes in the reaction chamber, and the cooled and condensed reaction products were collected in the water. By the end of the week, the water was a deep red due to the organic compounds. The water was removed and analyzed and showed the presence of amino acids.

## 30.3 LIFE IN THE SOLAR SYSTEM

Could any other planet be inhabited? Distances from the Sun and the resulting temperatures immediately restrict the possibilities. Mercury certainly and Venus probably are far too hot, Jupiter and the more distant planets are far too cold. Mars is the obvious candidate and, as we recall, provided the motivation for Lowell to establish his observatory in Arizona.

In 1924 when there was a particularly close approach of Mars, David Todd organized a search for radio signals. The Army and Navy, among others, cooperated. The federal government endorsed an appeal for periods of radio quiet so that the search would have minimal interference. Considering the primitive state of radio technology at that time, it is not surprising that a number of listeners reported receiving strange signals, none clearly the result of intelligent transmission.

A *New York Times* editorial took note of these attempts, and gave as its opinion

> Nevertheless, it is well not to be too nearly sure that Mars is uninhabited. It would be a dreadful place to live, indeed, but so is many a city slum, so that's no final proof.

The same editorial also quoted Marconi's comment on these attempts to detect signals from another planet

> a fantastic absurdity.

Somewhat more surprising was the *Scientific American's* comment that

> For ourselves, the question of communication with Mars leaves us absolutely cold. Whatever the attempt might cost, from ten cents to twenty million dollars, we could very easily spend it in a better scientific cause.

Now, more than 50 years later and after the Viking landings, the canals have not materialized and neither have any radio signals. The strongest evidence for life on Mars lies only in the ambiguous interpretation of some of the Viking landers' experiments.

No organic molecules have been seen in any lunar samples, but some have been found in several meteorites. Contamination of meteorites on the ground before discovery can be a problem, but the detection of organic molecules has been confirmed. At the least, these results indicate that meteorites come from or have passed through regions in which the Miller-Urey or similar reactions occur, but the size and location of these regions and their significance for the origin of life remain open questions.

## 30.4 LIFE BEYOND THE SOLAR SYSTEM

By now it is clear that if living organisms are found anywhere else in the solar system, they will have to be at a very primitive level. We will have to look beyond the solar system for advanced forms of life. What is the likelihood that there are Earthlike planets with intelligent beings aboard, around other stars? With about 100 billion stars in the galaxy, surely *somewhere* the conditions should be congenial to life? We know, from radio observations of the ISM, that complex carbon-based molecules are present, but are there enough in the right places?

We might start by directing our attention to sunlike stars (of type G2) that are at stages of their evolution where planets could exist with physical conditions hospitable to life. Perhaps 10 percent of the stars in the galaxy fall into this group, but we have no evidence yet for the presence of any satellites. Even for a star as close as Barnard's, no low-mass companion has been definitely detected.

Other nearby stars have also been examined for evidence of faint or unseen companions, possibly large planets. In all cases, apparent perturbations in the stellar positions were at the very limits of the experimental techniques.

NASA's Project Orion in 1976 included a workshop to study the ways of searching for planetary systems. Telescopes now in use for astrometric work are all refractors built more than 65 years ago, and none was designed for the precision and stability needed for planetary searches. The workshop concluded that there were now better methods to search for possible planets than through improvements in ground-based astrometry. The preferred methods involve the detection of radiation from the planets, either visible light reflected from the parent star or IR radiation emitted by the planet's surface.

The Orion study did not include satellite-borne astrometry, which is free of effects due to the Earth's atmosphere but still prone to other optical problems. Perhaps improved parallax and proper motion measurements, such as will be conducted by the European collaboration's Hipparcos satellite, will produce the first clear evidence for other planets.

Attempts might be made to detect other civilizations or send out signals they might recognize. Before doing this, however, considerable planning is needed so that search strategies optimize the likelihood of success. For example, it would be foolish to try to communicate at wavelengths that are easily absorbed in the ISM.

Much of the present thinking about searches for extraterrestrial intelligence can be traced to an important 1959 article by Giuseppe Cocconi and Philip Morrison, then at Cornell University. In that article, "Searching for Interstellar Communications," we find for the first time a careful review in modern technical terms of the requirements of radio communication. They advocated using the 1420 MHz (21 cm) hydrogen line, as the spectral line that would most surely be universally recognized by intelligent civilizations, and they ended their brief paper with this observation

> The probability of success is difficult to estimate; but if we never search, the chance of success is zero.

Since 1960 several groups have studied this problem, paying great attention to such matters as radio antennas, amplifier sensitivity, and frequency selection. One recent study prepared for NASA and now published as a paperback, can be strongly recommended for those of you who wish to pursue this topic. It is listed in the Bibliography.

An initial and essentially practical step is the estimation of the number of possible civilizations. One might, of course, just go ahead and conduct a search on the basis of hope, but unless there seem to be reasonable chances of success, it will be hard to persuade scientists to devote their time and, equally important, to persuade the Congress to appropriate the funds that will be needed.

In 1961 a Cornell radio astronomer and colleague of Cocconi and Morrison, Frank Drake, devised a formula that connects the relevant factors in an attempt to put this calculation on a quantitative basis. Drake's formula for the number of civilizations that are at the stage where they could send out identifiable signals is

$$N = R_* f_g f_p n f_l f_i f_c L$$

Here $R_*$ is the average rate of star formation; $f_g$ is the fraction of stars that have conditions suitable for life support, and $f_p$ is the fraction of these stars that have suitable planets; $n$ is the number of suitable planets per star in such a system; $f_l$ is the fraction of the suitable planets in which life develops; $f_i$ is the fraction of these planets that produce intelligent life; $f_c$ is the fraction of intelligent communities that are at a communicative stage; and $L$ is the duration of that stage. You might well wish to add other factors or devise a different formula, but this one provides a good starting point.

Some of the factors in the Drake formula ($R_*$, $f_p$) can be reasonably estimated from our present astronomical knowledge. Other factors ($f_g$, $n$, $f_l$, $f_i$) involve biochemical considerations; and $f_c$ and $L$ would seem to be open to wide speculation.

Let us try to put numbers into the formula. For $R_*$, we know that our galaxy has about $10^{11}$ stars and an age of about $10^{10}$ years; an average star formation rate of ten per year is plausible. Since star formation did not start promptly and since some massive stars will have long since disappeared, this is a conservative estimate. We might confine our attention to sunlike stars, that is, near type G2, and set $f_g = 0.1$. Of these stars, perhaps $f_p = 0.1$ is a fair guess for the fraction with planets and $n = 2$ an average for the number of suitable planets. An optimistic estimate for the chance of life evolving sets $f_l = 1$, and we guess the chance of intelligent life to develop to be $f_i = 0.5$. We might estimate that the duration of intelligent life, $L$, is a million years, with 500 million years to get to that stage and another 500 million years of deterioration, so that

$$f_c = \frac{10^6 \text{ years of intelligent life}}{10^9 \text{ years total life}} = 10^{-3}$$

We thus find

$$N \sim (10)(.1)(.1)(2)(1)(0.5)(10^{-3})(10^6) = 100$$

for the number of civilizations with which we might now communicate.

Obviously, the result is very sensitive to the values assumed for the different factors. More generous estimates easily lead to a value of one million for *N*, and conservative values for some of the factors soon reduce *N* to a value of ten or less. The final result is highly subjective, but the formula is an interesting example of an attempt to handle an apparently shapeless problem. Just because it uses a formula and numbers, the Drake formula should not be considered in the same way as the many other formulas we have encountered. Rather, it provides a useful way of focusing the discussion and identifying the various factors.

The next steps depend on the value of *N* just derived and thus to a considerable extent one's prejudices. A large value for *N* can be used to justify searches for life, and a small value of *N* could be taken to indicate that a successful search is so improbable as not to justify the time, effort, and expense.

## 30.5 PROJECTS THUS FAR

What physical options are open to us to reach these civilizations, and what has been done? Space travel is out of the question for the foreseeable future. Depending on one's evaluation of the Drake formula, the nearest civilization might be 100 LY or more away. The technology does not yet exist to propel space ships at even half the speed of light. Even at the speed of probes such as Voyager, such a journey would take a million years. The unfolding of a voyage lasting many generations belongs to science fiction, but not yet to science.

A more realistic and already available avenue of communication is provided by radio waves that suffer minimal interstellar absorption. The technology at hand is sufficient for us to be able to send a detectable radio signal across the galaxy. With the range of wavelengths available, how does one decide which one to use? Considering the cosmic radio noise that will be "seen" in all parts of the galaxy and the intrinsic limitations of radio detector sensitivity, a band of frequencies between 1400 and 1720 Mhz seems best. (The corresponding wavelengths are 12 to 20 cm). This spectral band lies between the lines of hydrogen (H) and the OH radical, components of water ($H_2O$), and has been termed the *water hole* (Fig. 30.3). Just as some wild animals gather at the water hole each day, so intelligent beings might meet in this part of the spectrum for their communication.

As far back as 1960 in Project Ozma, Drake used the world's largest radio dish, the 1000 ft reflector at Arecibo, Puerto Rico to try to detect signals from two nearby stars, Epsilon Eridani and Tau Ceti. Ozma II continued for some years after 1972 with Ben Zuckerman (University of Maryland) and Patrick Palmer (University of Chicago) systematically examining about 600 sunlike stars. Table 30.1 lists a selection of the many searches that have by now been undertaken.

The merit of this type of search for extraterrestrial intelligence is cheapness. The limitations are the amount of time available on the radio telescopes, in competition with other scheduled astronomical observations, and the smallest detectable signal which depends on the collecting area of the dish. An ambitious project (Cyclops) was designed in 1971 (Fig. 30.4). This would have been devoted to SETI and used more than 1000 radio telescope dishes, spread over about 10 square mi. Another project considered briefly would have involved a network of radio telescopes on the lunar surface, making use of craters much as the Arecibo telescope makes use of some of the hills in Puerto Rico. With a projected cost of 5 billion 1971 dollars for the Earth-based system, the project never progressed beyond the drawings and feasibility studies.

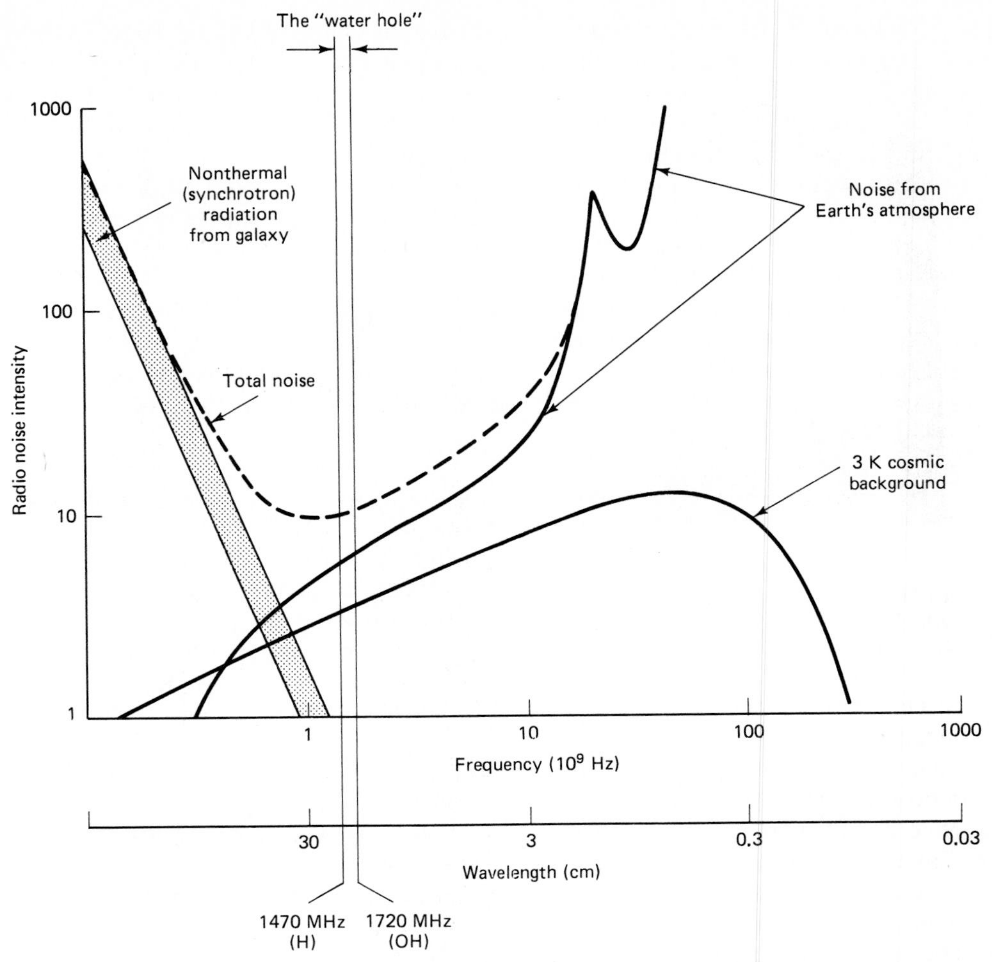


**Figure 30.3** The water hole. The best wavelengths for searching for extraterrestrial signals are between the spectral lines of H and OH. At other wavelengths, there will be far greater background noise, originating either in the Earth's atmosphere or from electrons in our galaxy (the synchrotron radiation) (Adapted from a NASA diagram).

*Table 30.1* SEARCHES FOR EXTRATERRESTIAL INTELLIGENCE: A SELECTION

| *Investigator* | *Telescope* | *Dates* | *Frequency or Wavelength* | *Targets* |
|---|---|---|---|---|
| Drake | N.R.A.O. | 1960 | 1420 MHz | Epsilon Eridani; Tau Ceti |
| Troitsky | Gorky, U.S.S.R. | 1968 | 21 cm and 30 cm | 12 nearby sunlike stars |
| Verschuur | N.R.A.O. | 1972 | 1420 MHz | 9 nearby stars |
| Zuckerman, Palmer | N.R.A.O. 91 m | 1972–1976 | 1420 MHz | 674 nearby sunlike stars |
| Kardashev | Eurasian network, I.C.R. | 1972 | several | pulsed signals from entire sky |
| Dixon, Kraus | Ohio State Univ. 53 m | 1973 | 1420 MHz | area search |
| Bridle, Feldman | Algonquin Park 46 m | 1974 | 23 GHz | 70 nearby stars |
| Drake, Sagan | Arecibo 305 m | 1975–1976 | 1420, 1653, 2300 MHz | several nearby galaxies |
| Tarter, Cuzzi, Black, Clark | N.R.A.O. 91 m | 1977 | 18 cm | 200 nearby stars |
| Horowitz | Arecibo | 1978 | 21 cm | 185 nearby stars |
| Cohen | Arecibo | 1978 | 6 cm | 25 globular clusters |
| Witteborn | Mt. Lemon | 1980 | 8–13 micron (IR) | 20 stars |
| Biraud, Tarter | Nancay | 1981 | 6 cm | 300 stars |

(a) (b)

**Figure 30.4** Early and ambitious designs for radio telescopes for SETI. (a) An array of over 1000 radio dishes on Earth, (b) utilization of Moon craters, much as the Arecibo radio antenna is housed in the hollow between hills (NASA/Ames).

The various SETI projects were attempts to detect signals from other civilizations. We can gain some useful insights for the planning of further searches by asking the inverse question: Are there regular activities on Earth that produce signals that could be detected by distant observers? If they were to detect such signals, might they deduce that there was a civilization on Earth, or would the signals from Earth appear only as a noisy jumble?

Radio, television, and radar are widely used on Earth. Some of the signals are of low power, but the total power radiated is large. Only a minute fraction of the energy radiated by radio and television stations is intercepted by our antennas. The rest is absorbed in the atmosphere or leaks away from the Earth. **Eavesdropping** makes use of this leakage of signals that are not narrowly beamed but are transmitted in a wide range of directions. Some of the signals at frequencies above 20 MHz might be detectable in this way far from Earth.

The total power radiated by all 9000 FM radio stations on Earth is about 300 Megawatts or $3 \times 10^{15}$ ergs/sec. The 2000 strongest televisions stations transmit a total of about $10^{16}$ ergs/sec. If distant civilizations possessed a technology as sophisticated as present technology on Earth, these signals could be detected from a distance of only about 2 LY, half the distance to the nearest star.

Supposing the power quadrupled to detectable levels over the next few years; what sort of signals might an observer detect? A study was carried out by W. T. Sullivan and his colleagues at the University of Washington. They took into account the geographical distribution of transmitters, concentrated mostly in North America, western Europe, and Japan, and they calculated the way in which signals from the Earth would appear to change as the Earth rotates. The results are shown in Fig. 30.5.

Our distant observer would note a 24 hr periodicity, with marked fluctuations as the concentrations of stations along the U.S. east and west coasts rose and set with the Earth's rotation. These eavesdropping observations would allow our observer to measure the rate of rotation of the Earth. Doppler shifts in the detected frequency could be used to track the Earth's orbital movement. Variations in signal strength could be correlated with solar ac-

**Figure 30.5** Eavesdropping. Radio signals that would be detected by an observer located at Barnard's star, 1.8 pc away. The large peaks correspond to the concentrations of radio and television transmitters, *Rise* and *set* refer to the appearance and disappearance of each region as the Earth rotates (W. T. Sullivan III, University of Washington, and *Science*).

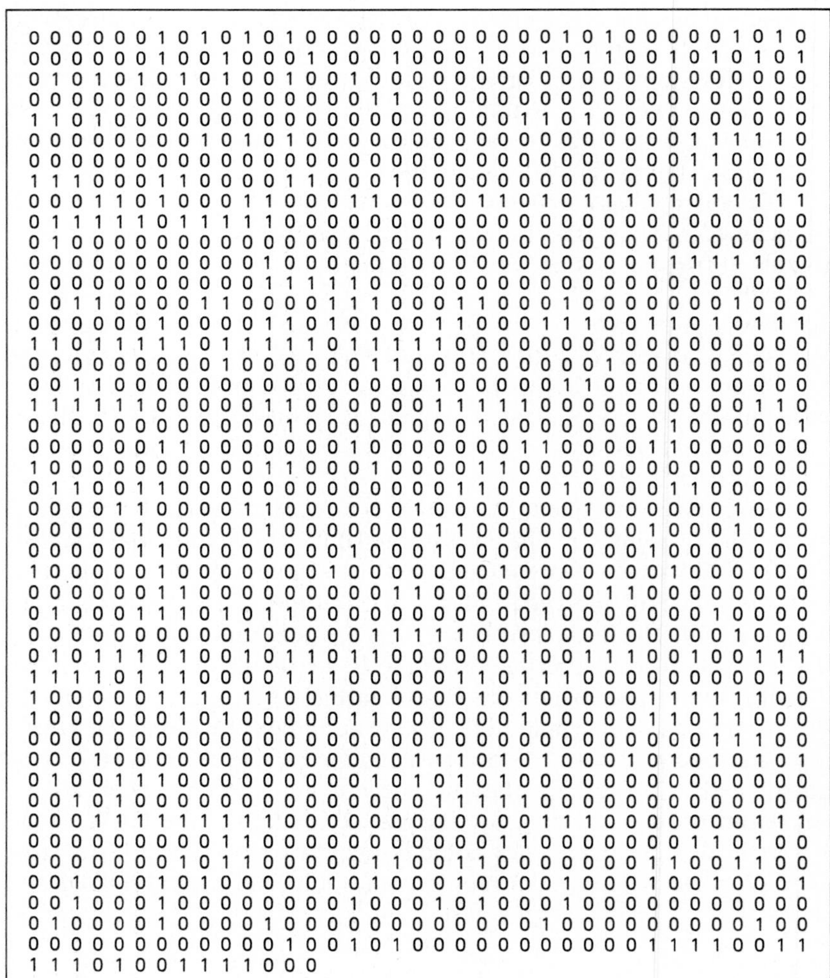

```
0 0 0 0 0 0 1 0 1 0 1 0 1 0 0 0 0 0 0 0 0 0 0 0 0 1 0 1 0 0 0 0 0 1 0 1 0
0 0 0 0 0 0 1 0 0 1 0 0 0 1 0 0 0 1 0 0 0 1 0 0 1 0 1 1 0 0 1 0 1 0 1 0 1
0 1 0 1 0 1 0 1 0 1 0 0 1 0 0 1 0 0 0 0 0 0 0 0 0 0 0 0 0 0 0 0 0 0 0 0 0
0 0 0 0 0 0 0 0 0 0 0 0 0 0 0 0 1 1 0 0 0 0 0 0 0 0 0 0 0 0 0 0 0 0 0 0 0
1 1 0 1 0 0 0 0 0 0 0 0 0 0 0 0 0 0 0 0 0 0 0 1 1 0 1 0 0 0 0 0 0 0 0 0 0
0 0 0 0 0 0 0 0 1 0 1 0 1 0 0 0 0 0 0 0 0 0 0 0 0 0 0 0 0 0 0 1 1 1 1 1 0
0 0 0 0 0 0 0 0 0 0 0 0 0 0 0 0 0 0 0 0 0 0 0 0 0 0 0 0 0 0 0 1 1 0 0 0 0
1 1 1 0 0 0 1 1 0 0 0 0 1 1 0 0 0 1 0 0 0 0 0 0 0 0 0 0 0 0 0 1 1 0 0 1 0
0 0 0 1 1 0 1 0 0 0 1 1 0 0 0 1 1 0 0 0 0 1 1 0 1 0 1 1 1 1 1 0 1 1 1 1 1
0 1 1 1 1 1 0 1 1 1 1 1 0 0 0 0 0 0 0 0 0 0 0 0 0 0 0 0 0 0 0 0 0 0 0 0 0
0 1 0 0 0 0 0 0 0 0 0 0 0 0 0 0 0 0 0 1 0 0 0 0 0 0 0 0 0 0 0 0 0 0 0 0 0
0 0 0 0 0 0 0 0 0 0 0 1 0 0 0 0 0 0 0 0 0 0 0 0 0 0 0 0 0 1 1 1 1 1 1 0 0
0 0 0 0 0 0 0 0 0 0 0 1 1 1 1 1 0 0 0 0 0 0 0 0 0 0 0 0 0 0 0 0 0 0 0 0 0
0 0 1 1 0 0 0 0 1 1 0 0 0 0 1 1 1 0 0 0 1 1 0 0 0 1 0 0 0 0 0 0 0 1 0 0 0
0 0 0 0 0 0 1 0 0 0 0 1 1 0 1 0 0 0 0 1 1 0 0 0 1 1 1 0 0 1 1 0 1 0 1 1 1
1 1 0 1 1 1 1 1 0 1 1 1 1 1 0 1 1 1 1 1 0 0 0 0 0 0 0 0 0 0 0 0 0 0 0 0 0
0 0 0 0 0 0 0 0 0 1 0 0 0 0 0 0 1 1 0 0 0 0 0 0 0 0 0 1 0 0 0 0 0 0 0 0 0
0 0 1 1 0 0 0 0 0 0 0 0 0 0 0 0 0 0 0 1 0 0 0 0 0 1 1 0 0 0 0 0 0 0 0 0 0
1 1 1 1 1 1 0 0 0 0 0 1 1 0 0 0 0 0 0 1 1 1 1 1 0 0 0 0 0 0 0 0 0 0 1 1 0
0 0 0 0 0 0 0 0 0 0 0 0 1 0 0 0 0 0 0 0 0 1 0 0 0 0 0 0 0 0 1 0 0 0 0 0 1
0 0 0 0 0 0 1 1 0 0 0 0 0 0 0 1 0 0 0 0 0 0 0 1 1 0 0 0 0 1 1 0 0 0 0 0 0
1 0 0 0 0 0 0 0 0 0 0 1 1 0 0 0 1 0 0 0 0 1 1 0 0 0 0 0 0 0 0 0 0 0 0 0 0
0 1 1 0 0 1 1 0 0 0 0 0 0 0 0 0 0 0 0 0 1 1 0 0 0 1 0 0 0 0 1 1 0 0 0 0 0
0 0 0 0 1 1 0 0 0 0 1 1 0 0 0 0 0 0 1 0 0 0 0 0 0 0 1 0 0 0 0 0 0 1 0 0 0
0 0 0 0 0 1 0 0 0 0 0 1 0 0 0 0 0 0 0 1 1 0 0 0 0 0 0 0 0 1 0 0 0 1 0 0 0
0 0 0 0 0 1 1 0 0 0 0 0 0 0 0 1 0 0 0 1 0 0 0 0 0 0 0 0 0 1 0 0 0 0 0 0 0
1 0 0 0 0 0 1 0 0 0 0 0 0 0 1 0 0 0 0 0 0 0 1 0 0 0 0 0 0 0 1 0 0 0 0 0 0
0 0 0 0 0 0 1 1 0 0 0 0 0 0 0 0 0 1 1 0 0 0 0 0 0 0 0 1 1 0 0 0 0 0 0 0 0
0 1 0 0 0 1 1 1 0 1 0 1 1 0 0 0 0 0 0 0 0 0 0 0 1 0 0 0 0 0 0 0 1 0 0 0 0
0 0 0 0 0 0 0 0 0 1 0 0 0 0 0 0 1 1 1 1 1 0 0 0 0 0 0 0 0 0 0 0 0 1 0 0 0
0 1 0 1 1 1 0 1 0 0 1 0 1 1 0 1 1 0 0 0 0 0 0 1 0 0 1 1 1 0 0 1 0 0 1 1 1
1 1 1 1 0 1 1 1 0 0 0 0 1 1 1 0 0 0 0 0 1 1 0 1 1 1 0 0 0 0 0 0 0 0 0 1 0
1 0 0 0 0 0 1 1 1 0 1 1 0 0 1 0 0 0 0 0 0 1 0 1 0 0 0 0 0 1 1 1 1 1 1 0 0
1 0 0 0 0 0 0 1 0 1 0 0 0 0 0 1 1 0 0 0 0 0 0 1 0 0 0 0 0 1 1 0 1 1 0 0 0
0 0 0 0 0 0 0 0 0 0 0 0 0 0 0 0 0 0 0 0 0 0 0 0 0 0 0 0 0 0 0 0 1 1 1 0 0
0 0 0 1 0 0 0 0 0 0 0 0 0 0 0 0 0 0 1 1 1 0 1 0 1 0 0 0 1 0 1 0 1 0 1 0 1
0 1 0 0 1 1 1 0 0 0 0 0 0 0 0 0 1 0 1 0 1 0 1 0 0 0 0 0 0 0 0 0 0 0 0 0 0
0 0 1 0 1 0 0 0 0 0 0 0 0 0 0 0 0 0 0 1 1 1 1 1 0 0 0 0 0 0 0 0 0 0 0 0 0
0 0 0 1 1 1 1 1 1 1 1 1 0 0 0 0 0 0 0 0 0 0 0 0 1 1 1 0 0 0 0 0 0 0 1 1 1
0 0 0 0 0 0 0 0 0 1 1 0 0 0 0 0 0 0 0 0 0 0 1 1 0 0 0 0 0 0 0 1 1 0 1 0 0
0 0 0 0 0 0 0 1 0 1 1 0 0 0 0 0 1 1 0 0 1 1 0 0 0 0 0 0 0 1 1 0 0 1 1 0 0
0 0 1 0 0 0 1 0 1 0 0 0 0 0 1 0 1 0 0 0 1 0 0 0 0 1 0 0 0 1 0 0 1 0 0 0 1
0 0 1 0 0 0 1 0 0 0 0 0 0 0 0 1 0 0 0 1 0 1 0 0 0 1 0 0 0 0 0 0 0 0 0 0 0
0 1 0 0 0 0 1 0 0 0 0 1 0 0 0 0 0 0 0 0 0 0 0 0 1 0 0 0 0 0 0 0 0 0 1 0 0
0 0 0 0 0 0 0 0 0 0 0 0 1 0 0 1 0 1 0 0 0 0 0 0 0 0 0 0 0 1 1 1 1 0 0 1 1
1 1 1 0 1 0 0 1 1 1 1 0 0 0
```

**Figure 30.6** Digitized message sent out from the Arecibo antenna in the direction of the globular cluster M13. The message was in binary form consisting of a sequence 1679 characters, representing 0s and 1s. (Each character was represented by one of two different radio frequencies.) To decode the message, it must be broken into seventy-three consecutive groups of twenty-three characters to display the pattern of 0s and 1s. The pattern is then made into a diagram by filling in the 1s and leaving the 0s blank as shown in Fig. 30.7 (Arecibo Observatory).

tivity (but the observer might not know that it was the solar wind that was causing this).

Some of the radars used by the Department of Defense are extremely powerful, but they are focused into narrow beams that sweep across the sky to detect incoming missiles or planes. These narrow beams could be detected at much greater distances than could the wide-beam radio or television, but a distant observer would detect only irregular blips and not necessarily at the same frequency each time. It would be hard to discover that these were man-made and not natural, such as from lightning.

Whereas SETI is passive, CETI is active. It is the deliberate communication with extraterrestrial intelligence. We have the ability to beam messages across the galaxy. What kind of message will be intelligible to the presumed audience? This problem is similar to that encountered in the translation of Egyptian hieroglyphics or the Cretan Linear B, but in those cases a Rosetta stone and a record of ancient languages were available. Without this sort of guide, some ancient scripts have not yet been deciphered.

Messages sent in digital form (a sequence of zeros and ones) are subject to less distortion than is the usual analog signal that is used for radio and television. (The improvement in digital phonograph records over analog is obvious at first hearing.) Unscrambling a digital message would consist of displaying the 0s and 1s as open and filled squares in an array to show a pattern. The resulting pattern should carry a message whose meaning would be universally recognizable. Much ingenuity went into the selection of the material in the message shown (Figs. 30.6 and 30.7) which was sent from the Arecibo radio telescope towards the globular cluster M13, 8 kpc away

**Figure 30.7** In this form, the pattern represents a number of very basic scientific facts that an intelligent civilization might be expected to recognize (Arecibo Observatory).

**Figure 30.8** M13, the globular cluster towards which the Arecibo message was beamed. This cluster is 8 kpc distant, and has an angular diameter of 4.8 arc min. It was discovered by Halley in 1714. Messier in 1764 was unable to see individual stars because of the poor quality of his telescope. The H-R diagram for this cluster shows a large number of giants that have evolved from the main sequence, and the cluster's age is estimated to be about 10 billion years (Yerkes Observatory).

(Fig. 30.8). Transmission of the message took 169 sec, at a radiated power of $3 \times 10^{12}$ watt. Interestingly, the selected frequency was at 2380 MHz, rather than in the water hole. The message will take 24,000 years to reach any listeners so that the soonest we can expect a response will be in 48,000 years. Of course, for us to receive a return signal requires that our own signal be recognized as a message and not thought to be only noise. Incorrect translation might produce an incoherent pattern or, worse, be taken as an insult, perhaps even as a declaration of war. We should also recall what happened when Galileo sent a coded message to Kepler (Chap. 6).

## 30.6 SUMMARY

While there are some theological reasons for believing that life on Earth is unique, it is hard to escape the scientific inference that life probably exists in other parts of the galaxy as well. Making contact with other civilizations presents some severe technical difficulties. Whether such efforts are desirable, worthwhile or cost-effective is a question that produces very personal responses. Some astronomers are actively engaged, others think it is a waste of time and money.

Some astronomers who took part in earlier searches seem now to be much more critical. Indeed, in 1979 a conference was held at the University of Maryland, with the topic "Where are they?" The discussions focused on problems raised by the persistent failures of all searches. A line of argument is this: If there are so many intelligent civilizations, some should be very advanced. If so, why have they not overcome the problems of extended space flight and colonized as far as the solar system? Why have they left no traces? After all, it took only 12 years of the space age from the first Sputnik to the landing of the men on the Moon.

Research could settle the question in one way: through positive results. Failure to detect life or to record intelligible messages, however, will not rule out the possibility that life does actually exist somewhere.

If the decision on further projects rested only on scientific grounds, then it seems most likely that research would continue. The national financial climate will probably continue to force the scientific community into hard decisions in which choices will have to be made, for example, between SETI or an additional shuttle flight. Many political, nonscientific interests will surely also enter. SETI did, however, receive a favorable recommendation for a total funding of twenty million dollars in the 1982 report of the Astronomy Survey Committee of the National Academy of Sciences:

**Plate 49** The Pleiades, one of the nearest open clusters. The wispy glowing regions are dust clouds, reradiating light from the young hot stars (Palomar Observatory photograph).

**Plate 50** Emission nebula (M16, NGC 6611) is about 8 pc across optically, but radio observations show that the nebula extends about three times further (KPNO).

**Plate 51** Eta Carinae nebula, NGC 3372, contains many very hot young stars. At a distance of 2.5 kpc, the gas and dust in this nebula radiates energy absorbed from the hot stars, one of which exploded spectacularly in 1843 and was for a while the second brightest star in the sky (CTIO).

**Plate 52** The Crab Nebula, remnant of the 1054 A.D. supernova, located about 2 kpc from Earth. Expanding filaments of gas show colors that indicate the types of atoms present, and the general glow comes from very-high-speed electrons following the spiral paths in the nebula (Palomar Observatory photograph).

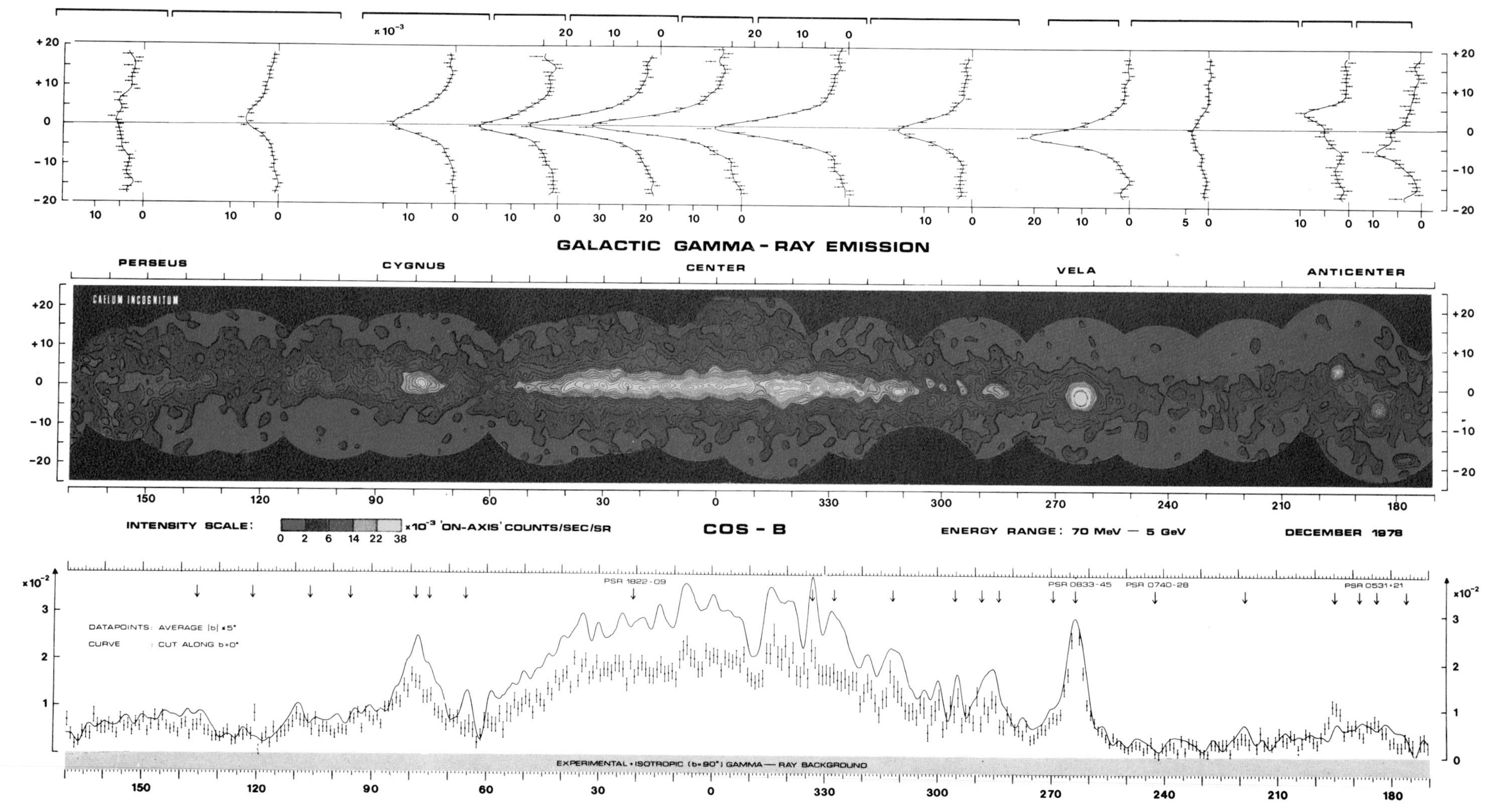


**Plate 53** Map of the Milky Way galaxy, based on gamma-ray observations with the COS-B satellite (Mayer-Hasselwander, et al., Max-Planck Institut für Astrofysik, and *Astronomy and Astrophysics*).

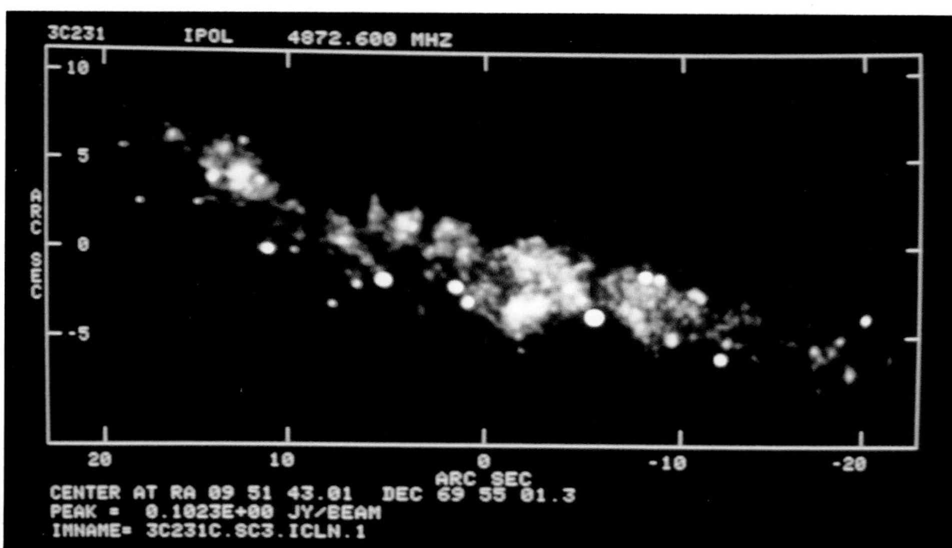


**Plate 54** Map of M82, based on observations with the VLA at 6 cm wavelength. Shown here is the central 700 pc of the galaxy which has a total extent of 8000 pc. Hot spots and jetlike features can be seen (NRAO).

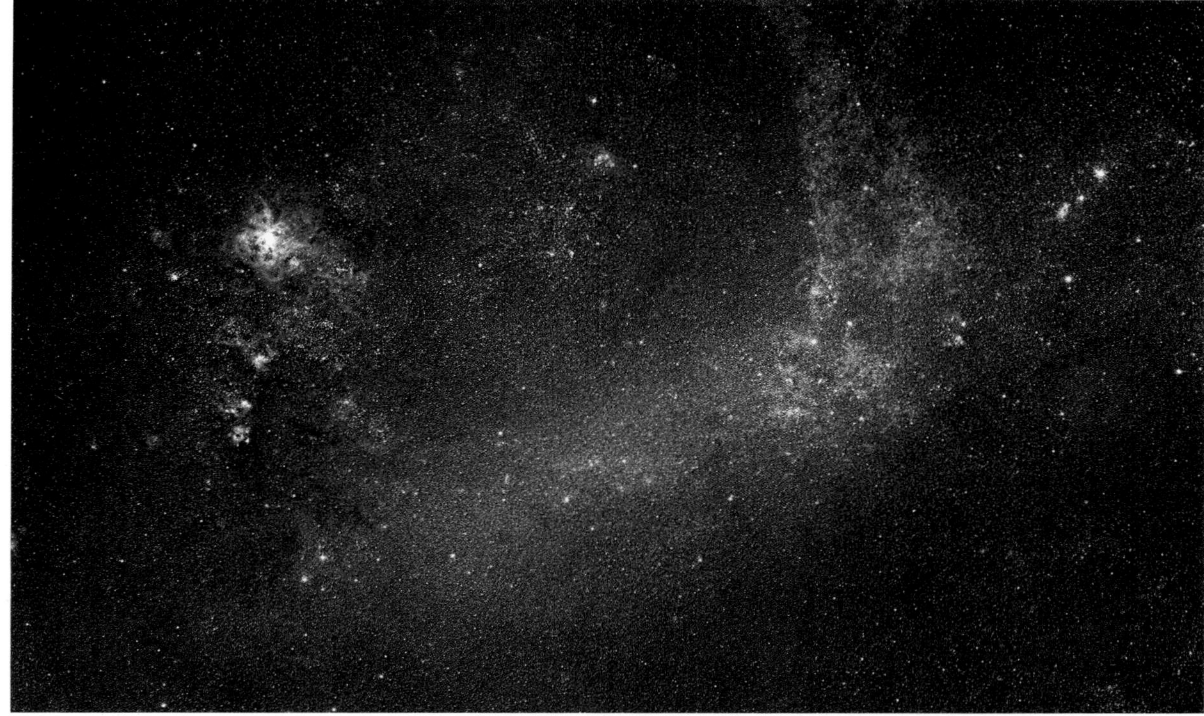

**Plate 55** The Large and Small Magellanic clouds, members of the Local Group of galaxies. The galaxies are not visible from northern observatories (CTIO).

**Plate 56** The Trifid Nebula, (M30, NGC 6514), an emission nebula heated by its hot *O* stars, with dense dust lanes appearing as dark rifts (KPNO).

**Plate 57** The Andromeda Nebula (M31, NGC 224), a member of the Local Group and thought to resemble the Milky Way galaxy in its spiral structure (Palomar Observatory photograph).

**Plate 58** Quasar 3C273. During the 900 sec exposure, the light from the quasar was blocked, and a very sensitive charge coupled device (CCD) was used to produce the image of the light from the galaxy containing the quasar. The jet can be seen clearly as well as the fainter outer regions of the galaxy [J.A. Tyson, Bell Laboratories].

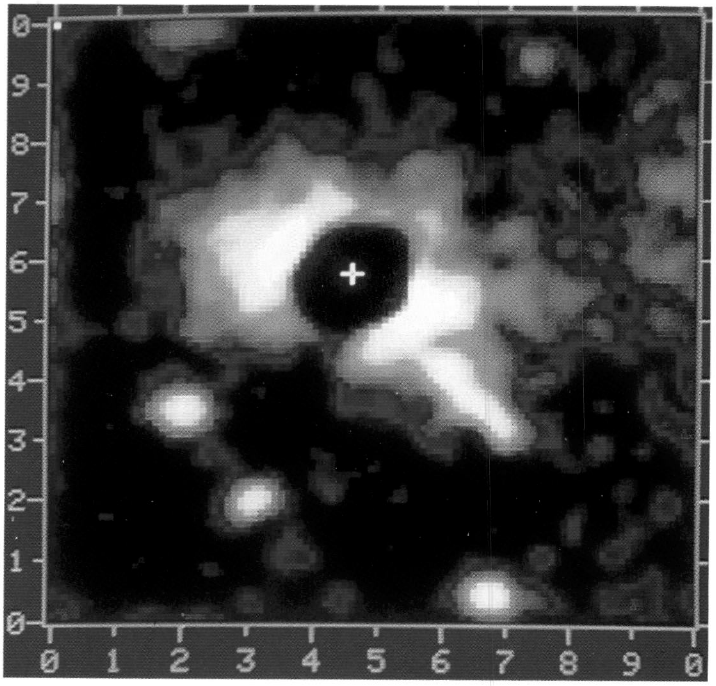

**Plate 59** One of the 34m radio telescopes that comprise the NASA Deep Space Network for receiving signals from satellites and space probes. It has been proposed that this network be used in the program that attempts to detect signals from extraterrestrial civilizations (Jill C. Tarter and NASA/Ames).

**Plate 60** Microwave antennas used on high-altitude airplane for measurement of the anisotropy of the 3 K background radiation (G. F. Smoot, Lawrence Berkeley Laboratory).

An astronomical search for Extraterrestrial Intelligence (SETI) supported at a modest level, undertaken as a long-term rather than as a short-term project, and open to the participation of the general scientific community.

## CHAPTER REVIEW

**Conditions for life as we know it:** chemical composition, temperature range.

**Origin:** formation of molecules in primitive atmosphere on Earth.

— Miller-Urey experiment.

**Life in the solar system:** on Mars—no signs

— elsewhere: conditions probably not favorable.

**Life beyond the solar system:** CETI communication, and SETI search.

— Drake equation: attempt to quantify estimates of likelihood of life existing.

**Projects:** OZMA and others—searches at various frequencies.

— message transmitted to M13 globular cluster.

## NEW TERMS

**CETI**
**Drake equation**
**eavesdropping**
**exobiology**
**Miller-Urey experiment**
**SETI**
**water hole**

## PROBLEMS

**1.** List the methods that have been used to date to search for life beyond the Earth. What different assumptions did these experiments make about the kinds of life they tried to detect?

**2.** Why is the Miller-Urey experiment important? What important fact is *not* shown by that experiment?

**3.** If you observed the Earth from Mars, what evidence would you be able to obtain for the existence of intelligent life?

**4.** Why are wavelengths around 21 cm thought to be best for extraterrestrial signals?

**5.** What is the Drake equation? On the basis of what you have learned in this course and your own best guesses for some quantities, compute a value for $N$.

**6.** Suppose that planets are found around a star that is four times as luminous as the Sun. At what distance from that star (in AU) might you find a planet that had the same temperature as we have on Earth, assuming the same atmosphere as we have?

**7.** Suppose that superluminal speeds are indeed possible, even for space travel. How long might it take for travelers to arrive at Earth from the nearest star, if they could travel at ten times the speed of light?

**8.** Consider the content of the message transmitted from Arecibo. Do you think that it might be understood? Are there other items that you would have used in addition to or instead of any of the actual sections?

**9.** The total radio/television power transmitted from Earth is about $10^{16}$ ergs/sec. What will be the intensity of the signal (in erg/cm$^2$ sec) at the nearest star?

**10.** The message transmitted from Arecibo contained 1679 characters, intended to be displayed in seventy-three rows of twenty-three characters each. Suppose that the pattern is interpreted instead as twenty-three rows of seventy-three characters. How will the pattern appear? Are there other ways of setting out the 1679 characters that seem to make sense?

**11.** If eavesdroppers do indeed detect signals from the Earth, how large a Doppler shift in wavelength will they be able to observe for signals from the edges of the Earth (due to the Earth's rotation)?

**12.** What is the difference between CETI and SETI?

**13.** Why do biological molecules have hydrogen, carbon, nitrogen, and oxygen as most of their constituent atoms?

**14.** Why is stellar evolution necessary for life?

**15.** Suppose that there was intelligent life on the surface of Venus despite the very high temperature. What wavelengths would have to be used for signals from Earth in order for the Venusians to be able to receive them?

CHAPTER

31

# Cosmology

Through the preceding chapters, we have gradually worked our way from the local to the most distant, from the planets and other occupants of the solar system out to galaxies and quasars far beyond. We are now in a position to formulate answers to the sweeping questions of **cosmology**: What if any was the origin of our universe? How if at all has it evolved? How will it change in the future?

One does not need modern astrophysics in order to pose these questions. Indeed, cosmological speculation is very old, and the history of cosmology tells us about social change at the same time that it chronicles the accumulation of astronomical knowledge. We do not have space here to review the long and fascinating history of cosmological ideas, although we have caught glimpses earlier when, for example, we traced the progress of the Copernican revolution that shifted the center of the universe away from the Earth. What distinguishes modern cosmology from those of earlier times is the wealth of observational data with which cosmological theories can now be confronted. As a result, cosmology has been transformed from unconfined speculation to a science.

Earlier theories could be invented without the embarrassment of being testable; today, any theory must meet some critical observational tests. This still leaves plenty of room for ingenuity, and some of the major questions remain unanswered, but we think we are moving in the right direction, (as no doubt, people have always felt).

## 31.1 OLBERS' PARADOX

Two popular books on astronomy published about 100 years ago show how far astronomy had by then advanced and how it has since moved. *Astronomy* was written by Simon Newcomb (Professor of Mathematics at John Hopkins University) and Edward Holden (Director of the Lick Observatory in California) in 1879. They pointed to the loss of energy by the Sun and the cooling of the Earth and correctly took these as indications of change, that conditions must have been different in the past and will presumably be different in the future. An evolving universe was accepted, but further detail could not be filled in. The *Story of the Heavens* by Sir Robert Ball went through many editions; evolution was accepted, but in 1900 he could still state that

> the actual steps of the process by which the primeval nebula became transformed into the solar system seem to lie beyond reach of discovery.

What is striking about both of these books is not their recognition of cosmic change but their omission of any mention of one of the most important cosmological observations: The sky is dark at night. This common observation would seem to be a most unlikely source for cosmological insight. Even to ask the question, "Why is the sky dark at night?" seems to invite a trivial response.

This innocuous question was discussed by several early scientists including Halley, but it seems to have had its first serious consideration by Heinrich Olbers in an article published in 1823, "On the transparency of space." Olbers was a noted astronomer in Bremen, already famed for his discovery of two of the minor planets, Pallas (1802) and Vesta (1807). He accepted the idea of an infinite and unchanging universe, homogeneously filled with stars that had, on the average, about the same luminosity as the Sun, and spaced at roughly equal distances apart.

Olbers then reasoned that the sky should always be bright, for it would be impossible to find a viewing direction that did not point to some star (Fig. 31.1). If every line of sight ended in a star, then all parts of the sky should be as bright—and clearly this was not the case. Olbers' resolution of this paradox was to assume that some light was absorbed in the interstellar space by material for which there at that time was no other evidence. Olbers then proceeded to show that the sky would be dark at night because stars farther than 30,000 times the distance to the Sun would be too faint to be seen.

**Figure 31.1** Olbers' paradox. (a) If the universe is infinite in extent and in age and is filled with stars, then every line of sight will end on a star, no matter in which direction we look. The sky will, therefore, always be bright. (b) If the universe has a finite age, the light from very distant stars will not yet have been able to reach us.

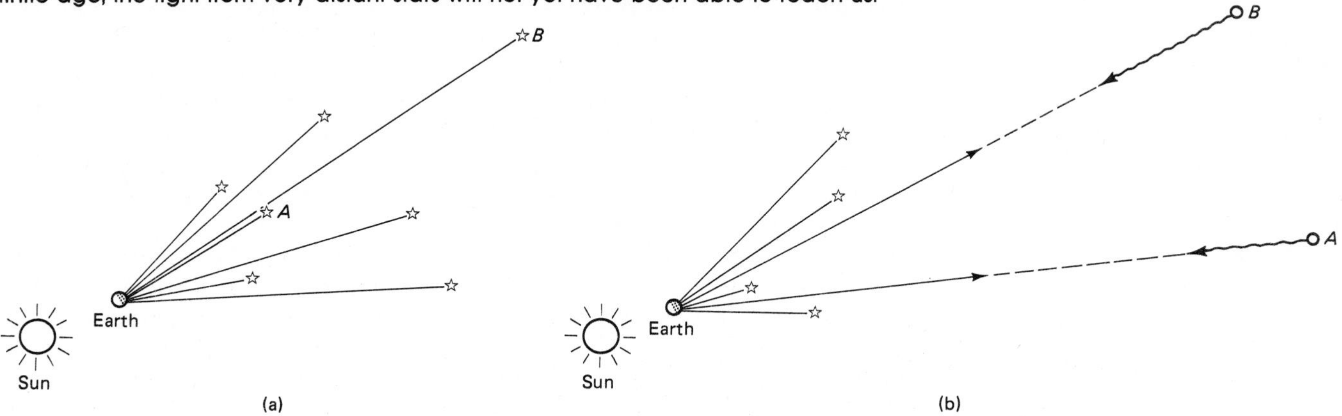

Olbers' paradox attracted little attention and was soon forgotten. Excellent astronomy books published as recently as 1965 contain no mention of the paradox. It was revived by Herman Bondi in a classic article in 1948 and then in his book on cosmology in 1952. At the time that Olbers was writing, nothing was known of the ISM, luminosities, or radiation from hot bodies, so let us reexamine the problem in modern terms.

If the stars (or collections of stars, in the galaxies) do extend out to infinity with about the same number of stars in each unit volume (on the average), then, Olbers concluded, the sky should be bright. But today we cannot expect interstellar absorption to solve our problem. We know that there is an ISM, but it is too thin to do what Olbers wanted. If the ISM did absorb as much starlight as Olbers assumed, it would heat up until it too was glowing and we would still have a bright sky. Instead, the resolution can involve several factors: As we look to greater distances, we are also looking further back in time, so it might be that more remote objects were less luminous than those nearer in distance and time. Further, we now know, distant galaxies are receding from us, and their red-shifted radiation reaches us with reduced energy, contributing less to the sky brightness. And, finally, as pointed out recently by E. R. Harrison of the University of Massachusetts, if the universe is not infinitely old, there will be distant regions from which light will not yet have been able to reach us (Fig. 31.1). In total, therefore, we can understand the darkness of the night sky if our universe is not static: Distant parts must be receding at increasing speeds. Further, it cannot have existed in its present form for an infinite time, for if it had our sky should be brighter.

## 31.2 BASIC OBSERVATIONS AND ASSUMPTIONS

Modern cosmology rests on an important observation, an important theory and some necessary assumptions. First, the observation: as was seen in chapter 28, distant galaxies are receding from us. The universe is definitely expanding and no static model can any longer be considered. Second, the theory: Einstein's General Theory of Relativity, dating from 1915, has revolutionized our views of space, time, and gravity. Support for the theory has come from observations of the precession of the perihelion of Mercury, the deflection of starlight and from the gravitational redshift. Theories of the large-scale structure of the universe must, therefore, include general relativity.

We assume that the laws and constants of physics have had the same form and the same values at all times and places. Newton had made this assumption for the universe that was accessible to him, and we now extend it considerably. There has been research into the question whether the gravitational constant $G$ has changed, and it has been possible to show that any changes must be smaller than about one part in $10^{11}$ per year. Other constants, such as the masses of the proton and electron and the electric charges they carry, also appear to have remained constant to at least that degree. We therefore assume at present that no changes have occurred but one must keep an open mind.

The basic astronomical assumptions of cosmology have been incorporated into the **cosmological principle** (Fig. 31.2). This principle asserts that all parts of the universe have the same average density of matter and radiation, and that observations made of the regions accessible to us also describe (on the average) the inaccessible regions. The principle also asserts that there is no preferred or special place in the universe—that we would see the same general features no matter where in the universe we would happen to be. Of course, if we went to the Andromeda galaxy we would be able to look back at the Milky Way and see different galaxies in different directions. What the principle asserts is that we will find about the same

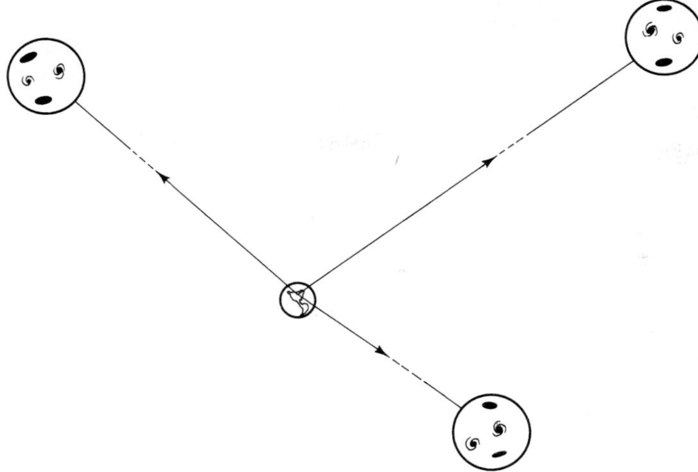

**Figure 31.2** The Cosmological Principle: No matter which direction we look, the universe will appear to have about the same average appearance and density of matter.

number and types of galaxies per cubic Mpc anywhere in the universe. The principle includes the expansion—that, wherever we might be, we would see galaxies moving from us, always following the same Hubble relation.

This idea always poses problems of visualization. Surely common sense tells us that galaxies receding from us must appear to be approaching someone else? We can try to picture this by using two well-worn analogies illustrated in Fig. 31.3.

In the dough of raisin bread, the raisins initially have some sort of random distribution. As the dough rises, the raisins spread out so that the shape of the distribution remains unchanged even though the distances are increasing. Raisin *A* will find the same Hubble Law for raisins as will raisin *B*. What about edge effects? Surely raisin *C* will know it is at the edge of the bread? At this point, visualization is harder, and we draw on another analogy. Try to imagine yourself on the surface of a sphere, with no world anywhere except on the surface of the sphere. As the sphere expands, all points on the surface move steadily apart (Fig. 31.3b) but their mutual pattern is retained. In this model, light travels parallel to the sphere's surface so that seeing to greater distances involves receiving light that has followed a curved path. The distance to which you can see will be set by the age of this universe—how long has the light had in which to travel to you? On a very large sphere, the light will not have been able to travel all the way around. But no matter how far (back) you can see and no matter where on the sphere you stand, the general view will be the same. General relativity tells us that space is curved over very large distances, but it does not tell us how to imagine this. Each analogy is useful in some limited way, but don't push any of them too far.

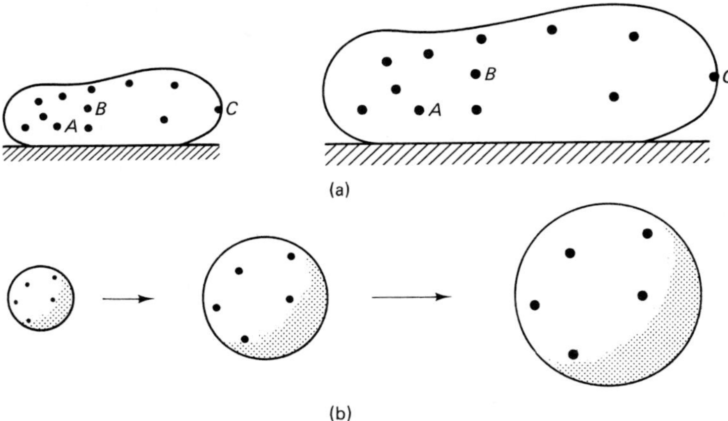


**Figure 31.3** Cosmological Principle and the expansion of the universe. (a) As a raisin bread expands, the average distance between raisins increases, but the general pattern of distances between raisins is preserved. (b) On the surface of an expanding sphere, the distances between various points will increase, but the pattern remains the same.

## 31.3 COSMOLOGICAL MODELS

After the publication of his *Special Theory of Relativity* with its revolutionary ideas of space and time (1905), Einstein went on to investigate the effect of gravity on light (1911), predicting the deflection of starlight, which was actually observed during the 1919 solar eclipse. In 1916, he published the first of several papers on the general theory of relativity and began to examine the large-scale behavior of the universe in the light of his newly formulated theory of gravity. Einstein found it necessary to modify his complicated equations to describe the universe that was then thought to be static. At that time, the general expansion was not known.

After the discovery of the galactic redshifts and their interpretation as indicators of an expanding universe, cosmological models remained mostly concerned with the large-scale movements. During the 1930s, the steady advance in nuclear physics provided additional input and, as we saw earlier, understanding the source of the energy in the Sun came from identification of the chains of nuclear reactions.

Since 1945 two different types of cosmological models have attempted to include both the expansion and the nuclear aspects. They have done this in different ways, and they have alternated in their appearance of success. The first is called the hot **Big Bang** model, for it considered the universe to have evolved explosively from an earlier dense and hot epoch; the second is the **Steady State** model, which does not assume an explosive origin (Fig. 31.4).

In the Big Bang model, as the expansion continues, the contents of the universe get more and more spread out but the same total amount of matter is present at all times. The starting point was in the Big Bang; the future is less certain. We do not have enough data to know whether the expansion will continue indefinitely or whether it will one day stop to be followed by a contraction back to a dense stage (Fig. 31.5). The progress of the Big Bang has been explored in great detail, and it is now widely accepted as our best present theory. In the next section, we shall give it the space it deserves.

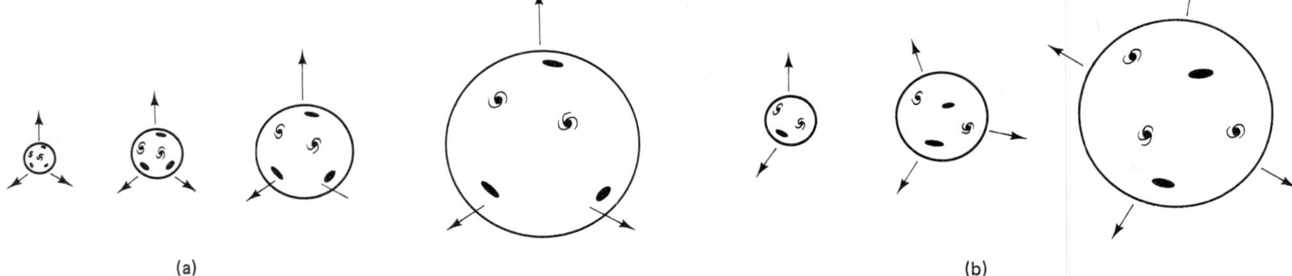


**Figure 31.4** (a) The Big Bang model: As the universe expands and the distances between galaxies increase, the average density of matter decreases. (b) Steady state model: Unlike the situation in the Big Bang model, the average density of matter is kept constant through the spontaneous creation of new atoms. From this newly created matter, new galaxies form, and the appearance of an average volume of the universe will thus appear the same at all times as well as in all regions.

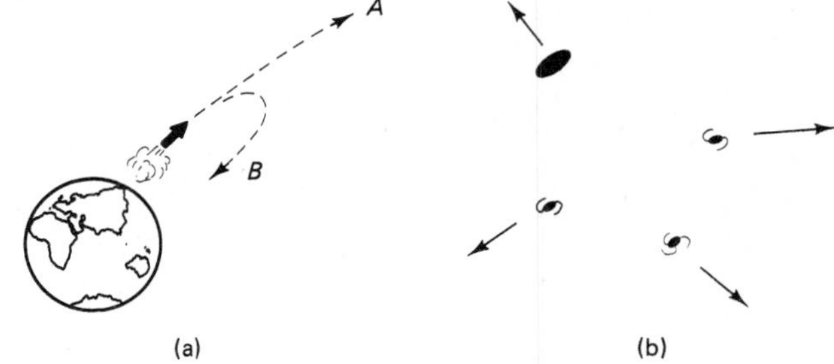


**Figure 31.5** (a) If a rocket fired from the Earth is given a high enough speed, it will be able to escape from the Earth. If, however, the rocket's speed is less than 11.2 km/sec, the gravitational pull of the Earth will bring the rocket back again (*B*). (b) Galaxies are observed to be moving apart with high speeds. If there is not sufficient mass in the universe, then the expansion will continue forever. If there is sufficient mass then the expansion may slow and then reverse. It is not yet known with any certainty which path the universe will follow, but the evidence favors continued expansion.

## BOX 31.1
*The Hubble Constant and the Age of the Universe*

The Hubble constant was defined by the velocity-distance relation (Fig. 31.6).

$$v = HD$$

where $v$ km/sec is the recession speed of a galaxy that is a distance $D$ Mpc away from us (see Chap. 28).

The time taken to travel that distance $D$ would be $T = D/v = D/(HD) = 1/H$. The reciprocal of the Hubble constant thus gives us a measure of the time that could have elapsed since distant galaxies were adjacent to our own. This is the Hubble time.

Distant galaxy
Recession speed v (determined from redshift)
Distance D, from v and the Hubble relation
Earth

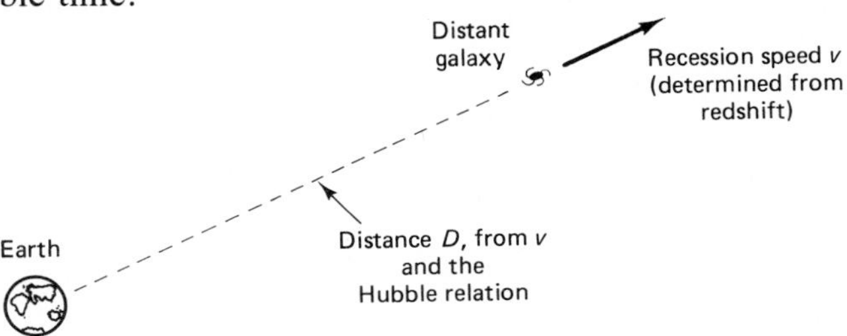

**Figure 31.6** The recession speeds of distant galaxies are proportional to their distances. This relation is described by the Hubble relation: $v = HD$.

**Figure 31.7** Redshifted spectra of galaxies at different distances. The optical spectra show the redshifts increasing with distance. For the most distant galaxy shown in this diagram, IC 173, the recession speed is 13,920 km/sec. Also shown are the profiles of the 21 cm hydrogen line. The redshift of this line can be used to determine the speed of recession (Vera C. Rubin, Carnegie Institution of Washington).

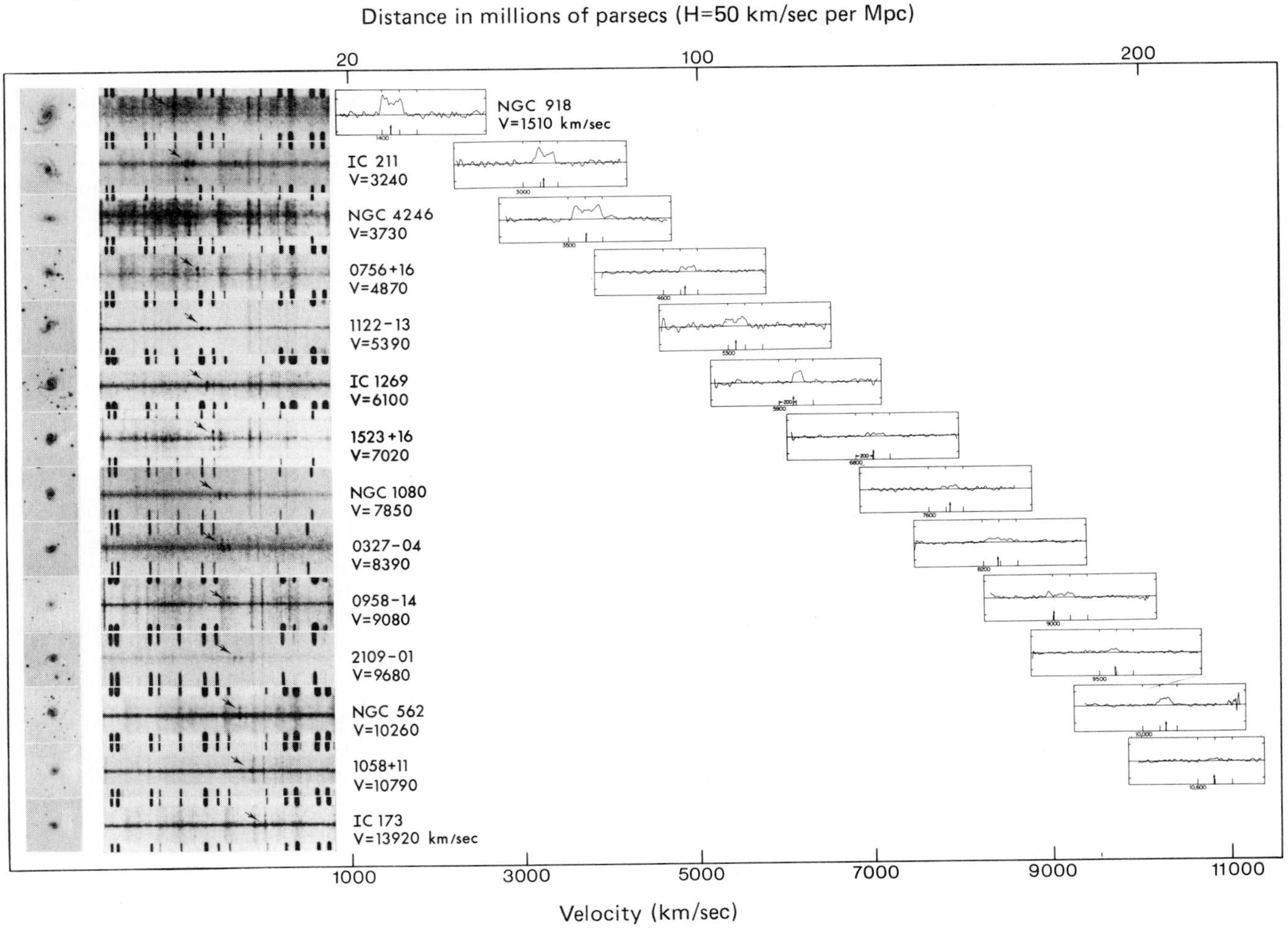

Distance in millions of parsecs (H=50 km/sec per Mpc)
20
100
200
NGC 918 V=1510 km/sec
IC 211 V=3240
NGC 4246 V=3730
0756+16 V=4870
1122-13 V=5390
IC 1269 V=6100
1523+16 V=7020
NGC 1080 V=7850
0327-04 V=8390
0958-14 V=9080
2109-01 V=9680
NGC 562 V=10260
1058+11 V=10790
IC 173 V=13920 km/sec
1000
3000
5000
7000
9000
11000
Velocity (km/sec)

For the Big Bang model, this is therefore the maximum age of the universe, the time during which the expansion would have proceeded at its present rate so that galaxies are at their present distances apart.

Hubble's first (1929) estimate for the value of the constant was 530 km/sec/Mpc. Later measurements reduced this somewhat, but the major change came in 1952 when Walter Baade announced the result of his recalibration, based on his discovery that the Cepheids fell into two classes. Baade's value was 180 km/sec/Mpc.

The present value of the Hubble constant is about 75 km/sec/Mpc (Fig. 31.7). It is convenient to change the units: multiply by $10^5$ to convert km/sec to cm/sec, and multiply by $3.086 \times 10^{24}$ to convert the Mpc to cm. Then

$$H = \frac{75 \text{ km/sec}}{\text{Mpc}} = \frac{75 \times 10^5 \text{ cm/sec}}{3.086 \times 10^{24} \text{ cm}} = 2.43 \times 10^{-18} \text{ sec}^{-1}$$

We then obtain

$$T = 1/H = 1/(2.43 \times 10^{-18}) \text{ sec}$$
$$= 4.11 \times 10^{17} \text{ sec} = 13 \text{ billion years.}$$

If Hubble's first value of $H$ were to be used, we would calculate the age of the universe to be only 1.8 billion years; Baade's value for $H$ makes the universe older, 5.4 billion years.

If the galaxies had lower speeds in the past, then they would have taken longer to cover the distances and the age should be greater. Conversely if the galaxies had actually traveled faster in the past but had slowed down to the present speeds, then the time age is less than the Hubble time.

Two early and serious prolems confronted the Big Bang theory. Around 1950 the high value then measured for the Hubble constant led to an age for the universe that was slightly shorter than the ages of the oldest globular clusters. Those ages had been derived from analyses of the H-R diagrams for stars in the clusters (Chap. 26). In addition, it had been thought that in the early stages of the Big Bang, all nuclei would be those of hydrogen and that some would fuse to produce the variety of heavier nuclei that we now find in nature. Computations were now showing that this would not work—there had to be some other site for the formation of the elements, even those as light as carbon, nitrogen, and oxygen.

Attempts to avoid these dilemmas took two routes. One led to the consideration of continued nucleosynthesis in stellar interiors and explosions, as described in the classic article by Burbidge, Burbidge, Fowler, and Hoyle in 1957, which showed how lighter nuclei could be and probably were fused to produce the variety and abundances of the nuclei observed in nature. The other path led to the Big Bang's major competitor, the **Steady State** theory, set out in 1948 in papers by Herman Bondi and Tom Gold and (in a somewhat different approach) by Fred Hoyle, all of Cambridge University. Their starting point was also the observed expansion but they extended the cosmological principle by assuming that new matter was spontaneously created in space to make up for the decrease in density that would otherwise result from the continued expansion. In this way new stars and galaxies would gradually form from the new matter and replace those that had since moved away. The universe would not only appear the same, on the average, at all places but also at all times. Bondi and Gold termed this the **Perfect Cosmological Principle**.

In the Steady State model, the universe has no beginning and no end. It expands forever. The expansion has been well documented, but what about the spontaneous creation of new atoms? The required rate of creation of new atoms is so low—about 1 hydrogen atom per cubic centimeter every $10^{15}$ year, or about 20 $M_\odot$ per million years in our galaxy—that it would not be directly observable. Creating atoms requires energy, and the idea of suddenly producing mass (energy) out of nothing runs counter to the principle of conservation of mass-energy that has proved of so much use in physics. But if we ask whether this principle has ever been tested to this degree the answer is "no." In recent years it has been found that some laws of physics do not hold exactly. In certain circumstances small departures produce measurable effects on a nuclear scale. Objections to the Steady State theory need to be based on more than an instinctive dislike for spontaneous creation of matter. The correct question to ask is: How do predictions from the Steady State theory compare with observations? For many years, until 1965, no clear-cut choice could be made between the Steady State and Big Bang theories, but a discovery in that year has effectively eliminated the Steady State theory and added overwhelming support to the Big Bang model.

The Steady State theory should not now be lightly dismissed. It is an excellent example of the ability of the scientific community to deal seriously with a radical theory that is ultimately retained or rejected, not on the basis of its initial appeal but rather on the outcome of its confrontation with observation.

There is an interesting parallel with Einstein's Special theory of relativity. The assumptions and consequences run so contrary to common sense, that they still attract considerable hostility, often articulate but usually misinformed. Unlike the Steady State theory, the Special Theory of relativity has had extensive confirmation, and it is an essential ingredient in many calculations.

## 31.4 BIG BANG THEORY

**Origins** Einstein's cosmological equations are too complex to be solved in general, but solutions for special cases can be found, as was done by the Dutch mathematician, Willem de Sitter in 1917. Other solutions were devised in 1920 by Alexander Friedman, a Soviet mathematician, but the first to produce solutions after the discovery of the redshifts was Abbe Georges Lemaitre of Belgium in 1927. Lemaitre suggested that our observed universe had an origin in a very hot and dense phase that expanded explosively. The modern version of the Big Bang can probably be dated to the work of George Gamow and his colleagues in the late 1940s. Gamow introduced nuclear physics into what had been a dynamical model only, and as already noted, he encountered problems in understanding the production of nuclei heavier than hydrogen. Gamow predicted that there should be some residual radiation from the Big Bang. We shall return to both of these points, but first we shall review the current version of the Big Bang, to see how its predictions compare with the observations. The main framework is generally accepted. Different versions are possible, especially for the very early stages. What we shall describe is a plausible model, but not necessarily the last word.

**General outline** The present rate of expansion is an obvious fact. Has the rate of the expansion been different in the past? What will happen in the future? (Fig. 31.8). The answers depend on the amount of matter in the universe, and the situation can be compared to the problem of escape velocity on Earth (Chap. 11).

If the total mass of the universe is large enough, then the gravitational attraction between galaxies will gradually slow them down so that the expansion will ultimately stop and the motions will be reversed. The galaxies will then be approaching rather than receding from each other. This will continue until they are jammed back together again as they presumably were at the start. (This has been called the Big Crunch.) With this model (the

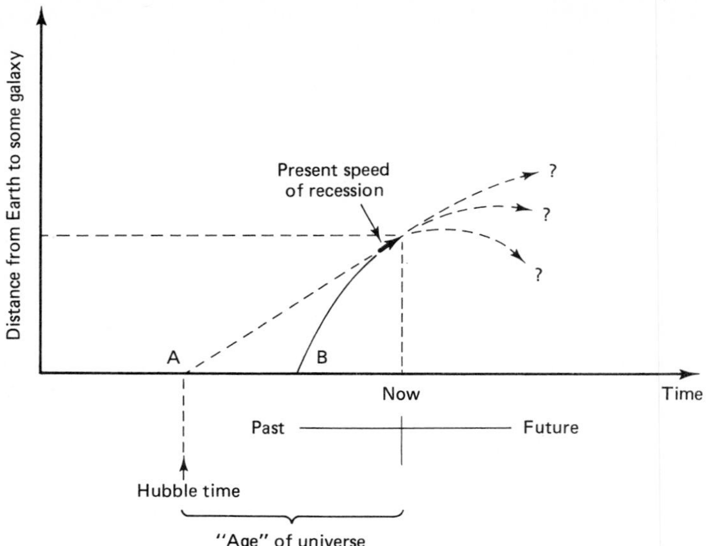


**Figure 31.8** The distance between the Earth and any galaxy is now increasing. How this distance has increased in the past and how it will change in the future are not yet known. If the relative speed between the galaxies did not change with time, then a plot of distance versus time would be a straight line. Extrapolating back from the present known speed would then indicate how far in the past the galaxies had been jammed together. This is the Hubble time, calculated from the Hubble constant. However, gravitational attraction between galaxies has slowed the expansion rate so that the track that starts at time *B* is more realistic. Projections into the future depend on knowledge of the average density of matter in the universe and are uncertain.

**closed** universe), it is possible that after such a collapse, a new expansion will start and there will be an oscillating universe, our present expansion being just one in a series of recurring bounces.

Alternately, there might be too little matter in the universe to force a reversal so that the expansion will be slowed but not halted. This is the open model of the universe. *Total mass in the universe* can be translated into *average density* × *total volume*. The critical boundary between the **closed** and **open** alternatives can be specified in terms of the average density. The critical density is $3 \times 10^{-29}$ g/cm$^3$. If the average density exceeds this value, the universe will be closed; if the average density is lower, the universe is open and the expansion will continue. The critical density is greater than has so far been observed. Finding the average density of the universe is not easy. We can count galaxies and determine their masses, but any cold intergalactic matter is much harder to detect. Black holes, in or especially outside galaxies could add appreciably to the mass and thus to the average density. Direct observational checks have so far not located enough *missing mass* to exceed the critical density. We therefore think that the universe is open and that it will continue to expand forever.

An independent check is provided by the redshifts of distant galaxies. If the expansion is slowing down, then in the distant past the expansion must have been faster, and the redshifts of distant galaxies would appear higher than predicted by the Hubble relation. In Fig. 31.9 we display the redshifts for distant galaxies and the curves to be expected for open and closed versions of our universe. The data, mostly from Sandage and Tammann at Mt. Palomar, support the open version.

How old is our universe? We know from radioactivity measurements that the solar system was formed about 4.6 billion years ago. The oldest globular clusters are about 8 to 16 billion years old. The Hubble constant provides another marker: $H$ is a measure of the rate of expansion so that $1/H$ is a measure of how long the universe has been expanding if it did so continually at the present rate. (Box 31.1, Fig. 31.8). The present value of $H$ is about 75 km/sec/Mpc or $2.43 \times 10^{-18}$ sec$^{-1}$. The 'age' is thus $1/H = 1/(2.43 \times 10^{-18})$ sec = 13 billion years. A value of $H$ larger than 100 km/sec/Mpc would reduce this Hubble time to below 10 billion years. Such a short time would be a severe embarrassment to the theory, for the universe cannot be younger than its components. Each different method of estimating age can be criticized, but a value in the range 10 to 15 billion years agrees with most estimates.

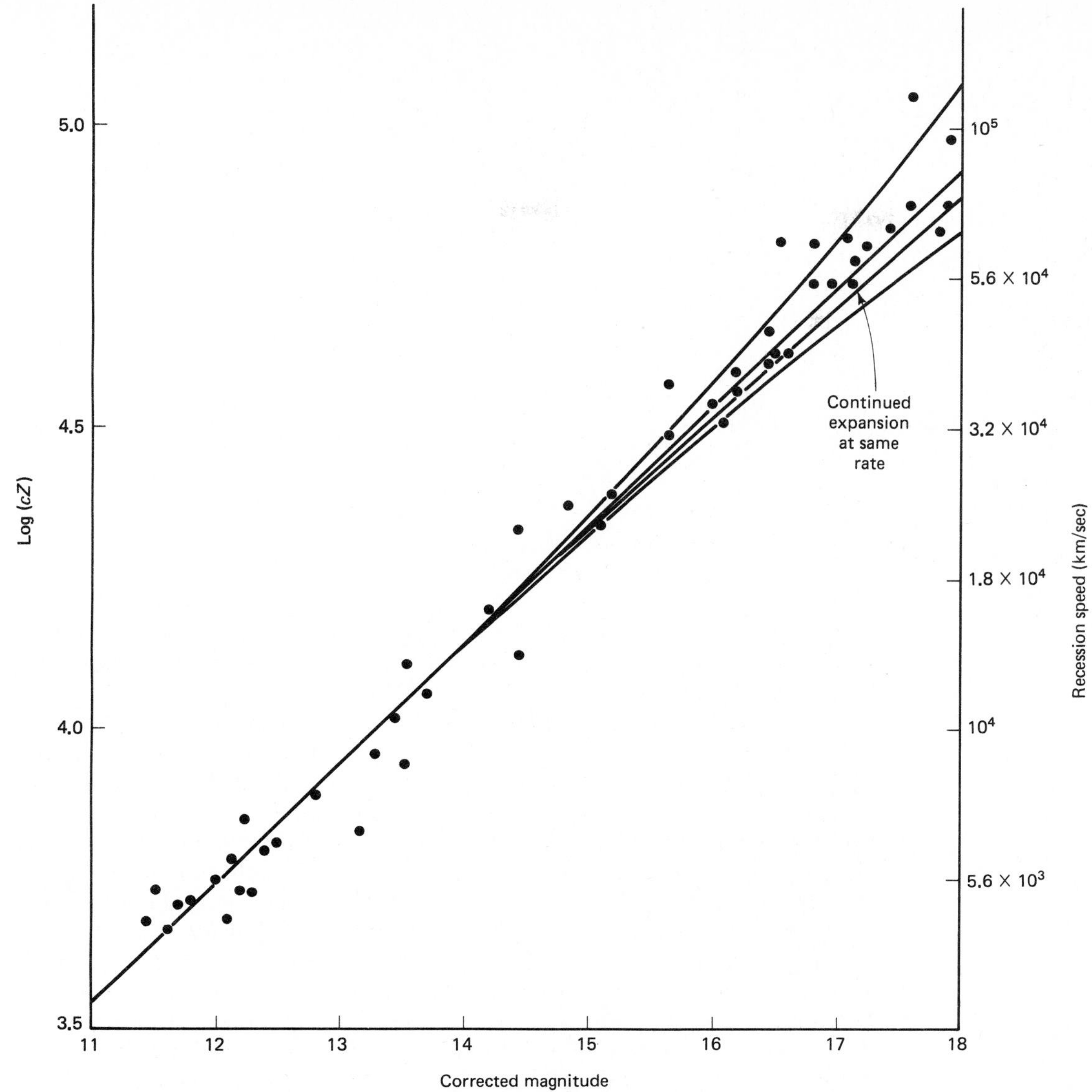


**Figure 31.9** The Hubble diagram, plotting redshift against apparent magnitude for clusters of galaxies. The redshift is plotted as log ($cZ$), where $c$ is the speed of light and $Z$ the redshift; thus $cZ$ is the recession speed in km/sec, and $\log cZ = 5$ corresponds to $10^5$ km/sec. The magnitudes have been corrected for various effects and are indicators of distance, since the faintest galaxies will be at the greatest distances. For distances as great as those corresponding to $16^m$, the points lie close to a straight line, represented by the Hubble relation, showing recession speed proportional to distance. In principle the behavior of the data points for greater distances can indicate how the universe will behave in the future. If the points group along a line that bends up (to the left) away from the continuation of the straight line, this implies that velocities were greater in the past (i.e., at great distances) and the universe is slowing down. If the points were to lie far enough to the left, then the slowing would be great enough to indicate that the universe is closed, and the expansion will ultimately reverse. The data do not yet permit a clear choice to be made. (Data from Kristian, Sandage, and Westphal, *Astrophysical Journal*, Vol. 221.)

Residual radiation At the start of the Big Bang, in what has been called the cosmic fireball (by analogy with the fireball seen in nuclear bomb explosions), the temperature must have been unimaginably high and the compact universe filled with intense radiation. As the expansion proceeded, the contents cooled (Fig. 31.10). The radiation, filling all of the expanding volume, must also have cooled, that is, it lost energy. The spectrum of thermal radiation is described by the Planck radiation formula. In the early very hot stages, the peak wavelength must have been at very short wavelengths.

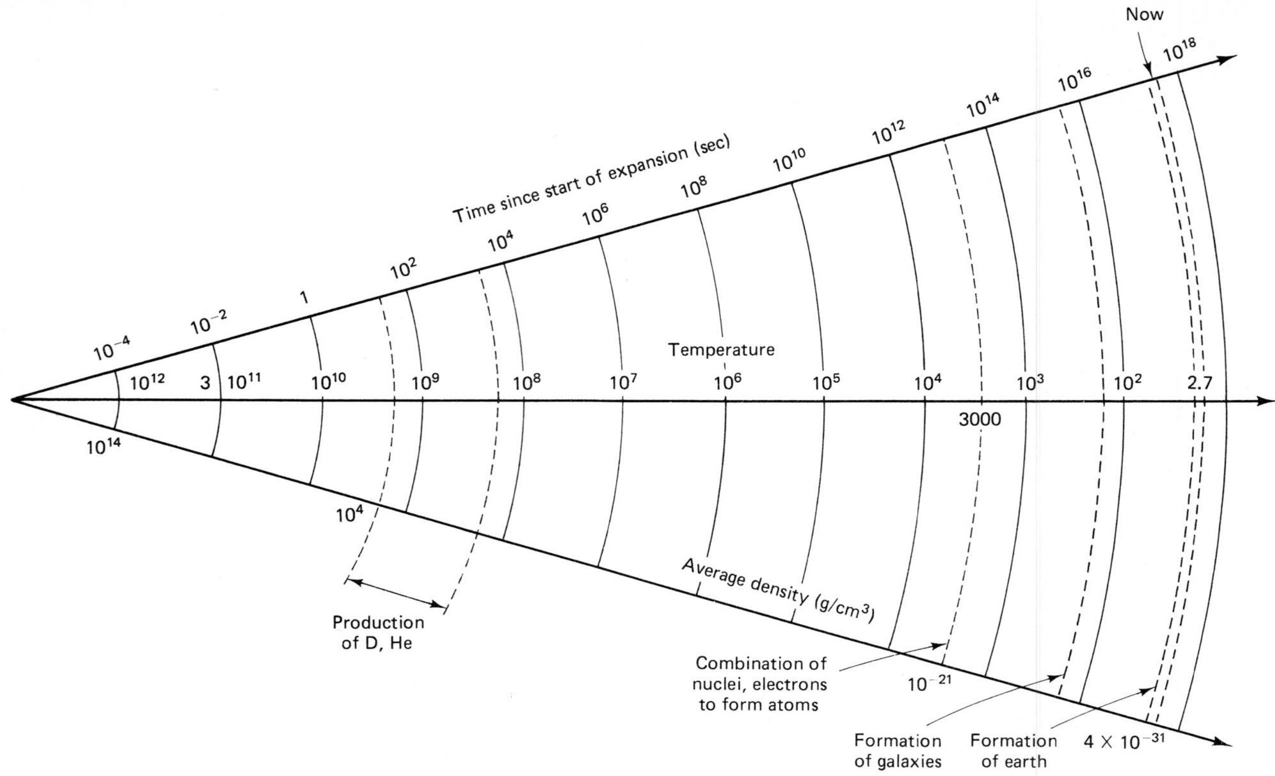


**Figure 31.10** Progress of the Big Bang. Since the start of the expansion, the temperature has decreased to the present 2.7 K, and the average density has dropped to the present $4 \times 10^{-31}$ g/cm³. The density and temperature at different times have controlled the physical processes that can occur. Nuclear collisions that produced deuterium and helium could occur only before the first 2000 sec, for after that time temperature would be too low. Electrons and nuclei will be able to combine to form atoms only when temperatures have dropped to around 3000 K, and that would not have occurred until about a few hundred thousand years.

If the universe has actually evolved along the lines described in the Big Bang model, what should the present radiation temperature now be, around 10 billion years after the start? The peak of the temperature must have shifted to longer wavelengths because of the cooling. We can also think of this as a Doppler shift, much as we see in galaxies. The radiation that reaches us today must have come from very far back in the past and, as a result, be highly redshifted by the time we detect it. The first calculation was carried out in 1946 by George Gamow who predicted that there should be a residual thermal radiation corresponding to a temperature of about 5 K. Recall that $\lambda_{max} T \sim 0.3$ cm degree; at this low temperature, the spectrum should peak at 0.6 cm wavelength, well beyond observational possibilities in 1946.

The presence of this cooler radiation now seems a natural consequence of the theory; it is strange that there were few attempts to try to look for the radiation at other wavelengths. In 1946 Robert Dicke (then at M. I. T.) carried out surveys at wavelengths of 1.00, 1.25, and 1.50 cm. Finding no signals above noise, he set 20 K as an upper limit to the temperature of any residual radiation. There seems to have been no experimental attempt to pursue this, perhaps because by then the Steady State theory was favored. In any event the later discovery of the residual radiation came as a complete surprise.

In 1965 Arno A. Penzias and Robert W. Wilson of the Bell Telephone Laboratories in New Jersey were engaged in research into the galactic synchrotron radiation and were calibrating their radio antenna system. This required identification of all noncosmic sources of radio noise ("static"). Penzias and Wilson found that even after making allowance for all known sources of noise, they were still left with a signal that did not depend on the direction in which their antenna was pointing. The noise intensity (at their wavelength of 7.35 cm) corresponded to what one would expect if the Earth was completely surrounded by thermal radiation of temperature 3.5 K. Penzias and Wilson's paper in *Astrophysical Journal* carries the title "A Measurement of Excess Antenna Temperature at 4080 Mc/s."

Unknown to Penzias and Wilson, Dicke and his associates (P. J. E. Peebles, Peter Roll, and David Wilkinson), now at Princeton, had returned to this search and had calculated that the residual radiation should be observable at centimeter wavelengths. They were building a radio telescope to conduct such a search when Penzias and Wilson called to tell of their puzzling discovery. The probable identity of the residual fireball radiation was recognized.

Measurement at a single wavelength does not identify a spectrum, however, and further observations were quickly made at several wavelengths. Roll and Wilkinson found a temperature of 2.7 K at 3.2 cm. Within a few years, observations were made at wavelengths from 3.3 mm to 73.5 cm; an average temperature of 2.7 K was found. Interesting confirmation came from a very different approach. In chapter 27, we described the investigation of the ISM by means of the absorption lines in stellar spectra. Lines due to CN, CH, and $CH^+$ had long been noted; with Penzias' and Wilson's discovery, these lines were reexamined. It was found that the absorption lines served not only to identify these molecules but also to indicate that those molecules had to be at a temperature of about 3 K, presumably being kept "warm" by the residual fireball radiation.

For a thermal temperature of 2.7 K, the spectrum will peak at a wavelength of ~0.3/2.7 or 0.1 cm, in the far-IR which is heavily absorbed in the Earth's atmosphere. Attempts to check this feature on some early rocket flights gave discordant results (around 8 K), probably because the on-board telescope was picking up some signal from the atmosphere. Very careful observations carried out on high-altitude balloons by groups at the University of California in Berkeley and Queen Mary College (London) have generally confirmed the thermal shape of the spectrum near the expected peak (Fig. 31.11). These balloon-borne experiments are extremely difficult. Cooling of the equipment to less than 2 K was needed and was accomplished by

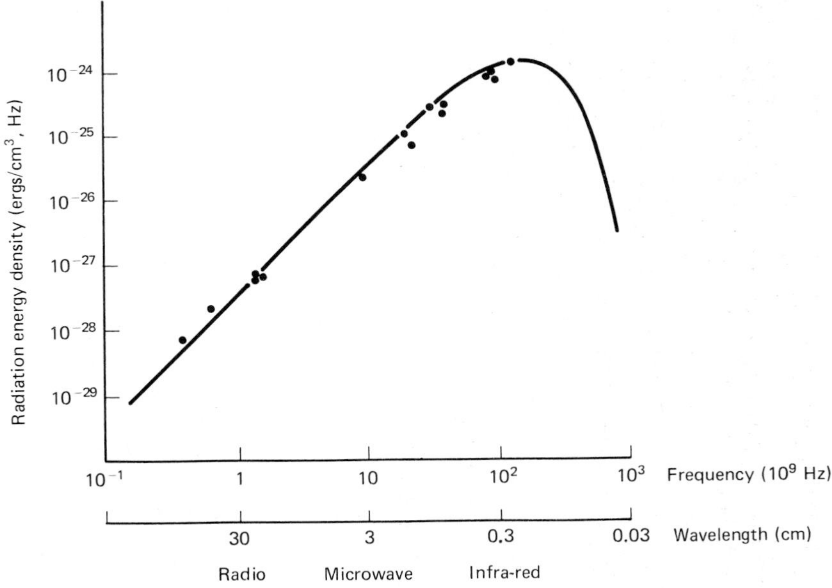


**Figure 31.11** Observed spectrum of the cosmic microwave radiation, believed to be the remnant radiation from the Big Bang. Measurements have been made at wavelengths from the radio to the IR, and the data are very well fitted by a Planck spectrum for 2.7 K temperature.

using liquid helium (Fig. 31.12). Radiation from even the small amount of overlying atmosphere must be subtracted. While there is a general agreement on the temperature, the Berkeley group has found some indications for small departures from a pure thermal spectrum and further measurements are being undertaken.

The isotropy of this residual radiation has been checked. Its intensity is almost the same from all directions, to about one part in a thousand. Recent measurements at high altitudes have shown small departures from exact isotropy and this can be interpreted as indicating a small velocity of our galaxy through the radiation (Fig. 31.13). Further observations are planned to be made from the Cosmic Background Explorer (COBE) satellite.

The discovery of the 3 K background radiation (as it is sometimes known) ranks in importance with the discovery of galactic redshifts. The Steady State theory cannot explain this radiation, whereas the Big Bang includes it in a natural way.

**The earliest stages of the Big Bang** Both the Steady State and Big Bang theories seemed very speculative; the discovery of the 3 K radiation was a remarkable confirmation of one theory. What else does the theory predict, and are there other tests we can apply? We can start by following the progress of matter (atoms, nuclei, particles) through the early stages of the universe

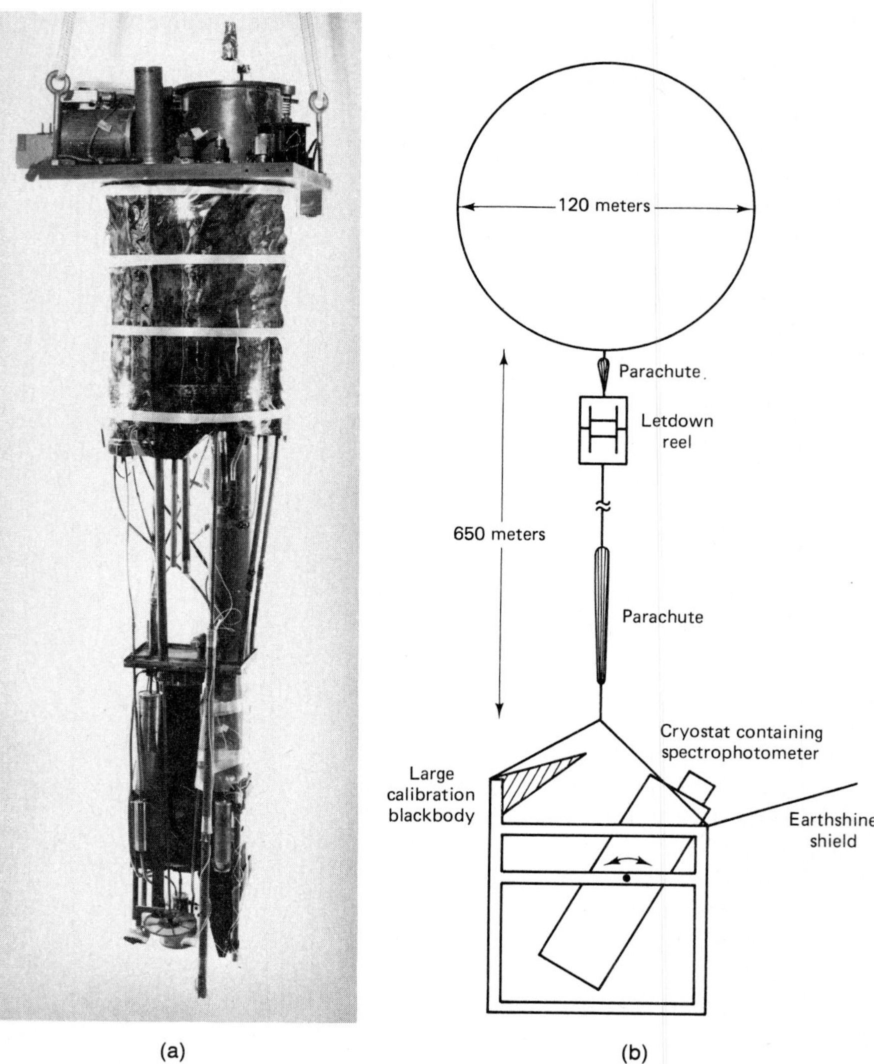


**Figure 31.12** Balloon-borne experiment to measure the cosmic microwave radiation at far-IR wavelengths. Liquid helium was used to keep the detector and optics cooled to less than 2K. The equipment was suspended below a balloon that floated at an altitude of 39 km and could be winched down so that the relatively hot balloon would not itself be in the field of view (Paul L. Richards, University of California, Berkeley).

**Figure 31.13** Microwave horn antennas used on high-altitude U2 airplane for measuring the departures from isotropy of the cosmic background radiation at a frequency of 54 GHz. Signals were measured in two directions 180° apart in azimuth, and the differences showed the presence of a very small anisotropy, about one part in a thousand in intensity (George F. Smoot, University of California, Berkeley).

and we find that although the expansion proceeds steadily, its changing temperature and density at different times are accompanied by important changes in the physical processes taking place.

For the first 1/10,000 sec after the start, the temperature is above $10^{12}$ K. Most of the energy will be in radiation; there may be a few protons and neutrons. The matter and radiation are dense, and the photons each have so much energy that many can convert into pairs of particles and antiparticles. (The existence of antiparticles was predicted by P. A. M. Dirac in 1928 from quantum mechanics calculations. The antiproton has the same mass as the regular proton, but it has a negative electric charge. Particles and antiparticles annihilate when they collide, producing photons and mesons. The mesons are about one-seventh the mass of the proton.) During this first phase, the density is greater than $10^{14}$ g/cm$^3$; we know very little about the behavior of particles at such densities and the calculations are correspondingly insecure.

As the expansion continues, the temperature drops so that individual photons no longer have sufficient energy to produce pairs of protons and antiprotons, but pairs of electrons and positrons can still be formed. Protons and electrons can combine to produce neutrons. This stage continues for about 24 sec until the temperature drops to a relatively low $5 \times 10^9$ K. The photons now have too little energy to produce any particles; annihilations continue between protons and antiprotons, and between electrons and positrons. The density is now too low to keep the protons and electrons jammed together as neutrons (as in neutron stars) and the free neutrons decay back into protons and electrons.

While the expansion continues, the temperature drops. Collisions between protons continue so that some deuterons will be formed. Further collisions between the deuterons and protons will lead to the production of $He^3$ and $He^4$, with traces of a few slightly heavier nuclei. As the expansion continues, the average distance between nuclei increases and the rate of colliding will drop. The production of deuterons and helium will, therefore, shut off after some time. Calculations show that, by the time these fusion reactions stop, about 25 to 30 percent of the mass should be made up by helium. This helium will be found in galaxies and stars and in the ISM (Fig. 31.14).

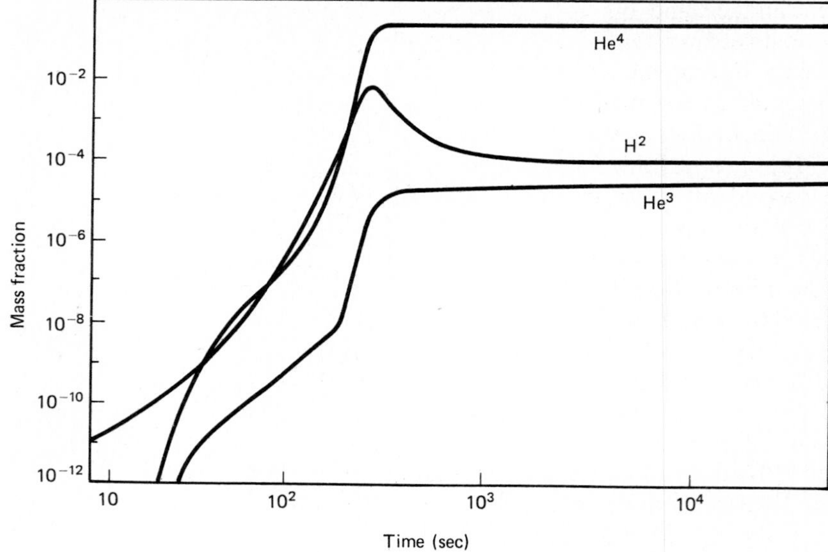


**Figure 31.14** The production of deuterium ($H^2$) and the two isotopes of helium ($He^3$ and $He^4$) during the early stages of the Big Bang. These curves, adapted from those computed by Robert V. Wagoner (Stanford University *Astrophysical Journal*, Vol. 179 (1973)) show how the abundance of these nuclei changed during the first $10^4$ sec of the expansion. The times during expansion are shown along the lower axis.

From the analysis of stellar spectra, we know that helium now in the stars comprises around 25 percent of the mass. On the other hand, calculation of the production of helium by the fusion of hydrogen in stellar cores has shown that far too little is produced to explain the observed amount of helium. The discrepancy is serious: Stellar nucleosynthesis can account for at most 10 percent of the known helium, and it was recognized that some different origin had to be found for the other 90 percent. The Big Bang theory provides an understanding of the helium abundance.

The deuterium abundance today is also an indicator of conditions in the early Big Bang. As with the helium, most of the deuterium existing today comes from the Big Bang, not from stellar nucleosynthesis. Production of deuterium was governed by the density in the early stages of the Big Bang. The relation between the density in those early stages and the average density of universal matter today is similar to that between the radiation in the early stages and the residual radiation today. Thus, if we can determine the deuterium abundance today, we can work our way through this chain of calculations to arrive at the value of the average density of universal matter today. Observations of interstellar deuterium from the Copernicus satellite (OAO-2) lead to a value for the present average density of matter in the universe of $4 \times 10^{-31}$ g/cm$^3$. This is too low to close the universe.

After about 34 min of expansion, the temperature will have dropped to $3 \times 10^8$ K, and the production of helium and deuterium stopped. The expansion continues without any significant number of nuclear reactions. After a few hundred thousand years the temperature will be down to around 3000 K. This is cool enough that electrons can be captured and retained by nuclei to form atoms. There may also be some unevenness in the distribution of the matter; local dense spots can grow to form galaxies. The expansion continues until today the radiation temperature is 2.7 K.

## 31.5 SOME REMAINING PUZZLES

The choice of cosmological model might seem to be settled by the discovery of the 3 K radiation; however, some puzzles remain. What is the value of the Hubble constant? Its measured value has changed markedly over the years, and a final value does not yet seem to have been found. By 1936 Hubble was reporting a value of 526 km/sec/Mpc, based on the use of Cepheids as distance indicators. Hubble's pioneering work has been continued by Alan Sandage and his colleagues at Mt. Palomar. Sandage and Tammann's most recent value is 50 ± 4 km/sec/Mpc. Robert Kirshner and John Kwan using supernovae as luminosity standards have found a value for $H$ in the range 45 to 80 km/sec/Mpc.

An independent new method has been developed by Marc Aronson, Jeremy Mould, and John Huchra using the broadening of the hydrogen 21 cm line caused by galactic rotation. (This broadening is greater than the thermal broadening discussed in Chap. 27.) This rotation can be used to compute the mass of a galaxy. Infrared measurements were used to estimate the luminosities of the galaxies, and a galactic mass-luminosity relation thus constructed (as was done earlier with stars). Calibration of the galactic mass-luminosity relation effectively calibrates some galaxies as luminosity standards to be used in distance estimation. Aronson and his colleagues found $H = 94 \pm 4$ km/sec/Mpc for galaxies at distances between 40 and 60 Mpc, and a value of $H = 65 \pm 4$ for the nearby Virgo cluster (15.7 Mpc).

The difficulties involved in the measurements of $H$ show up in a recent article that gave a critical review of values reported by several different research groups: 50 to 100, 77, 53, 95 ± 15, 100 ± 15, and 75 ± 17 km/sec/Mpc. Some of these values came from the application of a single method, some were averages based on the careful weighing of results of different methods. Clearly, the value seems to be somewhere in the range 50 to 100 km/sec/Mpc, but this wide range leaves an equally wide range in the corresponding ages attributed to the universe.

What seems to be involved is a statistical point. Even though we may be able to estimate the statistical errors in assigning a "±" to each value of $H$, the systematic errors may be extremely hard to locate and the best protection against systematic errors lies in having agreement between results from quite different methods. A disagreement of this kind between the age estimates derived from quite different measurements can be productive. We should recall the serious discrepancy between the newly discovered age of the Earth (deduced from measurements of radioactivity) and the then inadequate sources of energy known for the Sun. Out of such a disagreement we can often find the way to a much better understanding.

Quasars continue to cause doubts. If the dissenters prove correct, then at least some quasars may not follow the usual Hubble relation. If quasars defect, why should not other types of objects also?

Our ability to probe back in time is presently limited by our lack of knowledge of the behavior of particles and radiation at the enormous densities for which we have no direct laboratory experience. As a result, the behavior of the Big Bang before $10^{-4}$ sec is still very uncertain. However, this should not affect our predictions for the future course of evolution for here we are starting from present-day observations and a better understanding.

As progress has been made in the theory of elementary particles, there has been speculation that the proton itself might be unstable, that it might undergo radioactive decay with a very long lifetime. It is known that the proton lifetime must be at least $10^{30}$ years, and experiments are now underway to check this limit. If the proton is found to be unstable, this might resurrect the problem of spontaneous creation in a new form of Steady State theory, with protons both appearing and decaying.

There are still other problems. We need more accurate values for the universal abundance of helium and deuterium. How and when did galaxies form? The agreement on a qualitative model of turbulence and instabilities is a long way from a satisfactory quantitative theory. What is the true significance of the small anisotropy in the 3 K radiation? Suggested causes include the effect of gravitational radiation; a general rotation of the universe; a Hubble expansion that is slightly different in different directions; and patchiness in the density of the universe at an earlier stage. Neither the effect nor its interpretation is yet settled.

One thing we can safely conclude is that cosmology will continue to provide stimulating surprises.

## CHAPTER REVIEW

**Olbers' paradox:** Why is the sky dark at night?

— because the universe has finite age and is expanding.

**Observations:** expansion of the universe, as shown by redshifts of distant galaxies.

**Cosmological Principal:** assumes the universe is the same everywhere, on the average (in large enough regions).

**Perfect Cosmological Principle:** as for Cosmological Principle but with the added assumption that the universe has the same average appearance at all times.

**Steady State model:** assumes spontaneous creation of new matter to replace matter that has moved away in expansion.

**Big Bang model:** assumed expansion in a giant explosion.

— high-temperature radiation in the fireball has by now cooled to around 3 K.

— in early stages of expansion, nuclear reactions produce deuterium and helium.

**Cosmic Background radiation:** discovered in 1965, has temperature of 2.7 K, as expected from Big Bang, but cannot be explained by Steady State model.

**Further evolution:** will the universe continue to expand, or will expansion slow down and perhaps stop, then reverse?

— present data indicates universe is open, that expansion will continue, because there is not enough matter in universe to reverse the expansion.

## NEW TERMS

**antiparticles**
**Big Bang**
**closed universe**
**cosmic background radiation**
**Cosmological Principle**
**Hubble time**
**Olbers' paradox**
**open universe**
**Perfect Cosmological Principle**
**Steady State Theory**
**three degree radiation**

## PROBLEMS

**1.** Which single observational fact is the starting point for all cosmological theories, and what is the evidence for that fact?

**2.** Most atomic nuclei are thought to have been formed in reactions in stars. Which nuclei were probably not formed in this way? How and when were they formed?

**3.** What does it mean to say that the universe is open or closed? What evidence do we have, and what is the present view—open or closed?

**4.** What are the major differences between the Steady State and Big Bang cosmological theories?

**5.** The Hubble time based on a Hubble constant of 50 km/sec/Mpc is 20 billion years. The age of old globular clusters is around 13 billion years. Suppose new measurements lead to a value of the Hubble constant of 100 km/sec/Mpc. What problem would this pose?

**6.** What determines whether the universe is open or closed? Which astronomical quantity could allow us to decide between the options?

**7.** The Steady State theory required the spontaneous creation of 1 hydrogen atom per $cm^3$ every $10^{15}$ years. What total mass per year does this amount to for the solar system? Assume that the volume of the solar system is that of a sphere of 40 AU radius. (This is surely an overestimate.)

**8.** Why is the abundance of helium an indicator of conditions in the early universe?

**9.** What evidence is there that the universe is expanding?

**10.** What single observation effectively rules out the Steady State cosmological theory but is a natural part of the Big Bang theory?

**11.** At what wavelength does the 3 K background radiation have its maximum intensity? In what part of the spectrum does this lie? Can this be measured from ground level or must the observations be carried out from high altitude?

**12.** What is the cosmological principle?

**13.** What is the Hubble time, and what is its value?

**14.** The Hubble constant is close to 75 km/sec/Mpc.

a. What is the corresponding Hubble time?
b. What is the recession speed of a galaxy that is 20 Mpc distant?
c. For the galaxy described in (b), what will be the redshift in its spectrum as observed at Earth?

# APPENDIX A

# Glossary

**aberration.** a. Apparent displacement of a celestial object from its true position, caused by the Earth's orbital motion. b. Image defects produced by a lens or mirror. See chromatic aberration.

**absolute magnitude.** Magnitude that a star would appear to have if it were at a distance of 10 pc from the Sun.

**absolute zero.** Lowest point on the absolute or Kelvin temperature scale, equivalent to −273°C.

**absorption lines.** Dark lines superimposed on a continuous spectrum.

**acceleration.** Change of speed (increase or decrease); change in direction of motion without necessarily a change in speed.

**acceleration of gravity.** Actual value of acceleration due to gravity, for example at the surface of a planet or moon. Usually denoted by $g$. For example, on the Earth's surface, $g$ is close to 980 cm/sec$^2$.

**accretion.** Addition of mass to a body, usually by gravitational attraction. For example, in the early stages of star formation, matter accretes from the surrounding gas and dust cloud to form the star; residual matter may accrete to form planets.

**achondrite.** A type of stony meteorite with no chondrules.

**achromat.** Without chromatic (color) aberration. Usually applied to lenses that have been corrected so that two or more wavelengths will be brought to a common focus.

**active galaxy.** A galaxy with nonthermal spectrum and a large luminosity.

**active Sun.** Variation from average solar activity. Often intense activity in a small region, or a stage of wider-spread activity. Examples are sun-spots, flares, prominences.

**albedo.** Fraction of incident sunlight reflected by a planet or minor planet.

**alpha particle.** Nucleus of a helium atom. Emitted by many radioactive nuclei such as uranium and radium. Also an important component of the cosmic radiation.

**alt-azimuth.** Type of telescope mount, in which one axis is vertical and the other horizontal. Used in the Multiple Mirror Telescope (MMT).

**altitude.** a. Distance above the surface of a planet, used in describing an atmosphere or a spacecraft orbit. b. Angle of elevation above the horizon, for the line of sight to a celestial object.

**angstrom.** Unit of length, $10^{-8}$ cm, often used to describe wavelengths of visible light.

**angular diameter.** Apparent size of an object, measured as the angle across its diameter.

**angular momentum.** Momentum of a object rotating about an axis or in orbit about a fixed point. See momentum.

**angular resolution.** Smallest angle between two adjacent objects that can be seen to be separate (resolved) by an optical instrument; a measure of the detail that can be seen.

**annular eclipse.** Eclipse of the sun in which a ring of the Sun remains visible around the Moon.

**anorthosite.** Type of basaltic lunar rock, composed mostly of oxides of silicon, aluminum, and calcium. Abundant on lunar highlands.

**aperture.** Diameter of the objective lens or mirror of a telescope.

**aperture synthesis.** Simultaneous use of several small radio telescopes to achieve the angular resolution otherwise attainable only with a larger single telescope.

**aphelion.** In a noncircular orbit around the sun, the aphelion is the position where a planet is farthest from the sun. For a noncircular orbit around the Earth, the apogee is the position where a satellite is farthest from the Earth.

**Apollo spacecraft.** Spacecraft employed for lunar exploration. Apollo 11, 12, 14, 15, 16, and 17 landed on the Moon and later returned to Earth.

**apparent magnitude.** The brightness of an astronomical object, as observed on Earth and referred to the appearance of some objects chosen as standards. The scale of magnitudes is defined so that a difference of 5 magnitudes corresponds to a ratio of 100 in observed radiation intensity.

**archaeoastronomy.** Study of evidence for astronomical knowledge in prehistoric and ancient times. See megalithic astronomy.

**asteroid.** Older name for minor planet. Object in orbit around the Sun, intermediate in size between meteoroids and planets. Sizes range from under 1 kilometer to several hundreds of kilometers.

**asteroid belt.** The region of the solar system in which most asteroids have their orbits, between those of Mars and Jupiter.

**astrolabe.** Circular type of calculating device, used especially by Arab astronomers around A. D. 1000. Used for predicting positions of a small number of stars. Can also be used for sighting and thus measuring angles.

**astrology.** A system in which the positions of the Sun, Moon, and planets are supposed to exert an influence on events on Earth. Originally a part of astronomy, astrology is today without scientific content.

**astrometric binary.** Binary star system in which the presence of one of the stars is inferred from the observed motion of its companion.

**astrometry.** Part of astronomy concerned with the precise measurement of positions and motions of celestial objects.

**astronomical unit (AU).** Semimajor axis of the Earth's orbit around the Sun; 149.6 million km, $1.496 \times 10^{13}$ cm.

**atom.** Smallest particle of a chemical element. Consists of a nucleus and surrounding electrons. Typical sizes are in the range $10^{-7}$–$10^{-8}$ cm.

**atomic clock.** A clock whose regulation is based on the intrinsic ability of certain atoms to absorb and emit particular frequencies of radiation.

**atomic mass unit (amu).** Scale for atomic masses, defined with the mass of the most abundant carbon isotope being 12. 1 amu $= 1.67 \times 10^{-24}$ g.

**atomic number.** The number of protons in the nucleus of a particular atom. The atomic number of hydrogen is 1, of helium 2, carbon 6.

**aurora.** Light emitted in the upper atmosphere in the far northern and southern latitudes. Produced by collisions of high-speed particles with atmospheric atoms and molecules, and strongly influenced by the Earth's magnetic field and solar activity.

**azimuth.** Angle measured clockwise around the horizon, from 0° towards the north, through 90° to the east, 180° to the south, 270° to the west, and back to 360° due north.

**Balmer lines.** Absorption or emission lines in the spectrum of atomic hydrogen, produced by the transition of atoms between the first excited energy level and higher levels.

**bands (in spectra).** Absorption or emission lines that are so plentiful and so close together that they are not seen separately but rather as a continuous band; seen in the spectra of molecules and radicals.

**barred spiral galaxy.** Spiral galaxy with spiral arms connected to a spoke or bar that extends out from the galactic center. See spiral galaxy.

**basalt.** Type of rock produced by the cooling and solidification of molten lava.

**belts.** Dark bands seen on Jupiter and Saturn.

**beta particle.** An electron or positron emitted by some radioactive atoms.

**Big Bang theory.** Cosmological theory that assumes a universal expansion starting from an explosion in a very dense and compact stage. Some big bang theories predict continued expansion, others a slowing expansion to be followed by a reversal leading to a contraction.

**billion.** 1000 million ($10^9$), as used in the U.S., but one million million, as used in Great Britain. To avoid confusion, scientific usage now denotes $10^9$ by the prefix giga and $10^{12}$ by terra.

**binary star.** Double star, with the two stars in orbit around one another.

**binding energy.** Energy released when light atoms fuse to form a heavier atom, for example hydrogen atoms joining to form helium; energy required in order to break a heavier atom into its smaller components.

**black body.** Hypothetical surface that absorbs all radiation falling on it and then emits that radiation. Resulting radiation spectrum has distinctive shape, described by Planck's radiation law.

**black dwarf.** Final state of stellar evolution, when a star has used up all of its energy resources and can no longer radiate.

**black hole.** Body that is so massive and so compact that no light can leave its surface. Suggested by General Theory of Relativity, but not yet definitely observed although some candidates might have been detected in binary stars that emit X-rays.

**Bode's Law.** Numerical sequence that closely approximates the relative sizes of the orbits of the planets through Uranus, and the location of the asteroid belt. No known scientific basis for this law.

**Bohr atom model.** Model for the atom proposed by Niels Bohr. Similarities to solar system construction: electrons in orbit around a central nucleus, but with only certain orbits (and associated energies) permitted. Energy absorbed or emitted when electron goes from one permitted orbit to another.

**breccia.** Type of rock composed of compacted fragments of other rocks; commonly produced by meteorite impacts on the lunar surface.

**burster.** Type of celestial object that emits X- or gamma rays in sharp bursts, usually very irregularly.

Thought to indicate uneven transfer of matter from one star to another (compact) star in a close binary.

**canali.** Description applied by Giovanni Schiaparelli to markings he observed on Mars.

**carbon cycle (CNO cycle).** Sequence of nuclear reactions during which four hydrogen nuclei join to form one helium nucleus. In these reactions, carbon, nitrogen, and oxygen nuclei participate but are not used up. The CNO cycle is an important source of energy in stars that are much larger than the Sun.

**Cassegrain.** Inventor of a type of reflecting telescope that now bears his name. In this design, light from the concave primary mirror is reflected a second time by a convex mirror, through a hole cut in the primary, to produce an image at the Cassegrain focus.

**Cassini's division.** Most prominent gap in the ring structure around Saturn.

**CCD.** Charged-couple device. An electronic device for measuring light intensity, far more sensitive than photographic materials.

**celestial equator.** Great circle formed on the celestial sphere by the extension out of the Earth's equatorial plane.

**celestial sphere.** Apparent spherical surface, centered on the Earth to which the stars seem to be fixed.

**center of mass.** Mean position of the masses that comprise a system or larger body; for two bodies, the center of mass is a point on the line joining them. For a binary star system, the motion of each star can be computed about the center of mass.

**Cepheid variable.** A type of star with regular variation in its brightness; named for Delta Cephei, the first star recognized of this type. These stars play an important role in the determination of the scale of distances to galaxies.

**Chandrasekhar limit.** Theoretical maximum mass for a white dwarf star, about 1.4 solar masses.

**chondrite.** A type of meteorite that has stony appearance and contains small spherical particles, chondrules.

**chondrule.** Small round body (generally less than 1 mm) found in meteorites.

**chromatic aberration.** Lens defect which produces images of different colors at slightly different places. See achromat.

**chromosphere.** A region of the solar atmosphere between the bright photosphere and the more extensive corona. Hard to observe because of its relative faintness.

**circumpolar stars.** Stars that are close enough to the celestial pole that they do not rise and set each night but can be seen to move around the pole in a circular path. Latitude of observer determines which stars appear circumpolar.

**Clouds of Magellan.** Two irregularly shaped galaxies, one large and the other small, relatively near to our Milky Way galaxy. Visible by eye from the southern hemisphere. Usually denoted as LMC, SMC.

**cluster.** A group of apparently near-by objects: clusters of stars or galaxies. Star clusters are open or globular.

**CNO cycle.** See carbon cycle.

**color excess.** Change in color index due to absorption of starlight by interstellar dust.

**color index.** Difference between magnitudes of a star or other object when measured in two different wavelength regions. The CI is an indicator of temperature and is used for stellar classification.

**coma.** Large gaseous region surrounding the nucleus of a comet, together making up the head.

**comet.** Small body in the solar system, in orbit around the Sun. Some of its frozen material vaporizes during the closer parts of its approach to the Sun, to produce the characteristic tail, behind the bright head.

**compound.** Chemical combination of two or more dissimilar types of atoms.

**conduction.** Transfer of energy from atom to atom without any bulk motion. See convection.

**conjunction.** Closest apparent approach of two celestial objects. Planetary conjunctions were once considered important omens for events on Earth.

**constellation.** A group of stars that seemed to suggest the shape of some god, person, animal, or object. Now a term used to designate a region of the sky.

**continuous spectrum.** Radiation spectrum that displays a smooth variation of intensity as the wavelength changes, as opposed to a line spectrum that derives from sharply defined energies of atoms.

**convection.** Transfer of energy by bulk motion of material containing the energy. See conduction.

**core.** Central region of a planet. The Earth's core is liquid, possibly with a small, solid inner core.

**corona.** Outer (high-temperature) region of the solar atmosphere.

**coronagraph.** Instrument that blocks out the bright light from the solar photosphere and so makes it possible to observe the chromosphere and corona.

**coronal hole.** Region of the corona that appears dark against the rest of the corona, as a result of locally lower density and temperature.

**cosmic rays.** Atomic nuclei and electrons that travel in space at very high speeds. Most cosmic rays that are detected on Earth come from distant parts of the galaxy, but some come from the Sun, especially during its active phases.

**cosmogony.** A term sometimes used to describe the study of the origin of the universe, but more frequently now used in the restricted sense of the origin of the solar system. See cosmology.

**cosmological principle.** Assumption that the properties of all large regions of the universe are the same, at the same time. According to this principle, the universe appears generally the same in all directions of observation.

**cosmology.** The study of the origin and large-scale features of the universe. See cosmogony.

**Crab nebula.** Remnant of the supernova observed in A. D. 1054, now observable as an expanding and tangled cloud of gas, with a pulsar at the center.

**deceleration parameter.** Quantity that enters in cosmological models and describes the rate at which the universal expansion is changing.

**declination.** Angular distance of an object north or south of the celestial equator, measured in degrees. Thus the north celestial pole has a declination of +90°N.

**deferent.** One of the circles used by Ptolmey in his geometrical model of the solar system.

**degenerate electron gas.** Gas of electrons at very high pressure which does not behave like the "perfect" low-density gases. Quantum theory is needed for description of degenerate gas.

**density.** A measure of compactness: mass of an object divided by its volume, usually expressed in $g/cm^3$.

**density-wave theory.** A traveling wave in the galaxy that has been suggested to explain spiral structure and the initiation of star formation.

**deuterium.** Isotope of hydrogen. Each deuterium atom has one proton and one neutron in its nucleus, which is termed a deuteron.

**diffraction.** Behavior of light that passes close to the edge of an opaque object or through an aperture. The emerging light shows a pattern of bright and dark regions. Well explained by the wave theory of light.

**diffraction grating.** A device that spreads a multicolored beam of light into component colors (wavelengths) so that they can be seen separately. Composed of many closely spaced lines, on a transparent film or on a mirror; each line diffracts the light falling on it, and the light from many lines is focused to produce the spectrum.

**disk.** Equatorial flattened region of the Milky Way galaxy, containing most of the stars.

**dispersion.** Separation of radiation of different wavelengths because they travel with different speeds. In transparent substances such as glass, this causes the refractive index to depend on wavelength, and different colors are refracted by different amounts. (See chromatic aberration). Radio waves traveling from distant objects are dispersed by the interstellar medium and arrive at Earth at different times.

**Doppler effect, shift.** Apparent change in wavelength observed when the observer and the source of the radiation are moving relative to one another. The size of the wavelength change increases with increasing relative speed.

**Drake equation or formula.** Formula that relates the number of intelligent civilizations in the galaxy to various quantities such as the number of stars with planetary systems, the duration of civilized intelligent life, etc.

**dwarf.** Main-sequence star of low luminosity. See also white dwarf.

**dyne.** Unit of force in metric system that uses grams for masses and centimeters for length.

**eccentric.** Off-center; used by Ptolemy in his attempt to have a model that preserved uniform motions around circles, by placing the Earth away from the center of the presumed planetary orbits.

**eccentricity.** A measure of the shape of an ellipse: the ratio of the distance between foci to the major axis. For a circle, the eccentricity is zero. A long, narrow ellipse will have a value of eccentricity close to 1.0.

**eclipse.** Blocking of light from one body by another that passes in front of it. Eclipse can be total or partial.

**eclipse path.** Narrow path on the Earth's surface traced by the Moon's shadow during a solar eclipse.

**eclipsing binary star.** Binary star whose mutual orbit is viewed almost edge-on so that the light observed is regularly decreased each time one star eclipses the other.

**ecliptic.** Path that the Sun appears to follow, against the stars on the celestial sphere, during the course of a year.

**ecliptic plane.** Plane defined by the Earth's orbit around the Sun.

**electromagnetic radiation.** Radiation that carries energy through regular variations in its electric and magnetic force. Includes radio waves, infrared and ultraviolet radiation, visible light, X- and gamma rays.

**electromagnetic spectrum.** Radiations of different wavelengths; see electromagnetic radiation.

**electric field.** A region in which an electric charge exerts a force, for example on other electric charges.

**electron.** Constituent of atoms, in orbit around atomic nucleus. Electrons can also be separated from atoms and made to travel at high speeds. Mass is $9.1 \times 10^{-28}$ g, much smaller than that of atom. See beta particles, cosmic rays.

**electron volt.** Unit of energy. One electron volt is $1.6 \times 10^{-12}$ erg. Larger units are keV (thousand electron volts) and MeV (million electron volts).

**element.** Substance that cannot be broken into simpler substances, using only chemical methods. A chemical element consists of atoms that are all alike;

different elements have different types of atoms.

**ellipse.** Type of closed curve whose shape is specified in terms of its distance from one or two points. A circle is a special form of ellipse. In appearance, an ellipse is oval-shaped. See eccentricity.

**elliptical galaxy.** Type of galaxy that appears to have an elliptical shape and contains no visible interstellar material.

**elongation.** Angle between the directions of a planet and the Sun, as viewed from the Earth.

**emission line.** Bright line of a spectrum, with the wavelength defined by the energy levels of the atoms or molecule from which the radiation is emitted.

**emission nebula.** A cloud of interstellar gas that glows because of ultraviolet radiation absorbed from a nearby hot star.

**emission spectrum.** Spectrum that consists of emission lines.

**energy.** Physical property possessed by a body or system, often associated with motion. Thus kinetic energy is energy of motion, and thermal energy is internal energy due to molecular or atomic motion, such as in a gas. Energy is consumed when a force is exerted, such as to accelerate a body.

**energy level.** In an atom, molecule, or ion, the internal energy may have only certain values (levels) that are specified by quantum theory. Changes between energy levels can absorb or emit energy as radiation.

**energy spectrum.** The range of energies of a collection of high-speed particles, such as molecules in a gas or the cosmic-ray particles. Often expressed in mathematical form.

**ephemeris.** Table of predicted position of various celestial bodies at specified times.

**epicycle.** One of the circular components in Ptolemy's model of the solar system. Each planet was supposed to travel around a circle (the epicycle), and the center of this circle itself moved in a circular path (the deferent) around the Earth.

**equant.** Position defined by Ptolemy as the point about which planetary motion was supposed to proceed with uniform angular speed.

**equation of state.** Relation between pressure, temperature, and density of a gas.

**equatorial mount.** Type of telescope mount that permits easy compensation for the Earth's rotation and thus prolonged viewing of a given object. This is achieved by having one of the supporting axes of the telescope mount parallel to the Earth's axis and the other at right angles to it.

**equinox.** Two days each year when the Sun is above and below the horizon for equal lengths of time.

**erg.** Unit of energy, in the metric system of units with mass measured in grams and length centimeters.

**errors.** Unavoidable or unnoticed inaccuracies in a measurement. Statistical errors are those that tend to be as often high as low, hence balance, on the average. Systematic errors are those that tend to be high or low (such as may occur by using an inaccurate scale), and so do not balance.

**escape velocity.** Minimum speed that must be given to a body in order that it can travel away from its starting point and not be restrained by gravity to following a closed orbit. Often employed in relation to escape of spacecraft from Earth or molecules from a planetary atmosphere.

**excited, excitation.** Terms applied to atoms or molecules that have internal energies above the lowest allowed amounts. Deexcitation may be achieved through emission of the additional energy as radiation.

**exobiology.** Study of life that might exist away from the Earth.

**extinction.** Reduction in the intensity of light from a distant object, for example by absorption or scattering in the Earth's atmosphere or by interstellar dust.

**extragalactic.** Beyond the Milky Way galaxy.

**eyepiece.** Lens or system of lenses used in a telescope or other optical instrument, to enlarge the image formed by the objective (lens or mirror).

**Faraday rotation.** Polarized radiation has its plane of vibration rotated when passing through an ionized gas in a magnetic field.

**fission.** Splitting of a heavy atomic nucleus into two or more lighter fragments with the release of energy.

**flare, solar.** Rapid release of energy from a small region of the solar surface, observed as a sudden and localized brightening. Solar flares can emit cosmic rays as well as electromagnetic radiation.

**flare star.** Star that sometimes brightens very suddenly, with no regular repetition.

**flash spectrum.** Spectrum of the chromosphere. Hard to observe because of faintness; can be detected just before or after totality in solar eclipse.

**flux.** Rate of energy flow, expressed for example as $ergs/cm.^2$ sec.

**focal length.** Distance of image from lens or mirror where light from a distant object is brought to a focus. Focal length depends on curvature of lens or mirror surfaces and (for a lens) on the type of glass.

**focal ratio.** Ratio of focal length of lens or mirror to its diameter. For a camera, the focal ratio is often termed the *f*-number.

**focus.** Point where radiation converges after passing through a lens or after being reflected from a mirror: or point at the center of a circle or one of two special points within an ellipse.

**force.** Agent that produces a change in the motion of a body, either changing its speed or the direction of motion.

**Fraunhofer lines.** (Dark) absorption lines observed in solar or stellar spectrum.

**frequency.** In a wave motion or vibration, the number of vibrations per second.

**fringes.** Interference fringes: pattern of light and dark, produced by combining two or more waves. Closely related to diffraction.

**full moon.** Observed once each month, when the Earth lies between the Sun and Moon, and the fully illuminated disk of the Moon can be seen.

**fusion.** Joining of lighter atomic nuclei to form a heavier nucleus, with release of energy. Thus four hydrogen nuclei fuse to produce helium.

**galactic cluster.** Cluster of stars with an open (not too closely packed) appearance, as opposed to globular cluster.

**galactic equator.** Central plane of the Milky Way galaxy.

**galactic latitude and longitude.** System of coordinates useful in specifying the location of objects with respect to the galactic equator (for latitude) and the direction toward the galactic center (for longitude).

**galaxy.** Large number of stars with their interstellar gas and dust, grouped into a region that is well separated from other galaxies. (Star clusters occur on a scale much smaller than that of galaxies.)

**Galilean satellites.** Four satellites of Jupiter discovered by Galileo who first named them Medicean satellites, after Cosimo de Medici, his patron.

**gamma rays.** Shortest wavelength electromagnetic radiation.

**gauss.** Unit of magnetic field strength. On the Earth's surface, the magnetic field has a strength of about 1 gauss.

**Gegenschein.** Dim and hazy glow in the sky, seen in the direction opposite to the Sun, probably caused by sunlight reflected from many small dust particles.

**geomagnetic field.** Magnetic field around the Earth, probably produced by electric currents within the Earth. Long used with compasses for navigation. Geomagnetic field well above Earth's surface controls the behavior of electrically charged particles in the Van Allen belts.

**giant.** Star with very large luminosity and radius, much more luminous than main-sequence stars with same surface temperature.

**gibbous moon.** Phase of the Moon during which more than half of the disk appears illuminated.

**globular cluster.** Tight cluster of stars that is circular and symmetrical in appearance, as opposed to open or galactic clusters.

**granulation.** Terrazzo-like appearance of solar photosphere when observed at high magnification.

**gravitational constant, G.** Numerical constant in Newton's Law of Gravitation, $6.67 \times 10^{-8}$ dyn. $cm^2/g^2$ in metric units.

**gravitational energy.** Energy released by a system that contracts to smaller dimensions.

**gravitational lens.** Action of a large and massive object such as a galaxy that brings to focus the light from a more distant object. Focusing is produced by gravitational bending of light.

**gravitational redshift.** Change (increase) in wavelength of radiation that is emitted from the surface of a massive object.

**gravitational radiation waves.** Energy emitted in rapid movement of large masses, such as collapse of large star or by rapid motion of stars in binary system.

**greenhouse effect.** Heating of a planetary surface and lower atmosphere, produced by selective transmission and absorption of radiation that depends on wavelength. Thought to be responsible for the high temperature on the surface of Venus.

**H.** See Hubble constant.

**HI.** Atoms of neutral hydrogen, thus HI region.

**HII.** Atoms of ionized hydrogen, thus HII region; usually found near very hot and luminous stars.

**Hα.** Longest wavelength line in Balmar spectrum of hydrogen. Wavelength is 6563 Å, in red end of visible spectrum.

**Hβ.** Second-longest wavelength line in Balmer spectrum of hydrogen, at 4861 Å.

**Half-life.** Characteristic time for a radioactive species of atoms or particles; time during which half of the original number of particles decay into less massive particles.

**halo (of galaxy).** Region of a spiral galaxy that extends away from the flattened disk, and contains few stars, some globular clusters, making a roughly spherical overall shape for the galaxy.

**head (of a comet).** Bright part of a comet, containing the small nucleus and its surrounding coma.

**heliocentric.** With the Sun at the center, as in the Copernican model for the solar system.

**Helmholtz-Kelvin contraction.** Early phase in evolution of a star, when the original extensive gas and dust cloud contracts to starlike dimensions, with the release of gravitational energy.

**Hertz (abbreviated Hz).** Unit of frequency (of vibration or variation). One cycle or vibration per second, formerly designated as c/s or cps.

**Hertzsprung-Russell diagram.** Graph showing absolute magnitude or luminosity plotted against temperature or color index for individual stars.

**horizon system.** Astronomical coordinate system in which one angle (azimuth) is measured from north around a horizontal circle, and the other angle (altitude) is measured above or below the horizon circle.

**hour angle.** Astronomical coordinate, corresponding roughly to longitude on Earth and measured along the celestial equator.

**H-R diagram.** See Hertzsprung-Russell diagram.

**Hubble constant.** Ratio of recession speed of a distant galaxy to its distance from Earth, usually measured in km/sec/Mpc. Value is in the range of 50–100 km/sec/Mpc, but not yet precisely determined.

**Hubble law.** Observed relation between recession velocity and distance for galaxies: $v = HD$, where $H$ is the Hubble constant.

**hydrostatic equilibrium.** Condition in a star or planet in which the weight of overlying matter is supported by pressure exerted by matter closer to the center.

**igneous rock.** Rock that has crystallized from molten lava.

**image.** Radiation from an object that has been focused (by a lens or mirror) to create the appearance of the original object. Image may be enlarged or reduced in size, compared to original object, depending on lens or mirror system used.

**image intensifier or tube.** Electronic device that produces a bright secondary image from an originally faint image focused by a telescope.

**inclination.** Angle between the plane of an orbit and some reference plane. For example, the angle between the orbital plane of a planet and the ecliptic plane.

**Index Catalog (IC).** A supplement to Dreyer's New General Catalog (NGC) of galaxies and clusters of stars.

**inferior planet.** Mercury and Venus, those planets whose orbits lie between the Earth's orbit and the Sun.

**infrared radiation.** Radiations having wavelengths longer than visible but shorter than radio. Near-infrared starts at wavelengths around 7000 Å and far-infrared wavelengths extend to 1 mm.

**interference.** Combination of two or more waves that results in addition at some places and cancellation at others. When light waves interfere, a pattern of light and dark lines can be produced.

**interferometer.** Device to produce interference patterns (fringes). Laboratory interferometer can be used to measure wavelengths, and a stellar interferometer can be used to measure angular diameter of a star.

**interplanetary medium.** Gas and dust that exists in the space between planets and other bodies of the solar system.

**interstellar dust.** Small, solid particles or grains (typically $10^{-4}$–$10^{-5}$ cm diameter), probably mostly silicates and graphite.

**interstellar medium (ISM).** Gas and dust in the space between stars.

**interstellar lines.** Absorption lines seen in the spectra of stars, produced by interstellar gas.

**interstellar reddening.** Relative reduction of the intensity of the shorter (blue) wavelengths of a spectrum, compared to the longer (red) wavelengths, caused by absorption and scattering of light by interstellar dust.

**inverse square law.** A relation between two quantities in which one decreases in proportion to the square of the inverse of the other. The gravitational force and the intensity of light both show an inverse square distance dependence, that is the force and the intensity vary as $1/R^2$, where $R$ is the distance from the source to the point in question.

**ion.** An atom that has a net electric charge through the addition or loss of one or more electrons.

**ionosphere.** Outer region of the Earth's atmosphere where many of the atoms have been ionized by the absorption of solar ultraviolet radiation.

**ionization.** The process of removing one or more electrons from an atom or molecule.

**irregular galaxy.** Galaxy that has no symmetrical or spiral shape.

**isotope.** Atoms of the same chemical element that have different numbers of neutrons in their nuclei, but the same numbers of protons.

**isotropic.** Having the same properties in all directions. Usually applied to the large-scale appearance of the universe and assumed in some cosmological models.

**Jovian planet.** Any of the large outer planets: Jupiter, Saturn, Uranus, and Neptune.

**Keplerian.** An orbit that follows Kepler's Laws.

**kiloparsec (kpc).** 1000 parsecs or $3.09 \times 10^{21}$ cm. Unit of distance between objects in a galaxy.

**kinetic energy.** A form of energy possessed by moving objects.

**laser.** Optical device that produces a light beam of high intensity and very narrow band of wavelengths. This light beam can be so well focused that it spreads out very little even over as large distance as that of the Moon from the Earth, and so can be used to monitor changes in that distance.

**latitude.** Coordinate used to measure (in degrees) the angular distance of a point or celestial object above or below an equator.

**law (of science).** Relation that is found to hold between two or more quantities that describe some natural phenomenon and can be used to make quantitative predictions of future behavior of a system. Thus Newton's Laws of Motion, Snell's Law of refraction, Murphy's Law.

**libration.** Phenomenon relating to the mutual positions of Earth and Moon that allows us, from Earth, to see slightly more than one half of the Moon's surface by making observations over an extended period.

**light curve.** Graph showing the variation with time of a star's apparent magnitude or brightness.

**light year.** Distance that light travels in 1 year; 9.46 $\times$ $10^{17}$ cm.

**limb.** Edge of a bright object (Sun or planet) as viewed from Earth.

**line (spectral).** Well-defined feature in many spectra, resulting from absorption or emission of energy by an atom or molecule, between allowed energy levels.

**lithosphere.** Upper layer of the Earth, in which continental drift (plate tectonics) takes place.

**Local Group.** Group of galaxies closest to the Milky Way galaxy; includes the Large and Small Magellanic Clouds, and Andromeda nebula.

**logarithm.** Mathematical quantity by which one quantity is related to another in terms of powers of 10. Logarithms are used in converting from apparent to absolute magnitude.

**longitude.** Coordinate used to specify the position of a point or direction around (or parallel to) an equator. Measured in degrees or in units of time (hours, minutes), depending on system being described.

**luminosity.** Rate at which radiant energy is emitted by a star or other object, usually expressed in erg/sec.

**luminosity class.** Classification of stars on the H-R diagram within the same spectral class.

**lunar.** Of the Moon: thus lunar eclipse, lunar samples.

**Lyman series.** Group of lines in the hydrogen spectrum, related by having the hydrogen atoms initially or finally at the same (lowest) energy level. These lines occur in the ultraviolet and for astronomical objects cannot be observed from the Earth's surface because of atmospheric absorption.

**Magellanic clouds.** See Clouds of Magellan.

**magnetic field.** Region surrounding a magnet or electric current, in which magnetic force can be detected. In such a region, high-speed electrically charged particles will generally move along curved paths and radiate energy (see synchrotron radiation).

**magnetic pole.** One of the two regions on Earth to which a compass needle will point. Poles also exist on magnets, and the magnetic fields of some electric currents can have an equivalent behavior.

**magnetosphere.** Region surrounding star or planet (including Earth) in which a magnetic field exists.

**magnification.** Apparent enlargement by optical means as with a microscope or telescope.

**magnitude.** Scale for describing brightness of a celestial object; see apparent magnitude, absolute magnitude.

**main sequence.** Band on the H-R diagram, populated by stars that derive their radiant energy from the fusion of hydrogen to helium in their cores. Along the main sequence, the more massive stars appear hot and bright, and the less massive appear relatively cool and less bright.

**major axis (of ellipse).** The longest diameter of an ellipse. See also minor axis. Relation between major and minor axes defines eccentricity of an ellipse.

**mantle.** Region of the Earth between the crust and the core.

**mare (plural: maria).** (Latin) name applied to areas on the Moon's surface that seemed sealike when first viewed from Earth through a telescope.

**mascon.** Apparent concentration of mass beneath the lunar surface, detected through effects on orbits of spacecraft. First thought to result from buried asteroid, mascons are now explained by the higher density of mare material than surrounding highlands.

**maser.** Device that can amplify radio signals or emit a beam of radio waves within a very narrow wavelength range. Underlying physical mechanism is similar to that for lasers.

**mass.** Total amount of material in a body, that determines the size of the gravitational force between bodies at any specified separation. Mass also determines the way in which a body responds to the forces exerted on it.

**mass detect.** Loss of mass that occurs during fusion of light nuclei to form a heavier nucleus. Energy released is related to mass loss by Einstein's formula $E = mc^2$.

**mass-luminosity relation.** Relation discovered between mass and luminosity for main sequence stars, now understood in terms of theory of stellar structure.

**Maunder minimum.** Period between midseventeenth and early eighteenth centuries, when very few sunspots were observed.

**mean.** Average.

**mechanics.** Study of the behavior or motion of bodies under the influence of forces. Within mechanics, kinematics deals with the description of motion without regard to causes, while dynamics includes forces, masses, and energy. Statics is concerned with equilibrium of bodies that are subject to the simultaneous action of two or more forces.

**megalithic astronomy.** Study of ancient structures such as Stonehenge in which large stones have been arranged in a pattern in order to identify astronomical alignments, usually on the Sun and Moon. See archeoastronomy.

**megaparsec (Mpc).** One million parsec; $3.086 \times 10^{24}$ cm.

**meridian.** Great circle, on the celestial sphere or the Earth, that passes through both north and south poles and an observer's zenith or location.

**Messier's catalog.** Catalog of about 100 objects of nonstellar appearance, compiled by Charles Messier in 1787. Objects are often known by their Messier listings, such as M31 for the Andromeda nebula.

**metal.** Term usually applied to a substance that is a good conductor of electricity (copper, silver). Astronomically, *metal* refers to any atoms heavier than helium.

**meteor.** Glowing trail in the upper atmosphere, produced by meteoroid burning up as it moves at high speed.

**meteor shower.** Numerous meteors seen in short time span as the Earth moves through a cloud of meteoroids, probably remnants of a comet and still following the comet's orbit.

**meteorite.** Remnant of meteoroid that has been partially eroded in passage through the Earth's atmosphere before hitting the surface. Term now also applied to similar bodies that collide with the surfaces of the other planets and their satellites, producing craters.

**meteoroid.** Large rock (but much smaller than minor planet) moving in an orbit in the solar system. Meteoroids that enter in the Earth's atmosphere are termed meteors or meteorites, depending on their behavior.

**metric system.** System of scientific units based on meters or centimeters, grams or kilograms, and seconds, in contrast to the English units of pounds and feet.

**micrometeorite.** Very small meteoroid particle. Some that enter the Earth's atmosphere will slow down gradually rather than burn up, and settle slowly down or can be found as dustlike particles in the stratosphere. Those that collide with the lunar surface or spacecraft can be detected by the microscopic pits they produce.

**micron.** 1/1000 of a millimeter, 1/25,000 of an inch.

**Milky Way.** Bright band that stretches across the sky, produced by large number of stars and other bright objects that lie near the equatorial plane of our galaxy.

**Milky Way galaxy.** Concentration of stars, gaseous nebulas, interstellar gas and dust in which the Sun and solar system are located, well separated from other concentrations.

**minor axis.** Smallest diameter of an ellipse. See also major diameter.

**minor planet.** See asteroid.

**model.** Comprehensive description, that can be based on both theory and observation, for a physical system or object. Thus there are models for the interiors of stars and planets, for planetary atmospheres, and for the formation and evolution of the solar system. Models, like theories, are subject to revision depending on new observations or calculations.

**molecule.** Combination of two or more atoms, behaving as a single object under many circumstances. The component atoms can be of the same type or different. Properties of a molecule can differ greatly from those of its constituent atoms.

**momentum.** A property of a moving object, the product of mass and velocity. A quantity that enters into the calculation of the behavior of a body or system under the influence of forces. Angular momentum is a related quantity for objects rotating about an axis or in orbit.

**monochromatic.** Having only one wavelength, such as light from a laser.

**nebula.** Object with nonstellar appearance. Objects originally labeled as nebulae are now known to include galaxies (Andromeda nebula), clouds of gas and dust (Orion nebula), and supernova remnants (Crab nebula).

**neutrino.** Subatomic particle that has no mass or electric charge. Appears as a product of some nuclear reactions or from the decay of some radioactive nuclei or particles.

**neutron.** Subatomic particle with mass closely similar to that of the proton ($1.67 \times 10^{-24}$ g) but carrying no electric charge. A constituent of all atomic nuclei except hydrogen.

**neutron star.** Star composed of neutrons except for a very thin surface layer of atoms. Neutron stars have masses similar to the Sun but dimensions not much larger than the Earth, and, as a result, have very high densities.

**New General Catalog (NGC).** Comprehensive listing of star clusters, nebulae and galaxies published in 1888 by J. L. E. Dreyer, director of Armagh Observatory, Ireland. Later supplemented by two Index Catalogs (IC).

**new moon.** Phase of the moon when its motion brings it between the Earth and Sun, and thus appears to us not to be illuminated.

**Newtonian telescope.** Type of reflecting telescope devised by Newton in which a small flat mirror deflects the light from the primary mirror to a focus outside the telescope tube. Thus Newtonian focus.

**Newton's Laws.** Three laws of motion and the law of gravitation, set out by Isaac Newton in the seventeenth century and sufficiently accurate that they are still used, for example, for the calculation of orbits of planets and the trajectories of spacecraft. Included as special cases in Einstein's Theory of Relativity.

**nonthermal radiation.** Radiation that does not come

from hot objects or regions and so cannot be described by Planck's radiation laws. Example: synchroton radiation from high-speed electrons.

**nova.** Abbreviation from nova stella, Latin for new star, literally meaning the sudden appearance of a star where none had previously been known. Term now applied to sudden large brightening of a star, followed by a less rapid decrease in brightness.

**nuclear.** Referring to some property of a nucleus.

**nuclear reaction.** Process in which two nuclei collide and change identities, often with the release of energy and the emission of additional particles.

**nucleon.** Collective term for protons and neutrons, that is, those basic particles that constitute atomic nuclei.

**nucleosynthesis.** Fusion of lighter nuclei into heavier nuclei, as inside stars but also (to some extent) in the explosion of nuclear weapons.

**nucleus.** a. of an atom: the central part, composed of neutrons and protons; b. of a comet: the central region within the head; c. of a galaxy, central region.

**objective.** Main lens or mirror of a telescope or microscope, that collects the light from a distant object and brings it to an image at the prime focus.

**obliquity (of the ecliptic).** Angle between the plane of the ecliptic and the equatorial plane of the Earth, about 23.5°.

**occultation.** Eclipse of a planet or star behind the Moon or one of the planets.

**Olbers' paradox.** Contrast between the observation that the night sky is generally dark, and the result of a calculation that would lead to the conclusion that the sky should be as bright by night as by day.

**Oort cloud.** Suggested reservoir of comets, located in a spherical region around the Sun and about 50,000 AU in radius.

**opacity.** Property of a gas that describes its ability to absorb radiation and thus reduce its transmission. Opacity of a gas depends on its composition, density, and on the wavelength of the radiation.

**open cluster.** Galactic cluster of stars in which the individual stars can be seen, located within the spiral arm or disk of the galaxy.

**open universe.** Cosmological model in which the universe expands forever.

**opposition.** Planetary position when it is on the opposite side of the Earth from the Sun.

**optical binary.** Pair of stars that appear to be related because they seem close together, but are actually situated at very different distances from the Earth.

**optical depth.** A measure of the fraction of radiation that can emerge from an absorbing region, a combination of opacity and physical thickness.

**orbit.** Path traced out by one object around another.

**organic.** Term referring to molecules or processes relating to living object. Organic molecules are all based on the combination of carbon atoms, with one or more of hydrogen, nitrogen and oxygen and small numbers of other atoms.

**parabola.** Shape of a geometric curve. Can be obtained with algebra or considered as an ellipse with eccentricity equal to 1. Many comets have orbits that are closely parabolic.

**paraboloid.** Shape of a surface obtained by rotating a parabola about its axis, and used for the design of mirrors in many reflecting telescopes.

**parallax.** Visual effect in which an object seems to move against its background when viewed from different positions. Parallax is the term applied to a method used for measuring stellar distances. Values of stellar parallax are usually expressed as angles, in seconds of arc.

**parsec.** Distance of an object from Earth when it has a parallax of 1 arc sec. One parsec is equivalent to $3.086 \times 10^{18}$ cm. Larger units are kilo- (1000) and mega (one million) parsec.

**Paschen series.** Related series of spectral lines in the hydrogen spectrum, located in the near-infrared.

**penumbra.** Part of a shadow from which some of the illuminating object can still be seen.

**perfect cosmological principle.** Expanded form of the cosmological principle in which it is also assumed that the universe has the same large-scale average appearance in all places at all times.

**perfect radiator.** Idealized *black body*: one that absorbs all the radiation falling on it and then reemits that radiation. The resulting radiation spectrum is described by Planck's radiation law.

**periastron.** Position of closest approach between two stars in a binary system. Similar phenomenon to perigee, perihelion.

**perigee.** Point in an Earth satellite's orbit where it is closest to the center of the Earth.

**perihelion.** Place in an orbit around the Sun which is closest to the center of the Sun. A term most mostly frequently encountered in describing cometary orbits.

**period.** Time interval between regular repetitions in motion. For example, the period of a pendulum is the time taken for one swing; the period of a planet is the time taken to travel one complete orbit.

**period-luminosity relation.** Relation found to hold between the periods and luminosities of variable stars.

**perturbation.** Usually small deviation of an object from the trajectory it was following, caused by the introduction of a force that does not control the main motion. For example, cometary orbits are influenced mostly by the Sun, but are perturbed when the comets pass close to the large planets.

**phases of the moon.** Cycle of variations in the Moon's appearance, produced by the changing Sun-Moon-Earth angle through each month. The result is a regular cycle of changes in the Moon's brightness and apparent shape.

**photocell, photomultiplier.** Names given to devices that use a sensitive surface from which electrons are ejected by the absorption of light. Those electrons constitute a minute current that can be amplified so that very weak light intensities can be detected.

**photography.** Detection of light through the use of small crystals of silver bromide, suspended in a thin film of transparent gelatin. Chemical processing (developing) results in dark silver grains forming wherever the light was absorbed.

**photometry.** Measurement of the intensity of light.

**photon.** Particlelike unit of energy in electromagnetic radiation. The energy of a photon is proportional to the frequency of the radiation.

**photosphere.** Bright apparent surface of a star from which most of the radiant energy is emitted. Not a solid surface but rather a region within the stellar (or solar) atmosphere.

**plage.** Bright part of the solar surface when viewed in monochromatic light such as H$\alpha$.

**Planck's constant, *h*.** Constant of proportionality between the energy of a photon and the frequency of the associated wave. Numerical value is $6.63 \times 10^{-27}$ erg sec.

**Planck's Radiation Law.** Formula derived by Max Planck in 1900, relating the radiation spectrum of a perfect radiator to its temperature. Spectra of radiation from many astronomical objects are well described by Planck's Law.

**planet.** Satellites in orbit around a star. In the solar system, nine large planets and many small (minor) planets are known.

**planetary nebula.** Relatively thin shell of gas, blown off from and illuminated by a very hot star. Expands at high speed as it disperses into the interstellar medium.

**planetesimal.** Small bodies, probably formed during the early stage of the evolution of the solar system. Present planets probably formed through the collision and sticking together of planetesimals.

**plasma.** Very hot gas of ionized (electrically charged) particles.

**plate tectonics.** Movements of segments (plates) of the Earth's crust, leading to changing positions and appearance of the continents. Also known as continental drift.

**polar axis.** Axis of an equatorial telescope mounting that is parallel to the Earth's rotation axis. (The other axis of the telescope mounting will be at right angles to this.)

**polarization.** Selection or confinement of electromagnetic radiation so that all of the planes of the electric vibrations are parallel to one another. If this condition is not fully met, the radiation may be termed partially polarized.

**Polaroid.** Registered trade name for transparent material that polarizes the light it transmits.

**Populations (of stars).** Classes of stars, first suggested by Walter Baade. Population I stars occur in spiral arms in the galaxy, have relatively high metal abundances and are generally younger than Population II stars which occur in the galactic halo and have very low metal abundances.

**positron.** Particle with the same mass as the electron but with a positive rather than a negative charge. Emitted in some nuclear reactions and some radioactive decays.

**precession.** a. of the Earth: slow, toplike motion of the Earth's rotation axis, caused by the gravitational effects of Sun and Moon on the Earth's nonspherical shape. A result of this precession is a slow drift in the apparent positions of the stars and other celestial bodies. b. of a planet's orbit around the Sun: slow rotation of the major axis of a planetary orbit. Observed first for the orbit of Mercury and fully explained by the General Theory of Relativity.

**primary cosmic rays.** Cosmic-ray particles arriving at the top of the Earth's atmosphere, from the Sun and more distant objects. Secondary cosmic rays are particles produced by collisions of the primary cosmic rays with constituents of the atmosphere. Cosmic rays detected at sea level are almost entirely secondaries (or from later generations of collisions).

**prime focus.** Focus produced by the primary mirror or lens alone.

**primordial radiation.** Radiation detectable today, coming almost evenly from all directions and with a spectrum corresponding to a temperature of 2.7 K, probably the cool remnant radiation from an initial very hot stage in the evolution of the universe. Sometimes called cosmic microwave radiation or 3 K cosmic background.

**prominence.** Flamelike hot gas visible at the solar limb.

**proper motion.** Regular movement of a star against background stars, expressed as an angular speed in arc seconds per century.

**proton.** Nucleus of the hydrogen atom, with positive electric charge and a mass of $1.67 \times 10^{-24}$ gm. Protons occur in every type of nucleus; the number of protons in a nucleus is termed the atomic number.

**proton-proton (p-p) chain.** Sequence of nuclear reactions that can occur at temperatures above about 10 million K, and in which four protons fuse to produce one helium nucleus, with the release of energy. This reaction chain is the major source of the energy within the Sun.

**pulsar.** Object that emits pulses of radiation with extreme regularity. Pulse periods for different objects range between 1.5 thousandths of a sec and about 5 sec. Pulsars are thought to be rotating neutron stars, extremely compact objects of great density.

**quasar.** Object that appears starlike but is actually extra-galactic, moving away from us at high speed. Distance, deduced from velocity-distance relation, is very large. Extremely luminous.

**RR Lyrae variable.** Class of variable stars with periods less than 1 day.

**radar.** Technique in which radio signals are sent out and their reflections detected, to determine distances and speeds of moving objects.

**radial velocity.** Component of velocity along the line of sight to a star (as opposed to component of velocity across the line of sight; see transverse velocity).

**radiation.** Method of conveying energy without needing the help of matter (as in conduction and convection). Electromagnetic radiation travels by the rapid alternation of electric and magnetic forces.

**radiation pressure.** Pressure that is exerted by radiation on a body or surface on which it falls.

**radioactive dating.** Method for determining the ages of rocks and other materials from the measurement of abundances of radioactive atoms and their decay products.

**radioactivity.** Property of some atomic nuclei which spontaneously change (decay) by emitting one or more particles and become less massive nuclei. Each type of radioactive nucleus has a characteristic time for this to occur—the half-life.

**radiocarbon.** Radioactive isotope of carbon ($C^{14}$) widely used for dating organic remains such as wood and bone.

**ray.** On the surface of the Moon: bright streak of debris sometimes seen near a crater.

**red giant.** Large star with relatively low temperature but high luminosity; a stage in stellar evolution after a star has left the main sequence.

**reddening.** Alteration of a spectrum of light that has passed through a dusty region, produced by the preferential scattering and absorption of the shorter wavelength (blue) light, leaving the red light less affected.

**redshift.** Shift of wavelengths to longer (redder) values, caused by either a relative velocity of separation of source and detector, or else by a gravitational field.

**reflecting telescope.** Type of telescope in which the objective is a concave mirror.

**reflection nebula.** Interstellar cloud of gas and dust that is seen by the light which it reflects from nearby stars.

**refracting telescope, refractor.** Type of telescope in which the objective is a lens, usually an achromat.

**refraction.** Bending of light and other electromagnetic radiation in passing from one transparent medium to another.

**refractive index.** Of a medium: the numerical value that defines that medium's refractive power. If we know the refractive indexes of two media, we can calculate how much a light ray will be bent (refracted) in going from one medium to the other.

**refractory material.** Material that requires high temperatures before evaporating. Example: rocks, minerals.

**relativistic.** Term applied to situations where the theory of relativity must be used. For example, very-high-speed electrons, or the collisions between photons and electrons.

**relativity.** Theories formulated by Einstein. The Special Theory (1905) and the General Theory (1916) include Newton's Laws, but are much broader in their application. The Special Theory deals with relative motion and steady speeds, and the General Theory includes gravitation and accelerations.

**resolving power.** Of an optical system: the finest detail that can be separately seen, usually stated as an angle.

**retrograde motion or rotation.** Apparent backward (westward) motion of a planet, as seen against the stars (as opposed to the regular or prograde motion in an easterly direction). Also applied to the direction of rotation of a planet (such as Venus) about its own axis in the opposite direction to that of the other planets.

**revolution.** Movement of one body in an orbit around another.

**right ascension (R.A.).** Coordinate for measuring celestial longitude along the celestial equator; measured in hours and minutes.

**rille.** Depression on the lunar surface, usually with well-defined edges.

**Roche's limit.** Distance from one body within which gravitational forces would break up a second body. See tidal force.

**rotation.** Movement (spin) of a body about an axis that passes through that body. Distinct from revolution, which is motion in an orbit about some point or other body.

**r-process.** Sequence of nuclear reactions in which neutrons are captured to form heavier nuclei, and the rate of reactions is fast; r-process reactions can produce heavy nuclei such as uranium and thorium.

**Rydberg constant.** Quantity that appears in a formula which can be used to calculate the wavelengths of lines in atomic spectra. Discovered empirically by Rydberg, but now derived from knowledge of the structure of atoms.

**saros.** Cycle of eclipses, every 18 years 11 days. These eclipses will not be seen repeatedly at the same place on Earth.

**satellite.** Body that revolves in orbit around another body. Planets are satellites of the Sun, the Moon is a satellite of the Earth, and artificial satellites have been sent into orbit around the Earth, Moon, Mars, and Venus.

**scarp.** Steep surface exposed by the slide of material to a lower level, as on the lunar surface.

**Schmidt telescope.** Type of reflecting telescope that uses a spherical primary mirror and a thin correcting lens across the full aperture.

**Schwarzschild radius.** Radius of a spherical surface around a black hole from which no light can emerge because of the gravitational effect of the compact mass.

**secondary.** Not the primary: For example, some telescopes use a secondary mirror in addition to the primary (objective); secondary cosmic rays are produced by the collisions of primary cosmic rays in the Earth's atmosphere.

**seeing.** Effect of the Earth's atmosphere on the angular resolution attainable through a telescope. Usually expressed in arc seconds. Good seeing conditions permit resolution better than 1 arc sec.

**seismic waves.** Elastic vibrations that travel through the Earth, as a result of earthquakes. Similar waves have been detected on the lunar surface.

**seismology.** The study of earthquakes and the Earth's interior using observations of seismic waves.

**semimajor axis.** Half of the longest diameter or major axis, as of an ellipse.

**Seyfert galaxy.** Type of spiral galaxy with a bright nucleus whose spectrum shows emission lines.

**shock wave.** Expanding pressure wave that moves with the speed of sound and is produced by an object that moves faster than sound. For example, the "boom" you can hear when a supersonic plane passes.

**sidereal.** Related to the stars. Thus sidereal day, month, period, year: lengths of time intervals specified by motion of some object relative to the stars, as opposed to apparent lengths of those time intervals that will depend on the Earth's own movements.

**significant figures.** Number of meaningful digits in a physical quantity, determined by the accuracy of the measurements on which it is based.

**Small Magellanic Cloud (SMC).** See Clouds of Magellan.

**solar activity.** Variable phenomena observed on the Sun. Some (such as the sunspot cycle) will be fairly regular, but individual spots, prominences will not display any regularity.

**solar constant.** A measure of the average amount of energy radiated by the Sun and incident on the top of the Earth's atmosphere. The value is $1.37 \times 10^6$ ergs/cm$^2$ sec.

**solar nebula.** Extended cloud of gas and dust from which the Sun, planets and other bodies of the solar system are thought to have formed.

**solar wind.** Flow of high-speed electrically charged particles from the solar corona, outward through the solar system.

**solstice.** Extreme positions reached by the Sun, north and south of the celestial equator. When the Sun is at these positions, the shadows it casts on the Earth's surface are longest or shortest (depending on whether you are making this observation in the northern or southern hemisphere).

**space motion (velocity).** Velocity of a star with respect to the Sun. This motion has two components—radial (along the line between it and the Sun) and transverse (at right angles to the radial line).

**spectral class (type).** Classification of stars by their spectral features, into groups designated by *O*, *B*, *A*, etc. This grouping also corresponds to surface temperatures.

**spectrogram.** Photograph of spectrum.

**spectrograph.** Instrument for dispersing light into a spectrum and then photographing it.

**spectrophotometry.** Measurement of the intensity of light in various parts of a spectrum.

**spectroscope.** Instrument for viewing a spectrum. Usually contains a prism or grating that disperses the light.

**spectroscopic binary.** Type of binary star whose components are not seen separately, but whose spectrum shows periodic wavelength shifts that can be interpreted as the results of orbital motion.

**spectrum.** The spread or range of wavelengths in the radiation emitted by some body or region. The type of spectrum depends on the physical processes involved in the emission of the radiation.

**spicule.** Glowing jet of gas, seen rising in the solar chromosphere.

**spiral galaxy.** Type of galaxy that has a flattened shape, with arms of stars and gas and dust extending either from the nucleus or from short bars that are themselves connected to the nucleus.

**s-process.** Sequence of nuclear reactions in which neutrons are captured to form heavier nuclei, and the rate of reaction is slow; s-process reactions produce hardly any nuclei heavier than lead and bismuth.

**star cluster.** Group of stars within a galaxy, either very closely packed (in globular clusters) or further apart (in open clusters).

**Steady State theory.** Cosmological theory that assumes continuous creation of matter to compensate for steady expansion of the universe, to keep the average density of matter constant.

**Stefan's Law.** Formula relating the energy radiated per second from a hot surface, to the fourth power of that temperature.

**stellar model.** Theoretical description of the structure of a star: the way in which its density, temperature and other physical properties vary with distance from the center and the way in which energy is produced and radiated.

**stellar parallax.** Angle (usually in arc seconds) between lines from a star to the Sun and Earth when they are precisely 1 AU apart. Distance to the star is then (1/parallax), in parsec, where 1 parsec = $3.086 \times 10^{18}$ cm.

**stony meteorite.** Type of meteorite that is made mostly of stony material. See chondrite.

**stratosphere.** One of the upper layers of the Earth's atmosphere, above the troposphere that contains most of the weather, and below the ionosphere.

**Strömgren sphere.** Spherical region of ionized gas around a very hot star.

**summer solstice.** Extreme northern position of the Sun on the celestial sphere, north of the celestial equator (summer refers to the northern hemisphere).

**sunspot.** Area that appears dark on the solar disc because the sunspot has a temperature somewhat lower than its surroundings.

**sunspot cycle.** 11-year periodicity in the number of sunspots.

**supergiant.** Most luminous class of stars.

**supergranulation.** Pattern of bright and dark areas seen on the solar photosphere, caused by convective motion.

**supernova.** Explosive brightening of a star in which an outer layer is blown off and the star may be $10^8$ times (or more) as luminous as the Sun for some months.

**surface gravity.** Acceleration due to gravity on the surface of a body, such as a planet or satellite.

**synchrotron radiation.** Radiation emitted by relativistic electrons as they move along spiral paths in magnetic fields. Synchrotron radiation may be anywhere in the electromagnetic spectrum, from radio waves to the ultraviolet, depending on the electron energy and the magnetic field strength.

**synodic period.** Time interval between successive repetitions of the same apparent positions of the Sun, Earth, and a body. Thus synodic period for a planet, or synodic month for the Moon. See also sidereal.

**temperature (absolute or Kelvin).** Temperature measured on a scale that has its zero at about −273 C. On this scale, water freezes at 273 K and boils at 373 K.

**temperature (Celsius).** Often termed centigrade because this scale of temperature has 100 degrees between the freezing and boiling points of water.

**temperature (effective).** Temperature of a perfect (black body) radiator which would radiate exactly the same amount of energy as a given body whose surface is not necessarily black.

**temperature (Fahrenheit).** Temperature scale in which water freezes at 32 F and boils at 212 F. Not widely used for scientific work (see temperature, Celsius or absolute). Fahrenheit temperatures are used in the definition of BTU, a unit of heat energy widely used in the U.S. for domestic heaters and air-conditioners.

**terminator.** Line between bright and dark areas on the Moon or a planet, marking the boundary of the area illuminated by the Sun at that time.

**terrestrial planet.** Mercury, Venus, Earth, or Mars.

**thermal energy.** Internal energy in a body or substance, associated with the motion of the constituent atoms or molecules.

**thermal radiation.** Radiation emitted by a body or gas as a result of its temperature. The thermal radiation from a perfect radiator will follow Planck's Radiation Law.

**thermonuclear reactions.** Nuclear reactions occurring at high temperatures, such as fusion reactions in which hydrogen nuclei join to form helium.

**tidal force.** Gravitational force that is stronger on one side of a body than on the other. On the Earth, the oceans can respond to tidal forces and move relatively easily. On a completely solid object, the tidal force can produce a deformation that might even break the body. See Roche's limit.

**total eclipse.** Eclipse of the Sun in which the Moon completely hides the solar photosphere, or an eclipse of the Moon in which it passes completely into the umbra behind the Earth.

**transit.** Passage of a celestial body across the meridian, or an instrument used to observe transits. Also, passage of one body in front of another (without eclipsing it): for example, transits of Mercury across the face of the Sun.

**transverse velocity.** Component of velocity at right angles to the line of sight to a star (as opposed to component of velocity along the line of sight; see radial velocity).

**transverse wave.** Wave in which the fundamental vibrations are at right angles to the direction of travel of the wave. Examples: light, radio waves.

**trillion.** Usually now $10^{12}$ (one million millions). Rarely used scientifically.

**triple-alpha process.** Sequence of the nuclear reactions that can take place at temperatures above about $10^8$ K, in which three helium nuclei (alpha particles) are joined to form a carbon nucleus.

**Trojan asteroids.** Asteroids that are in solar orbit, in the same orbit as Jupiter but precede or follow the planet by 60°.

**Tropics.** Parallels of latitude that are 23½° north (Cancer) and south (Capricorn). The Sun appears directly overhead at each tropic at the appropriate solstice.

**troposphere.** Layer of the Earth's atmosphere, from sea level to about 10–15 km altitude, containing most of the weather.

**U, B, V magnitudes.** Stellar magnitudes as measured through filters that restrict observation to band of wavelengths in the ultraviolet ($U$), blue ($B$) and green-yellow ($V$) parts of the visible spectrum.

**ultraviolet radiation.** Electromagnetic radiation having wavelengths shorter than the visible violet, but longer than X-rays; approximate range of wavelengths is from 4000 angstroms down to around 100 angstroms. Strongly absorbed in the Earth's atmosphere.

**umbra.** Central part of a shadow.

**universe.** All matter, radiation and the space that contains them.

**Van Allen belts.** Regions around the Earth where the Earth's magnetic field confines high-speed electrically charged particles, mostly protons and electrons.

**variable star.** Star whose luminosity changes. This designation will include stars with explosive changes (novae and supernovae) as well as cyclic changes (Cepheids and RR Lyrae).

**velocity.** Speed in a designated direction. The rate at which a body changes its position is usually designated as *speed*, when the direction of motion is not considered. *Velocity* implies a definite direction.

**visual binary star.** Binary star whose two components can be seen (by telescope) to be separate.

**VLA.** Very Large Array; array of twenty-seven radio telescopes in New Mexico, one of the systems of radio telescopes operated by the National Radio Astronomy Observatory (NRAO).

**VLBI.** Very long baseline interferometry; use of radio telescopes that are thousands of kilometers apart, to observe the same celestial object and through combining their observations achieve very high angular resolution.

**volatile material.** Material that evaporates at relatively low temperatures. Examples: water, methane.

**watt.** Unit of power, the rate at which energy is consumed or radiated. Numerically, 10 million ergs/sec. Larger units are kilowatt (1000 watts) and megawatt (1 million watts). Named for an eighteenth century scientist, James Watt. Unrelated to prominent politician.

**wavelength.** Distance between successive crests or troughs of a wave.

**weight.** Gravitational force of attraction. Related to mass via $W = M g$, where $M$ is the mass of an object and $g$ is the local acceleration due to gravity, such as on the surface of a planet.

**white dwarf.** Star that is less massive than 1.4 times the solar mass, that has consumed almost all of its nuclear fuel and has contracted to a size not much larger than the Earth. Characterized by high surface temperature but small luminosity, as compared to the Sun. White dwarfs fall below the main sequence in the H-R diagram.

**Wien's Law.** Formula that was discovered empirically but can now be deduced from Planck's Radiation Law. Wien's Law relates the temperature of a black surface to the wavelength at which it emits the highest intensity radiation.

**winter solstice.** Extreme southern position of the Sun on the celestial sphere, south of the celestial equator (winter refers to the northern hemisphere).

**world.** Term sometimes used as synonym for *universe*, but not generally used in astronomy. Sometimes refers only to Earth, or to relatively confined environment.

**X-rays.** Electromagnetic radiation with wavelengths shorter than ultraviolet. Very short wavelength X-rays overlap the gamma-ray part of the spectrum.

**year.** Time taken by the Earth to complete one revolution of its orbit around the Sun.

**ZAMS.** Zero age main sequence; main sequence on the H-R diagram, occupied by stars that are no longer contracting but are now obtaining all their energy from nuclear reactions. As these stars age, their composition will change and they will move from the ZAMS.

**Zeeman effect.** Splitting of spectral lines emitted by atoms in magnetic fields.

**zenith.** Point on the sky directly overhead.

**zenith distance.** Angle in the meridian between the zenith and the direction to a celestial body. The sum of an object's altitude and zenith distance is 90°.

**zodiac.** Band on the sky, centered on the ecliptic, and about 18° wide, through which the Sun, Moon and planets appear to move through the course of each year.

**zodiacal light.** Faint glow seen at night near the ecliptic, probably sunlight reflected by interplanetary dust.

**zones.** Light-colored bands seen on Jupiter and Saturn.

APPENDIX B

# Mathematical Notes

## SCIENTIFIC NOTATION: POWERS OF TEN

Astronomy requires us to deal with some numbers that are very small and others that are very large. We have to carry out calculations that involve the dimensions of atoms and the wavelengths of light, and also the dimensions of galaxies. It is therefore a considerable convenience to use the scientific notation in which many zeros, before or after a decimal, are combined and expressed as a power of ten. Thus 100 can be written as $10^2$, 200 as $2.00 \times 10^2$ and 5 million = 5,000,000 as $5 \times 10^6$.

To express a large number in powers of ten, move the decimal to leave only one numeral to the left of the decimal, and count how many places the decimal had to be moved. Thus 517857.6 will be written as $5.178576 \times 10^5$, since the decimal was moved five places.

A similar rule holds for numbers less than one. Move the decimal to the right, and list the number of places moved as a negative power of ten. Thus 0.00112 becomes $1.12 \times 10^{-3}$.

Multiplication and division of large or small numbers is made considerably easier by using scientific notation, with the added rule that powers of ten are added when multiplying, and subtracted when dividing. For example:

$$\begin{aligned} 300 \times 2000 &= (3 \times 10^2) \times (2 \times 10^3) \\ &= (3 \times 2) \times (10^2 \times 10^3) \\ &= 6 \times 10^5 \end{aligned}$$

and

$$\begin{aligned} (6 \times 10^{27})/(4 \times 10^{25}) &= (6/4) \times (10^{27}/10^{25}) \\ &= 1.5 \times 10^2 \end{aligned}$$

Square roots require the power of ten to be halved:

$$\begin{aligned} 490{,}000 &= \sqrt{49 \times 10^4} = \sqrt{49} \times \sqrt{10^4} \\ &= 7 \times 10^2 \end{aligned}$$

## LOGARITHMS

Before the introduction of computers, logarithms were used for many calculations, but just about the only place where we shall encounter them is in the relation between apparent magnitude, absolute magnitude, and distance. The logarithm of a number is the power to which 10 must be raised to produce the desired number. Thus $10^2 = 100$, and the logarithm of 100 is 2, or $\log 100 = 2$. This will work also with the fractional powers: $10^{0.5}$ is the square root of 10, or 3.16. Thus $\log 3.16 = 0.5$. Tables of logarithms are available, and some hand calculators have the logarithm function built in.

APPENDIX

# Units

The different quantities encountered in astronomy are measured in different units. The metric system is generally used, but in some cases English units have been given because these are more familiar. In the metric system, the units are the meter for length, the gram for mass and the second for time. Larger and small multiples are often more convenient than these basic units. Angles are measured in degrees, temperatures are also designated in degrees.

## LENGTH

1 km = 1 kilometer = 1000 m = 0.6214 mile

1 m = 1 meter = 1.0914 yards = 39.37 inches

1 cm = 1 centimeter = 0.01 m = 0.3937 inches

1 mm = 1 millimeter = 0.001 m = 0.03937 inches

1 μ = 1 micron = $10^{-6}$ m = $10^{-4}$ cm

1 Å = 1 Angstrom = $10^{-10}$ m = $10^{-8}$ cm

1 mile = 1.6093 km

1 inch = 2.5400 cm

## MASS

1 metric ton = $10^6$ g = 1000 kg = 2204.6 lb (almost 1 ton)

1 kg = 1 kilogram = 1000 g

1 g = 1 gram

1 lb = 453.6 g

1 oz = 28.35 g

## ANGLES

1° = 60′ = 60 minutes

1′ = 60″ = 60 seconds

To distinguish minutes and seconds of angle from minutes and seconds of time, angles are usually designated as arc minute and arc second.

Circle: circumference of a circle = $2\pi R$, where $R$ is the radius

$\pi$ = 3.1415926

= 355/113 approximately (this is a better approximation than 22/7)

There is a simple relation between the diameter ($d$) of an object and the angular diameter ($\theta$) that it appears to have when viewed from a distance $D$:

$$\theta = 2.05 \times 10^5 \, (d/D) \text{ arc sec.}$$

This relation is accurate enough for our purposes for $\theta$ less than about 20°.

## TEMPERATURE

For most scientific use the Celsius or centigrade scale is used, rather than the more familiar Fahrenheit. On the Celsius scale, the freezing point of water is defined as 0° and the boiling point at 100°, both at atmospheric pressure.

On the Fahrenheit scale, the freezing point of water is 32°F, and the boiling point is 212°F. Conversion between Celsius and Fahrenheit follows simple rules:

$$C = 5/9 \, (F - 32)$$

$$F = 32 + 9/5 \, C$$

For many scientific purposes, the absolute or Kelvin scale is preferred. On this scale, the freezing point of water is 273, the boiling point is 373, and the temperatures are designated as K. Thus K = C + 273.

APPENDIX

# Basic Physical and Astronomical Constants

## PHYSICAL CONSTANTS

| | | |
|---|---|---|
| Speed of light | $c$ | = 2.99 792 458 cm/sec |
| Constant of gravitation | $G$ | = $6.672 \times 10^{-8}$ cm$^3$/g sec$^2$ |
| Planck's constant | $h$ | = $6.6262 \times 10^{-27}$ erg sec |
| Boltzmann's constant | $k$ | = $1.3806 \times 10^{-16}$ erg/kelvin |
| Stefan's constant | $\sigma$ | = $5.66956 \times 10^{-5}$ erg/cm$^2$ deg$^4$ s |
| Wien displacement constant | $\lambda_{max}T$ | = 0.289789 cm K |
| Mass of hydrogen atom | $m_H$ | = $1.6735 \times 10^{-24}$ g |
| Mass of neutron | $m_n$ | = $1.6749 \times 10^{-24}$ g |
| Mass of proton | $m_p$ | = $1.6726 \times 10^{-24}$ g |
| Mass of electron | $m_e$ | = $9.1096 \times 10^{-28}$ g |
| Rydberg's constant | $R$ | = $1.09677 \times 10^5$/cm |

## ASTRONOMICAL CONSTANTS

| | | |
|---|---|---|
| Astronomical unit | 1 AU | = $1.495\ 978\ 70 \times 10^{13}$ cm |
| Parsec | pc | = 206,264.806 AU |
| | | = 3.261 LY |
| | | = $3.085678 \times 10^{18}$ cm |
| Light year | LY | = $9.460530 \times 10^{17}$ cm |
| Tropical year (1900)—equinox to equinox | | = 365.24219878 days |
| 1 day | | = 86400 sec |
| Sidereal year | | = 365.256366 days |
| | | = $3.155815 \times 10^7$ sec |
| Mass of Sun | $M_\odot$ | = $1.9891 \times 10^{33}$ g |
| Radius of Sun | $R_\odot$ | = $6.96 \times 10^{10}$ cm |
| Luminosity of Sun | $L_\odot$ | = $3.827 \times 10^{33}$ erg/sec |
| Mass of Earth | $M_\oplus$ | = $5.9742 \times 10^{27}$ g |
| Equatorial radius of Earth | $R_\oplus$ | = 6378.140 km |
| Mean distance center of Earth to center of Moon | | = 384,403 km |
| Radius of Moon | $R_M$ | = 1738 km |
| Mass of Moon | $M_M$ | = $7.35 \times 10^{25}$ g |
| Solar constant | $S$ | = $1.37 \times 10^6$ ergs/cm$^2$ sec |

APPENDIX

# The Planets: Orbital Properties

| Name | Semimajor Axis (AU) | Semimajor Axis ($10^6$ km) | Sidereal Period (years) | Sidereal Period (days) | Synodic Period (days) | Eccentricity | Mean Orbital Speed (km/sec) | Inclination to Ecliptic |
|---|---|---|---|---|---|---|---|---|
| Mercury | 0.3871 | 57.9 | 0.24084 | 87.96 | 115.9 | 0.206 | 47.9 | 7°00′ |
| Venus | 0.7233 | 108.2 | 0.61521 | 224.70 | 584.0 | 0.007 | 35.0 | 3°24′ |
| Earth | 1 | 149.6 | 1.00004 | 365.26 | — | 0.017 | 29.8 | 0°00′ |
| Mars | 1.5237 | 227.9 | 1.88089 | 686.98 | 779.9 | 0.093 | 24.1 | 1°51′ |
| Jupiter | 5.2028 | 778.3 | 11.86 | 4333 | 398.9 | 0.048 | 13.1 | 1°18′ |
| Saturn | 9.5388 | 1427.0 | 29.46 | | 378.1 | 0.056 | 9.6 | 2°29′ |
| Uranus | 19.191 | 2871 | 84.07 | | 369.7 | 0.046 | 6.8 | 0°48′ |
| Neptune | 30.061 | 4497 | 164.81 | | 367.5 | 0.010 | 5.4 | 1°46′ |
| Pluto | 39.529 | 5913 | 248.53 | | 366.7 | 0.248 | 4.7 | 17°09′ |

APPENDIX

# The Planets: Physical Properties

| Name | Equatorial Radius (km) | Equatorial Radius ÷ Earth | Mass ÷ Earth | Surface Gravity (Earth = 1) | Sidereal Rotation Period (days) | Escape Velocity (km/sec) | Visual Magnitude at Brightest | Inclination of Equator to Orbit |
|---|---|---|---|---|---|---|---|---|
| Mercury | 2439 | 0.3824 | 0.0553 | 0.38 | 58.6 | 4.2 | −1.9 | 7.0° |
| Venus | 6052 | 0.9489 | 0.8150 | 0.89 | 243R[a] | 10.3 | −4.4 | 177.4° |
| Earth | 6378 | 1 | 1 | 1 | 0.9973 | 11.2 | — | 23.4° |
| Mars | 3397 | 0.5236 | 0.1074 | 0.38 | 1.026 | 5.0 | −2.0 | 25.2° |
| Jupiter | 70850 | 11.11 | 317.8 | 2.54 | 0.41 | 61 | −2.7 | 3.1° |
| Saturn | 60000 | 9.41 | 95.1 | 1.07 | 0.43 | 37 | +0.7 | 26.7° |
| Uranus | 25900 | 4.06 | 14.6 | 0.88 | 0.65R | 22 | +5.5 | 97.9° |
| Neptune | 24750 | 3.88 | 17.2 | 1.14 | 0.77 | 25 | +7.8 | 29° |
| Pluto | ~1800 | 0.28 | 0.0022 | 0.028 | 6.39 | ~1 | +15 | ? |

[a] R indicates retrograde rotation.

APPENDIX

# Satellites of the Planets

| Planet | # | Satellite Name | Semimajor Axis of Orbit (km) | Sidereal Period (days) | Diameter (km) | Mass (g) | Density (g/cm³) |
|---|---|---|---|---|---|---|---|
| Earth | | Moon | $3.84 \times 10^5$ | 27.322 | 3476 | $7.35 \times 10^{25}$ | 3.34 |
| Mars | | Phobos | $9.4 \times 10^3$ | 0.3189 | 28 × 22 × 20 | $9.6 \times 10^{18}$ | 2.0 |
| | | Deimos | $2.35 \times 10^4$ | 1.2624 | 15 × 12 × 11 | $2.0 \times 10^{18}$ | 1.9 |
| Jupiter | 16 | Metis | $1.2796 \times 10^5$ | 0.295 | 40 | | |
| | 15 | Adrastea | $1.2898 \times 10^5$ | 0.298 | 24 | | |
| | 5 | Amalthea | $1.813 \times 10^5$ | 0.498 | 270 | | |
| | 14 | Thebe | $2.22 \times 10^5$ | 0.675 | 110 | | |
| | 1 | Io | $4.22 \times 10^5$ | 1.769 | 3638 | $8.92 \times 10^{25}$ | 3.55 |
| | 2 | Europa | $6.71 \times 10^5$ | 3.551 | 3138 | $4.87 \times 10^{25}$ | 3.04 |
| | 3 | Ganymede | $1.07 \times 10^6$ | 7.155 | 5262 | $1.49 \times 10^{26}$ | 1.93 |
| | 4 | Callisto | $1.88 \times 10^6$ | 16.689 | 4800 | $1.08 \times 10^{26}$ | 1.83 |
| | 13 | Leda | $1.11 \times 10^7$ | 238.7 | 10 | | |
| | 6 | Himalia | $1.15 \times 10^7$ | 250.6 | 180 | | |
| | 10 | Lysithia | $1.17 \times 10^7$ | 259.2 | 20 | | |
| | 7 | Elara | $1.17 \times 10^7$ | 259.7 | 80 | | |
| | 12 | Ananke | $2.12 \times 10^7$ | 631 | 20 | | |
| | 11 | Carme | $2.26 \times 10^7$ | 692 | 30 | | |
| | 8 | Pasiphae | $2.35 \times 10^7$ | 735 | 40 | | |
| | 9 | Sinope | $2.37 \times 10^7$ | 758 | 28 | | |
| Saturn | 15 | Atlas | $1.3767 \times 10^5$ | 0.602 | 40 × 20 | | |
| | 14 | 1980 S27 | $1.3935 \times 10^5$ | 0.613 | 140 × 100 × 80 | | |
| | 13 | 1980 S26 | $1.4170 \times 10^5$ | 0.629 | 110 × 90 × 70 | | |
| | 10 | Janus | $1.5142 \times 10^5$ | 0.694 | 220 × 190 × 160 | | |
| | 11 | Epimetheus | $1.5147 \times 10^5$ | 0.695 | 140 × 120 × 100 | | |
| | 1 | Mimas | $1.8554 \times 10^5$ | 0.942 | 392 | $4.6 \times 10^{22}$ | 1.44 |
| | 2 | Enceladus | $2.3804 \times 10^5$ | 1.370 | 500 | $8.4 \times 10^{22}$ | 1.2 |
| | 16 | 1980 S13 | $2.9467 \times 10^5$ | 1.888 | 34 × 28 × 26 | | |
| | 17 | 1980 S25 | $2.9467 \times 10^5$ | 1.888 | 34 × 22 × 22 | | |
| | 3 | Tethys | $2.9467 \times 10^5$ | 1.888 | 1060 | $7.6 \times 10^{23}$ | 1.21 |
| | 4 | Dione | $3.7742 \times 10^5$ | 2.737 | 1120 | $1.05 \times 10^{24}$ | 1.43 |
| | 12 | 1980 S6 | $3.7811 \times 10^5$ | 2.739 | 36 × 32 × 30 | | |
| | 5 | Rhea | $5.2710 \times 10^5$ | 4.518 | 1530 | $2.5 \times 10^{24}$ | 1.33 |
| | 6 | Titan | $1.222 \times 10^6$ | 15.945 | 5150 | $1.35 \times 10^{26}$ | 1.88 |
| | 7 | Hyperion | $1.481 \times 10^6$ | 21.277 | 410 × 260 × 220 | | |
| | 8 | Iapetis | $3.561 \times 10^6$ | 79.331 | 1460 | $1.88 \times 10^{24}$ | 1.16 |
| | 9 | Phoebe | $1.295 \times 10^7$ | 550.45 | 220 | | |
| Uranus | 5 | Miranda | $1.298 \times 10^5$ | 1.414 | 300 | | |
| | 1 | Ariel | $1.908 \times 10^5$ | 2.520 | 800 | | |
| | 2 | Umbriel | $2.658 \times 10^5$ | 4.144 | 550 | | |
| | 3 | Titania | $4.361 \times 10^5$ | 8.706 | 1000 | | |
| | 4 | Oberon | $5.831 \times 10^5$ | 13.46 | 900 | | |
| Neptune | | Triton | $3.553 \times 10^5$ | 5.877 | 3600 | | |
| | | Nereid | $5.511 \times 10^6$ | 359.9 | 300 | | |
| Pluto | | Charon | $1.97 \times 10^4$ | 6.39 | ? | | |

Dimensions of many of the small satellites are not yet well established.

APPENDIX

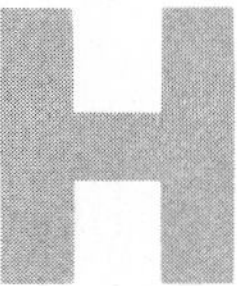

# The Nearest Stars

| Star | Parallax, $\pi$ (") | Distance (pc) | Proper Motion, $\mu$ ("/year) | Visual Magnitude, m | Absolute Magnitude $M_v$ | Spectral Type |
|---|---|---|---|---|---|---|
| Sun | | | | −26.8 | +4.85 | G2 |
| $\alpha$ Cen A | 0.760 | 1.32 | 3.68 | 0.1 | 4.4 | G2 |
| B | | | | 1.5 | 5.7 | K5 |
| C | | | | 11.0 | 15.5 | M5 |
| Barnard's | .552 | 1.81 | 10.30 | 9.5 | 13.3 | M5 |
| Wolf 359 | .431 | 2.32 | 4.84 | 15.3 | 16.7 | M6 |
| Lalande 21185 | .402 | 2.49 | 4.78 | 7.5 | 10.5 | M2 |
| Sirius A | .377 | 2.65 | 1.32 | −1.5 | 1.4 | A1 |
| B | | | | 7.2 | 11.6 | wd |
| Luyten 726-8 A | .365 | 2.74 | 3.35 | 12.5 | 15.3 | M6 |
| B (UV Ceti) | | | | 13.0 | 15.8 | M6 |
| Ross 154 | .345 | 2.90 | 0.74 | 10.6 | 13.3 | M5 |
| Ross 248 | .317 | 3.15 | 1.82 | 12.2 | 14.8 | M6 |
| $\epsilon$ Eri | .305 | 3.28 | 0.97 | 3.7 | 6.2 | K2 |
| Luyten 789-6 | .302 | 3.31 | 3.27 | 12.2 | 14.6 | M6 |
| Ross 128 | .301 | 3.32 | 1.40 | 11.1 | 13.5 | M5 |
| 61 Cyg A | .292 | 3.42 | 5.22 | 5.2 | 7.6 | K5 |
| B | | | | 6.0 | 8.4 | K7 |
| $\epsilon$ Ind | .291 | 3.44 | 4.67 | 4.7 | 7.0 | K5 |
| Procyon A | .287 | 3.48 | 1.25 | 0.3 | 2.7 | F5 |
| B | | | | 10.8 | 13.0 | wd |
| $\Sigma$ 2398 A | .284 | 3.52 | 2.29 | 8.9 | 11.2 | M3.5 |
| B | | | | 9.7 | 12.0 | M4 |
| BD + 43°44 A | .282 | 3.55 | 2.91 | 8.1 | 10.3 | M1 |
| B | | | | 11.0 | 13.3 | M6 |
| Lacaille 9352 | .279 | 3.58 | 6.78 | 7.4 | 9.6 | M2 |
| $\tau$ Ceti | .273 | 3.66 | 1.92 | 3.5 | 5.7 | G8 |
| BD+5°1668 | .266 | 3.76 | 3.73 | 9.8 | 11.9 | M4 |
| L725-32 | .262 | 3.82 | 1.31 | 11.5 | 13.7 | M5 |
| Lacaille 8760 | .260 | 3.85 | 3.46 | 6.7 | 8.8 | M1 |
| Kapteyn's | .256 | 3.91 | 8.79 | 8.8 | 10.9 | M0 |
| Kruger 60 A | .254 | 3.94 | 0.87 | 9.7 | 11.9 | M4 |
| B | | | | 11.2 | 13.3 | M6 |

APPENDIX

# The Brightest Stars

| Star | Apparent Magnitude | Absolute Magnitude | B − V | Spectral Type | Distance (pc) | Proper Motion ("/year) | Position (1980) Right Ascension | Declination |
|---|---|---|---|---|---|---|---|---|
| Archenar | 0.51 | −1.0 | −0.16 | B3 V | 36 | 0.10 | $01^h37^m.0$ | −57°20′ |
| Aldebaran | 0.86 | −0.2 | +1.52 | K5 III | 21 | 0.20 | 04 34.8 | +16 28 |
| Rigel | 0.14 | −6.8 | −0.04 | B8 Ia | 276 | 0.00 | 05 13.6 | −08 13 |
| Capella | 0.05 | −0.7 | +0.80 | G8 III | 14 | 0.44 | 05 15.2 | +45 59 |
| Betelgeuse | 0.41 | −5.5 | +1.87 | M2 Iab | 159 | 0.03 | 05 54.0 | +07 24 |
| Canopus | −0.72 | −3.1 | +0.16 | FO Ib-II | 30 | 0.025 | 06 23.5 | −52 41 |
| Sirius | −1.47 | +1.4 | +0.01 | A1 V | 2.7 | 1.32 | 06 44.2 | −16 42 |
| Procyon | 0.37 | +2.6 | +0.41 | F5 IV−V | 3.5 | 1.25 | 07 38.2 | +05 17 |
| Pollux | 1.16 | +0.8 | +1.02 | KO III | 11 | 0.62 | 07 44.1 | +28 05 |
| α Crucis A | 1.39 | −4.0 | −0.25 | B0.5 IV | 114 | 0.04 | 12 25.4 | −62 59 |
| B | 1.86 | −3.5 | −0.25 | B1 V | 114 | 0.04 | 12 25.4 | −62 59 |
| β Crucis | 1.28 | −4.6 | −0.25 | B0.5 III | 150 | 0.05 | 12 46.6 | −59 35 |
| Spica | 0.91 | −3.6 | −0.24 | B1 V | 67 | 0.05 | 13 24.1 | −11 03 |
| Hadar | 0.62 | −4.1 | −0.23 | B1 III | 150 | 0.04 | 12 02.4 | −60 16 |
| Arcturus | −0.06 | −0.3 | +1.23 | K2 III | 11 | 2.28 | 14 14.8 | +19 17 |
| α Centauri A | 0.01 | +4.4 | +0.68 | G2 V | 1.3 | 3.68 | 14 38.4 | −60 46 |
| B | 1.40 | +5.7 | +0.73 | K4 V | 1.3 | 3.68 | 14.38.4 | −60 46 |
| Antares | 0.92 | −4.5 | +1.84 | M1 Ib | 160 | 0.03 | 16 28.2 | −26 23 |
| Vega | 0.04 | +0.5 | 0.00 | A0 V | 8 | 0.34 | 18 36.2 | +38 46 |
| Altair | 0.77 | +2.2 | +0.22 | A7 IV−V | 5 | 0.66 | 19 49.8 | +08 49 |
| Deneb | 1.26 | −6.9 | +0.09 | A2 Ia | 491 | 0.00 | 20 40.7 | +45 12 |
| Fomalhaut | 1.15 | +2.0 | +0.10 | A3 V | 7 | 0.37 | 22 56.5 | −29 44 |

APPENDIX

# The Local Group of Galaxies

| Galaxy | Type | Apparent Magnitude, $m_V$ | Distance (kpc) | Diameter (kpc) | Radial Velocity[a] (km/sec) | Mass $M/M_\odot$ |
|---|---|---|---|---|---|---|
| Milky Way | Sb | — | — | 30 | — | $4 \times 10^{11}$ |
| LMC | Irr I | 0.9 | 48 | 10 | +276 | $2.5 \times 10^{10}$ |
| SMC | Irr I | 2.5 | 56 | 8 | +168 | — |
| Ursa Major system | E4 dw | — | 70 | 1 | — | — |
| Sculptor system | E3 dw | 8.0 | 83 | 2.2 | — | $2\text{–}4 \times 10^6$ |
| Draco system | E2 dw | — | 100 | 1.4 | — | — |
| Carina system | E3 dw | — | 170 | 1.5 | — | — |
| Fornax system | E3 dw | 8.4 | 250 | 4.5 | +39 | $1.2\text{–}2 \times 10^7$ |
| Leo I system | E4 dw | 12 | 280 | 1.5 | — | — |
| Leo II system | E0 dw | — | 230 | 1.6 | — | $1.1 \times 10^6$ |
| NGC 6822 | Irr I | 8.9 | 460 | 2.7 | −32 | — |
| NGC 147 | E6 | 9.7 | 570 | 3 | — | — |
| NGC 185 | E2 | 9.4 | 570 | 2.3 | −305 | — |
| NGC 205 | E5 | 8.2 | 680 | 5 | −239 | — |
| NGC 221 (M32) | E3 | 8.2 | 680 | 2.4 | −214 | — |
| IC 1613 | Irr I | 9.6 | 680 | 5 | −238 | — |
| Andromeda (M31) | Sb | 3.5 | 680 | 40 | −266 | $3 \times 10^{11}$ |
| And I | E0 dw | 14 | 680 | 0.5 | — | — |
| And II | E0 dw | 14 | 680 | 0.7 | — | — |
| And III | E3 dw | 14 | 680 | 0.9 | — | — |
| NGC 598 (M33) | Sc | 5.8 | 720 | 17 | −189 | $8 \times 10^9$ |

[a] Radial velocities are designated positive if the galaxy is receding from us, negative if approaching.

APPENDIX

# The Messier Catalog[a]

| M | NGC or IC | Apparent Magnitude | Description | Illustration |
|---|---|---|---|---|
| 1 | 1952 | 8.4 | "Crab" nebula, remains of SN 1054 | Fig. 25.9 |
| 2 | 7089 | 6.4 | Globular cluster | Fig. 26.7 |
| 3 | 5272 | 6.3 | Globular cluster | Fig. 24.19 |
| 4 | 6121 | 6.5 | Globular cluster | |
| 5 | 5904 | 6.1 | Globular cluster | |
| 6 | 6405 | 5.5 | Open cluster | |
| 7 | 6475 | 3.3 | Open cluster | |
| 8 | 6523 | 5.1 | "Lagoon" nebula | |
| 9 | 6333 | 8.0 | Globular cluster | |
| 10 | 6254 | 6.7 | Globular cluster | |
| 11 | 6705 | 6.8 | Open cluster | |
| 12 | 6218 | 6.6 | Globular cluster | |
| 13 | 6205 | 5.9 | Globular cluster | Fig. 30.8 |
| 14 | 6402 | 8.0 | Globular cluster | |
| 15 | 7078 | 6.4 | Globular cluster | |
| 16 | 6611 | 6.6 | Open cluster with nebulosity | |
| 17 | 6618 | 7.5 | "Swan" or "Omega" nebula | |
| 18 | 6613 | 7.2 | Open cluster | |
| 19 | 6273 | 6.9 | Globular cluster | |
| 20 | 6514 | 8.5 | "Trifid" nebula | |
| 21 | 6531 | 6.5 | Open cluster | |
| 22 | 6656 | 5.6 | Globular cluster | |
| 23 | 6494 | 5.9 | Open cluster | |
| 24 | 6603 | 4.6 | Open cluster | |
| 25 | 4725 IC | 6.2 | Open cluster | |
| 26 | 6694 | 9.3 | Open cluster | |
| 27 | 6853 | 8.2 | "Dumbbell" planetary nebula | |
| 28 | 6626 | 7.6 | Globular cluster | |
| 29 | 6913 | 8.0 | Open cluster | |
| 30 | 7099 | 7.7 | Globular cluster | |
| 31 | 224 | 3.5 | Andromeda galaxy | Fig. 28.1 |
| 32 | 221 | 8.2 | Elliptical galaxy; companion to M31 | Fig. 28.1 |
| 33 | 598 | 5.8 | Spiral galaxy | |
| 34 | 1039 | 5.8 | Open cluster | |
| 35 | 2168 | 5.6 | Open cluster | |
| 36 | 1960 | 6.5 | Open cluster | |
| 37 | 2099 | 6.2 | Open cluster | |
| 38 | 1912 | 7.0 | Open cluster | |
| 39 | 7092 | 5.3 | Open cluster | |
| 40 | | | Close double star | |
| 41 | 2287 | 5.0 | Loose open cluster | |
| 42 | 1976 | 4 | Orion nebula | Fig. 24.2 |
| 43 | 1982 | 9 | Northeast portion of Orion nebula | |
| 44 | 2632 | 3.9 | Praesepe; open cluster | |
| 45 | | 1.6 | The Pleiades; open cluster | Fig. 24.17 |
| 46 | 2437 | 6.6 | Open cluster | |
| 47 | 2422 | 5 | Loose group of stars | |

| | | | | |
|---|---|---|---|---|
| 48 | 2548 | 6 | Open cluster | |
| 49 | 4472 | 8.5 | Elliptical galaxy | |
| 50 | 2323 | 6.3 | Loose open cluster | |
| 51 | 5194 | 8.4 | "Whirlpool" spiral galaxy | Fig. 28.3 |
| 52 | 7654 | 8.2 | Loose open cluster | |
| 53 | 5024 | 7.8 | Globular cluster | |
| 54 | 6715 | 7.8 | Globular cluster | |
| 55 | 6809 | 6.2 | Globular cluster | |
| 56 | 6779 | 8.7 | Globular cluster | |
| 57 | 6720 | 9.0 | "Ring" nebula; planetary nebula | Fig. 24.14 |
| 58 | 4579 | 9.9 | Spiral galaxy | |
| 59 | 4621 | 10.0 | Spiral galaxy | |
| 60 | 4649 | 9.0 | Elliptical galaxy | |
| 61 | 4303 | 9.6 | Spiral galaxy | |
| 62 | 6262 | 6.6 | Globular cluster | |
| 63 | 5055 | 8.9 | Spiral galaxy | |
| 64 | 4826 | 8.5 | Spiral galaxy | |
| 65 | 3623 | 9.4 | Spiral galaxy | |
| 66 | 3627 | 9.0 | Spiral galaxy: companion to M65 | |
| 67 | 2682 | 6.1 | Open cluster | Fig. 24.16 |
| 68 | 4590 | 8.2 | Globular cluster | |
| 69 | 6627 | 8.0 | Globular cluster | |
| 70 | 6681 | 8.1 | Globular cluster | |
| 71 | 6838 | 7.6 | Globular cluster | |
| 72 | 6981 | 9.3 | Globular cluster | |
| 73 | 6994 | 9.1 | Open cluster | |
| 74 | 628 | 9.3 | Spiral galaxy | |
| 75 | 6864 | 8.6 | Globular cluster | |
| 76 | 650 | 11.4 | Planetary nebula | |
| 77 | 1068 | 8.9 | Spiral galaxy | |
| 78 | 2068 | 8.3 | Small emission nebula | |
| 79 | 1904 | 7.5 | Globular cluster | |
| 80 | 6093 | 7.5 | Globular cluster | |
| 81 | 3031 | 7.0 | Spiral galaxy | Fig. 28.10 |
| 82 | 3034 | 8.4 | Irregular galaxy | Fig. 29.14 |
| 83 | 5236 | 7.6 | Spiral galaxy | |
| 84 | 4374 | 9.4 | Elliptical galaxy | Fig. 28.30 |
| 85 | 4382 | 9.3 | Elliptical galaxy | |
| 86 | 4406 | 9.2 | Elliptical galaxy | Fig. 28.30 |
| 87 | 4486 | 8.7 | Elliptical galaxy | Fig. 28.23 |
| 88 | 4501 | 9.5 | Spiral galaxy | |
| 89 | 4552 | 10.3 | Elliptical galaxy | |
| 90 | 4569 | 9.6 | Spiral galaxy | |
| 91 | | | | |
| 92 | 6341 | 6.4 | Globular cluster | |
| 93 | 2447 | 6.5 | Open cluster | |
| 94 | 4736 | 8.3 | Spiral galaxy | |
| 95 | 3351 | 9.8 | Barred spiral galaxy | |
| 96 | 3368 | 9.3 | Spiral galaxy | |
| 97 | 3587 | 11.1 | "Owl" nebula; planetary nebula | |
| 98 | 4192 | 10.2 | Spiral galaxy | |
| 99 | 4254 | 9.9 | Spiral galaxy | |
| 100 | 4321 | 9.4 | Spiral galaxy | |
| 101 | 5457 | 7.9 | Spiral galaxy | |
| 102 | 5866(?) | 10.5 | Spiral galaxy | |
| 103 | 581 | 6.9 | Open cluster | |
| 104 | 4594 | 8.3 | Spiral galaxy | Fig. 28.11 |
| 105 | 3379 | 9.7 | Elliptical galaxy | |
| 106 | 4258 | 8.4 | Spiral galaxy | |
| 107 | 6171 | 9.2 | Globular cluster | |
| 108 | 3556 | 10.5 | Spiral galaxy | |
| 109 | 3992 | 10.0 | Spiral galaxy | |
| 110 | 205 | 9.4 | Elliptical galaxy | |

[a] Messier's original catalog contained 103 entries, to which another 7 were added later. There is some dispute about the identification of M102, and M 91 has not been identified.

APPENDIX

L

# Quasars: A Selection

| Name | Apparent Magnitude, $m_v$ | $B - V$ | Redshift $\lambda/\lambda_0$ |
|---|---|---|---|
| 3C2 | 19.4 | +0.79 | 2.037 |
| 3C9 | 19.2 | +0.23 | 3.012 |
| 3C47 | 18.1 | +0.05 | 1.425 |
| 3C138 | 18.8 | +0.53 | 1.759 |
| 3C249.1 | 15.7 | −0.02 | 1.311 |
| PKS 1127–14 | 16.9 | +0.27 | 2.187 |
| 3C270.1 | 18.6 | +0.19 | 2.519 |
| 3C283 | 12.8 | +0.21 | 1.158 |
| 3C277.1 | 17.9 | −0.17 | 1.320 |

APPENDIX

# Pulsars: A Selection

| Name | Period (sec) | Slowdown Rate ($10^{-9}$ sec/day) | Distance (pc) |
|---|---|---|---|
| CP 0329 + 54 | 0.714 5187 | 0.177 | 500 |
| NP 0531 + 21 | 0.033 0976 | 36.5 | 1700 |
| CP 0809 + 74 | 1.282 2413 | 0.014 | 130 |
| PSR 0833 − 45 | 0.089 2098 | 10.8 | 400 |
| CP 0950 + 08 | 0.253 0650 | 0.02 | 60 |
| PSR 1749 − 28 | 0.562 5582 | 0.704 | 1000 |
| CP 1919 + 21 | 1.337 3011 | 0.117 | 250 |
| PSR 2045 − 16 | 1.961 5669 | 0.947 | 400 |

APPENDIX

# The Chemical Elements: Abundances in the Solar System

| Element | Symbol | Atomic Number (Z) | Number of Atoms per $10^{12}$ Hydrogen Atoms |
|---|---|---|---|
| Hydrogen | H | 1 | $1 \times 10^{12}$ |
| Helium | He | 2 | $6.7 \times 10^{10}$ |
| Lithium | Li | 3 | |
| Beryllium | Be | 4 | $2.6 \times 10^{3}$ |
| Boron | B | 5 | |
| Carbon | C | 6 | $1.1 \times 10^{7}$ |
| Nitrogen | N | 7 | $2.3 \times 10^{6}$ |
| Oxygen | O | 8 | $1.8 \times 10^{7}$ |
| Fluorine | F | 9 | $7.8 \times 10^{2}$ |
| Neon | Ne | 10 | $2.6 \times 10^{6}$ |
| Sodium | Na | 11 | $6.0 \times 10^{4}$ |
| Magnesium | Mg | 12 | $1.1 \times 10^{6}$ |
| Aluminum | Al | 13 | $8.5 \times 10^{6}$ |
| Silicon | Si | 14 | $1.0 \times 10^{6}$ |
| Phosphorous | P | 15 | |
| Sulfur | S | 16 | |
| Chlorine | Cl | 17 | |
| Argon | Ar(A) | 18 | |
| Potassium | K | 19 | $2.7 \times 10^{7}$ |
| Calcium | Ca | 20 | |
| Scandium | Sc | 21 | |
| Titanium | Ti | 22 | |
| Vanadium | V | 23 | |
| Chromium | Cr | 24 | |
| Manganese | Mn | 25 | |
| Iron | Fe | 26 | $3.4 \times 10^{7}$ |
| Cobalt | Co | 27 | |
| Nickel | Ni | 28 | |
| Copper | Cu | 29 | |
| Zinc | Zn | 30 | $2.0 \times 10^{6}$ |
| Gallium | Ga | 31 | |
| Germanium | Gr | 32 | |
| All other elements, including: | | $Z > 32$ | $7.8 \times 10^{3}$ |
| Lead | Pb | 82 | 98 |
| Thorium | Th | 90 | 1.7 |
| Uranium | U | 92 | 1.0 |

APPENDIX

# Further Readings: A Bibliography

There are many books and articles on astronomy that expand on what we have been able to cover and which can add to your enjoyment of this fascinating subject. The listings given here are not intended to be complete. Rather, I have tried to cover a wide range of topics. These include many recent publications, some older and more historical books, and a few personal commentaries by eminent astronomers. Some have been issued in several editions by different publishers. The selections have been assembled here in broad groups, but you will find that some books overlap a few categories.

## Popular magazines on astronomy

*Astronomy,* published by AstroMedia Corporation, 411 Mason St., Milwaukee, Wisconsin 53202

*The Griffith Observer,* published by the Griffith Observatory, 2800 East Observatory Road, Los Angeles, California 90027

*Mercury,* published by the Astronomical Society of the Pacific, 1290 24th Ave., San Francisco, California 94122

*Sky and Telescope,* published by Sky Publishing Corp., 49 Bay State Road, Cambridge, Massachusetts 02238

The May/June issue of Mercury contains an extensive listing of astronomy books published during each previous year.

## History of astronomy


A. Berry, *A Short History of Astronomy,* New York, Dover Publications Inc., 1968 reprint of 1908 original.

J. L. E. Dreyer, *A History of Astronomy from Thales to Kepler,* New York, Dover Publications Inc., 1956 reprint of 1906 original.

A. Pannekoek, *A History of Astronomy,* New York, Interscience Publishers, 1961.

S. Toulmin and J. Goodfield, *The Fabric of the Heavens,* New York, Harper and Row Publishers, 1961.


## Archaeoastronomy


K. Brecher and M. Feirtag, *Astronomy of the Ancients,* Cambridge, Massachusettts, M. I. T. Press, 1979.

D. H. Clark and F. R. Stephenson, *The Historical Supernovae,* Oxford, Pergamon Press, 1977.

J. Cornell, *The First Stargazers,* New York, Charles Scribner's Sons, 1981.

E. Hadingham, *Circles and Standing Stones,* New York, Walker and Co., 1975.

E. C. Krupp, *Echoes of the Ancient Skies,* New York, Harper and Row Publishers, 1983.

E. C. Krupp, *In Search of Ancient Astronomies,* Garden City, New York, Doubleday and Co., 1977.

A. Marshack, *The Roots of Civilization,* New York, McGraw-Hill Co., 1972.

F. R. Stephenson and D. H. Clark, *Applications of Early Astronomical Records,* Bristol, Adam Hilger Ltd., 1978.

J. Wood, *Sun, Moon and Standing Stones,* Oxford, Oxford University Press, 1978.


## Handbooks, References


C. W. Allen, *Astrophysical Quantities,* (3rd ed.) London, The Athlone Press, 1973.

K. R. Lang, *Astrophysical Formulae,* New York, Springer-Verlag, 1980.

Patrick Moore (ed), *Yearbook of Astronomy,* New York, W. W. Norton and Co., (annual).

Royal Astronomical Society of Canada, *The Observer's Handbook,* (annual).


## Observing


M. Chartrand, *Skyguide,* New York, Western, 1982.

W. Kals, *Stargazer's Bible,* Garden City, New York, Doubleday and Co., 1980.

J. H. Mallas and E. Kreimer, *The Messier Album,* Cambridge, Massachusetts, Sky Publishing Co., 1978.

M. E. MARTIN, *The Friendly Stars,* New York, Dover Publications, Inc., 1964 republication, with revisions by D. H. Menzel.

D. H. MENZEL, *A Field Guide to the Stars and Planets,* Boston, Houghton Mifflin Co., 1975.

A. P. NORTON, *Sky Atlas,* (17th ed), Cambridge, Massachusetts, Sky Publishing Corp., 1978.

P. C. SHERROD, *A Complete Manual of Amateur Astronomy,* Englewood Cliffs, New Jersey, Prentice-Hall Inc., 1981.

C. A. WHITNEY, *Whitney's Star Finder,* (2nd ed), New York, Alfred A. Knopf, Inc., 1977.

## Photographs and color slides

Write for catalogs of items currently available.

Hansen Planetarium, 1098 South 200 West, Salt Lake City, Utah 84101. (Catalog includes items from Kitt Peak National Observatory, Palomar Observatory, Royal Observatory (Edinburgh), U.S. Naval Observatory, Sacramento Peak Observatory and others.)

Lick Observatory, University of California, Santa Cruz, Santa Cruz, California 95140.

Palomar Observatory, California Institute of Technology, Bookstore, 1201 East California Blvd., Pasadena, California 91125.

Yerkes Observatory, (Photographic Dept.), Williams Bay, Wisconsin 53191.

## General

R. HANBURY BROWN, *Man and the Stars,* Oxford, Oxford University Press, 1978.

N. CALDER, *Violent Universe,* New York, The Viking Press, 1969.

J. CORNELL AND A. P. LIGHTMAN, *Revealing the Universe,* Cambridge, Massachusetts, M. I. T. Press, 1982.

M. GROSSER, *The Discovery of Neptune,* New York, Dover Publications Inc., 1979.

J. S. HEY, *The Radio Universe,* Oxford, Pergamon Press, 1971.

W. G. HOYT, *Lowell and Mars,* Tucson, Arizona, University of Arizona Press, 1976.

S. MITTON (ed), *The Cambridge Encyclopedia of Astronomy,* New York, Crow Publishers, 1977.

P. MOORE, *The New Guide to the Stars,* New York, W. W. Norton and Co., 1974.

C. RONAN, *Deep Space,* New York, Macmillan Publishing Co., 1982.

C. W. TOMBAUGH AND P. MOORE, *Out of the Darkness,* New York, Mentor Books, 1981.

A. J. WHYTE, *The Planet Pluto,* Oxford, Pergamon Press, 1980.

## More advanced readings

There are many excellent texts that are at about the same level as our book or slightly lower or higher (in terms of their math and physics content). Many are updated every few years. The next few recommendations are intended to guide you to somewhat more advanced treatments which are still not at the senior/graduate level.

E. v. P. SMITH AND K. C. JACOBS, *Introductory Astronomy and Astrophysics,* Philadelphia, W. B. Saunders Company, 1973.

F. S. SHU, *The Physical Universe,* Mill Valley, California, University Science Books, 1982.

A. UNSOLD AND B. BASCHER, *The New Cosmos,* (3rd ed), New York, Springer-Verlag, 1983.

The next three sections provide more specialized but not excessively detailed references.

## The solar system

G. BRIGGS AND F. TAYLOR, *The Cambridge Photographic Atlas of the Planets,* Cambridge, U.K., Cambridge University Press, 1982.

J. K. BEATTY, B. O'LEARY, A. CHAIKIN, *The New Solar System* (2nd ed), Cambridge, U.K., Cambridge University Press, 1982.

N. CALDER, *The Comet is Coming,* New York, The Viking Press, 1981.

C. R. CHAPMAN, *Planets of Rock and Ice,* New York, Charles Scribner's Sons, 1982.

B. FRENCH, *The Moon Book,* New York, Penguin Books, 1977.

W. K. HARTMANN, *Moons and Planets*, Belmont, California, Wadsworth Publishing Co., Inc., 1973.

G. HUNT AND P. MOORE, *Jupiter,* New York, Royal Astronomical Society and Rand McNally and Co., 1981.

W. J. KAUFMANN III, *Exploration of the Solar System,* New York, Macmillan Publishing Co., Inc., 1978.

S. R. TAYLOR, *Lunar Science: A Post-Apollo View,* Oxford, Pergamon Press Inc., 1975.

F. L. WHIPPLE, *Orbiting the Sun,* Cambridge, Massachusetts, Harvard University Press, 1981.

## Stars

B. J. BOK AND P. F. BOK, *The Milky Way* (5th ed), Cambridge, Massachusetts, Harvard University Press, 1981.

K. FRAZIER, *Our Turbulent Sun,* Englewood Cliffs, New Jersey, Prentice-Hall Inc., 1982.

R. JASTROW, *Red Giants and White Dwarfs* (2nd ed), N.Y., W. W. Norton and Co., 1979.

R. W. Noyes, *The Sun, Our Star,* Cambridge, Massachusetts, Harvard University Press, 1982.

S. Mitton, *The Crab Nebula,* New York, Charles Scribner's Sons, 1978.

C. Payne-Gaposchkin, *Stars and Clusters,* Cambridge, Massachusetts, Harvard University Press, 1979.

H. Shipman, *Black Holes, Quasars and the Universe,* Boston, Houghton Mifflin Co., 1976.

## Galaxies and cosmology

R. Berendzen, R. Hart and D. Seeley, *Man Discovers the Galaxies,* New York, Science History Publications, 1976.

E. R. Harrison, *Cosmology,* Cambridge, U.K., Cambridge University Press, 1981.

E. Hubble, *The Realm of the Nebulae,* New York, Dover Publications Inc., 1958 reissue of 1938 original.

L. John (ed), *Cosmology Now,* London, British Broadcasting Corporation, 1973.

S. Mitton, *Exploring the Galaxies,* London, Faber, and Faber Ltd, 1976.

H. Shapley, *Galaxies,* Cambridge, Massachusetts, Harvard University Press, 1972 revision by P. W. Hodge.

J. Silk, *The Big Bang,* San Francisco, W. H. Freeman and Co., 1980.

S. Weinberg, *The First Three Minutes,* New York, Bantam Books, 1979.

## Life in the universe

E. Chaisson, *Cosmic Dawn,* Boston, Little Brown and Co., 1981.

G. Feinberg and R. Shapiro, *Life Beyond the Earth,* New York, Wm. Morrow and Co., Inc., 1980.

D. Goldsmith, *The Quest for Extraterrestrial Life,* Mill Valley, California, University Science Books, 1980.

D. Goldsmith and T. Owen, *The Search for Life in the Universe,* Reading, Massachusetts, Addison-Wesley, 1980.

M. Hart and B. Zuckerman, *Extraterrestrials—where are they?,* Oxford, Pergamon Press, 1982.

P. Morrison, J. Billingham and J. Wolfe (eds), *The Search for Extraterrestrial Intelligence,* New York, Dover Publications Inc., 1979. (This is also available as NASA Publication SP-419.)

R. Rood and J. Trefil, *Are We Alone?,* New York, Charles Scribner's Sons, 1981.

## Astronomical calculations with a pocket calculator

P. Duffett-Smith, *Practical Astronomy with your Pocket Calculator* (2nd ed), Cambridge, U.K., Cambridge University Press, 1981.

A. Jones, *Mathematical Astronomy with a Pocket Calculator,* New York, John Wiley and Sons, 1978.

## Scientific American

Almost every issue contains an article on an astronomical topic, and many have articles on scientific history. The graphics are always outstanding, but the level of the articles will sometimes be high. There are too many articles to list separately here. The *Scientific American* has published a cumulative index for 1948 to 1978.

The March/April 1981 issue of *Mercury* contains Michael Zeilik's comprehensive index to astronomy articles in the *Scientific American* for the years 1960 to 1981.

Several collections of reprinted articles have appeared, sometimes augmented by commentary or articles from other sources. These have been published by W. H. Freeman and Co., and now include:

*Comets* (1981), ed. by J. C. Brandt

*Cosmology +1* (1977), ed. by O. Gingerich

*The New Astronomy and Space Science Reader* (1977) ed. by J. C. Brandt and S. P. Maran

*New Frontiers in Astronomy* (1975), ed. by O. Gingerich

*The Planets* (1983), ed. by B. Murray

*The Solar System* (1975) (reprint of special September 1975 issue)

## NASA publications

NASA has produced many outstanding publications, at all levels of scientific detail. A catalog can be obtained from the Superintendent of Documents, U.S. Government Printing Office, Washington, D.C. 20402. Ask for the SP series. There is a good survey of these by Andrew Fraknoi in the Sept/Oct 1981 issue of *Mercury.*

# Index

Page numbers refer to entries in the body of the text; italicized numbers refer to figures or tables.

## A



## B

# C

# D



# E

# F



# G



# H

# I



# J

# K



# L



# M

# N



# O



# P

# Q



# R

## S

# T

# U



# V



# W

## Z

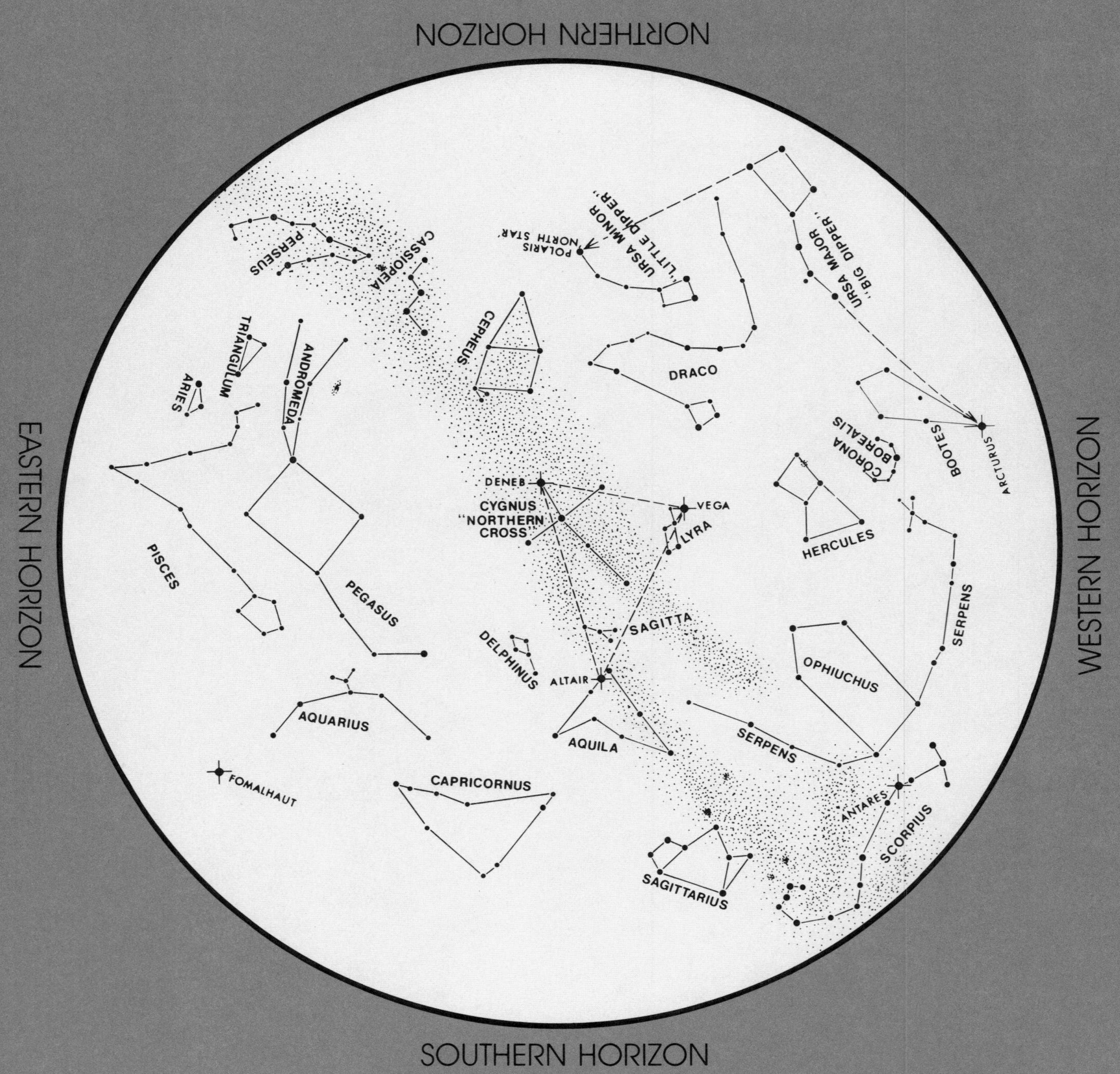


# THE NIGHT SKY IN SEPTEMBER